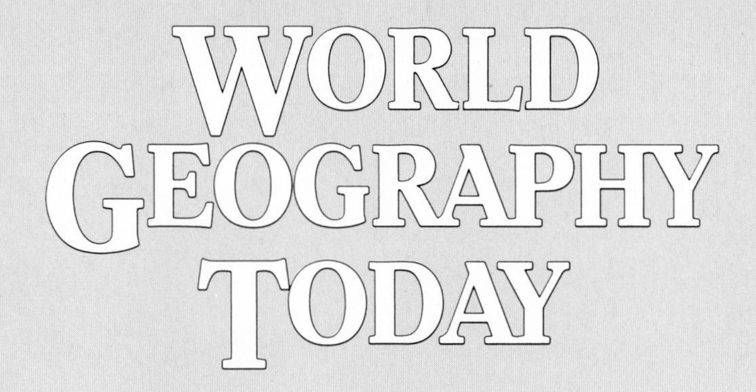

Annotated Teacher's Edition

Authors of the Teacher's Guide
Stephanie Abraham Hirsh
Director of Program and Staff Development
Richardson I.S.D.
Richardson, Texas

Karen Tindel Wiggins
Social Studies Consultant
Richardson I.S.D.
Richardson, Texas

HOLT, RINEHART AND WINSTON
AUSTIN NEW YORK SAN DIEGO CHICAGO TORONTO MONTREAL

Printed in the United States of America.

For permission to reprint copyrighted material, grateful acknowledgment is made to the following sources:

Boston University Press: From "Stanley's Despatches to the New York Herald 1871-1872, 1874-1877" edited by Norman R. Bennett. Copyright© 1970 by the Trustees of Boston University.

Century-Hutchinson, London: From *Wanderings in Tasmania* by George Porter. Published by Selwyn & Blount, Ltd., Autumn Books, 1934.

Grove Press, Inc.: From *Life and Death in Shanghai,* by Nien Cheng. Copyright© 1986 by Nien Cheng.

Her Majesty's Stationery Office: Excerpt "Margaret Thatcher's Speech Delivered to a Joint Session of the U.S. Congress, Washington D.C., February 20, 1985" in *Vital Speeches,* March 15, 1985, v. LI, no. 11.

Keefe, Bruyette & Woods: Adaptation of "Ten Biggest U.S. Bank Lenders to Mexico" by Keefe, Bruyette, & Woods. Titled "What Mexico Owes to U.S. Banks" from *The Wall Street Journal,* December 30, 1987. Published by Dow Jones & Company, Inc.

National Geographic Society: From "Brazil's Wild Frontier" in *National Geographic,* v. 152, no. 5, November 1977. Copyright© 1977 by National Geographic Society. From "Eruption in Colombia" and "Earthquake in Mexico" by Bart McDowell in *National Geographic,* v. 169, no. 5, May 1986. Copyright© 1986 by National Geographic Society.

Pantheon Books; a Division of Random House, Inc.: From *Kibbutz Makom: Report from an Israeli Kibbutz* by Amia Lieblich. Copyright© 1981 by Amia Lieblich.

Times Books, a Division of Random House, Inc.: From *The Russians* by Hedrick Smith. Copyright© 1976 by Hedrick Smith.

United States Department of State, Bureau of Public Affairs: "Trade and Economic Change" from *GIST,* Harriet Culley, editor. From "Obstacles to Investment and Economic Growth in Latin America" by Richard T. McCormack in *Current Policy,* no. 862.

Viking Penguin, Inc.: From *Life on the Mississippi* by Mark Twain. Copyright© 1948 by the Mark Twain Company.

J. Weston Walch: Map, "Gondwana Reassembled," after de Blij. Copyright© 1983 by J. Weston Walch.

ISBN 0-03-021379-7

890123456 036 987654321

TABLE OF CONTENTS

LIST OF MAPS

ATLAS

USING
WORLD GEOGRAPHY TODAY

World Geography Today is designed to help students gain the confidence and skills necessary to become independent learners. From the opening pages, the emphasis is placed on students' building a sound skills base and learning to use the book effectively.

TO THE STUDENT

World Geography Today is designed to provide you with an effective tool for learning about our world. It has standard features found in all good books. It also has some special features to make your reading and studying more effective and enjoyable. The following pages describe the design of *World Geography Today* and explain how the book is organized.

Organization of *World Geography Today*

It is helpful to think of *World Geography Today* as having three parts. The first part is called the front matter. As the name implies, this is material in the front of the book. Included in the front matter is the **Table of Contents**, which lists the titles of all units, chapters, and special-feature pages throughout the book. The pages you are reading now are part of the front matter, as is the Atlas. The **Atlas** is a compilation of maps to supplement your study of world geography.

The largest part of *World Geography Today* is devoted to units and chapters describing the study of geography and the world's major regions. The first unit begins on page 30. Each unit is divided into two or more chapters.

The third part of your textbook is called the back matter. It includes an **Appendix** with information in statistical form. It also includes a **Glossary**, which lists all the important vocabulary words that appear in blue type throughout the book. Last, the back matter includes an Index. The **Index** is your guide to the hundreds of topics discussed in the book, with page references for each topic.

Each unit opens with a two-page spread like the one shown here. Three photos depicting various aspects of geography are included. Objectives for the unit are included to help guide your reading and study. For Units 2 through 9, the area of the world discussed in that unit is displayed on the world map.

x

To the Student *acquaints students with the book's organizational format. Highlights of the text are explained in clear, concise language to ensure learning success.*

BASIC MAP AND GLOBE SKILLS

Before you begin your study of world geography, you must master some basic map and globe skills. Throughout this book, you will have the opportunity to improve these skills and build upon them.

Geographers are interested in answering such questions as these: Where is it? Why is it located there? What is it near? What influence does it have on its environment? To answer these and other questions, geographers use certain tools. The three major tools are globes, maps, and remote sensors.

The Globe

A Model of the Earth One of the best sources that geographers use to answer their questions is a globe. A globe is a scale model of the earth. It is especially useful for looking at the entire earth or at large areas of the earth's surface.

About 71 percent of the earth's surface is covered by water. The remainder is land. Geographers organize the earth's land surface into seven large masses of land, called continents. The continents are North America, South America, Europe, Asia, Africa, Australia, and Antarctica. Use a globe to locate the continents shown in Figure 1.

Notice on the globe that Europe is not really a separate continent, but part of the earth's largest landmass, called Eurasia. This landmass is traditionally divided into two continents, Europe and Asia. Landmasses smaller than continents that are completely surrounded by water are called islands. Locate some islands on the globe.

Like the land surface, the water surface of the earth is also organized into parts. It is divided into separate bodies of water. Study the globe to find the largest bodies of water, called oceans. The major oceans are the Pacific, the Atlantic, the Indian, and a much smaller ocean called the Arctic. There are other terms used to identify smaller bodies of water as well. Can you locate some of these on the globe?

Figure 1 The Continents

Lines on the Globe Look at Figure 2, the diagram of a globe. You will notice a pattern of lines that circle the earth in east-west and north-south directions. This pattern is called a grid. The two sets of lines that make up the grid do not actually exist on the surface of the earth. The purpose of these imaginary lines is to help us find the exact location of places on the earth. These lines are called latitude and longitude.

Lines of latitude are drawn in an east-west direction around the globe. Latitude is the measurement that tells you exactly how far north or south of the equator you are. The equator is the imaginary line that also circles the globe in an east-west direction. It is halfway between the North and South poles. In other words, latitude locates exactly where you are between the equator and either the North or the South Pole.

Because the lines of latitude are always parallel to the equator and to each other, they are called parallels of latitude. If you look at the diagram of the globe again, you will see that the parallels of latitude are labeled with the symbol °. This symbol is a unit of measurement called degrees. These parallels are also labeled with an *N* for *north* or an *S* for *south*. The letter *N* or *S* tells you whether a place is located north or south of the equator. Lines of latitude range from 0°, for locations on the equator, to 90° N or 90° S, for locations at the North or South Pole.

Latitude is only half of what you need to know in order to locate a place. There are many locations along the equator or along any parallel of latitude. How can you tell one place from another? To do this, you must use the lines of longitude. These lines circle the globe from pole to pole.

Just as we use the equator as a line to determine latitude, we use an imaginary line called the prime meridian to determine longitude. The prime meridian runs from the North Pole to the South Pole through Greenwich, England. Since lines of longitude are drawn in the same direction as the prime meridian, they are all called meridians. Imagine looking down at the globe from above the North Pole. No matter where on Earth a place is, you could draw a line that runs from pole to pole through your location.

Notice on the diagram of the globe that the meridians and parallels are similarly labeled. Each meridian indicates how many degrees of longitude a location is east or west of the prime meridian. A *W* for *west* and an *E* for *east* is added to the longitude degree. Lines of longitude range from 0° on the prime meridian to 180° at a meridian in the mid-Pacific Ocean.

By knowing the latitude and longitude of any place on Earth, you can find its exact location. No other place on Earth has this exact latitude and longitude.

Figure 2 Globe

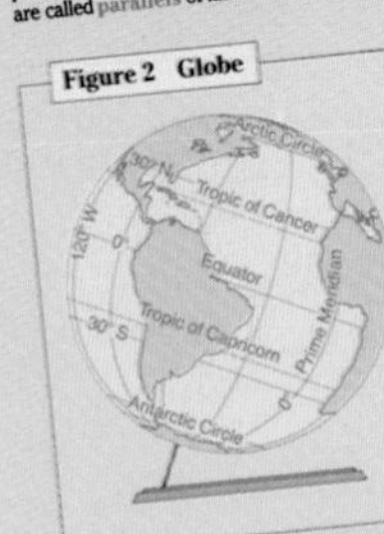

22

23

Preliminary **Basic Map and Globe Skills** *lessons teach and reinforce geographic skills such as reading map keys, determining latitude and longitude, and using map scales correctly.*

THE FUNCTIONAL ORGANIZATION PROVIDES AN EASY-TO-FOLLOW FRAMEWORK.

A stone fortress stands on an oasis at the base of the Atlas Mountains in Morocco. The Atlas Mountains extend for 1,500 miles (2,400 kilometers) across northwest Africa.

Islam, which is widely practiced in the Middle East and North Africa, has about 540 million followers. These Muslims are praying toward the holy city of Mecca.

Farmers along the coast and in the mountains of the Middle East raise a variety of livestock. Most people in this region are farmers. Sheep, which are raised for meat and wool, are traded and sold at this outdoor market in Morocco.

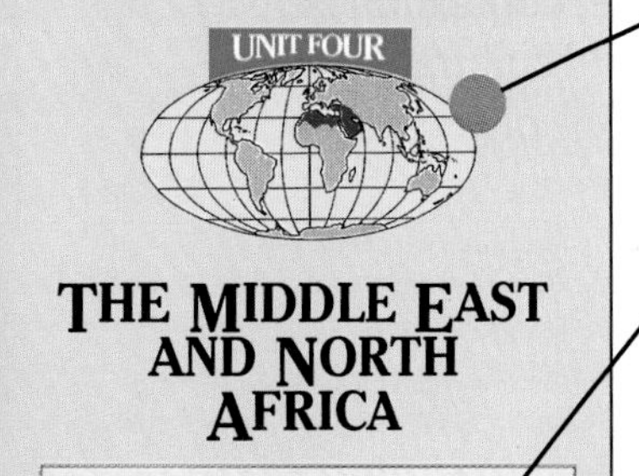

THE MIDDLE EAST AND NORTH AFRICA

OBJECTIVES

- *To trace the history and the development of religion in the Middle East and North Africa and to understand the important role of this region in today's world*
- *To describe how climate and landforms affect the way people live in the Middle East and North Africa*
- *To describe the economic, political, and social issues that the Middle East and North Africa face and how the region is trying to address those issues*
- *To describe the significance of the strategic location of this region*
- *To interpret a pie graph*

A global **locator map** *indicates the relative position of the region.*

Unit **Objectives** *help students set realistic goals for learning.*

Chapter headings provide an outline that helps students follow the sequence of ideas.

Each unit opens with stunning photographs representing economic, physical, and cultural aspects of the region's geography.

Chapter 21

The Desert Region

Muslims worship at Mecca's Grand Mosque.

Egypt's pyramids are monuments to the past.

The Middle East and North Africa are a crossroads where Asia, Europe, and Africa meet. The early foundations of Western civilization were laid in the river valleys of Egypt and Iraq. The ruins of great monuments, temples, and ancient cities can still be seen. Three major religions—Judaism, Christianity, and Islam—first appeared in this region. Today, the holy places of these religions draw pilgrims from throughout the world. Most of the people in the Middle East and North Africa are Arabs who speak Arabic and practice Islam. Other groups include Jews, Arab Christians, and non-Arabs who practice Islam.

The desert region has experienced much economic development and cultural change in a short period of time. Nationalism, religious conflict, and modernization are challenging traditional cultural values. In addition, a wealth of oil deposits has given parts of the region worldwide political importance.

GEOGRAPHY DICTIONARY

Judaism	arid
Christianity	oasis
Islam	exotic river
Muslim	Bedouins
monotheism	commodity
mosque	chromium

242

Historical Geography

Ancient Civilizations The world's first civilizations developed in the area known as the Fertile Crescent. The Fertile Crescent is a great arc of productive land that runs along the eastern shore of the Mediterranean Sea, through the plains along the Tigris (TIE-gruhs) and Euphrates (yu-FRAYT-eez) rivers to the Persian Gulf. The ancient name for the plains between the Tigris and Euphrates rivers is Mesopotamia (mehs-uh-puh-TAY-mee-uh). Large-scale irrigation was developed here long ago. Many of the plants and animals found throughout the world today were first domesticated, or tamed by humans, in the Fertile Crescent.

The period from about 3500 B.C. to 600 B.C. saw the rise and fall of great civilizations in the Middle East and North Africa. In Mesopotamia, the Sumerians developed writing. Later, the Assyrians developed iron weapons. In Egypt, the great pyramids, monuments to the pharaohs, were built along the Nile. Along the Mediterranean shore, Phoenician merchants established a great trading network. Judaism, the first religion centered around the belief in a single god, also appeared during this period within the Fertile Crescent in Palestine. Each culture left a part of its knowledge and way of life for the civilizations that followed.

After 600 B.C., most of the region was controlled by one great empire after another. These included the Persian, Macedonian (Greek), and Roman empires. The Roman Empire was divided during the fourth century A.D. into the Eastern Roman Empire and the Western Roman Empire. The eastern part was controlled from Constantinople, the site of present-day Istanbul, Turkey. The Turks eventually overthrew the Eastern Roman Empire. Germanic invaders gradually took over the western section. The Western Roman Empire collapsed in A.D. 476.

Christianity developed out of Judaism and spread during the Roman era. It is a religion founded on the teachings of Jesus Christ.

The ancient Egyptian mummy of King Tutankhamen was discovered, with many other artifacts, in 1922.

Christians believe that Jesus Christ is the son of God. The Roman leaders distrusted Jesus Christ and crucified him. Christians were persecuted by the Romans for many years. By the late fourth century, however, Christianity had become the official religion of the Roman Empire. During the ninth century, Christianity split into the Eastern Orthodox church and the Roman Catholic church. In the 1500s, another group of Christians, called Protestants, broke away from the Roman Catholic church. Today, Christianity is the dominant religion of the Western Hemisphere.

Rise of Islam Islam was founded by the prophet Muhammad (moh-HAM-uhd), who lived from about A.D. 570 to 632. Muhammad lived in Mecca, a trading city in western

243

Each chapter begins with two dramatic photographs representative of cultural and economic aspects of the region. An accompanying introduction provides insight into the historical, physical, economic, and cultural influences of the region on its inhabitants.

A list of key words provides opportunities for preteaching vocabulary.

Exciting photographs and illustrations provide realistic views of each area and its people.

UP-TO-DATE CARTOGRAPHY EQUIPS STUDENTS TO DEAL WITH TODAY'S HEADLINES.

Maps showing climate, population density, economics, history, and cultural patterns illustrate different facets of geography.

Western Europe

countries of Western Europe have the lowest population densities? the highest? Why?

113

Western Europe

A combined physical/ political map of the region introduces each unit.

Complete map keys and fully drawn lines of latitude and longitude facilitate understanding.

Geography Skills *lessons provide practice of essential skills and develop students' ability to use geographical information.*

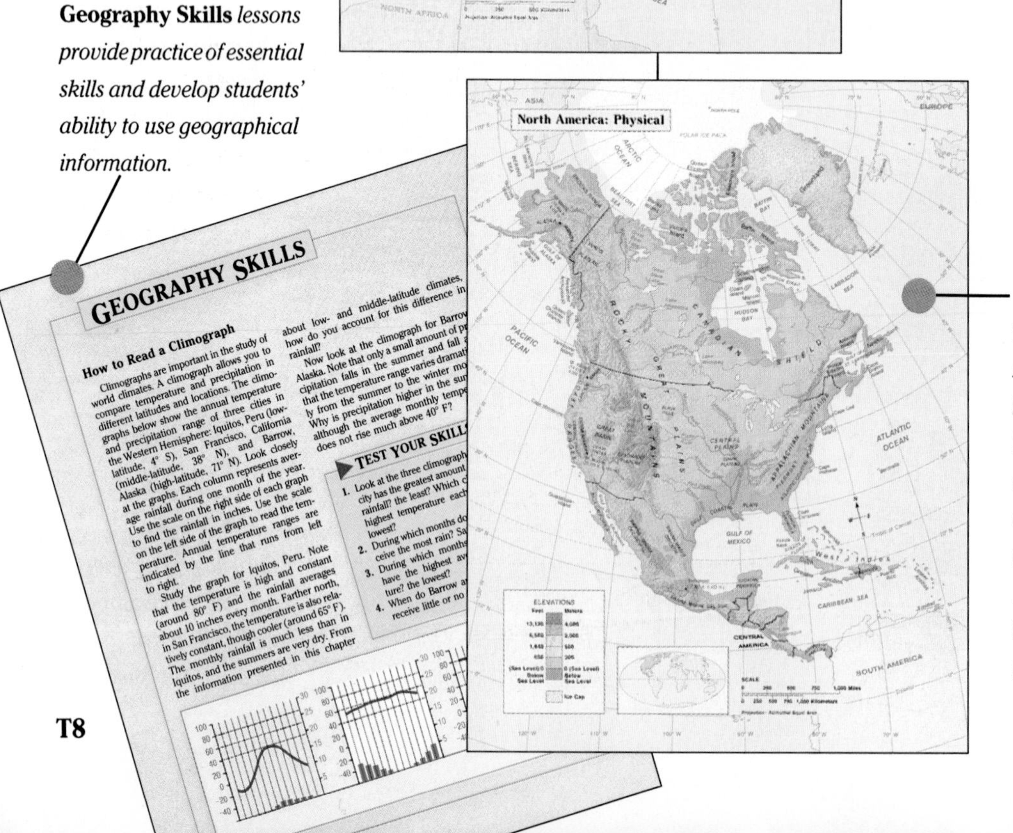

The new and expanded ***Atlas*** *provides comprehensive support for the text. Twenty-one pages of full-color maps illustrate political and physical aspects of the world and its regions. A reference section highlights statistical information for individual countries.*

SPECIAL ELEMENTS BOOST VOCABULARY, READING, AND COMPREHENSION SKILLS.

Content is organized around physical, economic, and cultural geography. This uniform format promotes ease in reading and comprehension.

Iceland's capital, Reykjavík, is visible on the far left of this Landsat photograph. Most of Iceland is covered by lava and volcanic ash.

Iceland

Physical Geography To the west of the Scandinavian Peninsula, in the middle of the North Atlantic Ocean, is the island country of Iceland. (See the map on page 133.) This island covers an area about the same size as the state of Kentucky. Iceland has the world's most northern capital, Reykjavík (RAYK-yuh-vik), located very near the Arctic Circle. Situated on the west coast, Reykjavík contains more than half of Iceland's population.

Iceland was originally settled by Vikings from Norway. Icelanders still speak a language very similar to that spoken by the Vikings. The Icelandic parliament is more than 1,000 years old, the oldest in the world. The country is situated on top of the volcanic Mid-Atlantic Ridge and has more than 200 volcanoes. Many of these volcanoes are active. Three-quarters of Iceland is mountainous and uninhabitable. Glacial ice covers about one-eighth of the high mountain interior. The coastline is extremely rugged and constantly battered by waves.

The name Iceland is misleading, since the majority of the island is not ice covered or extremely cold. Because of the island's location in the middle of the Atlantic, the ocean currents keep temperatures mild. The country has mainly a cool, maritime climate. It is windy, wet, and cloudy much of the year. Temperatures usually remain above freezing.

Less than one percent of Iceland supports trees. The ground is mainly barren volcanic lava or covered with grasses. Hay and potatoes are grown on the grasslands. Sheep and cattle are also grazed there. Wool is used mainly to manufacture sweaters, which are needed by the people in this cool climate.

Iceland's most important natural resource is the rich fishing waters that surround the island. Fishing and fish products are the main industries. Over 70 percent of the nation's export income comes from fishing. To protect its fishing resources, Iceland claims a 200-nautical-mile (320-kilometer) fishing limit around the island.

Because of its mountains and wet climate, Iceland produces much hydroelectricity. It also gets energy from the ground. As a result of volcanic activity just below the earth's surface, underground water rises all over the island as steam, [illegible]ng **geysers**. The word *geyser* comes fr[illegible] Icelandic term for a hot spring that shoots up fountains of hot water and steam into the air. In most Icelandic homes, hot water is piped from the boiling springs for heat. The hot springs are also used to heat greenhouses, where crops such as tomatoes, grapes, and cucumbers can be grown year-round. Geysers are tapped to generate electricity. This cheap electrical energy from geysers and waterpower has attracted the aluminum-refining industry to Iceland.

Despite its isolated location and barren land, Iceland has a high standard of living. The

136

Key terms are printed in blue and defined.

The Icelanders have adapted to their physical environment. Heat and electricity are generated by hot springs and geysers.

country has little crime, unemployment, or illiteracy. Icelanders have made great progress in using their country's natural resources to live comfortably in a challenging environment.

Finland

Physical Geography Finland lies across the Gulf of Bothnia from Sweden. It borders the Soviet Union in the east and Sweden and Norway in the far north. (See the map on page 133.) Finland's landforms show the effects of the last ice age, during which time the whole country was covered by thick sheets of ice. Today, Finland is made up mostly of a low, glaciated plateau with more than 60,000 lakes. A narrow coastal plain is found in the west along the Gulf of Bothnia and south along the Gulf of Finland. Several thousand small islands in the Gulf of Bothnia are also part of Finland. Finland's rivers play an important role in producing hydroelectricity and in providing local transportation.

Because the Gulf of Bothnia freezes during most winters, Finland has a more severe climate than the other Nordic countries, which are nearer to open seawater. Finland can be divided into three major climate regions. In the far south of the country is a humid continental climate. The central part of the country has a subarctic climate. To the far north is a tundra climate. In the humid continental climate region, the summer growing season is very short. Despite this, the Finns grow some crops. The subarctic climate region is covered mostly by evergreen forests. In the far north, where there is only sparse tundra, the main activity is reindeer herding, which is done by people known as the Lapps. Later in this chapter, you will read more about this tundra region and the Lapps who inhabit it.

Economic Geography Less than 10 percent of Finland is suitable for farming. Glaciers

137

In-text references give students helpful information to aid comprehension.

Agriculture In spite of successes in industrialization, agriculture in the Soviet Union still does not meet all of the people's needs, particularly for meat. Nearly a quarter of Soviet workers are employed in agriculture versus only 3 percent in the United States. The Soviet land area devoted to farming is also larger than that of the United States. However, the Soviet Union regularly imports large amounts of agricultural products, especially grains to feed cattle.

The difficulties of Soviet agriculture are caused partly by the climate. In most of the country, the growing season is short, and droughts are frequent. However, the organization of Soviet agriculture seems to be a main problem. Areas of comparable soil and climate in North America and Scandinavia are much more productive.

Soviet farmland is divided between state and collective farms. State farms, which occupy more than half of the farmland, operate like factories. The farmers receive pay for each hour they work. These farms average about 15,000 acres (6,000 hectares) in size. In contrast, the average American farm is about 350 acres (140 hectares) in size. (See Themes in Geography "*Kolkhozy* and *Sovkhozy*" on page 202.)

Collective farms take up the remaining farmland. Workers receive a share of the production and profits rather than wages. A collective farm, which averages about 7,700 acres (3,100 hectares), may have more than 400 families living and working on it.

Workers on both state and collective farms also have small plots of land on which to grow crops to sell for personal profit. They farm the plots in addition to working on government-owned farms. These private plots make up only 2 to 3 percent of all farmland in the Soviet Union. Carefully tended and highly productive, they provide most of the Soviet Union's fresh fruits and vegetables. Farmers are able to sell these products privately to people in the cities. Farmers do not seem to have enough incentive to make the state-controlled lands as productive.

Life in the Soviet Union

All Soviet citizens are expected to sacrifice and make communism the most important aspect in their lives. The Soviet government has made heavy industry and the production of military goods more important than providing consumer goods. As a result, consumer goods are expensive, and there is little variety available. Soviet people must frequently line up to buy goods at stores.

Generally, the Soviet people live in small apartments, have few cars, and have little opportunity to travel outside the USSR. This standard of living is better than the Soviet people have had in the past. Yet, it is well behind that of most of the United States and the countries of Western Europe.

Life in the Soviet Union requires that one obey what the government says. Soviet citizens have few personal freedoms. If a person disagrees with the Communist ideals or government policies, then a promotion or a good job may be difficult to get. At worst, the person could be sent to a prison labor camp in Siberia. While there are not as many people in

For the Record

Personally, I have known of people who stood in line 90 minutes to buy four pineapples . . . 3 and a half hours to buy three large heads of cabbage only to find the cabbages were gone as they approached the front of the line, 18 hours to sign up to purchase a rug at some later date, all through a freezing December night to register on a list for buying a car, and then waiting 18 more months for actual delivery. . . . Lines can run from a few yards long to half a block to nearly a mile, and usually they move at an excruciating creep. Some friends . . . watched and photographed a line that lasted two solid days and nights, four abreast and running all through an apartment development. They guessed there were 10,000–15,000 people, signing up to buy rugs, an opportunity that came only once a year in that entire section of Moscow.

From *The Russians* by Hedrick Smith.

201

reaches more than 500 feet (152 meters) above sea level.

The North European Plain is Western Europe's most important landform region. More than two-thirds of the population of Western Europe lives here. Nearly all of Europe's major rivers flow across this plain. Most of Western Europe's major agricultural and industrial centers are located here.

The third landform region is the Central Uplands. This ancient, hilly area of Western Europe consists of the Massif Central (ma-SEEF sehn-TRAHL) of France, the Jura (JUR-uh) Mountains on the Swiss-French border, the Black Forest and Bavarian Plateau of Germany, and the Ardennes (ahr-DEHN), a wooded plateau region in Belgium and Luxembourg. These hills and plateaus are forested and have important mineral deposits, such as coal and iron. Agriculture is quite productive in this region.

The Alpine Mountains comprise the most rugged and spectacular landform region of Western Europe. These young and active mountains are divided into several major ranges. The Alps are the highest and longest mountain range in Western Europe. They extend from southern France through Switzerland and Austria and across northern Italy. The highest peak, Mont Blanc (MOHN BLAHN), reaches to 15,771 feet (4,808 meters). Many peaks in the Alps reach more than 14,000 feet (4,268 meters). Because of their high elevations, the Alps receive much snow, and large glaciers are still present.

Another range in the Alpine Mou[illegible]gion is the Pyrenees (PIR-uh-neez). [illegible]e mountains form a **natural boundary** between Spain and France. When political borders are established by landforms, they are called natural boundaries. The highest peak reaches an elevation of 11,168 feet (3,404 meters). Several high mountain ranges are found in Spain, including the Sierra Nevada. South of the Alps are the Apennines (AP-uh-neenz) of Italy and the rugged Pindus (PIN-duhs)

The Pyrenees Mountains form a natural boundary between France and Spain. These are old, rugged landforms. What other European landforms form natural boundaries?

116

Phonetic respellings are provided for words with difficult pronunciations. A **Glossary** *listing also provides a pronunciation key and page references.*

Caption questions engage students in thinking critically about the content.

Primary sources create interest and relate additional information in support of the text.

UNIQUE FEATURES LINK THE STUDY OF GEOGRAPHY TO STUDENTS' LIVES.

Cities of the World introduces students to the concepts of urban geography and shows how these concepts apply to the unique development of each city.

This feature helps students to understand the common issues facing the cities of the world and to appreciate culturally diverse patterns of life.

Cities of the World

PARIS

Paris, the capital of France, is the ce
of the country's economic, political, and
tural life. It is known for its tree-lined s
and its beautiful palaces, cathedrals
monuments. Paris is the home of the
famous Eiffel Tower, Arc de Triomphe
of Triumph), Louvre museum, and
Dame cathedral.

The city of Paris was founded m
2,000 years ago on an island, called
Cité (EEL duh lah see-TAY), in the
the Seine River. This island was at
roads of water and land routes

▼ The Île de la Cité is the heart o

146

Have students locate Paris on the map on page 143.

Right Bank. Called the center of intellectual life, the Left Bank has traditionally been the home of artists, writers, universities, museums, and government offices. The Right Bank is the heart of the city's economic life. It has many cafés, shops, banks, foreign embassies, and small industries such as clothing design. In this century, there has been rapid ...th of suburban areas outside the center

Themes in Geography

THE PERSIAN GULF

Most of the world's oil lies beneath the earth in one small area of southwestern Asia: the Persian Gulf. Because oil has become necessary to modern life with its automobiles, aircraft, electric power, and plastics industries, it is not surprising that the Gulf region has attracted the world's attention.

The Persian Gulf is a body of water linked to the Indian Ocean by the narrow Strait of Hormuz. However, *Persian Gulf* also refers to the countries that surround the Gulf itself. The largest of these countries is the kingdom of Saudi Arabia, and the smallest is the tiny island state of Bahrain.

The Gulf countries may differ greatly in size and population, but they have much in common. To keep their economies running, they all rely heavily on money earned from exporting oil. The people of the region share a common religion, Islam. And almost the entire area has a dry, desert climate.

The largest customers for the oil produced in the Persian Gulf are the world's major industrial countries, particularly the countries of Western Europe, North America, and Japan. This means that a large amount of oil must be moved from the Gulf to the places where it is needed. Some of the oil is exported overland by pipeline to neighboring countries. Most, however, travels great distances via enormous tanker ships.

Region **It is because the Gulf states have so much in common—religion, climate, and dependence on oil—that we refer to the Persian Gulf as a region. Yet, despite their similarities, these countries also have important differences. The country that differs most from its neighbors is Iran. Unlike the other Gulf states, Iran is not an Arab country. Also, most Iranians belong to a branch of Islam that is different from the Islam practiced by most other Gulf residents. How do you think these differences might cause conflict between Iran and its neighbors?**

270

Movement **Most of the Persian Gulf's oil exports travel through the Strait of Hormuz. This makes the strait an extremely important area. If it were to be closed, for war or any other reason, the flow of oil to the industrialized world would be greatly reduced. This is why many world powers, including the United States and the Soviet Union, are concerned about what happens in the Gulf region. Study the world maps in the Atlas at the front of this book. Can you find any other narrow waterway in the world where the flow of traffic could easily be stopped?**

The Persian Gulf region has a long and impressive history that dates back to ancient times. The country called Iraq today was the site of ancient Mesopotamia. Mesopotamia was known as "the land between the two rivers," which are the Tigris and the Euphrates. Later, in another part of the Gulf, called Arabia (today called Saudi Arabia), the prophet Muhammad lived. There, he founded Islam, which is one of the world's major religions today.

Location **The Persian Gulf was an important part of the world long before oil became a valuable product. Since the time of ancient Mesopotamia, it has been one of the world's most important waterways. The Gulf's location has always encouraged trading. Long ago, Gulf traders sold food products from Mesopotamia and pearls from the Gulf; they bought porcelain wares and tea from India and China. These traders also traded products from the Gulf for African gold, spices, and slaves. What sorts of locations offer the best opportunities for trade? Why?**

271

Geographic concepts are promoted by focusing the content around five critical aspects of geography:

- location
- place
- region
- relationships within places
- movement

The consistent exploration of these five geographic themes enables students to develop universal points of reference for comparing and contrasting regions and cultures.

The special feature **Themes in Geography** links each region to the major geographic themes.

EXTENSIVE REVIEWS AID COMPREHENSION.

Each **Chapter Check** evaluates students' mastery of the content and skills covered in the chapter.

1. **Reviewing the Main Ideas** summarizes major ideas from the chapter.
2. **Building a Vocabulary** asks students to define key terms and answer questions pertaining to these words.
3. **Recalling and Reviewing** requires students to recall information and utilize critical-thinking skills.
4. **Using Geography Skills** focuses on activities using map and graph skills.

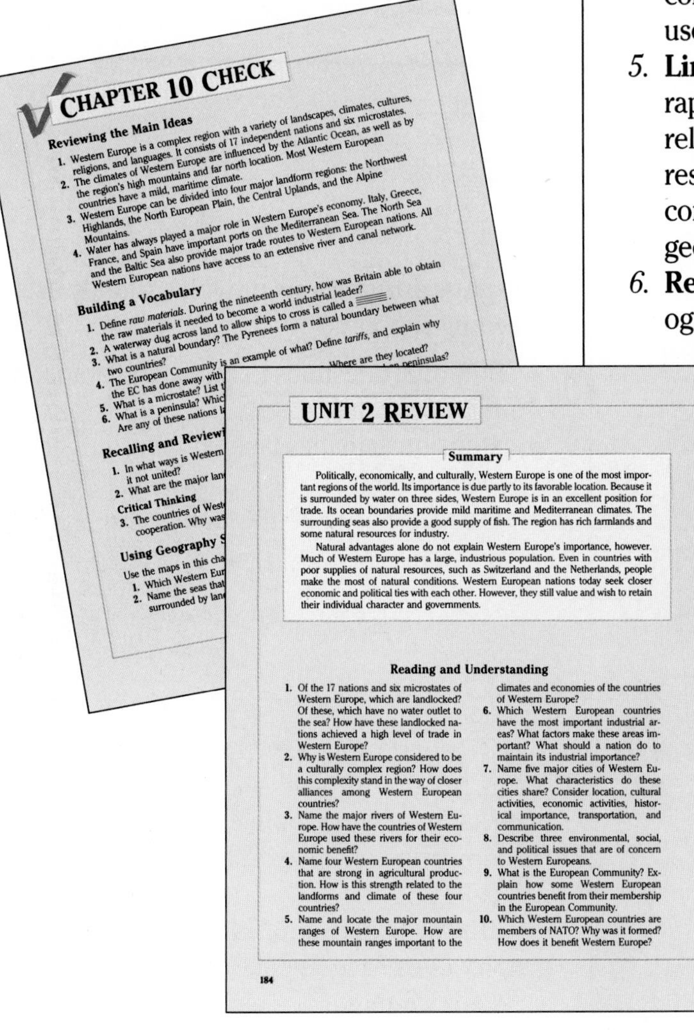

CHAPTER 10 CHECK

Reviewing the Main Ideas

1. Western Europe is a complex region with a variety of landscapes, climates, cultures, religions, and languages. It consists of 17 independent nations and six microstates.
2. The climates of Western Europe are influenced by the Atlantic Ocean, as well as by the region's high mountains and far north location. Most Western European countries have a mild, maritime climate.
3. Western Europe can be divided into four major landform regions: the Northwest Highlands, the North European Plain, the Central Uplands, and the Alpine Mountains.
4. Water has always played a major role in Western Europe's economy. Italy, Greece, France, and Spain have important ports on the Mediterranean Sea. The North Sea and the Baltic Sea also provide major trade routes to Western European nations. All Western European nations have access to an extensive river and canal network.

Building a Vocabulary

1. Define *raw materials*. During the nineteenth century, how was Britain able to obtain the raw materials it needed to become a world industrial leader?
2. A waterway dug across land to allow ships to cross is called a ______.
3. What is a natural boundary? The Pyrenees form a natural boundary between what two countries?
4. The European Community is an example of what? Define *tariffs*, and explain why the EC has done away with
5. What is a microstate? List t
6. What is a peninsula? Whic ... Where are they located? Are any of these nations l ... peninsulas?

Recalling and Reviewi

1. In what ways is Western ... it not united?
2. What are the major lan

Critical Thinking

3. The countries of West ... cooperation. Why was

Using Geography S

Use the maps in this cha
1. Which Western Eur
2. Name the seas that ... surrounded by lan

Each **Unit Review** measures students' newly acquired knowledge of geography.

1. **Summary** surveys the unit's major points.
2. **Reading and Understanding** provides questions covering the unit content. Recalling important facts and drawing conclusions are emphasized.
3. **Mastering Geography Skills** asks for visual interpretation of graphs and maps. Activities provide students with opportunities to use newly acquired geographic skills.
4. **Applying and Extending** applies learned concepts to new situations and requires the use of critical-thinking skills.
5. **Linking Geography and**...relates geography to such topics as history, economics, religion, culture, art, and literature. These research activities help students find connections and relationships between geography and other areas of life.
6. **Reading for Enrichment** provides a bibliography of source materials.

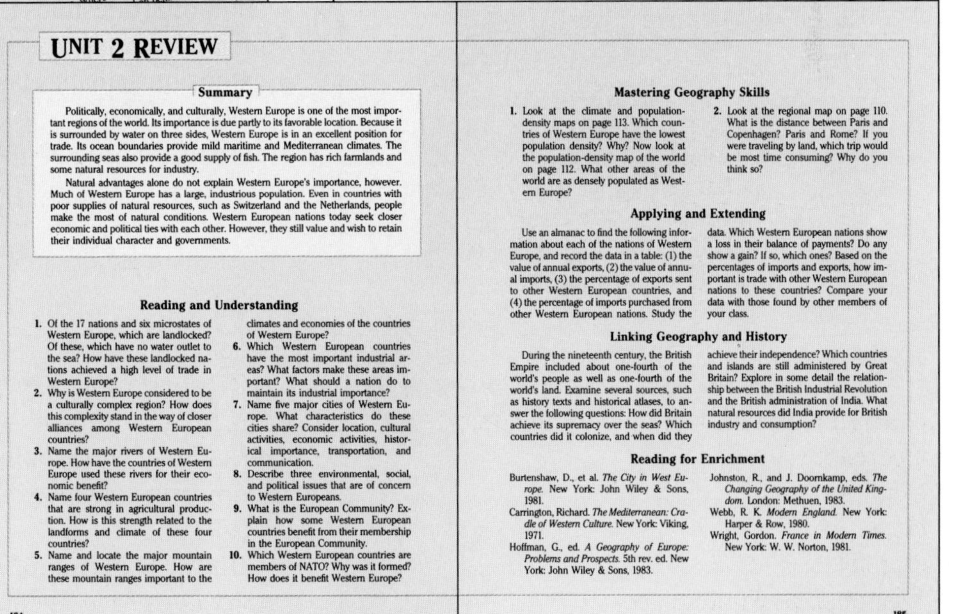

UNIT 2 REVIEW

Summary

Politically, economically, and culturally, Western Europe is one of the most important regions of the world. Its importance is due partly to its favorable location. Because it is surrounded by water on three sides, Western Europe is in an excellent position for trade. Its ocean boundaries provide mild maritime and Mediterranean climates. The surrounding seas also provide a good supply of fish. The region has rich farmlands and some natural resources for industry.

Natural advantages alone do not explain Western Europe's importance, however. Much of Western Europe has a large, industrious population. Even in countries with poor supplies of natural resources, such as Switzerland and the Netherlands, people make the most of natural conditions. Western European nations today seek closer economic and political ties with each other. However, they still value and wish to retain their individual character and governments.

Reading and Understanding

1. Of the 17 nations and six microstates of Western Europe, which are landlocked? Of these, which have no water outlet to the sea? How have these landlocked nations achieved a high level of trade in Western Europe?
2. Why is Western Europe considered to be a culturally complex region? How does this complexity stand in the way of closer alliances among Western European countries?
3. Name the major rivers of Western Europe. How have the countries of Western Europe used these rivers for their economic benefit?
4. Name four Western European countries that are strong in agricultural production. How is this strength related to the landforms and climate of these four countries?
5. Name and locate the major mountain ranges of Western Europe. How are these mountain ranges important to the climates and economies of the countries of Western Europe?
6. Which Western European countries have the most important industrial areas? What factors make these areas important? What should a nation do to maintain its industrial importance?
7. Name five major cities of Western Europe. What characteristics do these cities share? Consider location, cultural activities, economic activities, historical importance, transportation, and communication.
8. Describe three environmental, social, and political issues that are of concern to Western Europeans.
9. What is the European Community? Explain how some Western European countries benefit from their membership in the European Community.
10. Which Western European countries are members of NATO? Why was it formed? How does it benefit Western Europe?

184

Mastering Geography Skills

1. Look at the climate and population-density maps on page 113. Which countries of Western Europe have the lowest population density? Why? Now look at the population-density map of the world on page 112. What other areas of the world are as densely populated as Western Europe?
2. Look at the regional map on page 110. What is the distance between Paris and Copenhagen? Paris and Rome? If you were traveling by land, which trip would be most time consuming? Why do you think so?

Applying and Extending

Use an almanac to find the following information about each of the nations of Western Europe, and record the data in a table: (1) the value of annual exports, (2) the value of annual imports, (3) the percentage of exports sent to other Western European countries, and (4) the percentage of imports purchased from other Western European nations. Study the data. Which Western European nations show a loss in their balance of payments? Do any show a gain? If so, which ones? Based on the percentages of imports and exports, how important is trade with other Western European nations to these countries? Compare your data with those found by other members of your class.

Linking Geography and History

During the nineteenth century, the British Empire included about one-fourth of the world's people as well as one-fourth of the world's land. Examine several sources, such as history texts and historical atlases, to answer the following questions: How did Britain achieve its supremacy over the seas? Which countries did it colonize, and when did they achieve their independence? Which countries and islands are still administered by Great Britain? Explore in some detail the relationship between the British Industrial Revolution and the British administration of India. What natural resources did India provide for British industry and consumption?

Reading for Enrichment

Burtenshaw, D., et al. *The City in West Europe.* New York: John Wiley & Sons, 1981.
Carrington, Richard. *The Mediterranean: Cradle of Western Culture.* New York: Viking, 1971.
Hoffman, G., ed. *A Geography of Europe: Problems and Prospects.* 5th rev. ed. New York: John Wiley & Sons, 1983.
Johnston, R., and J. Doornkamp, eds. *The Changing Geography of the United Kingdom.* London: Methuen, 1983.
Webb, R. K. *Modern England.* New York: Harper & Row, 1980.
Wright, Gordon. *France in Modern Times.* New York: W. W. Norton, 1981.

185

THE COMPLETE SUPPORT PROGRAM PROVIDES MAXIMUM TEACHING EFFECTIVENESS.

The **Teacher's Guide**, bound into the front of the *Teacher's Edition*, gives practical suggestions for day-to-day teaching:

- an outline of chapter topics and suggestions for preteaching vocabulary
- a purpose and objectives for each lesson
- motivating teaching strategies including time-line activities
- strategies for evaluation, reteaching, and enrichment
- answers to **Chapter Check** and **Unit Review** questions

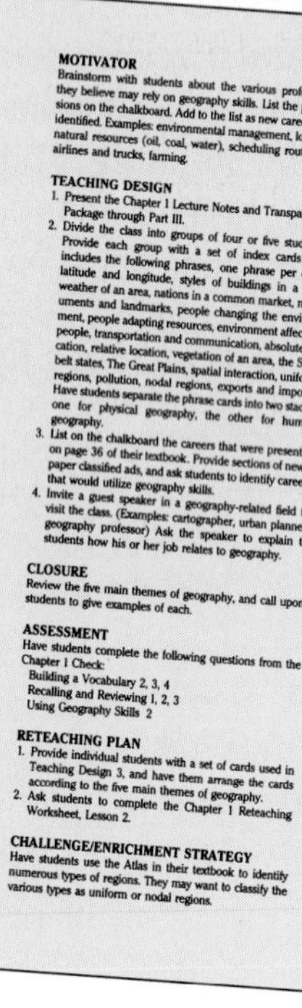

MOTIVATOR
Brainstorm with students about the various professions they believe may rely on geography skills. List the professions on the chalkboard. Add to the list as new careers are identified. Examples: environmental management, locating natural resources (oil, coal, water), scheduling routes for airlines and trucks, farming.

TEACHING DESIGN
1. Present the Chapter 1 Lecture Notes and Transparency Package through Part III.
2. Divide the class into groups of four or five students. Provide each group with a set of index cards that includes the following phrases, one phrase per card: latitude and longitude, styles of buildings in a city, weather of an area, nations in a common market, monuments and landmarks, people changing the environment, people adapting resources, environment affecting people, transportation and communication, absolute location, relative location, vegetation of an area, the Sunbelt states, The Great Plains, spatial interaction, uniform regions, pollution, nodal regions, exports and imports. Have students separate the phrase cards into two stacks, one for physical geography, the other for human geography.
3. List on the chalkboard the careers that were presented on page 36 of their textbook. Provide sections of newspaper classified ads, and ask students to identify careers that would utilize geography skills.
4. Invite a guest speaker in a geography-related field to visit the class. (Examples: cartographer, urban planner, geography professor) Ask the speaker to explain to students how his or her job relates to geography.

CLOSURE
Review the five main themes of geography, and call upon students to give examples of each.

ASSESSMENT
Have students complete the following questions from the Chapter 1 Check:
Building a Vocabulary 2, 3, 4
Recalling and Reviewing 1, 2, 3
Using Geography Skills 2

RETEACHING PLAN
1. Provide individual students with a set of cards used in Teaching Design 3, and have them arrange the cards according to the five main themes of geography.
2. Ask students to complete the Chapter 1 Reteaching Worksheet, Lesson 2.

CHALLENGE/ENRICHMENT STRATEGY
Have students use the Atlas in their textbook to identify numerous types of regions. They may want to classify the various types as uniform or nodal regions.

▶ LESSON 3
Latitude and Longitude

LESSON OBJECTIVES
The student should be able to
- define *absolute location*
- use latitude and longitude to determine the absolute location of various sites

PURPOSE
The study of latitude and longitude will acquaint students with the international method for locating places on the earth's surface.

MOTIVATOR
Lead a discussion identifying possible reasons for the development of the grid system of latitude and longitude. Give examples of how the system is used today.

TEACHING DESIGN
1. Have students reread the section "Lines on the Globe" on pages 22–23 in their textbook.
2. Present the Lecture Notes and Transparency Package, Part IV.
3. Using an appropriate wall map of the world, ask volunteers to locate sites using the following coordinates.

Coordinates	City
42° N 73° W	(New York, New York)
44° N 89° W	(Chicago, Illinois)
35° S 60° W	(Buenos Aires, Argentina)
14° N 16° W	(Dakar, Senegal)
7° N 60° W	(Georgetown, Guyana)

4. Call out coordinates, and direct students to name the corresponding site, using a map in their textbook.

CLOSURE
Ask students to summarize the important applications of latitude and longitude in today's world.

ASSESSMENT
1. Have students complete the following question from the Chapter 1 Check:
   Using Geography Skills 3
2. Have students complete the Chapter 1 Skills Worksheet.
3. Ask students to complete the Chapter 1 Review/Test.

RETEACHING PLAN
1. Using the Skills Worksheet, ask the students to verbalize the process of determining absolute location with given coordinates.
2. Ask students to complete the Chapter 1 Reteaching Worksheet, Lesson 3.

CHALLENGE/ENRICHMENT STRATEGY
1. Have students investigate the history of the development of the latitude and longitude system.

T21

Level A: Have students research a world-famous volcanic eruption. Ask them to write a "news report" of the event.

The Interior of the Earth

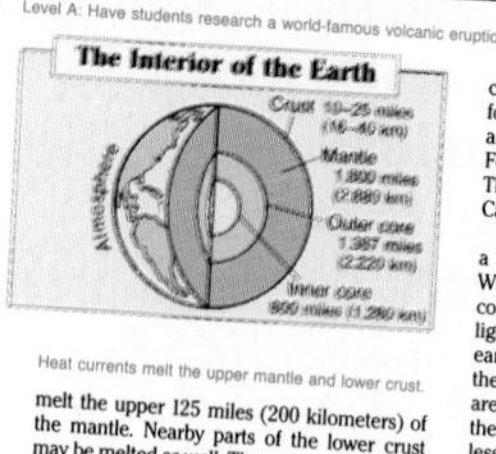

Heat currents melt the upper mantle and lower crust.

melt the upper 125 miles (200 kilometers) of the mantle. Nearby parts of the lower crust may be melted as well. The melted, liquid rock from within the earth that is brought to the earth's surface is called **lava**.

The crust above the rising heat currents usually lifts upward. Lava may break through and flow over the surface, filling valleys and building the land upward. A great mountain of lava, called a **volcano**, might be built. On the other hand, the crust above cool parts of the mantle usually sinks, creating broad depressions, or low spots, on the earth's surface.

Forces working within the earth not only create lava flows and volcanoes but also bend and break rock. When rock layers are bent, the result is a **fold**. When rock layers break and move apart, the result is a **fault**. The shock waves caused by the moving rock above a fault can create an **earthquake**. Whether rocks fold or fault depends on how hard the rocks are and how sudden and strong the forces are. Folding and faulting also cause mountains. The Rocky Mountains of the United States and Canada are results of these forces.

The weight and hardness of rocks also play a role in moving parts of the earth's crust. When melted by heat currents from the earth's core, heavier rocks tend to sink beneath the lighter rocks. Over the long history of the earth, the lightest rocks have risen to the top of the crust. They now form the continents. In an area of a continent undergoing rapid erosion, the weight of surface rock becomes less and less. As a result, this area will slowly rise. Similarly, areas being continually buried by sediments tend to sink as the load becomes heavier. Scientists think these forces that change the surface of the earth have been at work for millions of years.

Large Landforms and Plate Tectonics

Plates The shapes on the earth's surface are called **landforms**. Examples of landforms include hills, valleys, and nearly flat areas, called **plains**. Landforms vary in size from the smallest indentation made by a raindrop to the largest mountain range.

Folds and Faults

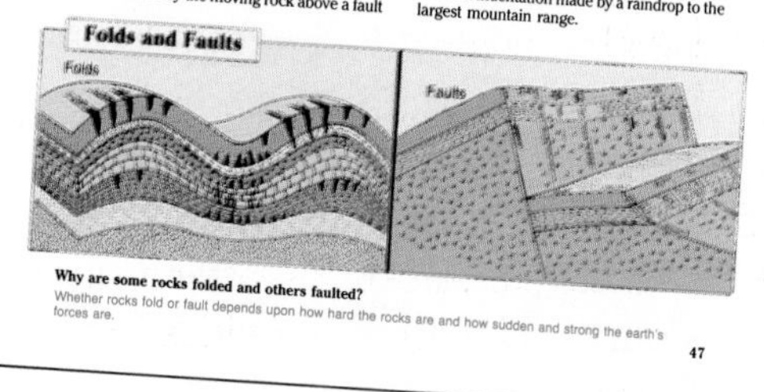

Why are some rocks folded and others faulted?
Whether rocks fold or fault depends upon how hard the rocks are and how sudden and strong the earth's forces are.

47

On-page annotations provide instantly useful teaching supports:

- additional questions to clarify and emphasize major concepts
- extension activities labeled according to student ability
- answers to caption questions

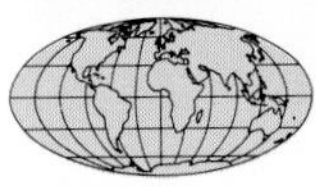

ANCILLARIES HELP ENSURE LEARNING SUCCESS.

TEACHER'S RESOURCE BINDER

The *Teacher's Resource Binder* includes materials to motivate and guide students for mastery learning.

1. The *Lecture Notes* present chapter content and historical background in outline form,and the transparencies correlate to each chapter.
2. The four-color *Transparencies,* at least one per chapter, provide maps, charts, and tables suitable for classroom discussion.
3. The *Workbook, Teacher's Edition* contains the reproduction of the Pupil's Edition with overprinted answers.
4. The *Test Book* provides Chapter and Unit Tests in a standardized-test format for easy grading. A complete Answer Key is supplied in the back.
5. The *Worksheets,* in blackline master format, provide four additional opportunities for practice and extension:
    - Skills
    - Reteaching
    - Critical Thinking
    - Challenge/Enrichment

WORKBOOK, PUPIL'S EDITION

This ancillary provides perforated worksheets that emphasize vocabulary as it relates to chapter content. The *Workbook* makes effective use of a variety of formats: puzzles, fill-in-the-blank, and matching.

COMPUTER TEST GENERATOR

This flexible, computerized program consists of a *User's Manual,* a *Test Generator Booklet* that provides a printout of each test, and a disk on which the original test questions are stored. Questions follow both an objective and short-answer format and are leveled according to student ability. Questions may be added or deleted, and items may be randomly reordered to create different test versions.

The *Computer Test Generator* is available for Apple® IIe/ IIc/ IIGS (with IIe configuration).

Apple® is a registered trademark of Apple Computer, Inc.

USING THE ANNOTATED TEACHER'S EDITION

The Annotated Teacher's Edition of *World Geography Today* contains three major features: a Teacher's Guide bound into the front of the student text, annotations printed in blue on the student pages, and a Bibliography. Each is designed to assist the teacher in planning and carrying out daily instruction in world geography.

ORGANIZATION OF THE TEACHER'S GUIDE

Bound into the front of the Annotated Teacher's Edition is a 192-page Teacher's Guide. The guide is organized by units and chapters to accompany the pupil text. Each chapter is divided into daily lesson plans that follow a consistent format. The organization of each unit of the Teacher's Guide is as follows:

UNITS

Each unit begins with the title and the pages of the Pupil's Edition covered in that unit, followed by an overview of the unit in paragraph form. The Unit Overview provides a brief summary of the unit's content.

The Unit Overview is followed by a Chapter Time Line. The time line lists the number and title of each chapter in the unit and a suggested number of days required to teach each one. A total number of days is provided for the entire unit. Please note that the time line is a suggestion only. You may wish to adjust time allocation to fit your curriculum and the needs of your students.

Following the time line is a list of Major Topics covered in the unit. These focus on the most important points of the unit.

This portion of the Teacher's Guide ends with Introducing the Unit, a suggested strategy or activity to use with students. The purpose of this section is to focus attention on the content and to help prepare students for new instruction.

CHAPTERS/CHAPTER PLANNING GUIDES

Following the general material covering the unit are specific materials designed to assist with teaching each of the chapters in that unit. These begin with the chapter title and number, followed by the relevant pages in the Pupil's Edition. Each of these divisions is followed by a Chapter Planning Guide. The Planning Guide is designed to help in the following ways:

- It provides a list of suggested lessons for that chapter.
- It lists the pupil pages covered by each lesson.
- It provides a key to supplementary materials available in the Teacher's Resource Binder for each lesson.

As with the unit materials, a suggested Time Allocation for the chapter is provided. This allocation corresponds to the number of lessons suggested for each chapter, with each lesson designed for one day of instruction. Again, you may want to adjust Time Allocation to fit your curriculum and the special needs of your students.

The Time Allocation is followed by a list of Chapter Topics. These topics address both content and skills that the students will encounter in each chapter of the Pupil's Edition.

Finally, the introductory material for each chapter ends with a section called Preteaching Chapter Vocabulary. This section includes definitions of the vocabulary terms introduced in the chapter and suggestions for acquainting students with the terms so that they will understand them when they encounter them in their reading. Note that these vocabulary terms appear in bold, blue type in the Pupil's Edition.

DAILY LESSON PLANS

The greatest part of the Teacher's Guide for *World Geography Today* is devoted to Daily Lesson Plans. These plans correspond to the Time Allocation and to the Chapter Planning Guide. Each lesson is clearly numbered and follows a clear, consistent format designed to provide maximum assistance to the teacher. The format of the Daily Lesson Plans is as follows:

Lesson Objectives	Closure
Purpose	Assessment
Motivator	Reteaching Plan
Teaching Design	Challenge/Enrichment Strategy

The Lesson Objectives expand and refine the objectives listed for the chapter as a whole. They provide a handy reference around which to plan instruction on a daily basis. These objectives can be used to clarify for students what they are expected to accomplish in the study of each chapter, thus helping them set goals for achievement.

Following the Lesson Objectives is a statement of Purpose. Because students learn best when they know what to look for and why it is important, the statement of Purpose is designed to give the students reasons they will be asked to study the material in each lesson.

In conjunction with the Purpose, the Motivator contains a suggested activity for getting started with the content for each lesson. These activities range from informal discussions to individual activities, but all are designed to arouse student interest in the material they are about to explore.

The main body of the lesson plans is contained in the section called Teaching Design. This section encompasses from two to six suggestions or strategies for teaching the content of the lesson. These suggestions are structured around the objectives established for the lesson, and care has been taken to include a variety of activities suitable for whole-group instruction, small-group instruction, and individual instruction. Additionally, the activities and strategies have been written bearing in mind the fact that students learn through different styles. Writing activities, reading and study exercises, drill, note-taking, problem-solving, and geography skills practice are all included.

The Closure section of each lesson plan contains suggestions for wrapping up the day's activities. This section is intended to pull together the content and to summarize before proceeding to the next lesson.

The next section of the Daily Lesson Plan provides for assessment of student progress. This section may include suggestions for informal assessment by the teacher on a whole-group basis as well as a key noting where the lesson content is reviewed in the Chapter Check of the Pupil's Edition. It also may guide the teacher to supplementary materials in the Teacher's Resource Binder.

The section for Assessment also notes when to provide the students with the test for each chapter. *World Geography Today* is organized into 55 short chapters and 10 units; therefore, you may want to use the chapter tests as additional review, rather than as formal tests. For this reason, each chapter test is called Chapter Review/Test. As there are 10 units, you may use the Unit Tests to serve as the formal tests for your course.

The Daily Lesson Plans in the *World Geography Today* Teacher's Guide are structured for mastery learning. A key component in this scheme is the Reteaching Plan. The Reteaching Plan, like the rest of the Teacher's Guide, follows a consistent format. It contains ideas for presenting the basic lesson concepts and content in a new way, built around the lesson objectives. It also notes the availability of Reteaching Worksheets in the Teacher's Resource Binder.

Finally, each lesson plan ends with Challenge/Enrichment Strategy, which includes several ideas for taking students beyond the content presented in the Pupil's Edition. The ideas were written to provide variety in order to allow you the best chance of presenting material suitable for all students.

ANSWERS TO CHAPTER CHECK AND UNIT REVIEW

The units and chapters in the Teacher's Guide follow the same organization as the Pupil's Edition. Because each chapter ends with a Chapter Check and each unit ends with a Unit Review, answers to these questions and activities are found in the same order. At the end of the last lesson plan in each chapter, you will find the Answers to the Chapter Check. Answers to Geography Skills questions are likewise found in their proper order. Complete answers are provided for every question, and suggested answers are provided for open-ended questions or exercises in which various students might arrive at different conclusions.

At the end of every chapter of the Teacher's Guide are the Unit Review/Test and the Answers. Like the introduction to each chapter, the Unit Review/Test provides a Planning Guide and a Time Allocation for teaching the Unit Review in the Pupil's Edition and is keyed to the Unit Test in the Teacher's Resource Binder. In the Answers to the Unit Review, answers are offered for most open-ended questions or exercises.

USING THE TEACHER'S EDITION ANNOTATIONS

Blue annotations are overprinted on the pupil pages for ease of use. These annotations appear on nearly every page, and they serve a number of useful functions.

1. Many of the annotations tie together the various parts of the text for the teacher. These annotations note the location of maps and other illustrations throughout the text. They also indicate where a topic under study has been encountered previously or where in the text it will be studied again. Such annotations allow reviewing, comparing, and contrasting material from one chapter to another.

2. A second group of annotations is designed to provide questions to ask the students as a way of providing review and reinforcement. Many of these are discussion ideas centered around concepts the students will encounter when they are reading the text.

3. A third set of annotations is targeted at the maps, photographs, and other illustrations in the pupil text. In every regional chapter, for example, an annotation notes on the first page where the map for that chapter can be found. There are additional annotations above or below the maps, referring to other maps that may be useful to enhance study of the topic or region. For example, there are annotations with every map of a subregion referring to the larger regional map of which that area is a part; thus, you can easily direct students to locate Great Britain and Ireland on the larger regional map of Western Europe. This cross reference allows students to see the subregion in the proper perspective: as a segment of a larger part of the world.

4. Next, there are annotations labeled "Level A" or "Level B" denoting their suitability for students of various learning styles and abilities. Those activities labeled Level A are suitable for students with the desire and ability to work on their own. Many of these activities call for independent or small-group research. Those activities labeled Level B are more suitable for students of average or lower ability. Most of these activities are short in duration and tied closely to the content of Geography Skills features in the text.

5. Finally, there is a group of annotations that alert you to resources available in the Teacher's Guide or in the Teacher's Resource Binder. These are found on the opening page of every chapter and on every Chapter Check and Unit Review

The lesson plans in the Teacher's Guide and the annotations are designed to complement each other. Together they contain a wealth of useful and practical information for the teacher. If, however, you should desire less structure than that provided in the Daily Lesson Plans, the Annotated Teacher's Edition can easily be adapted for your own special needs. You may wish to utilize only those suggestions in the lesson plans that fit the goals and objectives you have set for each chapter or the course as a whole. An alternative strategy might be to rely more heavily on the annotations for teaching suggestions and activities. The annotations alone provide numerous ideas for review, enrichment, practice, and assessment, with the added advantage of being leveled for students of varying abilities.

USING THE BIBLIOGRAPHY

The third major feature of the Annotated Teacher's Edition is the Bibliography. The Bibliography includes audiovisual materials as well as books. It is organized by unit for easy reference. Films, filmstrips, and recordings are offered on a variety of countries, regions, and topics.

Please note that the Bibliography for students, "Reading for Enrichment," is found in the Unit Review after each of the 10 units of the Pupil's Edition.

ADJUSTING THE SCOPE AND SEQUENCE

The Daily Lesson Plans in the Teacher's Guide have been written to provide a full year of instruction in the same sequence in which the Pupil's Edition is written; however, the scope and sequence can be adjusted in a number of ways to fit your curriculum needs. You could, for example, follow Unit 1 with a study of the United States and Canada if you wanted greater emphasis on that region of the world.

You also could easily adjust the scope and sequence to accommodate courses of less than one year in duration. This can be done by surveying selected regions of the world instead of studying all of them in depth. For example, after teaching Unit 1, you could assign only the first chapter of each unit to be surveyed, as the first chapter of each regional unit is an overview of the entire region. You could assign Chapter 10 if you wanted your students to survey Western Europe without studying individual countries or subregions. A course emphasizing Africa and Asia might look like this:

Unit 1	The Earth and Its People: all chapters
Unit 2	Western Europe: Chapter 10 only
Unit 3	The Soviet Union and Eastern Europe: Chapter 17 and Chapter 20
Unit 4	The Middle East and North Africa: all chapters
Unit 5	Sub-Saharan Africa: all chapters
Unit 6	The Orient: all chapters
Unit 7	The Pacific World: all chapters
Unit 8	Latin America: Chapter 37 only
Unit 9	The United States and Canada: Chapter 44 only
Unit 10	Sharing the World's Resources: optional

A course could likewise be designed in which students sur the region as a whole and then study only selected countries or subregions within the region. In such a case, students could be assigned Chapter 37 for an overview of Latin America followed by Chapter 38 — Mexico — and Chapter 42 — Brazil. Given the 10 units and 55 chapters in *World Geography Today*, the scope and sequence can be modified to fit many different situations.

BASIC MAP AND GLOBE SKILLS
Pages 22–29

PLANNING GUIDE		
Lesson	**Textbook**	**Teacher's Resource Binder**
1. Introducing the Globe	pp. 22–25	Content/Vocabulary Workbook Reteaching Worksheet
2. Reading Maps	pp. 25–29	Lesson Review/Test Reteaching Worksheet

TIME ALLOCATION: 2 days

LESSON TOPICS
- The Globe
- The Hemispheres
- Reading Maps
- Map Projections
- Great-Circle Routes
- Scales and Directional Indicators
- Using Map Skills
- Remote Sensors and Other Technology

PRETEACHING LESSON VOCABULARY
Write the following terms on the chalkboard:

globe	scales	directional indicator
continents	parallels	great-circle routes
oceans	hemispheres	map projections
latitude	meridians	prime meridian
longitude	equator	remote sensors
islands	degrees	compass rose
map	atlas	legend
Landsat	grid	

Conduct a quick preteaching survey with the class. Name each vocabulary word, and ask students to indicate by a show of hands the words that they feel certain they can define. Then place the students in groups, and ask them to separate the terms into the following categories: Words We Can Define and Words We Need to Learn More About. Use this information as a guide when selecting activities for each class.

▶ LESSON 1
Introducing the Globe

LESSON OBJECTIVES
The student should be able to
- define significant vocabulary terms
- describe geographic tools and methodologies
- observe for detail on globes
- locate and describe major landforms and features of the earth

PURPOSE
In this lesson, students will be introduced to the most basic concepts of physical geography. An understanding of the globe is a prerequisite to the study of physical geography and is essential to using maps successfully.

MOTIVATOR
Display a globe for the class to see. Indicate various points on the globe while asking the following questions. Have students write their answers on a sheet of paper.

1. What is the sphere called? (globe)
2. What are these landmasses called? (continents)
3. What percent of the earth is covered by water? (71)
4. What are the largest bodies of water called? (oceans)
5. What are the names of this landmass? (Europe and Asia)
6. Which ocean is this? (Atlantic)
7. Which ocean is this? (Pacific)
8. Which ocean is this? (Indian)
9. What is this line called? (prime meridian)
10. What is this line called? (equator)

Ask students to correct their own papers. You may wish to collect them to assess students' knowledge of globes.

TEACHING DESIGN
1. Divide the class into groups, and provide each group with a globe. Ask students to study the globe and list information on the globe that they would like to understand. After 10 minutes of group discussion, invite students to share their questions, and discuss the answers with the class as a whole.
2. Ask students to imagine how the early explorers found their way to new and exciting destinations. Review with students the advantages of using latitude and longitude to locate places.
3. Draw on the chalkboard a grid to represent your classroom. Have students indicate the coordinates at which to draw in key furniture, including desks. Provide students with a series of questions to answer using the completed classroom grid.
   Examples:
   a. At what intersection is the teacher's desk located?
   b. At what intersection is Mary seated?
   c. At what intersection is Bill seated?
   d. At what intersection is the classroom door?
   Discuss the similarities between the classroom grid and a map that utilizes latitude and longitude.
4. Review with students the location of latitude and longitude lines on the globe and the significance of the equator and the prime meridian. Indicate to students that they will have further opportunities for practice using latitude and longitude in Chapter 1, Lesson 3.
5. Ask students for the definition of the term *hemisphere.* (half a ball) Use a globe to point out the Northern, Southern, Eastern, and Western hemispheres. Ask students to list the continents and oceans that are included in each hemisphere.

CLOSURE

Invite several students to come to the front of the class, hold a globe, and identify as many features as they can in one minute. You may wish to hold a contest to determine who is able to recall the most places.

ASSESSMENT

Ask students to complete the Basic Map and Globe Skills Content/Vocabulary Worksheet in their Workbook.

RETEACHING PLAN

1. Ask students to complete the Basic Map and Globe Skills Reteaching Worksheet, Lesson 1.
2. Coordinate a map relay with several students to review the locations of the continents and oceans. Place students in rows, and provide each row with the list of continents and oceans. Send a world map down the row, and ask each student to locate and cross off one of the items on the list. Students should continue to pass the map until all places have been located. The first row to identify all places correctly is declared the winner.
3. Invite students to participate in a geography scavenger hunt. Encourage students to collect several newspapers and locate key terms related to maps and globes. Students should circle the words in red. Ask students to complete the following information for all the words they discover.

Word	Newspaper Article Title	Context for Use of Word

CHALLENGE/ENRICHMENT STRATEGY

1. Direct students to create a map that identifies the sources of the food they ate for lunch yesterday.
2. Place a world map in the center of a bulletin board. Invite students to bring in news articles about various countries in the world and place them beside the map, with yarn leading from the articles to the appropriate countries. At specified times, encourage students to summarize the articles.
3. Challenge students to discover why Australia is considered to be a continent rather than an island.
4. Provide students with the opportunity to investigate the types of navigation tools used by early sailors.
5. Select students to interview an airline pilot about techniques utilized for navigation in the skies, then report the interview to the class.

▶ LESSON 2
Reading Maps

LESSON OBJECTIVES

The student should be able to

- describe tools and methodologies used by geographers to study the earth
- define significant geographic terminology
- interpret information located on a map
- translate information from one medium to another

PURPOSE

In this lesson, students are asked to refresh their map skills. For many students, this will provide an important review of content they have studied in the past. For other students, it will serve as a positive reminder of important knowledge they need for success in this course. Remind students that learning to interpret a map is a skill they will use throughout the year and throughout their lives.

MOTIVATOR

Refer students to several maps in their textbook. Ask them to look for at least five similarities among all maps. Challenge them to identify more. Following five minutes of searching, write the similarities on the chalkboard. (The list should include scales, directions, legends, colors, symbols, and so on.) Indicate to students that the lesson focus will be on the skills they need to interpret this information.

TEACHING DESIGN

1. Bring several atlases to class, and give them to groups around the room. Ask each group to select a map to describe for the rest of the class. Encourage students to cover all the factors they listed as similarities in the Motivator. Check for accuracy as students locate and describe legends, scales, and directional indicators.
2. Refer students to the map "Great Britain and Ireland" on page 124 of their textbook. Ask them to practice using a legend to identify the following.
   a. The capitals
   b. Three places where fishing occurs
   c. Ways of making a living in Ireland
   d. The three countries that have industrial centers
   e. The symbol for iron
3. Refer students to the map "The South" on page 551 of their textbook. Ask students to practice using the scale to determine the following.
   a. Distance between El Paso, Texas, and New Orleans, Louisiana
   b. Distance between Dallas and Houston, Texas
   c. Distance between Little Rock, Arkansas, and Jackson, Mississippi
   d. City nearest the halfway point between Austin, Texas, and Oklahoma City, Oklahoma
4. Display for students the three types of map projections described in their textbook. As you examine the projections, create a list of the advantages and disadvantages of each type of projection.
5. Review the importance of remote sensors and Landsat photography. Challenge students to look through their textbook to locate the photographs that were made with this form of technology.

CLOSURE

Display a map from an atlas. Invite several students to stand before the class to identify all the key features reviewed in the lesson. Determine whether the students are able to identify the correct type of projection. Ask them to indicate the clues they used to make this identification. Invite several other students to do the same with other maps.

ASSESSMENT

Ask students to complete the Basic Map and Globe Skills Lesson Review/Test.

RETEACHING PLAN

1. Ask students to create a map of their neighborhood. The map should include a legend and a directional indicator. Ask students to identify all the directions they travel on their way to school.
2. Ask students to complete the Basic Map and Globe Skills Reteaching Worksheet, Lesson 2.
3. Instruct students to continue practicing determining the difference in distance between two sites as measured on a flat map and on a globe.
4. Provide students with a world atlas that contains many types of projections. Encourage students to look through the atlas for examples of the three types of map projections presented in their textbook.

CHALLENGE/ENRICHMENT STRATEGY

1. Encourage students to look through various news sources for maps. Students should bring the maps to school and share with the class how the maps are used. A bulletin board could be created to display the various maps used by the news media.
2. Provide students with a United States road map. Tell them to assume that they are planning a trip from New York City to Los Angeles. Students should list all the roads they will follow. They also should identify where they will stop each evening and the distance they will have traveled. Indicate to students that they will camp out on this trip, so evening stops must be planned for campgrounds.
3. Invite a local meteorologist to visit the class and discuss the use of satellites and remote sensors in weather forecasting. If this person is unavailable to visit the class, encourage students to interview the weather specialist and present to the class the results of the interview.
4. Invite a city official to visit the class and speak about how maps are used by the various departments of the city in their daily work.

UNIT 1 The Earth and Its People Pages 30–105

UNIT OVERVIEW

This unit provides the background necessary to help students understand the relationship between cultural geography and physical geography. Significant physical geography concepts are taught to ensure that students are able to apply them to the study of cultural regions.

CHAPTER TIME LINE

Chapter	Title	Time Allocation
1	The Nature of Geography	3 days
2	The Earth in Space	4 days
3	The Earth's Surface	4 days
4	The Earth's Atmosphere	3 days
5	Global Climates	2 days
6	The Water Planet	3 days
7	World Patterns of Vegetation and Soils	3 days
8	World Cultures	3 days
9	Economic Development	4 days
	Unit 1 Review	1 day
	Unit 1 Test	1 day
		31 days

MAJOR TOPICS

- Foundation for the Study of Geography
- Distinction Between Physical and Cultural Geography
- Core Themes in the Study of Geography
- Understanding the Concept of Culture
- Interrelationship of Land, Water, and Atmosphere

INTRODUCING THE UNIT

Provide all students with newspapers or magazines. Ask them each to locate and cut out one article related to the study of geography. Discuss their selections and, whenever possible, establish the link between cultural and physical geography. Invite students to hypothesize about the impact of geography on our lives.

CHAPTER 1 The Nature of Geography Pages 32–37

PLANNING GUIDE		
Lesson	**Textbook**	**Teacher's Resource Binder**
1. Definition of Geography	p. 33	Content/Vocabulary Workbook Reteaching Worksheet Critical Thinking Worksheet
2. Themes of Geography and Related Careers	pp. 33–36	Lecture Notes and Transparency Package Reteaching Worksheet
3. Latitude and Longitude		Lecture Notes and Transparency Package Skills Worksheet Chapter Review/Test Reteaching Worksheet Challenge Worksheet

TIME ALLOCATION: 3 days

CHAPTER TOPICS

- Definition of Geography
- Fundamental Themes of Geography
  - Location
  - Place
  - Relationships Within Places
  - Movement Across the Earth's Surface
  - Regions
- Careers in Geography

PRETEACHING CHAPTER VOCABULARY

Write the following terms on the chalkboard:

human geography	relative location
physical geography	spatial interaction
region	uniform region
regional geography	nodal region
absolute location	cartography

Ask students how many terms or parts of terms they already know. Discuss their ideas. Point out that the term *geography* comes from the Greek language, with the prefix *geo-*, meaning "earth," and *-graphy,* meaning "writing." The term means "the study of or writing about the earth." Ask students to identify the other term that contains *-graphy*. Point out that *carte* in French means "map." Distinguish between *cartography* and *cartographer*. Ask students to identify the key words in understanding absolute and relative location. Have a volunteer read the dictionary definitions of *absolute* and *relative* to determine the meanings of the two terms in the chapter. Point out that the prefix *uni-* means "consisting of only one." Note that a *node* is a knob or swelling and that *nodal* is the adjective form. Relate these meanings to the context of the terms in the chapter.

▶ LESSON 1

Chapter Vocabulary and Definition of Geography

LESSON OBJECTIVES

The student should be able to

- define *geography*
- distinguish between physical and human geography
- define the significant terms of the lesson

PURPOSE

The study of geography is important because students are linked to all the world's peoples. Students share resources, products, and ideas from around the world.

MOTIVATOR

Ask students what they expect to learn in their geography course. Discuss why they think the study of geography will be important in their lives. (learn how the world's peoples are linked, how events far away influence us)

TEACHING DESIGN

1. Through class discussion, create a word web on the chalkboard related to the term *geography*. Use questioning to identify broad categories for classifying the terms, leading students to an awareness of the major kinds of geography: human and physical.
2. Have students work with a partner to classify the vocabulary according to human and physical geography.

CLOSURE

Review the purpose of the lesson. Call on students to define *geography* in their own words and distinguish between physical and human geography.

ASSESSMENT

1. Have students complete the following questions from the Chapter 1 Check:
   Building a Vocabulary 1
   Using Geography Skills 1
2. Ask students to complete the Chapter 1 Content/Vocabulary Worksheet in their workbook.

RETEACHING PLAN

1. Ask students to complete the Chapter 1 Reteaching Worksheet, Lesson 1
2. Provide a series of pictures that represent human or physical geography. Have students categorize them and explain the rationale for their categorization.

CHALLENGE/ENRICHMENT STRATEGY

1. Have students locate newspaper articles related to issues of geography. Ask each student to prepare a statement summarizing the relationship of geography to a current news event.
2. Have students complete the Chapter 1 Critical Thinking Worksheet.

▶ LESSON 2

Themes of Geography and Related Careers

LESSON OBJECTIVES

The student should be able to

- list and explain the five themes of geography
- explain criteria for determining regions
- cite an example of how people are affected by the environment and vice versa
- describe three professions related to geography

PURPOSE

The study of the five main themes of geography provides a foundation for understanding and interpreting the information presented throughout the course. Information about careers in geography demonstrates the application of geography to daily life.

MOTIVATOR

Brainstorm with students about the various professions they believe may rely on geography skills. List the professions on the chalkboard. Add to the list as new careers are identified. Examples: environmental management, locating natural resources (oil, coal, water), scheduling routes for airlines and trucks, farming.

TEACHING DESIGN

1. Present the Chapter 1 Lecture Notes and Transparency Package through Part III.
2. Divide the class into groups of four or five students. Provide each group with a set of index cards that includes the following phrases, one phrase per card: latitude and longitude, styles of buildings in a city, weather of an area, nations in a common market, monuments and landmarks, people changing the environment, people adapting resources, environment affecting people, transportation and communication, absolute location, relative location, vegetation of an area, the Sunbelt states, The Great Plains, spatial interaction, uniform regions, pollution, nodal regions, exports and imports. Have students separate the phrase cards into two stacks, one for physical geography, the other for human geography.
3. List on the chalkboard the careers that were presented on page 36 of their textbook. Provide sections of newspaper classified ads, and ask students to identify careers that would utilize geography skills.
4. Invite a guest speaker in a geography-related field to visit the class. (Examples: cartographer, urban planner, geography professor) Ask the speaker to explain to students how his or her job relates to geography.

CLOSURE

Review the five main themes of geography, and call upon students to give examples of each.

ASSESSMENT

Have students complete the following questions from the Chapter 1 Check:

Building a Vocabulary 2, 3, 4
Recalling and Reviewing 1, 2, 3
Using Geography Skills 2

RETEACHING PLAN

1. Provide individual students with a set of cards used in Teaching Design 3, and have them arrange the cards according to the five main themes of geography.
2. Ask students to complete the Chapter 1 Reteaching Worksheet, Lesson 2.

CHALLENGE/ENRICHMENT STRATEGY

Have students use the Atlas in their textbook to identify numerous types of regions. They may want to classify the various types as uniform or nodal regions.

▶ LESSON 3
Latitude and Longitude

LESSON OBJECTIVES

The student should be able to

- define *absolute location*
- use latitude and longitude to determine the absolute location of various sites

PURPOSE

The study of latitude and longitude will acquaint students with the international method for locating places on the earth's surface.

MOTIVATOR

Lead a discussion identifying possible reasons for the development of the grid system of latitude and longitude. Give examples of how the system is used today.

TEACHING DESIGN

1. Have students reread the section "Lines on the Globe" on pages 22–23 in their textbook.
2. Present the Lecture Notes and Transparency Package, Part IV.
3. Using an appropriate wall map of the world, ask volunteers to locate sites using the following coordinates.

Coordinates	City
42° N 73° W	(New York, New York)
44° N 89° W	(Chicago, Illinois)
35° S 60° W	(Buenos Aires, Argentina)
14° N 16° W	(Dakar, Senegal)
7° N 60° W	(Georgetown, Guyana)

4. Call out coordinates, and direct students to name the corresponding site, using a map in their textbook.

CLOSURE

Ask students to summarize the important applications of latitude and longitude in today's world.

ASSESSMENT

1. Have students complete the following question from the Chapter 1 Check:
   Using Geography Skills 3
2. Have students complete the Chapter 1 Skills Worksheet.
3. Ask students to complete the Chapter 1 Review/Test.

RETEACHING PLAN

1. Using the Skills Worksheet, ask the students to verbalize the process of determining absolute location with given coordinates.
2. Ask students to complete the Chapter 1 Reteaching Worksheet, Lesson 3.

CHALLENGE/ENRICHMENT STRATEGY

1. Have students investigate the history of the development of the latitude and longitude system.

2. Ask students to investigate types of grid systems that are used in various professions or fields of study.
3. Have students complete the Chapter 1 Challenge Worksheet.

Chapter 1 Check, page 37, Answers

Building a Vocabulary

1. Physical geography is the study of the earth's natural features and how they vary. It includes the study of landforms, climate, water, and vegetation. Human geography is the study of the activities of people and their interaction with the environment. It includes economic and political geography, the geography of cities and farms, and the geography of religions and languages.
2. Answers will vary but might include an example of trade, travel, or manufacturing. Spatial interaction is the movement—of individuals, goods, ideas, and information—to a final destination or in interaction with others. It is important to understand the concept of movement in order to understand geographic change.
3. Absolute location is the exact site on the earth's surface where something is found. Relative location is the relationship of one place to another, either in terms of physical distance or in the form in which transportation and communication between two places is established.
4. A uniform region has one or more common features, such as language or government. A nodal region is defined by the spatial interaction around a place—the purposeful movement of people, ideas, and things. France is a uniform region because its people share a common language and government.

Recalling and Reviewing

1. The five basic themes of geography provide us with the tools of geographic inquiry. They enable us to ask questions about places on the earth and their relationships to the people who inhabit them. It is important to know what questions to ask to gain the appropriate foundation for good citizenship. The more we know, the more informed decisions and predictions we can make about our world.
2. Regions are manageable units of geographic study. By narrowing the scope of a region to its physical, cultural, economic, and political characteristics, it is possible for us to make better-informed judgments. Students should name uniform and nodal regions in their state.

Critical Thinking

3. Geographers use physical characteristics, such as climate and landforms, and human characteristics, such as ideas and activities (political, economic), to identify places. In describing their hometowns, students can consider these factors and also discover how the "place" has changed over a period of time.

Using Geography Skills

1. The maps that show elevation and landforms are concerned with physical geography; they show the shape of the land and the patterns of waterways. The political maps are concerned with human geography; they show political boundaries, countries, and cities. Some maps combine physical and political information.
2. The areas that might be more populated would be around the Great Lakes, the Mississippi and other major rivers, and the Pacific, Atlantic, and Gulf coastal regions. Communities flourish near moist and irrigated areas, which provide good soil for agriculture. Coastal regions also prosper because of trade.
3. On the political map, each country is a separate region. The physical map shows regions that are separated by landforms. Some of these are the British Isles, the Greek islands, the Alps, and the Iberian Peninsula.

CHAPTER 2 The Earth in Space Pages 38–43

PLANNING GUIDE

Lesson	Textbook	Teacher's Resource Binder
1. Universe, Galaxies, and Solar System	pp. 38–40	Content/Vocabulary Workbook Reteaching Worksheet Critical Thinking Worksheet
2. Earth's Rotation, Revolution, and Tilt	p. 40	Reteaching Worksheet Challenge Worksheet
3. Seasons and Their Effect on Us	pp. 41–42	Skills Worksheet Reteaching Worksheet
4. Seasons Throughout the World	pp. 41–42	Lecture Notes and Transparency Package Chapter Review/Test Reteaching Worksheet

TIME ALLOCATION: 4 days

CHAPTER TOPICS

- The Universe and Galaxies
- The Earth and the Solar System
- Earth's Rotation, Revolution, and Tilt
- Seasons, Solstice, and Equinox

PRETEACHING CHAPTER VOCABULARY

Write the following terms on the chalkboard:

universe	polar regions	solstice
galaxy	rotation	Tropic of Capricorn
solar system	revolution	Arctic Circle
planets	equinox	Antarctic Circle
tropics	seasons	Tropic of Cancer

Ask the students to complete a prediction chart as follows with the first 11 terms.

Terms	Predicted Meaning Before Reading	Actual Meaning After Reading
1. universe		

Take time to identify and locate the final four terms on a globe. Write the corresponding latitude on the chalkboard. Challenge students to search for the significance of these terms as they read the chapter.

▶ LESSON 1
Universe, Galaxies, and Solar System

LESSON OBJECTIVES
The student should be able to
- define significant chapter vocabulary
- distinguish among universe, galaxy, and solar system
- compare the size of the sun to the size of the earth

PURPOSE
The study of the nature of the universe allows students to develop a sense of perspective regarding the earth.

MOTIVATOR
Ask students to think of products (such as food, automobiles, and electronic equipment) named after objects in the solar system and in space. List their responses on the chalkboard. Ask them to hypothesize about the manufacturers' reasons for choosing such names for their products.

TEACHING DESIGN
1. Have students work in groups to create a diagram to represent the relationships among universe, galaxy, solar system, and Earth. Ask a spokesperson to present the diagram and explanation to the class. Summarize the discussion.
2. Lead a discussion asking students to predict how their daily lives might change if they moved to the North Pole or the equator.

CLOSURE
1. Summarize the differences among the universe, galaxy, solar system, sun, and Earth.
2. Ask students to recall the hypotheses they made in the Motivator and determine whether or not the hypotheses were logical.

ASSESSMENT
1. Have students complete the following questions from the Chapter 2 Check:
   Building a Vocabulary 1
   Recalling and Reviewing 1, 2
   Using Geography Skills 1
2. Ask students to complete the Chapter 2 Content/Vocabulary Worksheet in their Workbook.

RETEACHING PLAN
1. Ask students to complete the Reteaching Worksheet, Lesson 1.
2. Have students illustrate the relationships among the universe, galaxy, solar system, sun, and earth.

CHALLENGE/ENRICHMENT STRATEGY
1. Have students research and report on other planets in our solar system.
2. Have students complete the Chapter 2 Critical Thinking Worksheet.

▶ LESSON 2
Earth's Rotation, Revolution, and Tilt

LESSON OBJECTIVES
The student should be able to
- define significant vocabulary
- list three factors that control the amount of the sun's energy that falls on various parts of the earth
- describe the process and impact of the earth's rotation
- explain the process of the earth's revolution
- describe the impact of the tilt of the earth

PURPOSE
An understanding of the impact of the earth's rotation, revolution, and tilt prepares students to understand the cause of seasons. Ultimately, seasons significantly influence our way of life.

MOTIVATOR
Ask students to do the following: Think about the planet Earth. There are three things that cause it to be unique among other planets. These three factors control the amount of the sun's energy that falls on different parts of the earth. How many students can name these three items? (rotation, revolution, and tilt)

TEACHING DESIGN
1. Ask students to reread pages 39-40. Write Rotation, Revolution, and Tilt on the chalkboard. Ask students to summarize significant facts about each concept located in the chapter. Review the significant information.
   Rotation
   - equals one complete spin on the axis, which equals one day
   - is in an eastward direction so the sun appears to rise in the east and set in the west
   - makes it possible to expose all parts of earth's surface to the sun's energy
   - creates day and night

   Revolution
   - is a path around the sun that is completed once a year (every 365 days)
   - is another word for *orbit*
   - is one-fourth day longer than a calendar year, so an additional day is added to the calendar in leap years, which occurs every four years.

Tilt
- is at a 23½-degree angle to the sun
- is constant, so the earth's north polar axis always points to the North Star
- is fixed in relation to the stars but not to the sun
- varies as the earth moves around the sun
- causes different parts of the earth to receive varying amounts of the sun's energy

3. Use a globe and a flashlight to demonstrate for students the impact of the revolution and tilt on the movement of the earth.

CLOSURE
Call upon some of the students to use the flashlight and globe to discuss rotation, revolution, and tilt.

ASSESSMENT
Have students complete the following questions from the Chapter 2 Check:
Building a Vocabulary 2
Critical Thinking 3
Using Geography Skills 2

RETEACHING PLAN
1. Ask students to complete the Chapter 2 Reteaching Worksheet, Lesson 2.
2. Have students create an illustration to demonstrate the tilt and revolution of the earth.

CHALLENGE/ENRICHMENT STRATEGY
1. Have students complete the Chapter 2 Challenge Worksheet.
2. Ask students to discuss the impact of the following situations:
   a. The earth suddenly stops rotating.
   b. The earth begins to rotate in the opposite direction.
   c. The earth shifts from a 23½° tilt to a 46° tilt.
   d. The earth's revolution around the sun takes only 175 days.

► LESSON 3
Seasons and Their Effect on Us

LESSON OBJECTIVES
The student should be able to
- define significant vocabulary
- explain the cause of seasons

PURPOSE
Students should develop an understanding and appreciation of seasons in their location and of how the seasons influence their way of life.

MOTIVATOR
Ask students which is their favorite season and what they like most about that season. List their favorite activities for the season on the chalkboard. Indicate the importance of understanding and appreciating the cause of seasons.

TEACHING DESIGN
1. Review the following concepts with students:
   a. The earth *rotates* daily on its axis from west to east, so the apparent motion of the stars, including the sun, is from east to west.
   b. The earth *revolves* around the sun.
2. Discuss possible causes of seasons. Use the diagram below to demonstrate that seasons are not caused by the earth being closer to the sun but that the earth is in fact closer in the winter.

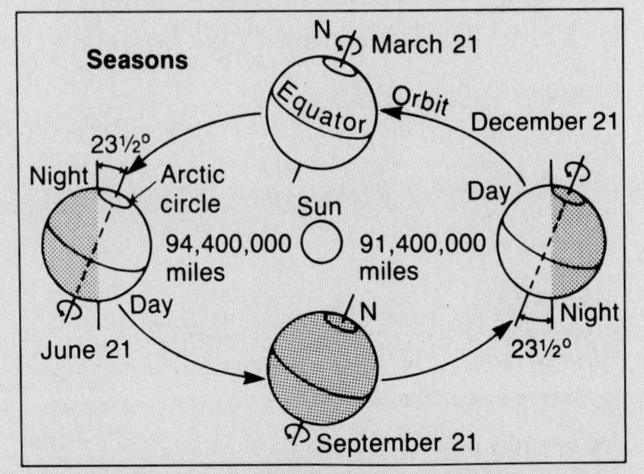

3. Ask students to examine the diagram for other possible causes. Explain that the seasons occur because the plane in which the earth revolves is not the same as the plane of the earth's equator; hence, the intensity and direction of the sun's energy differs as the earth rotates around the sun at a 23½-degree tilt.
4. Use a globe and a model for the sun to show which continents face the sun at different times of the year. For example, the Southern Hemisphere is inclined toward the sun in December; the Northern Hemisphere is inclined toward the sun in June.

CLOSURE
Ask students to review the two basic causes of seasons and explain their meanings. (tilt and revolution)

ASSESSMENT
1. Have students complete the following questions from the Chapter 2 Check:
   Building a Vocabulary 3
   Using Geography Skills 3
2. Have students complete the Chapter 2 Skills Worksheet.

RETEACHING PLAN
1. Ask students to create their own illustration to symbolize the cause of seasons. Have them label the Northern and Southern hemispheres.
2. Ask students to complete the Chapter 2 Reteaching Worksheet, Lesson 3.

CHALLENGE/ENRICHMENT STRATEGY

1. Ask students to investigate the definition of the following terms and their application to the discussion:
   ecliptic	obliquity
   celestial equator	zenith
2. Have students investigate ancient celebrations or rites associated with the solstices and/or equinoxes.

▶ LESSON 4
Seasons Throughout the World

LESSON OBJECTIVES

The student should be able to

- distinguish between equinox and solstice
- describe seasonal changes at different places on the earth

PURPOSE

Students should develop an understanding and appreciation of the ways in which seasons influence people's ways of life throughout the world.

MOTIVATOR

Review with students the cause of seasons. Ask students why they believe there is more sunlight in summer than in winter. Draw conclusions.

TEACHING DESIGN

1. Review significant concepts presented in Lesson 3.
2. Present the Chapter 2 Lecture Notes and Transparency Package.
3. Have students work with a partner and use their textbook to complete the following chart:

	Winter Solstice	Spring Equinox	Summer Solstice	Fall Equinox
What is the date?				
What season begins in the Northern Hemisphere?				
What season begins in the Southern Hemisphere?				
What occurs at the Tropic of Capricorn?				
What occurs at the Tropic of Cancer?				
What occurs at the Antarctic Circle?				
What occurs at the Arctic Circle?				

CLOSURE

1. Ask students to tell a classmate next to them the difference between equinox and solstice.
2. Review with the transparency from the Chapter 2 Transparency Package the movement of the earth and the cause of seasons.

ASSESSMENT

1. Have students complete the following questions from the Chapter 2 Check:
   Building a Vocabulary 4
2. Ask students to complete the Chapter 2 Review/Test.

RETEACHING PLAN

1. Direct students to complete the following chart:

Dates	What is the geographical term?	What season begins in the Northern Hemisphere?	Is there more or less sunlight than darkness?
September 21			
December 21			
March 21			
June 21			

2. Create an illustration to show the differences in seasons across the globe on the above four significant dates. Use it to explain these differences to students.
3. Ask students to complete the Chapter 2 Reteaching Worksheet, Lesson 4.

CHALLENGE/ENRICHMENT STRATEGY

1. Suggest that students research and report on whether or not seasons exist on other planets. Do other planets have a tilt?
2. Ask students to determine what we would observe if the axis of the earth were tilted more or less than 23½ degrees. What if the earth revolved faster or slower than it does now?

Chapter 2 Check, page 43, Answers

Building a Vocabulary

1. rotation; day
2. The spring and fall equinoxes. On each of these days, day and night are exactly 12 hours long. During the spring equinox, the Northern Hemisphere begins to warm. The Southern Hemisphere begins to warm after the fall equinox.
3. The solstice is when one or the other pole is tilted away from the sun more than at any other time of year. The equinox is when there is equal light and darkness everywhere on Earth.

4. Around the earth's poles. They are too far from the equator to receive direct heat energy from the sun's rays.

Recalling and Reviewing

1. The earth's rotation, revolution, and tilt.
2. Our solar system is part of the Milky Way galaxy.

Critical Thinking

3. No. The side of the earth away from the sun would be completely dark and extremely cold all the time, while the part facing the sun would be extremely hot.

Using Geography Skills

1. North polar region; south polar region.
2. Eastward.
3. 23° south of the equator in the Southern Hemisphere; Tropic of Capricorn.

CHAPTER 3 The Earth's Surface Pages 44–53

PLANNING GUIDE		
Lesson	Textbook	Teacher's Resource Binder
1. Forces at the Earth's Surface	pp. 44–46	Content/Vocabulary Workbook Reteaching Worksheet Critical Thinking Worksheet
2. Forces Within the Earth	pp. 46–47	Lecture Notes and Transparency Package Reteaching Worksheet Challenge Worksheet
3. Landforms	pp. 47–52	Reteaching Worksheet
4. Reading an Elevation Map	pp. 48–49	Skills Worksheet Chapter Review/Test

TIME ALLOCATION: 4 days

CHAPTER TOPICS

- Forces at the Earth's Surface
- Forces Within the Earth's Surface
- Landforms: Development and Impact
- Elevation Maps and Cross-Sectional Diagrams

PRETEACHING CHAPTER VOCABULARY

Write the following terms on the chalkboard:

erosion	lava	plate tectonics
rock weathering	volcano	midoceanic ridge
sediments	fold	relief
canyon	fault	plateau
glaciers	earthquake	floodplain
elevation	landforms	alluvial fan
sand dunes	plains	delta

There are photographs and illustrations throughout Chapter 3 that represent examples of some of these terms. Take students on a picture tour through the chapter and other pages in their textbook to locate examples of as many terms as possible. Call upon students to create definitions for each term, based upon the pictures. You may wish to bring in additional geography books and distribute them to selected students to locate additional examples.

▶ LESSON 1
Forces at the Earth's Surface

LESSON OBJECTIVES

The student should be able to

- define significant chapter vocabulary
- list four forces that cause changes in the earth's surface
- explain how rock weathering, water, glaciers, and wind change the surface of the earth

PURPOSE

Students need to be aware of the powerful impact of natural forces on the earth. Ultimately, their surroundings could be altered by one or more of these forces.

MOTIVATOR

Display for students pictures of the Grand Canyon. Ask students to hypothesize about how it came to exist. Discuss whether or not students believe that the Grand Canyon continues to change. Discuss why or why not. (rock weathering, water, wind)

TEACHING DESIGN

1. Display additional pictures to demonstrate the impact of erosion.
2. Review the definition of *erosion* (wearing away of the land through the force of water and wind), and ask students to cite examples they have witnessed.
3. Identify for students the four forces that cause changes in the earth's surface: rock weathering, water, glaciers, and wind.
4. Ask students to reread pages 44–46 and use the information to complete the following chart:

Definitions (key terms)	Rock Weathering	Water Erosion	Glaciers	Wind Erosion
Process (How does it work?)				
Impact (What will it eventually cause to happen?)				

5. Ask students to hypothesize about the answers to the following questions:
   a. Which form is most powerful?
   b. Which is quickest?
   c. Which causes the most change?

CLOSURE

Ask students to identify examples of erosion that exist in their area. What was the land like before the erosion?

ASSESSMENT

1. Have students complete the following questions from the Chapter 3 Check:
   Building a Vocabulary 3, 4
   Recalling and Reviewing 1, 2
   Using Geography Skills 3
2. Ask students to complete the Chapter 3 Content/Vocabulary Worksheet in their Workbook.

RETEACHING PLAN

1. Ask students to complete the Chapter 3 Reteaching Worksheet, Lesson 1.
2. Have students list significant terms to remember about each force presented in this section.

CHALLENGE/ENRICHMENT STRATEGY

1. Have students research the correct answers to the four questions in Teaching Design 5.
2. Ask students to complete the Chapter 3 Critical Thinking Worksheet.

▶ LESSON 2
Forces Within the Earth

LESSON OBJECTIVES

The student should be able to

- explain how forces within the earth keep the surface from becoming flat
- describe the theory of plate tectonics and name the three types of plate boundaries
- analyze the formation of continents

PURPOSE

This information provides students with an understanding of the development of the earth and reasons that students should expect the surface to continue to change.

MOTIVATOR

Read aloud to students the following eyewitness account of the Mexico City earthquake of September 19, 1985. [From *National Geographic*, Vol. 169, No. 5, May 1986]

> In his third-floor office in downtown Mexico City, pathologist José Hernández Cabañas bent over his morning work last September 19. At 7:18 he felt the eight-story building begin to shudder beneath him.
>
> Then the structure began to rock slowly back and forth in widening swings. Cabañas gripped a window frame. Looking out, he saw trees sway almost to the ground and thought, "This building can't hold up."
>
> He lurched to a stairwell door but found it jammed shut. He smashed it open with his shoulder. On the quaking steps he slipped, fell, and tumbled downward, pitching helplessly from landing to landing. Passing the second floor, he heard people screaming. . . .
>
> Cabañas landed hard near a ground-level exit door. A tremor shook it open and he started through, only to fall again as flying debris gashed his head. Covered with blood, almost senseless, he crawled desperately into the street. At that instant his building toppled over backward.

Ask students if any of them has ever experienced or heard about an earthquake. Have them describe what happened and tell how their description compares with the account read in class. Indicate to students that they will learn the causes of such natural phenomena in this lesson.

TEACHING DESIGN

1. Present the Chapter 3 Lecture Notes and Transparency Package.
2. Create through class discussion an information map to help students process the chapter information regarding changes from within the earth.
3. Utilize the diagram "Movement at Plate Boundaries" on page 50 of the textbook to discuss the theory of plate tectonics. Ask students to explain the various types of plate boundaries and the results of collisions at plate boundaries.
4. Provide groups of students with maps of the supercontinents, and ask them to identify the present-day continents included within each.

CLOSURE

1. Ask students to summarize key information about all forces that change the earth's surface.
2. Ask additional students to describe the plate tectonic theory in their own words.

ASSESSMENT

Have students complete the following questions from the Chapter 3 Check:
Building a Vocabulary 1, 2
Recalling and Reviewing 3, 5
Using Geography Skills 2

RETEACHING PLAN

Ask students to complete the Chapter 3 Reteaching Worksheet, Lesson 2.

CHALLENGE/ENRICHMENT STRATEGY

1. Have students complete the Chapter 3 Challenge Worksheet.

2. Have students create a model to demonstrate the impact of movement at plate boundaries.
3. Ask students to create a three-dimensional representation of the core of the earth.

▶ LESSON 3
Landforms

LESSON OBJECTIVES

The student should be able to

- distinguish between primary and secondary landforms
- categorize landforms by how they were formed

PURPOSE

Students must possess an understanding of landforms prior to studying the relationship between landforms and climate. In choosing a place to live in the future, students will perceive how landforms could affect their way of life.

MOTIVATOR

Turn students' attention to the picture of the island volcano Surtsey on page 51 of their textbook. Remind students that Surtsey did not exist prior to 1963. Ask: Remembering the discussion in Lesson 2, what do you think is the probable cause for the appearance of this island? What will cause the island to continue to change?

TEACHING DESIGN

1. Display pictures of landforms for the class, and ask students to classify them as either primary or secondary landforms. Review the characteristics discussed in the chapter for determining primary and secondary landforms. Conclude with the students that as a landform appears, forces such as erosion begin to work upon it; hence, there are very few primary landforms.
2. Discuss with students the other ways to classify landforms according to composition: (a) rocks with a thin layer of weathered sediments and soil at the surface and (b) sediments deposited by water, wind, or ice. Have students reclassify the pictures into these two categories.
3. Provide students with the following list of landforms: islands, plateaus, hills, mountains, plains, valleys. Ask students to draw or to cut and paste pictures from magazines to illustrate each term.

CLOSURE

Call upon students to identify pictures of landforms and classify them into the two categories in the chapter.

ASSESSMENT

Have students complete the following questions from the Chapter 3 Check:

Building a Vocabulary 5
Recalling and Reviewing 4
Using Geography Skills 1

RETEACHING PLAN

1. Ask students to complete the Chapter 3 Reteaching Worksheet, Lesson 3.
2. Create diagrams of landforms, and ask students to label them.

CHALLENGE/ENRICHMENT STRATEGY

1. Ask students to locate pictures of primary landforms immediately following their formation. (Example: Surtsey, 1963.)
2. Have students create a bulletin board about landforms.
3. Ask students to locate landforms in their region, decide how they were formed, and label them by type.

▶ LESSON 4
Reading an Elevation Map

LESSON OBJECTIVES

The student should be able to

- interpret an elevation map
- understand a cross-sectional diagram

PURPOSE

Reading an elevation map is a skill that students may apply in many situations. Students who ski will use a type of elevation map to help route courses. Students who take car trips through varying levels should be concerned with elevation.

MOTIVATOR

Ask students to identify places they have visited or would like to visit with high and low elevations. Ask them if they know the exact elevation for these sites. (For example, they might indicate that the ocean is at sea level.) Ask students where they could look to determine who in the class has been to the highest elevation. (on an elevation map)

TEACHING DESIGN

1. Have students read the Geography Skills feature "How to Read an Elevation Map and a Cross-Sectional Diagram" on pages 48–49 of their textbook.
2. Ask students to form small groups and prepare answers to the questions at the end of the Skills feature.
3. Bring to class several different atlases to present and discuss different types of elevation maps and cross-sectional diagrams.
4. Using the Atlas in the front of their textbook, students should determine the approximate elevations of the following cities: Moscow, USSR; Paris, France; Washington, D.C., U.S.A.; Cairo, Egypt; Melbourne, Australia; Brasília, Brazil; London, England.

CLOSURE

Have students investigate encyclopedias and other resource books to find various elevation and cross-sectional diagrams.

ASSESSMENT

1. Have students complete the Chapter 3 Skills Worksheet.
2. Ask students to complete the Chapter 3 Review/Test.

RETEACHING PLAN

Suggest that students work in pairs to create a fictional map and cross-sectional diagram. Examples of subjects students may choose are an eight-layer cake and a three-story house.

CHALLENGE/ENRICHMENT STRATEGY

1. Ask students to research and report on the highest places on Earth, indicating where humans have and have not been.
2. Have students research and report on the impact of higher elevations on traveling, cooking, and the human respiratory system.
3. Have students write a "You Are There" television program based upon their research of an actual attempt to climb Mount Everest.

Geography Skill, pages 48-49, Answers

How to Read an Elevation Map and a Cross-Sectional Diagram

1. Dark orange represents highest elevation; dark green represents lowest elevation.
2. About 300 miles (500 km).
3. About 2.5 miles (4 km).
4. A cross-sectional diagram makes it easier to see the shapes of the landforms as they actually appear. The shape of a cross-sectional diagram is not accurate, but rather it shows where the land rises and falls. Vertical and horizontal distances on a cross-sectional diagram are exaggerated in order to show the relief of the landforms more dramatically. If the vertical scale were not exaggerated, even tall mountains would appear as small bumps on a cross-sectional diagram.
5. Puno: about 2.5 miles (4 km); Arica: less than 1 mile (1.6 km). Mountains: about 4 or 5 miles (6 or 8 km).

Chapter 3 Check, page 53, Answers

Building a Vocabulary

1. Folds occur when rock layers are bent. Faults occur when rock layers break and move apart.
2. The midoceanic ridge is a chain of mountains on the ocean floor. It is formed at the boundary where plates move apart and heat currents force the crust to be lifted up. It is 40,000 miles (64,000 km) long.
3. Relief
4. A very deep, narrow valley with steep sides. It is formed by water erosion.
5. An alluvial fan is a landform created by sediment eroded from a mountain and deposited by a stream as it enters a plain along the base of a mountain. A floodplain is level ground built by sediment deposited by a river or stream. A delta is the deposit of mud and sand at the river's mouth. All three landforms are formed by sediment deposits.
6. Lava is melted, liquid rock from within the earth that is brought to the earth's surface.

Recalling and Reviewing

1. Wind, water, ice, and rock weathering cause erosion. Wind and water carry soil to different places. Rock weathering breaks rock into particles of sand or mud. Rocks are also decayed by frozen water that pushes rocks apart.
2. Sheet glaciers, or ice sheets, and mountain glaciers. A sheet glacier covers a large area of land and flows outward from great domes of ice where snow accumulates. Mountain glaciers are smaller and more common. They are found in the high mountain valleys of the world.
3. Lava, volcanoes, and earthquakes continue to lift up the earth's surface and tear it apart.
4. Primary landforms are masses of rock raised by volcanic eruptions and other uplifting forces. Secondary landforms are formed by erosion of primary landforms by wind, water, ice, and rock weathering.
5. The theory of plate tectonics states that the earth's surface is made up of rigid, moving plates. The plates float upon the upper mantle. Along the edges, or boundaries, of the plates, forces inside the earth cause volcanoes, faults, and folds at the surface. The three types of plate boundaries are those where plates move away from each other, those where plates crash into each other, and those where plates slide under or over each other. The continents were formed by movement at plate boundaries. Continental landforms were made by volcanic eruptions and continental plates crashing into each other. They are generally found along plate boundaries.

Critical Thinking

6. Answers will vary but might include landforms such as mountain ranges, canyons, plains, and deltas. Students should explain how they were formed, using the theory of plate tectonics. Human interaction with the environment can be seen in such activities as farming, mining, and construction.

Using Geography Skills

1. Plateaus, canyons, deserts, plains, valleys, hills, and mountains are examples of erosion. Floodplains, alluvial fans, and deltas are examples of sediment deposits.
2. In the first part of the diagram, Plates 1 and 2 are moving away from each other. As a result, the sea floor spreads, and lava moves upward from below to fill the opening. This creates the midoceanic ridge. In the second part of the diagram, Plates 1 and 2 are colliding. The result is the formation of mountains where the continents collide.
3. Answers will vary but might include examples of erosion such as gullies, ravines, canyons, and beaches.

CHAPTER 4 The Earth's Atmosphere Pages 54–61

PLANNING GUIDE		
Lesson	Textbook	Teacher's Resource Binder
1. Global Energy Balance, Pressure Systems, and Wind Systems	pp. 54–56	Content/Vocabulary Workbook Reteaching Worksheet Critical Thinking Worksheet
2. Types of Winds, Moisture in the Air, and Storms	pp. 56–59	Lecture Notes and Transparency Package Reteaching Worksheet Challenge Worksheet
3. Temperature Effect and Orographic Effect	p. 60	Chapter Review/Test Reteaching Worksheet Skills Worksheet

TIME ALLOCATION: 4 days

CHAPTER TOPICS

- Global Energy Balance
- Atmospheric Exchange Systems
- Elevation and Climate

PRETEACHING CHAPTER VOCABULARY

Write the following terms on the chalkboard:

weather	polar easterlies	hurricanes or typhoons
climate	front	
air pressure	evaporation	tornadoes
trade winds	humidity	orographic effect
doldrums	condensation	
westerlies	precipitation	

Distinguish *weather* from *climate*. (Weather is the condition at any given time; climate is the weather conditions in an area over a long period of time.)

Use the following visual to elicit suggestions for definitions of the following terms:

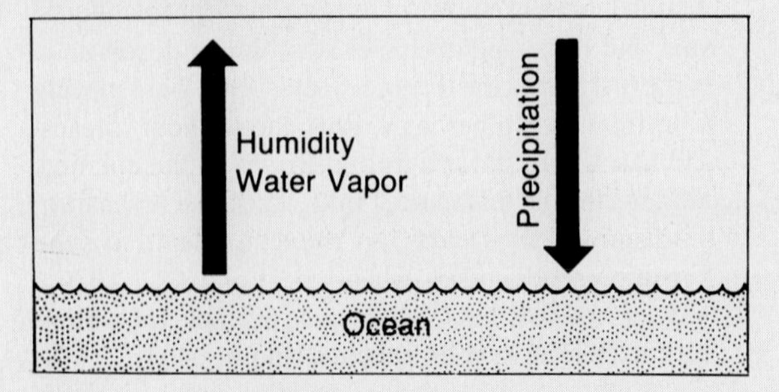

Discuss the similarities and differences among *hurricanes, typhoons,* and *tornadoes*.

Ask students to think about a meaning for *air pressure*. Have them consider the individual words. What definitions can they suggest? Check the answer at the end of Lesson 1.

Indicate to students that all the following terms relate to *wind*. With this word as a context clue, what conclusions can they draw? Words include *trade winds, doldrums, westerlies, polar easterlies,* and *front*. Discuss the accuracy of their suggestions.

Have students compare the definitions of *evaporation* and *condensation* in their textbook glossary. Indicate to students that the term *orographic effect* comes from the combination of *oro-*, meaning "mountain," *graph,* meaning "writing," and *-ic,* meaning "pertaining to." What definition can they create?

▶ LESSON 1
Global Energy Balance, Pressure Systems, and Wind Systems

LESSON OBJECTIVES

The student should be able to

- define significant vocabulary
- explain how the earth maintains an energy balance
- describe the cause of air-pressure differences
- describe the movement of wind patterns across the globe

PURPOSE

This lesson will provide students with a very basic introduction to information provided on the daily weather forecasts. Students will learn why they should take note when they hear the forecaster indicate, for example, that the air pressure is falling.

MOTIVATOR

Ask students what they listen for on the daily weather forecast. What are their specific concerns? Create a list of terms used by weather forecasters that the students would like to understand. Indicate your intent to discuss these concerns throughout their study of Chapter 4.

TEACHING DESIGN

1. Conduct a discussion based on the following questions:
   a. What is the sun's energy transformed into when it reaches the earth? (heat)
   b. What prevents the energy from escaping from the earth? (atmosphere)
   c. How is the energy balance maintained across the earth? (Some energy escapes to space as heat. Other energy is transferred to and from the polar regions and tropics.)
   d. What causes air-pressure differences? (unequal heating of earth's surface as air masses with differing tem-

peratures move back and forth up and down the earth)

2. Have students work with a partner to complete the following chart.

	Cause	Effect
Low-pressure areas	(Air is warmed, expands, becomes lighter, and rises.)	(unstable air causing clouds, rain storms)
High-pressure areas	(Cold air is dense, heavy, and stable.)	(clear, calm weather)

3. Have students work in groups to discuss what occurs under the following conditions:
   a. Cold air meets warm air.
   b. Air from the sea meets air over the land at noon.
   c. Air from the sea meets air over the land at midnight.
4. Use a globe to demonstrate how warm air rises from the equator and settles at about 30° N and S. Cold air rises from the poles and settles at about 60° N and S. Cold air meeting warm air at 60° N and S causes unstable weather conditions.
5. Have students compare their understanding of air pressure with the definitions they suggested in Preteaching Chapter Vocabulary.

CLOSURE

Have students look on a world map for 60° N and S. Are these areas known for fluctuations in weather conditions?

ASSESSMENT

1. Have students complete the following questions from the Chapter 4 Check:
   Building a Vocabulary 1
   Recalling and Reviewing 1, 2
2. Ask students to complete the Chapter 4 Content/Vocabulary Worksheet in their Workbook.

RETEACHING PLAN

1. Ask students to complete the Chapter 4 Reteaching Worksheet, Lesson 1.
2. Assist students with the preparation of their own diagrams to demonstrate the movement of air across the earth as described in Teaching Design 4.
3. Provide students with cards containing the descriptions of pressure areas listed in Teaching Design 2, and ask them to match these to the corresponding area.

CHALLENGE/ENRICHMENT STRATEGY

1. Suggest that students investigate the storms typical of low-pressure areas in the United States and rank the storms that have occurred in the last five years according to the damage caused.
2. Have students complete the Chapter 4 Critical Thinking Worksheet.
3. Choose students to investigate and report the reasons for naming hurricanes and not naming other storms. How are names chosen for hurricanes?
4. Ask a student to report on the Galveston, Texas, hurricane of 1900.

▶ LESSON 2

Types of Wind, Moisture in the Air, and Storms

LESSON OBJECTIVES

The student should be able to

- describe prevailing winds across the globe
- list the effects of the oceans on climate and weather
- list the steps that lead to the occurrence of precipitation
- discuss the different types of storms and their causes

PURPOSE

Students need to know how to listen to a weather report and interpret the information necessary to prepare for their daily activities.

MOTIVATOR

Bring to class examples of weather information published in the daily newspaper. Discuss the different diagrams and charts, and determine how much the students can interpret. Indicate that the class will be looking at additional information helpful to understanding the forecast and the information published about it.

TEACHING DESIGN

1. Present the Chapter 4 Lecture Notes and Transparency Package.
2. Write the chart below on the chalkboard, and have students complete it. Discuss it with the class.

	Low Latitudes	Middle Latitudes	High Latitudes
Prevailing winds	(trade winds)	(westerlies)	(polar easterlies)
Latitude	(0°–30°)	(30°–60°)	(60°–90°)
Direction of wind current in Northern Hemisphere	(from northeast)	(from west)	
Direction of wind current in Southern Hemisphere	(from southeast)	(from west)	

3. Discuss three effects of oceans on climate and weather.
   a. Transfer heat from warm tropical regions to cold polar regions
   b. Provide oxygen to the air
   c. Produce moisture in the atmosphere

4. Ask students to create a graphic display describing the steps that lead to rain and/or a storm.

CLOSURE
1. Allow some students to present and explain their graphic displays on precipitation to the class.
2. Review with the class the movement of air across the globe.

ASSESSMENT
Have students complete the following questions from the Chapter 4 Check:
- Building a Vocabulary 2, 4, 5
- Recalling and Reviewing 3, 4
- Using Geography Skills 1

RETEACHING PLAN
1. Prepare cards with the information contained in Teaching Design 2. Ask students to organize the information according to latitudes.
2. Ask students to complete the Chapter 4 Reteaching Worksheet, Lesson 2.

CHALLENGE/ENRICHMENT STRATEGY
1. Have students complete the Chapter 4 Challenge Worksheet.
2. Ask students to present an explanation of symbols used on weather maps by meteorologists.
3. Have students investigate weather satellites and their contributions to weather reporting. How does accurate information contribute to safe travel?
4. Select a student to research and report on storm-tracking systems.

▶ LESSON 3
Temperature Effect and Orographic Effect

LESSON OBJECTIVES
The student should be able to
- describe the two effects of elevation upon climate
- distinguish between leeward and windward sides of mountains

PURPOSE
From this lesson, students will learn why people must adapt to their climate regions.

MOTIVATOR
Bring to class magazine photographs of a ski resort and a desert city. Discuss the length of the ski season and how long a desert will go without rain. Determine whether students can see any common factor (geographic feature) that influences these extreme differences in climate. Tell students they will learn the answer in the lesson.

TEACHING DESIGN
1. Ask students to review page 60 to describe the two effects that elevation has upon climate.
2. Discuss temperature effect and orographic effect.
3. Draw a mountain on the chalkboard. Label the leeward and windward sides. Indicate the direction of movement of moisture.

CLOSURE
Ask students to summarize how they believe their local climate affects their way of life.

ASSESSMENT
1. Have students complete the following questions from the Chapter 4 Check:
   - Building a Vocabulary 3
   - Critical Thinking 5
   - Using Geography Skills 2
2. Ask students to complete the Chapter 4 Review/Test.

RETEACHING PLAN
1. Direct students to create a diagram to demonstrate the temperature and orographic effects on climate.
2. Ask students to complete the Chapter 4 Reteaching Worksheet, Lesson 3.

CHALLENGE/ENRICHMENT STRATEGY
Have students complete the Chapter 4 Skills Worksheet.

Chapter 4 Check, page 61, Answers

Building a Vocabulary
1. Air pressure is the measurement of the weight of the air. Differences in air pressure are caused by unequal heating of the earth's surface. When air is warmed, it expands, becomes lighter, and rises, creating low-pressure areas that cause clouds and precipitation. Cool air tends to be heavy and sinks. This sinking action produces high-pressure areas and clear, calm weather.
2. Precipitation.
3. When moist air flowing from the ocean meets a mountain barrier, it is forced to rise. As it rises, it cools, and rain or snow are the result. The side of the mountain facing the wind (the windward side) receives the most moisture. The leeward side of the mountain receives less moist air, and the climate is drier.
4. The polar front is created; stormy weather.
5. Doldrums are areas with no strong winds. They are found between the trade winds and the equatorial zone (between the equator and 30° north and south).

Recalling and Reviewing
1. All of the sun's energy absorbed by the earth eventually leaves the earth as lost heat, giving the earth an energy balance.

2. The oceans absorb most of the sun's energy. The major wind belts move across the surface of the ocean. They set the ocean currents in motion. Warm currents reduce water temperature by carrying warm air from low latitudes into cooler high latitudes. Cool currents return cool water from higher latitudes to be rewarmed in lower latitudes. Thus, ocean currents move heat. Plants in the ocean produce oxygen for air. The oceans are also the most important source of moisture. Most water vapor in the air is evaporated from oceans. This vapor stores the heat energy that brings changes in the earth's atmosphere.
3. Most water vapor comes from the oceans. The rest comes from lakes, rivers, plants, and soil. When the temperature rises, the air can hold more water vapor. This is called humidity. When the temperature falls, water vapor condenses and can be seen in the form of clouds, fog, and precipitation.
4. Hurricanes or typhoons, tornadoes, and thunderstorms. Thunderstorms are the most common storms.

Critical Thinking

5. Answers will vary. Wind factors should include types of winds. Water factors might include such conditions as ocean currents, humidity, rainstorms, or hurricanes.

Using Geography Skills

1. Between 0° and 30° north and south. In the Northern Hemisphere, they flow from the northeast; in the Southern Hemisphere, from the southeast.
2. When the land becomes warmer during the day, warm air rises, creating low pressure. Cool air from the higher pressure over the sea blows inland. At night, the land cools more rapidly than the sea. The cool land breeze flows from the land (high-pressure area) to the sea (low-pressure area).

CHAPTER 5 Global Climates Pages 62–71

PLANNING GUIDE

Lesson	Textbook	Teacher's Resource Binder
1. Climate Types	pp. 62–69	Lecture Notes and Transparency Package Content/Vocabulary Workbook Reteaching Worksheet Challenge Worksheet
2. Climate and Human Activity	p. 69	Critical Thinking Worksheet Chapter Review/Test Reteaching Worksheet Skills Worksheet

TIME ALLOCATION: 2 days

CHAPTER TOPICS

- Climate Types
- World Climate Regions
- Climographs
- Climate and Human Activity

PRETEACHING CHAPTER VOCABULARY

Write the following terms on the chalkboard:
monsoon permafrost
Locate pictures in the textbook illustrating these terms. Tell students that these are both conditions related to climate.

Identify *monsoon* as a wind that blows part of the year from one direction and that blows the rest of the year from the opposite direction. Ask students to hypothesize about why monsoon winds occur in the Indian Ocean and southern Asia.

Ask students what they believe *permafrost* means. Where do they think it is found? Have them find the definition in the textbook Glossary and then compare it to the definition in a dictionary. Discuss the similarities and differences.

▶ LESSON 1 Climate Types

LESSON OBJECTIVES

The students should be able to

- define significant vocabulary
- discriminate between world climate regions
- identify factors that cause changes in the landscape
- interpret a climograph

PURPOSE

Students will learn about the climate regions in the world and develop an understanding of the reasons for the location of all types of cities.

MOTIVATOR

Ask students to name places they would like to visit. List these places on the chalkboard. Ask them what type of weather they expect to find in each. How do they know this? Discuss different sources they could use to locate the climate of these places.

TEACHING DESIGN

1. Present the Chapter 5 Lecture Notes and Transparency Package.
2. Ask students to use the map "World Climate Regions" on pages 66–67 of their textbook to reclassify the world climate regions into the following categories:
   a. Warm/Cold Climates
   b. Wet/Dry Climates
   c. Continental/Maritime Climates

3. Ask students to describe the types of clothing and shelter probably typical of each climate region. Discuss their conclusions.
4. Write on the chalkboard a list of 10 to 20 major world cities, and ask students to identify the climate region in which each is located. Select various students to verify the answers in a world atlas or an encyclopedia. If any contrasts exist, have students explain the cause.
5. Have students read the Geography Skills feature "How to Read a Climograph" on page 68 of their textbook and complete the questions.

CLOSURE
Have students write a paragraph titled "Where Do I Live?" describing a fictitious person and his or her way of life in a particular climate region. Read several examples to the class, and allow students to identify the regions.

ASSESSMENT
1. Have students complete the following questions from the Chapter 5 Check:
   Building a Vocabulary 1, 2
   Recalling and Reviewing 1, 2, 3, 4
2. Ask students to complete the Chapter 5 Content/Vocabulary Worksheet in their Workbook.

RETEACHING PLAN
1. Collect the "Where Do I Live?" paragraphs from the Closure, and ask students to complete additional ones.
2. Ask students to complete the Chapter 5 Reteaching Worksheet, Lesson 1.

CHALLENGE/ENRICHMENT STRATEGY
1. Ask students to create a climograph for their own city and a city in an equivalent latitude in the Eastern Hemisphere.
2. Have students investigate how meteorologists use a climograph. Invite a meteorologist to class to discuss his or her skills, training, and application of the concepts studied in this chapter.
3. Ask students to locate five cities whose climate is not typical of the surrounding climate region. Have them write a paragraph to explain the cause.
4. Have students complete the Chapter 5 Challenge Worksheet.

▶ LESSON 2
Climate and Human Activity

LESSON OBJECTIVES
The student should be able to
- discriminate between world climate regions
- describe ways in which people adapt to climate

PURPOSE
Students are completing their study of climate and weather. They now have the skills to interpret a weather report and use the information to plan for an event or activity. This lesson will reinforce the close relationship between human activity and climate.

MOTIVATOR
Assign students to meet in groups. Ask each group to prepare three generalizations about the relationship between climate and human activity, then ask each group to select a spokesperson. Have each spokesperson present the generalizations. Write these on poster board and save for use in the Closure.

TEACHING DESIGN
1. Direct students to review the description of highland climates. Identify several famous mountains and mountain ranges, and ask students to compare the climate at the bottom of the mountain to the climate at the top of the mountain. (Mountains may include the Rockies, Mount Everest, the Andes, the Alps.)
2. Ask students to work in groups to identify all the items they used this week to protect themselves from the weather. Have the groups identify activities they were and were not able to accomplish because of weather conditions. Discuss how this situation might differ six months from now.
3. Have students complete the Chapter 5 Critical Thinking Worksheet.
4. Have students create a chart showing 10 cities that are popular because of their climate and 10 cities that are popular in spite of their climate.

CLOSURE
1. Ask students to write a paragraph describing the relationship between human activity and climate.
2. Return to the list of generalizations created in the Motivator. Determine which statements remain accurate. Explain why others do not.

ASSESSMENT
1. Have students complete the following questions from the Chapter 5 Check:
   Critical Thinking 5
   Using Geography Skills 1, 2
2. Ask students to complete the Chapter 5 Review/Test.

RETEACHING PLAN
1. Ask students to complete the Chapter 5 Reteaching Worksheet, Lesson 2.
2. Direct students to create a word web from the phrase below. Students should try to include all the key terms and ideas they have studied.

CHALLENGE/ENRICHMENT STRATEGY

1. Have students complete the Chapter 5 Skills Worksheet.
2. Select several cities in all parts of the globe, and ask students to investigate and report on human adaptation to climate in these cities.
3. Ask students to locate 10 cities that are growing because of an advantageous climate.
4. Choose students to investigate and report on the most recent ice age and predictions regarding future climate periods.

Geography Skill, page 68, Answers

How to Read a Climograph

1. Iquitos. Barrow. Iquitos. Barrow.
2. Iquitos: March and December. San Francisco: January and December.
3. Barrow: highest—June, July, August; lowest—January, February, March. San Francisco: highest—June, July, August, September; lowest—November, December, January, February. Iquitos: highest—November, December, January, February; lowest—June, July, August.
4. Barrow receives no precipitation during January, February, March, April, and May and little precipitation during December. San Francisco receives no precipitation during June, July, and August and little precipitation during May and September.

Chapter 5 Check, page 71, Answers

Building a Vocabulary

1. A monsoon is a change in the winds that brings wet and dry seasons. Monsoons occur in Asia, especially India. During the summer months, moist air flows into Asia from the warm ocean, bringing heavy rains. During the winter, dry air flows off the continent, bringing dry conditions to the area.
2. Permafrost is permanently frozen ground. It is found in the tundra climate region.

Recalling and Reviewing

1. Humid continental and subarctic. There are no large land areas in the Southern Hemisphere at these latitudes.
2. Since the equator is constantly being heated by the sun's rays, warm air is always rising in the tropics. This continuous uplift of warm air causes almost daily thunderstorms and heavy rainfall.
3. Continental climates have extreme temperatures—hot in the summer and cold in the winter. Maritime climates are rainy; temperatures are fairly moderate all year. Continental climates are found inland or on the eastern coasts of continents; maritime climates are generally found on the west coasts of continents. Students may choose the continental desert, the continental steppe, or the humid continental climate to contrast with the Mediterranean or maritime climate.
4. Tropical and subtropical deserts are found on the west coasts of continents. Examples are the Sahara (northern Africa), Mojave Desert (North America), and Atacama Desert (South America).

Critical Thinking

5. Answers will vary but might include clothing, food, and activities.

Using Geography Skills

1. Northwest coast of North America, southern Chile and Argentina, parts of Brazil; northwestern Europe, southeastern tip of Africa, southeastern Australia, New Zealand. The maritime climate is generally found on the west coasts of continents in the middle latitudes. Their climates are therefore influenced by the oceans.
2. Tropical and subtropical desert in the middle latitudes. Tundra in the high latitudes.
3. The warm currents warm the east coast, the Gulf of Mexico, and the Gulf of Alaska. The cold currents cool the islands north of Canada and the Pacific west coast. The humid subtropical region is located in the southeastern third of the United States.

CHAPTER 6 The Water Planet Pages 72–77

PLANNING GUIDE

Lesson	Textbook	Teacher's Resource Binder
1. The Hydrologic Cycle	pp. 72–74	Lecture Notes and Transparency Package Content/Vocabulary Workbook Reteaching Worksheet
2. The Oceans	pp. 74–75	Skills Worksheet Reteaching Worksheet Challenge Worksheet
3. Rivers, Lakes, and Groundwater; People and Water	pp. 75–76	Critical Thinking Worksheet Chapter Review/Test Reteaching Worksheet

TIME ALLOCATION: 3 days

CHAPTER TOPICS

- Definition of the Hydrosphere
- The Hydrologic Cycle
- The Oceans
- Rivers and Lakes
- Groundwater
- People and Water

PRETEACHING CHAPTER VOCABULARY

Write the following terms on the chalkboard:

hydrosphere	continental shelf	groundwater
hydrologic cycle	headwaters	water table
transpiration	tributary	aquifer
evapotranspiration	estuary	irrigation

Have students complete the following chart to rate their knowledge of the terms before reading the selection. Have students rate their knowledge again after studying the chapter.

Words	Words I Can Define	Words I Have Seen or Heard	New Words

Advise students to pay careful attention to words they note under "New Words" when they encounter them in the text. Point out that *hydro* is a Greek term meaning "water" and that *aqua* is a Latin term meaning "water."

▶ LESSON 1
The Hydrologic Cycle

LESSON OBJECTIVES

The student should be able to

- define the significant terms of the chapter
- describe the hydrosphere
- explain the process of the hydrologic cycle

PURPOSE

The study of the water of the earth is important because it is an essential resource. Students must understand its importance in order to make wise use of this resource.

MOTIVATOR

Divide the class into groups of three or four students, and ask them to brainstorm to develop a list of all the places where water is found on the earth. One student in each group should list the group's ideas on a sheet of paper. Write the ideas on the chalkboard as each recorder reads from the list. Point out any water on the earth that students fail to mention. Ask why water is important to our planet.

TEACHING DESIGN

1. Present the Chapter 6 Lecture Notes and Transparency Package.
2. Direct students to create a visual representation of the hydrologic cycle on a continent and on an ocean. Each student should explain his or her visual to a partner, who notes the vocabulary terms of the lesson that are used in the explanation.

CLOSURE

After requesting that all students think through the cycle's phases to themselves, ask volunteers to state the phases of the hydrologic cycle.

ASSESSMENT

1. Have students complete the following questions from the Chapter 6 Check:
   Building a Vocabulary 1, 5
   Using Geography Skills 1
2. Ask students to complete the Chapter 6 Content/Vocabulary Worksheet in their Workbook.

RETEACHING PLAN

1. Ask students to complete the Chapter 6 Reteaching Worksheet, Lesson 1.
2. Provide small groups of students with a set of cards containing phases of the hydrologic cycle. Students should take turns arranging the cards in sequence and explaining the arrangement to others in the group.
3. Have students create a word web that represents the sources of water in the hydrosphere.

CHALLENGE/ENRICHMENT STRATEGY

1. Ask students to collect newspaper articles that deal with the subject of water.
2. Invite a guest speaker from the water department to present to the class information regarding sources of water and its uses in your area.

▶ LESSON 2
The Oceans

LESSON OBJECTIVES

The student should be able to

- locate major bodies of water
- distinguish among types of water on the earth's surface
- identify and describe major features of the ocean floor.

PURPOSE

The study of marine geography is important in order for students to develop an understanding of the resources from our oceans.

MOTIVATOR

Divide the class into groups equal to the number of globes available in the classroom. Ask each group to estimate the percentage of the earth's surface that is covered by oceans, seas, and gulfs. Ask each group to suggest an explanation for the earth's designation as the "blue planet."

TEACHING DESIGN

1. Ask students to locate the following bodies of water on individual world maps: Pacific, Atlantic, Indian, and Arctic oceans; Mediterranean, Coral, and Caribbean seas.

Students should list the bodies of water by size. Ask students to share their rankings.

2. Using maps or globes showing the ocean floor, ask students to identify the following features and describe their characteristics: midoceanic ridge, Mariana Trench, and continental shelf. Students should be prepared to explain why the continental shelf is wider in the Atlantic than in the Pacific.
3. Ask students to reread pages 74–75 in their textbook and complete the following chart.

	Seawater and Its Characteristics
Elements	
Water temperature	
Water pressure	
Dynamic system	
Plant life	
Animal life	

CLOSURE
Write the following outline on the chalkboard or on a transparency. Review the lesson's main ideas by calling on students to complete the outline aloud.

I. Oceans
  A. Pacific
  B.
  C.
II. Seas
  A. Mediterranean
  B.
  C.
III. Features of ocean floor
  A. Continental shelf
  B.
  C.
IV. Characteristics of seawater
  A. Dissolved elements
  B.
  C.
  D.
  E.
  F. Most marine life in shallow areas

ASSESSMENT
Have students complete the following questions from the Chapter 6 Check:

Building a Vocabulary 2
Recalling and Reviewing 1, 2
Using Geography Skills 2

RETEACHING PLAN
1. Ask students to complete the Chapter 6 Reteaching Worksheet, Lesson 2.
2. Have students complete the Chapter 6 Skills Worksheet.

CHALLENGE/ENRICHMENT STRATEGY
1. Ask students to study a world map or globe to note the relationship of land and water ratios to population densities in each hemisphere. Have students write a summary of their findings.
2. Suggest that students use an atlas to locate a map or maps of the ocean floor to note significant landforms. Have students report their findings to the class.
3. Have students complete the Chapter 6 Challenge Worksheet.

▶ LESSON 3

Rivers, Lakes, and Groundwater; People and Water

LESSON OBJECTIVES
The student should be able to
- distinguish among headwaters, rivers, tributaries, estuaries, and lakes
- describe the formation and properties of groundwater, water tables, and aquifers
- explain the necessity of preserving our water resources

PURPOSE
The study of our water resources in rivers, lakes, and groundwater will help students appreciate the value and necessity of conserving these resources for current and future generations.

MOTIVATOR
Divide the class into groups of four or five students. Ask each group to brainstorm to develop a list of ideas related to two topics: (1) the uses of water and (2) the necessity of preserving water. Time the groups on each topic, allowing two minutes per topic. Record each group's ideas as volunteers share group contributions.

TEACHING DESIGN
1. Using a wall map, point out the major bodies of water included in the lesson: headwaters, rivers, tributaries, estuaries, lakes. Ask small groups of students to use an appropriate map to identify these bodies of water in an assigned country, state, or continent.
2. Direct students to create a cross-sectional diagram illustrating locations of groundwater. Have students include a legend noting water table, well, and aquifer.
3. Ask students to create an individual water-use log for one day. They should note the time of day and the way they used water. After completing the log, students should write a paragraph with the following topic sentence: Water is one of the most valuable resources in our daily lives.
4. Have students complete the Chapter 6 Critical Thinking Worksheet.

CLOSURE

Divide the class into four groups, and ask each group to present a one-minute summary of the main ideas of one lesson objective.

ASSESSMENT

1. Have students complete the following questions from the Chapter 6 Check:
   Building a Vocabulary 3, 4
   Recalling and Reviewing 3, 4
   Critical Thinking 5
2. Ask students to complete the Chapter 6 Review/Test.

RETEACHING PLAN

1. Ask students to complete the Chapter 6 Reteaching Worksheet, Lesson 3.
2. Direct students to create illustrations or cut out pictures to show the different types of water bodies.

CHALLENGE/ENRICHMENT STRATEGY

1. Ask students to research the major transportation networks, using waterways for a selected hemisphere. Students should draw their findings on a map.
2. Suggest that students research and report on ways in which people in modern society use their water supply to improve their way of life.

Chapter 6 Check, page 77, Answers

Building a Vocabulary

1. The hydrosphere contains all the water of the earth. The hydrologic cycle is the circulation of water from one part of the hydrosphere to another, among the atmosphere, oceans, continents, ice sheets, and all living things.
2. The shallowest part of the ocean and the part of the sea floor that slopes from the continents.
3. Headwaters are the first streams to form from precipitation on hills and mountains. Tributaries are small streams that join larger streams. Estuaries are where rivers meet seawater.
4. water table
5. Evaporation is the process by which water is changed from liquid to gas. Transpiration is the evaporation of water through the leaves of plants. Evapotranspiration is the combined evaporation of water from the ground and transpiration by plants.
6. The watering of land through pipes, ditches, or canals. Because without water, neither industry nor agriculture can be productive.

Recalling and Reviewing

1. The Mariana Trench. The continental shelf.
2. Seawater is made of all the materials it dissolves, especially salt. It contains every element known on Earth. It is slow to heat up and to cool down, and it is 800 times denser than air.
3. Rivers bring fresh water and minerals to the coast. Estuaries are rich in fish and shellfish.
4. Precipitation. Water sinks into the ground and seeps downward until the ground is saturated. The top of this saturated zone is the water table. The water table follows the slope of landforms, and its depth varies. Groundwater flows into a rock layer called an aquifer; water is stored here and moves from space to space.

Critical Thinking

5. Answers will vary but might include reduced food production, limited electrical power, health and pollution problems, and atmospheric changes.

Using Geography Skills

1. Water evaporates from the oceans, rivers, and lakes and from plants. As it rises and cools, it condenses into moisture, which falls into the oceans and onto the continents in the forms of precipitation. The precipitation that falls onto the continents becomes ice, runoff that eventually returns to the oceans, and groundwater.
2. Marine life flourishes on the continental shelf because of the rich food supply. Because the ocean is shallow here, the sunlight warms the water and allows plant and animal life to flourish.

CHAPTER 7 World Patterns of Vegetation and Soils Pages 78–85

PLANNING GUIDE		
Lesson	**Textbook**	**Teacher's Resource Binder**
1. Plant Communities	pp. 78–81	Lecture Notes and Transparency Package Content/Vocabulary Workbook Reteaching Worksheet
2. Soil and Biomes	pp. 81–84	Critical Thinking Worksheet Reteaching Worksheet
3. Biomes and Changes in the Biosphere	p. 84	Skills Worksheet Chapter Review/Test Reteaching Worksheet Challenge Worksheet

TIME ALLOCATION: 3 days

CHAPTER TOPICS

- Plants as a Major Resource
- Plant Communities and Succession
- Formation and Characteristics of Soil
- Basic Biomes
- People and the Biosphere

PRETEACHING CHAPTER VOCABULARY

Write the following terms on the chalkboard:

biosphere	humus	coniferous forest
food chain	soil horizons	savanna
plant community	leaching	prairie
plant succession	biome	steppe
climax community	deciduous forest	

Ask students to list any terms or parts of terms that they can already define. Discuss the terms, calling on volunteers to provide definitions, and direct students to add to their list terms that they can define after the class discussion. The following are suggestions for stimulating discussion.

Ask students if someone can define the prefix *bio-*, which is found in the terms *biosphere* and *biome.* (*Bio-* is a Greek term that means "life" or "living things.") Point out that *sphere* in this context refers to an environment in which things exist. Discuss how this context could relate to the context of the term in geometry. Point out that *-ome* comes from the field of botany and means "a group." Ask students to suggest final definitions for the terms and compare them to the definitions in the glossary. Students may suggest other words that are related to the word parts *bio-*, *sphere,* and *-ome.*

A key to decoding *climax community* is to understand the ecological context of the words. In ecology (the science of relationships between organisms and their environment), *climax* refers to the stage of development in which a community of organisms becomes stable and begins to perpetuate itself. *Community* refers to a group of plants or animals living in a region under similar conditions. Ask students to consider a definition of the term related to plants.

Succession refers to a sequence or order. Ask students to suggest a definition of *plant succession. Horizon* refers to a layer of soil as seen in a cross section of the land. *Leaching* occurs when rain washes essential nutrients downward and out of topsoil, depleting its fertility. Ask students how leached soil compares to alluvial soil. *Deciduous* comes from a Latin word that means "to fall off." *Coniferous* comes from a Latin word that means "cone-bearing." Suggest examples of trees that lose their leaves at the end of a growing season and those that bear cones and remain evergreen. Then ask students to define the terms *deciduous forest* and *coniferous forest.* Indicate that a *steppe* is a dry, short grassland.

▶ LESSON 1
Plant Communities

LESSON OBJECTIVES

The student should be able to

- define the significant terms of the chapter
- identify the importance of the earth's plant resources in sustaining life on Earth
- identify the stages of a food chain
- explain the concept of plant communities
- provide an example to illustrate plant succession
- identify examples of human actions and natural events that can affect plant succession

PURPOSE

The study of this lesson will enable students to develop an appreciation of the role of plants in sustaining animal and human life on earth.

MOTIVATOR

Ask students to list the foods that they ate during the previous day. They should classify the foods by source: plant or animal. For foods they ate from an animal source, they should note the source of that animal's food.

Example:

Foods I Ate	Food Source	Food Source
Sandwich:		
bread	plant	
lettuce	plant	
tomato	plant	
roast beef	animal	plant

As students review their lists, ask them to write a statement summarizing the importance of plants as a source of life. Explain that in this lesson, they will learn more about the key role of plants in their lives.

TEACHING DESIGN

1. Use the Lecture Notes and Transparency Package to introduce the main ideas of the lesson. Lead a class discussion to create a food-chain diagram on the chalkboard. Point out that the states of the food-chain cycle can be compared to another cycle studied: the hydrologic cycle. Ask students to suggest points of comparison. Point out that both cycles use the materials of the earth in an efficient and continuous manner.
2. After discussing plant succession related to the illustration "Forest Succession After a Fire" on page 80 of their textbook, ask students to think of two examples of plant succession in their own community. One example should represent a natural sequence resulting in a climax community; the other should represent a sequence that is changed by human or natural events. Students may want to consider areas such as farmland, ranch land, lawns, gardens, and flower beds. Students should share their examples with classmates.

CLOSURE

Write the following terms on the chalkboard: *biosphere, food chain, plant communities, plant succession.* Ask students to write a paragraph summarizing the importance of plants as a source of life for animals and people. The vocabulary terms should be utilized in the paragraphs. Call on volunteers to share their paragraphs.

ASSESSMENT

1. Have students complete the following questions from the Chapter 7 Check:
   Building a Vocabulary 1, 2
   Recalling and Reviewing 1
2. Ask students to complete the Chapter 7 Content/Vocabulary Worksheet in their Workbook.

RETEACHING PLAN

1. Ask students to complete the Chapter 7 Reteaching Worksheet, Lesson 1.
2. Provide students with the list of terms, and ask them to cut a sheet of paper into four parts. One term should be written on each section of the paper, with the appropriate definition on the other side. Students should check their definitions with a partner, then take turns asking each other the meanings of the terms.
3. Divide students into pairs, and ask half of the class to find photographs in their textbook of a forest (page 79) and the other half to find a desert (page 84) to explain the concept of plant communities. Students should address the following ideas in their explanations: effect of climate on plant communities; needs of plants for sunlight, water, and nutrients; and plant interdependence within each community. Each student should present his or her explanation to a student who used a different photograph.

CHALLENGE/ENRICHMENT STRATEGY

1. Ask students to collect colorful pictures of various plant communities to display in the classroom.
2. Divide students into groups of three or four, and have them create a visual on a transparency to represent a "plant factory." Components of the factory should include sun, sun's rays, air, carbon dioxide, plants, soil, water, nutrients in the soil, and oxygen. A representative of each group should explain the group's visual representation of how plants produce oxygen and food for sustaining animal and human life.

▶ LESSON 2
Soil and Biomes

LESSON OBJECTIVES

The student should be able to
- identify major factors of soil formation
- describe the characteristics of soil horizons
- identify the five basic biomes of the world
- describe the characteristics of four types of forests in a forest biome

PURPOSE

The study of soil and biomes helps students develop an understanding of the relationships among climate, soil, and natural vegetation. Knowledge of this interaction is a key to understanding the impact of the environment on ways of life in a region.

MOTIVATOR

Display pictures of three different types of biomes, and ask students to serve as photo analysts. Students should view the pictures, then suggest hypotheses regarding the climate, types of vegetation, and types of animals one could expect to find in the areas shown. After students have discussed each picture, explain that the plant and animal community of an area is referred to as a *biome*. Ask students to predict what factor might cause the major differences in the vegetation of different biomes.

TEACHING DESIGN

1. Ask students to review the material in the section "Soil Types" on pages 81–82 of their textbook in order to complete the information map shown below. Students should begin with the box labeled "Soil."

What factors control the development of soil types?

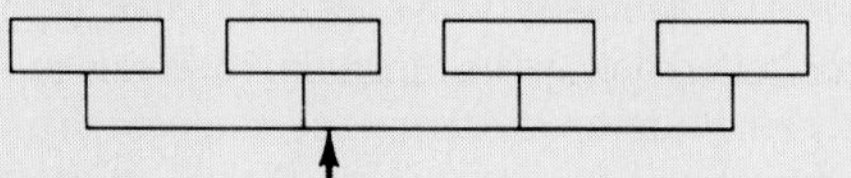

What is the most important factor in the development of soil types?

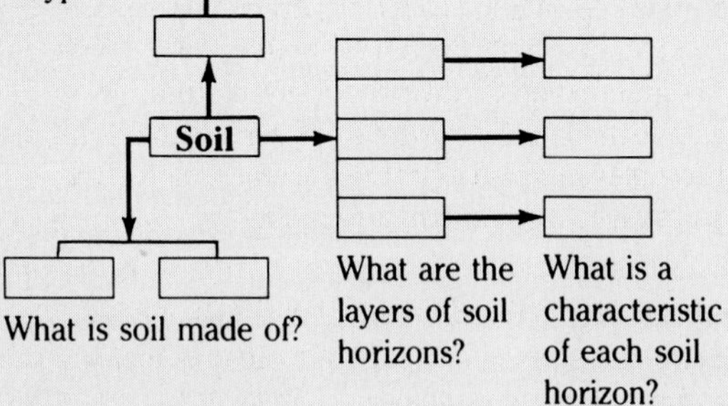

Each student should share the completed map with another student. Create a complete class information map on the chalkboard or on a transparency so that students may check their work.

2. Review the term *biome* and the fact that differences in biomes correspond to differences in climate. Have students complete the following chart, using information from Chapters 5 and 7. (Information for the climate heading is found in Chapter 5.) The chart for each biome should be written on a separate sheet of paper or on an index card.

Forest Biome			
Type of Community	**Location**	**Climate**	**Type of Plants**
Tropical rain forest			
Middle-latitude forest			
Boreal forest			
Mediterranean scrub forest			

Discuss the chart after students complete their work. Ask students to point out an example of each type of forest community on a wall map of the world.

CLOSURE

1. Direct a question-answer session to review the main ideas of the lesson's objectives.
2. Have students complete the Chapter 7 Critical Thinking Worksheet.

ASSESSMENT

Have students complete the following questions from the Chapter 7 Check:

Building a Vocabulary 3, 4, 5
Using Geography Skills 1, 2

RETEACHING PLAN

1. Ask students to complete the Chapter 7 Reteaching Worksheet, Lesson 2.
2. Provide students with index cards on which to copy the information map in Teaching Design 2. After reviewing their completed maps, have students arrange the cards in the order they choose. Students should then explain the information on their cards to a partner. The partner should then do the same. Each listener should check the presenter's explanation for accuracy, using the completed information map as a guide.

CHALLENGE/ENRICHMENT STRATEGY

1. Have students collect magazine pictures that represent the different types of biomes. Display the pictures in the classroom.
2. Ask students to contact a county agricultural agent or the owner of a plant nursery to investigate the types of soil in the local community. Students should report their findings in class.

▶ LESSON 3
Biomes and Changes in the Biosphere

LESSON OBJECTIVES

The student should be able to

- describe the characteristics of savanna, grassland, desert, and tundra biomes and of barren regions
- identify natural and human actions that can change the biosphere
- explain the importance of preserving natural resources

PURPOSE

Through study of the basic biomes and changes that can occur as the result of human or natural causes, students should begin to develop an awareness of ecological balance and the need to preserve natural resources for the future survival of people and nature.

MOTIVATOR

Divide students into groups of four or five. List the five basic biomes of the world on the chalkboard, and review the vegetation of each. Assign each group a different biome, and tell students to imagine that they live within this type of biome. Each group should determine how they would live in this area, what occupations they would have, what problems they might encounter, and what changes they would make in the natural conditions of the area. Each group should discuss its assigned biome. After each group shares information, ask if the changes in the natural conditions would be beneficial or damaging to the balance of life in the area.

TEACHING DESIGN

1. Direct students to complete a chart in which they show the characteristics of the savanna, grassland, desert, and tundra biomes and of barren regions. After students complete their charts, create a master chart on the chalkboard or on a transparency so that students may check their work.
2. Divide students into groups of four or five. Ask each group to study a different picture in the textbook representing a biome and to explain the characteristics of the biome. A representative from each group should then explain the biome to the class.
3. Have students brainstorm as a class to develop a list of ways that natural events and human actions can cause changes in the biosphere. After listing the ideas on the chalkboard, select one natural event and one human action, and predict the effects of each in the form of a futures' projection, as shown below. Add to the diagram to include as many positive and negative effects as needed. After the diagram is completed, the students should prepare a generalization about the importance of preserving natural resources.

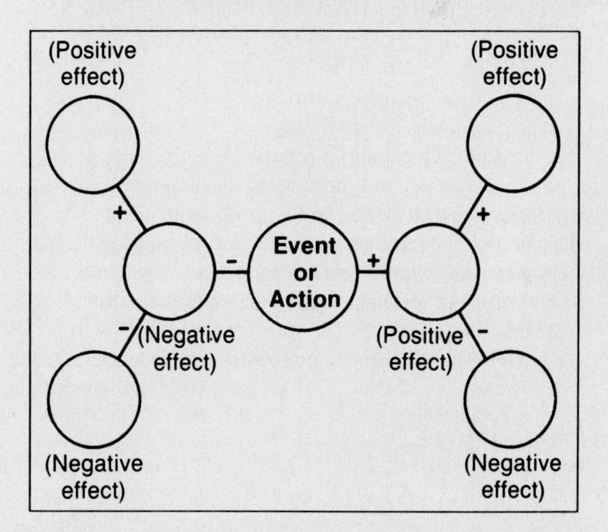

4. Have students complete the Chapter 7 Skills Worksheet.

CLOSURE

Direct students to turn to the map "The World's Basic Biomes" on page 82 of their textbook. Call on volunteers to state the characteristics of each biome on the map.

ASSESSMENT

1. Have students complete the following questions from the Chapter 7 Check:
   Recalling and Reviewing 2
   Critical Thinking 3, 4, 5
   Using Geography Skills 3
2. Ask students to complete the Chapter 7 Review/Test.

RETEACHING PLAN

1. Ask students to complete the Chapter 7 Reteaching Worksheet, Lesson 3.
2. Direct students to work in small groups to describe the characteristics of various biomes depicted in illustrations on pages 78–84 in their textbook. Students should use the pictures as a reference source.

CHALLENGE/ENRICHMENT STRATEGY

1. Ask students to review their charts to determine the biome with the greatest potential for varied agricultural production. Students should provide reasons for their answer.
2. Ask students to create a list of 20 countries and identify the biome(s) found in each. Students should predict types of economic activity found in each country, using encyclopedias to verify their answers.
3. Have students complete the Chapter 7 Challenge Worksheet.

Chapter 7 Check, page 85, Answers

Building a Vocabulary

1. All living things and the areas they inhabit on Earth. Yes.
2. The food chain is a series of stages in which energy is passed along through living things. Some animals eat plants; these animals in turn are eaten by other, usually larger, animals. At each stage of the food chain, the number of living things is reduced. Energy is lost as each higher level eats the production of the level below it. Humans are at the top of the food chain.
3. When the necessary nutrients for plant growth are washed downward out of the topsoil by heavy rainfall. As the result of leaching, nutrients cannot be reached by plant roots.
4. A plant and animal community that covers a large area of the world's surface. Climate influences the type and quantity of vegetation, as well as the water supply; therefore, it has a direct impact on the kind of life each biome can support.
5. deciduous; coniferous
6. Plant succession occurs when one group of plants replaces another. The plants that are able to grow after a forest fire, for instance, differ from the plants that can grow in a fully recovered forest. The process of plant succession can eventually lead to the formation of a climax community.

Recalling and Reviewing

1. A climax community occurs only after a long process of plant succession. A climax community can be destroyed by natural events such as forest fires, storms, volcanic eruptions, or plant disease. Human activities.
2. Desert biome plants survive by using very little water or by storing water. Polar regions are called barren regions because their year-round ice and snow prohibit the growth of plant life.

Critical Thinking

3. Answers will vary but should include specific vegetation such as grass and weeds, mosses and ferns, land or water plants, flowering or flowerless plants, and seed or seedless plants.
4. Answers will vary but should include major characteristics of two or more biomes.
5. Answers will vary but should include details about climate, growing season, and type of soil in the area.

Using Geography Skills

1. Middle-latitude forest.
2. Eastern half of the United States.
3. Barren regions and desert and tundra biomes.

CHAPTER 8 World Cultures Pages 86–93

PLANNING GUIDE		
Lesson	**Textbook**	**Teacher's Resource Binder**
1. Cultural Geography, Culture Regions, and Culture Change	pp. 86–88	Lecture Notes and Transparency Package Content/Vocabulary Workbook Reteaching Worksheet
2. People on the Land	pp. 88–90	Lecture Notes and Transparency Package Skills Worksheet Reteaching Worksheet Challenge Worksheet
3. Culture and World Events; Cultural Geography in the Future	pp. 90–92	Critical Thinking Worksheet Chapter Review/Test Reteaching Worksheet

TIME ALLOCATION: 3 days

CHAPTER TOPICS

- Definition of Cultural Geography
- Culture Regions
- Culture Change
- Culture and World Events
- Cultural Geography in the Future

PRETEACHING CHAPTER VOCABULARY

Write the following terms on the chalkboard:

cultural geography	domestication
culture	subsistence agriculture
culture trait	urbanization
culture region	industrialization
innovation	nationalism
diffusion	totalitarian government
acculturation	democratic government

Have students complete the following chart to rate their knowledge of the terms before reading the chapter.

Words	Words I Can Define	Words I Have Seen or Heard	New Words

Advise students to pay careful attention to words they note under "New Words" when they encounter them in their textbook.

Point out that six of the terms include the word or forms of the word *culture,* which comes from a Latin word that means "to cultivate" or "to grow." Ask students to suggest how this meaning relates to the meaning used in the textbook. (all the features of a society's way of life)

Point out the prefix *agri-* in *subsistence agriculture*. It comes from a Greek word that means "field," "earth," or "soil." *Subsistence* comes from a Latin word that means "to continue to exist" or "to maintain." Ask a volunteer to look up the definition of *subsistence agriculture* in the glossary. Ask others to relate the definition in the glossary to the definition of the Latin word.

The term *urban* comes from a Latin word that means "city." Ask students to suggest an antonym for *urban*. (*rural*)

Ask students to identify the root words in *nationalism* (*nation*) and *totalitarian* (*total*). *Total* in this context means "complete" or "absolute." Point out that *democratic* comes from the Greek word *demos,* meaning "people."

▶ LESSON 1

Cultural Geography, Culture Regions, and Culture Change

LESSON OBJECTIVES

The student should be able to

- define significant chapter vocabulary
- identify culture traits
- explain the concept of culture regions
- explain the process of culture change related to innovation, diffusion, and acculturation
- identify examples of innovations in his or her culture

PURPOSE

By studying culture and culture change, students learn the similarities and differences in the ways people live. Classification by culture regions assists students in understanding their world.

MOTIVATOR

Create a word web on a transparency with *culture* as the main idea. Ask students to identify words related to this term and add them to the web. Retain the web for use in the Closure.

TEACHING DESIGN

1. Present the Chapter 8 Lecture Notes and Transparency Package to introduce the concepts of cultural geography, culture regions, and culture change. Use the transparency to illustrate culture diffusion, exemplified by the spread of Christianity.
2. Direct students to turn to the map "World Religions" on page 91 of their textbooks and think about different culture traits, as well as the religions, that are shared in various regions. Ask students to suggest possible divisions of the world according to culture regions. (political boundaries, languages, religions, and so on)
3. Suggest that students work in small groups to discuss the problem of culture change, with each group assigned one of the questions listed below.
   a. What is an innovation?
   b. What are modern examples of innovation?
   c. What are examples of innovation in history?
   d. What could cause an innovation to be accepted or rejected by a culture group?
   e. What are some innovations our culture has borrowed from other cultures?
   f. What is cultural diffusion?
   g. How do geographers trace cultural diffusion? How does diffusion occur?
   h. How does acculturation occur?
   i. What is an example of acculturation?

Call on a member of each group to answer the question.

CLOSURE

Create a new word web for *culture* similar to the one used in the Motivator, with students adding words and ideas from the lesson. Compare the two word webs to note similarities and differences.

ASSESSMENT

1. Have students complete the following questions from the Chapter 8 Check:
   Building a Vocabulary 1, 2, 3

Recalling and Reviewing 1, 2
Using Geography Skills 1

2. Have students complete the Chapter 8 Content/Vocabulary Worksheet in their Workbook.

RETEACHING PLAN

1. Ask students to complete the Chapter 8 Reteaching Worksheet, Lesson 1.
2. Create a matching activity with the key terms and their definitions. Give each pair of students a set of cards, each card having either a term or a definition on it. Each pair of students should display the entire set of cards and take turns matching the terms with the definitions.

CHALLENGE/ENRICHMENT STRATEGY

1. Have students use a world atlas to investigate regions of the world that share religions, languages, political systems, and economic systems.
2. Direct students to investigate the diffusion of a language or a religion from the beginning of its known history to the present. Students should report their findings in class, using maps, time lines, and other visual aids.
3. Ask students to consider how modern communication, transportation, and trade affect cultural diffusion, then write a paragraph explaining their views. Students may want to consider whether or not cultural diversity is increasing or decreasing.

► LESSON 2
People on the Land

LESSON OBJECTIVES

The student should be able to

- identify three innovations that have affected the world's cultural geography
- trace the development of how people have met the need for food, shelter, and clothing
- distinguish between subsistence and commercial agriculture
- explain the relationship between urbanization and cultural diffusion
- identify culture changes that have resulted from industrialization

PURPOSE

In this lesson, students will gain a clearer understanding of the relationship of people to the land on which they live. Students also will learn the importance of the effects of innovation on people's ways of life.

MOTIVATOR

Ask students to consider the following situation:

You have been transported by means of a time machine to an area uninhabited by people. What problems must you solve in this new environment in order to survive?

After students discuss their solutions, ask volunteers to make a generalization about people's basic needs. Tell students that they will learn in this lesson how people have met these needs throughout history.

TEACHING DESIGN

1. Review basic human needs as related to the Motivator. Use the Chapter 8 Lecture Notes and Transparency Package to trace the major innovations that changed human culture.
2. Suggest that students work in groups of three to explain why productivity is more limited in subsistence farming than in commercial agriculture. One person in each group should record the group's ideas. Have students take notes as a representative from each group presents the ideas to the class.
3. Ask students to work in small groups to create a visual that would represent the key role that cities or urban centers play in cultural diffusion. The students' visuals should be discussed and displayed in the classroom.
4. Lead a class discussion of the culture changes that have occurred as the result of industrialization. You may want to create a web or diagram to represent students' ideas.

CLOSURE

Ask students to list in note form the main ideas of the lesson. Call on students to share their ideas.

ASSESSMENT

1. Have students complete the following questions from the Chapter 8 Check:
   Building a Vocabulary 4, 5, 6
   Recalling and Reviewing 3, 4
2. Have students complete the Chapter 8 Skills Worksheet.

RETEACHING PLAN

1. Ask students to complete the Chapter 8 Reteaching Worksheet, Lesson 2.
2. Have students complete the following questions, using their textbooks for reference.

People on the Land

a. The three basic needs that all people have are _____, _____, and _____.
b. The three innovations through history that have changed how people meet their needs are _____, _____, and _____.
c. The two types of agriculture are _____ and _____.
d. The growth of towns and cities is called _____.
e. Cities play a major role in the creation of most _____.
f. The growth of the method of production using factories, workers, and machinery is called _____.

CHALLENGE/ENRICHMENT STRATEGY

1. Ask students to investigate scientific and technological advances that are improving agricultural production.
2. Invite an urban geographer or a city planner to class to discuss careers in his or her field.
3. Ask students to investigate modern methods of production for a selected product. (Examples: automobiles, aircraft, computers, radios, clothing, beverages.) Students may wish to compare production methods in the United States with those in another country.
4. Have students complete the Chapter 8 Challenge Worksheet.

▶ LESSON 3

Culture and World Events, and Cultural Geography in the Future

LESSON OBJECTIVES

The student should be able to

- identify culture traits that have related to conflicts among people and nations
- explain the relationship between culture traits and conflict
- identify examples of culture conflicts in past and current times
- identify the advantages of respecting cultural diversity among people of the world

PURPOSE

This lesson contains important concepts that promote cultural understanding and tolerance of people with different cultures. Tolerance and appreciation of cultural diversity are keys to promoting world peace.

MOTIVATOR

Provide students with copies of daily and/or Sunday editions of a major newspaper. Divide the class into groups, and ask students to circle the headlines of news stories about conflicts at state, national, and international levels. One student in each group should list what appears to be the source, or underlying cause, of each conflict. Ask a representative from each group to share the group's information, and record it on the chalkboard. Ask students to review the class' list and suggest categories for classifying the sources of conflict. Ask students whether or not there is a relationship between conflict and culture traits.

TEACHING DESIGN

1. Provide the following chart design on the chalkboard, and ask students to complete it, working in groups of three. For each culture trait, students should list a conflict in the past and one in the present that has occurred because of cultural differences.

Culture and World Events		
Culture Traits	**Conflict in the Past**	**Conflict in the Present**
Religion		
Nationalism		
Traditional and modern values		
Politics		
Economics		

Discuss each culture trait and its relationship to conflicts in the past and present. Ask students to propose solutions to the problems of culture conflicts and to suggest advantages of cultures that are different from theirs.

2. Have students complete the Chapter 8 Critical Thinking Worksheet.

CLOSURE

Ask students to think about differences in culture traits and suggest solutions to world conflicts that stem from cultural diversity.

ASSESSMENT

1. Have students complete the following question from the Chapter 8 Check:
   Critical Thinking 5, 6
2. Ask students to complete the Chapter 8 Review/Test.

RETEACHING PLAN

1. Ask students to complete the Chapter 8 Reteaching Worksheet, Lesson 3.
2. Have students complete the following information map to review the lesson.

Culture and World Events
What culture traits are sources of conflict among people?
What is an example of a conflict related to differences in each culture trait?

CHALLENGE/ENRICHMENT STRATEGY

1. Direct students to investigate the religious beliefs of Jews, Muslims, and Christians, noting similarities and differences. Ask students to make a generalization regarding religious conflicts in the Middle East.
2. Have students use an atlas to research the distribution patterns of the world's resources such as oil, coal, iron, fresh water, and arable land. Students could map or

chart the distribution. They should report their findings in class and suggest possible world problems that could result from unequal distribution of resources.

3. Direct students to react to the following statement by writing a paragraph:

   The knowledge of a group's language is the key to unlocking the mysteries of that group's culture.

Chapter 8 Check, page 93, Answers

Building a Vocabulary

1. Acculturation occurs when one culture changes as it meets another. Diffusion occurs when an innovation or culture trait moves through a society.
2. A culture region is an area where there are shared culture traits. Examples include Japan, the southern United States, and Scandinavia.
3. A thing that a group of people normally does. Examples can include dress, patterns of communication, religious and social traditions, and architecture. Those seen every day will vary.
4. Urbanization.
5. industrialization
6. Studying cultural geography is important because many geographic problems remain to be solved and because understanding the differences and similarities among different groups of people may help resolve conflicts in the future.

Recalling and Reviewing

1. Food, clothing, and shelter are the basic needs that all people have. No, these needs have not always been met in the same way. Hunting and gathering, agriculture, urbanization, and industrialization, developed in that order, are how individuals and then societies have adapted and learned to meet their basic needs.
2. When people grow food mainly for themselves, they practice subsistence agriculture; their farm tools are simple, and they use animals as the main source of power. In commercial agriculture, farmers grow food and other crops to sell to others for profit. Machinery is used, and large farms and fewer workers are typical.
3. Communication of innovations and other culture traits occurs more rapidly in cities. New ideas are more quickly tested and adopted, or discarded. Because of transportation, cities receive more input from other cultures and influence other cultures as well.
4. Answers will vary but should focus on the acquisition of food, clothing, and shelter.

Critical Thinking

5. Answers will vary but might include conflicts in Lebanon, between Israel and Arab countries, between the United States and the Soviet Union, and between ethnic groups in individual African nations, such as Nigeria, Zaire, and Kenya.
6. Answers will vary but might include innovations in technology, medicine, architecture, and language.

Using Geography Skills

1. Answers will vary. To be exposed to innovations and various culture traits, one would have to go first to an urban center. In Australia, the urban centers are located in the coastal region.

CHAPTER 9 Economic Development Pages 94–103

PLANNING GUIDE		
Lesson	**Textbook**	**Teacher's Resource Binder**
1. Economic Geography	pp. 94–95	Content/Vocabulary Workbook Reteaching Worksheet
2. Wealth and Poverty	pp. 96–97	Lecture Notes and Transparency Package Reteaching Worksheet Challenge Worksheet
3. Population Growth and Economic Development	pp. 97–102	Reteaching Worksheet
4. Using a Population Pyramid	p. 98	Skills Worksheet Chapter Review/Test Reteaching Worksheet Critical Thinking Worksheet

TIME ALLOCATION: 4 days

CHAPTER TOPICS

- Economic Geography
- Wealth and Poverty
- Population Growth
- Economic Development
- Population Pyramids

PRETEACHING CHAPTER VOCABULARY

Write the following terms on the chalkboard:

economic geography
primary economic activities
secondary economic activities
tertiary economic activities
developed countries
less developed countries (LDCs)
illiteracy
gross national product (GNP)
capitalism
free enterprise
command economy
communism
population geography
birthrate
death rate
multinational company

Use root, compound, and common words to develop an understanding of the following terms:

multinational company	*multi-* means "many" *nation-* means "nation"
population geography	*populus* means "people"
command economy	(Greek) *oikonomia* means "household manager"

Ask students to read in their textbook Glossary the definitions of any terms they cannot define.

▶ LESSON 1
Economic Geography

LESSON OBJECTIVES
The student should be able to
- define significant vocabulary
- distinguish among primary, secondary, and tertiary economic activities
- explain economic geography

PURPOSE
When they are adults, students will find that the place where they live may influence their ability to be successful in their chosen career. Students should study this section carefully to learn how geography and economics are related, as the relationship can affect their future.

MOTIVATOR
Ask students to contribute to a word web on the chalkboard based upon the phrase *Careers and Jobs*. Ask students to suggest different methods of classifying and reclassifying the information on the chalkboard. Save the word web for use in Teaching Design 3.

TEACHING DESIGN
1. Introduce the concept of scarcity to the class in the following manner: Indicate to students that everyone has unlimited wants. Demonstrate this concept by asking students to list all their wants. Conclude that the list could be infinite. Ask students why individuals do not have all they want. Conclude that the world's resources necessary for satisfying all the wants are limited. Tell students that the reason for scarcity is unlimited wants and limited resources. Direct students to create a graphic display to represent this concept.
2. Announce to students that one definition of *economics* is "the study of how people use scarce resources to satisfy needs and wants." Discuss what they think *economic geography* means.
3. Present the three categories of economic activities introduced in the chapter. (primary, secondary, and tertiary) Ask students to work in groups and classify the careers and jobs on the chalkboard (from the word web in the Motivator) into the three categories.

CLOSURE
Direct students to tell a nearby classmate a definition of *economic geography* and an example of each of the three types of economic activities.

ASSESSMENT
1. Have students complete the following questions from the Chapter 9 Check:
   Building a Vocabulary 1, 4
   Critical Thinking 3
2. Ask students to complete the Chapter 9 Content/Vocabulary Worksheet in their Workbook.

RETEACHING PLAN
1. Have students copy the following chart and complete it with examples.

Economic Geography	Examples
Earth's resources	
Ways to earn a living	
Distribution of products	

2. Ask students to complete the Chapter 9 Reteaching Worksheet, Lesson 1.
3. Select students to locate and classify pictures from magazines as examples of economic activities. Then have them arrange the pictures on the classroom bulletin board.

CHALLENGE/ENRICHMENT STRATEGY
1. Have students select 10 major cities of the world and identify their major economic activities. Classify the activities as primary, secondary, or tertiary. Tell students that they should look for conclusions to be drawn.
2. Ask students to locate newspaper articles related to economic geography and to describe the relationship.

▶ LESSON 2
Wealth and Poverty

LESSON OBJECTIVES
The student should be able to
- classify countries according to their economic system
- describe the differences between developed and less developed countries

PURPOSE
This lesson will give students an appreciation for the free enterprise system of the United States.

MOTIVATOR
Write the following schematic on the chalkboard:

←——————————————→
Less Developed **Developed**

Ask students to name and place various countries on the line. Discuss their reasons for the placements. Save the diagram for use in the Closure.

TEACHING DESIGN

1. Present the Chapter 9 Lecture Notes and Transparency Package.
2. Review the differences among free enterprise (market), command, and traditional economies. Ask students to contribute to the completion of the chart below.

Who Decides...	Free Enterprise	Command	Traditional
What to produce?	(market, or consumers)	(government)	(family)
How much to produce?	(supply and demand, or consumers)	(government)	(family)
For whom to produce?	(consumers)	(government)	(family)

4. Select a group of 20 countries. Distribute almanacs and encyclopedias. Decide whether the economies of the countries are free enterprise, command, or traditional. Classify the countries as developed and less developed. Rank the countries by GNP and per capita income.

CLOSURE

Return to the schematic from the Motivator. Ask students to review their placements and decide whether they wish to move any countries. Have them discuss their reasons.

ASSESSMENT

Have students complete the following questions from the Chapter 9 Check:

Building a Vocabulary 2, 3
Recalling and Reviewing 1

RETEACHING PLAN

1. Have students read an introductory chapter in a basic economics book. Direct them to prepare their own outline of the information presented.
2. Ask students to complete the Chapter 9 Reteaching Worksheet, Lesson 2.

CHALLENGE/ENRICHMENT STRATEGY

1. Have students complete the Chapter 9 Challenge Worksheet.
2. Ask students to list all the countries that appear in the newspaper in one day. Have them classify these according to the level of development. Continue the project for a week. What conclusions can the students draw?

▶ LESSON 3

Population Growth and Economic Development

LESSON OBJECTIVES

The student should be able to

- define significant vocabulary
- indicate the level of population growth in the world
- describe ways for countries to improve their economic conditions

PURPOSE

This lesson is important for helping students understand the impact of population growth and its relationship to economic development. This is a topic that all adults are expected to be able to discuss.

MOTIVATOR

Create a word web on the chalkboard around the phrase *Population Growth*. After writing numerous words on the chalkboard, rate each word with a + or −, depending on the class interpretation. Indicate to students that population growth causes many positive and negative effects.

TEACHING DESIGN

1. Review appropriate chapter vocabulary, and explain how these words affect governmental policies.
2. Have students write a paragraph describing what they perceive as the impact of population growth on the world.
3. Discuss with students alternatives for countries that desire to further their economic development as presented in the chapter.
4. Create a list of less developed countries. Ask students to investigate each country and select an appropriate alternative for economic growth. Ask them to decide which alternative from the chapter would help the country most. Students should defend their answers.
5. Have students read the Cities of the World feature "Urban Land Use" on pages 100–101 of their textbook. Discuss how the information in this feature applies to students' hometown.

CLOSURE

Direct students to write a generalization about population and economic development. Read several to the class.

ASSESSMENT

Have students complete the following question from the Chapter 9 Check:

Recalling and Reviewing 2

RETEACHING PLAN

1. Ask students to list at least three ways to improve the conditions of less developed countries.
2. Ask students to complete the Chapter 9 Reteaching Worksheet, Lesson 3.

CHALLENGE/ENRICHMENT STRATEGY

Instruct students to select a less developed nation. Have them research the country and submit in writing a mock history of its economic development as if it were written in the year 2100.

▶ LESSON 4

Using a Population Pyramid

LESSON OBJECTIVES

The student should be able to

- define *population pyramid*
- interpret a population pyramid
- analyze the impact of population patterns on the world

PURPOSE

Like Lesson 3, this lesson is important for helping students understand the impact of population growth and its relationship to economic development.

MOTIVATOR

Review the basic content of Lesson 3. Ask students what they know about the Egyptian pyramids. Show some pictures of pyramids. Ask students what they think a population pyramid is and why a pyramid is chosen to represent this information.

TEACHING DESIGN

1. Have students read the Geography Skills feature "How to Use a Population Pyramid," on page 98 of their textbook.
2. Discuss with students the relationship between the Egyptian pyramids and the population pyramids.
3. Have students complete the Chapter 9 Skills Worksheet.
4. Bring to class other maps that provide information about population. Ask students how to interpret each map.

CLOSURE

Review with students how geographers use population pyramids.

ASSESSMENT

1. Have students complete the following questions from the Chapter 9 Check:
   Using Geography Skills 1, 2
2. Ask students to complete the Chapter 9 Review/Test.

RETEACHING PLAN

Ask students to complete the Chapter 9 Reteaching Worksheet, Lesson 4.

CHALLENGE/ENRICHMENT STRATEGY

1. Ask students to identify and complete the following chart on five multinational companies.

Company Name	Products Sold	Countries Located	Gross Sales	Charities Supported

2. Have students complete the Chapter 9 Critical Thinking Worksheet.

Geography Skills, page 98, Answers

How to Use a Population Pyramid

1. Soviet women: 2.5 percent; 4 percent. American women: 4 percent; 3.5 percent.
2. Brazil: 0–4; 75–79. 0+; United States: 25–29, 80+; Soviet Union: 20–24, 80+. In developed countries, the majority of the population is in its productive, working phase; in less developed countries, the majority of the population is young.
3. Most students will fall into the 10–14 or 15–19 age categories. They can determine the percentage of their age bracket by measuring the column against the horizontal axis.

Chapter 9 Check, page 103, Answers

Building a Vocabulary

1. Primary economic activities are generally controlled by physical geography; examples are agriculture, forestry, and mining. Secondary economic activities are food processing and manufacturing; examples are dairies, bakeries, and steel industries. Tertiary economic activities are service industries; examples are stores, restaurants, banking and insurance, government activities, education, and transportation.
2. A type of economy in which resources, businesses, and industries are owned by private individuals. Some examples are the United States, Canada, some countries of Western Europe, Japan, Israel, Australia, and New Zealand.
3. gross national product (GNP)
4. Developed countries have higher standards of living than less developed countries; most people live in cities and work in manufacturing and service industries; products and services are widely available. Less developed countries have little industrial development, and most people practice subsistence agriculture. LDCs usually sell their natural resources to and purchase manufactured goods from developed countries; illiteracy is widespread, and populations are expanding.
5. In a command economy, prices of goods and labor are set by the government. Per capita incomes are generally lower than in a free enterprise system, where supply and demand determine prices and wages. In a free enterprise system, people have freedom to operate private businesses for a profit.
6. Birthrate is the number of births per 1,000 people in a given year. Death rate is the number of deaths per 1,000 people in a given year. These concepts provide a way to measure and predict a country's economic development and population growth.

Recalling and Reviewing

1. The study of economic geography is important because it shows us how the earth's resources are used, how products are distributed, and how people earn their living. It is necessary to understand these things in order to

make predictions and suggestions about future world development.
2. Rapid population growth makes economic progress very difficult. Increased numbers of people place pressure upon the natural resources, which are scarce in many countries, causing standards of living, including health, education, and housing, to remain poor.

Critical Thinking

3. Answers will vary. Students should describe the natural resources in their area that have contributed to economic development. They can list agriculture, manufacturing, and banking as examples of primary, secondary, and tertiary economic activities.

Using Geography Skills

1. 6 billion; 6.5 billion.
2. The first dramatic rise in population growth occurred after 1000 (around 1650). This was the result of improved health standards and living conditions. Population growth has exploded during the last half of the twentieth century.

UNIT 1 Review/Test

PLANNING GUIDE		
Lesson	Textbook	Teacher's Resource Binder
1. Unit 1 Review	pp. 104–105	
2. Unit 1 Test		Unit 1 Test

TIME ALLOCATION: 2 days

▶ LESSON 1
Unit 1 Review

Have students read the Unit 1 Summary. Use the Unit 1 Objectives as the basis for review, and discuss any questions that students have on the Unit 1 material.

Have students complete the questions in the textbook under the following headings:
Reading and Understanding
Mastering Geography Skills

Have students read the textbook material under the following headings and complete the appropriate or selected activities:
Applying and Extending
Linking Geography and Economics

▶ LESSON 2
Unit 1 Test

Have students complete the Unit 1 Test from the Teacher's Resource Binder.

Unit 1 Review, pages 104-105, Answers

Reading and Understanding

1. Location, place, relationships within places, movement, and regions. Movement (spatial interaction) because it involves the transporting and use of various resources from different places.
2. If the earth did not rotate on its axis, the side of the earth facing the sun would always be extremely hot, while the side away from the sun would always be extremely cold and dark.
3. The humid tropical climate is much hotter than the tundra climate because the regions near the equator receive much more solar energy than those areas near the poles.
4. The maritime climate is influenced by oceans and has moderate temperatures all year. Because a continental climate has no oceans to moderate temperature it is characterized by great temperature extremes.
5. Solar energy, air pressure, the general earth circulation systems, oceans, and mountains can determine the amount of precipitation that a region receives.
6. Because of the hydrologic cycle, life on Earth is possible. Without rain, groundwater, and plants, human beings could not survive. The earth is called the "water planet" because it is the only planet in the solar system to have liquid water. Because Earth is the only planet with water, it is the only planet with life.
7. Forest, savanna, grassland, desert, and tundra. Climate accounts for the differences among the biomes.
8. Agriculture meant that people could live closer in larger villages; therefore, population increased.
9. If an innovation is forced upon one culture by another, conflict can arise. For example, the introduction of Western ways into Middle Eastern cultures has caused conflict between traditional and modern values.
10. Multinational companies can provide jobs and develop natural and human resources. Many less developed nations do not have the resources to start new factories. Multinationals contribute to industrial development and the training of new workers.

Mastering Geography Skills

1. Barrow, Alaska: The cold climate and short growing season limit the variety of vegetation. Agriculture is not productive in this area; therefore, population would also be limited. San Francisco, California: Vegetation and population patterns would be fairly consistent, given the mild climate and amount of precipitation. Iquitos, Peru: Because of the high temperature and precipitation, vegetation would be dense and varied; population might be dense if it were possible to clear away the abundance of vegetation to make room for agricultural projects.
2. The Andes Mountains were formed when two plates collided. They were then shaped by glacier erosion.

3. It seems likely that the population of developed countries will stabilize or decrease in the future. Less developed countries are likely to continue their trend of high birthrates and death rates.

Applying and Extending

The highest temperature ever recorded was 136° F (58° C) at El Azizia, Libya, in northern Africa on September 13, 1922. The lowest temperature ever recorded was −126.9° F (−88.3° C) at Soviet Antarctic Station, Vostok, on August 24, 1960. The highest temperature for the United States was 134° F (57° C) in Death Valley, California, on July 10, 1913. The lowest temperature was −80° F (−62° C) at Prospect Creek, Alaska, on January 23, 1971. Elevations: Death Valley is 282 feet (86 meters); Prospect Creek is 1,640 feet (500 meters) to 6,560 feet (2,000 meters). Elevation affects climate in several ways. The higher the elevation, the colder the climate becomes. Various types of vegetation grow at the various elevation levels.

Linking Geography and Economics

Answers will vary but might include the limited amount of resources for industrial development found in these types of countries. In many less developed countries, the climate is not suitable to large-scale agriculture. Many less developed countries also have landforms such as deserts or tropical rain forests, which hinder economic development.

UNIT 2
Western Europe
Pages 106–185

UNIT OVERVIEW

Western Europe, though less than one-half the size of the United States, is one of the most densely populated regions on Earth. It contains 17 nations, six microstates, and a variety of landscapes, climates, and languages. Industrial development and relatively high standards of living characterize this region that once dominated the world. Economic and political cooperation among Western European countries has become important to the future of the region.

CHAPTER TIME LINE

Chapter	Title	Time Allocation
10	The Geography of Western Europe	3 days
11	Great Britain and Ireland	2 days
12	The Nordic Countries	2 days
13	France and the Low Countries	3 days
14	West Germany and the Alpine Countries	2 days
15	Italy and Greece	2 days
16	Spain and Portugal	2 days
	Unit 2 Review	1 day
	Unit 2 Test	1 day
		18 days

MAJOR TOPICS

- Relationships Among Land, Climate, and Ways of Life
- Historical Development Fostering Economic, Social, and Cultural Interchange
- Western Europe's Rise to World Domination
- Current Issues Facing Nations and the Region

INTRODUCING THE UNIT

Use a wall map of the world to introduce the culture region of Western Europe. Point out the area, and name the countries of the region. Ask students if any of their ancestors came from any of the countries of Western Europe. Note the size of the area in comparison to the rest of the world, as well as its relatively high population density. Identify countries throughout the world that were once colonies or territories under the rule of Western European countries. Ask students what impact Western European colonization had on other culture groups in many places of the world. Explain that the influences of Western European culture are still seen far beyond the boundaries of this culture region.

CHAPTER 10 The Geography of Western Europe Pages 108–121

PLANNING GUIDE		
Lesson	**Textbook**	**Teacher's Resource Binder**
1. Historical Geography	pp. 108–114	Lecture Notes and Transparency Package Content/Vocabulary Workbook Reteaching Worksheet
2. Physical Geography, Climate Regions, and Landform Regions	pp. 114–117	Challenge Worksheet Skills Worksheet Reteaching Worksheet
3. Natural Resources, Economy, and Political Ties	pp. 117–120	Critical Thinking Worksheet Chapter Review/Test Reteaching Worksheet

TIME ALLOCATION: 3 days

CHAPTER TOPICS

- Historical Geography
- Physical Geography
- Natural Resources
- Economic and Political Ties

PRETEACHING CHAPTER VOCABULARY

Write the following terms on the chalkboard:

microstate
imperialism
natural boundary
landlocked
European Community(EC)

raw materials	canal	tariffs
alliance	hydroelectricity	terrorism
peninsula	economic association	

Lead a discussion of the definitions by asking volunteers to provide definitions and/or examples of the terms. For terms needing clarification, point out specific information that may assist students' retention of the definitions. After discussing the terms, ask students to turn to a partner and provide an example that relates to each term.

micro-	(small)
imperialism	(*Imperial* comes from a Latin word meaning "empire"; the suffix *-ism* here means "a policy or doctrine.")
raw	(in a natural condition; prior to refining or manufacturing)
alliances	(formal agreements between nations for a common cause) Refer students to related words, such as *ally* and *allied,* and their usage.
peninsula	Ask students to identify an example of a peninsula in the United States.
natural boundary	Synonyms for *boundary* are *limit, border.*
canal	(a human-made waterway for irrigation, shipping, or travel) Point out major canals, such as Erie, Suez, and Panama.
hydroelectricity	(In Chapter 5, students learned that *hydro-* means "water.")
economic association	(an organized group with a common economic interest or purpose)
European Community	Point out that a community is a group of people or nations with common interests.
tariffs	(comes from an Arabic term meaning "information or notification." Tariffs are lists or notices of duties, or taxes, on imported or exported goods charged by governments.)
terrorism	(The suffix *-ism* in this sense refers to an action.)
landlocked	(cut off from the sea)

▶ LESSON 1
Historical Geography

LESSON OBJECTIVES
The student should be able to
- define the significant terms of the chapter
- sequence events in history
- identify areas of conflict among the nations

PURPOSE
The study of Western Europe is essential in order for students to understand the cultural heritage of many groups in the United States.

MOTIVATOR
Make a transparency of a political outline map of Western Europe. Ask volunteers to identify the various countries and label them on the transparency. Use question marks to identify any unknown nations. Ask volunteers to point out the language or languages spoken in each nation. Ask students to suggest reasons for the existence of so many different languages and countries in an area that is less than one-half the size of the United States. Indicate that students will better understand the reasons after this lesson.

TEACHING DESIGN
1. Use the Chapter 10 Lecture Notes and Transparency Package to trace the historical development of Western Europe.
2. Divide students into groups, and ask each group to investigate major time periods in Western European history. Provide encyclopedias and other reference materials for students to use, along with their notes from the lecture. Have the groups use the following chart to guide their research.

Western European History

Dates	Events	Conflicts	People

3. Ask each group to report the key information to the class, using maps and time lines as visual aids.

CLOSURE
Name key people, dates, and events in Western European history, and ask all students to think about their significance. Then call upon volunteers to share their thoughts.

ASSESSMENT
1. Have students complete the following question from the Chapter 10 Check:
   Building a Vocabulary 1
2. Ask students to complete the Chapter 10 Content/Vocabulary Worksheet in their Workbook.

RETEACHING PLAN
1. Ask students to complete the Chapter 10 Reteaching Worksheet, Lesson 1.
2. Provide students with a time line containing key dates in Western European history, and ask them to work with a partner to label corresponding events.

CHALLENGE/ENRICHMENT STRATEGY
1. Have students use a historical atlas to make maps showing changes of boundaries and names in the development of nations in Western Europe.
2. Direct students to create a map showing empires of Western European nations during the age of colonization and imperialism. Students should also identify the dates of independence in each colony or possession and any lasting effects of European culture.

3. Ask students to create a map identifying the major languages spoken in Western Europe today. Related research could include tracing the history of the development of these languages.

▶ LESSON 2
Physical Geography, Climate Regions, and Landform Regions

LESSON OBJECTIVES
The student should be able to
- locate the countries of Western Europe
- identify the major factors influencing the climates of Western Europe and their effects
- locate and describe the major landforms
- explain the significance of bodies of water in the development of Western Europe
- identify major natural resources and give their uses

PURPOSE
The study of the physical geography of Western Europe will enable students to understand the impact of environment on the ways of life in a region.

MOTIVATOR
Provide students with a physical map of Western Europe showing features such as mountains, rivers, and elevation. The names of the countries, rivers, and bodies of water should *not* be shown. Ask students to work in groups to identify on the map possible locations of boundaries of countries, inland cities, and coastal cities. Have each group share information and compare locations. Ask students to compare their maps with actual physical/political maps of Western Europe and note similarities and differences.

TEACHING DESIGN
1. Have students practice locating the countries of Western Europe, using wall maps and world maps in their textbook as well as in atlases, encyclopedias, and desk maps. Divide students into groups of three, and assign to each group one of the following topics to research: Landform Regions, Climate Types, Bodies of Water, Physical Features, Political Boundaries, Natural Resources. The information of each group topic should be divided so that each person in the group has something to share. Each group should then share its information with the class, each member presenting a part of the group information. Each group should select the most significant facts and features about its topic. Students may wish to use maps, transparencies, charts, and such to complement their information.
2. Direct students to begin a chart of the nations of Western Europe that will be completed by the end of this unit of study. Students should use their textbook as a resource. The chart should contain these headings: Country, Capital, Major Waterways, Natural Resources, Type of Government, Landforms, Imports, Exports, Major Economic Activities, Mountains.
3. Homework: Assign the Chapter 10 Challenge Worksheet and the Chapter 10 Skills Worksheet.

CLOSURE
Write the six topics listed in Teaching Design 1 on the chalkboard. Ask students to think about one main idea related to each topic. Call upon volunteers to write key information under each topic, and ask the class to make additions or revisions to the information listed.

ASSESSMENT
Have students complete the following questions from the Chapter 10 Check:

Building a Vocabulary 2, 3
Recalling and Reviewing 2
Using Geography Skills 1, 2

RETEACHING PLAN
1. Ask students to complete the Chapter 10 Reteaching Worksheet, Lesson 2.
2. Have students work in groups to complete notes related to the following major topics of the lesson. Students may then drill each other on the information. Major topics are Countries of Western Europe, Major Influences on Climate, and Landform Regions.

CHALLENGE/ENRICHMENT STRATEGY
1. Ask students to study maps of Western Europe and its major waterways, then write a paragraph summarizing the importance of coastal harbors and navigable rivers to the economic development of this region.
2. Encourage students to collect travel brochures, posters, and postcards that pertain to famous Western European cities, coastal resorts, and mountain resorts for display in the classroom. Students may contact airlines, travel agents, consulates, or embassies for information.

▶ LESSON 3
Natural Resources, Economy, and Political Ties

LESSON OBJECTIVES
The student should be able to
- identify the importance of water for transportation and economic activity
- identify resources and explain their importance
- identify examples of economic and political ties that promote cooperation in Western Europe

PURPOSE
The study of the natural resources of Western Europe will enable students to understand the relationship between environment and the ways of life that develop in a region,

as well as the advantages of economic and political cooperation to a region's development.

MOTIVATOR

Show students pictures of major cities in Western Europe at the end of World War II and at the current time. Ask them to provide adjectives describing the cities at these two times. Note the date of the ending of the war and the span of years to the present. Ask students to suggest explanations for the tremendous rebuilding and recovery that have occurred in Western Europe since World War II. List the explanations on the chalkboard.

TEACHING DESIGN

1. Ask students to label the following on a map titled "Natural Resources of Western Europe": navigable rivers, bodies of water bordering land, forest areas, and mineral resources. The Atlas may be used as reference in addition to the textbook. Have students plot water routes that could be used for transporting goods.
2. Have students work in groups, using their maps developed in Teaching Design 1, to answer the following questions:
   a. What countries in Western Europe are landlocked?
   b. What bodies of water create coastlines?
   c. What major sea routes could Western European countries use to transport goods?
   d. What countries or microstates have frozen bodies of water nearby in winter months?
   e. What are the names of major forests?
   f. What mineral deposits are the most common?
   g. What industries have developed because of their use of natural resources?

   Lead a class discussion of the students' answers.
3. Write the following headings on the chalkboard:

Economic Cooperation		Political Cooperation	
Disadvantages	Advantages	Disadvantages	Advantages

   Ask students to brainstorm about possible advantages and disadvantages of economic and political cooperation among the nations of Western Europe and list them on the chalkboard. Have the class vote to determine if the majority of students would favor cooperation or non-cooperation in economic and political areas.
4. Have students review Chapter 10 in the textbook and list examples of economic and political cooperation in Western Europe since World War II.
5. Have students complete the Chapter 10 Critical Thinking Worksheet.

CLOSURE

1. Project the transparency of the information map from the Chapter 10 Lecture Notes and Transparency Package, and have students complete the chart.
2. Have students compare their explanations in the Motivator for Western Europe's recovery after World War II with the explanations that they can give now.

ASSESSMENT

1. Have students complete the following questions from the Chapter 10 Check:
   Building a Vocabulary 4, 5, 6
   Recalling and Reviewing 1
   Critical Thinking 3
2. Ask students to complete the Chapter 10 Review/Test.

RETEACHING PLAN

1. Ask students to complete the Chapter 10 Reteaching Worksheet, Lesson 3.
2. Reproduce the transparency from the Closure, and have students work in pairs to review the lesson by asking each other questions and by pointing out pertinent locations on the map "Western Europe" on page 110 in their textbook.

CHALLENGE/ENRICHMENT STRATEGY

1. Ask students to investigate the direction of flow of the major rivers in Western Europe. They should use atlases to check changes in elevation along rivers.
2. Have students research and report on United States assistance to the nations of Western Europe after World War II. Students also should note areas of economic, political, and military cooperation between the United States and countries of Western Europe today.
3. Have students investigate and report on the organization and activities of the European Community.

Geography Skills, pages 112–113, Answers

1. More than 500 persons per square mile (195 per square kilometer).
2. The Eastern Hemisphere.
3. Asia; Antarctica.

Chapter 10 Check, page 121, Answers

Building a Vocabulary

1. Materials in their natural state that are used to produce finished goods. From its overseas colonies.
2. canal
3. A political boundary established by landforms. Spain and France.
4. An economic association. Tariffs are taxes on imported goods; the EC has abolished them in order to open trade among member nations.
5. A very small country that usually depends upon its larger neighbors for military and economic assistance. Western Europe's microstates are Andorra (between France and Spain), Vatican City (in Rome), San Marino (in Italy), Monaco (on France's Mediterranean coast), Liechtenstein (between Switzerland and Austria), and Luxembourg (at the crossroads of France, Belgium, and West Germany).
6. A landform surrounded by water on three sides. Italy, Spain, Portugal, Norway, Sweden, and Denmark. No.

Recalling and Reviewing

1. The countries of Western Europe are cooperating with one another through the European Community and NATO. Not all the nations of Western Europe belong to these organizations.
2. Northwest Highlands: the hills in Ireland, the Pennine Mountains in England, the Scottish Highlands, the Plateau of Brittany, the Iberian Peninsula, the Kjølen Mountains in Norway and Sweden; North European Plain: Atlantic coast of Western Europe into the Soviet Union; Central Uplands: the Massif Central of France, the Jura Mountains on the Swiss-French border, the Black Forest and the Bavarian Plateau of Germany, and the Ardennes in Belgium and Luxembourg; Alpine Mountains: parts of France, West Germany, Switzerland, Italy, Austria, Spain, Greece.

Critical Thinking

3. Answers will vary but should include the desire to control certain land areas as motivation for conflict and the need to rebuild economically and politically after World War II as motivation for cooperation.

Using Geography Skills

1. Great Britain, Republic of Ireland, Iceland.
2. Baltic Sea, North Sea, Mediterranean Sea, Tyrrhenian Sea, Ionian Sea, Adriatic Sea, Aegean Sea; Baltic Sea; West Germany, Denmark, Sweden, Finland.

CHAPTER 11 Great Britain and Ireland Pages 122–129

PLANNING GUIDE		
Lesson	**Textbook**	**Teacher's Resource Binder**
1. Geographical and Historical Perspective	pp. 122–126	Lecture Notes and Transparency Package Content/Vocabulary Workbook Reteaching Worksheet
2. Economic Issues and Division of Ireland	pp. 126–128	Skills Worksheet Chapter Review/Test Reteaching Worksheet Critical Thinking Worksheet Challenge Worksheet

TIME ALLOCATION: 2 days

CHAPTER TOPICS

- Great Britain
- Ireland

PRETEACHING CHAPTER VOCABULARY

Write the following terms on the chalkboard:

lochs
constitutional monarchy
fossil fuels
nationalized
commonwealth
bog
peat

Discuss the terms, asking students to suggest definitions. Point out that *lochs* are long, narrow lakes created by sheet glaciers. Describe the process of their creation.

Ask students to identify root words in the term *constitutional monarchy*. Remind them that a constitution is a system of laws that prescribes the nature, functions, and limits of government. A monarch is a ruler of a nation, and a monarchy is a government ruled by a monarch. Thus a constitutional monarchy is a government in which the monarch's powers are limited to those granted by a constitution. Compare this type of government to that of the United States.

Ask students to provide the definition of *fossil fuels* from the two terms and to suggest examples of fuels that are derived from organisms of a past geologic age.

Point out that the suffix *-ize* means "to become or to make into." *Nationalize* means "to make privately owned business into government owned business."

Define the term *commonwealth* by analyzing the two words in it. *Common* means "public" or "community"; *wealth* refers to welfare or well-being. Point out that some states of the United States have *commonwealth* in their official names. (Kentucky, Virginia, Massachusetts, Pennsylvania)

Students may not be familiar with the term *peat*. Explain that it is partially carbonized vegetable matter, usually moss, that is found in *bogs*. It is frequently used in Great Britain and Ireland for fertilizer and fuel.

Ask students to select categories for classifying the vocabulary terms. (geography, government)

▶ LESSON 1
Geographical and Historical Perspective

LESSON OBJECTIVES

The student should be able to

- define the significant terms of the chapter
- locate the United Kingdom and the Republic of Ireland
- identify and describe the major landforms and features of Great Britain and Ireland
- describe the major economic activities in these nations
- sequence key historical events in the development of these nations
- locate major cities

PURPOSE

The study of the physical and historical setting of the United Kingdom and the Republic of Ireland will acquaint students with a small part of the world that has had a major influence on the development of culture in the United States as well as other areas in the world.

MOTIVATOR

Present the following scenario to students:

Your way of life today is the same as that of your parents, your grandparents, and your great-grandparents. You are either a worker involved in agriculture using hand tools and animal power or a worker involved in producing handcrafted goods in your home. Change then comes to your nation. New inventions bring mechanization, factories, and steam power. What impact will this industrial revolution have on the way of life of your country's people? Imagine all of the possibilities (mechanization, growth of cities, larger middle class, and so forth).

Direct students to brainstorm about the changes, and record them on the chalkboard. Explain that an industrial revolution began in Great Britain, changing its agricultural way of life to an industrial way of life. It is called a revolution because it produced many vast changes.

TEACHING DESIGN

1. Present the Chapter 11 Lecture Notes and Transparency Package to illustrate the physical and historical settings of Great Britain and Ireland.
2. Ask students to look at the map "Great Britain and Ireland" on page 124 of their textbook and list the natural resources and economic activities on a sheet of paper. Beside each, they should note the locations where the resource or activity is found. Ask students to use the map and their listing to answer the following questions:
   a. Which are the most important economic activities?
   b. Where are most steel factories located?
   c. Which areas are most suitable for farming?
3. Direct students to work in groups of three to create a sequence of industrial development in Britain and Ireland from the late 1700s to the present. Students may create a visual to illustrate the developments graphically. Students' information should include factors that made the Industrial Revolution possible, early industries, countries competing with Great Britain, industries nationalized to curb industrial decline, problems of industrial decline, and new industries. Choose volunteers to share their information with classmates.

CLOSURE

Instruct students to work with a partner to review the main ideas of the lesson by beginning with the Western Europe chart for the United Kingdom and the Republic of Ireland. Indicate that students should use their notes and information in the text as references.

ASSESSMENT

1. Have students complete the following questions from the Chapter 11 Check:
   Building a Vocabulary 1, 2, 3
   Recalling and Reviewing 1, 2
   Critical Thinking 5
   Using Geography Skills 1, 2
2. Ask students to complete the Chapter 11 Content/Vocabulary Worksheet in their Workbook.

RETEACHING PLAN

1. Ask students to complete the Chapter 11 Reteaching Worksheet, Lesson 1.
2. Divide students into two teams, with one person serving as a game leader and scorekeeper. The game leader should select headings from the Western Europe chart, and a student from each team must provide correct information in order to score a point. The team with the highest number of points at the end of Round 1 gets to select the game leader for Round 2.

CHALLENGE/ENRICHMENT STRATEGY

1. Ask students to report on the inventions that made factory production possible during the Industrial Revolution in Great Britain.
2. Have students compare Great Britain's role in world trade in the nineteenth century to its role today. What changes has the nation made to compete in world markets?
3. Select volunteers to serve "high tea" to classmates while explaining this custom and the traditional foods and beverages served.

▶ LESSON 2
Economic Issues and Division of Ireland

LESSON OBJECTIVES

The student should be able to

- identify economic, social, and cultural interchange
- identify major cities and describe their functions
- describe the major economic activities of Ireland

PURPOSE

The study of economic, political, and religious issues in Great Britain and Ireland will create students' awareness of the complexity of problems facing nations in their own part of the world.

MOTIVATOR

Ask students to review facts related to the United States' colonial period as part of the British Empire. Instruct students to identify aspects of the United States' culture that can be traced to its English heritage. (language, religion, values, representative government, literature) Explain that they will learn more about this part of the world, which had a great influence on the development of our nation.

TEACHING DESIGN

1. Show students photographs or slides of London along with a city map, and have them locate famous areas and

sites on the map as they are discussed. Point out the importance of the Thames River to the development of London as a major urban center. Compare London's problems with those of urban areas in the United States.
2. Review the history of the expansion of the British Empire, English domination of the seas, and the later independence of colonial possessions.
3. Ask students to work in groups to brainstorm about possible solutions of the economic problems identified on page 128 of their textbook. Each group should consider one problem and be prepared to present solutions.
4. Trace the development of events leading to the division of Ireland into Northern Ireland and the Republic of Ireland, and the religious conflict occurring in Northern Ireland. Identify major cities and economic activities of the two divisions. Discuss the problems facing these areas and how people are attempting to solve their problems. Ask students to suggest solutions.
5. Have students complete the Chapter 11 Skills Worksheet.

CLOSURE

Have students work with a partner to complete the entries for the United Kingdom and Republic of Ireland on their Western Europe chart.

ASSESSMENT

1. Have students complete the following questions from the Chapter 11 Check:
   Building a Vocabulary 4, 5, 6
   Recalling and Reviewing 3, 4
2. Ask students to complete the Chapter 11 Review/Test.

RETEACHING PLAN

1. Ask students to complete the Chapter 11 Reteaching Worksheet, Lesson 2.
2. Divide students into groups to prepare fact cards for the information presented in this lesson.

CHALLENGE/ENRICHMENT STRATEGY

1. Have students complete the Chapter 11 Critical Thinking Worksheet.
2. Have students complete the Chapter 11 Challenge Worksheet.
3. Ask students to investigate the activities of the Irish Republican Army and the British government's efforts to control this group.
4. Choose volunteers to research and report on the process of manufacturing lead crystal, such as Waterford, and bone china, such as Wedgwood or Royal Doulton.

Chapter 11 Check, page 129, Answers

Building a Vocabulary

1. To symbolize unity, history, and patriotism.
2. nationalized. Steel, shipbuilding.
3. An association of self-governing states with similar backgrounds that are united by a common loyalty.
4. Decayed vegetable matter; peat is found in bog areas.
5. Long, deep lakes carved by glaciers. Highland Britain.
6. Coal, oil, and natural gas.

Recalling and Reviewing

1. Lowland Britain: level, fertile soil, in southeastern Britain; highland Britain: rocky, glaciated landscape, low but rugged mountains, in northwest Britain.
2. London has become a trade center because it is close to the English Channel, which facilitates trade with Western Europe.
3. Economic cooperation and special trade arrangements.
4. The northern part of Ireland was populated by Scottish and English Protestants while the rest of the island remained Roman Catholic. This religious difference has divided the island into two political units.

Critical Thinking

5. During the ice ages, sheet glaciers covered much of the British Isles. As the glaciers advanced, they shaved off topsoil, leaving behind a rocky landscape with only a thin layer of good soil.

Using Geography Skills

1. The English Channel.
2. Coal and oil.

CHAPTER 12 The Nordic Countries Pages 130–139

PLANNING GUIDE		
Lesson	Textbook	Teacher's Resource Binder
1. Norway and Sweden	pp. 130–134	Lecture Notes and Transparency Package Skills Worksheet Content/Vocabulary Workbook Reteaching Worksheet
2. Denmark, Iceland, and Finland	pp. 134–138	Critical Thinking Worksheet Chapter Review/Test Reteaching Worksheet Challenge Worksheet

TIME ALLOCATION: 2 days

CHAPTER TOPICS

- Norway
- Sweden
- Denmark and Greenland
- Iceland
- Finland and Lapland

PRETEACHING CHAPTER VOCABULARY

Write the following terms on the chalkboard:

fiord	neutrality
socialism	uninhabitable
cooperative	geyser

Ask students to turn to the pictures of a fiord on page 131 and geysers on page 137. Describe the geologic action that has caused the development of these two features. Ask students where they could find other examples of fiords (coasts of New Zealand, Alaska) and geysers (Wyoming, in Yellowstone National Park).

Remind students that the suffix *-ism* refers to a doctrine, theory, or system. *Socialism* is the economic system in which the government possesses both the political power and the means to produce and distribute goods. Explain that socialism is in effect in some of the Nordic countries, and provide examples. (government ownership of utilities, railroads, airlines, major industries)

The noun *cooperative* refers to an enterprise that is collectively owned and operated for mutual benefit. In some Nordic countries, cooperative dairies have been formed, so that small farmers have a means of functioning as a larger unit. Point out related words: *cooperating* means "working together toward a common purpose"; *cooperative* people are people who work together.

Neutrality is a nation's policy of not taking sides in international affairs. Ask students to describe situations in which they have played a neutral role.

Ask a volunteer to analyze the term *uninhabitable* by its prefix, root, and suffix. Remind students that the suffix *-able* means "capable of." The prefix *un-* means "not." *Inhabit* means "to reside in." Thus, *uninhabitable* means "an area not capable of being resided in." Point out that an *inhabitant* is a permanent resident of an area.

▶ LESSON 1
Norway and Sweden

LESSON OBJECTIVES

The student should be able to

- define the significant terms of the chapter
- locate the nations of Scandinavia
- identify and describe the major landforms, features, and climate of Norway and Sweden
- identify glaciation as a force that has altered the features of the earth's crust
- locate natural resources and give their uses
- explain the impact of environment on ways of life
- identify economic activities in these nations
- locate major cities and identify their functions
- identify the importance of water as a resource

PURPOSE

The study of Norway and Sweden should help students realize the importance of people as a nation's key resource. Although they do not have abundant natural resources, Norway and Sweden are prosperous nations with high standards of living.

MOTIVATOR

To groups of students, provide a physical map of the Nordic countries showing latitude, longitude, mountain ranges, rivers, and surrounding bodies of water. (Use the transparency from the Chapter 12 Lecture Notes and Transparency Package as a master.) Students should work together to predict the answers to the following questions.

a. What effects do the northern latitude locations, ocean currents, and prevailing winds have upon this area?
b. What is the climate in this area?
c. How suitable is the area for farming?
d. Where are cities located?
e. What economic activities are important in the area?

Record each group's predictions on a transparency, and check to see whose predictions are closest to the information presented in the textbook.

TEACHING DESIGN

1. Use the Lecture Notes and Transparency Package to introduce the Nordic countries before focusing upon Norway and Sweden in this lesson. Students should take notes and locate sites on a map of the area.
2. Provide students with a physical/political map of Norway and Sweden, and ask them to label boundaries, bodies of water, landforms, and major cities. Students should show natural resources and industries on the map, using symbols and a map key. Place the transparency of predictions from the Motivator on the overhead projector, and lead a discussion comparing students' predictions with the information on their maps.
3. Have students complete the Chapter 12 Skills Worksheet.

CLOSURE

1. Ask students to think about the geography, resources, and economic activities of Norway and Sweden. Write these three topics on the chalkboard. Ask each student to confer with a neighbor and name, for each country, one main idea related to the three topics.

ASSESSMENT

1. Have students respond aloud to the following questions from the Chapter 12 Check:
   Building a Vocabulary 1, 3
   Recalling and Reviewing 1, 2
2. Have students complete the Western Europe chart for Norway and Sweden.
3. Ask students to complete the Chapter 12 Content/Vocabulary Worksheet in their Workbook.

RETEACHING PLAN

1. Ask students to complete the Chapter 12 Reteaching Worksheet, Lesson 1.

2. Have students create review cards for Norway and Sweden that are cut in the shapes of those countries. On each country's cards, students should list key ideas and descriptions that they should remember. Students may switch cards with others to check the information.

CHALLENGE/ENRICHMENT STRATEGY

1. Ask students to collect colored pictures or postcards depicting scenes found in Scandinavian countries. Use pictures to create a visual display in the classroom.
2. Allow interested students to investigate and report on popular sports in the Scandinavian countries.
3. Have students compare the welfare system, tax rate, and standard of living in Sweden to welfare programs, tax rate, and standard of living in the United States.

▶ LESSON 2
Denmark, Iceland, and Finland

LESSON OBJECTIVES

The student should be able to

- identify and describe the major landforms, features, and climate of Denmark, Iceland, and Finland
- locate natural resources and give their uses
- explain the impact of environment on ways of life
- identify economic activities in these nations
- locate major cities and identify their functions

PURPOSE

The study of these nations will help students understand the critical factor of people as one of the most important resources in the development of a nation.

MOTIVATOR

Ask students to consider the following statements of fact related to a "mystery" country. Each statement should be read, allowing time for students to consider the impact of each statement on the development of the country. Students should then comment on each statement.

a. Less than 10 percent of the land is suitable for farming.
b. One-fourth of the nation is lakes and swamps.
c. Forests cover 60 percent of the country.
d. The country has access to the sea.
e. The coastline has many good harbors.
f. Rivers flow swiftly down the glacial hills of the south.
g. Three climate regions exist: humid continental, subarctic, and tundra.

After students have commented on the impact of each statement, ask them to rate this country's possibilities for development in the following areas by placing a check in the appropriate column.

	Poor	Fair	Good
Agriculture			
Industry			
Urban centers			
Lumber			
Varied economy			

Read the following factual statements about the "mystery" country and compare them to the students' ratings:

a. This country has developed a productive and varied economy.
b. The people of this country enjoy a high standard of living.

Explain that students will be learning about this country during today's lesson. They should seek explanations for the final two factual statements in their study.

TEACHING DESIGN

1. Using a large wall map, call on volunteers to locate the countries, major bodies of water, and neighboring countries. Ask students what effect Finland's location near a major world power would have on its policies with the USSR and nations of Western Europe.
2. Have students work in groups to review the information about the three countries and the self-governing province of Greenland. Students should complete the features checklist comparing the similarities and differences among the nations, then compare their checklist to one from another group.

Features	Denmark	Iceland	Finland
Smallest country			
Contains world's largest island			
Contains world's most northern capital			
People originated from Central Asia			
Landforms show effects of glaciation			
Contains many islands			
High living standard			
Policy of neutrality			
Limited resources			
Fishing industry			
Hydroelectric power			
Uses geysers to generate electricity			
Dairy and livestock production			
Most densely populated country			
Extensive social welfare program			

3. Instruct students to complete their Western Europe charts for Denmark, Iceland, and Finland.
4. Homework: Have students complete the Chapter 12 Critical Thinking Worksheet.

CLOSURE

Ask volunteers to share their information on the three countries from their Western Europe charts. Have students make revisions or additions as necessary. Ask which was the "mystery" country discussed in the Motivator.

ASSESSMENT

1. Have students complete the following questions from the Chapter 12 Check:
   - Building a Vocabulary 2, 4, 5, 6
   - Recalling and Reviewing 3
   - Critical Thinking 4
   - Using Geography Skills 1
2. Ask students to complete the Chapter 12 Review/Test.

RETEACHING PLAN

1. Ask students to complete the Chapter 12 Reteaching Worksheet, Lesson 2.
2. Divide students into small groups, and provide each group with a list of the three countries and their capitals, major bodies of water, resources, and products. On a large outline map of the region, students should label on the appropriate country each item on the list. Resources and products could be labeled by using picture symbols. Display the review maps in the classroom.

CHALLENGE/ENRICHMENT STRATEGY

1. Choose students to investigate the Scandinavian lumber industry and create visuals illustrating the manufacturing process and the resulting products.
2. Have students write a narration for a television travel documentary on Nordic countries. What places will be shown to give viewers a realistic perspective of the land and the people of this area?
3. Have students complete the Chapter 12 Challenge Worksheet.

Chapter 12 Check, page 139, Answers

Building a Vocabulary

1. A business organization that is owned by and operated for the mutual benefit of its members. Cooperatives benefit businesses and farmers because those who cannot individually afford to transport products can pool their efforts to distribute goods to market.
2. Most of the people have settled along the coast.
3. A narrow inlet of the sea between high, rocky banks. Norway's fiords were formed by glaciers that extended down to the coast.
4. A hot spring that shoots hot water and steam into the air. Iceland.
5. An economic system in which the government owns and controls the means of producing goods. Sweden.
6. Sweden maintains a military force, in spite of its policy of neutrality, because it is concerned about potential Soviet military threats.

Recalling and Reviewing

1. Because the interior of the country is colder and mountainous; also, many Norwegians earn their living from the sea through the merchant marine, fishing, and oil fields in the North Sea.
2. Forestry, farming, mining, and manufacturing. The Swedes have highly efficient farms and some of the highest grain yields in the world.
3. Because of poor soil and a short growing season, the Finns are limited to livestock raising and dairy farming. Rivers provide an important source of hydroelectric power. Abundant forests have made lumbering and wood products important industries.

Critical Thinking

4. Answers will vary, but students should mention the importance of skis for transportation. Among the other sports that could be mentioned are hunting, fishing, horse racing, and auto racing.

Using Geography Skills

1. Subarctic. Fur trapping, mining.

CHAPTER 13 France and the Low Countries Pages 140–155

PLANNING GUIDE		
Lesson	**Textbook**	**Teacher's Resource Binder**
1. Physical Geography	pp. 140–144	Content/Vocabulary Workbook Reteaching Worksheet Critical Thinking Worksheet
2. Urban Geography and Culture	pp. 144–148	Lecture Notes and Transparency Package Reteaching Worksheet Challenge Worksheet
3. The Low Countries	pp. 148–154	Skills Worksheet Chapter Review/Test Reteaching Worksheet

TIME ALLOCATION: 3 days

CHAPTER TOPICS

- France
- Monaco
- The Netherlands
- Belgium
- Luxembourg

PRETEACHING CHAPTER VOCABULARY

Write the following terms on the chalkboard:

vineyard polders
dikes bilingual

Pronounce each word. Students may be familiar with the word *vineyard,* but point out that this term refers to land that is planted with grapevines. Note the relationship between the word *vine* and terms that mean "wine" in various languages: *vino* in Spanish and Italian, *vin* in French.

Explain the meanings of *dikes* and *polders*. Ask a volunteer to analyze the term *bilingual* to determine its meaning. (The word comes from Latin, with the prefix *bi-* meaning "two" and *-lingual* referring to "tongue," which in this case means "language." A bilingual person can speak two languages fluently.)

▶ LESSON 1

Physical Geography

LESSON OBJECTIVES

The student should be able to

- define the significant terms of the chapter
- identify major landforms and features of France
- describe the physical setting of France
- identify and locate significant bodies of water
- describe the agricultural base of France

PURPOSE

By studying the physical geography of France, students will learn more about the impact of geography on that country's way of life.

MOTIVATOR

Write the word *France* on the chalkboard. Ask students to work in groups of five to generate a list of words that come to mind when they think of France. After someone in each group has recorded the group's ideas, ask a representative to read the list to classmates. Record each new word about France on an overhead transparency. Tell the students that they will learn more about France in this lesson and that the class will review the words after studying this country.

TEACHING DESIGN

1. Geographic Setting
   a. Use a wall map of Western Europe to note the physical location of France. Point out that France is larger than the states of Oregon and Nevada, but smaller than Texas. Note that France is the largest of the Western European countries.
   b. Ask students to study the economic map "France and the Low Countries" on page 143 of their textbook and locate surrounding countries, bodies of water, rivers, landforms, and major cities. Point out that France has the greatest variety of landforms in Western Europe: mountains, hills, plateaus, plains, and coastlines. Ask students how this diversity of landforms could affect ways of life in France.
   c. Have students write the names of the regions (North European Plain, Northwest Highlands, Central Uplands, Alpine Mountains) on a sheet of paper and list the major characteristics of each.
   d. Ask students to name the important bodies of water forming French coastlines and to locate them on a map. Ask students to suggest why these bodies of water are significant to the nation.
2. Ask students to label a map of France with climates, indicating major characteristics of each in a map key.
3. Invite students to work in groups to determine the best areas of France for farming, using the information they have learned about landforms, rivers, and climate. After each group has determined areas suitable for farming, ask them to use their textbook to add products to the climate map completed in Teaching Design 2. Map symbols should be used with a map key.
4. Have students continue work on the chart of Western Europe that they began in Chapter 10.

CLOSURE

1. Ask volunteers to read the information that pertains to France from their chart of Western Europe. Other students should check their entries and make necessary revisions.
2. Have students review the list of words they created in the Motivator and compare them to the list they would make after studying France in this lesson.

ASSESSMENT

1. Have students complete the following questions from the Chapter 13 Check:
   Building a Vocabulary 3
   Recalling and Reviewing 1, 2
   Using Geography Skills 1
2. Ask students to complete the Chapter 13 Content/ Vocabulary Worksheet in their Workbook.

RETEACHING PLAN

1. Ask students to complete the Chapter 13 Reteaching Worksheet, Lesson 1.
2. Have students work with a partner to review the information from the Western Europe chart under the following headings: Landforms, Mountains, Waterways, Natural Resources.

CHALLENGE/ENRICHMENT STRATEGY

1. Direct a group of students to plan a grand tour of France and present the travel itinerary and description of sites.
2. Have interested students report on one of the following industries in France: perfume, china, fashion, wine.
3. Ask students to investigate and report on famous French historical figures.
4. Have students complete the Chapter 13 Critical Thinking Worksheet.

▶ LESSON 2
Urban Geography and Culture

LESSON OBJECTIVES
The student should be able to
- identify major cities, their locations, and their functions
- identify famous figures representing French contributions to civilization and history

PURPOSE
The study of urban geography and French culture will provide students with a better understanding of the contributions of this national group to world culture.

MOTIVATOR
Show slides of Paris, pass around color photographs, or display posters in the classroom. Ask students to give their impressions of the famous "city of lights." Point out that Paris remains the population, political, economic, and cultural center of France.

TEACHING DESIGN
1. Present the Chapter 13 Lecture Notes and Transparency Package to highlight the content on the role of France in the world community. Have students take notes.
2. Have students read the Cities of the World feature "Paris" on pages 146–147 in their textbooks and discuss the advantages and disadvantages of living there.
3. Divide students into small groups, and assign one of the following to each group to prepare:
   a. A chamber of commerce visitor's brochure of the following cities: Paris (city plan and famous buildings, monuments, art and music center, high-fashion center, transportation center), Le Havre, Cherbourg, Marseilles, Lyons, Bordeaux, Toulouse, Monte Carlo. Students should focus on reasons for the location of a city and what functions each serves in the country.
   b. A book cover, museum brochure, or record album cover to represent information about one of the following people: Molière, Voltaire, Balzac, Camus, Hugo, Rodin, Monet, Manet, Gauguin, Debussy, Berlioz, Renoir, Descartes, Pascal, Montaigne.

   Provide encyclopedias and selected copies of *National Geographic* as references. A spokesperson from each group could present a report on the city or person to the class. The classroom could be arranged to simulate a sidewalk café for the presentations, with students providing French food.

CLOSURE
Ask students to turn to a classmate and share three new ideas they learned about French cities and/or culture.

ASSESSMENT
Have students complete the following question from the Chapter 13 Check:
Recalling and Reviewing 3

RETEACHING PLAN
1. Ask students to complete the Chapter 13 Reteaching Worksheet, Lesson 2.
2. Have students prepare fact cards about cities and people from the lesson. Pairs of students should take turns quizzing each other on the information.

CHALLENGE/ENRICHMENT STRATEGY
1. Ask students to investigate the history related to the national flag, holiday, anthem, and motto of France and create a visual to explain their findings.
2. Select students to investigate and report on English words that have been adopted by the French people and French words that are part of the English language.
3. Have students complete the Chapter 13 Challenge Worksheet.

▶ LESSON 3
The Low Countries

LESSON OBJECTIVES
The student should be able to
- locate the Low Countries on a map of Western Europe and identify their capitals and major cities
- describe the landforms of the area
- describe major economic activities
- identify the action to reclaim land from below sea level

PURPOSE
As students study these countries, they will learn about how people can have a dramatic impact on the environment and about the contributions to world culture made by people of the Low Countries.

MOTIVATOR
Show students a map of Western Europe, pointing out the Netherlands, Belgium, and Luxembourg and the countries in close proximity. Point out the distances in miles from major cities to other countries. Ask students to recall the languages spoken in neighboring countries and to explain why it would be advantageous for people in the Low Countries to speak more than one language. Inquire if anyone knows the languages spoken in these three countries.

TEACHING DESIGN
1. Ask students to complete a physical/political map of the Low Countries, using atlases for reference. Differences in elevation should be shaded in and designated on a map key. Students should also note rivers and bodies of water. Ask students to use their maps to determine the areas that are below sea level. Have students read the Themes in Geography feature "Polders of the Netherlands" on pages 150–151 in order to answer the following questions: How has the Netherlands reclaimed land that is below sea level? How are floods prevented? How were windmills used? Call on a volunteer to explain why these countries are known as the Benelux countries.

2. Have students complete their Western Europe charts for the Netherlands, Belgium, and Luxembourg.
3. Suggest that students work with a partner to complete the Chapter 13 Skills Worksheet.

CLOSURE

Ask volunteers to present the information on their Western Europe charts for the three countries. A review game could be played if time permits.

ASSESSMENT

1. Have students complete the following questions from the Chapter 13 Check:
   Building a Vocabulary 1, 2
   Recalling and Reviewing 4, 5
   Critical Thinking 6
   Using Geography Skills 2
2. Ask students to complete the Chapter 13 Review/Test.

RETEACHING PLAN

1. Ask students to complete the Chapter 13 Reteaching Worksheet, Lesson 3.
2. Suggest that students work in small groups to view a map of the Netherlands, Belgium, and Luxembourg and answer the following questions:
   a. Where is each country located? (Give exact and relative locations.)
   b. What and where are the capitals of each country?
   c. What is the climate of the area?
   d. What land features are important?
   e. What are the major economic activities?

CHALLENGE/ENRICHMENT STRATEGY

1. Select students to research and report on the history of the Benelux countries during and after World War II.
2. Have students compare the types of currency used in the three countries and investigate their values relative to the U.S. dollar.

Chapter 13 Check, page 155, Answers

Building a Vocabulary

1. Walls built around an area to be drained, intended to keep out water. After the dikes were built, the Dutch built canals to remove the water, using windmills to operate the pumps. Polders have increased the acreage available for agriculture in the Netherlands.
2. bilingual. Trilingual.
3. France is a world leader in the production of wines, and vineyards are used to grow grapes, which are used to produce these wines.

Recalling and Reviewing

1. The North European Plain: lowland area, level land, and fertile soil; the Central Uplands: the Massif Central, Ardennes, Vosges, and Jura mountain ranges and the island of Corsica; the Alpine Mountains: the Pyrenees, the Alps, and the French Riviera.
2. It faces four bodies of water: the North Sea, the English Channel, the Atlantic Ocean, and the Mediterranean Sea; it has large, navigable rivers and a system of canals that connects the major rivers.
3. On the French Riviera on the Mediterranean Sea. Tourism, small industries, and the sale of postage stamps.
4. Netherlands: High sand dune area, low coastal plain, higher hilly area with forest and pasture; Belgium: coastal region, a central plain, the Ardennes Plateau; Luxembourg: the Ardennes Plateau.
5. In Belgium, farming is limited to growing fruits, vegetables, sugar beets, and some grains. Luxembourg is mainly hilly and forested, and the soil is not very productive for agriculture; as a result, industry has become more important than agriculture in the two countries.

Critical Thinking

6. Answers will vary, but students should mention economic benefits such as a single currency and pooling of resources. Possible problems are language and cultural differences as well as political differences.

Using Geography Skills

1. Along the Loire River east of Nantes, in Burgundy by the Saône River north of Lyons, northeast of Paris near the Meuse River, and along France's southern coast between the Massif Central and the Mediterranean Sea. Maritime and Mediterranean climates.
2. Generally more than 500 persons per square mile (195 per square kilometer). France's population density averages between 125 and 500 persons per square mile (49 and 195 per square kilometer).

CHAPTER 14 West Germany and the Alpine Countries Pages 156–165

PLANNING GUIDE		
Lesson	Textbook	Teacher's Resource Binder
1. West Germany	pp. 156–161	Lecture Notes and Transparency Package Critical Thinking Worksheet Content/Vocabulary Workbook Reteaching Worksheet
2. Switzerland, Austria, and Liechtenstein	pp. 161–164	Skills Worksheet Chapter/Review Test Reteaching Worksheet Challenge Worksheet

TIME ALLOCATION: 2 days

CHAPTER TOPICS

- West Germany
- Switzerland
- Austria
- Liechtenstein

PRETEACHING CHAPTER VOCABULARY

Write the following terms on the chalkboard:

dialect	confederation	arable
loess	foehn	republic
enclave	tree line	basin

Read and discuss the following definitions, then ask questions to guide the discussion.

dialect (regional variety of a major language)
loess (soil deposited by wind)
enclave (a part of one country completely within the boundaries of another)
confederation (group of states joined together for a common purpose) Ask students to name another confederation.
foehn (warm, dry wind that blows down the sides of mountains)
tree line (elevation beyond which trees do not grow)
arable (suitable for growing crops) Where is the nearest arable land?
republic (form of government in which citizens elect representatives) Ask students if they can name a republic.
basin (a broad depression in the earth's surface)

Ask students to divide the terms into two categories. Solicit suggestions for names of categories. Choose the best, and write it on the chalkboard.

▶ LESSON 1
West Germany

LESSON OBJECTIVES

The student should be able to

- define significant vocabulary terms
- describe key events in the history of West Germany
- locate important places in West Germany
- identify key resources and industries
- describe the political system of West Germany

MOTIVATOR

Write the following numbers on the chalkboard:

1,800 (small kingdoms once ruled Germany)
4 (countries once divided Berlin)
20 (years to rebuild after World War II)
3 (regions divide West Germany)
60 (percent of Berlin destroyed in World War II)
100 (miles, length of Berlin wall)

Have students work in groups. Tell the groups to scan pages 157–161 in their textbooks to locate the significance of each number as it relates to West Germany. At the end of 10 minutes, discuss the findings of each group.

TEACHING DESIGN

1. Present the Chapter 14 Lecture Notes and Transparency Package to provide a historical overview of West Germany and the Alpine Countries.
2. Refer students to the chart developed during the study of Chapter 10 regarding key events in the history of Europe. Review with students the major events that had an impact on West Germany.
3. Review with students the location of the following key places: West Germany, Rhine River, Alps, North Sea, Danube River, Hamburg, Austria, Bonn, Elbe River, Frankfurt, Weser River.
4. Direct students to work with a partner to complete the West Germany section of their Western Europe chart.
5. Ask students to continue to work with a partner to create a map showing the major exports, imports, and trading partners of West Germany.
6. Have students complete the Chapter 14 Critical Thinking Worksheet.

CLOSURE

Determine with the class three major ideas to remember about West Germany. Have students write these in their notebooks.

ASSESSMENT

1. Ask students to complete the Chapter 14 Content/Vocabulary Worksheet in their Workbook.
2. Have students complete the following questions from the Chapter 14 Check:
   Building a Vocabulary 1, 2, 3
   Recalling and Reviewing 1, 2, 4
   Using Geography Skills 2

RETEACHING PLAN

1. Direct students to label each event in the following chart according to the category.

Germany or West Germany	Independent/ Controlled by Another Country/ Controls Other Countries	Conflict or Peace	Economy: Stable or Unstable
Before 1871			
1871–1914			
World War I			
1918–1939			
World War II			
1949–Today			

2. Ask students to complete the Chapter 14 Reteaching Worksheet, Lesson 1.

CHALLENGE/ENRICHMENT STRATEGY

1. Invite students to bring objects made in West Germany from home. Students may also bring German food.
2. Ask students to investigate and report on the escapes over the Berlin Wall from East Berlin to West Berlin.
3. Choose volunteers to investigate the Holocaust during World War II.

▶ LESSON 2
Switzerland, Austria, and Liechtenstein

LESSON OBJECTIVES

The student should be able to

- describe key events in the history of Switzerland and Austria
- locate important sites in Austria, Switzerland, and Liechtenstein
- identify key resources and industries
- describe the relationship between Switzerland and the rest of the world

PURPOSE

This lesson will identify for students a country that maintains a standard of living that is higher than that of the United States. This will enable students to recognize the value of learning about other countries in order to achieve higher goals for our own nation.

MOTIVATOR

Read the following statement from the textbook: "To a great extent, the rise of Switzerland to world importance is the result of its convenient geographic location at the center of Europe." Locate Switzerland on a wall map of Western Europe. Discuss with students possible interpretations of this statement, and ask them to decide whether they agree or disagree.

TEACHING DESIGN

1. After writing the following on the chalkboard, direct students to sit with a partner and locate these places on maps in their textbooks: Switzerland, Jura Mountains, Alps, Rhine River, Mittelland, Lake Geneva, Lake Constance, Bern, Zurich, Geneva, Austria, Danube River, Vienna, Graz, Innsbruck, Salzburg, Liechtenstein.
2. Show the transparency from the Chapter 14 Lecture Notes and Transparency Package, and have students work with a partner to identify the country that each statement in the chart describes. (Some statements may apply to more than one country.)
3. Review with students the major industries of Austria and Switzerland, and write them on the chalkboard. Ask students to create a brochure to solicit new business or more visitors for each country.
4. Have students complete the Chapter 14 Skills Worksheet.

CLOSURE

Select several students to present their brochures to the class. Have all students make a list of the significant points brought out about each country.

ASSESSMENT

1. Have students complete the following questions from the Chapter 14 Check:
   Building a Vocabulary 4, 5, 6, 7
   Recalling and Reviewing 3
   Critical Thinking 5
   Using Geography Skills 1, 3
2. Ask students to complete the Chapter 14 Review/Test.

RETEACHING PLAN

1. Write the key facts and places studied in the lesson on separate index cards. Have students separate the cards into three piles to represent the three countries. Ask students to perform this activity several times.
2. Ask students to complete the Chapter 14 Reteaching Worksheet, Lesson 2.

CHALLENGE/ENRICHMENT STRATEGY

1. Invite students to bring to class several examples of products made in Switzerland and Austria.
2. Select volunteers to investigate the steps involved in opening a Swiss bank account. Have them report their findings to the class.
3. Have students complete the Chapter 14 Challenge Worksheet.

Chapter 14 Check, page 165, Answers

Building a Vocabulary

1. Dust-sized particles of soil deposited by the wind. South and east of the North German Plain, extending into Poland and the Soviet Union.
2. A warm, dry wind that blows down the side of a mountain. It keeps temperatures mild in winter. Yes; most Swiss farms are in the Mittelland, or Swiss Plateau, the region that is warmed by the foehn.
3. A government in which citizens elect the representatives and head of state. The United States and West Germany.
4. Land that is suitable for growing crops. No, because the climate is too cold above the tree line.
5. Confederation.
6. A part of one country completely within the boundaries of another. West Berlin, which is a West German city, is an enclave because it is surrounded by East Germany.
7. A regional variety of a major language.

Recalling and Reviewing

1. Germany is divided into West Germany and East Germany. For purposes of temporary occupation, it was separated after World War II into zones under the jurisdiction of the Allies. The zones of the Western Allies

eventually became West Germany, and the Soviet zone eventually became East Germany.

2. It provides inexpensive and convenient transportation, which is necessary for the country's industrial success.
3. All three countries share the Alps and have many fertile lowland valleys.
4. A stable government and membership in international economic and political organizations.

Critical Thinking

5. Answers will vary, but generally, the answer is yes. West Germany and the Alpine Countries have abundant natural resources, including water resources, good soil, and attractive scenery. Students should list natural resources for each country and explain how these resources have enabled each country to develop economically.

Using Geography Skills

1. West Germany, France, Italy, Austria, Liechtenstein. The language spoken might vary according to the border that Switzerland shares with another country. For example, French may be spoken in the Jura Mountains and Mittelland regions, German in the area around Lake Constance and Zurich, and Italian in the southern Alpine region.
2. Essen, Saarbrücken, Bielefeld; all are industrial cities.
3. West Germany, Liechtenstein, Switzerland, Italy, Yugoslavia, Hungary, Czechoslovakia. Austria is vulnerable because it has few natural boundaries that could provide it with a defense against invasion.

CHAPTER 15 Italy and Greece Pages 166–175

PLANNING GUIDE		
Lesson	Textbook	Teacher's Resource Binder
1. Italy	pp. 166–172	Lecture Notes and Transparency Package Content/Vocabulary Workbook Reteaching Worksheet Critical Thinking Worksheet
2. Greece	pp. 172–174	Challenge Worksheet Skills Worksheet Chapter Review/Test Reteaching Worksheet

TIME ALLOCATION: 2 days

CHAPTER TOPICS

- Italy
- Greece

PRETEACHING CHAPTER VOCABULARY

Write the following terms on the chalkboard:

strait — coalition government
balance of trade — isthmus
alluvial soils

Ask students which of the terms do not relate to physical geography. Show students a map with a strait and an isthmus identified. Ask students to help develop a definition. Then ask them to locate another isthmus and another strait in the Western Hemisphere.

Ask students what they think *alluvial soils* might be. Relate the definition of alluvial soils to a delta. Determine a definition.

Present the definitions of *balance of trade* and *coalition government*. Discuss possible reasons for the many failures of coalition governments in Italy.

▶ LESSON 1
Italy

LESSON OBJECTIVES

The student should be able to

- define significant vocabulary terms
- identify key events in the history of Italy
- explain how the geography of Italy contributes to its development
- locate significant places in Italy
- describe contributions of Italian culture to the world

PURPOSE

This lesson gives students the opportunity to learn the basis of much of their cultural heritage. It is important that world citizens know the past and present contributions of Italy to the world community.

MOTIVATOR

Write the following words on the chalkboard, and ask students to identify contributions made in each category to Western civilization by the Roman Empire.

Art — Architecture — Government

TEACHING DESIGN

1. Use the Chapter 15 Lecture Notes and Transparency Package to introduce the countries.
2. Divide the class into groups of three. Assign each group one of the following topics, and give them 20 minutes to conduct research and prepare a presentation for the rest of the class. Provide several resource books in the classroom, and encourage students to use the wall maps.

Topics	Questions
Natural Regions Northern Italy Central Italy Southern Italy Vatican City, San Marino, and Malta	1. What are three important facts about your topic? 2. What are significant dates, people, places, and products?

3. Allow each group 3 to 5 minutes in which to present their findings. Grade the presentation on the following

characteristics: (1) selection of important facts, (2) organization, and (3) presentation. Indicate to students the criteria for grades before they begin the research. Remind students to take notes during the presentations. If necessary, reports may continue during Lesson 2.
4. Homework: Have students complete information on Greece and Italy on their Western Europe chart.

CLOSURE
1. Ask students to compare their notes with a those of a student nearby to verify that they provide an accurate representation of the content presented.
2. Call on several students to present their notes in each category.

ASSESSMENT
1. Ask students to complete the Chapter 15 Content/Vocabulary Worksheet in their Workbook.
2. Have students complete the following questions from the Chapter 15 Check:
   Building a Vocabulary 1, 2, 3
   Recalling and Reviewing 2, 3
   Using Geography Skills 4

RETEACHING PLAN
1. Ask students to complete the Chapter 15 Reteaching Worksheet, Lesson 1.
2. Direct students to use the *World Book* or another appropriate encyclopedia to complete an outline on Italy. The outline should contain the following headings: History, Physical Geography, Political Organization, Economy, and Social/Cultural Activities.

CHALLENGE/ENRICHMENT STRATEGY
1. Have students complete the Chapter 15 Critical Thinking Worksheet.
2. Select students to investigate and bring pictures of the work of Italian artists from any period in history.
3. Play a music selection from a famous Italian composer.
4. Invite students to have a food fair to taste a variety of homemade Italian recipes.

▶ LESSON 2
Greece

LESSON OBJECTIVES
The student should be able to
- identify key events in the history of Greece
- explain how the physical geography of Greece contributes to its development
- locate significant places in Greece
- describe contributions of Greek culture to the world

PURPOSE
Like the lesson on Italy, this lesson gives students the opportunity to learn the basis of much of their cultural heritage. It is important that world citizens know the past and present contributions of Greece to the world community.

MOTIVATOR
Bring to class several pictures that represent contributions made by the Greeks to Western civilization. Include pictures of the Olympics, elections, the Acropolis, the Parthenon, and Greek sculpture. Use the pictures to emphasize the impact of the early Greek culture on our way of life.

TEACHING DESIGN
1. Begin class by finishing any oral reports remaining to be presented from Lesson 1.
2. Use the classroom wall map to review the location of the following important sites: Balkan Peninsula, Greece, Ionian Sea, Aegean Sea, Mediterranean Sea, Black Sea, Crete, Pindus Mountains, Rhodope Mountains, Mount Olympus. Discuss why Greece is said to hold a strategic location in the world.
3. Create a class chart on the chalkboard to summarize the key factors listed by the various groups.
4. Have students complete the Chapter 15 Challenge Worksheet.

CLOSURE
1. Using the chart written on the chalkboard during Teaching Design 3, have students make a review sheet.
2. Have students answer aloud the following questions from the Chapter 15 Check:
   Building a Vocabulary 4
   Recalling and Reviewing 1, 4
   Critical Thinking 5
   Using Geography Skills 1, 2, 3

ASSESSMENT
1. Have students complete the Chapter 15 Skills Worksheet.
2. Ask students to complete the Chapter 15 Review/Test.

RETEACHING PLAN
1. Ask students to complete the Chapter 15 Reteaching Worksheet, Lesson 2.
2. Transcribe the information written on the chalkboard onto index cards. Instruct students to match the cards to the categories and countries, then identify whether they indicate a similarity or a difference.

CHALLENGE/ENRICHMENT STRATEGY
1. Choose several students to investigate Greek and Roman mythology, then present an oral report.
2. Invite students to bring Greek foods to class and play Greek music.
3. Select volunteers to investigate the national flag, motto, holidays, folk costumes, and anthem of Greece.
4. Ask several students to research and report on famous ancient Greeks.

Chapter 15 Check, page 175, Answers

Building a Vocabulary

1. An isthmus is a narrow neck of land that connects larger land areas; a strait is a narrow body of water between two land areas. Isthmus: neck of land separating the Peloponnesian Peninsula from southern Greece; strait: the Strait of Gibraltar.
2. Fertile soils deposited by rivers. They are found in Italy's Po Valley. Fertile soil is needed for agriculture.
3. A government in which several political parties join together to run the country. Because each has been successful only for a short period of time.
4. The difference between the value of the goods a country exports and the value of the goods it imports. Nearly all of Italy's energy resources, especially coal, oil, and natural gas, must be imported.

Recalling and Reviewing

1. Greek ideas form the base of the Western democratic style of government. Roman culture spread from Italy throughout Western Europe during the Roman Empire. Greek and Italian ideas about art, architecture, literature, and science spread throughout Europe and influenced the development of Western culture.
2. Economic disparity between north and south, poverty, and political instability. Membership in the European Community has helped solve some of these problems by providing job opportunities for Italian workers in other EC countries.
3. Northern Italy: the Alps, the Apennines curve from the Alps in the northwest down the center of the peninsula toward the south, the Po Valley, a rich agricultural region, between the Alps and the Apennines, major rivers; Southern Italy: dry, volcanic region with poor soils and no major rivers.
4. Southern Greece: agriculture, tourism, and fishing; the Greek Islands: agriculture and tourism; central Greece: agriculture, mining, industry, and tourism; northern Greece: livestock raising, forestry, agriculture, and mining.

Critical Thinking

5. Answers will vary but should consider Greece's strategic location on Mediterranean sea routes. Because Greece is bordered by Communist countries of the Soviet bloc to the north and east, NATO would provide a defense system for Greece against Communist aggression. Greece controls the Aegean Sea, which is part of the Mediterranean Sea. This control prevents Soviet naval activity and blocks Soviet access to the Mediterranean Sea.

Using Geography Skills

1. Italy: Sardinia, Elba, Sicily; Greece: Crete, Corfu, Lesbos, Chios, Rhodes, Dodecanes Islands, Cyclades Islands, Ionian Islands.
2. Through the Malta Channel or the Strait of Messina to the Ionian Sea, through the Aegean Sea, the Dardanelles, and the Sea of Marmara.
3. Athens: 38° N, 23° E.
4. Those pertaining to vineyards, livestock, orchards, and olives.

CHAPTER 16 Spain and Portugal Pages 176–183

PLANNING GUIDE		
Lesson	**Textbook**	**Teacher's Resource Binder**
1. History, Geography, and Agriculture	pp. 176–179	Lecture Notes and Transparency Package Skills Worksheet Reteaching Worksheet Challenge Worksheet
2. Resources, Economy and Culture	pp. 179–182	Content/Vocabulary Workbook Critical Thinking Worksheet Chapter Review/Test Reteaching Worksheet

TIME ALLOCATION: 2 days

CHAPTER TOPICS

- Iberian History
- Spain
- Portugal
- Andorra and Gibraltar

PRETEACHING CHAPTER VOCABULARY

Write the following terms on the chalkboard:

navigable plaza autonomy

Ask students to suggest meanings of the words, then to check their answers in their textbook Glossary. Provide the following additional information to facilitate students' understanding of the terms.

The root *navis* is the Latin word for "ship"; the suffix *-able* means "capable of"; the verb *navigate* means "to plan, record, or control the course or position of a ship." Point out related words such as *navigation, navigator*.

The word *plaza* is from a Latin term meaning "a broad courtyard or street." Remind students that Spanish, like French and Italian, is a Romance language, based upon Latin. In Spain, plazas are public squares or open areas in towns or cities. Ask students what the equivalent of a plaza would be in a small American town. (town square)

The word *autonomy* comes from a Greek word and refers to the condition of being self-governing. A country has autonomy or is autonomous if it governs itself without intervention from another country. Ask students if the United States is an autonomous nation.

▶ LESSON 1
History, Geography, and Agriculture

LESSON OBJECTIVES
The student should be able to
- define the significant terms of the lesson
- locate the nations of the Iberian Peninsula
- identify and describe the major landforms, features, and climate of the Iberian Peninsula
- sequence key events in the history of Spain
- identify the agricultural base in the Iberian Peninsula according to climate and landforms

PURPOSE
The study of Spain and Portugal will provide examples of the relationship of people to the environment. People adapt to the geographic setting to meet basic needs, and they expand beyond their immediate environment by using ingenuity, creativity, and determination.

MOTIVATOR
Write the following terms on the chalkboard:

ranch	Los Angeles	Santa Fe	mosquito
sombrero	San Antonio	gazpacho	cork
Sierra Nevada	Colorado	plaza	flamenco
El Paso	Florida	Corpus Christi	siesta

Ask students to provide their meanings and origins. Explain to students that these words and names of places are all derived from Spanish. Point out that Spain expanded into the Americas and left lasting influences. Remind students that Spain, like other countries of Europe, was once a major power in the world. Explain that students will learn about Spain's place in the world community in this lesson.

TEACHING DESIGN
1. Use the Chapter 16 Lecture Notes and Transparency Package to present the history and geography of the Iberian Peninsula. Write key dates on the chalkboard, and instruct students to take notes on significant events presented in the lecture. Students should also note sites and regions on the map "Spain and Portugal" on page 178 of their textbook.
2. On the chalkboard, provide students with a list of the agricultural products of Spain and Portugal. Ask students to work in groups to predict in what areas these agricultural goods are grown, based upon the information in their maps on landforms, rivers, and climate types. Students should then compare their predictions of the locations for agricultural production with information in their textbook.
3. Have students complete the Chapter 16 Skills Worksheet.

CLOSURE
1. Ask representatives from each group from Teaching Design 2 to share their knowledge of agricultural production in Spain and Portugal.
2. Direct students to turn to a nearby student and review the historical events associated with the dates listed on the chalkboard.

ASSESSMENT
1. Have students complete the following questions from the Chapter 16 Check:
   Building a Vocabulary 1, 2
   Recalling and Reviewing 1, 2, 3
   Critical Thinking 6
   Using Geography Skills 1, 2
2. Ask students to complete the information on Spain and Portugal in their Western Europe chart.

RETEACHING PLAN
1. Ask students to complete the Chapter 16 Reteaching Worksheet, Lesson 1.
2. Direct students to work in small groups to locate on the map "Spain and Portugal" on page 178 in their textbook the rivers, landforms, islands, bodies of water, and neighboring countries of Spain and Portugal. One student in each group should call out locations while the other students in the group point to the sites.

CHALLENGE/ENRICHMENT STRATEGY
1. Have students complete the Chapter 16 Challenge Worksheet.
2. Select students to investigate the lives and contributions of one of the following people, then report to the class: Prince Henry the Navigator, Bartholomeu Dias, Ferdinand Magellan, Phillip II, King Juan Carlos, Vasco da Gama, Pedro Cabral, Ferdinand and Isabella, Franco.
3. Ask students to locate the areas of the world that were once part of the colonial possessions of Spain and Portugal. Students also should list the places in the world today where Spanish or Portuguese is spoken, then share this information with the class.

▶ LESSON 2
Resources, Economy, and Culture

LESSON OBJECTIVES
The student should be able to
- identify the natural resources of the Iberian Peninsula
- identify the major economic activities in the region
- locate urban centers and identify their functions
- identify the cultural contributions of Spain and Portugal

PURPOSE
The study of natural resources and related economic activities in Spain and Portugal will assist students in understanding the relationship between the two. Students also will see examples of the development of urban centers for processing, manufacturing, and marketing of products.

MOTIVATOR
Ask students to name the natural resources that are found in the area where they live. List these on the chalkboard. Ask what products, industries, or economic activities are related to these resources. List these beside each resource. Ask students to make a generalization reflecting the relationship of economic activities to the location of natural resources in an area. Tell students that they will study the natural resources of the Iberian Peninsula in this lesson.

TEACHING DESIGN
1. Direct students to work in groups to list the natural resources found in Spain and Portugal. Each group should use encyclopedias and/or dictionaries to determine possible industries or products derived from the resources. A spokesperson from each group should share the information with the class.
2. Provide encyclopedias and books on Spain and Portugal for students to use as references. Each student should work with a partner to select four scenes of cities, famous sites, or cultural events in one of the countries and write a caption as it would appear on the back of a postcard sold in the country. You may wish to provide postcard-size index cards for students. Students should describe what is pictured on the postcard as well as provide the caption.
3. Assign students to complete the Chapter 16 Content/Vocabulary Worksheet in their Workbook.

CLOSURE
Have volunteers share their postcards with the class. Ask: What are the major cities? Where are they located? What functions of the cities have made them urban centers? What industries are linked to these urban centers?

ASSESSMENT
1. Have students answer the following questions from the Chapter 16 Check:
   Building a Vocabulary 3
   Recalling and Reviewing 4, 5
   Using Geography Skills 3, 4
2. Have students complete the Chapter 16 Critical Thinking Worksheet.
3. Ask students to complete the Chapter 16 Review/Test.

RETEACHING PLAN
1. Ask students to complete the Chapter 16 Reteaching Worksheet, Lesson 2.
2. Direct students to work with a partner to review the information in their Western Europe chart on Spain and Portugal. Ask them to focus their attention on these headings: Natural Resources, Major Economic Activities, Exports, and Imports.

CHALLENGE/ENRICHMENT STRATEGY
1. Select volunteers to investigate and report on writers and artists from the Iberian Peninsula.
2. Choose students to investigate special dances (bolero, flamenco, fandango), festivals, or events (bullfights) that are identified with the Iberian culture.
3. Ask students to investigate the political conflicts associated with the area of Gibraltar. Ask what is meant by the expression "steady as the Rock of Gibraltar."

Chapter 16 check, page 183, Answers

Building a Vocabulary
1. A body of water that is deep enough and wide enough for ships to pass through; navigable rivers are important to trade, and the Guadalquivir is one of Spain's few navigable rivers.
2. The plaza is a public square and the center of most village activity. It is often the site of a marketplace and a gathering place for people.
3. The right to local self-government. Language differences among various regional groups have contributed to the drive toward autonomy.

Recalling and Reviewing
1. A group of Muslims from Africa (the Moors) invaded and ruled parts of the Iberian Peninsula for more than 700 years, spreading Islamic art, architecture, agriculture, and science over the peninsula.
2. They are used in two of Spain's major industries, the wine and olive-oil industries.
3. Both countries were ruled by dictators; they both have become democracies since the 1970s.
4. Fertile soil, fish, iron, mercury, uranium, copper, tungsten, zinc, lead. Spain must import many essential resources such as coal, oil, and timber because of its limited energy resources. Few Spaniards have been trained as scientists and engineers, and a great disparity between rich and poor still exists.
5. Trade deficit, the need to import energy resources, limited agricultural production, high unemployment, and political instability. Increase farmland, extend irrigation and power projects, promote industrial production, membership in the European Community.

Critical Thinking
6. Specific locations will vary, but students should choose an area that has the continental steppe climate and vegetation characteristic of the American West. The area around the middle of the Ebro River and the area between the Cantabrian and Guadarrama mountains are two possible answers.

Using Geography Skills
1. About 50 miles (80 kilometers). Because it is connected with the ocean by the Guadalquivir River, which is navigable.
2. About 15 miles (24 kilometers). The distance is so short that control of Gibraltar could mean control of access to the Mediterranean from the Atlantic.

3. Oporto. Portugal.
4. About 41° N, 4° W. Advantage: Madrid's location is central; disadvantages: could be cut off from other areas of Spain by mountains and rivers, not a good location for international trade.

UNIT 2 Review/Test

PLANNING GUIDE		
Lesson	**Textbook**	**Teacher's Resource Binder**
1. Unit 2 Review	pp. 184–185	
2. Unit 2 Test		Unit 2 Test

TIME ALLOCATION: 2 days

▶ LESSON 1
Unit 2 Review

Have students read the Unit 2 Summary. Use the Unit 2 Objectives as the basis for review, and discuss any questions that students have on the Unit 2 material.

Have students complete the questions in the textbook under the following headings:
Reading and Understanding
Mastering Geography Skills

Have students read the textbook material under the following headings and complete the appropriate or selected activities:
Applying and Extending
Linking Geography and History

▶ LESSON 2
Unit 2 Test

Have students complete the Unit 2 Test from the Teacher's Resource Binder.

Unit 2 Review, pages 184–185, Answers

Reading and Understanding

1. Luxembourg, Andorra, Switzerland, Austria, Liechtenstein. All have a water access to the sea by use of the various rivers of Western Europe and through a series of canals; Luxembourg: through its iron and steel industry, foreign investment, and banking; Andorra: through tourism; Switzerland: through its central location, industry, tourism, banking, and location of international organizations; Austria: through industry and tourism; Liechtenstein: through small-scale industry, being headquarters for many foreign corporations, tourism, and selling of postage stamps.
2. Because of its various languages and political systems. Most Western European nations, although seeking closer economic and political ties, are anxious to retain their individual national characters.
3. Rhine, Danube, Elbe, Moselle, Seine, Meuse, Saône, Ebro, Seine, Po, Rhône, Loire, and other rivers connected by canals. The rivers of Western Europe have provided easy and efficient water transportation for trade and hydroelectric power for industry.
4. France, Denmark, the Netherlands, and West Germany. These have a mild maritime climate and are located on the fertile North European Plain.
5. Alps, Pyrenees, Jura, Sierra Nevada, Kjølen, Apennines, Pennines, Vosjes, Pindus. They serve as borders for some countries and provide a major source of water power and tourism; the mountains, especially the Alps and Pyrenees, also create a milder Mediterranean climate in the southern part of Western Europe.
6. France, Great Britain, West Germany, the Netherlands, Belgium, Luxembourg, and Italy. Water transportation, supplies of coal and other natural resources. Answers will vary but should include lower unemployment, lower inflation, development of new and more efficient sources of energy, modernized agriculture, and maintaining a balance of trade.
7. London, Paris, Milan, Amsterdam, and Rome. All have diverse populations, are located on major waterways near important mineral deposits, and have rich cultural histories.
8. Answers will vary but might include the debate over nuclear power, the influx of people from former Asian and African colonies into urban areas, unemployment, and inflation.
9. The economic association of Western Europe. Ireland: foreign investment; Italy: Italians can get jobs in other EC nations, and Italy can obtain necessary resources for industry; Spain and Portugal: hope for increased economic development and trade; Greece: improvement and development of its economic situation.
10. Great Britain, Norway, France, Spain, Portugal, Greece, Italy, West Germany, the Netherlands, Luxembourg, Ireland, and Belgium. It was formed as an alliance for the common defense of Western Europe. NATO provides its members with a strong military defense against outside aggression.

Mastering Geography Skills

1. Switzerland, Austria, Norway, Sweden, Finland. Switzerland and Austria are dominated by highland climates; the Scandinavian countries have the subarctic and tundra climates. East and south Asia, parts of Africa, South America, east and west coast of North America, some coastal regions of Australia.
2. About 600 miles (960 kilometers); about 700 miles (1,120 kilometers). The trip from Paris to Rome would be more time consuming because you would need to cross several mountain ranges.

Applying and Extending
Answers will vary.

Linking Geography and History
Students should explain the history of the British Empire, especially in terms of its relationship with India.

UNIT 3 The Soviet Union and Eastern Europe Pages 186–239

UNIT OVERVIEW
In 1917, the Soviet Union became the world's first Communist country. Today, it is one of the world's most powerful nations. This unit provides a background for this development as well as a current look at the USSR and adjacent Communist-bloc countries.

CHAPTER TIME LINE

Chapter	Title	Time Allocation
17	Soviet Union	6 days
18	Western Soviet Union	3 days
19	Eastern Soviet Union	2 days
20	Eastern Europe	2 days
	Unit 3 Review	1 day
	Unit 3 Test	1 day
		15 days

MAJOR TOPICS
- Factors That Have Contributed to the Soviet Union's Becoming a Major World Power
- Relationship Between Land, Climate, and Ways of Life
- Economic and Political Ties Between the USSR and Eastern Europe
- Relationship Between Climate and Population Density
- Current Issues

INTRODUCING THE UNIT
Show students several pictures from the Soviet Union and Eastern Europe. Discuss with students what they would like to learn about this region. Write their answers on paper, and refer back to them at the end of the unit.

CHAPTER 17 The Soviet Union Pages 188–205

PLANNING GUIDE		
Lesson	**Textbook**	**Teacher's Resource Binder**
1. Physical Geography	pp. 194–198	Lecture Notes and Transparency Package Reteaching Worksheet
2. History to World War II	pp. 188–192	Content/Vocabulary Workbook Reteaching Worksheet
3. History Since World War II; Political Organization	pp. 192–200	Reteaching Worksheet Critical Thinking Worksheet
4. Economy	pp. 200–201	Reteaching Worksheet Challenge Worksheet
5. Life in the USSR	pp. 201–204	Reteaching Worksheet
6. The Soviet Union Today; Map Skills	pp. 196–204	Skills Worksheet Chapter Review/Test Reteaching Worksheet

TIME ALLOCATION: 6 days

CHAPTER TOPICS
- Historical Geography
- Physical Geography
- Peoples of the Soviet Union
- Soviet Political Organization
- Economic Geography
- Life in the Soviet Union
- The Soviet Union and World Affairs

PRETEACHING CHAPTER VOCABULARY
Write the following terms on the chalkboard:

portage	soviet	collective farm	autarky
tsar	ethnic group	taiga	state farm

Create a prereading assessment chart as follows.

Words	I Have Seen Before	Predicted Meaning	Definition

Discuss words that the students have studied previously. See whether the class can identify the words that are types of economic systems. Which other words relate to economic activities? Which words probably relate to the government of the Soviet Union? Following the discussion, have students use the Glossary in their textbook to ensure that they have accurate definitions.

▶ LESSON 1
Physical Geography

LESSON OBJECTIVES
The student should be able to
- define significant vocabulary
- locate significant sites in the Soviet Union
- compare the size of the United States with the size of the Soviet Union
- describe the relationships between the land regions of the USSR and the climate regions

PURPOSE
This lesson will provide students with an overview of the land and geography of the USSR. The study of the physical geography is critical to an understanding of the position held by the nation as a world superpower.

MOTIVATOR
Create a word cluster around the term *Soviet Union*. Solicit contributions from all students. Display the cluster in the classroom on a sheet of poster board. Have students refer to the poster throughout the study of the Soviet Union. Highlight the correct impressions as they are studied in the textbook and through class discussion.

TEACHING DESIGN
1. Call on individual students to approach a classroom wall map and locate the following sites discussed in the reading: USSR, Sea of Japan, Tian Shan, Aral Sea, Ob River, Arctic Ocean, Pacific Ocean, Ural Mountains, Dnieper River, Yenisei River, Crimean Peninsula, Sea of Okhotsk, Finland, Kamchatka Peninsula, Caucasus Mountains, Don River, Lena River, Bering Sea, East Siberian Sea, Caspian Sea, Volga River, Black Sea.
2. Use the Lecture Notes and Transparency Package to introduce the chapter.
3. Review with students the vegetation of the climate regions that encompass the land of the USSR. Refer to the climate map on page 197. Have students work with a partner to create a chart that will compare the land regions of the Soviet Union to the climate regions and vegetation. Prior to the students' beginning work, indicate that they should expect overlaps between regions and climates.

CLOSURE
Locate several regions from the chapter on a world map or a map of the Soviet Union. Ask students to use their charts to identify the type of climate and vegetation prevalent in each region. Ask students to predict economic activities that might occur in the regions.

ASSESSMENT
Have students complete the following questions from the Chapter 17 Check:
Building a Vocabulary 4, 5
Recalling and Reviewing 3
Using Geography Skills 1, 2

RETEACHING PLAN
Ask students to complete the Chapter 17 Reteaching Worksheet, Lesson 1.

CHALLENGE/ENRICHMENT STRATEGY
1. Ask students to report on the significance of the rivers in the USSR to the development of the country.
2. Allow interested students to investigate the "Russian Riviera" discussed in the chapter.

▶ LESSON 2
History to World War II

LESSON OBJECTIVES
The student should be able to
- locate significant sites in the Soviet Union
- identify important dates in the development of the Soviet Union
- recognize key people in the history of the Soviet Union

PURPOSE
This lesson will provide students with an understanding of the key events in the history of the Soviet Union. This study will enable students to comprehend the reasons for the current political system in the Soviet Union.

MOTIVATOR
Begin this lesson by creating a futures wheel on the chalkboard around the assumption that the Soviet Union adopts a democratic government. Discuss the potential advantages and disadvantages.

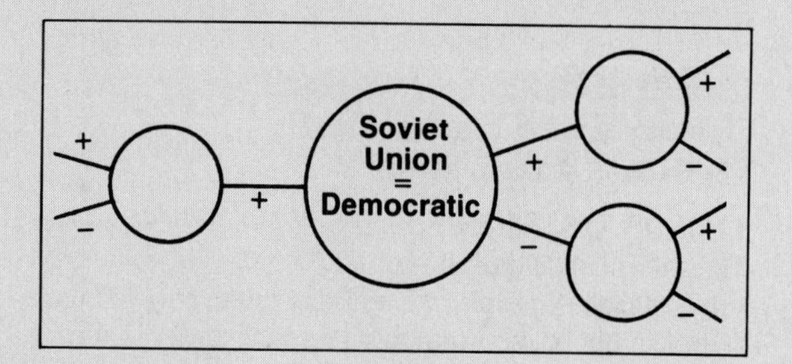

TEACHING DESIGN

1. List the following significant dates on the chalkboard.

1237	1547	1796	1920
1240	1637	1867	1922
1490s	1703	1917	1929

   Have students locate the significant dates in the reading selection and identify the key events associated with each date. Give students time to find the answers before calling on them to explain the event. Write the key ideas on a transparency. Have students take notes.
2. Locate the following places on a large classroom world map: Kiev, China and India, Moscow, Alaska, Mediterranean and Baltic Seas, Leningrad, Black Sea. Direct students to work in discussion groups to identify a date associated with the location. Call on a group spokesperson to explain the relationship.

CLOSURE

1. Call on volunteers to use their time line to "tell a story" about the history of the Soviet Union. Provide each individual with a different time period. (Example: 1237–1547) When the time period is complete, call on another student to continue the story.

ASSESSMENT

1. Ask students to complete the Chapter 17 Content/Vocabulary Worksheet in their Workbook.
2. Have students complete the following questions from the Chapter 17 Check:
   Building a Vocabulary 1
   Recalling and Reviewing 1

RETEACHING PLAN

1. Have students research the following individuals in an encyclopedia and summarize in a paragraph their contributions to Russian history: Ivan III, Ivan IV, Peter the Great, Catherine the Great, Lenin, Stalin.
2. Ask students to complete the Chapter 17 Reteaching Worksheet, Lesson 2.

CHALLENGE/ENRICHMENT STRATEGY

1. Invite a group of students to host a "You Are There" television program and interview selected leaders from Russian history. Award extra points for creativity.
2. Select students to investigate and report their answers to the following questions.
   a. Why was Ivan IV called "Ivan the Terrible?"
   b. Why did Russia sell Alaska to the United States?
   c. Who was Rasputin?

▶ LESSON 3
History Since World War II; Political Organization

LESSON OBJECTIVES

The student should be able to

- recognize key people in the history of the Soviet Union
- explain the political organization of the Soviet Union

PURPOSE

In this lesson, a continuation of Lesson 2, students will continue learning the chronology of key events in the history of the Soviet Union. This study will enable them to understand the basis for the "cold war" between the USSR and the United States.

MOTIVATOR

Invite several students to the front of the class to identify a key person or date from Lesson 2. Ask each individual to tell something he or she remembers about the person or date. Encourage students to continue to build an oral history of the Soviet Union.

TEACHING DESIGN

1. Ask students to identify the negative and positive results of World War II for the Soviet Union. Discuss which result was most important to the government. Identify possible reasons.
2. Write the following on the chalkboard:

1945	Gorbachev
1953	Khrushchev
1964	Stalin
1985	Brezhnev

   Ask students to match the year with the leader.
3. Have students work with a partner to create a diagram that represents the organization and operation of the political system in the Soviet Union. Be sure that students include the following terms in their charts: Soviet Socialist Republics, RSFSR, soviets, education, agriculture, finance, Supreme Soviet, Congress, Central Committee, Politburo, Communist party.
4. Each pair should then begin a chart (see sample below), using information from this section regarding similarities to and differences from the United States. Students will continue this chart in Lesson 4.

Similar to U.S.	**Different from U.S.**

CLOSURE

Invite several students to address the class and present the diagram they developed in Teaching Design 4 to describe the political organization of the Soviet Union.

ASSESSMENT

Have students complete the following questions from the Chapter 17 Check:
Building a Vocabulary 2
Recalling and Reviewing 4

RETEACHING PLAN

Ask students to complete the Chapter 17 Reteaching Worksheet, Lesson 3.

CHALLENGE/ENRICHMENT STRATEGY

1. Have students complete the Chapter 17 Critical Thinking Worksheet.
2. Ask several students to investigate and report on cold war encounters between the United States and the USSR, including the Cuban Missile Crisis.

▶ LESSON 4
Economy

LESSON OBJECTIVES

The student should be able to

- define significant vocabulary terms
- identify the major resources of the Soviet Union
- analyze the impact of the environment on ways of life
- describe the economic activities of the Soviet Union
- determine the agricultural base of the Soviet Union

PURPOSE

Students will continue to explore how the geography of an area affects its economy. In this lesson, students will expand their understanding of the key differences between the Soviet Union and the United States.

MOTIVATOR

Write the following chart on the chalkboard. Call on students to review the information.

Communism: Who Decides	
What to produce?	
How much to produce?	
For whom to produce?	

TEACHING DESIGN

1. Place students in small groups. Have each group search the chapter for information to complete the following:

USSR	
Natural Resources	Products

2. Have students reread page 194 on the policies of Gorbachev. Ask students to identify characteristics of capitalism "between the lines" of his new policies.
3. Ask students to read the Themes in Geography feature "Kolkhozy and Sovkhozy" on pages 202–203 in their textbook, then write a paragraph in response to the following question: Why are American farms more productive than collective and state farms of the Soviet Union?

CLOSURE

Return to the charts in which students listed differences and similarities between the United States and the Soviet Union. Ask for additional statements to add to the charts.

ASSESSMENT

Have students complete the following questions from the Chapter 17 Check:

Recalling and Reviewing 2
Critical Thinking 5, 6

RETEACHING PLAN

1. Provide students with a map to locate the key natural resources and vegetation regions of the Soviet Union.
2. Ask students to complete the Chapter 17 Reteaching Worksheet, Lesson 4.

CHALLENGE/ENRICHMENT STRATEGY

1. Have students complete the Chapter 17 Challenge Worksheet.
2. Choose students to investigate and report on the status of imports and exports of the Soviet Union. Questions to be answered are: Who are their major trading partners? Do they have a trade deficit or surplus? Why are so few consumer goods imported?

▶ LESSON 5
Life in the USSR

LESSON OBJECTIVES

The student should be able to

- describe the culture of the Soviet Union
- identify ethnic groups of the Soviet Union

PURPOSE

This lesson will provide students with an understanding of the way of life of Soviet citizens. It will instill in students a greater appreciation for the American way of life.

MOTIVATOR

Ask students to imagine life as a teenager in the Soviet Union. Based upon their reading, what are some of the images they would create? Discuss how life for teenagers in the Soviet Union is different from life for teenagers in the United States.

TEACHING DESIGN

1. Refer students to the map "Ethnic Groups in the Soviet Union" on page 193. Write the following questions on the chalkboard or on a transparency, and have students write the answers.
   a. Which ethnic group covers the largest percentage of the country?
   b. Which is the second largest ethnic group?
   c. Approximately how many ethnic groups live in the Soviet Union?
   d. Which ethnic group lives nearest Kiev? Moscow?
   e. Based upon the reading, define the characteristics that are usually associated with each ethnic group.
2. Place students in groups of three or four to develop responses to the following questions.
   a. How is the ethnic composition of the USSR similar to

and different from that of the United States?

b. What is the greatest problem in the USSR caused by the large number of ethnic groups?

c. What are secondary problems?

Lead a class discussion on the students' responses.

CLOSURE
Call upon several students to read their letters from Teaching Design 3 to the class.

ASSESSMENT
Have students complete the following questions from the Chapter 17 Check:

Building a Vocabulary 3

RETEACHING PLAN
1. List on the chalkboard the following five cultural institutions: religion, economy, politics, language, family. Ask students to write a paragraph describing each cultural aspect as it is in the Soviet Union.
2. Ask students to complete the Chapter 17 Reteaching Worksheet, Lesson 5.

CHALLENGE/ENRICHMENT STRATEGY
1. Select students to investigate and report on one of the following topics as it relates to the USSR: Music, Soviet Jews, Afghanistan, The KGB.
2. Assume that a new leader has taken control in the Soviet Union. What actions should he or she take to establish Soviet national unity?

▶ LESSON 6
The Soviet Union Today; Map Skills

LESSON OBJECTIVES
The student should be able to
- define the basis for the cold war
- describe the current relationship between the Soviet Union and the United States
- interpret information provided in a climate map and a population-density map

PURPOSE
This lesson will provide students with an understanding of the role of the Soviet Union in the world today. Students need this information if they are to make informed decisions in selecting future United States policymakers.

MOTIVATOR
Provide newspapers for all students. Have each student locate and read an article concerning a Communist country. Select students to present an overview of the articles they read. Solicit generalizations about the USSR and its relationship to the world today.

TEACHING DESIGN
1. Write the following countries on the chalkboard: Hungary, Czechoslovakia, Poland, Angola, Cuba, Afghanistan, Ethiopia, China. Point out these countries on a world map. Ask students to answer the following questions: What is the relationship between the Soviet Union and each of these countries? What events may have caused this relationship to evolve?
2. Discuss the issue of nuclear war. Ask students why much of the world is opposed to nuclear war and what various nations are doing to prevent a nuclear war.
3. Ask each student to reread the Geography Skills feature "How to Compare a Climate Map to a Population-Density Map" on pages 196-197, then complete the questions.
4. Have students return to the word cluster they created in the Lesson 1 Motivator to determine which student impressions of the USSR were most accurate.

CLOSURE
1. Call upon several students to announce a generalization regarding population density and climate.
2. Have students complete the Chapter 17 Skills Worksheet.

ASSESSMENT
Ask students to complete the Chapter 17 Review/Test.

RETEACHING PLAN
Ask students to complete the Chapter 17 Reteaching Worksheet, Lesson 6.

CHALLENGE/ENRICHMENT STRATEGY
1. Ask students if they can name a defector from the Soviet Union. Direct students to prepare a letter to the United States government as if it were from a potential Soviet defector. Tell them that the letter should state reasons for seeking political asylum.
2. Have selected students research and act out for the class the most recent summit conference and the major items under discussion.
3. Select students to investigate and report on the issue of spies in the United States and the Soviet Union.

Geography Skills, Page 196, Answers
1. Dark blue; pale yellow. Dark green; mauve; gray-green. Mild climate regions are more densely populated than cold climate regions.
2. Eastern Europe.
3. This region has a continental desert climate, which is not conducive to agriculture or settlement.
4. South of the Caucasus Mountains. Highland.
5. Peninsulas and islands in the Arctic Ocean. 75° to 80° N, 50° to 180° E.

Chapter 17 Check, Page 205, Answers

Building a Vocabulary
1. A governing council of a republic. Political control is held by the Communist party leaders in Moscow, but

each republic has its own soviet, or special regional governments, created for the many ethnic groups.

2. A policy to develop self-sufficiency in industrial production. Because it has a wealth of natural resources, including oil, natural gas, coal, hydroelectric power, iron ore, gold, timber, copper, chromium, and manganese.
3. A group of people who share the same customs and language. More than 150.
4. In the subarctic region. The taiga provides the Soviet Union with a large supply of products, and some agriculture is possible in this area during the short summer.
5. Low land areas between lakes or rivers where boats and freight can be carried across. Because travel was easiest by water, many Russian cities were located at stream junctions and portages.

Recalling and Reviewing

1. Stalin introduced a Five Year Plan that outlined production goals for a five-year period. He also collectivized more than half of Soviet farmland; put all factories under government control; set production quotas; and built new mines, steel foundries, railroads, canals, and industrial cities.
2. On a state farm, workers receive wages, but the farm is operated by the government. On a collective farm, workers receive lower wages than those on a state farm, but they receive a share of the production and the profits.
3. They provide a network of water transportation that is cheaper than land transportation; rivers also provide hydroelectric power. Major rivers are the Dnieper, Don, Volga, Ob, Yenisei, and Lena.
4. The governments of the Soviet Union's 15 republics are controlled by the central government in Moscow. Real governing power in the Soviet Union rests with the Communist party; party leaders make all important decisions. The Congress is the party's highest authority and elects the Central Committee, which determines who will be members of important policy committees. The Politburo is the most important committee and is the center of party power.

Critical Thinking

5. In most of the country, the growing season is short, and droughts are frequent. The collective and state farms are not as productive as comparable privately owned farms in other countries and private plots in the USSR. Answers will vary as to whether or not the problem can be solved.
6. Answers will vary, but students may mention the advantages of meeting national goals and the disadvantages of having a limited group make complex decisions that affect large numbers of people.

Using Geography Skills

1. The Volga, Irtysh, Ob, Yenisei, Amur, and Ussuri rivers and the Ural, Sayan, and Yablonoi mountains. The humid continental and subarctic climate regions.
2. The Caspian Sea and the Black Sea.

CHAPTER 18 Western Soviet Union Pages 206–215

PLANNING GUIDE		
Lesson	**Textbook**	**Teacher's Resource Binder**
1. Industrial Development; European Russia	pp. 206–212	Content/Vocabulary Workbook Reteaching Worksheet Critical Thinking Worksheet
2. Western Soviet Republics	pp. 212–214	Reteaching Worksheet Challenge Worksheet
3. Language and Culture	pp. 210–211	Lecture Notes and Transparency Package Chapter Review/Test Reteaching Worksheet

TIME ALLOCATION: 3 days

CHAPTER TOPICS

- Industrial Location
- European Russia
- The Ukraine
- Moldavia
- Byelorussia
- The Baltic Republics
- The Caucasus

PRETEACHING CHAPTER VOCABULARY

Write the following terms on the chalkboard:

heartland   periphery   infrastructure

Point out to students the following information.

heartland (An important central region, especially one regarded as vital to a nation's economy or defense.)

periphery (*Peri-* comes from the Latin word for "around." A periphery is the outermost part or region within a precise boundary.)

infrastructure (*Infra-* means "below, beneath basic facilities"; hence, an *infrastructure* is the equipment and services needed for the growth and functioning of a country or community.)

▶ LESSON 1

Industrial Development; European Russia

LESSON OBJECTIVES

The student should be able to

- define significant vocabulary
- identify key places in the western Soviet Union
- assess Soviet industrial development
- describe the process used to develop industry
- name key facts about European Russia

PURPOSE

Students will learn the results of the attempt by the Soviet government to relocate citizens and establish new industries. This will provide them with an example of factors that are considered in urban planning.

MOTIVATOR

Ask students to think about a business they may wish to own someday. Then tell them to imagine that the United States government is trying to motivate people to move to an unsettled and unoccupied territory in Alaska. The government will provide all persons who agree to relocate with the money necessary to open their businesses. This sounds like an offer too good to be true. Write the following questions on the chalkboard: After careful consideration, what are the questions you want answered by the government? What services does the government need to provide if you are to resettle? Have students discuss the problem.

TEACHING DESIGN

1. Draw the following flow chart of Soviet industrial development on the chalkboard.

1861	1928	Today
Cause:	Cause:	Cause:
Effect:	Effect:	Effect:

   Have students review pages 207–209 in their textbook to locate information to complete the chart. Complete the chart on the chalkboard as you call on students to give their answers.
2. Ask students to review the four goals of the Soviet Union for starting new industries. Have students work in groups to answer the following questions for each goal.
   a. How could its achievement help the country?
   b. What could the government do to ensure the achievement of the goal?

   Have students discuss the responses as a class.
3. Have students read the Cities of the World feature "Moscow" on pages 210–211 and note the highlights.
4. Provide students with the following chart, and ask them to complete the first row across.

USSR Republics	Largest Cities	Resources	Industries	Major Products
RSFSR				
Ukraine				
Moldavia				
Byelorussia				
Estonia, Latvia, and Lithuania				
Georgia, Armenia, and Azerbaijan				

CLOSURE

1. Review with students the four goals of the Soviet Union for industrial development. Name three goals, and ask students to identify the missing one.
2. Discuss the first row of information completed on the Republics chart.

ASSESSMENT

1. Have students complete the following questions from the Chapter 18 Check:
   Building a Vocabulary 1
   Recalling and Reviewing 1, 2, 3, 4
   Critical Thinking 5
2. Ask students to complete the Chapter 18 Content/Vocabulary Worksheet in their Workbook.

RETEACHING PLAN

Ask students to complete the Chapter 18 Reteaching Worksheet, Lesson 1.

CHALLENGE/ENRICHMENT STRATEGY

1. Have students complete the Chapter 18 Critical Thinking Worksheet.
2. List several cities described in the chapter. Ask students to complete the following chart for determining a location for a new government clothing plant.

New Location to Be	
Positive	Negative

Decision

____

▶ LESSON 2
Western Soviet Republics

LESSON OBJECTIVES

The student should be able to
- identify key places in the western Soviet Union
- name key facts about European Russia

PURPOSE

Students will now have an understanding of the regional organization of the USSR, further extending their knowledge of this superpower nation.

MOTIVATOR

Ask the class the following question: What geographical-political divisions does the Soviet Union have that compare to those of the United States? Lead students to the conclusion that both nations have political subdivisions; in the United States, they are called states, whereas in the USSR, they are called republics. On an unlabeled map of the western Soviet Union, locate and label the 15 republics.

TEACHING DESIGN

1. Ask students to return to the chart they began in Lesson 1, Teaching Design 4. Have them complete it for the remaining republics, using the abbreviation N/A for information not available in their textbook.
2. Ask students to work in groups, using their chart to answer the following questions.
   a. Which republics are primarily agricultural?
   b. Which republics produce steel?
   c. Which republics have a history and culture that support independence?
   d. Which republics have the three largest cities?
   e. Which republics have access to a warm-water port?
   f. Which republics have access to a river allowing transportation of goods?
   g. Which republics have an economy supported by manufacturing?

CLOSURE

1. Discuss the answers to the preceding questions with the entire class. Have one student at a time read and answer a question.
2. Review the locations of the Soviet republics discussed in this chapter.

ASSESSMENT

Have students complete the following questions from the Chapter 18 Check:

Building a Vocabulary 2, 3
Using Geography Skills 1, 2, 3

RETEACHING PLAN

Ask students to complete the Chapter 18 Reteaching Worksheet, Lesson 2.

CHALLENGE/ENRICHMENT STRATEGY

1. Have students complete the Chapter 18 Challenge Worksheet.
2. Select various students to conduct a talk show, inviting "guests" from the major cities in the western Soviet Union. Indicate that each "guest" should discuss why the audience should plan a trip to his or her city.

▶ LESSON 3
Language and Culture

LESSON OBJECTIVES

The student should be able to

- determine the cultural interchange among regions
- explain criteria for the various regional boundaries
- explain the importance of Moscow as the capital

PURPOSE

Students will learn the value of knowing more than one language as they examine the impact of language barriers in the Soviet Union.

MOTIVATOR

Ask students how many of them have enjoyed watching the Olympic Games in the past. Ask students to reflect on various types of uniforms worn by the various nationalities. Determine whether any are able to recall the uniforms worn by the competitors from the Soviet Union. Ask students if they remember the letters written across the Soviet uniforms. Write the letters (CCCP) on the chalkboard, and ask whether anyone knows their meaning. Leave the letters on the chalkboard for use in the Closure.

TEACHING DESIGN

1. Lead a class discussion on the implications of the Soviet government's having to deal with many different languages in the USSR.
2. Place the students in discussion groups to address the following questions about the information presented.
   a. Which groups do you think the Soviet government permits to maintain their individual identities? Why do you suppose the government allows these differences to exist?
   b. How do differences among ethnic groups in the Soviet Union compare to differences in the United States? Identify some similarities and differences.
   c. The study of Russian is now mandatory in all Soviet schools. What is the reason for this policy?
3. Discuss the answers developed in the groups. Ask students to add to their notes when you indicate that some valuable information has been offered.

CLOSURE

Return to the chalkboard where you wrote CCCP. Indicate for the students that the letters are from the Cyrillic alphabet and stand for SSSR. Now, ask the students what they think the letters represent. Indicate to the students that the first *S* stands for *soyuz,* which is the Russian word for "union." Note that in 1975, the Soviet spacecraft *Soyuz* docked with the American spacecraft *Apollo,* and the crews visited each other during this union in space. Refer students to the chart "The Roman and Cyrillic Alphabets" on page 236 of their textbook.

ASSESSMENT

Ask students to complete the Chapter 18 Review/Test.

RETEACHING PLAN

1. Have students read in encyclopedias about the ethnic diversity in the Soviet Union. Have them list the 10 most significant facts they learned from the readings.
2. Ask students to complete the Chapter 18 Reteaching Worksheet, Lesson 3.

CHALLENGE/ENRICHMENT STRATEGY

1. Ask selected students to create a map to demonstrate the population of the 10 largest cities in the Soviet Union. Ask them to identify the primary nationality or ethnic group associated with each city.

2. Choose students to investigate and report orally on a comparison between bilingual education in the Soviet Union and the United States.

Chapter 18 Check, page 215, Answers

Building a Vocabulary

1. A nation's basic residential and transportation facilities. Often, infrastructure must be built before new industries can be started, and thus, it is the necessary foundation of an industrialized country.
2. The core, or innermost region, of a country. Because this region has the most factories and the most productive agriculture in the Soviet Union.
3. Because there are few people and little economic development in the periphery of the Soviet Union; more than two-thirds of the Soviet people live in the heartland, which is less than one-third of the country's land.

Recalling and Reviewing

1. Moscow region: factories for smaller consumer items and advanced-technology industries; the Ukraine: heavy industry; along the middle Volga from Kazan to Volgograd: machine building; across the southern Urals: industries based on iron and petroleum.
2. Because the region is very rich in oil, iron ore, copper, lead, zinc, tungsten, gold, and manganese.
3. It is one of the greatest livestock areas in the country; it also produces citrus fruits, tea, tobacco, cotton, grains, and hay and is a source of oil, manganese, coal, lead, zinc, and copper.
4. The Ukraine is the most productive farming region in the USSR, producing large amounts of grain. It contains rich deposits of coal, iron ore, and manganese. It is a center for the production of iron, steel, and related products. Dams on the Dnieper River supply electricity to industries in the region.

Critical Thinking

5. Answers will vary but should include the vast transportation network that connects the industrial regions to the areas where natural resources are located. Answers will vary for the second part of the question but should include possible political and economic cooperation between Western Europe and the USSR.

Using Geography Skills

1. Leningrad is located at approximately 60° N, 30° E.
2. Georgian, Armenian, Tadzhik, Ukrainian, Turkmen, Uzbek, Kazakh, Azerbaijan, Estonian, Latvian, Lithuanian, Kirghiz, Byelorussian, and Moldavian.
3. Primarily rivers such as the Volga, the Don, and the Dnieper, and the Ural Mountains.

CHAPTER 19 Eastern Soviet Union Pages 216–223

PLANNING GUIDE		
Lesson	**Textbook**	**Teacher's Resource Binder**
1. Physical Geography	pp. 216–220	Content/Vocabulary Workbook Lecture Notes and Transparency Package Reteaching Worksheet Critical Thinking Worksheet
2. Economic Development	pp. 218–222	Skills Worksheet Chapter Review/Test Reteaching Worksheet Challenge Worksheet

TIME ALLOCATION: 2 days

CHAPTER TOPICS

- Soviet Central Asia
- Siberia

PRETEACHING CHAPTER VOCABULARY

Write the following terms on the chalkboard:

dryland agriculture habitation fog

Ask students to suggest what *dryland agriculture* may mean. Explain that the term refers to farming that depends upon rainfall to water crops. Discuss the problems associated with this method. Identify alternatives practiced in the United States. Discuss the meaning of the word *fog*. Indicate that a habitation fog is caused by a combination of warm air, dust, and water vapor.

▶ LESSON 1 Physical Geography

LESSON OBJECTIVES

The student should be able to

- define significant vocabulary
- locate key places in the eastern Soviet Union
- discuss the influence of landforms on ways of life

PURPOSE

Students will further develop an understanding of the ways an environment can influence an individual's life and a country's development.

MOTIVATOR

Draw the following chart on a sheet of poster board.

Siberia		
Student Beliefs	Fact	Opinion
1.		

Ask students to indicate what they believe to be true about Siberia. Write their beliefs on the poster board. Tell students that by the end of the lesson, they should be able to determine which of the statements are fact and which are opinion.

TEACHING DESIGN

1. Present the Chapter 19 Lecture Notes and Transparency Package.
2. Review locations of the following places. Make tags to put on a wall map to highlight these places. Give students time to prepare to identify the significance of each place as it relates to this region of the USSR.
   Soviet Central Asia: Ural Mountains, Caspian Sea, China, Iran, Afghanistan, Indian Ocean, Atlantic Ocean, Syr Darya (river), Amu Darya (river).
   Siberia: Pacific Ocean, Arctic Ocean, Mongolia, China, Ob River, Yenisei River, Lena River, Trans-Siberian Railroad, Vladivostok, Sea of Japan, BAM, Lake Baikal, Novosibirsk.
   Soviet Far East: Khabarovsk, Amur River, Sakhalin Island, Kuril Islands.
3. Direct students to take notes from Chapters 18 and 19 in their textbooks to complete the chart below.

	Western Soviet Union	Soviet Central Asia	Siberia	Soviet Far East
Climates				
Landforms				
Size				
Resources				

Place students in groups to prepare one key statement to summarize the comparisons for each characteristic.

CLOSURE

1. Assign a spokesperson for each group. Identify a characteristic from the chart in Teaching Design 3. Ask each spokesperson to read the summary statement. Provide an overall summary for each category.
2. Compare what students have learned about Siberia to the beliefs they had before the lesson by referring to the chart in the Motivator.

ASSESSMENT

1. Have students complete the following questions from the Chapter 19 Check:
   Building a Vocabulary 1, 2
   Recalling and Reviewing 3
   Critical Thinking 6
   Using Geography Skills 1
2. Ask students to complete the Chapter 19 Content/Vocabulary Worksheet in their Workbook.

RETEACHING PLAN

Ask students to complete the Chapter 19 Reteaching Worksheet, Lesson 1.

CHALLENGE/ENRICHMENT STRATEGY

1. Have students complete the Chapter 19 Critical Thinking Worksheet.
2. Ask students to read the book *One Day in the Life of Ivan Denisovich* by Alexander Solzhenitsyn. Conduct a discussion about the Siberian way of life in the story.

▶ LESSON 2
Economic Development

LESSON OBJECTIVES

The student should be able to

- identify key places in the eastern Soviet Union
- explain the impact of landforms on economic development of the eastern Soviet Union
- discuss the problems associated with the development of the eastern Soviet Union
- name major resources of the eastern Soviet Union

PURPOSE

Students will learn about the contributions made to the world market by the Soviet Union, thereby gaining a better understanding of the world market in general.

MOTIVATOR

Write the following numbers on the chalkboard. Indicate to students that each number can be found somewhere in the chapter. Place the students in groups, and allot them 10 minutes. Provide recognition to the group that identifies the most correct responses.

2/5	1/3	1/8	75 percent
5	5,700	1/10	5,000
.5 million	2,000	1/2	

TEACHING DESIGN

1. Compare the development of Soviet Central Asia to the economic development of the western part of the United States. Ask students to brainstorm about similarities and differences. Select a spokesperson to present group ideas. Summarize them on the chalkboard.
2. Ask students to complete the following features checklist for this region of the USSR.

Features	Soviet Central Asia	Siberia	Soviet Far East
Coal			
Farming			
Fishing			
Forests			
Furs			
Industry			
Iron			
Minerals			
Natural gas			
Oil			

Students should place a check mark beside every applicable characteristic for each region.

3. Have students work with a partner. Ask each pair to develop a five-point plan for establishing a new business in Soviet Asia. They should select a business from the column on the left and a location from the column on the right. Each pair's five-point plan should support their decision and identify their exact location. Students should use their features chart as a reference.

Industry	Location
Fishing boat sales	Soviet Central Asia
Steel plant	Siberia
Plastics plant	Soviet Far East
Clothing manufacturing plant	
Bread factory	

4. Have students complete the Chapter 19 Skills Worksheet.

CLOSURE

Invite students to present their five-point plans to the class. List the decisions on the chalkboard, and try to reach a class consensus.

ASSESSMENT

1. Have students complete the following questions from the Chapter 19 Check:
   Recalling and Reviewing 1, 2, 4, 5
   Using Geography Skills 2
2. Ask students to complete the Chapter 19 Reteaching Worksheet, Lesson 2.
3. Ask students to complete the Chapter 19 Review/Test.

RETEACHING PLAN

1. Ask students to complete the following chart to identify the significant factors presented in the lesson.

	Economic Resources	Ways of Life
Soviet Central Asia		
Siberia		
Soviet Far East		

2. Ask students to complete the Chapter 19 Reteaching Worksheet, Lesson 2.

CHALLENGE/ENRICHMENT STRATEGY

1. Have students complete the Chapter 19 Challenge Worksheet.
2. Select students to investigate and report to the class on the origin of the bear as a symbol for the Soviet Union in political cartoons.

Chapter 19 Check, page 223, Answers

Building a Vocabulary

1. A method of farming that depends upon rainfall to water the crops; in the northwestern parts of the Kazakh SSR, where there are huge areas of steppe. The goal is to use modern methods and machinery to increase grain production on huge state farms.
2. A winter fog caused by the warmth, dust, and water vapor that cities create in the atmosphere. They are typically found hanging over cities in Siberia.

Recalling and Reviewing

1. Because of the cold, dry, harsh climate. To make use of the vast natural resources in the area.
2. Through irrigation, modern farming methods, and machinery. Coal, copper, lead, zinc, nickel, and uranium.
3. West Siberian Lowland: taiga in the north is swamp in the summer and frozen in the winter; steppe covers the south; Central Siberian Plateau: a rugged region of thick forests; Central Siberian Plateau: a rugged region of thick forests; Eastern Highlands: a mountainous area. The Eastern Highlands region is least developed.
4. People are reluctant to move here because of the harsh climate and rugged terrain, which make building costs very high.
5. Rich coal deposits have made this region a center of industry and transportation. With further development of Siberian resources, the Kuznetsk Basin could become one of the Soviet Union's greatest industrial areas.

Critical Thinking

6. Soviet planners have not been able to decide what to do with the area because it can never be self-sufficient in food, it has only low-grade deposits of the iron ore and coal needed for industry, and it is far from all potential markets. The USSR's rivalry with China may lead it to develop this region.

Using Geography Skills

1. Fishing, farming, industry; Russian.
2. Kirghiz SSR, Kazakh SSR. Zero to 25 persons per square mile (0 to 10 per square kilometer). Because of the high elevation and rough terrain.

CHAPTER 20 Eastern Europe Pages 224–237

PLANNING GUIDE		
Lesson	**Textbook**	**Teacher's Resource Binder**
1. History and Physical Geography	pp. 224–227	Lecture Notes and Transparency Package Content/Vocabulary Workbook Reteaching Worksheet
2. Economic Activity and Culture	pp. 227–236	Challenge Worksheet Chapter Review/Test Reteaching Worksheet Critical Thinking Worksheet

TIME ALLOCATION: 2 days

CHAPTER TOPICS

- Historical Geography
- Physical Geography
- Economic Geography
- East Germany
- Poland
- Czechoslovakia
- Hungary
- Romania
- Bulgaria
- Yugoslavia
- Albania

PRETEACHING CHAPTER VOCABULARY

Write the following terms on the chalkboard:

COMECON antimony
martial law foodstuff

Ask students to review the definition of each term in the textbook Glossary. Invite the class to give examples of martial law, antimony, and foodstuff.

▶ LESSON 1
History and Geography

LESSON OBJECTIVES

The student should be able to

- define significant vocabulary
- locate the eight countries of Eastern Europe
- describe the economic interchange among countries in this region
- sequence historical data concerning Eastern Europe

PURPOSE

In this lesson, students will gain perspective on the role of communism in countries other than the USSR. This will enable students to recognize that communism exists in many forms throughout the world.

MOTIVATOR

Provide the students with the following scenario:

Imagine that your family has been transferred to a new country to live. In this new country, many languages are spoken. None of the languages is similar to your own. How does this make you feel? What will you do?

TEACHING DESIGN

1. Present the Chapter 20 Lecture Notes and Transparency Package. Ask students to take notes.
2. Write the following events on the chalkboard, and ask students to place the events in sequence in the history of Eastern Europe.
   a. World War II
   b. Unsuccessful Polish revolt for independence
   c. Occupation by Russian armies throughout Eastern Europe
   d. A period involving control by other nations, including German, Austrian, and Turkish empires
   e. Czechoslovakian rebellion for independence
   f. A period of independence for Eastern Europe
   g. Hungarian rebellion for independence
   h. World War I
3. Ask the class as a group to prepare a summary paragraph to describe the nature of Soviet control in Eastern Europe. Discuss possible reasons for the amount of conflict that has occurred in this region.
4. Lead a class discussion on COMECON. Discuss why it exists and how it compares to the European Community. Provide students with the diagram below, and ask them to contribute ideas. Write five of the generalizations on the chalkboard.

COMECON	
Similarities to EC	Differences from EC

5. Discuss the impact of language barriers in the Eastern European countries. Identify disadvantages and advantages to individual governments and to the USSR.

CLOSURE

1. Ask students to work with a classmate to review the names of the Eastern European countries and capitals.
2. Summarize the history of this region.

ASSESSMENT

1. Ask students to complete the Chapter 20 Content/Vocabulary Worksheet in their Workbook.
2. Have students complete the following questions from the Chapter 20 Check:
   Building a Vocabulary 1

Recalling and Reviewing 1
Using Geography Skills 1

RETEACHING PLAN
Ask students to complete the Chapter 20 Reteaching Worksheet, Lesson 1.

CHALLENGE/ENRICHMENT STRATEGY
1. Direct students to research the Marshall Plan and its impact on European development after World War II.
2. Ask students to use a map of the world to create a drawing that represents the trade that takes place between Eastern European countries and the rest of the world.

▶ LESSON 2
Economic Activity and Culture

LESSON OBJECTIVES
The student should be able to
- identify major resources available in Eastern Europe
- describe significant cultural, economic, and political characteristics of Eastern European countries
- locate the countries of Eastern Europe

PURPOSE
This lesson will acquaint students with the people and land of Eastern Europe. Students will increase their prediction skills and their knowledge of the world's people.

MOTIVATOR
Refer students to the climate map "The Soviet Union and Eastern Europe" on page 197 of their textbook. Ask them to use the map to predict answers to the following questions about Eastern Europe: Where might you expect to find farming? Where might you expect to find mining? Where might you expect to find forestry?

TEACHING DESIGN
1. Refer students to the map "Eastern Europe and the Balkan Peninsula" on page 226 of their textbook. Lead a class discussion to determine whether or not the predictions they developed in the Motivator were accurate.
2. Each of the eight countries in Eastern Europe has a distinct personality. Ask students to review the information about each country in the chapter, then choose a study partner. Provide students with large index cards. Have each pair write eight paragraphs, each of which describes an individual living in a different country of Eastern Europe. Students should look for information that makes each country unique.

   Read to the class the following example:

   My name is Karl, and I am 16 years old. Although it doesn't make the government happy, my family and I practice Roman Catholicism. We are proud that the Pope is from our country. My father used to belong to Solidarity; however, the government threatened to hold him in jail, so he resigned. Now he continues to work in an automobile plant. Where do you think I live? (Poland)

   Invite several students to read their paragraphs to the class. Build a chart on the chalkboard, listing the key facts that students used to distinguish the countries. Ask students to compare the class list to their individual lists.
3. Have students complete the Chapter 20 Challenge Worksheet.

CLOSURE
Ask remaining students to read their paragraphs, and use the class list of information to help students practice identifying each country.

ASSESSMENT
1. Have students complete the following questions from the Chapter 20 Check:
   Building a Vocabulary 2, 3, 4
   Recalling and Reviewing 2, 3, 4, 5
   Critical Thinking 6
   Using Geography Skills 2
2. Ask students to complete the Chapter 20 Review/Test.

RETEACHING PLAN
1. Ask Students to complete the Chapter 20 Reteaching Worksheet, Lesson 2.
2. Refer back to student questions you listed in the Introduction to Unit 3. Ask students to determine whether or not they have learned the answers to their questions.

CHALLENGE/ENRICHMENT STRATEGY
1. Ask students to place the Eastern European countries on a schematic as follows to represent each country's ties with the USSR.

   ←——————————————————→

Close Political Ties with USSR	Distant Political Ties with USSR
2. Have students complete the Chapter 20 Critical Thinking Worksheet.

Chapter 20 Check, page 237, Answers

Building a Vocabulary
1. COMECON, the Council for Mutual Economic Assistance, is an organization that promotes economic cooperation among the countries of Eastern Europe and the Soviet Union. After World War II, when it was founded, COMECON encouraged each country to specialize in those products that it could make best, but recently the organization has allowed the production of a wider assortment of goods. Goods produced in COMECON countries are made available with low import duties to other COMECON countries, thereby encouraging interregional trade.

2. Law administered by the military forces of a country when it is believed that the civilian government is unable to keep public order. It was imposed on Poland in 1980 because of Solidarity (trade union) strikes and anti-Soviet activities.
3. A silvery white element that is used in medicine and to make other metals. Yugoslavia.
4. A substance with food value, particularly the raw material of food. Corn is an example of a foodstuff.

Recalling and Reviewing

1. Eastern Europe came under Soviet domination during World War II when the area was caught between the Soviet Union and Germany; at the end of the war, Russian armies were found throughout the region, Soviet-style governments and cultural institutions were installed, and Communist parties were placed in power. Yugoslavia and Albania are most independent, partly because of their greater distance from the Soviet Union, and partly because of their political leaders.
2. Warm, moist climate and fertile soil provide good farming conditions; major deposits of coal, iron ore, zinc, and lead are important for industrial development.
3. Romania has insisted on the right of each Communist country to handle its own affairs without Soviet interference, and it has maintained an independent course in managing its relations with other countries. Bulgaria is a loyal supporter of the USSR's foreign policy; Czechoslovakia has a Western European outlook and a history of industrial prosperity, and therefore it has never been attracted to Soviet communism.
4. Yugoslavia consists of may different ethnic groups, languages, landform regions, climate zones, and levels of economic development; the religions and ways of life of the people vary also.
5. Albania has few natural resources. Little industry has been developed, and Albania is culturally isolated from the rest of Europe.

Critical Thinking

6. Answers will vary but should mention problems in communicating, lack of common goals, and identity with the ethnic group rather than with the nation. They should also mention the advantages provided by the ideas, values, and ways of life contributed by all the ethnic groups.

Using Geography Skills

1. Czechoslovakia and Hungary. Both have river outlets to the sea; Czechoslovakia has the Elbe River, which provides easy access to Western Europe and the North Sea; Hungary has the Danube River, which is connected to the Rhine River by canal and is an outlet to the Black Sea; therefore, being landlocked is not a problem.
2. Elbe, Danube and its tributaries. The rivers may encourage trade and exchange of ideas between the two areas.

UNIT 3 Review/Test

PLANNING GUIDE		
Lesson	Textbook	Teacher's Resource Binder
1. Unit 3 Review	pp. 238–239	
2. Unit 3 Test		Unit 3 Test

TIME ALLOCATION: 2 days

► LESSON 1
Unit 3 Review

Have students read the Unit 3 Summary. Use the Unit 3 Objectives as the basis for review, and discuss any questions that students have on the Unit 3 material.

Have students complete the questions in the textbook under the following headings:
Reading and Understanding
Mastering Geography Skills

Have students read the textbook material under the following headings and complete the appropriate or selected activities:
Applying and Extending
Linking Geography and Economics

► LESSON 2
Unit 3 Test

Have students complete the Unit 3 Test from the Teacher's Resource Binder.

Unit 3 Review, pages 238–239, Answers

Reading and Understanding

1. The collapse of Mongol rule allowed Russian fur trappers, hunters, and pioneers to expand the tsar's empire eastward into Siberia.
2. Government agencies act to establish production goals, identify new areas for development, and determine what goods will be available. The government controls nearly all aspects of the Soviet people's lives.
3. The governments of the republics are controlled by the central government in Moscow. The soviet of each republic is responsible for local matters in education, agriculture, and finance, but real governing power rests with the Communist party.
4. Oil, natural gas, coal, hydroelectric power, iron ore, gold, timber, copper, chromium, manganese, and ores of nearly all other industrial metals also are available. These resources have enabled the Soviet Union to become one of the world's most industrialized nations.

5. Autarky has enabled the Soviet Union to become almost self-sufficient by importing few goods from other countries. Advantages: the Soviet Union can provide almost all of its basic needs through this policy; disadvantages: the Soviet people have had to do without some goods rather than import them.
6. Because the roads connecting cities in the Soviet Union are poor, the goods produced in the Soviet industrial regions are carried primarily by rail and water. Railroads connect all parts of the western Soviet Union, and canals and barges link the great rivers of the region, thus providing a good transportation network for bringing raw materials to industrial centers.
7. Soviet agriculture is a major problem; it is still not providing enough food for the nation's people; therefore, the Soviet Union must import large amounts of food regularly. By allowing workers on state farms and collective farms to have small plots of land for growing crops; the workers can sell the crops grown on these plots for personal profit; consequently, these plots are highly productive.
8. Through the economic organization COMECON. Most Eastern European countries have tried to develop their own styles of government; however, the Soviet Union has been able to stop such developments with invasion or interference.
9. Crop production in some areas of Eastern Europe is low. Farmers have not always produced enough food on collective farms, and agriculture in some areas is still practiced with little modern technology. In Poland, crop production is improving because of modern farming methods.
10. Romania, Bulgaria, and Albania. Romania: most large mineral resources are undeveloped; poor transportation in rural areas. Bulgaria: most manufacturing is geared for the local market. Albania: few factories and few natural resources.
11. Ethnic group.

Mastering Geography Skills

1. Zero to 25 persons per square mile (0 to 10 persons per square kilometer).
2. More than 500 persons per square mile (195 persons per square kilometer); Leningrad.
3. Generally between 25 and 125 persons per square mile (10 and 49 persons per square kilometer). Answers to the second part of the question will vary.

Applying and Extending

Answers will vary for the first part. A warm-water port was acquired at Vladivostok in 1860 when Russia forced a weak China to give up its lands north of the Amur River. Virtually all of present-day Soviet Union was acquired by some form of conflict or by force with the exception of the area bounded on the north by the Gulf of Finland, on the west by the Baltic states, on the south by the Dniester River, and on the northeast by the Volga River.

Linking Geography and Economics

Answers will vary.

UNIT 4
The Middle East and North Africa
Pages 240–289

UNIT OVERVIEW

The region of the Middle East and North Africa is a crossroads where the continents of Asia, Europe, and Africa meet. Its strategic location has made it a focal point in world affairs. Three of the world's great religions developed in this region. Water and oil are the two key resources that transform deserts into rich farmland and poverty into wealth. Agriculture remains a chief occupation, even though the area contains the world's largest arid region. Clashes occur between traditional and modern ways of life, between religious groups, and between Arab nations and Israel. Modernization, economic development, and cultural change are challenging traditional values and ways of life. To a large extent, world peace and economic development depend upon political stability in this region.

CHAPTER TIME LINE

Chapter	Title	Time Allocation
21	The Desert Region	3 days
22	The Countries of the Eastern Mediterranean	3 days
23	Iraq, Iran, and the Arabian Peninsula	2 days
24	North Africa: Egypt and the Arab West	3 days
	Unit 4 Review	1 day
	Unit 4 Test	1 day
		13 days

MAJOR TOPICS

- Climate and Ways of Life
- Water and Population
- Economic, Social, and Cultural Interchange
- Key Resources
- Economic Development and Political Stability
- Modernization and Traditional Values
- Current Issues

INTRODUCING THE UNIT

Use a large wall map of the world to introduce this culture region to the class. Ask students to suggest reasons that this region can be called a crossroads. Explore land and water routes that connect this region to Europe, Asia, and Africa. Ask volunteers to share their knowledge about the early civilizations of this region and its most famous places. Challenge students to think of current events that illustrate the region's importance in world affairs.

CHAPTER 21 The Desert Region Pages 242–251

PLANNING GUIDE		
Lesson	**Textbook**	**Teacher's Resource Binder**
1. Landforms and Climate	pp. 246–248	Content/Vocabulary Workbook Reteaching Worksheet
2. Historical Geography	pp. 242–246	Lecture Notes and Transparency Package Skills Worksheet Reteaching Worksheet
3. Resources, Industry, and World Events	pp. 248–250	Critical Thinking Worksheet Chapter Review/Test Reteaching Worksheet Challenge Worksheet

TIME ALLOCATION: 3 days

CHAPTER TOPICS

- Historical Geography
- Physical Geography
- Economic Geography
- The Desert Region and World Events

PRETEACHING CHAPTER VOCABULARY

List the following terms on the chalkboard:

Judaism	monotheism	exotic river
Christianity	mosque	Bedouins
Islam	arid	commodity
Muslim	oasis	chromium

Pronounce each word for students, and ask them to define the terms that they know: place a check mark (√) beside these terms. Call upon volunteers to use the textbook Glossary to find the definitions of any remaining terms. Point out two of the terms that relate to religion: *monotheism* and *mosque*. Write the term *polytheism* above the term *monotheism*. Ask volunteers to analyze the parts of these words.

▶ LESSON 1

Landforms and Climate

LESSON OBJECTIVES

The student should be able to

- define the significant terms of the lesson
- locate the countries of the region
- locate and describe major landforms of the region
- explain the significance of water in this region
- explain the causes of population patterns
- explain the reasons for the sites of major cities

PURPOSE

The study of the physical geography of the desert region will help students understand the influence of environment on ways of life in the Middle East and North Africa.

MOTIVATOR

Ask students to turn to the map "The Middle East and North Africa" on page 244 in their textbook and work with a partner to locate as many of the following geographic features as possible in a five-minute period.

Feature	Name	Country or Countries
river	(Moulouya River)	(Morocco)
sea	(Black Sea)	(Turkey)
mountains		
desert		
gulf		
island		
lake		
oasis		
strait		

TEACHING DESIGN

1. In preparation for each class, ask a volunteer to create large outline maps of each country of the Middle East and North Africa by projecting a transparency of a political map onto butcher paper taped to the wall. One map of the entire region should be projected in order that all countries are drawn to the same scale. After each country's outline is drawn onto the paper, ask a student to cut out the countries.

   Display the country cutouts in the classroom, and assign two or three students to each country. Write the following assignment for each country on the chalkboard.

   a. Illustrate your country on the cutout model.
   b. Note important physical features.
   c. Note important cities.
   d. Note farming areas and products.
   e. Use symbols to note the location of important mineral resources and industries.
   f. Use colors or stripes to note climate regions.
   g. Use colors or dots to note populated areas.
   h. Present your country to the class in a three-minute time period, using the illustrated cutout map.
   i. Use the textbook, encyclopedias, and atlases as references.
   j. Watch for current information on your country in the media, and collect articles to bring to class. Find out the names of current political leaders of your country and try to find pictures of them.

   Provide each student with the following chart, and announce that as the report on each country is presented in class, all students should complete the information on their chart. Information and countries should be added throughout the study of the unit.

Country	Significance of Location	Important Physical Features	Climate Features	Major Cities/Capital	Areas of High Population Density	Agricultural Products	Resources/Industries
Turkey							

2. Have students read the chapter and begin their country assignment.

CLOSURE

Ask volunteers to name various physical features found in the Middle East and North Africa and to state generalizations regarding the geography of the region.

ASSESSMENT

1. Have students complete the following questions from the Chapter 21 Check:
   Building a Vocabulary 2, 4, 6
   Recalling and Reviewing 1
   Using Geography Skills 2
2. Ask students to complete the Chapter 21 Content/Vocabulary Worksheet in their Workbook.

RETEACHING PLAN

1. Ask students to complete the Chapter 21 Reteaching Worksheet, Lesson 1.
2. Direct students to use the map "The Middle East and North Africa" on page 244 of their textbook to review the countries of this region, their capitals, and major geographic features.

CHALLENGE/ENRICHMENT STRATEGY

1. Ask students to investigate the way of life of the Bedouins in past and present times.
2. Select students to investigate and report on the various expeditions in search of the source of the Nile River.
3. Choose students to research and report on the building and operating procedure of the Suez Canal.

▶ LESSON 2
Historical Geography

LESSON OBJECTIVES

The student should be able to

- identify key time periods, people, and events in the history of the Middle East and North Africa
- identify cultural interchange in the region

PURPOSE

The study of this region should help students understand and appreciate the cultural contributions of the desert people to world civilization.

MOTIVATOR

Using a large wall map of the world, direct students to point out the area of the Middle East and North Africa. Ask students to think about reasons that part of this area is called the Middle East. Ask the following questions to stimulate discussion: If this area is known as the Middle East, what area of the world would be considered the Far East? From which part of the world would this area be considered the Middle East?

TEACHING DESIGN

1. Introduce the Chapter 21 Lecture Notes and Transparency Package to trace the history of this region.
2. Create a culture/religion web that depicts the many cultures of this region. Students should label either a culture group or a religion in each circle, using their textbook as a resource.

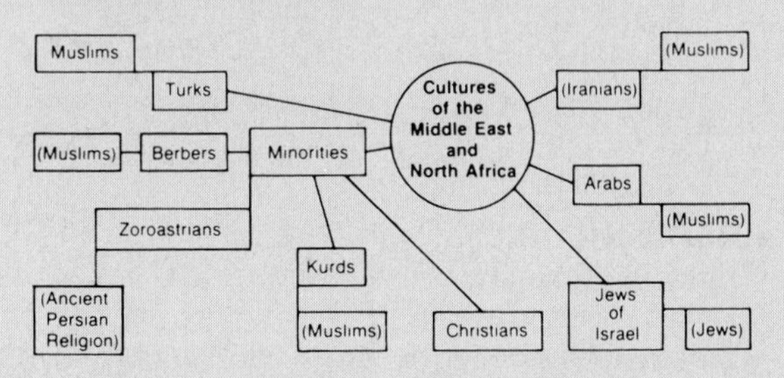

Note: the answers in parentheses should be left blank.

CLOSURE

Ask students to think about a question they might ask that would review important topics in the lesson. Allow time for students to think of a question. Call on a volunteer to pose a question and call on someone to answer, beginning a "chain reaction" of questions and answers in the class.

ASSESSMENT

1. Have students complete the following questions from the Chapter 21 Check:
   Building a Vocabulary 1, 3
   Recalling and Reviewing 2, 3
   Critical Thinking 7
2. Have students complete the Chapter 21 Skills Worksheet.

RETEACHING PLAN

Ask students to complete the Chapter 21 Reteaching Worksheet, Lesson 2.

CHALLENGE/ENRICHMENT STRATEGY

1. Choose students to write a report on the Crusades and results of these religious wars in the Middle East.
2. Ask students to create a comparative chart of the major religions that began in this region and to write a paragraph summarizing the similarities and differences among the religions.

▶ LESSON 3
Resources, Industry, and World Events

LESSON OBJECTIVES
The student should be able to
- describe the agricultural base of the region
- identify major natural resources and their uses
- identify the major economic activities in the region

PURPOSE
The study of resources, industry, and world events will enable students to understand the importance of the desert region to the rest of the world.

MOTIVATOR
Write the following headings on the chalkboard: Agriculture, Mineral Resources, and World Events. Ask students to brainstorm, without looking in their textbook, to determine what information they would expect to find under these headings in the chapter. Remind them that they already have a few clues from their study of the geography of the Middle East and North Africa and from their knowledge of world events.

TEACHING DESIGN
1. Using a wall map of the region, call on volunteers to locate the areas that would have suitable climate, landforms, and water for agriculture. Students should explain why their locations are suitable. They should then determine which areas would be more suitable for commercial rather than for subsistence agriculture.
2. Ask students to work with a partner to study the section "Agriculture" on pages 248–249 in their textbook, then list all of the examples of agricultural products on a sheet of paper. Each pair should review the products and suggest categories for grouping them.

CLOSURE
Assign each row of students a topic from the lesson, and give them three minutes to write a generalization related to the topic and the region.

ASSESSMENT
1. Have students complete the following questions from the Chapter 21 Check:
   Building a Vocabulary 5
   Recalling and Reviewing 4, 5
   Critical Thinking 6
   Using Geography Skills 1
2. Have students complete the Chapter 21 Critical Thinking Worksheet.
3. Ask students to complete the Chapter 21 Review/Test.

RETEACHING PLAN
1. Ask students to complete the Chapter 21 Reteaching Worksheet, Lesson 3.
2. Distribute four index cards to each student, then direct them to prepare review cards by writing a heading on each card: Agriculture, Mineral Resources, Regional Issues, and World Issues. Students should use information from their textbook to list data that will be important to remember from the lesson.

CHALLENGE/ENRICHMENT STRATEGY
1. Have students complete the Chapter 21 Challenge Worksheet.
2. Select students to investigate and report on the organization and decisions of OPEC.

Geography Skills, page 267, Answers

Test Your Skills
1. The oil reserves of the world (percent of total).
2. Middle East; USSR, China and Eastern Europe; Latin America; Africa; the United States and Canada; Western Europe; Asia, Australia, and the Pacific. The Middle East; Asia, Australia, and the Pacific.
3. 54.5 percent; 12.0 percent; 3.5 percent.
4. It shows that more than half of the world's oil reserves are located in the Middle East.

Chapter 21 Check, Page 251, Answers

Building a Vocabulary
1. The belief in one god. Islam, Judaism, and Christianity; students may choose two of these.
2. Because it begins in a humid region and crosses into a desert region.
3. A mosque.
4. A place in the desert where there are springs and wells. Oases are found beneath dry riverbeds in deserts; they are stopping places for traders and their goods.
5. A valuable economic good, such as a product of agriculture or mining. Oil.
6. Dry. Cities and farmlands are generally located along river valleys and wherever else water is available.

Recalling and Reviewing
1. Rugged low mountains, deserts, and coastal plains.
2. An arc of productive land that runs along the eastern shore of the Mediterranean Sea, through the plains along the Tigris and Euphrates rivers to the Persian Gulf. In this region, the world's first civilizations developed, early uses of large-scale irrigation were developed, many plants and animals were first domesticated, and several of the world's great religions were founded.
3. Judaism
4. A religion founded on the teachings of Jesus Christ; Eastern Orthodox, Roman Catholic, Protestant.
5. Muhammad. Muslims.

Critical Thinking
6. Around the Persian Gulf, in the Algerian and Libyan sections of the Sahara, and in northern Mesopotamia.

Oil is an important export commodity; the countries of the desert region have few resources other than oil that can be exported to the developed countries. Money from oil exports can be used to support economic development, but. Western ideas and influences brought into the region have caused conflict with traditional values. Many oil-producing countries have used oil as a political and economic weapon in international relations

7. Answers will vary but should include diet, politics, culture, and social relations. Answers to the second part of the question also will vary.

Using Geography Skills

1. From the Atlantic Ocean through the Strait of Gibraltar to the Mediterranean Sea, through the Suez Canal to the Red Sea, and from the Gulf of Aden to the Arabian Sea. Egypt, which controls the Sinai Peninsula, also controls the Suez Canal and its use.
2. Tropical and subtropical desert. Except for the Nile valley, most people live in the regions with Mediterranean, wet and dry tropical/subtropical, and continental steppe climates. They are most productive for agriculture.

CHAPTER 22 The Countries of the Eastern Mediterranean Pages 252–261

PLANNING GUIDE

Lesson	Textbook	Teacher's Resource Binder
1. Turkey and Cyprus	pp. 252–255	Lecture Notes and Transparency Package Content/Vocabulary Workbook Skills Worksheet Reteaching Worksheet
2. Israel	pp. 255–258	Reteaching Worksheet Challenge Worksheet
3. Syria, Lebanon, and Jordan	pp. 258–260	Critical Thinking Worksheet Chapter Review/Test Reteaching Worksheet

TIME ALLOCATION: 3 days

CHAPTER TOPICS

- Turkey
- Cyprus
- Israel
- Syria
- Lebanon
- Jordan

PRETEACHING CHAPTER VOCABULARY

Write the following terms on the chalkboard:

Zionism kibbutz minaret
Holocaust guerrilla

Provide the following explanations of specific terms. Ask students to identify the root (*Zion*) and suffix (*-ism*) in the term *Zionism*. Point out that *Zion* refers to the Jewish people, Israel, or the Jewish homeland and that the term comes from a Hebrew word. Students should recall that the suffix *-ism* means "a practice or process." See if students can then suggest a meaning for *Zionism*.

Point out the following clues to help students understand the term *Holocaust*: The prefix *holo-* comes from a Greek word meaning "whole or entire," and *caust* comes from a Greek word meaning "burned." Explain that when capitalized in a historical context, the term refers to the widespread destruction of Jews by Nazis during World War II.

A *kibbutz* is a collective farm in modern Israel. Point out that the term is derived from a Hebrew word meaning "gathering." Ask students to note other countries they have studied that have collective farms.

Ask students if they have heard of guerrilla warfare. Call on volunteers to describe that type of warfare. Point out that the term *guerrilla* comes from the Spanish word *guerra*, which means "war." Explain that guerrillas seek to isolate the forces of the opposition by organizing members of the local population and by using tactics such as sudden attack and harassment.

Have students look again at the photograph of mosques on page 254 of their textbook. Point out that the tall towers are called *minarets*. Ask students to suggest how these towers might be used. (to call Muslims to prayer)

▶ LESSON 1
Turkey and Cyprus

LESSON OBJECTIVES

The student should be able to

- define the significant terms of the chapter
- identify the climates of Turkey and Cyprus
- identify physical features of Turkey and Cyprus
- explain the significance of water to these nations
- identify major natural resources and their uses
- describe the agricultural base of these nations
- describe major economic activities in these nations
- explain the reasons for the location of cities

PURPOSE

The study of Turkey and Cyprus will enable students to understand the concept of strategic location and the importance of such a location to other nations in the world.

MOTIVATOR

Using a large wall map, point out the locations of Turkey and Cyprus. Ask students the following questions: Why do the United States and the Soviet Union both have an interest in Turkey? What waterways near Turkey should be important to the Soviet Union?

TEACHING DESIGN

1. Ask students to look at the maps "The Middle East and North Africa" on page 244 and "Eastern Mediterranean" on page 255 of their textbook. Ask the following questions:

a. What bodies of water border Turkey?
b. What continents is Turkey part of?
c. What body of water surrounds Cyprus?
d. What countries border Turkey?
e. What resources are found in Turkey and Cyprus?
f. What factors might influence the climates?

2. Discuss the land and climate of Turkey. Ask students to point out the areas that would be suitable for farming.
3. Present the Chapter 22 Lecture Notes and Transparency Package. Have students take notes.
4. Call on students who were assigned Turkey and Cyprus in the Teaching Design 1, Lesson 1, Chapter 21 to present their illustrated map cutouts and related information to the class. Remind all students to listen carefully in order to complete their charts. Have students post the illustrated cutouts in the classroom, making sure that they are placed in their proper locations. Post current news articles above their cutouts, and attach string from the country cutout to the news articles.

CLOSURE
Ask each pair of students to present its TV news status report on Turkey and Cyprus. Have students check for similarities and differences in reports.

ASSESSMENT
1. Have students complete the following questions from the Chapter 22 Check:
   Recalling and Reviewing 1, 2
2. Ask students to complete the Chapter 22 Content/Vocabulary Worksheet in their Workbook.
3. Have students complete the Chapter 22 Skills Worksheet.

RETEACHING PLAN
Ask students to complete the Chapter 22 Reteaching Worksheet, Lesson 1.

CHALLENGE/ENRICHMENT STRATEGY
1. Ask students to pretend they are applying for a job as a tour guide in the city of Istanbul. Have them prepare their tour of the city and present the information in class.
2. Have students research the special handcrafted goods produced in Turkey.

▶ LESSON 2
Israel

LESSON OBJECTIVES
The student should be able to
- identify significant physical features of Israel
- explain the economic development of Israel
- describe the agricultural base of the country
- arrange historical events in sequential order
- identify the major economic activities of Israel

PURPOSE
The study of Israel will enable students to realize the complexity of problems in the Middle East, where there appear to be no simple solutions that are acceptable to all parties.

MOTIVATOR
Have students read the primary source feature "For the Record" on page 257 of their textbook. Ask students what perceptions they draw from the article about kibbutz life.

TEACHING DESIGN
1. Call on students who were assigned Israel in the Teaching Design 1, Lesson 1, Chapter 21 to present their illustrated map and related information to the class. Other students should take notes and complete their charts. Post the map in the classroom, along with any news articles about Israel that students have collected.
2. Lead a discussion about the problems facing Israel today. As students identify problems, list them on the chalkboard. Have students work in groups to find solutions that would be acceptable to Israelis and to their Arab neighbors.

CLOSURE
Ask a representative from each group to present its solutions to the problems that were listed on the chalkboard. Discuss the recommendations.

ASSESSMENT
Have students complete the following questions from the Chapter 22 Check:
Building a Vocabulary 1, 2, 3, 4
Recalling and Reviewing 3

RETEACHING PLAN
Ask students to complete the Chapter 22 Reteaching Worksheet,

CHALLENGE/ENRICHMENT STRATEGY
1. Have students complete the Chapter 22 Challenge Worksheet.
2. Encourage students to investigate the United States' policies toward Israel since its creation in 1948.

▶ LESSON 3
Syria, Lebanon, and Jordan

LESSON OBJECTIVES
The student should be able to
- identify the climates of Syria, Lebanon, and Jordan
- locate important physical features of these nations
- explain the importance of water to these nations
- describe the agricultural base of these nations
- identify the major economic activities of these nations

PURPOSE
The study of Syria, Lebanon, and Jordan will assist students in developing an understanding of this part of the world.

MOTIVATOR
Invite volunteers to read aloud from the *World Book* (or other encyclopedia) paragraphs describing the governments of Jordan, Lebanon, and Syria. Have other students

take notes from the readings. Call on students to identify the similarities and differences between these countries' two types of government (constitutional monarchy and republic), and record the main ideas on the chalkboard.

TEACHING DESIGN

1. Ask students who were assigned Syria, Lebanon, and Jordan in Teaching Design 1, Lesson 1, Chapter 21 to present their illustrated maps and related information to the class. Other students should take notes and complete their charts. Post in the classroom the illustrated maps and any current news articles students have collected about Syria, Lebanon, or Jordan. Have students check the developing map of the region to note if any bodies of water that serve as boundaries should be added to make the wall map more accurate.
2. Have students work in groups of three to complete the Chapter 22 Critical Thinking Worksheet.

CLOSURE

Each group of students should report the steps that they developed in the Chapter 22 Critical Thinking Worksheet to solving problems in the countries in the Middle East.

ASSESSMENT

1. Have students complete the following questions from the Chapter 22 Check:
   Building a Vocabulary 5
   Recalling and Reviewing 4, 5
   Critical Thinking 6
   Using Geography Skills 1
2. Ask students to complete the Chapter 22 Review/Test.

RETEACHING PLAN

Ask students to complete the Chapter 22 Reteaching Worksheet, Lesson 3.

CHALLENGE/ENRICHMENT STRATEGY

1. Ask students to investigate both sides of the Israeli-Arab conflict and list the major arguments of each side on a chart. Have students post their charts after presenting the information aloud in turn to classmates. The class could suggest solutions to end the conflict.
2. Suggest that students investigate the backgrounds of the leaders of Syria, Lebanon, and Jordan, then present profiles of these leaders to the class.

Chapter 22 Check, page 261, Answers

Building a Vocabulary

1. The belief in a Jewish state in Palestine. In Europe.
2. The persecution of the Jews by the Nazis during World War II; the word *Holocaust* means "widespread destruction." Many nations, sympathetic to the Jews, supported the idea of creating the modern nation of Israel.
3. Kibbutz.
4. guerrillas
5. The tower of a mosque. Damascus.

Recalling and Reviewing

1. In the early 1920s, a far-reaching revolution modernized Turkish society, and Turkey was opened up to Western ideas. As Islam lost its status as state religion, the government took over many schools controlled by religious groups, substituted the Roman alphabet for the Arabic alphabet, modified its Islamic laws according to the European code, mandated the wearing of European-style clothing, and recognized women's rights. Turkey has built factories, developed mineral deposits, and built roads and railroads.
2. Three-quarters of the people in Cyprus consider themselves Greek, while most of the rest of the people consider themselves Turks, causing a bitter dispute that has resulted in riots and civil wars. There are few Cypriot nationalists to provide a sense of unity in the country. As a result of this conflict, the economy of Cyprus has not developed.
3. Jews have lived in Palestine since ancient times, but at the beginning of this century, the area was dominated by Islamic Arabs. The Zionist movement among Jews in Europe embraced the idea of Palestine as a Jewish state to which Jews from all over the world could move; after World War II, a Zionist state, Israel, was established in Palestine by the United Nations. The Arabs of Palestine and other Arab countries felt that Israel was founded on Arab land and that Israelis denied Arabs their essential rights. As a result, war erupted, and many Palestinian Arabs fled to nearby states to live in refugee camps; their displacement is a basic issue in the conflict.
4. Because the region's Mediterranean climate allows for productive agriculture.
5. Because Palestinian refugees far outnumber the original inhabitants and because the country lacks many natural resources that could support economic development. Jordan is struggling to irrigate more land and to develop a base for large-scale industry in the cities.

Critical Thinking

6. Answers will vary but should include the facts that the economies of the Arab nations that took Palestinian refugees cannot absorb the added population and that the presence of Palestinian refugees in these countries has increased tension in the region. Answers to the second question will vary, but the most popular solution is to create a Palestinian state, or states, in the occupied West Bank of Jordan or in the Gaza Strip.

Using Geography Skills

1. In 1967, Israel gained the Sinai Peninsula, the Gaza Strip, the West Bank of the Jordan River, East Jerusalem, and the Golan Heights. The Sinai has since been returned to Egypt. The West Bank originally belonged to Jordan and contained some of its best agricultural land as well as the site of its second most important city, East Jerusalem. Jordan wants the land back, and its relations with Israel are poor.

CHAPTER 23 Iraq, Iran, and the Arabian Peninsula Pages 262–275

PLANNING GUIDE		
Lesson	**Textbook**	**Teacher's Resource Binder**
1. Physical and Cultural Geography	pp. 262–274	Lecture Notes and Transparency Package Content/Vocabulary Workbook Reteaching Worksheet Challenge Worksheet
2. Current Events	pp. 262–274	Skills Worksheet Chapter Review/Test Reteaching Worksheet Critical Thinking Worksheet

TIME ALLOCATION: 2 days

CHAPTER TOPICS

- Iraq
- Iran
- Saudi Arabia
- Small Arabian States

PRETEACHING CHAPTER VOCABULARY

Write the following terms on the chalkboard:

fallow theocracy desalinization

Pronounce the words for the class. Inform students that the term *fallow* refers to the practice of leaving land idle for a growing season, rather than cultivating it. Ask students to suggest synonyms and antonyms for the term. Ask students why they think farmers would leave land fallow instead of planting a crop.

Ask students to analyze the terms *theocracy* and *desalinization,* looking for prefixes, root words, and suffixes. They should recall from their study of the term *monotheism* that *theo-* refers to a god. The suffix *-cracy* comes from a Greek word meaning "rule by."

The prefix *de-* means "to remove" something. Point out that *salin* in the term *desalinization* refers to salt. The suffix *-ation* refers to an action or process. Thus, desalinization is the process of removing salt from water. Ask students why this process would be important in the Middle East.

▶ LESSON 1
Physical and Cultural Geography

LESSON OBJECTIVES

The student should be able to

- define the significant terms of the chapter
- identify the major factors influencing the climates of Iraq, Iran, and the Arabian Peninsula
- locate important physical features of these areas
- explain the importance of water to these nations
- describe the agricultural base of these nations
- describe major economic activities in these areas

PURPOSE

The study of Iraq, Iran, and the Arabian Peninsula in this lesson will enable students to understand the impact that religion has on lives of people in this region.

MOTIVATOR

Have students take notes as you read aloud the description of the Five Pillars of Islam from the *Encyclopædia Britannica* (see "Islam," Volume 9). Show the transparency from the Chapter 21 Transparency Package charting the Five Pillars of Islam, and ask students to compare their notes to the chart. Ask students to suggest how being a Muslim might affect one's daily activities.

TEACHING DESIGN

1. Present the Chapter 23 Lecture Notes and Transparency Package. Direct students to take notes.
2. Call on students who were assigned Iraq, Iran, Saudi Arabia, and the small Arabian states in Teaching Design 1, Lesson 1, Chapter 21 to present their illustrated maps and related information to the class. Remind other students to take notes to complete their charts. Post in the classroom the students' illustrated maps of these countries and any current news articles they have collected.

CLOSURE

Ask students to think of similarities and differences among the nations studied in this lesson. Call on volunteers, and list their ideas on the chalkboard below the headings Similarities and Differences. Determine which similarities were affected by the common religion.

ASSESSMENT

1. Have students complete the following questions from the Chapter 23 Check:
   Building a Vocabulary 1, 2
   Recalling and Reviewing 1, 2
   Using Geography Skills 3, 4
2. Ask students to complete the Chapter 23 Content/ Vocabulary Worksheet in their Workbook.

RETEACHING PLAN

Ask students to complete the Chapter 23 Reteaching Worksheet, Lesson 1.

CHALLENGE/ENRICHMENT STRATEGY

1. Have students complete the Chapter 23 Challenge Worksheet.
2. Select students to compare the role of women in Muslim countries to that of women in the United States. (Source: *National Geographic,* October 1987)

▶ LESSON 2
Current Events

LESSON OBJECTIVES

The student should be able to

- describe major economic activities in the region

- identify the economic interchange among regions
- explain the economic importance of oil resources

PURPOSE

The study of this region will enable students to review people's critical role in using and conserving resources.

MOTIVATOR

Ask volunteers to go to the illustrated maps and news articles posted in the classroom and read the headlines of articles related to Iraq, Iran, Saudi Arabia, and the small Arabian states. Ask students if any events in this part of the world affect the United States, Europe, the USSR, or Japan.

TEACHING DESIGN

1. Have students read the feature "The Persian Gulf" on pages 270-271 of their textbook and locate on a wall map the countries surrounding the Persian Gulf.
2. Have students read the Geography Skills lesson "How to Use a Pie Graph" on page 267 of their textbook and to answer the questions aloud with a nearby classmate. Discuss how oil has changed the way of life in this region. Ask students what regional and world conflicts have been created by the oil industry. Discuss why this region does not need vast amounts of oil.
3. Ask students to refer to the pie graphs "Oil Reserves of the World" on page 267 and "Oil Reserves of the Middle East" on page 289 of their textbook. Have them complete a map locating the areas of oil reserves and the routes of pipelines throughout the region. Students also may refer to the map "Iraq, Iran, and the Arabian Peninsula" on page 265.

CLOSURE

Call on students to answer aloud the following questions from the Chapter 23 Check:

Building a Vocabulary 3
Recalling and Reviewing 3
Critical Thinking 4, 5
Using Geography Skills 1, 2

ASSESSMENT

1. Have students complete the Chapter 23 Skills Worksheet.
2. Ask students to complete the Chapter 23 Review/Test.

RETEACHING PLAN

1. Ask students to complete the Chapter 23 Reteaching Worksheet, Lesson 2.
2. Have students create a radio news broadcast describing a current situation in Iran, Iraq, or the Arabian Peninsula.

CHALLENGE/ENRICHMENT STRATEGY

1. Have students complete the Chapter 23 Critical Thinking Worksheet.
2. Suggest that students investigate who earns the wealth from oil produced in this region. Students may report how this wealth has changed ways of life in the area.
3. Select students to research and report on the architecture of the Middle East.

Chapter 23 Check, page 275, Answers

Building a Vocabulary

1. Land that is intentionally not planted, in order to conserve moisture and nutrients. To keep land productive.
2. A country governed by religious law. Iran. Theocracy affects educational opportunities, military structure, and economic growth.
3. A process that removes the salts from seawater and from salty water from wells. Desalinization has provided water for dairy farming, wheat farming, and industry.

Recalling and Reviewing

1. Where there is either water or oil; the most important area of settlement is in the western mountains, where the holy cities of Mecca and Medina are located.
2. The two Yemens do not have important oil deposits as do the other small states of the Arabian Peninsula.
3. Rapid economic development in these states has followed the discovery of oil.

Critical Thinking

4. Iraq has been supported by Egypt and other wealthy oil states in its war against Iran; its income has been diverted to military spending. Iran's oil reserves are not expected to last long, and its economy has been disrupted. Tremendous economic potential exists for Iraq because of its agricultural and oil resources.
5. Islam replaced the westernized Shah of Iran with Ayatollah Khomeini, who returned Iran to a conservative Muslim theocracy.

Using Geography Skills

1. By tanker from the Persian Gulf through the Strait of Hormuz to the Gulf of Oman into the Arabian Sea, through the Gulf of Aden and the Red Sea, then through the Suez Canal to the Mediterranean. If the tankers were not able to go through the Suez Canal, they would have to go around the southern tip of Africa, which would be very expensive.
2. At Jidda. It is a populated area with a work force and a local market, and it is on a major shipping route.
3. Iran. Iraq.
4. About 21° N, 39° E. North. About 225 miles (360 km).

CHAPTER 24 North Africa: Egypt and the Arab West Pages 276–287

PLANNING GUIDE		
Lesson	**Textbook**	**Teacher's Resource Binder**
1. The Sahara and the Nile	pp. 276–283	Content/Vocabulary Workbook Reteaching Worksheet
2. Egypt and Its Capital, Cairo	pp. 278–283	Skills Worksheet Reteaching Worksheet Challenge Worksheet
3. Libya, Tunisia, Algeria, and Morocco	pp. 283–286	Lecture Notes and Transparency Package Critical Thinking Worksheet Chapter Review/Test Reteaching Worksheet

TIME ALLOCATION: 3 days

CHAPTER TOPICS

- The Sahara
- Egypt
- Libya
- Tunisia
- Algeria
- Morocco

PRETEACHING CHAPTER VOCABULARY

Write the following terms on the chalkboard:

erg	silt	capital
reg	cash crop	
fossil groundwater	free port	

Pronounce the terms for students. Explain that two terms relate to formations in the Sahara. (*erg* and *reg*) Ask students to find the definitions in the reading and to use words or phrases that would describe the other terms.

▶ LESSON 1
The Sahara and the Nile

LESSON OBJECTIVES

The student should be able to

- define the significant terms of the lesson
- identify the major factors influencing the climate of the region and their effects
- identify significant physical features of North Africa
- explain the significance of water sources to the area

PURPOSE

This lesson will help students understand the relationship of environment to ways of life.

MOTIVATOR

Write the term *Sahara* on the chalkboard and pronounce it. Ask students to write on a sheet of paper the first three things that come to mind when they hear the term. Call on volunteers to share their ideas. Ask a student to record these ideas on a transparency. Follow the same procedure with the term *Nile River*. Explain that in this lesson, students will learn about these two features.

TEACHING DESIGN

1. After students have read the sections in their textbook on the Sahara and the Nile, ask them to create a features list, using the following sample.

Sahara
(desert)
(steep cliffs)
(sand dunes)

   Suggest that the features they list should be the most significant attributes of this vast desert. After students complete their list about the Sahara, have them make a similar list about the Nile.
2. Call on students to explain the information on their features list of the Sahara and to point out on a wall map the factors causing the desert. Ask volunteers to state how this desert environment influences ways of life.
3. Using a world wall map and a pointer, trace the course of the Nile River from its sources to the Mediterranean Sea. Ask a volunteer to use a large sheet of tracing paper to trace the length of the river on the wall map. Then, transpose the drawing onto a wall map of the United States, from east to west, to show the length of the river relative to the width of the United States. Ask volunteers to state how this river affects ways of life in its vicinity.

CLOSURE

Repeat the activity from the Motivator, then compare the Sahara and Nile transparencies made before the lesson with the transparencies made after the lesson.

ASSESSMENT

1. Have students complete the following questions from the Chapter 24 Check:
   Building a Vocabulary 1, 3, 4
2. Ask students to complete the Chapter 24 Content/Vocabulary Worksheet in their Workbook.

RETEACHING PLAN

Ask students to complete the Chapter 24 Reteaching Worksheet, Lesson 1.

CHALLENGE/ENRICHMENT STRATEGY

1. Select students to research life in the Sahara and report their findings to classmates. The report could include information about oases, caravan routes, temperatures, vegetation, and methods of survival.
2. Ask students to investigate ancient civilizations that developed along the Nile and their contributions to the history of the world.

▶ LESSON 2
Egypt and Its Capital, Cairo

LESSON OBJECTIVES
The student should be able to
- describe the agricultural base of regions
- identify economic activities in Egypt
- identify causes of population patterns
- identify patterns of urban growth in Cairo
- describe urban environmental issues

PURPOSE
The study of Egypt and Cairo will further students' understanding of people's adaptation to the environment.

MOTIVATOR
Refer students to pictures of Cairo in their textbook; call their attention to photographs depicting the old and new sections of the city. Ask students to state advantages and disadvantages of city life in Cairo. Ask how these compare to advantages and disadvantages of living in metropolitan areas studied previously. Explain that students will learn more about this ancient city in this lesson.

TEACHING DESIGN
1. Call on students who were assigned Egypt in Teaching Design 1, Lesson 1, Chapter 21 to present their illustrated map and related information to the class. Remind all students to complete the section on Egypt in their charts. Post in the classroom the students' illustrated map of Egypt and any current news articles about Egypt.
2. Divide the class in half; assign one half the topic Basin Irrigation and the other half the topic Perennial Irrigation. Students should be prepared to describe the assigned irrigation process and explain the advantages and disadvantages of each process. Allow time for study before calling on students.
3. Write the following chart on the chalkboard:

Egypt			
Resources		**Industries**	**Problems of Economic Development**
oil	⟶		1.
manganese	⟶		2.
iron ore	⟶		3.
cotton	⟶		4.
gypsum	⟶		5.
phosphate	⟶		
nitrate salts	⟶		

   Ask students to read the chalkboard carefully before reviewing the information in their textbook or their charts. Have students identify industries that have been developed in Egypt, based on resources that are available, and list problems of economic development.
4. Using the overhead transparency of the information map of Cairo, ask students to complete the map, using the special feature on Cairo. Call on volunteers to fill in the appropriate information on the transparency.

CLOSURE
Write the following topics on the chalkboard: Irrigation, Agricultural Production, Industrial Development, and Urban Problems (Cairo). Ask students to think of a statement that would summarize information from the lesson related to each topic. Allow students time to consider each topic before calling on individuals to share their statements.

ASSESSMENT
1. Have students complete the following questions from the Chapter 24 Check:
   Building a Vocabulary 2
   Recalling and Reviewing 1
   Critical Thinking 5
2. Have students complete the Chapter 24 Skills Worksheet.

RETEACHING PLAN
Ask students to complete the Chapter 24 Reteaching Worksheet, Lesson 2.

CHALLENGE/ENRICHMENT STRATEGY
1. Have students complete the Chapter 24 Challenge Worksheet.
2. Ask students to prepare a visitors' guide to Cairo. This brochure should explain Cairo's history as well as the characteristics of the city's old and new sections.

▶ LESSON 3
Libya, Tunisia, Algeria, and Morocco

LESSON OBJECTIVES
The student should be able to
- locate and describe major landforms
- locate these countries and their capitals
- identify natural resources and their uses
- identify the major economic activities
- describe areas suitable for agricultural development
- identify causes of population patterns
- explain the importance of water to the region

PURPOSE
The study of Libya, Tunisia, Algeria, and Morocco will further help students understand the relationship between the physical setting of a region and its ways of life.

MOTIVATOR
Provide groups of students with desk-size maps of North Africa, and ask them to label the following countries and their capitals: Libya, Tunisia, Algeria, and Morocco. Announce that each group is competing for prizes that will be won by the group members with the most accurate map. Students cannot use any reference source to complete their maps. Call time after a suitable interval, then ask a representative from each group to check its map while a volunteer points to the locations of the countries and their capitals on a wall map. Award appropriate prizes to the group with the most accurate map.

TEACHING DESIGN

1. Present the Chapter 24 Lecture Notes and Transparency Package to give an overview of the region.
2. Call on the students who were assigned these countries in Teaching Design 1, Lesson 1, Chapter 21 to present their illustrated maps and related information to the class. As students complete their charts of these nations, post in the classroom the illustrated maps and any current news articles that students have collected.

CLOSURE

Have students respond orally to the following questions from the Chapter 24 Check:
Building a Vocabulary 5, 6
Recalling and Reviewing 2, 3, 4
Using Geography Skills 1, 2

ASSESSMENT

1. Have students complete the Chapter 24 Critical Thinking Worksheet.
2. Ask students to complete the Chapter 24 Review/Test.

RETEACHING PLAN

1. Ask students to complete the Chapter 24 Reteaching Worksheet, Lesson 3.
2. Have students review information about the countries in this lesson by covering up various sections of their charts and trying to state major facts from each section.

CHALLENGE/ENRICHMENT STRATEGY

1. Ask students to investigate the French Foreign Legion (Légion Étrangère) and report on its past and present activities in North Africa.
2. Ask students to write an essay explaining the importance of water in North Africa.

Chapter 24 Check, page 287, Answers

Building a Vocabulary

1. An erg is a sea of high, shifting sand; a reg is a gravel-covered plain that has been swept free of dust and sand by the wind. In the Sahara.
2. A crop produced solely for direct sale in a market. Cotton.
3. A sediment that is coarser than clay, yet finer than sand. Silt is rich and fertile, deposited on the banks of a river by flooding; this has been important to Egyptian agriculture because the lands along the Nile have maintained their rich soils through the centuries.
4. Water that has been stored beneath the desert. The climate of the Sahara was almost humid years ago.
5. A port with almost no taxes on goods unloaded there. Tangier.
6. capital. Money invested in a savings account.

Recalling and Reviewing

1. The Nile River has deposited fertile soil for farming in the Nile valley and delta; it saved Egypt from being a desert wasteland. Benefits of Aswān High Dam: perennial irrigation of the Nile valley and the generation of electricity for industry. Problems: fertile soil is no longer deposited by annual flooding, so fertilizers must be used; nutrients once released into the Mediterranean for use by fish are no longer released, and there has been an increase in a disease produced by small organisms carried by snails.
2. Libya has a relatively small population and rich oil reserves. Political instability.
3. More than one-quarter of Tunisia's land is suitable for agriculture. Tunisia has a broadly based economy of small farms worked by their owners, and the ideals of European democracy are appreciated. The country has a good transportation system and several good ports. Women have voting rights, and a national school system educates nearly all children. Tunisia is a model of stability and social progress in the region.
4. Atlas Mountains, Sahara, and the coastal lowlands.

Critical Thinking

5. The amount of arable land is too small to support the existing population; hence, the country is said to be overpopulated. Answers will vary.

Using Geography Skills

1. Cairo, Egypt; Tripoli, Libya; Algiers, Algeria; Rabat, Morocco; Tunis, Tunisia. All are located on the coastal lowlands, all are port cities, and all are located in the Mediterranean climate region.
2. The Strait of Gibraltar lies between Spain and Morocco. No, the other route is the Suez Canal.

UNIT 4
Review/Test

PLANNING GUIDE		
Lesson	**Textbook**	**Teacher's Resource Binder**
1. Unit 4 Review	pp. 288–289	
2. Unit 4 Test		Unit 4 Test

TIME ALLOCATION: 2 days

▶ LESSON 1
Unit 4 Review

Have students read the Unit 4 Summary. Use the Unit 4 Objectives as the basis for review, and discuss any questions that students have on the Unit 4 material.

Have students complete the questions in the textbook under the following headings:
Reading and Understanding
Mastering Geography Skills

Have students read the textbook material under the following headings and complete the appropriate or selected activities:

Applying and Extending
Linking Geography and Religion

▶ LESSON 2

Unit 4 Test

Have students complete the Unit 4 Test from the Teacher's Resource Binder.

Unit 4 Review, pages 288–289, Answers

Reading and Understanding

1. The foundations of Western civilization were laid in the region, and three of the world's major religions—Judaism, Christianity, and Islam—were founded there. Today, oil wealth has given the region political and economic importance in the world. Forces of nationalism, religious conflict, and modernization make the region a focus of world conflict and policy decisions.
2. Because it allowed such productive agriculture, the Nile encouraged the development of civilization along its banks. Today, though basin irrigation has been replaced by perennial irrigation through the use of dams, the Nile is still important to Egypt; most of Egypt's people live near the Nile.
3. Perennial irrigation: a river is dammed, and water from the lake created is piped or taken by canals to fields for irrigation. Irrigation by tunnels: water is collected at the foot of mountains and taken by tunnels to the fields. Basin irrigation: basins were constructed to catch the water and silt from the annual flooding of the river. Benefits: perennial irrigation can irrigate large areas of farmland and provide electricity for industries and homes; tunnels provide water to irrigate the dry but good soil of Iran; basin irrigation enabled large areas of land far from the Nile to be irrigated; pipeline irrigation has enabled desert areas in Israel to be irrigated by the Jordan River, thus encouraging settlement and development of the Negev Desert. Problems: basin irrigation allowed the production of only one crop per year; the problems of the other methods are salinization , the need for constant maintenance and repair, and the increased incidence of a disease carried by snails.
4. Oil, tobacco, hazelnuts, cotton, citrus fruits, grapes, olives, copper, asbestos, coffee, hides, figs, dates, phospates, salt and chromium. They provide income.
5. Turkey, Israel, Saudi Arabia, Tunisia, Libya, and the oil-rich states of the Persian Gulf are the most economically developed; the least developed are Oman, the Yemen Arab Republic, the People's Democratic Republic of Yemen, Syria, Lebanon, and Jordan. The most economically developed countries are, with the exceptions of Israel, Turkey, and Tunisia, those with the income from oil reserves; some of the least developed countries have political conflict, which hinders economic growth.
6. Turkey, Iran, and Israel. Turkey and Iran have the Islamic religion in common with the other Arab countries. Israel is a nation established by Jewish immigrants from Europe, North Africa, and other areas of the world; it also has a large Arab population. Turkey has adopted aspects of European culture and modernized to a greater extent than the Arab countries. In Iran, the language is Persian rather than Arabic, and the dominant sect of Islam is different from that in the Arab countries. All have problems of economic development and limited arable land.
7. Much of Tunisia's architecture, its language, its form of goverment, its democratic traditions, and its infrastructure all have been influenced by Tunisia's relationship to France. Algeria, another former French colony, has been less fortunate economically since its independence; many Algerians work in France and send part of their earnings home.
8. Most people in this region settle near river valleys or oases. Some settlements have survived by storing rainwater. Future settlement is possible but not likely.
9. All Muslims are expected to accept the will of Allah and Muhammad, pray five times daily, give part of their wealth to the poor, fast during Ramadan, and make at least one pilgrimage to Mecca in a lifetime.
10. Nationalism has been expressed by various independence movements; conflict between Islamic and Western cultures; Greek and Turkish ethnic conflicts in Cyprus; the Arab, Palestinian, and Israeli conflict; the Kurdish independence movement in Turkey, Iraq, and Iran; the Iranian revolution against Western influences and culture and the expulsion of the Shah; the radical policies of Libya toward other nations; the Iraq-Iran conflict over territory and religion; and the conflict between the Yemen Arab Republic and the People's Democractic Republic of Yemen.

Mastering Geography Skills

1. This pie graph represents the oil reserves of the Middle East; the pie graph in Chapter 23 represents the oil reserves of the world.
2. Forty-five percent; Saudi Arabia can wield tremendous economic and political power.
3. Industry, agricultural land, livestock, minerals; answers to the second part of the question will vary.

Applying and Extending

Carafe: A glass bottle for serving water or wine at a table; originally, to dip; from Arabic, Spanish, Italian, French. *Bazaar:* A market usually consisting of an area lined with shops or stalls; originally, an Oriental market; from Persian, Turkish, Italian. *Muslin:* A sturdy, plain-weave cotton fabric; from Arabic, Italian, French. *Kosher:* Conforming to Jewish dietary laws, or proper, correct, permissible; from Hebrew, Yiddish. *Caravan:* A company of travelers journeying together; originally, a single file of pack animals; from Persian, Italian, French. *Algebra:* An area of mathematics in which symbols represent numbers; originally, reunification or bond-setting; from Arabic, Medieval Latin. *Salaam:* A respectful or ceremonial greeting; originally, peace; from Arabic. *Alcove:* A recess connected to or forming part of a room; originally, a vault; from Arabic, Spanish, French. *Jar:* A glass or earthenware container with a wide mouth and usually no handles; originally, a large earthenware vase; from Arabic, French.

These words illustrate cultural diffusion because they were picked up by contact with cultures of the Middle East, became part of the vocabularies of European and Asian countries, and have become part of our language. With some words, the meaning has changed through its use in a culture different from the original.

Linking Geography and Religion

Answers will vary.

UNIT 5
Sub-Saharan Africa
Pages 290–353

UNIT OVERVIEW

Sub-Saharan Africa is an area of rich tradition and potential. The history and economic development of the region are closely related to its unique environment. From the vast deserts, rain forests, and grasslands arose many great civilizations of the past. Many ethnic groups have their roots in these civilizations. Europeans brought changes to this area of the world. The nations of sub-Saharan Africa are still striving to overcome the problems of disease, illiteracy, and political uncertainties. Today, Africans seek to preserve the most valuable aspects of their traditional ways; they also seek to modernize and improve their ways of life.

CHAPTER TIME LINE

Chapter	Title	Time Allocation
25	Sub-Saharan Africa	4 days
26	West Africa	2 days
27	East Africa	2 days
28	Central Africa	2 days
29	Southern Africa	3 days
	Unit 5 Review	1 day
	Unit 5 Test	1 day
		15 days

MAJOR TOPICS

- Land, Climate, and Ways of Life
- Economic, Social, and Cultural Change
- Varying Goals and Growth Rates Among Nations
- Political Unrest and the Uncertain Future
- Challenges Facing the African Nations

INTRODUCING THE UNIT

Indicate to students that the sub-Saharan region of Africa today includes more than 40 countries. Show them the location on a world map. Ask what they believe they should learn about this part of the world. Provide students with several different editions of newspapers and magazines. Ask them to find articles about Africa. Discuss what students think are the key areas of interest as reported by the press. Ask students if this affects what they believe they should learn about this region.

CHAPTER 25 Sub-Saharan Africa Pages 290–303

PLANNING GUIDE		
Lesson	**Textbook**	**Teacher's Resource Binder**
1. Physical Geography	pp. 298–301	Content/Vocabulary Workbook Reteaching Worksheet
2. Landforms	pp. 300–301	Skills Worksheet Reteaching Worksheet
3. Historical Geography	pp. 292–298	Lecture Notes and Transparency Package Critical Thinking Worksheet Reteaching Worksheet
4. People and Ways of Life	pp. 301–302	Chapter Review/Test Reteaching Worksheet Challenge Worksheet

TIME ALLOCATION: 4 days

CHAPTER TOPICS

- Historical Geography
- Physical Geography
- Ways of Life
- Economic Geography
- Future Issues

PRETEACHING CHAPTER VOCABULARY

Write the following terms on the chalkboard:

plantation	Sahel
cash crop *	escarpment
subsistence agriculture *	rift
landlocked *	sorghum

* Review terms from previous chapters.

Locate photographs in the textbook that represent the following nouns.

plantation (p. 311)	Sahel (p. 308)
cash crop (p. 321)	escarpment (p. 340)
landlocked nation (p. 350)	rift (p. 317)

Refer students to each page, and ask them to guess, without referring to the caption, the noun associated with the picture. Then call on students to state a definition of each noun

▶ LESSON 1
Physical Geography

LESSON OBJECTIVES
The student should be able to
- define the significant terms of the chapter
- describe the physical setting of sub-Saharan Africa
- locate major nations and regions of sub-Saharan Africa
- analyze the impact of environment on population patterns in sub-Saharan Africa

PURPOSE
The study of sub-Saharan Africa is essential to students' understanding of the cultural heritage of many groups in the United States.

MOTIVATOR
Refer students to the statistical charts on sub-Saharan Africa in the back of their textbook. Instruct them to work in groups to discuss the following questions.
1. How many countries are in sub-Saharan Africa?
2. What is the largest country in terms of land area?
3. What is the smallest country in terms of land area?
4. What is the official language of this part of the world? other major languages?
5. What is the most common form of government?
6. Which country has the largest population?
7. Which country has the smallest population?

TEACHING DESIGN
1. Refer students to the climate and population-density maps "Sub-Saharan Africa" on page 299 of their textbook. Ask students to work with a partner and use the maps to complete the following chart.

Sub-Saharan Africa		
Climate Region	**Short Description**	**Population Density**

2. After students complete the chart, discuss the following questions with the class: What factors besides those in the chart contribute to population density? Which type of climate is most similar to our own? Where is it found?
3. Provide each student with an unlabeled map of sub-Saharan Africa, and have them label the following: Atlantic Ocean, Indian Ocean, Gulf of Guinea, Sahel, Namib Desert, Kalahari Desert, Cape of Good Hope. Ask students to keep the maps throughout the study of the unit and to label new places as they are studied.

CLOSURE
Show students a wall map of sub-Saharan Africa. Name the places that students located earlier on their maps, and invite them to locate each place for the class.

ASSESSMENT
1. Have students complete the following questions from the Chapter 25 Check:
   Building a Vocabulary 4
   Using Geography Skills 1
2. Ask students to complete the Chapter 25 Content/Vocabulary Worksheet in their Workbook.

RETEACHING PLAN
Ask students to complete the Chapter 25 Reteaching Worksheet, Lesson 1.

CHALLENGE/ENRICHMENT STRATEGY
1. Direct students to make a review chart with pictures to represent each climate region.
2. Choose several students to create a vegetation map for sub-Saharan Africa overlaid with the climate regions.
3. Ask students to begin to collect and read newspaper and newsmagazine articles about Africa.

▶ LESSON 2
Landforms

LESSON OBJECTIVES
The student should be able to
- locate and describe landforms in sub-Saharan Africa
- describe the physical forces that alter the landforms
- analyze the impact of environment on ways of life

PURPOSE
The study of landforms of sub-Saharan Africa will provide students with another review of the way landforms and the total environment affect the lives of people.

MOTIVATOR
Write on the chalkboard the following phrase that has been used to describe Africa: "The Sleeping Giant." Ask each student to write a paragraph to describe the possible reasons for the description. Invite several students to read their paragraphs to the class.

TEACHING DESIGN
1. Write the following sites on the chalkboard, and ask students to add them to their maps of sub-Saharan Africa: Atlas Mountains, Sudd, Sudan Basin, Nile River, Zaire Basin, Niger River, Zaire River, Kalahari Basin, Zambezi River, El Djouf basin, Red Sea, Chad Basin, and Lake Nyasa.
2. Ask students to use the information they have read in their textbooks to complete the following diagram.

Problems Faced Every Day	Unexpected Concerns

Brainstorm with students about possible solutions for everyday problems in sub-Saharan Africa.

CLOSURE
Ask students to review with a classmate the locations of important new sites that they have added to their maps of sub-Saharan Africa.

ASSESSMENT
1. Have students complete the following questions from the Chapter 25 Check:
   Building a Vocabulary 5
   Recalling and Reviewing 2
2. Ask students to complete the Chapter 25 Skills Worksheet.

RETEACHING PLAN
Ask students to complete the Chapter 25 Reteaching Worksheet, Lesson 2.

CHALLENGE/ENRICHMENT STRATEGY
1. Invite students to create a collage to represent the major landforms and bodies of water in sub-Saharan Africa.
2. Challenge students to locate vegetation or animals indigenous to sub-Saharan Africa.

▶ LESSON 3
Historical Geography

LESSON OBJECTIVES
The student should be able to
- interpret a historical map
- identify events in the history of sub-Saharan Africa
- analyze the impact of environment on development

PURPOSE
In this lesson, students will learn about the history and cultural heritage of Africa and about some of the contributions of the Africans to the American way of life.

MOTIVATOR
Discuss the difference between written history and oral history. Tell students that in this lesson they will review some of the significant time periods in the history of the African continent.

TEACHING DESIGN
1. Present the Chapter 25 Lecture Notes and Transparency Package. Have students take notes.
2. Ask students to work with a partner and use their textbook and encyclopedias to complete the following chart.

	Kingdom					
	Kush	Axum	Ghana	Mali	Songhai	Kanem and Bornu
Time Period						
Location						
Significance						

3. African history is often organized into three to five major time periods. Ask students to form groups and brainstorm about the possible reasons. Tell them they will use their textbook to focus on the three time periods in the following chart.

The Great Kingdoms About 1500	The Slave Trade About 1800	The Eve of Independence About 1956

   Ask students to use their textbook as a guide to list on the chart the most important information to remember about each time period.
4. Have students read the Geography Skills feature "How to Compare Historical Maps" on pages 296–297.

CLOSURE
Call on students to respond aloud to the "Test Your Skills" questions on page 296. Close with an oral review of the major events in the history of sub-Saharan Africa.

ASSESSMENT
Have students complete the following questions from the Chapter 25 Check:
   Building a Vocabulary 1, 2
   Recalling and Reviewing 1, 3
   Critical Thinking 6

RETEACHING PLAN
Ask students to complete the Chapter 25 Reteaching Worksheet, Lesson 3.

CHALLENGE/ENRICHMENT STRATEGY
1. Discuss the term *stereotype*. Ask a student to investigate and report on the old Hollywood image of Africa, comparing it to modern day Africa. Discuss how such portrayals affect people's ideas about places.
2. Ask students to complete the Chapter 25 Critical Thinking Worksheet.

▶ LESSON 4
People and Ways of Life

LESSON OBJECTIVES
The student should be able to
- analyze the impact of environment on ways of life
- explain the economic importance of various resources to the development of sub-Saharan Africa

PURPOSE
This lesson will provide students with an overview of the most serious issues facing this region of Africa. As citizens of a developed nation, students should be aware of issues.

MOTIVATOR
Read to the class a newspaper article describing any current situation in Africa that relates to the economy. Discuss the issues and possible solutions.

TEACHING DESIGN

1. Ask students to review pages 301–302 in their textbook, paying particular attention to the ways of life of Africans and the problems they often encounter.
2. Write the following diagram on the chalkboard, and ask students for ideas to complete the idea webs.

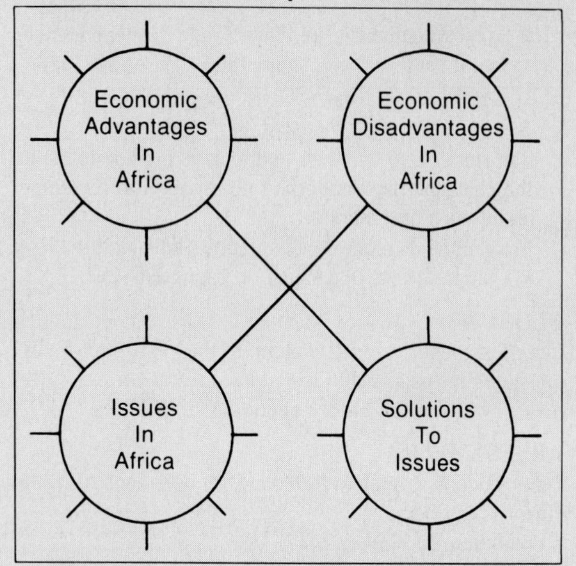

Lead a class discussion on the possible relationships among the four hubs of the diagram.

3. Have students work in groups, choosing to discuss either issues or solutions in sub-Saharan Africa. Have each group brainstorm about its topic and create a final list. Each group should take time to rank its list of issues or solutions in order of priority.

CLOSURE

Invite several group spokespersons to present their group rankings. Begin with issues, followed by solutions.

ASSESSMENT

1. Have students complete the following questions from the Chapter 25 Check:
   Building a Vocabulary 3
   Recalling and Reviewing 4, 5
   Using Geography Skills 2
2. Ask students to complete the Chapter 25 Review/Test.

RETEACHING PLAN

1. Ask students to complete the Chapter 25 Reteaching Worksheet, Lesson 4.
2. Ask students to review and outline the information under "Future Issues" on page 302.

CHALLENGE/ENRICHMENT STRATEGY

1. Ask students to complete the Chapter 25 Challenge Worksheet.
2. Select several students to create a map showing the region's predominant language groups.

Geography Skills, pages 296–297, Answers

1. Belgian Congo: Zaire; British East Africa: Kenya; Northern Rhodesia: Zambia; Southern Rhodesia: Zimbabwe; German Southwest Africa: Namibia; French Somaliland: Djibouti.
2. Ethiopia and Liberia. Two.
3. Cape Verde (1975) from Portugal; Guinea-Bissau (1974) from Portugal; São Tomé and Príncipe (1975) from Portugal; Cabinda (1975) from Portugal; Angola (1975) from Portugal; Mozambique (1975) from Portugal; Comoros (1975) from France; Seychelles (1976) from Great Britain; Djibouti (1977) from France.
4. Britain. Egypt, Anglo-Egyptian Sudan, British East Africa (Kenya), Uganda, British Somaliland, Nyasaland, Northern Rhodesia, Southern Rhodesia, Bechuanaland, Swaziland, Basutoland, Union of South Africa, Nigeria, Gold Coast, Sierra Leone, Gambia.

Chapter 25 Check, Page 303, Answers

Building a Vocabulary

1. A large farm that concentrates upon one cash crop. European colonists.
2. A steep slope capped by a plateau. Drakensberg.
3. The Sahel is a dry savanna region along the southern edge of the Sahara. It has been affected by a long period of drought since the late 1960s, and the desert is extending southward. Many people and farm animals have died from the droughts. Many more people have moved southward into the west savanna region, causing more crowded conditions there.
4. A grain that is one of the main crops grown in sub-Saharan Africa. Corn.
5. Eastern Rift Valley; Western Rift Valley.

Recalling and Reviewing

1. Most of the peoples of the region kept an oral history rather than a written record.
2. The El Djouf, Chad, Sudan, Zaire, and Kalahari basins. Three of the basins are along the southern edge of the Sahara—the El Djouf Basin to the west, the Chad Basin in the middle, and the Sudan Basin to the east. The Zaire Basin is in central Africa, and the Kalahari Basin is in southern Africa.
3. Most African nations are quite small and support large and growing populations. They have had trouble making economic progress, a problem that is complicated by difficulties in communication among ethnic groups.
4. Mining, farming, herding, and manufacturing of consumer goods and processed foods.
5. Geographers believe that Africa was once part of the supercontinent Gondwanaland.

Critical Thinking

6. Answers will vary. Students should consider the development of cities, agriculture, and political and economic

institutions. It is possible that the countries would cooperate better if they had not been politically and culturally fractured by the colonial powers.

Using Geography Skills

1. Least densely populated: several areas scattered along 20° N are not populated. The desert climate helps explain this. Most densely populated: the highlands of Ethiopia, around Lake Victoria, the Nile Valley in Sudan, along the western coast of Ghana, Benin, Togo, and Nigeria, eastern South Africa, and around the Niger River. Factors: fertile soil and wet and dry tropical/subtropical and maritime climates.
2. Chad, Niger, Mali, Burkina Faso, Central African Republic, Uganda, Rwanda, Burundi, Zaire, Zambia, Malawi, Zimbabwe, Botswana, Lesotho, Swaziland. Only Mali, Niger, Uganda, Zaire, and Zambia have large rivers running through them. In order to transport goods to the coast for export, these countries must trade overland.

CHAPTER 26 West Africa Pages 304–315

PLANNING GUIDE		
Lesson	**Textbook**	**Teacher's Resource Binder**
1. Physical Geography and History	pp. 304–307	Lecture Notes and Transparency Package Content/Vocabulary Workbook Reteaching Worksheet Critical Thinking Worksheet
2. Economic Geography; Countries of West Africa	pp. 307–314	Skills Worksheet Chapter Review/Test Reteaching Worksheet Challenge Worksheet

TIME ALLOCATION: 2 days

CHAPTER TOPICS

- Physical Geography
- Ways of Life
- Economic Development
- Sahel Countries
- Atlantic Coast Countries
- Guinea Coast Countries

PRETEACHING CHAPTER VOCABULARY

Write the following terms on the chalkboard:

desertification	rutile	columbite
cassava	coup	trust territory
animism		

Refer students to their textbook Glossary definition of *desertification*. How does it relate to the root word? Indicate that a starch taken from the root of the *cassava* helps make tapioca pudding. Ask students what they think *animism* is. Define it as a traditional belief in nature and spiritual beings. *Rutile* and *columbite* are minerals. Discuss the relationship that could exist between a *coup* and *trust territory*.

▶ LESSON 1
Physical Geography and History

LESSON OBJECTIVES

The student should be able to

- locate and identify key countries of West Africa
- identify events in the history of these countries
- describe ways of life in the region

PURPOSE

As world citizens, students should have some knowledge about all parts of the world. This lesson will introduce them to a part of the world that they may know little about.

MOTIVATOR

Refer students to the map "West Africa" on page 307 of their textbook. Ask students to work with a partner to locate answers to the following questions: How many countries are included in this region? What are the three largest countries? Which is the smallest?

TEACHING DESIGN

1. Ask students to complete the following diagram showing which region each country fits.

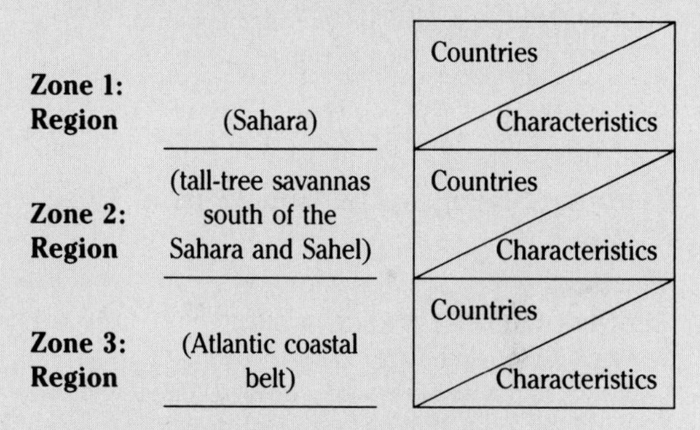

2. Refer students to the following features checklist, and ask them to complete it with a classmate.

	Features												
Countries	Farming	Cattle	Fishing	Forests	Industry	Iron ore	Minerals	Oil	Plantation crops	Sahel	Atlantic coast	Landlocked	Guinea coast
1.													
15.													

Ask students to write three generalizations to describe each region of West Africa.

3. Present the Chapter 26 Lecture Notes and Transparency Package. Have students take notes.

CLOSURE

Invite several students to read the generalizations that they wrote to describe the regions of this area. Use each generalization to initiate a summary of an important characteristic of West Africa.

ASSESSMENT

1. Have students complete the following questions from the Chapter 26 Check:
   Building a Vocabulary 1, 2, 3
   Recalling and Reviewing 1
   Critical Thinking 4
   Using Geography Skills 1, 2, 3
2. Ask students to complete the Chapter 26 Content/ Vocabulary Worksheet in their Workbook.

RETEACHING PLAN

Ask students to complete the Chapter 26 Reteaching Worksheet, Lesson 1.

CHALLENGE/ENRICHMENT STRATEGY

1. Have students complete the Chapter 26 Critical Thinking Worksheet.
2. Ask students to research the flags of West Africa and answer the following questions about each of them: What are its meanings? Is there any comparison with the flag of the former colonial empire?

▶ LESSON 2

Economic Geography; Countries of West Africa

LESSON OBJECTIVES

The student should be able to

- locate major natural resources of West Africa
- review the location of major nations of West Africa
- analyze the impact of the environment on ways of life
- describe major economic activities of West Africa

PURPOSE

This lesson will help students understand the challenges of the future development of West Africa and its relationship to the economies of these nations.

MOTIVATOR

As students enter the classroom, play recorded music by a group from a country in West Africa. (Example: King Sunny Adé of Nigeria.) Have students listen to one song from the record or tape. Ask students to list three or four adjectives describing the images that the music brings to mind. Invite volunteers to share their images, and list them on the chalkboard. Indicate to students that as they study this lesson, they should compare their adjectives with adjectives in their textbook that apply to the country.

TEACHING DESIGN

1. Divide students into approximately 10 groups. Assign one of the following countries to each group: Mauritania, Chad, Ivory Coast, Mali, Senegal, Ghana, Niger, Liberia, Nigeria, and Cameroon. Explain to students that they have two assignments:
   a. They should review the information about their countries in their textbook and in any of the classroom resources. Provide each group with an envelope containing five index cards. The students should select five important facts to review about their country and write the facts on their cards.
   b. Each group also should write a one-paragraph plan for the future development of their assigned country. Each group should consider the economic resources of the country and the past colonial ruler.
2. Have students play the following game as they circulate their envelopes: Each group should remove only one card at a time and use only as many clues as necessary to identify the country. Act as scorekeeper, adding points for every clue used. The team that correctly identifies the most countries using the fewest number of clues (having the fewest number of points) is the winner.
3. Have students complete the Chapter 26 Skills Worksheet.

CLOSURE

Invite several groups to share their plans for the future development of their country. Write similarities among plans on the chalkboard. Prepare a closing generalization.

ASSESSMENT

1. Have students complete the following questions from the Chapter 26 Check:
   Building a Vocabulary 4, 5
   Recalling and Reviewing 2, 3
   Critical Thinking 5
2. Ask students to complete the Chapter 26 Review/Test.

RETEACHING PLAN

1. Use the envelopes developed in several classes as a source of information for students to review. They can challenge each other with cards from different classes.
2. Ask students to complete the Chapter 26 Reteaching Worksheet, Lesson 2.

CHALLENGE/ENRICHMENT STRATEGY

1. Have students complete the Chapter 26 Challenge Worksheet.
2. Ask students to create an illustrated map representing two key facts about each country in West Africa.

Chapter 26 Check, page 315, Answers

Building a Vocabulary

1. The expansion of the desert caused by overgrazing and people cutting the few trees for wood. Sahel region.

2. Animism.
3. Rutile is a source of titanium and is found in Liberia and Sierra Leone; columbite is a metal used to harden stainless steel and is found in Nigeria.
4. A trust territory is placed under the control of another nation until it can govern itself. Cameroon.
5. The overthrow of an existing government by a small group. Benin.

Recalling and Reviewing

1. The arid Sahara in the north, where rainfall varies from year to year and where in some areas there is no rain; tall-tree savannas and forests, where the rainfall is more reliable but where people are subject to disease carried by tsetse flies; coastal belt, which contains rain forests and a humid tropical climate and where rain occurs almost every day. Most people live near the coastal region, where food is more plentiful, agriculture is more productive, and trading is more efficient.
2. Crops failed, and many animals and people have died.
3. Ivory Coast, because of its political stability and good relations with France; Ghana, because of its income from tropical products such as cacao, mineral resources, and some mineral-related industries; Nigeria, because of its large population and broad resource base: good farmland, minerals, oil deposits, and well-educated population; Gabon, because of its tropical rain forests, minerals, small population, good educational opportunities and health facilities, and developed roads and railways.

Critical Thinking

4. The British ruled through local leaders, left decisions to the Africans, and built an extensive infrastructure. The French invested heavily in one or two cities in each colony. The Germans invested in cities, roads, and railways. The Portuguese never fully developed their colonies. Positive effects: improved health care, new farming methods, resource development, and improved communications, transportation, and education. Negative effects: disturbance of African life, loss of profits from resources, lack of trained leaders, and lack of foreign investment because of social unrest and warfare.
5. Answers will vary but should include initiating economic associations.

Using Geography Skills

1. Chad, Nigeria, Benin, Ivory Coast, Sierra Leone, Mali, and Senegal. Humid tropical and wet and dry tropical/subtropical.
2. Along the coast in Senegal, Guinea, Ghana, and Cameroon, and in Nigeria on the Niger River and on the border with Niger. Answers will vary, but students should note that there is very little industry in the region.
3. Senegal River: into the Atlantic; Gambia River: into the Atlantic; Benue River: into the Niger River; Niger River: into the Gulf of Guinea.

CHAPTER 27 East Africa
Pages 316–325

PLANNING GUIDE		
Lesson	**Textbook**	**Teacher's Resource Binder**
1. Physical Geography and History	pp. 316–318	Lecture Notes and Transparency Package Content/Vocabulary Workbook Reteaching Worksheet Critical Thinking Worksheet
2. Economic Geography; Countries	pp. 318–324	Chapter Review/Test Reteaching Worksheet Challenge Worksheet

TIME ALLOCATION: 2 days

CHAPTER TOPICS

- Environments and Cultures
- Kenya
- Tanzania
- Uganda
- Sudan
- Ethiopia
- Somalia
- Djibouti

PRETEACHING CHAPTER VOCABULARY

Write the following terms on the chalkboard:
Swahili gum arabic sisal millet

Ask students how many of them are familiar with the term *Swahili.* Have students read aloud the definition in their textbook Glossary as you write it on the chalkboard.

Ask students to hypothesize about what might be a good category title for the three remaining terms. Have them look up the three words in their textbook Glossary.

▶ LESSON 1
Physical Geography and History

LESSON OBJECTIVES

The student should be able to
- define significant vocabulary terms
- locate and identify key countries of East Africa
- identify important historical events in these countries
- describe resources located in this region

PURPOSE

In this lesson, students will learn why East Africa is considered to have been the central point for the beginning of world civilization.

MOTIVATOR

In Chapter 25, students read about the idea that Africa was once the center of a supercontinent called Gondwanaland.

Show students the following diagram.

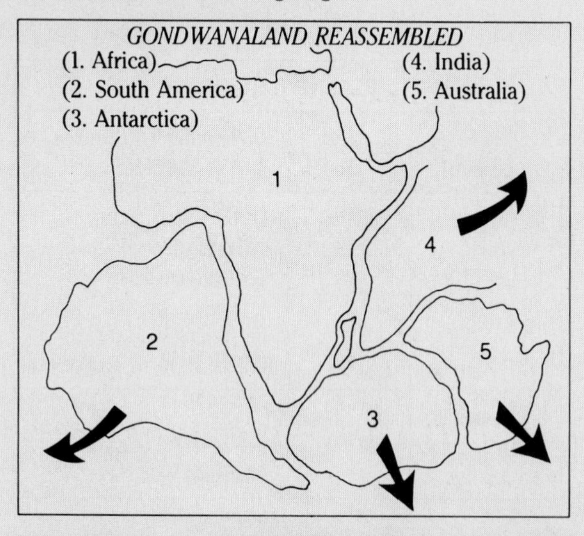

Ask students to identify the continents represented by the numbers. Tell students that according to the theory of continental drift, there once was a great landmass on earth that broke apart to form continents. Gondwanaland once may have been a part of a single huge landmass called Pangaea. Pangaea first divided into Gondwanaland and Laurasia. According to this theory, a clue to the original joining of these two continents is that the Atlas Mountains of Africa may have been part of the Alps of Europe. Another clue appears to be the similarity between the outline of the east coast of South America and the west coast of Africa. Challenge students to investigate the theory of continental drift in a reference source to identify additional clues.

TEACHING DESIGN

1. Ask students to refer to their maps of Africa and add the following countries and sites: Mount Kenya, Mount Elgon, Mount Kilimanjaro, Lake Victoria, Djibouti, Kenya, Uganda, Tanzania, Red Sea, Horn of Africa, Sudan, Ethiopia, and Somalia.
2. Present the Chapter 27 Lecture Notes and Transparency Package as an overview of the region. Have students take notes. Ask students to add information from their notes for each country they named on their map.

CLOSURE

Invite several students to share information that they have included in their assignment. Indicate to students that additional time will be given to finish in the next lesson.

ASSESSMENT

1. Have students complete the following questions from the Chapter 27 Check:
   Building a Vocabulary 2, 3, 4
   Using Geography Skills 1, 2, 3
2. Ask students to complete the Chapter 27 Content/Vocabulary Worksheet in their Workbook.

RETEACHING PLAN

1. Ask students to complete the Chapter 27 Reteaching Worksheet, Lesson 1.
2. Refer students to the *World Book* or another encyclopedia, and ask them to find information to complete the following chart on each East African country.

Country	Land	Resources	Government	People	Future
Kenya					

CHALLENGE/ENRICHMENT STRATEGY

1. Invite interested students to plan a photographic safari to Kenya.
2. Visit a local zoo. Have students make a list of the animals there that are native to East Africa.
3. Have students complete the Chapter 27 Critical Thinking Worksheet.

▶ LESSON 2

Economic Geography; Countries

LESSON OBJECTIVES

The student should be able to

- locate and identify key countries of East Africa
- describe resources located in this region
- describe major economic activities

PURPOSE

This lesson will increase students' knowledge of the East African countries, which are frequently in the news.

MOTIVATOR

Divide students into work groups. On the chalkboard, write Critical Thinking questions 5 and 6 from page 325 in their textbook. Encourage each group to review the chapter in order to answer one of the questions. After several minutes, call upon several groups to report their answers.

TEACHING DESIGN

1. Allow students to complete the information maps begun in the preceding lesson. Invite several students to share their information on various countries. Ask all students to review their maps for any information to add.
2. Ask students to continue to work in groups. Tell them that you will shortly begin making important announcements regarding the United States' involvement with East Africa. Explain that each group must prepare a response to each announcement. This response should include reasons for the group decision. Read the following announcements aloud to students, and allow them time to prepare their responses and reasons.
   a. Your group has been awarded a free one-week trip to East Africa. Where will you plan to visit? Why?
   b. The government of the United States wants your recommendation regarding a location for a new food-processing plant in East Africa. Where should it be built? Why?

c. Your group has decided to spend one month in the Peace Corps in East Africa. In which country will you work? Why?

CLOSURE

Invite several groups to report their decisions to the class. Review with students the map of East Africa and the unique characteristics of this region.

ASSESSMENT

1. Choose students to answer aloud the following questions from the Chapter 27 Check:
   Building a Vocabulary 1
   Recalling and Reviewing 1, 2, 3, 4
3. Ask students to complete the Chapter 27 Review/Test.

RETEACHING PLAN

Ask students to complete the Chapter 27 Reteaching Worksheet, Lesson 2.

CHALLENGE/ENRICHMENT STRATEGY

1. Encourage interested students to read about Olduvai Gorge and the evidence of early settlement that Louis and Mary Leakey have discovered there.
2. Have students complete the Chapter 27 Challenge Worksheet.

Chapter 27 Check, page 325, Answers

Building a Vocabulary

1. Swahili. By Arabic.
2. No, sisal is a cash crop. Kenya, Tanzania.
3. A product of the acacia tree. The production of medicine and candies.
4. A grass grown for its small seeds, which are edible.

Recalling and Reviewing

1. The Eastern Rift is drier and has plains areas; the Western Rift is more humid and is filled with lakes.
2. It has helped the economy of Kenya and has encouraged the government of Kenya to provide a stable government. Problems generally concern conservation of natural environments and wildlife when they come into conflict with economic development and human needs.
3. Nilotic peoples live near the Nile River on the plains of Sudan and the highlands to the south; Kushites live from the Red Sea coast through the Horn of Africa; the Bantu peoples live to the south in Kenya and Tanzania.
4. It contains one major ethnic group.

Critical Thinking

5. Swahili illustrates culture diffusion. Answers to the second part of the question will vary but might involve a discussion of colonial influence in the region.
6. Answers will vary but may include some of the following. Kenya: may soon be overpopulated; limited farmland; unemployment; unskilled laborers; few exportable mineral resources. Tanzania: few fertile soils; limited technical support to increase productivity; tsetse fly infestation; small deposits of minerals; cultural conflict between Arabs and Bantus. Uganda: rebuilding the country after Amin's rule; political instability; lack of foreign investment; and ethnic conflict. Kenya must use its human and agricultural resources more fully to provide better opportunities for all. Uganda must try to build a more stable government so that its resources can be fully developed. Tanzania might seek technical support to increase agricultural productivity.

Using Geography Skills

1. Between 10° and 15° N, 33° E. Because it lies between the Blue Nile and the White Nile, which provide water.
2. In eastern Kenya and eastern Ethiopia. Highlands and a wet and dry tropical/subtropical climate are good for growing coffee.
3. Lakes: Victoria, Albert, Kioga, Kivu, Tanganyika, Nyasa, Turkana; rivers: Blue Nile, White Nile, Atbara, Pangani. Burundi, Tanzania, Uganda, Kenya, Sudan, Ethiopia. Lake Victoria. On a plateau between the Eastern and Western rifts.

CHAPTER 28 Central Africa Pages 326–339

PLANNING GUIDE		
Lesson	Textbook	Teacher's Resource Binder
1. Physical Geography and History	pp. 326–328	Lecture Notes and Transparency Package Content/Vocabulary Workbook Reteaching Worksheet Skills Worksheet
2. Countries and Resources	pp. 328–338	Critical Thinking Worksheet Chapter Review/Test Reteaching Worksheet Challenge Worksheet

TIME ALLOCATION: 2 days

CHAPTER TOPICS

- Physical Geography
- Zaire
- The People's Republic of the Congo
- Central African Republic
- Rwanda and Burundi
- Zimbabwe
- Zambia
- Malawi
- Mozambique
- Angola
- Indian Ocean Islands

PRETEACHING CHAPTER VOCABULARY

Write the following terms on the chalkboard:

multinational companies*	periodic market	copra
subsistence farmers*	embargo	
escarpment*	vanadium	*Review terms

Ask students to use the context clues in the sentences below to help define the terms and complete the sentences.

______ is a *multinational company* because ______.
A *subsistence farmer* usually eats whatever ______.
An *escarpment* is a unique landform because ______.
A nation may suffer from an *embargo* of its products because ______.
You might expect to be able to buy ______ at a *periodic market*.
Vanadium added to iron produces ______.
______ is obtained from *copra*.

▶ LESSON 1
Physical Geography and History

LESSON OBJECTIVES
The student should be able to
- identify significant vocabulary terms
- locate major nations of the world
- describe the physical setting of Central Africa
- locate major landforms of Central Africa

PURPOSE
In this lesson, students will continue their journey through Africa, learning about the Central African nations and practicing their map skills.

MOTIVATOR
Bring recent editions of various newspapers to class. Distribute them, and ask students to look for articles related to Africa. Invite several students to summarize the articles they found. Create a word web to demonstrate the issues focused on in the newspapers.

TEACHING DESIGN
1. Present the Chapter 28 Lecture Notes and Transparency Package. Have students take notes.
2. Ask students to take out their maps of Africa and label the following sites: Zaire River, Namib Desert, Mitumba Mountains, and Zambezi River. Have any of these sites been labeled previously?
3. Have students add the following countries to their maps: Zaire, People's Republic of the Congo, Zambia, Central African Republic, Malawi, Rwanda, Burundi, Zimbabwe, Mozambique, Angola, and the Indian Ocean Islands.
4. Announce the following assignment to the class: Each student will work with a partner. Each pair of students will be assigned a country or countries from the list in Teaching Design 3. The students should spend time today researching answers to the following seven questions regarding their assigned country: Where is the country located? What are the land and climate like? What are the country's greatest resources? What are the country's greatest concerns? What might be the country's plan for future development? In what physical region is the country located? How do most people of the country live? Students should plan an interview format for sharing their information with the class.

CLOSURE
Summarize the purpose of the lesson and the need to expand our awareness of the world. Select several students to share one important new idea they have learned.

ASSESSMENT
1. Have students answer the following questions aloud from the Chapter 28 Check:
   Building a Vocabulary 1, 2, 3, 4
   Using Geography Skills 1, 2, 3
2. Ask students to complete the Chapter 28 Content/Vocabulary Worksheet in their Workbook.

RETEACHING PLAN
Ask students to complete the Chapter 28 Reteaching Worksheet, Lesson 1.

CHALLENGE/ENRICHMENT STRATEGY
1. Have students complete the Chapter 28 Skills Worksheet.
2. Invite students to investigate multinational companies with holdings in Central Africa. Who are they? What are the business opportunities there?

▶ LESSON 2
Countries and Resources

LESSON OBJECTIVES
The student should be able to
- identify resources of Central African nations
- describe major economic activities of Central Africa
- analyze the relationship between Central African environment and ways of life
- locate major nations of Central Africa

PURPOSE
This lesson will help students see potential methods by which people can use their environment to improve their future ways of life. Students will also gain practice in the important skills of taking notes and classifying.

MOTIVATOR
Acknowledge that students have been studying Africa for a long time. Tell them that you are interested in knowing if any of the students' perceptions about Africa have changed as a result of their study. Invite students to share old and new perceptions.

TEACHING DESIGN
1. Prior to calling upon the first pair of students to make their presentation prepared in Lesson 1, Teaching De-

sign 4, ask all students to create the following chart, adding boxes as necessary.

Country	Land	Climate	Resources	Concerns	Future
Zaire					

Have students use the chart to take notes as students present their interviews.

2. Invite students to present their research begun in Lesson 1, Teaching Design 4. Keep to your maximum time limit to ensure the completion of the project.
3. Have students complete the Chapter 28 Critical Thinking Worksheet.

CLOSURE

Refer students to the charts they created in Teaching Design 1. Use the following questions to close this lesson and chapter.

1. How would you characterize the overall physical geography of Central Africa?
2. What problems and concerns are found throughout the countries of Central Africa?

ASSESSMENT

1. Have students complete the following questions from the Chapter 28 Check:
   Recalling and Reviewing 1, 2, 3, 4
   Critical Thinking 5
   Using Geography Skills 4, 5
2. Ask students to complete the Chapter 28 Review/Test.

RETEACHING PLAN

Ask students to complete the Chapter 28 Reteaching Worksheet, Lesson 2.

CHALLENGE/ENRICHMENT STRATEGY

1. Choose volunteers to investigate art as an important feature of African culture.
2. Allow students to stage a pretend periodic market in the classroom, bringing objects to sell and trade. Students should devise methods for displaying their wares.
3. Have students complete the Chapter 28 Challenge Worksheet.

Chapter 28 Check, page 339, Answers

Building a Vocabulary

1. An open-air trading market held regularly at a crossroads or town. It resembles commercial agriculture on a small scale; many farmers have fields for cash and food crops, which they trade at the periodic market.
2. Embargo. Yes, restrictions on economic trade can seriously damage a country's economy. An embargo was ordered against Rhodesia in 1965 by the United Nations; this pressure, combined with successes of guerrilla fighters, brought black majority rule to Rhodesia.
3. A gray mineral added with iron to make strong steel.
4. The dried meat of the coconut from which coconut oil is obtained. In Mozambique.

Recalling and Reviewing

1. The river system, humid climate, and deposits of copper, iron ore, oil, manganese, gold, cobalt, tin, and industrial diamonds make Zaire a rich country. There are not enough people to fill technical and professional jobs; Kinshasa has grown so rapidly that there are not enough houses or jobs for people moving there; the country's various ethnic groups view each other with suspicion.
2. Zambia, Zimbabwe, and Malawi. Zimbabwe is still in a transition period after independence; civil war and economic pressure have left its once prosperous economy in ruins, and it still must resolve many ethnic conflicts. Zambia made the change to independence with ease, although it lost one of its most important trading partners (Zimbabwe) during the civil war; its economic potential is great, although it has been affected by problems associated with its large copper resources, and agricultural development has been uneven. Malawi has few resources, limited economic development, and slow industrialization; because it is landlocked, Malawi is too far from markets to be a major exporter of cash crops, and its government has been unable to take a strong position against white-ruled South Africa, as many of its people work there.
3. The Portuguese colonists made few long-term investments. Poverty, illiteracy, unemployment, and poor health facilities are common. Since independence, Angola's political instability and military conflict have prevented its economic development.
4. Parts of Central Africa's river systems contain great waterfalls over which ships cannot travel.

Critical Thinking

5. Answers will vary but should include the fact that roads and railroads are necessary for economic development.

Using Geography Skills

1. Central African Republic, Zaire, Zambia, Malawi, Rwanda, Burundi, and Zimbabwe. None.
2. Zambezi River. People's Republic of the Congo: Zaire River; Zaire: Zaire River, Lakes Tanganyika, Edward, Kiva; Angola: Zaire River and some tributaries; Burundi: Lake Tanganyika; Malawi: Lake Nyasa; Mozambique: Lake Nyasa.
3. Zaire River, Limpopo River, Zambezi River.
4. Throughout the Zaire Basin in Zaire. Rain forests.
5. Approximately 13° S, 20° E.

CHAPTER 29 Southern Africa Pages 340–351

PLANNING GUIDE		
Lesson	Textbook	Teacher's Resource Binder
1. Physical Geography and History	pp. 340–343	Lecture Notes and Transparency Package Skills Worksheet Content/Vocabulary Workbook Reteaching Worksheet
2. Contemporary South Africa	pp. 343–348	Critical Thinking Worksheet Reteaching Worksheet Challenge Worksheet
3. Comparing Nations of Sub-Saharan Africa	pp. 348–350	Chapter Review/Test Reteaching Worksheet

TIME ALLOCATION: 3 days

CHAPTER TOPICS
- Republic of South Africa
- Namibia
- Botswana
- Lesotho
- Swaziland

PRETEACHING CHAPTER VOCABULARY
Write the following terms on the chalkboard:
veld apartheid sanctions exile

Ask students to complete a word inventory for each term, using the following chart.

Terms	Terms I Have Seen Before	Terms I Think I Can Define	Definition
veld			

Ask students what clues they can recognize from the root word of *apartheid*. Does the prefix in *exile* offer a clue to its definition?

▶ LESSON 1
Physical Geography and History

LESSON OBJECTIVES
The student should be able to
- identify significant geography terms
- locate major nations of the world
- locate major landforms found in Southern Africa
- explain the historical development of Southern Africa

PURPOSE
Southern Africa and the United States' involvement with the nations of this region are subjects that often appear in the news. In order to comprehend the current issues, students must have an understanding of the region's historical development.

MOTIVATOR
Southern Africa may be one of the African regions most familiar to the students. Ask students to contribute to a word web using the term *Southern Africa*. Display the word web in the classroom to use as a reference as students study this region.

TEACHING DESIGN
1. Refer students to the map "Southern Africa" on page 343 in their textbook. After they have located the countries of Southern Africa, ask students to locate the following sites: Namib Desert, Vaal River, Kalahari Desert, Cape of Good Hope, Capetown, Kimberley, Durban, Pretoria, Drakensberg Escarpment, Bloemfontein, and Orange River. As students locate the places on the map, challenge them to create time lines showing the significance of each site in the history of this region.
2. Refer students again to the map "Southern Africa" on page 343 in their textbook; have them compare it to the climate and population maps of sub-Saharan Africa on page 299. Ask students to hypothesize, based upon the information presented in the three maps, about why the greatest population centers of Southern Africa exist where they do. Ask students to clarify their reasons.
3. Present the Chapter 29 Lecture Notes and Transparency Package. Have students take notes.
4. Have students complete the Chapter 29 Skills Worksheet.

CLOSURE
Ask several students to present their time lines to the class. Encourage other students to check their time lines and fill in any important events not mentioned.

ASSESSMENT
1. Have students complete the following questions from the Chapter 29 Check:
   Building a Vocabulary 1
   Recalling and Reviewing 1, 2
   Using Geography Skills 3
2. Ask students to complete the Chapter 29 Content/Vocabulary Worksheet in their Workbook.

RETEACHING PLAN
1. Ask students to complete the following time line:
   1652
   1702 (50 years after 1652)
   1806
   1868
   1899–1902
   1910
   1948
2. Ask students to complete the Chapter 29 Reteaching Worksheet, Lesson 1.

CHALLENGE/ENRICHMENT STRATEGY

1. Have students write editorials regarding any of the major events in the history of the Republic of South Africa.
2. Encourage students to read any of the current articles on South Africa in magazines and newspapers. Have students use the *Reader's Guide to Periodical Literature* to locate appropriate articles to read.

▶ LESSON 2

Contemporary South Africa

LESSON OBJECTIVES

The student should be able to

- describe major economic activities of South Africa
- describe interchanges among the ethnic groups
- support the basic values of American society
- develop criteria for making judgments
- explain the system of apartheid in South Africa

PURPOSE

In this lesson, students will examine the policy of apartheid. Students will learn about the impact this policy has on South Africa and the world.

MOTIVATOR

Have students read the Cities of the World feature "Johannesburg" on pages 344–345 in their textbook. Ask the class for their reaction to the feature. Indicate that they will spend this class period studying about apartheid.

TEACHING DESIGN

1. Introduce students to the concept of apartheid. Ask them to offer their own definitions of the policy. In order to help students understand the policy from both of its perspectives, lead a class discussion to complete the following chart.

Whites		Blacks	
Positive	Negative	Positive	Negative

2. South Africa is the most highly developed nation on the African continent. Ask students to use the information in the chapter to support this contention by locating 5 to 10 statements. Summarize the major ideas.
3. Have students complete the Chapter 29 Critical Thinking Worksheet.

CLOSURE

Lead a class discussion on possible plans for abolishing apartheid. List the student's ideas on the chalkboard. Review the factors that could prevent the South African government from proceeding with a peaceful plan.

ASSESSMENT

Have students answer aloud the following questions from the Chapter 29 Check:

Building a Vocabulary 2, 3, 4
Recalling and Reviewing 3, 4
Critical Thinking 5
Using Geography Skills 1, 2

RETEACHING PLAN

1. Ask students to verbalize how they might feel if they were among the privileged few or a segregated group in South Africa.
2. Ask students to complete the Chapter 29 Reteaching Worksheet, Lesson 2.

CHALLENGE/ENRICHMENT STRATEGY

1. Ask interested students to investigate and report on the life of Nelson Mandela.
2. Have students complete the Chapter 29 Challenge Worksheet.
3. Ask students to write letters to the editor of a local newspaper expressing their views of the apartheid system.

▶ LESSON 3

Comparing Nations of Sub-Saharan Africa

LESSON OBJECTIVES

The student should be able to

- draw comparisons regarding regional differences and similarities in sub-Saharan Africa
- identify the nations of sub-Saharan Africa
- identify the major economic resources
- describe the major ways of life in sub-Saharan Africa
- describe the agricultural base of sub-Saharan Africa

PURPOSE

By this time, students will have spent almost three weeks studying the sub-Saharan region of the African continent. This lesson will help them to synthesize the information.

MOTIVATOR

Write the following statements on the chalkboard:

1. There are no metropolitan cities in sub-Saharan Africa.
2. All black Africans have the same ethnic heritage.
3. Sub-Saharan Africa is poor in resources.
4. The future of the sub-Saharan African nations relies on the discovery of valuable minerals to support their technological growth.
5. All sub-Saharan African countries have obtained their independence from their colonial rulers.

Ask students how many of these statements they may have believed to be true prior to their study of Africa. If they say they would have counted several as true, ask them to identify the information they have learned in this unit to refute the false statements.

TEACHING DESIGN

1. Divide the class into groups of four or five. Ask each group to complete a portion of the following chart. For some classes, it may be desirable to assign a certain area of Africa and to research each topic. For more advanced classes, it may be valuable to assign a particular topic and ask students to find an answer for each region.

	West Africa	East Africa	Central Africa	Southern Africa
Physical regions				
Climate				
Resources				
People				
Cities				
Ways of life				
Major challenges				
Best solutions				

2. Provide the class with time to complete their charts. Following this work time, lead a class discussion of the following questions: Which regions of sub-Saharan Africa are most similar? Why? Which regions of sub-Saharan Africa are most different? Why? What are the most important resources of sub-Saharan Africa?
3. Create a class list of all the challenges for this area of the world. Ask students to return to their groups and prioritize their lists from 1, for most important, to 10, for least important. Then ask students for possible solutions.
4. Have students complete the Chapter 29 Challenge Worksheet.

CLOSURE

Give each group the opportunity to present its list, and create a class priority ranking. Provide a summary for the chapter, using generalizations.

ASSESSMENT

Ask students to complete the Chapter 29 Review/Test.

RETEACHING PLAN

1. Allow students who have mastered the objectives of this lesson to assist students who have not, by reviewing the content and ideas presented in the lesson.
2. Ask students to complete the Chapter 29 Reteaching Worksheet, Lesson 3.

CHALLENGE/ENRICHMENT STRATEGY

1. Select volunteers to create a collage to represent the contributions of sub-Saharan Africa to the world.
2. Have students create a pie graph to represent the sub-Saharan African portion of the world's mineral supplies.

Chapter 29 Check, page 351, Answers

Building a Vocabulary

1. Velds. Highveld, bushveld, and lowveld.
2. The South African policy of separation of the races. The opening of the economy to all people is essential to the country's prosperity. Because of apartheid, the threat of unrest and civil war is increasing.
3. Embargoes placed on products. Sanctions have been used by developed countries to try to force South Africa to stop the policy of apartheid.
4. An exile.

Recalling and Reviewing

1. The coastal areas and coastal sides of mountains are humid because of the rain-bearing winds that come off the ocean. The mountains keep the rain-bearing winds from reaching the interior plains and plateaus, so these areas are often cooler and drier than the coastal areas.
2. Under the Homeland policy, the Republic of South Africa has divided the country into white and black regions. The black regions have been subdivided into territories based on different African and ethnic groups. The government has decided that these territories, called "homelands," should be independent. Urban black South Africans living in white cities have been assigned citizenship in the homeland of their ethnic group, so they have become foreigners in the cities where they live and work and, as foreigners, they have only limited civil rights.
3. In the homelands, most agriculture is subsistence level. The primary crops are maize and sorghum; goats and cattle are herded. Overgrazing and soil erosion are major problems; food and manufactured goods must be imported because of the shortage of workers. Agriculture in white South Africa is modern and productive.
4. In the wetter eastern part of the country. Because much of the country occupies the Kalahari Desert.

Critical Thinking

5. Answers will vary but should include discussion of political reform and civil rights.

Using Geography Skills

1. Near Capetown and in the north near Johannesburg.
2. Mining, fishing, farming, and livestock raising.
3. Maseru, Mbabane. About 350 miles (560 km.). About 175 miles (280 km.) and about 100 miles (160 km.), respectively.

UNIT 5
Review/Test

PLANNING GUIDE		
Lesson	Textbook	Teacher's Resource Binder
1. Unit 5 Review	pp. 352–353	
2. Unit 5 Test		Unit 5 Test

TIME ALLOCATION: 2 days

▶ LESSON 1
Unit 5 Review

Have students read the Unit 5 Summary. Use the Unit 5 Objectives as the basis for review, and discuss any questions that students have on the Unit 5 material.

Have students complete the questions in the textbook under the following headings:

Reading and Understanding
Mastering Geography Skills

Have students read the textbook material under the following headings and complete the appropriate or selected activities:

Applying and Extending
Linking Geography and Economics

▶ LESSON 2
Unit 5 Test

Have students complete the Unit 5 Test from the Teacher's Resource Binder.

Unit 5 Review, pages 352–353, Answers

Reading and Understanding

1. Because inland travel was made difficult by the hostile tribes, rugged terrain, and disease.
2. Internal ethnic conflicts have taken time, energy, and resources that might otherwise have been used for economic development. In some cases, conflict has destroyed important roads and businesses.
3. Gambia, Cape Verde, Gabon, Lesotho, São Tomé and Príncipe, Ghana, Togo, Botswana, Swaziland, and Zimbabwe. These countries are limited as to the kinds of crops they can grow.
4. The savannas are important as the location of most of Africa's game parks, which attract tourists; they also are the areas where most Africans live. Africans use the savannas for herding and agriculture.
5. Sub-Saharan Africa's potential for agricultural development is limited because of the infertile, unproductive soils that are characteristic of many areas. Because sub-Saharan Africa does have substantial deposits of minerals such as gold, copper, chromium, manganese, cobalt, lead, zinc, and uranium, it has the potential for industrial development. Oil is the only resource needed for industrialization that the region does not have in great supply.
6. Overpopulation has resulted in overgrazing and soil erosion, migration of people to urban areas, unemployment, overcrowding, inadequate housing, poor health care and nutrition, and the growth of slums.
7. Barriers to cooperation in the region are the language and cultural differences among countries and ethnic groups, the long history of suspicion among ethnic groups, the colonial influence on trade patterns, and the alignment of countries into Soviet and Western-oriented blocs. The Organization of African Unity has helped member countries solve problems involving disputes between members, threats to the independence of members, foreign investments, and economic and social problems related to modernization. Cooperation is important because the countries are now competing for the same markets.
8. It is a developed country with abundant natural resources and a broad-based economy. Its dominant political group is the white minority.
9. The European colonial powers established political boundaries in order to settle disputes among themselves. The boundaries were drawn without regard for Africa's landforms, climate regions, or cultures. They divided ethnic groups and tribes and cut through agricultural lands. Many landlocked states, which have problems importing and exporting goods, were created. Some countries have only one climate region, limiting the types of crops that can be grown. Others have small populations and are too small to diversify their economies.
10. Sahara.

Mastering Geography Skills

1. Songhai. France.
2. Kush. Great Britain.
3. Tanzania, Burundi, and Rwanda. No. Germany was forced to give up these colonies after its defeat in World War I; these colonies were transferred to Great Britain as trust territories until independence.

Applying and Extending

Answers will vary, but students should consider the impact of the slave trade upon local cultures and upon the economy of the countries that imported slaves.

Linking Geography and Economics

Answers will vary, but students should note that developing countries tend to have a higher percentage of rural population, a higher percentage of people engaged in agriculture, a lower per capita income, a higher rate of illiteracy, and a higher rate of population growth than economically developed countries.

UNIT 6
The Orient
Pages 354–417

UNIT OVERVIEW

The Orient is the term used to refer to the Asiatic countries and islands. The term is derived from Latin and means "the direction from which the sun rises." This vast region, also called the East or the Far East, is an area of great diversity. In it are some of the most populated nations of the world. These nations face the challenge of providing adequate food supplies for their people. Traditional agriculture remains a major activity, but modernization, industrialization, and urbanization are bringing changes to the ancient cultures of this region.

CHAPTER TIME LINE

Chapter	Title	Time Allocation
30	The Oriental Realm	4 days
31	India and South Asia	3 days
32	China	4 days
33	Japan and Korea	4 days
34	Southeast Asia	4 days
	Unit 6 Review	1 day
	Unit 6 Test	1 day
		21 days

MAJOR TOPICS

- Relationship among Land, Climate, and Ways of Life
- Historical Development's Effect on Economic, Social, Political, and Cultural Change
- Distinctive Characteristics of Cultures
- Relationship among Resources, Political Organization, and Economic Development
- Current Issues Facing Nations of the Orient

INTRODUCING THE UNIT

Suggest that students study briefly the photographs in Unit 6 of their textbook, pages 354–416. Ask students to suggest similarities and differences among the pictures. Using a large wall map of the world, point out the Orient. Compare the size of this region to that of regions previously studied. Ask students to think about an appropriate generalization about the Orient. Call on volunteers to share their ideas. Stress the fact that a region of such vast size is likely to be a region of diversity.

CHAPTER 30 The Oriental Realm
Pages 356–367

PLANNING GUIDE		
Lesson	**Textbook**	**Teacher's Resource Binder**
1. Landforms	pp. 356–358	Content/Vocabulary Workbook Reteaching Worksheet
2. Climate and Resources	pp. 358–362	Skills Worksheet Reteaching Worksheet
3. Resources and People	pp. 362–366	Critical Thinking Worksheet Reteaching Worksheet
4. History and Religion	pp. 362–364	Lecture Notes and Transparency Package Chapter Review/Test Reteaching Worksheet Challenge Worksheet

TIME ALLOCATION: 4 days

CHAPTER TOPICS

- Physical Geography
- Cultural Geography
- Economic and Political Development

PRETEACHING CHAPTER VOCABULARY

Write the following terms on the chalkboard:

paddy aquaculture Hinduism Buddhism

Ask students in which context they have heard the word *paddy*. Explain that a paddy in the context of this chapter is a field covered with water for growing rice.

Ask students if they know the definition of *agriculture*. What do they assume is the definition of *aquaculture*? (Aquaculture is water or sea farming.)

Ask students to recall what the suffix *-ism* indicates. Tell students that they will learn the difference between Hinduism and Buddhism in this chapter.

► LESSON 1
Landforms

LESSON OBJECTIVES

The student should be able to

- define significant terms of the chapter
- locate and describe major landforms of the Orient
- describe the physical forces that have caused changes in the surface of the earth
- locate the major nations of the Orient

PURPOSE

In this lesson, students will be introduced to countries and landforms of the Orient, where more than 50 percent of the world's population lives.

MOTIVATOR

On a sheet of poster board, create a word web around the term *The Orient*. Ask students to name all the words and phrases that come to mind when they hear this term. Create a separate poster for each class. Ask students in each class to review the word web and determine whether it has raised questions about this region to which they would like to know the answers. List several of their questions below the word web on the poster board. Inform students that this unit of study will give them the opportunity to learn the answers to their questions.

TEACHING DESIGN

1. Refer students to the regional map "The Orient" on page 359 of their textbook to identify the location of the countries of this region. Indicate to students that *Orient* is a Western term meaning "in the direction of the sun." Ask students why Westerners would have named this area the Orient. What might be another appropriate name for this region? Write the students' suggestions on the chalkboard.
2. Provide each student with an oversized map of the Orient. Have them identify the following physical features on their maps: India, Pakistan, Bangladesh, Nepal, Bhutan, Afghanistan, Sri Lanka, China, Japan, North Korea, South Korea, Vietnam, Laos, Burma, Kampuchea, Thailand, Kalimantan (Borneo), Philippines, Malaysia, Hong Kong; Indian Ocean, Pacific Ocean; Pamir Knot, Himalayas, Mount Everest, Kunlun Mountains, Altay Mountains, Qin Ling Mountains, Tian Shan, Hindu Kush, Khyber Pass; Tibet, Mongolia; Gobi Desert, Thar Desert, Taklimakan Desert; Huang He, Chang River, Xi River, Mekong River, Irrawaddy River, Chao Phraya River, Ganges River, Brahmaputra River, Indus River; Sea of Japan, Yellow Sea, East China Sea, South China Sea.
3. Indicate to students that their study of the Orient will be diverse and that they will help with the presentation of content. On the chalkboard, write the following topics: Government, Economy, Religion, Family and Culture, Physical Setting, and Current Issues. Divide the class into six groups, and assign each group one of these six topics.

   Tell students that each group will become expert on one part of the Orient. Each group will research information on its topic for every country in the region. Inform students that on assigned days, they will give oral reports on their topic, covering each of the Oriental countries. Ask each group to limit its speech to three minutes and to write the information that is most important to remember on a sheet of poster board or on an overhead transparency. Answers to the following questions should be included in their report: What is the capital of the country? What are the important landforms and bodies of water, and how do these affect the lives of the people? What is the climate, and how does it affect the lives of the people? Allow students the remaining class time to begin their research.

CLOSURE

Review with students the three most important facts about the mountains and plateaus of the Orient: (1) they form a natural political boundary between India, China, and the USSR; (2) they provide natural frontiers for defense; and (3) they provide the source of rivers, whose rich alluvial soils aid in the production of crops to feed the population.

ASSESSMENT

1. Ask students to complete the Chapter 30 Content/Vocabulary Worksheet in their Workbook.
2. Have students complete the following questions from the Chapter 30 Check:
   Building a Vocabulary 1, 2
   Recalling and Reviewing 1, 2

RETEACHING PLAN

1. On mimeographed sheets of paper, create a second version of the map assigned in Teaching Design 2; this time, however, use numbers to indicate the various places identified for study. Allow students to use the maps to review the locations.
2. Ask students to complete the Chapter 30 Reteaching Worksheet, Lesson 1.

CHALLENGE/ENRICHMENT STRATEGY

1. Begin a bulletin board about the Orient. Encourage students to bring to class articles they find in newspapers and magazines for the next four weeks.
2. Invite each student to choose a book about the Orient to read, then prepare a book jacket to represent his or her feelings about the book. The jacket should include a short summary on the inside cover.

▶ LESSON 2
Climate and Resources

LESSON OBJECTIVES

The student should be able to
- draw inferences from data provided
- describe physical forces that alter the earth's features
- analyze the impact of environment on ways of life
- explain the causes of population patterns
- analyze the location of cities

PURPOSE

In this lesson, students will review the many factors that influence where people live. The lesson also will review the relationships between people and their environment.

MOTIVATOR

Encourage students to reflect on the information they studied in Lesson 1. Write the following terms on the chalkboard: mountains, rivers, oceans, plateaus. Ask students to discuss the negative and positive influences that people can expect if they live near these landforms and bodies of water. Ask students to work in groups of three or four, and have a recorder keep notes for the group. After their discussion, give each group a physical map of the Orient. Ask the groups to identify the places that they believe have the densest population. Then have them compare their ideas with the population-density map "The Orient" on page 362 of their textbook. Discuss their degree of accuracy. Review other factors that influence population distribution.

TEACHING DESIGN

1. Refer students to the climate map "The Orient" on page 360 of their textbook. Ask them to identify from the map the climates of the Orient. Write these climates on the chalkboard. Inform students that using this information, along with information in the chapter text, they are to work with a partner to complete the following chart.

Orient/Climate Regions	**(Subarctic)**				
Which landforms are nearby?					
Which bodies of water are nearby?					
What is the daily weather like?					
How would you describe population density?					
What effect does it have on ways of life?					
What are the primary resources of the region?					

2. Call students' attention to the tremendous storms that occur in this region every year. The word *monsoon* comes from the Arabic word *mausim,* which means "seasonal winds." What else does the textbook tell about these winds?
3. Have students prepare a one-page essay summarizing the general relationship between climate and ways of life in this region.
4. As students complete their essays, allow them any remaining class time to continue preparation of their reports; remind students that presentations begin with the study of Chapter 31, "India and South Asia."

CLOSURE

Invite several students to read their essays from Teaching Design 3 for the class. Summarize the key ideas.

ASSESSMENT

1. Have students complete the Chapter 30 Skills Worksheet.
2. Ask students to complete the following questions from the Chapter 30 Check:
   Recalling and Reviewing 3, 4
   Using Geography Skills 1, 2, 3

RETEACHING PLAN

Ask students to complete the Chapter 30 Reteaching Worksheet, Lesson 2.

CHALLENGE/ENRICHMENT STRATEGY

1. Challenge students to create a collage that represents the diverse effects of environment on various parts of people's lives. Display the collages in the classroom.
2. Invite students to bring from home items that were made in the Orient. Create a display of these items in the classroom. (Examples: chopsticks, cooking utensils, toys, electronic equipment, fans, artwork, food products)

▶ LESSON 3
Resources and People

LESSON OBJECTIVES

The student should be able to

- locate and describe major landforms of the Orient
- locate the major natural resources of the Orient
- understand the criteria for determining regions
- analyze the impact of the environment on ways of life
- explain the cause of population patterns
- determine kinds and sources of energy

PURPOSE

In this lesson, students will review the role of the environment in shaping the way people live.

MOTIVATOR

Throughout the chapter, students read about the many contrasts within the Orient. In this lesson, focus their attention on the similarities. On the chalkboard, create a list of reasons for calling this part of the world a culture region.

TEACHING DESIGN

1. Display the Chapter 30 transparency on an overhead projector. Call on students to provide answers to complete the chart.
2. Have students read the Geography Skills feature "How to Read a Line Graph" on page 365 of their textbook. Call on volunteers to answer related questions.
3. Use the following information map to review the relationship between climate and ways of life.

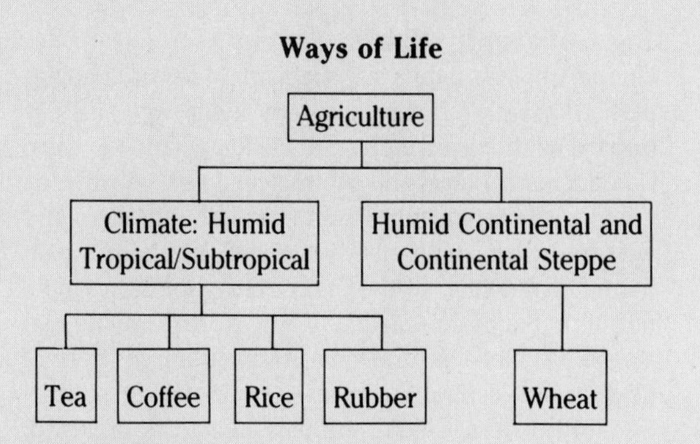

CLOSURE
Call on students to dictate the major climate regions of the Orient and the major crops in each, and write their answers on the chalkboard.

ASSESSMENT
Have students complete the Chapter 30 Critical Thinking Worksheet.

RETEACHING PLAN
Ask students to complete the Chapter 30 Reteaching Worksheet, Lesson 3.

CHALLENGE/ENRICHMENT STRATEGY
1. Challenge students to investigate why many of the Oriental countries use English as their business language.
2. Have students complete the Chapter 30 Challenge Worksheet.

▶ LESSON 4
History and Religion

LESSON OBJECTIVES
The student should be able to
- sequence and categorize historical events
- review characteristics of major world religions

PURPOSE
In this lesson, students will discover the major religions of the Orient and the significance of this culture region in the history of the world.

MOTIVATOR
Ask students to identify significant world events that have occurred in the Orient. Create a time line on the chalkboard as dictated by students. Make the time line large enough to allow for addition of events during the lesson.

TEACHING DESIGN
1. Present the Lecture Notes and Transparency Package to introduce students to the history of the Orient. Have students take notes.
2. Ask students to complete the following chart to summarize the history of this region.

Colonizing Nations	Which countries in the Orient were colonies?	When did the countries became independent?
Great Britain		
The Netherlands		
France		
Portugal		
United States		
Japan		

   Ask students if any parts of the Orient are still colonies. Discuss with students what they think will be the future of colonies in the Orient.
3. Use the time line on the chalkboard to summarize this section with the entire class.
4. Call on students to name all the religions mentioned in the text, and write them on the chalkboard. Ask students to work in groups of three to begin locating the answers to complete the following chart. Inform students that they will complete their charts as they study Unit 6.

	Hinduism	Islam	Buddhism	Sikhism	Taoism	Confucianism	Shintoism	Christianity
Name for followers								
Countries where followers are in majority								
Countries where followers are in minority								
Main ideas of the religion								

CLOSURE
Discuss with the class the similarities and differences they found in their research of the religions of the Orient. Write these on the chalkboard or on a transparency.

ASSESSMENT
1. Have students complete the following questions from the Chapter 30 Check:
   Building a Vocabulary 3
   Critical Thinking 6
2. Ask students to complete the Chapter 30 Review/Test.

RETEACHING PLAN

1. Ask students to complete the Chapter 30 Reteaching Worksheet, Lesson 4.
2. Have students use the *World Book* or another encyclopedia to outline the history of the Orient.

CHALLENGE/ENRICHMENT STRATEGY

1. Invite students to investigate the Soviet occupation of Afghanistan or the Chinese position in Tibet.
2. Challenge students to create a television talk show featuring the founders of the religions in this chapter.

Geography Skills, page 365, Answers

Test Your Skills

1. About 650 million; about 350 million. China: nearly 1,500 million (1.5 billion); India: about 1,200 million (1.2 billion).
2. 2000. 1970.
3. The growth rates are similar. Both curves rise at the same angle over time.
4. A line graph can provide a quick visual overview of data and how that data changes over time. It is useful for comparing population growth from more than one country over time.

Chapter 30 Check, Page 367, Answers

Building a Vocabulary

1. Paddies. Rice.
2. Water and sea farming. Many nations in the Orient depend upon freshwater fish and seafood for food. Fish and shellfish are commercially farmed.
3. Hindus worship many gods and believe that enlightenment is the result of good conduct; they also believe in reincarnation. Buddhists seek enlightenment through meditation. Hinduism.

Recalling and Reviewing

1. They serve as natural boundaries between India, China, and the USSR, provide natural frontiers for defense, affect Asia's climates, and are the source of major rivers.
2. The rivers influence where people live, depositing rich alluvial soils necessary for agriculture. They also are used for irrigation and transportation.
3. Humid tropical, wet and dry tropical/subtropical, humid subtropical, humid continental, tropical and subtropical desert, continental desert, continental steppe, subarctic, and highland. Population is most heavily concentrated in the humid tropical, wet and dry tropical/subtropical, and humid continental climate regions.
4. Fertile alluvial and volcanic soil, tin, tungsten, iron ore, coal, oil, hydroelectric power, fish, and forests.
5. Political instability, especially in Kampuchea, Laos, China, Vietnam, the Philippines, and South Korea. Overpopulation; some nations have successfully slowed population growth, but doing so may have a negative effect on economic growth.

Critical Thinking

6. Answers will vary. Example: China.

Using Geography Skills

1. Humid tropical: Indonesia, Vietnam, Malaysia, Bangladesh, Burma, Sri Lanka, western India; wet and dry tropical/subtropical: India, Sri Lanka, Thailand, Laos, Kampuchea; tropical and subtropical desert: northwestern India, Pakistan; humid subtropical: southeastern and western China, southern South Korea, southern Japan; humid continental: northeastern China, northern Japan, northern South Korea and North Korea; continental steppe: Mongolia, central India, Afghanistan; continental desert: north-central China, southern Mongolia; subarctic: northeastern China, northern Mongolia; highland: western China (Tibet), Nepal, Bhutan, Borneo, Irian Jaya.
2. Sea of Japan, Pacific Ocean, Yellow Sea, East China Sea, South China Sea, Celebes Sea, Arafura Sea, Java Sea, Gulf of Siam, Strait of Malacca, Bay of Bengal, Indian Ocean, Arabian Sea.
3. These countries are in the low to middle latitudes and therefore are subject to tropical storms. Typhoons follow erratic paths and can move from the low latitudes into the middle latitudes. In the low latitudes, the typhoons tend to move from east to west with the trade winds (into southern China). In middle latitudes, the typhoons tend to move eastward (across Taiwan, South Korea, and southern Japan).
4. Java; Bali.

CHAPTER 31 India and South Asia Pages 368–381

PLANNING GUIDE		
Lesson	Textbook	Teacher's Resource Binder
1. History and Geography	pp. 368–372	Lecture Notes and Transparency Package Content/Vocabulary Workbook Reteaching Worksheet
2. Resources and Ways of Life	pp. 372–377	Skills Worksheet Critical Thinking Worksheet Reteaching Worksheet
3. Issues and Neighbors	pp. 377–380	Challenge Worksheet Chapter Review/Test Reteaching Worksheet

TIME ALLOCATION: 3 days

CHAPTER TOPICS

- Historical Geography
- Physical Geography
- Economic Geography
- India
- Pakistan
- Afghanistan
- Nepal and Bhutan
- Bangladesh
- Sri Lanka
- The Maldives

PRETEACHING CHAPTER VOCABULARY

Write the following terms on the chalkboard:

subcontinent	caste system
Sikhism	cottage industries
tropical cyclone	nonalignment
storm surge	buffer state
deforestation	sultanate
Green Revolution	

Indicate to students that these terms may be classified by the following categories: geography, economics, and politics. Work with the students to correctly identify the appropriate placement of these terms.

▶ LESSON 1
History and Geography

LESSON OBJECTIVES

The student should be able to

- define significant chapter vocabulary
- describe physical forces that alter the earth's crust
- describe the physical setting of selected regions
- place historical events in sequential order

PURPOSE

In this lesson about India and the countries of South Asia, students will learn to identify places of world significance in this region. They also will understand how events in various parts of the world affect the development of other parts.

MOTIVATOR

Display a transparency of a physical map of India and the countries of South Asia. Ask students to work with a partner to see how many different countries and significant features they remember from their study of Chapter 30. Ask students to review page 51 in their textbook regarding the theory of Gondwanaland. Call upon several students to explain to the class what they have read. Point out the relationship to what the class learned in Unit 5.

TEACHING DESIGN

1. Present the Chapter 31 Lecture Notes and Transparency Package. Have students take notes.
2. On the chalkboard, write the following list of significant physical features of South Asia: Bay of Bengal, Arabian Sea, India, Pakistan, Bangladesh, Nepal, Bhutan, Afghanistan, Sri Lanka, Indus River, Ganges River, Brahmaputra River, Thar Desert, New Delhi, Madras, Bombay, Calcutta. Ask students to return to their partners to help locate these places on the map "India and South Asia" on page 371 of their textbook.
3. Refer students to the location of India on a wall map of the world. Ask students to use their prediction skills to suggest characteristics of India's climate, based upon their knowledge of its location. Review the significance of the monsoon as presented in Chapter 30. Refer students to the map "World Climate Regions" on pages 66–67 as you discuss the flow and impact of monsoon winds. Discuss the difference between cyclones and storm surges. Determine whether or not the students' predictions were accurate.
4. Call on the groups who were assigned the topics Physical Setting and Government in Teaching Design 3, Lesson 1, Chapter 30 to present their reports to the class on India, Pakistan, Bangladesh, Nepal, Bhutan, Afghanistan, and Sri Lanka. Remind all students to take notes and determine answers to the questions identified in each report. Check to make sure that students identify the correct answers.

CLOSURE

Name key dates and events in the history of India. Call on students to identify the significance of each event as presented in the Teaching Design 1 lecture.

ASSESSMENT

1. Have students complete the following questions from the Chapter 31 Check:
   Building a Vocabulary 1, 5
   Recalling and Reviewing 1, 3
   Using Geography Skills 1, 2
2. Ask students to complete the Chapter 31 Content/Vocabulary Worksheet in their Workbook.

RETEACHING PLAN

1. Ask students to complete the Chapter 31 Reteaching Worksheet, Lesson 1.
2. Provide students with unlabeled maps of India and South Asia. Have students label these maps with the significant physical features listed on the chalkboard in Teaching Design 2.

CHALLENGE/ENRICHMENT STRATEGY

1. Invite pairs of students to research the life of Gandhi and present an interview with Ghandi to the class.
2. Select volunteers to research and report on the Taj Mahal. Encourage them to explain the reason it is called one of the world's most beautiful buildings.

▶ LESSON 2
Resources and Ways of Life

LESSON OBJECTIVES

The student should be able to

- describe the physical setting of the regions of India
- identify major natural resources and their uses
- understand the criteria for determining regions
- analyze the impact of the environment on ways of life
- explain the cause of population patterns
- describe the agricultural base of India

PURPOSE

In this lesson, students will develop an understanding of how regions are defined and an appreciation of how traditional cultures can coexist with modern cultures.

MOTIVATOR

Read to students the following sentence: "The Indus River flows southward mainly through Pakistan and is the 'lifeblood' of that desert and mountain nation." Discuss with students the meaning of *lifeblood* in this context. How does the river support and sustain life for the people of Pakistan?

TEACHING DESIGN

1. Provide students with physical maps of India. Indicate to them that India is normally divided into three physical regions. Ask them to analyze their maps to predict the approximate locations of these regions.

   Refer students to the population-density map "The Orient" on page 362, and to the economic map "India and South Asia" on page 371. Have students compare their predictions to these textbook maps.
2. Draw the following schematic on the chalkboard:

   ←——————————————→
   Traditional   Modern

   Divide students into work groups of four or five, and assign a spokesperson for each group. Each group should identify the characteristics of Indian ways of life and list them under the traditional side or the modern side. Ask students if there are any characteristics that should be listed in the middle. If so, ask students to be prepared to discuss their reasons. Direct spokespersons to prepare a presentation of their group's diagram.
3. Call on students who were assigned the topics Economy and Family and Culture in Teaching Design 3, Lesson 1, Chapter 30. Have all students take notes.
4. Have students complete the Chapter 31 Skills Worksheet.

CLOSURE

Invite the spokespersons from each group in Teaching Design 2 to present their diagrams to the class. Display the diagrams in the class to use as a source of review.

ASSESSMENT

1. Have students complete the following questions from the Chapter 31 Check:
   Building a Vocabulary 3, 6
   Recalling and Reviewing 2
   Critical Thinking 4
2. Have students complete the Chapter 31 Critical Thinking Worksheet.

RETEACHING PLAN

Ask students to complete the Chapter 31 Reteaching Worksheet, Lesson 2.

CHALLENGE/ENRICHMENT STRATEGY

1. Invite students to investigate monsoons and storm surges, then write and present a television newscast describing a recent storm in India.
2. Ask students to create a brochure for tourists, describing places to see and things to do in major cities of India.

▶ LESSON 3
Issues and Neighbors

LESSON OBJECTIVES

The student should be able to

- describe the social and cultural interchange among the people of India and South Asia
- determine the economic importance of water and other resources to the development of nations

PURPOSE

The students will review in this lesson some of the issues faced by South Asia, and they will practice formulating solutions to problems.

MOTIVATOR

Provide the class with unlabeled political maps of India and South Asia. Conduct a friendly competition to see who can identify the most countries on this map. Indicate to students that they will learn significant information about India and South Asia in this lesson. Ask students what they think are the greatest concerns of these countries. Discuss what they believe to be similarities and differences among them.

TEACHING DESIGN

1. Refer students to their religion charts begun in Teaching Design 4, Lesson 4, Chapter 30. Have them fill in their charts for Hinduism, Islam, Buddhism, and Sikhism.
2. Call on students to present the information in their religion charts. Discuss the answers, making sure that all students correct their charts. Then, have students complete the following chart, checking the appropriate religions of India and the South Asian countries.

Country	Hinduism	Buddhism	Sikhism	Islam
India				
Pakistan				
Afghanistan				
Nepal				
Bhutan				
Bangladesh				
Sri Lanka				
Maldives				

   Discuss the result of the religious differences in South Asia. Ask students to consider why the conflict in these countries is so serious while different religions coexist peacefully in other nations.
3. Request that the groups who were assigned the topic Current Issues in Teaching Design 3, Lesson 1, Chapter 30 give their reports. Ask all students to take notes.
4. Have students complete the Chapter 31 Challenge Worksheet.

CLOSURE

Ask students: What do you want to remember about this part of the world? In the discussion that follows, guide students to an understanding of the South Asian people and appreciation for students' own way of life.

ASSESSMENT

1. Have students complete the following questions from the Chapter 31 Check:
   Building a Vocabulary 2, 4, 7
   Critical Thinking 4
2. Ask students to complete the Chapter 31 Review/Test.

RETEACHING PLAN

1. Ask students to complete the Chapter 31 Reteaching Worksheet, Lesson 3.
2. Have students create a diagram that represents the 10 most important issues related to this region.

CHALLENGE/ENRICHMENT STRATEGY

1. Invite a guest who has lived in or visited India or South Asia to speak to the class.
2. Allow students to bring Indian food to the class for all to taste and/or bring souvenirs for all to view.

Chapter 31 Check, page 381, Answers

Building a Vocabulary

1. A large landmass that is smaller than a continent. The Indian subcontinent broke away from Gondwanaland and collided with Eurasia, causing the sea floor beneath both landmasses to compress and lift, forming mountains. India continued to press into Eurasia, pushing the mountains higher to form the Himalayas.
2. Hinduism and Islam.
3. A rise in sea level caused by a storm approaching the coast. Bangladesh has been severely flooded, and thousands of lives have been lost.
4. The introduction of new types of grain that produce high yields. It has not been effective; natural disasters and population increases reduced the gains that had been made, and larger amounts of fertilizer and water were needed for the new grains.
5. A social system that divides a society into major classes, or castes, based upon occupation. Brahmins, Kshatriyas, Vaisyas, Sudras, untouchables. Not very effective, especially in the rural areas.
6. A foreign policy of not siding with major world powers. India has close ties with the Soviet Union and good relations with the United States.
7. A buffer state.

Recalling and Reviewing

1. A mountainous area to the north formed by the Himalayas, an alluvial plain south of the mountains around the Ganges River called the Indo-Gangetic Plain, and the Deccan Plateau covering the southern two-thirds of the peninsula. The mountain area guards India's northern border and is home for nomadic herders and some farmers. The alluvial plains form the agricultural and cultural center of the country. Most of India's mineral resources are in the plateau region.
2. Most of the good land is already cultivated. Farms are small, and scattered plots make farming less efficient. Crops are often destroyed by floods, droughts, or locusts, and new high-yield grains require larger amounts of fertilizer and water, which are in short supply.
3. India provided natural resources and a market to Britain.

Critical Thinking

4. Answers will vary.

Using Geography Skills

1. Kaveri, Krishna, Godavari, Tapti, Narmada, Yamuna, Ganges, Brahmaputra, Hooghly, Ravi, Indus. All but Ravi, Indus, Tapti, and Narmada flow into the Bay of Bengal.
2. In the Ganges River valley, in the Indus River valley, in narrow plains along the Indian coast, and in the southern part of the Deccan Plateau. These areas are good for agriculture and industry.

CHAPTER 32 China
Pages 382–393

PLANNING GUIDE		
Lesson	**Textbook**	**Teacher's Resource Binder**
1. Regional Geography	pp. 387–391	Skills Worksheet Reteaching Worksheet
2. History and Culture	pp. 382–387	Lecture Notes and Transparency Package Content/Vocabulary Workbook Reteaching Worksheet
3. China Today	pp. 386–387	Challenge Worksheet Reteaching Worksheet
4. Taiwan and Mongolia	pp. 391–392	Critical Thinking Worksheet Chapter Review/Test Reteaching Worksheet

TIME ALLOCATION: 4 days

CHAPTER TOPICS

- Historical Geography
- Regional Geography
- Taiwan
- Mongolia

PRETEACHING CHAPTER VOCABULARY

Write the following terms on the chalkboard:

dynasty double cropping commune terracing

Provide the following background to help students understand the meanings of the vocabulary terms.

A *dynasty* is a succession of rulers from the same family. Point out that dynastic cycles are a major theme in Chinese history. Each dynasty started out strong and prosperous, but over time was weakened by rebellions from within and invasions from outside.

A *commune* is a small, often rural community where members with common interests work together. Point out that in China, communes are responsible for agriculture and light industry.

Double cropping is planting and harvesting two crops during a single growing season. Ask students what they think the advantage of double cropping is.

Terracing is creating a raised bank of earth with sloping sides and a flat top. The word is derived from the French term *terrasse,* meaning "a pile of earth."

▶ LESSON 1
Regional Geography

LESSON OBJECTIVES
The student should be able to
- define the significant terms of the chapter
- identify the climate of China
- identify significant physical features of China
- describe the agricultural base
- identify economic activities in China
- identify the natural resources of China and their uses
- explain the causes of population patterns in China

PURPOSE
The study of the regional geography of China will help students understand this country's ways of life and economy.

MOTIVATOR
Call on volunteers to locate China and the continental United States on a wall map of the world. Ask students to study the wall map to identify similarities and differences in locations and features of these two nations. Write the two categories on the chalkboard. After students have sufficient time to make the comparisons, call on volunteers to share theirs. After writing their ideas on the chalkboard, ask students to suggest categories for grouping the items.

TEACHING DESIGN
1. Ask students who were assigned the topics Physical Settings and Economy in Teaching Design 3, Lesson 1, Chapter 30 to present their reports to the class. Remind students to take notes as they listen.
2. Divide students into groups of three. Ask each group to study the maps "China, Mongolia, and Taiwan" on page 385 of their textbook and "The Orient" on page 359 to answer the following questions:
   - What are the major rivers of China?
   - Why are the crops grown in northern China different from those grown in southern China?
   - Which part of China produces grains primarily?
   - What landforms are found in northern China?
   - What landforms are found in Tibet?
   - Why is western China sparsely populated?
   - What are the major mineral resources of China?
   - What are the major industries of China?

   Call on various groups to share their answers.

CLOSURE
On the chalkboard, write the names of the four regions of China. Through class discussion, write generalizations for each region about physical features, climate, economic activity, and population.

ASSESSMENT
1. Have students complete the following questions from the Chapter 32 Check:
   Building a Vocabulary 3,4
   Recalling and Reviewing 1, 2, 3
   Using Geography Skills 1, 2
2. Have students complete the Chapter 32 Skills Worksheet.

RETEACHING PLAN
Ask students to complete the Chapter 32 Reteaching Worksheet, Lesson 1.

CHALLENGE/ENRICHMENT STRATEGY
1. Suggest that students compare irrigation practices in Egypt and China.
2. Choose students to prepare a visitor's guide to selected cities in China.

▶ LESSON 2
History and Culture

LESSON OBJECTIVES
The student should be able to
- identify and sequence key events in Chinese history
- identify examples of economic, social, and cultural interchange among regions and countries
- identify areas of dispute among nations

PURPOSE
The study of major events in the history of China will help students develop a perspective for understanding the culture of this vast nation.

MOTIVATOR
Ask students to study the map "China, Mongolia, and Taiwan" on page 385 of their textbook. Ask students to choose a partner to discuss possible reasons that the Chinese civilization developed independently from other civilizations. After students have had time for discussion, call on volunteers to share their explanations.

TEACHING DESIGN
1. Present the Chapter 32 Lecture Notes and Transparency Package to trace the historical development of China. Have students take notes.

2. Divide students into groups of three or four, and ask each group to complete a short written assignment about the following historical scenarios. Assign one scenario to each group, giving some groups the same ones. Provide reference books, encyclopedias, and appropriate newsmagazines for assistance. Allow 15 minutes for students to discuss scenarios and to write.
   a. You are the leader of the Ch'in dynasty in the third century B.C. In order to protect your empire from invasions, you have decided to build a wall. To do so will require great numbers of laborers. To recruit laborers, write a decree explaining why it is necessary to build this protective wall for the country.
   b. The year is 1921; you are Sun Yat-sen. Your country has suffered at the hands of Western nations. You have organized the Nationalist party and have just been elected president of China. Write a speech to revive the nation's spirit, stating your future goals.
   c. You are Mao Zedong in 1957. You must outline for the Communist Central Committee your Great Leap Forward program and describe your future plans.

CLOSURE

Call on the groups in the chronological order of the scenarios to present their written assignments to the class. One member of each group should read the scenario, one member should write a date and character on the chalkboard, and one member should read the group's work.

ASSESSMENT

1. Have students complete the following question from the Chapter 32 Check:
   Building a Vocabulary 1
2. Ask students to complete the Chapter 32 Content/Vocabulary Worksheet in their Workbook.

RETEACHING PLAN

1. Ask students to complete the Chapter 32 Reteaching Worksheet, Lesson 2.
2. Provide every student with a set of shuffled cards, each card containing a major event in Chinese history. Students should arrange the cards in sequence.

CHALLENGE/ENRICHMENT STRATEGY

1. Choose students to research and report on the teachings of the Chinese philosophers Confucius and Lao-tzu.
2. Ask students to investigate the art, literature, architecture, or music of China and to report their findings.

▶ LESSON 3
China Today

LESSON OBJECTIVES

The student should be able to

- describe major economic activities in China
- identify examples of economic, social, and cultural interchange among regions and countries

PURPOSE

The study of this lesson will help students understand the importance of nations' cooperating to achieve common goals while retaining their cultural differences.

MOTIVATOR

Project on a transparency the following passage from the Department of State's Bureau of Public Affairs.

> China is now our 16th largest trading partner, and we are China's third largest, with $8 billion two-way trade in 1986. Major Chinese exports to the U.S. have been textiles, apparel, and petroleum products. Major U.S. exports to China have included railway equipment, aircraft, logs and lumber, and computers. Agricultural products, which once accounted for the bulk of U.S. exports, are now around 2% of the total. American investment in China is estimated at $1.5 billion and has been focused on energy exploration, electronics, textiles, food processing, hotels, and construction. More than 140 U.S. firms have invested in China, and many maintain resident offices in Beijing.

Ask students why it is in the interest of the United States to have economic and political relations with China. List students' reasons on the chalkboard, and ask the class to rank them in order of importance.

TEACHING DESIGN

1. Ask students who were assigned the topics Religion and Family and Culture in Teaching Design 3, Lesson 1, Chapter 30 to present their reports to the class. All students should take notes and answer the questions related to each topic.
2. Review the main features of China's economic system, then ask students who were assigned the topic Government in Teaching Design 3, Lesson 1, Chapter 30 to present their report. After students have answered the related questions, lead a discussion about the programs of economic development.
3. Ask students who were assigned the topic Current Issues in Teaching Design 3, Lesson 1, Chapter 30 to present their report and answer the related questions.
4. Have students read the primary source feature "For the Record" on page 386 in their textbook. Lead a class discussion about the treatment of political prisoners in Communist China.

CLOSURE

Write the following topics on the chalkboard: Religion and Family, Culture, Government, Economy, and Current Issues. Ask students to turn to a nearby classmate and develop a summary statement about each topic. Call on volunteers to share their summaries.

ASSESSMENT

1. Have students complete the following question from the Chapter 32 Check:
   Critical Thinking 6
2. Have students complete the Chapter 32 Challenge Worksheet.

RETEACHING PLAN
Ask students to complete the Chapter 32 Reteaching Worksheet, Lesson 3.

CHALLENGE/ENRICHMENT STRATEGY
1. Choose students to investigate Chinese cuisine. They may wish to invite a chef from a Chinese restaurant to present a cooking demonstration.
2. Challenge students to research and report on the history of the silk industry in China.

▶ LESSON 4
Taiwan and Mongolia

LESSON OBJECTIVES
The student should be able to
- identify physical features of Taiwan and Mongolia
- identify key events in the history of these nations
- describe economic activities of these nations
- identify the agricultural base of these nations
- locate nations and cities

PURPOSE
This lesson will give students the background necessary for understanding the current world status of Taiwan and Mongolia and the transition planned for these countries.

MOTIVATOR
Write the following names of countries on the chalkboard: People's Republic of China, Republic of China. Ask students to share their knowledge about these two Chinas in terms of government, economic systems, political leaders, and place in world affairs. Explain that students will learn more about the Republic of China in this lesson.

TEACHING DESIGN
1. On a wall map of the Orient, locate Taiwan and Mongolia. Ask students to use the map to discuss the location of these two nations in terms of nearby nations and bodies of water. Students also should note physical features that would affect ways of life in these two nations.
2. Refer students to encyclopedias, world almanacs, and history texts to investigate the history of Taiwan and Mongolia. Each student should work with a partner and organize the information they collect according to the categories Date, People, and Events. After allowing students adequate time to prepare, call on volunteers to share their findings. Record their information on the chalkboard in the form of a chart for each country.
3. Ask students to work in groups to research political and economic features, urban centers, and foreign policies of Taiwan and Mongolia. Each group should report its findings in the form of entries in a world almanac.

CLOSURE
Call on various groups to present their world almanac entries on Taiwan and Mongolia. All students should then write a summary statement about each country.

ASSESSMENT
1. Have students complete the following questions from the Chapter 32 Check:
   Recalling and Reviewing 4, 5
2. Have students complete the Chapter 32 Critical Thinking Worksheet.
3. Ask students to complete the Chapter 32 Review/Test.

RETEACHING PLAN
Ask students to complete the Chapter 32 Reteaching Worksheet, Lesson 4.

CHALLENGE/ENRICHMENT STRATEGY
1. Suggest that students create dioramas, collages, or mobiles to portray the history of Taiwan or Mongolia.
2. Ask students to report on the People's Republic of China's policies towards Taiwan and Mongolia.

Chapter 32 Check, page 393, Answers

Building a Vocabulary
1. A government ruled by a family whose power has lasted for a long period of history. Because of the great era of the Han dynasty, which ruled for about 400 years.
2. A commune is a cooperative group of farmers who pool their labor. Communes were organized after the Communists came to power in 1949 in order to produce more food for China's population. Mao Zedong.
3. Two crops are raised in one year on the same land. In the north, three crops can be grown. Yes. China is the world's leading rice producer and exporter.
4. Terracing.

Recalling and Reviewing
1. In northern China. It continues to be the center of culture and political power.
2. Oil, coal, iron ore, lead, zinc, and manganese.
3. The humid subtropical climate and alluvial lowlands have made southern China a prime agricultural area. A variety of crops can be grown in this region. The rainfall and mild climate make rice cultivation a major economic activity. Trade and industry have been encouraged by the navigable rivers. Unlike humid southern China, northern China suffers from insufficient and uncertain rainfall. Northern China has humid continental and continental steppe climates, which are appropriate for growing grains such as wheat, millet, and sorghum, but serious droughts sometimes occur in this region.
4. Taiwan not only produces enough food to feed its large population but also exports some food products. It is a major industrial nation, exporting textiles, metals, wood products, electronic equipment, ships, and steel. Much of the population is urbanized, and the people enjoy a relatively high standard of living.
5. It depends upon the Soviet Union for economic aid.

Critical Thinking

6. Answers will vary, but students should mention the effects of communism upon the traditional family structure in China. At one time, the oldest male was the dominant member of the family, whereas under communism, women are given equal status. The Communist government has tried to slow population growth.

Using Geography Skills.

1. In the valleys of the Huang He and the Chang River, in northeast China. Because of the fertile soil, which supports agriculture, and because of the industrial development and trade made possible in part by the rivers.
2. Xi River. Farming, industry, mining, and fishing.

CHAPTER 33 Japan and Korea Pages 394–405

PLANNING GUIDE		
Lesson	**Textbook**	**Teacher's Resource Binder**
1. History and Culture of Japan	pp. 394–396	Lecture Notes and Transparency Package Content/Vocabulary Workbook Reteaching Worksheet
2. Geography of Japan	pp. 396–399	Reteaching Worksheet
3. Economy of Japan	pp. 399–401	Skills Worksheet Critical Thinking Worksheet
4. North and South Korea	pp. 401–404	Lecture Notes and Transparency Package Chapter Review/Test Reteaching Worksheet Challenge Worksheet

TIME ALLOCATION: 4 days

CHAPTER TOPICS

• Japan • North and South Korea

PRETEACHING CHAPTER VOCABULARY

Write the following terms on the chalkboard:

tsunami balance of payments
subsidy export economy

Ask students to suggest possible categories under which three of these words could be grouped. (economics, trade) Explain that *tsunami* is the term for large sea waves caused by earthquakes and volcanic eruptions.

If students are unfamiliar with the economic terms, provide the following background information to assist them.

subsidy (monetary assistance granted by a government to a person or group) Ask students whether any groups in the United States receive subsidies from the government.

balance of payments (a summary of a nation's business transactions with other countries, including imports and exports, or payments and receipts) Ask students whether a nation would want a balance of payments.

export economy (an economy of a nation that focuses on producing goods for sale or trade to other nations) Ask students to define *exports*. Ask why nations sell goods to other countries.

▶ LESSON 1
History and Culture of Japan

LESSON OBJECTIVES

The student should be able to

• define the significant terms of the chapter
• identify in sequence key events in Japan's history
• identify examples of social and cultural interchange
• identify areas of dispute among nations

PURPOSE

The study of Japan's history and culture will provide students with a framework for understanding the people of this highly industrialized and complex nation.

MOTIVATOR

Explain to students that according to ancient Japanese mythology, the islands of Japan were created by the god Izanagi and the goddess Izanami. Izanami gave birth to the gods of fire, metal, earth, and agriculture. From water falling from the left eye of Izanagi, the sun goddess Amaterasu, ancestress of Japan's imperial family, was born. Izanagi gave the sun goddess a jewel from a necklace and told her to govern heaven. Compare this legend of the divine origin of the emperor to the idea of divine rulers in Europe. Ask students what effect belief in this legend could have on the people of Japan and their ideas about their country. Point out that the symbols of the imperial family today (iron sword, bronze mirror, curved jewel) represent the goddess of the Sun. Explain to students that they will learn about the history and people of Japan in this lesson.

TEACHING DESIGN

1. Present Part I of the Chapter 33 Lecture Notes and Transparency Package to introduce Japan. Advise students to take precise notes, as they will be asked to use them to make prediction statements in Lesson 2.
2. Call on students who were assigned Religion, Family and Culture, and Government in Teaching Design 3, Lesson 1, Chapter 30 to present their reports to the class. Remind students to take notes.

CLOSURE

After allowing students the opportunity to review the questions related to the topics presented in the reports, call on volunteers to answer the questions aloud.

ASSESSMENT

1. Have students complete the following question from the Chapter 33 Check:
   Recalling and Reviewing 2
2. Ask students to complete the Chapter 33 Content/Vocabulary Worksheet in their Workbook.

RETEACHING PLAN

1. Ask students to complete the Chapter 33 Reteaching Worksheet, Lesson 1.
2. Direct students to use their notes from the lecture to review the history of Japan.

CHALLENGE/ENRICHMENT STRATEGY

1. Have students research and report on the educational system of Japan.
2. Direct students to investigate Japanese housing and report their findings, using visual aids.
3. Ask a group of students to explain and demonstrate a Japanese tea ceremony.

▶ LESSON 2
Geography of Japan

LESSON OBJECTIVES

The student should be able to

- locate important physical features of Japan
- locate Japan on various maps
- describe physical forces that alter the earth's crust
- describe the agricultural base of Japan
- describe economic activities related to resources

PURPOSE

In this lesson, students are introduced to the physical geography and resources of Japan. Students will continue to focus on the interaction between people and environment.

MOTIVATOR

Refer students to the map "Japan and Korea" on page 397 of their textbook, as well as the map "Asia: Physical" in their textbook Atlas, page 6. Have students review their maps from the Chapter 33 Content/Vocabulary Worksheet. On a world map, compare the sizes of Japan and California. Compare the latitudes of the Japanese islands to the latitudes of the east coast of the United States. Ask students to predict the climate of the Japanese islands and to determine which area of Japan has a longer growing season, north or south.

TEACHING DESIGN

1. Refer students to the notes they took during the Lesson 1 lecture. Divide students into groups of three. Ask groups to discuss the information from the Lecture Notes in terms of its impact on the development of Japan. Each group should then write a prediction statement for each of the following topics: Land Surface, Climate, Vegetation, Agriculture, Industry, Population Density.
2. While students remain in their groups, call on those who were assigned the topic Physical Setting of Japan in Teaching Design 3, Lesson 1, Chapter 30 to present their reports. Remind all students to take notes and to determine answers to the questions identified for the report.
3. Direct student groups to compare their prediction statements about Land Surface, Climate, and Vegetation to the information presented in the report and the information in their textbook. Allow groups to make any necessary revisions of their prediction statements.

CLOSURE

Ask students to present their group's prediction statements about Land Surface, Climate, and Vegetation. Have the class dictate a paragraph containing the main ideas related to these three topics, and write it on the chalkboard.

ASSESSMENT

Have students complete the following questions from the Chapter 33 Check:
Recalling and Reviewing 1
Using Geography Skills 1

RETEACHING PLAN

1. Ask students to complete the Chapter 33 Reteaching Worksheet, Lesson 2.
2. Direct students to label a map of Japan, including important physical features and cities.

CHALLENGE/ENRICHMENT STRATEGY

1. Allow students to bring traditional Japanese foods to class for classmates to sample, or ask a chef of a Japanese restaurant to visit the class and discuss traditional Japanese foods and methods of preparation.
2. Ask students to report on sports and other popular forms of entertainment in Japan.

▶ LESSON 3
Economy of Japan

LESSON OBJECTIVES

The student should be able to

- describe major economic activities in Japan
- identify causes of population patterns in Japan
- explain the importance of natural resources to Japan
- identify the location of cities and note their functions

PURPOSE

The study of Japan's economic development will help students understand the importance of people as a major resource in a nation with limited land and limited natural resources for industrialization.

MOTIVATOR

Refer students to their prediction statements from the previous lesson. Ask volunteers to read their statements for the categories of Agriculture, Industry, and Population Density. As students share their statements, write the main ideas in note form on the chalkboard.

TEACHING DESIGN

1. Ask students who were assigned the topics Economy and Current Issues in Teaching Design 3, Lesson 1,

Chapter 30 to present their report to the class. Remind students to take notes.

2. Project the following chart on a transparency, and lead a class discussion by calling on students to provide information to fill in the appropriate spaces. Point out that it is appropriate to leave some spaces blank.

Japan: Economic Development				
Activity	**Disadvantages**	**Advantages**	**Products**	**Issues**
Agriculture				
Fishing				
Aquaculture				
Forestry				
Industry				

3. Ask students to review the chart and the information in the textbook to suggest reasons for Japan's development into the third largest industrial nation of the world with the highest economic growth rate.
4. Refer students to the map "Japan and Korea" on page 397 and the Cities of the World feature "Tokyo" on pages 402–403; ask them to list reasons for the growth of urban centers in the Tokyo Bay area, the Kansai region, and the Nagoya area. Each student should compare his or her list for each center with a partner.

CLOSURE
Refer students to the notes from their prediction statements on Agriculture, Industry, and Population Density. Ask the class to decide whether these predictions were accurate or inaccurate. Have volunteers make revisions.

ASSESSMENT
1. Have students complete the following questions from the Chapter 33 Check:
   Building a Vocabulary 1, 2, 3
   Recalling and Reviewing 3
   Critical Thinking 5
   Using Geography Skills 2
2. Have students complete the Chapter 33 Skills Worksheet.

RETEACHING PLAN
Ask students to complete the Chapter 33 Reteaching Worksheet, Lesson 3.

CHALLENGE/ENRICHMENT STRATEGY
1. Have students complete the Chapter 33 Critical Thinking Worksheet.
2. Select students to investigate worker-management relations in Japanese businesses.

▶ LESSON 4
North and South Korea

LESSON OBJECTIVES
The student should be able to
- describe the physical setting of North and South Korea
- locate North and South Korea on maps
- describe economic activities related to resources
- sequence events in history
- identify areas of dispute among nations

PURPOSE
The study of North and South Korea will offer students the opportunity to compare development between nations with different governmental and economic policies.

MOTIVATOR
Ask students to describe North and South Korea without looking at maps for references. Record students' descriptions on the chalkboard, and ask students to turn to page 397 in their textbook to review the map "Japan and Korea." Ask whether they wish to revise any of their descriptions.

TEACHING DESIGN
1. Use Part II of the Chapter 33 Lecture Notes and Transparency Package to introduce North and South Korea.
2. Call on students who were assigned the topics Physical Setting, Government, Economy, Religion, Family and Culture, and Current Issues in Teaching Design 3, Lesson 1, Chapter 30 to give their reports to the class. Remind all students to take notes during the presentations and determine answers to the questions.
3. Make available reference sources (world almanac, atlases, encyclopedias), and ask students to complete the chart below comparing North and South Korea.

	North Korea	South Korea
Natural resources		
Agricultural products		
Industrial products		
Gross National Product		
Per capita income		
Major trading nation		

Lead a class discussion based on the information in the chart. Ask students: Which nation is more productive? What factors have hampered North and South Korean development since 1950?

CLOSURE
Refer students to the activity in the Motivator and the notes on the chalkboard. Ask students to add to or revise their original descriptions of North and South Korea.

ASSESSMENT
1. Have students complete the following questions from the Chapter 23 Check:
   Building a Vocabulary 4
   Recalling and Reviewing 4
   Critical Thinking 6
2. Have students complete the Chapter 33 Review/Test.

RETEACHING PLAN

1. Ask students to complete the Chapter 33 Reteaching Worksheet, Lesson 4.
2. Have students create a product map of North and South Korea, using symbols and a map legend.

CHALLENGE/ENRICHMENT STRATEGY

1. Ask students to research and report on major events of the Korean War and the extent of the United States' involvement.
2. Have students complete the Chapter 33 Challenge Worksheet.

Chapter 33 Check, page 405, Answers

Building a Vocabulary

1. Large sea waves caused by earthquakes and volcanic eruptions.
2. Financial support given by a government. Because the Japanese farmer is protected by high subsidies for rice, for example, the Japanese consumer pays more for rice, but foreign competition is kept out.
3. Japan has a huge trade surplus. Japan could import more products.
4. Most of a country's products are for export to other nations. South Korea. Electronics, steel, automobile, shipbuilding, and textile industries.

Recalling and Reviewing

1. Japan is made up of four large islands and many smaller ones, and more than 70 percent of Japan is mountainous. Most Japanese live along the narrow coastal plains, where Japan's urban and agricultural areas are found.
2. Buddhism, writing system, system of rule by imperial dynasties before the eighth century.
3. Because the Japanese share the same language and culture, communication and the setting of common goals is possible. Japanese society is efficient, as Japanese people generally are well educated and have a strong work ethic. Japanese industry is newer and more efficiently run than many factories in the United States and Western Europe. Because Japanese workers take part in decision making in their companies, they are loyal to their employers. The Japanese government works closely with industry and provides financial aid. The Japanese have been able to copy Western products, upgrade them, and sell them at lower prices. They also have taken the lead in new technological fields.
4. South Korea. The nation is self-sufficient in food production, is an industrial nation, and has an urbanized middle class and political and religious freedoms.

Critical Thinking

5. Because Japan has been successful. It is possible, but Japan is successful in part because of unique aspects of Japanese culture.
6. Answers will vary. The beginning of trade and communication might make reunification possible.

Using Geography Skills

1. Sapporo: Hokkaido; Hiroshima: Honshu; Nagasaki: Kyushu; Kyoto: Honshu.
2. Industrial areas: Kitakyushu, Osaka, Sakai, Kobe, Hiroshima, Kyoto, Nagoya, Yokosuka, Yokohama, Kawasaki, Tokyo, Funabashi, Chiba, Sapporo, Okayama. Farming areas: same as industrial cities, as well as: Sasebo, Kagoshima, Kumamoto, Sendai, Akita, Hachinohe.

CHAPTER 34 Southeast Asia Pages 406–417

PLANNING GUIDE		
Lesson	**Textbook**	**Teacher's Resource Binder**
1. History and Landforms	pp. 406–407	Lecture Notes and Transparency Package Content/Vocabulary Workbook Reteaching Worksheet
2. Climate and Resources	pp. 407–409	Skills Worksheet Reteaching Worksheet Critical Thinking Worksheet
3. Country Profiles	pp. 410–416	Chapter Review/Test Reteaching Worksheet Challenge Worksheet

TIME ALLOCATION: 4 days

CHAPTER TOPICS

- Historical Geography
- Physical Geography
- Economic Geography
- Burma
- Thailand
- Kampuchea
- Laos
- Vietnam
- Malaysia
- Singapore and Brunei
- Indonesia
- The Philippines

PRETEACHING CHAPTER VOCABULARY

Write the following terms on the chalkboard:
slash-and-burn farming archipelago
intensive agriculture

Ask students to use context clues and root-word clues to determine the definitions of *slash-and-burn farming* and *intensive agriculture*. Make certain that students are able to distinguish between these two. Locate an *archipelago* on a map, and ask students to create their own definition.

► LESSON 1
History and Landforms

LESSON OBJECTIVES

The student should be able to

- define significant terms of the chapter
- locate important physical features of this region

- locate the nations of this region
- understand criteria for determining physical regions
- place historical events in a sequential order

PURPOSE

The study of Southeast Asia will help students understand the importance of this region to the world and the reasons for its past involvement in wars.

MOTIVATOR

Refer students to the following statement on page 406 of their textbook: "Southeast Asia is also strategically located between the Indian and Pacific oceans." Refer students to a wall map of the world, and ask them to compare the strategic location of these countries to that of the Middle East. Ask students for similarities and differences between the value of the location of each of these regions. Indicate that this knowledge provides a framework for understanding the world's concerns about Southeast Asia.

TEACHING DESIGN

1. Before class, prepare cutout shapes of each country in Southeast Asia. Refer students to the map "Southeast Asia" on page 409 of their textbook. Ask them to take three minutes to memorize the location and names of the countries in this region. As students are studying, distribute the cutouts in envelopes to various students around the classroom. At the end of three minutes, call on a volunteer who has an envelope. Explain that each individual with an envelope is expected to come to the chalkboard, identify the country in the envelope, and indicate where it is located in relation to all other countries in Southeast Asia. The countries should fit together at the conclusion of the exercise.
2. Provide all students with a physical map of Southeast Asia. Announce to students that this region of the world is usually divided into three physical regions. Explain that it is their task to determine where these three regions probably occur. Give each group approximately five minutes to discuss their ideas. Then call on each group to share their hypotheses. Write all the hypotheses on the chalkboard, and attempt to reach a class consensus. Finally, refer all students to page 407 of their textbook to assess the accuracy of their decisions.
3. Use the Lecture Notes and Transparency Package to present the cultural and physical geography of Southeast Asia. Have students take notes.
4. Call upon students who were assigned the topics Family and Culture and Physical Setting in Teaching Design 3, Lesson 1, Chapter 30 to make their presentations.

CLOSURE

Refer students to the discussion in Chapter 30 of religious systems of the Orient, pages 362–363. Call out each Southeast Asian country, and ask students to identify the predominant religion of that country.

ASSESSMENT

1. Have students complete the following questions from the Chapter 34 Check:
   Recalling and Reviewing 1
   Critical Thinking 5
   Using Geography Skills 1
2. Ask students to complete the Chapter 34 Content/Vocabulary Worksheet in their Workbook.

RETEACHING PLAN

Ask students to complete the Chapter 34 Reteaching Worksheet, Lesson 1.

CHALLENGE/ENRICHMENT STRATEGY

1. Invite students to create a diagram that represents how Southeast Asian languages, religions, art, architecture, and foods reflect the influences of their neighbors.
2. Challenge students to create a map that represents all the groups who came to settle in Southeast Asia and where they originated.

▶ LESSON 2
Climate and Resources

LESSON OBJECTIVES

The student should be able to

- describe the physical setting of Southeast Asia
- locate the major natural resources and give their uses
- analyze the impact of environment on ways of life
- describe the agricultural base of this region
- examine conservation concerns in the region

PURPOSE

In this lesson, students will learn of the influence that environment has on people's ways of life in Southeast Asia.

MOTIVATOR

Point out to students that a portion of Southeast Asia lies within what is sometimes referred to as the "Ring of Fire." Point out this area on a wall map. Ask students to discuss with a classmate possible reasons for this name. Then ask students to share their ideas. Discuss the correct response with the class. Ask students to identify other places where people live near active volcanoes and earthquakes. (Japan, Mexico) Indicate to students that this is another example of how the environment can influence people's way of living. Discuss how the constant threat of earthquakes and volcanic eruptions may have affected the way Southeast Asians live.

TEACHING DESIGN

1. Pose the following questions to the class: Near what important geographic feature of the world is Southeast Asia located? (equator) Judging from its location, what are the climate regions you would expect to find in this area? Ask students to list, with a partner, the advantages and disadvantages of living within a humid tropical climate region. Discuss their ideas.
2. Ask students to identify the three types of agriculture that occur in Southeast Asia. List these three types on

the chalkboard. Ask each student to create a cluster diagram to represent the characteristics of each type.

3. Ask students to identify any evidence in the chapter that indicates a conservation problem in Southeast Asia. Discuss why conservation appears to be as great an issue in less developed nations as in developed nations.
4. Call upon students assigned the topics Government and Economy in Teaching Design 3, Lesson 1, Chapter 30 to make their presentations. Have students take notes.

CLOSURE
Ask students to prepare summaries for this lesson under the headings Climate, Environment, and Conservation.

ASSESSMENT
1. Have students complete the following questions from the Chapter 34 Check:
   Building a Vocabulary 1, 2
   Recalling and Reviewing 2
   Using Geography Skills 2, 3
2. Have students complete the Chapter 34 Skills Worksheet.

RETEACHING PLAN
Ask students to complete the Chapter 34 Reteaching Worksheet, Lesson 2.

CHALLENGE/ENRICHMENT STRATEGY
1. Challenge students to prepare a travel brochure about Southeast Asia, highlighting the points of interest.
2. Have students complete the Chapter 34 Critical Thinking Worksheet.

▶ LESSON 3
Country Profiles

LESSON OBJECTIVES
The student should be able to
- locate major natural resources of Southeast Asia
- locate the major nations of this region
- describe the major economic activities of this region
- explain the economic importance of each country

PURPOSE
In this lesson, students will gain further understanding of the significance of Southeast Asia to the rest of the world.

MOTIVATOR
Divide students into groups of five. Give the groups 10 minutes to brainstorm about all the reasons that Southeast Asia is significant in relation to the world. Bring in resource materials and newspapers to help students identify reasons. Remind students also to use their textbook. At the end of the brainstorming session, create a composite list.

TEACHING DESIGN
1. List the countries of Southeast Asia on the chalkboard. Divide students into enough groups to ensure that all countries are covered. Ask each group to keep its assigned country a secret. Indicate to students that their first task is to take the name of their country and use the letters in its name to write fact clues describing it, as in the following example.
   B uddhists with a rich culture live here.
   U rban development is in Rangoon and Mandalay.
   R angoon is the capital.
   M ountainous western Southeast Asia is its location.
   A ll business is controlled by the government.
2. Explain to students that a game will proceed as follows: Group 1 will be selected to begin, but all groups will participate. Group 2 should call out a letter of the alphabet. If the letter is included in the name of Group 1's country, Group 1 is obligated to read its clue for the letter. If Group 2 correctly identifies Group 1's country, Group 2 wins points for all the letters in the name of the country, even those that have not been identified. If Group 2 is unable to identify Group 1's country, the next group (Group 3) gets a turn. When a country is identified, the identifying group is invited to share clues about its country. Proceed with the game until all countries are identified, and announce the winning group.
3. Call upon the groups who were assigned the topics Current Issues and Religion in Teaching Design 3, Lesson 1, Chapter 30 to make their presentations to the class.

CLOSURE
Call out the names of the countries of Southeast Asia, and call upon individuals to name significant facts they remember from the lesson.

ASSESSMENT
1. Have students complete the following questions from the Chapter 34 Check:
   Recalling and Reviewing 3, 4
   Critical Thinking 6
   Using Geography Skills 4, 5
2. Ask students to complete the Chapter 34 Review/Test.

RETEACHING PLAN
1. Suggest that students use their textbook to create a diagram that represents one significant fact they would like to remember about each country in Southeast Asia.
2. Ask students to complete the Chapter 34 Reteaching Worksheet, Lesson 3.

CHALLENGE/ENRICHMENT STRATEGY
1. Ask students to create a chart to represent the economic system of each country located in Southeast Asia.
2. Have students complete the Chapter 34 Challenge Worksheet.

Chapter 34 Check, page 417, Answers

Building a Vocabulary
1. In slash-and-burn farming, forests are burned, and crops are planted and then shifted to another location when the soil nutrients are depleted. Intensive agriculture, which requires much human labor, is most efficient. Yes. Slash-and-burn farming occurs in areas where population density is low, whereas intensive agriculture occurs where population density is high.

2. A large group of islands. Indonesia. Indonesia's 13,500 islands extend 3,000 miles (4,800 kilometers) between the Indian and Pacific Oceans.

Recalling and Reviewing

1. Mountains of the north: steep, though not very high, and covered with dense forests. The ranges fan out from the Himalayas and the Plateau of Tibet and extend southward, dividing Southeast Asia's northern valleys. Plains and plateaus of the center: the plains are flat, and the plateaus are low. Volcanic islands to the south and east: this region comprises thousands of volcanic islands, some of which have one volcano, while others have hundreds. Some of these islands have rich volcanic soil, but earthquakes and volcanic eruptions are common.
2. It provides crops for export. Coconuts (for copra), coffee, oil palms, sugarcane, and rubber.
3. Conflict in these countries has interfered with agriculture and has destroyed much existing industry, particularly in Vietnam. The conflict has taken time and resources that might have been used for economic development. All three countries are now faced with the problem of rebuilding their economies.
4. Conflicts over land ownership, rebel movements of Muslims and Communist guerrillas, and establishment of democratic government.

Critical Thinking

5. Answers will vary.
6. Answers will vary.

Using Geography Skills

1. Song Koi (Red River), Mekong, Nu, Irrawaddy, Chao Phraya, Chindwin. Mekong. Chao Phraya.
2. Through the Strait of Malacca between Sumatra and the Malay Peninsula, north to the South China Sea.
3. Forests: Thailand, Burma, Malaysia, Indonesia, Laos. Rice: Burma, Vietnam, the Philippines, Thailand.
4. Burma (between the Nu and Irrawaddy rivers), Gulf of Siam, Indonesia.
5. Luzon, Mindanao. Luzon.

UNIT 6
Review/Test

PLANNING GUIDE		
Lesson	**Textbook**	**Teacher's Resource Binder**
1. Unit 6 Review	pp. 418–419	
2. Unit 6 Test		Unit 6 Test

TIME ALLOCATION: 2 days

▶ LESSON 1
Unit 6 Review

Have students read the Unit 6 Summary. Use the Unit 6 Objectives as the basis for review, and discuss any questions that students have on the Unit 6 material.

Have students complete the questions in the textbook under the following headings:
Reading and Understanding
Mastering Geography Skills

Have students read the textbook material under the following headings and complete the appropriate or selected activities:
Applying and Extending
Linking Geography and Art

▶ LESSON 2
Unit 6 Test

Have students complete the Unit 6 Test from the Teacher's Resource Binder.

Unit 6 Review, pages 418–419, Answers

Reading and Understanding

1. The volcanic mountains have deposited rich volcanic soils. These active volcanoes, however, bring destructive eruptions and earthquakes. Many of the nonvolcanic islands are covered with thick rain forests, which provide valuable wood resources; however, because the mountain areas are not agriculturally productive, food must be produced in the lowland areas.
2. The summer monsoon usually brings rain, and the winter monsoon usually brings cool and dry weather, except to Japan and the Philippines.
3. Food supply has not been able to keep pace with population growth, and most of the best land is already under cultivation. Food imports use resources that might better be used in economic development. Countries in the region have attempted to limit the birthrate and have begun to plant high-yield grains as a result of the Green Revolution. Some countries, such as China, have combined small, inefficient plots into large communes.
4. Japan, Taiwan, Singapore, India, South Korea, and the Philippines, as well as the colony of Hong Kong.
5. Most developed: Japan and Singapore; these countries have been aided by good location for trade, the work ethic of the people, and government sponsorship of economic development. Least: Afghanistan, Bangladesh, Bhutan, and Nepal; these countries have been hindered by landforms and climate that make productive agriculture difficult, isolation from the overpopulation.
forces of modernization, and, in the case of Bangladesh,
6. Malaysia has large deposits of tin, which it mines and exports to pay for food imports; Brunei has large depos-

its of petroleum and natural gas, which it exports; Indonesia has large deposits of oil, tin, bauxite, and natural gas, which it exports; Burma has oil, lead, zinc, silver, and gemstones, but they are not fully developed; China has tungsten, oil, coal, iron ore, lead, and manganese, which it has used as the basis for industrialization.

7. Political instability has caused people to flee their countries and become refugees. Agriculture has been disrupted, and industry has been damaged by conflict, particularly in Vietnam. Political instability of the area has delayed economic development.

Mastering Geography Skills

1. Vertical: imports and exports in billions of U.S. dollars; horizontal: U.S. trade with Japan from 1968 to 1984.
2. 1968: about $3 billion; 1984: about $23 billion.
3. From 1976 to 1980, and from 1980 to 1984.
4. The United States has a trade imbalance with Japan because it imports much more than it exports, whereas Japan exports much more than it imports; this imbalance causes problems for both countries.

UNIT 7
The Pacific World
Pages 420–447

UNIT OVERVIEW

Australia, New Zealand, and the islands of the South Pacific were among the last regions of the world to be colonized. Australia is a modern, industrialized nation. New Zealand and the Pacific Islands remain heavily agricultural. All countries have played significant roles in the history of the world and today maintain trade and political relationships with important world nations.

CHAPTER TIME LINE

Chapter	Title	Time Allocation
35	Australia	4 days
36	New Zealand and the Pacific Islands	4 days
	Unit 7 Review	1 day
	Unit 7 Test	1 day
		10 days

MAJOR TOPICS

- Relationship Between Geography and History
- Impact of Geography on Economic Development
- Significance of the Region to the World Market
- Key Issues Facing the Nations of the Pacific World

INTRODUCING THE UNIT

Ask students to refer to the definition of the word *pacific* in the dictionary. Discuss the meaning, and brainstorm about possible reasons that this name was given to this ocean and group of islands. Relate it to the idea that this area was among the last parts of the world to be settled.

CHAPTER 35 Australia
Pages 422–433

PLANNING GUIDE		
Lesson	**Textbook**	**Teacher's Resource Binder**
1. Historical Geography and Landform Regions	pp. 422–424	Lecture Notes and Transparency Package Content/Vocabulary Workbook Reteaching Worksheet
2. Biogeography and Climate Regions	pp. 424–427	Reteaching Worksheet Critical Thinking Worksheet
3. Economic Geography	pp. 427–432	Skills Worksheet Reteaching Worksheet Challenge Worksheet
4. Urban Geography, Issues, and Sydney	pp. 430–432	Chapter Review/Test Reteaching Worksheet

TIME ALLOCATION: 4 days

CHAPTER TOPICS

- Historical Geography
- Physical Geography
- Economic Geography
- Urban Geography
- Future Issues

PRETEACHING CHAPTER VOCABULARY

Write the following terms on the chalkboard:

biogeography, reef, artesian well, marsupial, coral, hinterland

biogeography Ask students to guess a definition for this term, based upon their knowledge of the prefix *bio-* and the word *geography*.

marsupial Show several pictures of marsupials. Have students note common characteristics.

reef Display several pictures of reefs, and ask students to identify their characteristics.

coral Show students examples or pictures of various kinds of coral. Ask students to identify their characteristics.

artesian well Refer students to the definition of this term in their textbook Glossary.

hinterland — Refer students to the Glossary definition of *hinterland,* and ask them to compare it to synonyms commonly used in the United States. (*interior, outskirts*)

▶ LESSON 1

Historical Geography and Landform Regions

LESSON OBJECTIVES

The student should be able to

- define significant terms of the chapter
- identify in sequence key events in history
- locate and describe the major landforms of Australia

PURPOSE

This lesson will allow students to study an industrialized nation that exists outside the Northern Hemisphere and that has developed despite a lack of arable land.

MOTIVATOR

Use the Chapter 35 transparency of a physical map of Australia, and make a large cutout of Australia on poster board. Ask students to study the transparency for five minutes. Following the study period, host a contest to see who is able to remember the most places on the map of Australia. Label the sites correctly on the large cutout. Then ask students to identify questions they have about Australia. Write their questions on the back of the poster-board cutout.

TEACHING DESIGN

1. Present the Chapter 35 Lecture Notes and Transparency Package. Have students take notes.
2. Draw the following chart on the chalkboard.

	British Colonization of the United States	British Colonization of Australia
Purpose		
Time period		
Settlement pattern		
Results		

   Lead a class discussion comparing the British colonization of Australia with that of the United States.
3. Refer students to the map "Australia" on page 428 of their textbook, and ask them to practice locating the following sites: Mount Kosciusko, Darling Range, Great Australian Bight, Gulf of Carpentaria, Macdonnell Ranges, Hamersley Range, Great Artesian Basin, Lake Eyre, Musgrave Ranges, Flinders Range, Kimberley Range, Australian Alps, Great Dividing Range, Gulf of St. Vincent, Spencer Gulf.
4. Display a wall map of Australia. Direct students to listen carefully as you read aloud descriptions of the three major landform regions. Have students work in groups of three and, on the basis of the clues, determine the boundaries for the landform regions.

   Compare the divisions made by the students to the actual divisions on the economic map "Australia" on page 428 of their textbook. Discuss reasons for the discrepancies and similarities between them.

CLOSURE

Ask several students to point out on a wall map of Australia the three basic landform regions. Ask them to identify any significant landforms they recall from the lesson.

ASSESSMENT

1. Have students complete the following questions from the Chapter 35 Check:
   Recalling and Reviewing 1
   Using Geography Skills 2, 3
2. Ask students to complete the Chapter 35 Content/Vocabulary Worksheet in their Workbook.

RETEACHING PLAN

1. Ask students to complete the Chapter 35 Reteaching Worksheet, Lesson 1.
2. Provide students with a time line containing key dates in Australian history, and ask them to work with a partner to label the events that correspond to each date.

CHALLENGE/ENRICHMENT STRATEGY

1. Challenge students to investigate the legal differences between states and territories in Australia as they relate to Great Britain.
2. Encourage students to discover the reasons for the influx of Europeans to Australia in the 1860s.

▶ LESSON 2

Biogeography and Climate Regions

LESSON OBJECTIVES

The student should be able to

- identify the major factors influencing the climates of Australia and their effects
- explain the effect of bodies of water on development

PURPOSE

This part of the study of Australia will identify for students some of the unique qualities of the region, as well as characteristics that have developed as a result of Australia's isolation from the rest of the world.

MOTIVATOR

Refer students to the climate map "Australia and the Southwest Pacific" on page 427 in their textbook, and ask them to predict where the majority of Australia's population lives. Have students state the geographical clues that helped them make their predictions. Have them compare their predictions to the population-density map "Australia and the Southwest Pacific" on page 429.

TEACHING DESIGN

1. Write on the chalkboard: Exotic Animals, Native Animals and Plants. Ask students to review the information in the chapter to identify examples of each category. Discuss the distinction between native and exotic.
2. Ask students to fill in the blanks in the sentence below. Because Australia is in the (Southern) Hemisphere, its seasons are (the opposite) of those in the United States. Discuss the possible implications of this statement for trade between the two nations.
3. Ask students to answer the following questions pertaining to climate regions found in Australia. Encourage students to work with a partner. Involve them in a class discussion to ensure that the correct answers were discovered.
   a. In what region will you find a hot climate with desert shrubs and scarcely any rain?
   b. In what climate region will you find very hot summers, very cold winters, light rainfall, grazing land, grasses, shrubs, and trees along streams?
   c. In what climate region will you find very wet and very dry seasons with tall, rough grass and adequate trees for raising cattle?
   d. In what region will you find a rainy, hot climate with tall trees, vines, and plantations?
   e. Where will you find rainy, moderate temperatures, mixed forests, and land suitable for wheat farming?
   f. Where will you find mild winters with rain and long, dry, sunny summers?

CLOSURE

Refer students to the map "World Climate Regions" on page 66-67 and to the Atlas map "The World: Political" on pages 4-5 of their textbook. Ask them to locate cities in the rest of the world with climates that are comparable to cities in Australia.

ASSESSMENT

Have students complete the following questions from the Chapter 35 Check:
Building a Vocabulary 1, 2, 3, 4
Recalling and Reviewing 3

RETEACHING PLAN

1. Ask students to complete the Chapter 35 Reteaching Worksheet, Lesson 2.
2. Direct students to create a map of Australia that represents its various climate regions and life in each.

CHALLENGE/ENRICHMENT STRATEGY

1. Ask students to select five cities in Australia and prepare a weather report for each city for each of the following days: September 1, January 1, April 1, and July 1.
2. Have students complete the Chapter 35 Critical Thinking Worksheet.

▶ LESSON 3

Economic Geography

LESSON OBJECTIVES

The student should be able to

- describe the characteristics of an industrialized nation
- describe the major economic activities in Australia
- list the economic resources available in Australia

PURPOSE

This lesson will enable students to understand the significance and benefits of living in an industrialized nation.

MOTIVATOR

Assign students to groups of three. Ask groups to create a list of characteristics of the world's most industrialized nations. Have a representative from each group present its list, and create a master list on the chalkboard.

TEACHING DESIGN

1. Refer students to the master list that they created in the Motivator. Ask them to classify the list into three categories: Economic, Political, and Cultural. Discuss what it means to live in an industrialized nation.
2. Ask students to use their textbook and additional resources to create a diagram representing the export-import relationships among Australia, the United States, the Soviet Union, China, and the Middle East.
3. Ask students to work in groups to identify significant facts that they believe are important to remember about Australia, using the following chart.

Agriculture	Ranching	Mining	Services

Discuss the advantages and disadvantages of each way of earning a living in Australia. Identify which are increasing in importance and which are declining.

CLOSURE

Select representatives from each group to share important facts that they included on their charts. Create a class summary chart on the chalkboard.

ASSESSMENT

1. Have students complete the following questions from the Chapter 35 Check:
   Recalling and Reviewing 2
   Critical Thinking 5
2. Have students complete the Chapter 35 Skills Worksheet.

RETEACHING PLAN

Ask students to complete the Chapter 35 Reteaching Worksheet, Lesson 3.

CHALLENGE/ENRICHMENT STRATEGY

1. Have students complete the Chapter 35 Challenge Worksheet.
2. Challenge students to find out which American companies have offices in Australia.
3. Invite a former resident of Australia or someone who has recently visited there to address the class.

► LESSON 4
Urban Geography, Issues, and Sydney

LESSON OBJECTIVES

The student should be able to

- describe the relationship between Australia and the rest of the world
- locate and describe major landforms in Australia
- describe characteristics of Sydney, Australia

PURPOSE

This lesson will enable students to understand the significance and benefits of living in an industrialized nation.

MOTIVATOR

Prior to this lesson, locate advertisements that are designed to attract tourists to Australia. Display the advertisements for students, and ask them to suggest reasons that Australia may appeal to some people as a place to visit. List students' ideas on the chalkboard, and ask them to identify attractions in Australia that are similar to those in the United States and ones that are unique.

TEACHING DESIGN

1. Provide students with a map of Australia, and ask them to locate the states and their capitals. Then ask students to work with a partner, referring to their textbook, to answer the following questions.
   a. Which states have natural boundaries?
   b. Which states have the greatest population?
   c. Where does most of Australia's industry occur?
   d. What is the capital of Australia?
2. Provide resource books on Australia for the class. Ask students to select one of the state capitals and prepare a brochure describing a reason to visit their selected city. Invite students to share their brochures.

CLOSURE

Ask students to reflect on the future of Australia. Summarize the chapter by asking students to imagine Australia in the year 2000. Have them describe the positive and negative aspects of life; list responses on the chalkboard.

Check the back of the cardboard cutout of Australia for questions that students had in the Lesson 1 Motivator. Determine that all questions have been answered.

ASSESSMENT

1. Have students complete the following questions from the Chapter 35 Check:
   Building a Vocabulary 5
   Recalling and Reviewing 4
   Using Geography Skills 1
2. Ask students to complete the Chapter 35 Review/Test.

RETEACHING PLAN

1. Ask students to complete the Chapter 35 Reteaching Worksheet, Lesson 4.
2. Have students draw a map of Australia that includes important landforms, bodies of water, states, territories, and cities. Ask students to label all major sites.

CHALLENGE/ENRICHMENT STRATEGY

1. Challenge students to draw comparisons between the history of Sydney and that of three American cities.
2. Direct students to plan for a seven-day survival course in the Outback. Have them create a list of the 10 most important items to pack.
3. Invite students to investigate further and report on the state of the Aborigine culture in Australia today.

Chapter 35 Check, Page 433, Answers

Building a Vocabulary

1. The study of the geographic distribution of plants and animals. Because of its isolation from other continents.
2. Marsupials carry their young in pouches. Kangaroos and koalas.
3. A ridge of coral, rock, or sand lying at or near the surface of the water. Coral.
4. In Queensland. Rain falling on the mountains to the east enters an aquifer and flows westward into the desert.
5. hinterland

Recalling and Reviewing

1. The Western Plateau, the Central Plains, and the Eastern Highlands. The Western Plateau includes Western Australia, the Northern Territory, and most of South Australia. The Central Plains extends from the Gulf of Carpentaria in the north to the Great Australian Bight in the south. The Eastern Highlands, extends from Cape York to Victoria along Australia's east coast.
2. Sheep and cattle are raised in the Outback. Mining is also important.
3. The Great Dividing Range on the east coast is in the path of the moist southeast trade winds, preventing their reaching the leeward side of the range and the interior.
4. In urban areas. They have full citizenship, and the government has turned over one-third of the land in the Northern Territory to them.

Critical Thinking

5. The lack of water. Answers will vary.

Using Geography Skills

1. Along the southeast coast and the Great Australian Bight near Sydney, Melbourne, and Brisbane, and along the

southwest coast near Perth. Humid subtropical, maritime, Mediterranean.

2. States: Western Australia, Queensland, South Australia, Victoria, Tasmania, and New South Wales; territories: Northern Territory and Australian Capital Territory. Darwin: Northern Territory; Sydney: New South Wales; Perth: Western Australia.
3. Along Australia's northeastern coast. More than 1,250 miles (2,000 kilometers).

CHAPTER 36 New Zealand and the Pacific Islands Pages 434–445

PLANNING GUIDE		
Lesson	Textbook	Teacher's Resource Binder
1. Overview and Regions	p. 434	Content/Vocabulary Workbook Reteaching Worksheet
2. New Zealand	pp. 435–436	Reteaching Worksheet Critical Thinking Worksheet
3. Islands of the Pacific	pp. 437–444	Lecture Notes and Transparency Package Skills Worksheet Chapter Review/Test Reteaching Worksheet Challenge Worksheet

TIME ALLOCATION: 4 days

CHAPTER TOPICS

- New Zealand
- Islands of the Pacific

PRETEACHING CHAPTER VOCABULARY

Write the following terms on the chalkboard:

geothermal   continental island
oceanic island   atoll

Review the prefix and root of *geothermal. Geo-* relates to earth, and *thermal* relates to heat. Identify other areas of the world with geothermal energy. (Iceland) Brainstorm with students about possible differences between *oceanic islands* and *continental islands*. Indicate which of their ideas are accurate. Refer students to the diagram on page 442 in their textbook, which depicts the formation of an atoll. Ensure students' comprehension, and announce that they will learn more about atolls in this chapter.

▶ LESSON 1
Overview and Regions

LESSON OBJECTIVES

The student should be able to

- define significant vocabulary terms
- locate New Zealand and the Pacific islands
- describe the basis for identification of Pacific regions

PURPOSE

This lesson will reinforce students' acquaintance with the largest geographic feature on Earth and will expand their awareness of important places of the earth.

MOTIVATOR

Display a world map and a globe. Invite students to come to the map and globe and locate the Pacific Ocean, New Zealand, and the Pacific islands. Discuss with students the clues they use to remember the locations of important places in the world.

TEACHING DESIGN

1. Refer students to the map "Pacific Islands" on page 440 of their textbook. Ask students to locate these three regions: Melanesia, Micronesia, and Polynesia.
2. Lead a class discussion about the basis on which regional boundaries were established in the Pacific Ocean. Compare this rationale with those for boundaries previously studied in the chapters on Africa and Australia.
3. Divide the class into groups of three. Write one of the following topics on each of 10 index cards: New Zealand—Physical Geography; New Zealand—History and People; New Zealand—Economic Geography; New Zealand—Social and Political Issues; Islands of the Pacific—Physical Geography; Islands of the Pacific—History and People; Islands of the Pacific—Tonga, Western Samoa, and American Samoa; Islands of the Pacific—Melanesia; Islands of the Pacific—Micronesia; Islands of the Pacific—Polynesia. Provide one card to each group. Ask each group to review in their textbook the information describing their topic. Indicate that students are to prepare a three-minute presentation of the information, which they will give during Lesson 4.

CLOSURE

Ask students to review in their groups the location of the Pacific islands, New Zealand, and the regional boundaries.

ASSESSMENT

1. Have students complete the following questions from the Chapter 36 Check:
   Recalling and Reviewing 3
2. Ask students to complete the Chapter 36 Content/Vocabulary Worksheet in their Workbook.

RETEACHING PLAN

Ask students to complete the Chapter 36 Reteaching Worksheet, Lesson 1.

CHALLENGE/ENRICHMENT STRATEGY

1. Select students to report on how the introduction of farming, tourism, and military bases has affected the Pacific islands.
2. Invite students to research and share information about vacation sites in the Pacific.

New Zealand

LESSON OBJECTIVES

The student should be able to

- locate important geographic features of New Zealand
- describe the economic activity in New Zealand
- discuss the interchange between New Zealand and the rest of the world
- cite the major events in the history of New Zealand
- analyze population patterns of New Zealand

PURPOSE

In this lesson, students will practice their skills at analyzing the relationship between geographic features and population patterns.

MOTIVATOR

Refer students to the economic map "New Zealand" on page 436 of their textbook. Ask students to identify features that they believe would affect population-density patterns. List their ideas on the chalkboard. Refer students back to the map, and ask them to predict where the majority of people live in New Zealand. Have them check the validity of their ideas against the population-density map "Australia and the Southwest Pacific" on page 429.

TEACHING DESIGN

1. Draw the following chart on the chalkboard.

	New Zealand	
	North Island	**South Island**
Geography		
History		
Economy		
Culture		

   Ask students to recall the information presented in the chapter and to cite the main differences they studied. Then direct students to imagine the following situation: You have been given a major promotion offer at Worldwide Research Laboratories. The offer includes a transfer to an office in New Zealand. Prepare a short speech that you will present to your family in an attempt to convince them that you should accept this offer.
2. Invite students who prepared reports on New Zealand to present them to the class. Following the report of each group of students, ask remaining members of the class to write three questions about the presentations given.
3. After all presentations have been made, ask students to trade their questions with a nearby classmate. Give students time to answer the questions.

CLOSURE

Select several students to share one of their questions, and direct the rest of the class to respond with the answers.

ASSESSMENT

1. Have students complete the following questions from the Chapter 36 Check:
   Building a Vocabulary 1
   Recalling and Reviewing 1, 2
   Critical Thinking 5
   Using Geography Skills 1, 2

RETEACHING PLAN

1. Ask students to complete the Chapter 36 Reteaching Worksheet, Lesson 2.
2. Collect the questions that students developed in response to the presentations. Select the best questions on each topic, and give students a chance to respond.

CHALLENGE/ENRICHMENT STRATEGY

1. Have students complete the Chapter 36 Critical Thinking Worksheet.
2. Choose students to create a poster illustrating all the products that come from sheep.

▶ LESSON 3

Islands of the Pacific

LESSON OBJECTIVES

The student should be able to

- locate the major Pacific Islands
- describe the historical development
- discuss the economic life on the islands
- identify significant geographic features
- describe the relationship between the United States and the Pacific islands
- distinguish between oceanic and continental islands

PURPOSE

By the conclusion of this lesson, students will have knowledge of the unique relationship between the United States and many of the Pacific islands.

MOTIVATOR

Ask students to reflect on what they believe to be true about the Pacific islands. Using the words *Pacific islands* as a base, create a word web on the chalkboard to illustrate their ideas. Indicate to students that they should refer to the web throughout the lesson and see if any of their ideas change.

TEACHING DESIGN

1. Present the Lecture Notes and Transparency Package to develop the concepts of oceanic and continental islands. Mark the islands in chalk on a classroom wall map.
2. Refer students to the special feature "Life on an Atoll" on pages 442–443. Ask students how their views of the world would differ if they had grown up on an atoll.
3. Allow students to present their three-minute reports assigned in Lesson 1, Teaching Design 3. Ask students to

write three important ideas that they need to remember from each speech.
2. Have students complete the Chapter 36 Skills Worksheet.

CLOSURE
Invite several students to read their three statements of important points from each report topic.

ASSESSMENT
1. Have students complete the following questions from the Chapter 36 Check:
   Building a Vocabulary 2, 3
   Recalling and Reviewing 4
   Critical Thinking 6
2. Ask students to complete the Chapter 36 Review/Test.

RETEACHING PLAN
1. Ask students to complete the Chapter 36 Reteaching Worksheet, Lesson 3.
2. On several index cards, list characteristics of oceanic and continental islands, and have students practice placing them in the correct category.

CHALLENGE/ENRICHMENT STRATEGY
1. Have students complete the Chapter 36 Challenge Worksheet.
2. Ask students to write a fictitious page from a diary, describing one day on an atoll.
3. Choose volunteers to investigate and report on the naval campaign in the Pacific during World War II.

Geography Skills, pages 438–439, Answers
1. The prime meridian. 0°.
2. 7 A.M., Friday.
3. Four. Students should refer to the map and the text to answer the second part of the question.
4. 11 A.M. 8 A.M. 8 A.M. 1 P.M.

Chapter 36 Check, page 445, Answers

Building a Vocabulary
1. Heat escaping from the earth's interior, causing hot springs, geysers, and steam vents. North Island.
2. Oceanic islands are made of volcanic materials from the ocean floor; they are mountainous and forested, they have fresh water and good soils, and they can support human populations. Continental islands are made from sections of ancient continents. Oceanic: Tahiti, Hawaii, Guam; continental: New Guinea, New Caledonia.
3. A ring of small coral islands built up on a coral reef surrounding a shallow lagoon. No.

Recalling and Reviewing
1. New Zealand is a major exporter of lamb meat, and the second largest wool producer after Australia. Live sheep are exported, mainly to the Middle East.
2. Japan, Australia, and the United States.
3. Melanesia, Micronesia, and Polynesia.
4. If they are high islands, moist air rises over the mountains, causing the windward east side to be very wet. The leeward west side of the islands is much drier. Rainfall varies greatly over short distances on these islands.

Critical Thinking
5. Answers will vary. Dependence is generally not a problem as long as New Zealand maintains good relations with its trading partners.
6. Answers will vary, but students should mention the ocean as both an isolating factor and one that has prompted migration and the diffusion of cultures.

Using Geography Skills
1. Wellington. North Island. Industry, fishing, mining, and raising livestock.
2. On South Island.

UNIT 7 Review/Test

PLANNING GUIDE		
Lesson	Textbook	Teacher's Resource Binder
1. Unit 7 Review	pp. 446–447	
2. Unit 7 Test		Unit 7 Test

TIME ALLOCATION: 2 days

▶ LESSON 1
Unit 7 Review
Have students read the Unit 7 Summary. Use the Unit 7 Objectives as the basis for review, and discuss any questions that students have on the Unit 7 material.

Have students complete the questions in the textbook under the following headings:
Reading and Understanding
Mastering Geography Skills

Have students read the textbook material under the following headings and complete the appropriate or selected activities:
Applying and Extending
Linking Geography and Culture

▶ LESSON 2
Unit 7 Test
Have students complete the Unit 7 Test from the Teacher's Resource Binder.

Unit 7 Review, pages 446–447, Answers

Reading and Understanding

1. Many died from diseases brought by Europeans. Many Maoris died fighting the British.
2. There was not enough incentive before the 1860s for many Europeans to come to Australia and New Zealand. Gold rushes in both countries and the expansion of farmland in Australia encouraged more settlement.
3. The industries and many minerals among the coasts provide jobs. The coastal location is beneficial for trade, and the interior of Australia is too dry to support large numbers of people.
4. The Great Dividing Range keeps the moist winds from the ocean from reaching the leeward side of the mountains or the interior of the country; thus, climates to the east of the Great Dividing Range are moist, and climates to the west of the range are dry. The Southern Alps have the same effect; when the westerlies rise over the Southern Alps, they drop rain on the west coast of South Island, while the leeward side of the mountains remains dry.
5. The Murray-Darling system is Australia's only major river system. The Great Artesian Basin is the largest source of underground well water in Australia.
6. Sheep and cattle are important to both countries, and both produce a variety of food crops for local use. Australia, with its many climate regions, can raise a wider variety, including tropical and subtropical crops. Agriculture in much of Australia is hampered by lack of water, while New Zealand has a moist climate that is excellent for agriculture. Agriculture is important to the economy of each country.
7. Efficient farmers, small populations, many different climate regions.
8. Australia is rich in mineral resources.
9. North Island is volcanic. It is subject to volcanic eruptions and geothermal activity in the form of hot springs and geysers.
10. Europeans established plantation agriculture, which has provided a source of income. It also has drawn workers from other areas of the world.
11. Subsistence farming, fishing, plantation farming, agriculture, military bases, and tourism. These sources of income are not generally sufficient to meet the needs of the island populations. There is high unemployment and little industry.
12. (c) a coral island usually in the shape of a ring.
13. A high island is also an oceanic island and is made of volcanic material. High islands are mountainous and forested, have fresh water and good soils, and can support human populations. Low islands, which are generally just above sea level, are made of coral. They have few resources, limited fresh water, thin soils, few trees, and they support only small human populations.
14. Polynesia. The Maoris and other Polynesian groups make up New Zealand's ethnic minority.

Mastering Geography Skills

1. Three (including solar time).
2. Europe, Africa.
3. Back.
4. 9 A.M.

UNIT 8
Latin America

UNIT OVERVIEW

The vast region known as Latin America extends from the northern border of Mexico to the southern tip of South America and includes the islands in the Caribbean Sea. The region is called Latin America because the majority of its people speak languages that are derived from Latin—principally Spanish and Portuguese. The region is one of great variety in landforms, vegetation, and climate. The people reflect their ancestral history: Indian, Spanish, Portuguese, and African. Latin America's modern cosmopolitan cities coexist with small, traditional villages.

The region is characterized by a rapidly growing population, political and social change, and an expanding middle class. Many of the nations of Latin America are considered to be developing, but some resources remain undeveloped. With greater political stability and continued economic development, this region of ancient civilizations should continue to grow in world prominence.

CHAPTER TIME LINE

Chapter	Title	Time Allocation
37	Latin America	3 days
38	Mexico	3 days
39	Central America and the Caribbean	3 days
40	Northern South America	2 days
41	The Andean Countries	2 days
42	Brazil	2 days
43	Southern South America	3 days
	Unit 8 Review	1 day
	Unit 8 Test	1 day
		20 days

MAJOR TOPICS

- Relationship among Land, Climate, and Ways of Life
- Historical Development and Cultural Interchange
- Political Stability and Economic Development
- Current Issues Facing Nations of the Region

INTRODUCING THE UNIT

Use a wall map of the world to introduce Latin America. Identify the area of the region on the map, and compare it in size to North America, Africa, Europe, and China. Ask students to name places they have heard of in this region, such as cities and geographic features. Call on volunteers to locate these places on the map.

CHAPTER 37 Latin America Pages 450–459

PLANNING GUIDE		
Lesson	Textbook	Teacher's Resource Binder
1. Historical Geography	pp. 450–453	Lecture Notes and Transparency Package Content/Vocabulary Workbook Reteaching Worksheet
2. Physical Geography	pp. 453–456	Critical Thinking Worksheet Reteaching Worksheet
3. Economic Geography and Current Issues	pp. 456–458	Skills Worksheet Chapter Review/Test Reteaching Worksheet Challenge Worksheet

TIME ALLOCATION: 3 days

CHAPTER TOPICS

- Historical Geography
- Physical Geography
- Economic Geography
- Social and Political Issues

PRETEACHING CHAPTER VOCABULARY

Write the following terms on the chalkboard:

indigo	mulattos	llanos
mestizos	pampas	labor intensive

Pronounce the terms, and ask students to identify words that are Spanish in origin. (*indigo, mestizos, mulattos, pampas, llanos*) Point out that these terms have been incorporated into the English language. Call on volunteers to define words they know. Provide the following definitions.

indigo (a deep blue dye used to color cotton and wool)
mestizos (people of mixed Indian and European ancestry)
mulattos (people of mixed white and black ancestry) The singular form is *mulatto*.
pampas (extensive grass-covered plains)
llanos (open grassy plains)
labor intensive (refers to work that requires manual labor) Ask students to name synonyms for *intensive*. (exhaustive, concentrated)

▶ LESSON 1
Historical Geography

LESSON OBJECTIVES

The student should be able to

- define the significant terms of the chapter
- identify and sequence key events in history
- identify cultural interchange among nations

PURPOSE

The study of Latin America will help students understand neighboring people in the Western Hemisphere and the cultural heritage of many people in the United States.

MOTIVATOR

Refer students to a wall map of the Western Hemisphere. Call on volunteers to identify the general area of Latin America on the map. Ask students to name the languages spoken by the people of this region. (Students may readily identify Spanish but may not be familiar with other languages, such as French, Portuguese, English, Quechua, and Dutch.) Ask students who the original inhabitants of this region were. (Indians) Ask students to suggest reasons for the existence of many different languages in the region.

TEACHING DESIGN

1. Present the Chapter 37 Lecture Notes and Transparency Package. Have students take notes.
2. Divide students into groups according to the number of nations in Latin America. Announce that each group is to determine artifacts that would be representative of the geography, history, people, economy, and future of its assigned country. Students should then collect and/or create the artifacts, which could be authentic objects, original creations, pictures, news articles, and so on. These artifacts will become part of a "culture trunk" for each group's country. Each group will present each artifact "found" in the "trunk" to classmates and explain how it represents an important aspect of their assigned nation. Students may wish to decorate their trunk for the presentation. (Examples: for the geography of Mexico, students may read news articles about a devastating earthquake and show pictures of its effects. For Mexico's history, students may create a facsimile of a gold object dating from the Aztecs and summarize the Indian conquest by the conquistadores.) As each group presents a country, other students should take notes, using the chart below and adding boxes as necessary.

Latin America					
Country	Geography	History	People	Economy	Future

CLOSURE
Name key people, dates, and events in Latin American history, and ask students to think about their significance before calling upon volunteers to share their thoughts.

ASSESSMENT
1. Have students complete the following questions from the Chapter 37 Check:
   Building a Vocabulary 1, 2
   Recalling and Reviewing 1, 2
2. Ask students to complete the Chapter 37 Content/Vocabulary Worksheet in their Workbook.

RETEACHING PLAN
1. Ask students to complete the Chapter 37 Reteaching Worksheet, Lesson 1.
2. Provide students with a time line containing key dates in Latin American history; ask them to work with a partner to label the events that correspond to the dates.

CHALLENGE/ENRICHMENT STRATEGY
1. Direct students to create a map showing empires of Western European nations in Latin America during the ages of exploration and colonization. Students should identify lasting effects of European culture.
2. Select students to investigate and report on the accomplishments of the Mayas, Aztecs, and Incas.

▶ LESSON 2
Physical Geography

LESSON OBJECTIVES
The student should be able to
- locate the countries of Latin America
- locate and describe major landforms of Latin America
- identify the physical forces that alter the earth's crust
- identify the major factors influencing the climates
- explain the significance of water to development

PURPOSE
The study of the physical geography of Latin America will enable students to understand the effect of environment on the ways of life in this region.

MOTIVATOR
Provide students with the following chart:

Latin America			
Countries	I Have Heard of Before	I Know the General Location	I Can Label the Specific Location on a Map
Mountain peaks			
Mountain ranges			
Rivers			

Students should complete their knowledge inventory before and after studying the lesson, by listing the appropriate countries and landforms, then placing check marks in the appropriate spaces.

TEACHING DESIGN
1. Divide students into groups, and provide a globe and length of string to each group. Ask students to plot distances between the following locations and to note the longest and shortest distances within each group.
   Group 1: New York City (U.S.) to London (England), Buenos Aires (Argentina), Rio de Janeiro (Brazil), San Francisco (U.S.), Madrid (Spain), Mexico City (Mexico), San Juan (Puerto Rico), and Havana (Cuba).
   Group 2: Madrid (Spain) to Mexico City (Mexico), Panama City (Panama), Buenos Aires (Argentina), Santiago (Chile), and Lima (Peru).
   Group 3: Lisbon (Portugal) to Rio de Janeiro, Belém, and Manaus (Brazil).
   Group 4: Recife (Brazil) to Guayaquil (Ecuador), and Caracas (Venezuela) to Punta Arenas (Chile).
   After students complete their measurements, ask them to generalize about Latin America's location and size relative to the United States and to Europe.
2. Provide students with blank physical-political maps of Latin America. Divide students into groups of four, and assign one of the following topics to each group for studying and sharing with classmates: Landforms, Distinctive Physical Features, Climate Types, Vegetation Types, Bodies of Water and River Systems, Political Boundaries, and Major Cities and Capitals.

   Each group should learn its assigned topic, then travel from one group to another to teach that topic until every group has shared its topic with all others. Each group should divide its information so that each person has significant knowledge to share with classmates. Each group should ask three questions to test other groups' understanding of its lesson. Students may wish to use maps and pictures from various sources as visual aids. During the presentation, all students should make appropriate notations on their maps.

CLOSURE
Refer students to the knowledge-rating chart used in the Motivator, and ask students to rate their knowledge again.

ASSESSMENT
1. Have students complete the following questions from the Chapter 37 Check:
   Building a Vocabulary 2
   Using Geography Skills 1, 2, 3
2. Have students complete the Chapter 37 Critical Thinking Worksheet.

RETEACHING PLAN
Ask students to complete the Chapter 37 Reteaching Worksheet, Lesson 2.

CHALLENGE/ENRICHMENT STRATEGY
1. Invite students to imagine themselves as explorers on an expedition along the Amazon River, the Paraná River, or the Orinoco River and to write a diary entry describing the scenes they encounter.

2. Have students prepare a visual display to show the variety of climate regions and vegetation in Latin America.

▶ LESSON 3

Economic Geography and Current Issues

LESSON OBJECTIVES

The student should be able to

- identify natural resources and explain their uses
- describe major economic activities in a region
- identify the sources of energy in Latin America
- identify issues facing Latin American nations

PURPOSE

The study of the natural resources of Latin America will enable students to understand the relationship between environment and the ways of life in this region.

MOTIVATOR

Divide the class into groups of four, and provide every group with a copy of a current major newspaper or magazine, along with several highlighter pens. Have each group scan the material to locate articles about Latin American economy and current issues. Students should highlight the headlines of the articles. After allowing students time, call on groups to read their headlines. Make notes of the major topics on a transparency.

TEACHING DESIGN

1. Have students work in groups to create an outline, an information map, or an information web on the natural resources and farming systems of Latin America. Headings should include: Mineral Resources, Fuel Resources, Other Resources, Plantation Agriculture, Subsistence Agriculture, Mixed Agriculture, and Commercial Agriculture. Call on volunteers to share their information.
2. Lead a class discussion about current economic, social, and political issues in Latin America. On a map of Latin America, ask students to identify the countries that have experienced political strife (revolutions, coups, civil wars) in the twentieth century. Have students suggest the positive and negative results of these actions.

CLOSURE

Tell students to think about the future development of Latin American nations. Call on volunteers to name potential for and obstacles to development in this region.

ASSESSMENT

1. Have students complete the following questions from the Chapter 37 Check:
   Building a Vocabulary 4
   Recalling and Reviewing 3, 4, 5
   Critical Thinking 6
2. Have students complete the Chapter 37 Skills Worksheet.
3. Ask students to complete the Chapter 37 Review/Test.

RETEACHING PLAN

1. Ask students to complete the Chapter 37 Reteaching Worksheet, Lesson 3.
2. Have students create a collage that illustrates resources and products of Latin America.

CHALLENGE/ENRICHMENT STRATEGY

1. Have students complete the Chapter 37 Challenge Worksheet.
2. Ask students to report on the history of United States investment in Latin America.

Geography Skills, pages 457, Answers

1. Guatemala. Cuba. Mexico's population is about 10 times that of Cuba.
2. Along the Atlantic coast of Brazil.
3. For comparing climate, population density, historical information, and landforms.

Chapter 37 Check, Page 459, Answers

Building a Vocabulary

1. A plant used to make blue dye. Sugarcane, tobacco.
2. Mestizos are descended from Indians and Europeans; mulattos are descended from Europeans and Africans.
3. Extensive plains found in South America. Pampas are in Argentina; llanos are in Venezuela.
4. labor intensive. They can lessen economic gains.

Recalling and Reviewing

1. The Europeans brought diseases that killed many Indians. African slaves brought yellow fever and malaria.
2. Rich landowners, merchants, and generals replaced colonial governments. The economies remained unchanged.
3. Plantation agriculture: a single crop is grown for export; subsistence agriculture: food is produced mainly for home use; mixed agriculture: mixtures of Indian and European crops, combination of small farms and large estates, and less emphasis on subsistence agriculture; modern commercial agriculture: latest farming methods and mechanized equipment. Subsistence agriculture.
4. The Amazon system is located just south of the equator and south of the Guiana Highlands; the Orinoco system drains northern South America; the Paraná system begins in the highlands of eastern Brazil and the eastern slopes of the Andes and flows into the Río de la Plata.
5. According to the theory of plate tectonics, South America and Africa broke away from Gondwanaland more than 100 million years ago. South America moved westward, and old continental fragments from the Pacific region collided with the west coast, possibly forming the first Andes Mountains. A few million years ago, South America began to override the Pacific plate; mountains in the Andes have continued to be built as a result of folding, faulting, and volcanic activity.

Critical Thinking

6. Political instability has made economic and industrial development difficult. Latin America's population growth has put pressure on cities, and economic development in cities has not kept pace with migration.

Using Geography Skills

1. Humid tropical, wet and dry tropical/subtropical, tropical and subtropical desert, Mediterranean, humid subtropical, continental steppe, maritime, and highland. Latin America covers a large area north and south of the equator, extending through tropical, subtropical, and mild climate regions. The altitude of the mountains in the region produces additional climate variations.
2. The interior is densely forested, and although the climate would be suitable for agriculture, there is not a good transportation system to move goods to and from markets. Coastal regions provide trade opportunities.
3. The Amazon Basin, the Andes Mountains.
4. Colombia, Ecuador, Peru, Bolivia, Chile, Argentina.

CHAPTER 38 Mexico Pages 460–469

PLANNING GUIDE		
Lesson	**Textbook**	**Teacher's Resource Binder**
1. Physical Geography	p. 460	Lecture Notes and Transparency Package Content/Vocabulary Workbook Reteaching Worksheet
2. Economic Geography	pp. 461–465	Challenge Worksheet Reteaching Worksheet
3. Cultural Geography	pp. 465–468	Skills Worksheet Chapter Review/Test Reteaching Worksheet Critical Thinking Worksheet

TIME ALLOCATION: 3 days

CHAPTER TOPICS

- Physical Geography
- Economic Geography
- Mexico's Culture Regions
- Mexico's Future

PRETEACHING CHAPTER VOCABULARY

Write the following terms on the chalkboard:

hacienda plateau* peninsula*
isthmus* plain* land reform

*Review terms

Pronounce each term, and ask students to identify those that have already been introduced in previous chapters. (isthmus, plateau, plain, peninsula) Call on volunteers to state the most important attributes of each review term. Ask students to identify the term that is derived from Spanish. (*hacienda*) Ask students to raise their hands if they have heard of the term *hacienda*. Point out that haciendas are found in areas of the world that have a Spanish heritage. The term refers to a large estate, plantation, or ranch.

Ask students to analyze the term *land reform* by suggesting synonyms for the word *reform*. (improvement, correction) Explain that in Latin America, land reform usually refers to breaking up large estates owned by wealthy people.

▶ LESSON 1
Physical Geography

LESSON OBJECTIVES

The student should be able to

- define the significant terms of the chapter
- identify major landforms and features of Mexico
- describe the physical setting of Mexico
- identify major factors influencing the climate of Mexico
- identify the physical forces that alter the earth's crust

PURPOSE

The study of the physical geography of Mexico will enable students to note the relationship of environment to ways of life in this country.

MOTIVATOR

Write the word *Mexico* on the chalkboard. Ask students to work in groups of four to generate a list of images that come to mind when they hear this word. After three minutes, call on a member of each group to read its list to classmates. Record each new image on an overhead transparency, and ask students to categorize the terms.

TEACHING DESIGN

1. Use the Lecture Notes and Transparency Package to introduce Mexico. Have students take notes.
2. Divide students into groups of four, and ask each group to select an area of Mexico to live in, based upon the information they learned during the lecture. Allow groups time to discuss and select their area before reporting their selection and the reasons for making it.

CLOSURE

Review major features of Mexico by calling on volunteers to locate them on a wall map.

ASSESSMENT

1. Have students complete the following questions from the Chapter 38 Check:
   Recalling and Reviewing 3, 4
2. Ask students to complete the Chapter 38 Content/Vocabulary Worksheet in their Workbook.

RETEACHING PLAN

1. Ask students to complete the Chapter 38 Reteaching Worksheet, Lesson 1.
2. Provide students with a map of Mexico showing neighboring land and bodies of water, with numbers

representing various physical features, capital letters representing climate regions, and roman numerals representing surrounding countries and bodies of water. Students should work in pairs and take turns naming the corresponding features.

CHALLENGE/ENRICHMENT STRATEGY

1. Challenge students to plan a tour of Mexico that would interest geographers. Students should present a travel itinerary that includes a description of the sites.
2. Have students research and report on volcanic activity and earthquakes in Mexico during the last 50 years.

▶ LESSON 2
Economic Geography

LESSON OBJECTIVES

The student should be able to

- identify natural resources of Mexico and their uses
- analyze the impact of environment on ways of life
- describe the agricultural base of various regions
- describe major economic activities in Mexico
- describe the economic importance of water to the nation

PURPOSE

The study of economic geography in Mexico will give students a context for understanding the complex economic issues facing this neighboring country.

MOTIVATOR

Reproduce the following statistics or more recent data about Mexico's debts to North American banks. (Source: *The Wall Street Journal,* December 30, 1987.)

WHAT MEXICO OWES TO U.S. BANKS
Citicorp: $2.9 billion
BankAmerica Corp.: $2.41 billion
Manufacturers Hanover Corp.: $1.88 billion
Chemical New York Corp.: $1.73 billion
Chase Manhattan Corp.: $1.66 billion
Bankers Trust New York Corp.: $1.27 billion
J. P. Morgan & Co.: $1.14 billion
(SOURCE: Keefe, Bruyette & Woods Inc.)

Explain that Mexico, as well as other Latin American nations, has borrowed funds for its development from banks in the United States and other countries. Many Latin American countries have been unable to repay their loans in full because of economic problems. Ask students what happens to banks in the United States when Mexico is unable to repay its debts. Ask students to make a generalization regarding the relationship of Mexico's economic development to the United States' economic stability.

TEACHING DESIGN

1. Refer students to the map "Latin America" on page 455 in their textbook. Based upon Mexico's climate and landforms, volunteers should identify areas that would be suitable for farming and estimate the percentage of Mexico's land that could have crop production each year. Record this information on the chalkboard. Ask students to review the information in the text before completing the following chart on agricultural production in Mexico. Students should note the crops or livestock production for each category of agriculture.

Type of Agriculture	Crops	Livestock
Subsistence		
Commercial		
Plantation		

   Have students compare their information with a nearby classmate and circle the names of crops that are exported, then report the information from the chart.
2. Ask students to work with a partner to review the section on page 464 in their textbook under "Land Reform" and prepare to answer the following questions: What are the main characteristics of the hacienda and *ejidos* systems of land use? What are the advantages and disadvantages of each system of land use? After allowing five minutes for review and discussion, call on volunteers to answer the questions. Record the main ideas on the chalkboard.
3. Refer students to the map "Mexico" on page 465 of their textbook. Ask groups of students to rank these resources according to their probable economic value.

CLOSURE

Have students remain in their groups. Provide each group with a fact sheet containing economic information from the lesson. Ask each group to produce ideas for increasing productivity in Mexico and to determine a three-step plan for accomplishing this increase. Call on group representatives to report their plans.

ASSESSMENT

1. Have students complete the following questions from the Chapter 38 Check:
   Building a Vocabulary 1
   Recalling and Reviewing 1, 5
   Using Geography Skills 1, 3
2. Have students complete the Chapter 38 Challenge Worksheet.

RETEACHING PLAN

1. Ask students to complete the Chapter 38 Reteaching Worksheet, Lesson 2.
2. Have students create a product map of Mexico.

CHALLENGE/ENRICHMENT STRATEGY

1. Ask students to investigate the current status of immigration of Mexicans to the United States and to analyze immigration policies.
2. Choose volunteers to research cookbooks for authentic recipes and the special ingredients used to prepare foods in Mexico. Students should report their findings and bring food samples to share with classmates.

▶ LESSON 3
Cultural Geography

LESSON OBJECTIVES

The student should be able to

- analyze the location of cities
- describe examples of exchange among regions
- explain the causes of population patterns
- identify environmental issues in Mexico City

PURPOSE

In this lesson, students will have an opportunity to expand their knowledge of the diversity of Mexico's regions.

MOTIVATOR

Call on volunteers to define the term *region*. After reviewing the definition, have students name ways in which Mexico could be divided into regions. (vegetation, climate, location) List students' ideas on the chalkboard.

TEACHING DESIGN

1. Outline on the chalkboard the following information map, and ask students to work with partners to write three significant ideas that they should remember about each of the regions of Mexico.

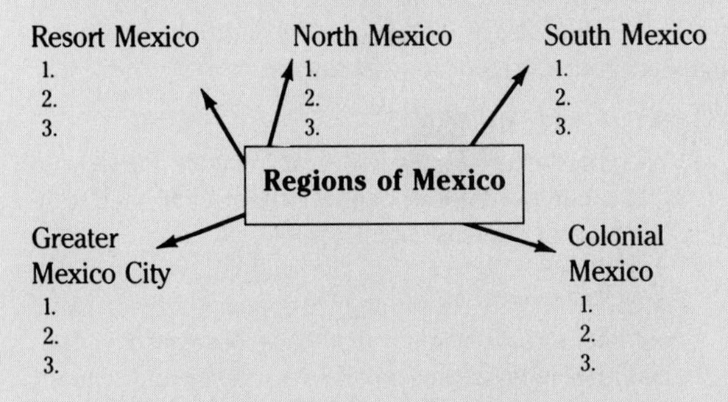

2. Have students read the Cities of the World feature "Mexico City" on pages 466–467 and the Themes in Geography feature "Mexico's Traditional Villages" on pages 462–463 of their textbook. Ask students to identify memorable ideas about ways of life in each setting. Discuss advantages and disadvantages of living in both settings. Ask students to name common factors of urban life in the major cities they have studied thus far.
3. Invite students to present their "Culture Trunk" reports on Mexico. Remind them to identify a significant fact for each topic and to write it on their charts.

CLOSURE

Call on volunteers to share key ideas they chose to write about Mexico on their charts.

ASSESSMENT

1. Have students complete the following questions from the Chapter 38 Check:
   Recalling and Reviewing 2
   Critical Thinking 6
   Using Geography Skills 2
2. Have students complete the Chapter 38 Skills Worksheet.
3. Ask students to complete the Chapter 38 Review/Test.

RETEACHING PLAN

1. Ask students to complete the Chapter 38 Reteaching Worksheet, Lesson 3.
2. Collect students' information maps on Mexico's regions, and use them to provide clues. Call on students to name the region that matches the clues.

CHALLENGE/ENRICHMENT STRATEGY

1. Have students complete the Chapter 38 Critical Thinking Worksheet.
2. Choose students to investigate and report on the handicrafts unique to each region of Mexico.

Chapter 38 Check, page 469, Answers

Building a Vocabulary

1. Attempts to break up large estates (haciendas). Fewer than 10 percent are landless today.
2. A large estate. Most haciendas have been replaced by cooperative villages (*ejidos*).

Recalling and Reviewing

1. Petroleum is Mexico's major mineral resource. The country's petroleum reserves are among the largest in the world, and more than a dozen cities are growing to support the oil business.
2. Dramatic population increases and a huge foreign debt are causing serious economic problems. Construction of new industries, schools, and housing cannot keep up with population growth. Income from export sales must go to buy food rather than build factories. Other export earnings must go to pay off foreign loans. Political unrest is another disruptive factor.
3. A central plateau called the Meseta Central. The Sierra Madre Oriental and the Sierra Madre Occidental to the east and west of the Meseta Central. The valley of Mexico south of the Meseta Central. The Isthmus of Tehuantepec and the Yucatán Peninsula in the south. Smaller mountain ranges, volcanoes, and fertile valleys continue southward into Central America. Greater Mexico City.
4. Mexico extends from the low to the middle latitudes. Elevation affects climate as well.
5. Cotton, wheat, alfalfa, fruits, vegetables, coffee, sugarcane, cacao, bananas, and pineapples.

Critical Thinking

6. Greater Mexico City: holds more than one-fourth of Mexico's population; Colonial Mexico: surrounds Mexico City and extends to both coasts; South Mexico: Mexico's poorest region, with subsistence agriculture and traditional villages; North Mexico: prosperous and modern, industry and commercial farming, closer ties with the United States; Resort Mexico: coastal ports and fishing villages visited by foreign tourists. Answers to the second part of the question will vary.

Using Geography Skills

1. Along the western and eastern coasts, on the Yucatán Peninsula, near Puebla and the Isthmus of Tehuantepec. Humid tropical and wet and dry tropical/subtropical.
2. 25 persons per square mile (10 per square kilometer); 125 persons per square mile (49 per square kilometer).
3. In the Bay of Campeche, on the Isthmus of Tehuantepec, in northeastern Mexico, and along the eastern coast. By boat.
4. Gulf of California.

CHAPTER 39 Central America and the Caribbean Pages 470–479

PLANNING GUIDE		
Lesson	Textbook	Teacher's Resource Binder
1. History and Culture; Central America	pp. 470–474	Lecture Notes and Transparency Package Content/Vocabulary Workbook Reteaching Worksheet
2. The Caribbean	pp. 474–478	Reteaching Worksheet Challenge Worksheet
3. Current Issues	pp. 470–478	Skills Worksheet Chapter Review/Test Reteaching Worksheet Critical Thinking Worksheet

TIME ALLOCATION: 3 days

CHAPTER TOPICS

- Central America
- The Islands

PRETEACHING CHAPTER VOCABULARY

Write the following terms on the chalkboard:

plantain voodoo allspice

plantain (a large tropical plant whose fruit resembles the banana; the fruit is a major food source in tropical areas) If possible, purchase a plantain and a banana and show them in class. Explain that plantains are usually cooked before eating.

voodoo (originated in Africa and is characterized by belief that good and evil play an important part in daily life)

allspice (a mildly pungent and aromatic spice prepared from the berries of a tree of the myrtle family) Bring a sample of ground allspice to class, and allow students to smell and/or taste it.

▶ LESSON 1
History and Geography

LESSON OBJECTIVES

The student should be able to

- define the significant terms of the chapter
- locate and describe major landforms and features
- identify the major factors influencing the climate
- identify the physical factors that alter the earth's crust
- identify major historical events of the region

PURPOSE

The study of the physical geography of Central America and the Caribbean will enable students to understand the interaction of physical and cultural geography.

MOTIVATOR

Provide students with a map of the region that identifies only two countries in Central America and four countries in the Caribbean. Give students a list of all the countries in the region, and allow them three minutes to label countries they can locate on their maps without research. At the end of the time allotment, have students check their labels against the map "Central America and the Caribbean" on page 472 in their textbook. Ask students how they would rate their knowledge of the countries' locations within this region: poor, fair, good, or superior.

TEACHING DESIGN

1. Use the Lecture Notes and Transparency Package to present an overview of Central America and the Caribbean. Have students take notes.
2. Ask students to suggest the cultural changes that occurred in this region after Europeans arrived. Point out the foreign nations that have controlled various countries in this region and the reasons for their interest in the area. Point out that the United States has sent troops into various nations of this region to protect United States interests. Ask students to discuss the effects of the long period of colonialism.
3. Ask students to present their "Culture Trunk" reports on the countries of Central America. Remind students to identify a significant fact for each topic on their charts.

CLOSURE

Ask each student to compare his or her chart information with another student's. Call on volunteers to share the most significant information about the seven countries.

ASSESSMENT

1. Have students complete the following questions from the Chapter 39 Check:
   Recalling and Reviewing 1, 2
   Using Geography Skills 1, 2, 3
2. Ask students to complete the Chapter 39 Content/Vocabulary Worksheet in their Workbook.

RETEACHING PLAN

Ask students to complete the Chapter 39 Reteaching Worksheet, Lesson 1.

CHALLENGE/ENRICHMENT STRATEGY

1. Ask students to write an essay explaining why Central America and the Caribbean have long been considered strategic locations in the eyes of other nations.
2. Challenge students to research the conflict in Nicaragua between the Sandinistas and the *contras* and to prepare a time line showing major events in that conflict.

▶ LESSON 2
The Caribbean

LESSON OBJECTIVES

The student should be able to

- identify major economic activities in the region
- describe the economic importance of resources
- identify major historical events of the region

PURPOSE

This lesson's focus on the Caribbean will increase students' knowledge of the land and people of this region.

MOTIVATOR

Collect travel brochures describing tourist sites on islands throughout the Caribbean. Divide students into five groups, and distribute brochures to each group. Have students peruse the brochures, then pass them to another group. After all groups have had an opportunity to view the brochures, call on volunteers to share their perceptions of this region. Record them on the chalkboard.

TEACHING DESIGN

1. Ask students to study the map "Central America and the Caribbean" on page 472 of their textbook, noting similarities and differences among the nations. Have students work in groups of three to complete a resource/produce chart for all countries of the region. Students should place a check mark in appropriate boxes for each country, using their textbook as a source. Resource/product headings should include Oil, Natural Gas, Bauxite, Iron Ore, Metals, Bananas, Coffee, Sugarcane, Tobacco, Cotton, Fish, Produce, and Livestock. After students complete their charts, ask them to rank the countries from 1 to 16, with 1 designating the country with the greatest number of checks. Call on volunteers to share their first five countries or islands.
2. Ask students to present their "Culture Trunk" reports on the countries and islands of the Caribbean. Remind students to identify a significant fact for each topic.

CLOSURE

Call on students to present an information clue from their charts or checklist for a country in the Caribbean, and ask volunteers to name the appropriate country.

ASSESSMENT

1. Have students complete the following questions from the Chapter 39 Check:
   Building a Vocabulary 1, 2, 3
   Recalling and Reviewing 3, 4

RETEACHING PLAN

Ask students to complete the Chapter 39 Reteaching Worksheet, Lesson 2.

CHALLENGE/ENRICHMENT STRATEGY

1. Have students complete the Chapter 39 Challenge Worksheet.
2. Ask students to invite a travel agent to visit the class and describe travel opportunities in the Caribbean.

▶ LESSON 3
Current Issues

LESSON OBJECTIVES

The student should be able to

- identify major activities of the region
- explain the importance of resources to development
- apply problem-solving skills

PURPOSE

This lesson will give students an overview of the complex issues facing the nations of this region and of the difficulty in finding feasible plans of action to address these issues.

MOTIVATOR

Reproduce the following passage on an overhead transparency, and project it for students to read.

Following is an address by Richard T. McCormack, Ambassador to the Organization of American States, before the second Conference of the Great Cities of the Americas, San Juan, Puerto Rico, June 20, 1986.

The obstacles to investment and growth in Latin America deserve the great attention they are receiving. Investment is particularly important because without investment no new jobs will be created. Without new jobs, there will be no economic growth. Without economic growth, political stability will gradually erode, and the precious democracies that all of us have worked so hard with Latin America to bring about may lapse into military dictatorships or the further spread of communist tyranny.

TEACHING DESIGN

1. Direct students to review Chapter 39 to create a list of major issues facing the people of Central America and Caribbean nations. Call on volunteers to share their lists, and record the issues on the chalkboard.
2. Provide reference materials for students. Refer students to the Atlas in the front of their textbook and to the

statistical charts in the back of their textbook as additional resources. Ask students to work in their "Culture Trunk" groups to create a statistical chart for their assigned country in Central America or the Caribbean. Headings should include Population, GNP, Per Capita Income, Literacy Rate, Infant Mortality Rate, Value of Exports, Value of Imports, Population Density, and Percentage of Unemployment. Create a completed master chart on an overhead transparency. Ask students to use the chart information to determine the countries with the highest and lowest standards of living.

3. Students should work in their groups to select one critical issue in their assigned country, to brainstorm about plans of action, and to create a plan for addressing that issue. A group recorder should write the group's plan on a sheet of paper, using the following format.

Country ______________ Issue ______________
Planned Actions *Resources Needed to Carry Out Plan*
1. a.

Students may select their issue from their textbook, their "Culture Trunk" report, or their statistical charts.

CLOSURE

Call on representatives from each group to share their plans of action with classmates. Allow students to vote for those plans that they think are the most feasible. Then have students compare the maps they labeled for the Lesson 1 Motivator to their present knowledge.

ASSESSMENT

1. Have students complete the following question from the Chapter 39 Check:
   Critical Thinking 5
2. Have students complete the Chapter 39 Skills Worksheet.
3. Ask students to complete the Chapter 39 Review/Test.

RETEACHING PLAN

Ask students to complete the Chapter 39 Reteaching Worksheet, Lesson 3.

CHALLENGE/ENRICHMENT STRATEGY

1. Have students complete the Chapter 39 Critical Thinking Worksheet.
2. Ask students to devise a plan to promote economic cooperation in Central America and the Caribbean and to benefit all participating nations. Students may wish to review economic associations such as the EC.

Chapter 39 Check, page 479, Answers

Building a Vocabulary

1. A type of banana usually used for cooking. Haiti.
2. A Haitian version of African animism. Hispaniola.
3. Jamaica.

Recalling and Reviewing

1. Humid tropical plains along the Caribbean coast. Inland mountains, where most of the people live.
2. Competition for markets for their export crops, dramatic population increases, unemployment, lack of land, underdevelopment. Costa Rica has more peaceful democratic traditions, a growing middle class, and a more even distribution of wealth. Most of its people are of primarily Spanish descent. It has the highest standard of living in Latin America.
3. Both countries produce sugarcane and coffee. Both have had periods of political unrest and dictatorship. Haiti is primarily French speaking, while the Dominican Republic is Spanish speaking. The Dominican Republic's economy is better developed than Haiti's, and it has made more advances in education, public housing, and health than Haiti. It also has been more successful at attracting foreign investment than Haiti.
4. Farming, tourism, food processing, manufacturing.

Critical Thinking

5. Answers will vary, but students should mention limited resources and population growth as factors. Answers to the second part of the question will vary.

Using Geography Skills

1. Pacific Ocean, Caribbean Sea. The canal is built on the narrowest stretch of land in Latin America. It shortens the distance that ships need to travel from the west to the east coasts of the United States.
2. At approximately 10° N, 85° W.
3. Cuba. About 90 miles (144 kilometers).

CHAPTER 40 Northern South America Pages 480–487

PLANNING GUIDE		
Lesson	**Textbook**	**Teacher's Resource Binder**
1. History and Culture	pp. 480–486	Lecture Notes and Transparency Package Skills Worksheet Reteaching Worksheet Critical Thinking Worksheet
2. Present and Future	pp. 480–486	Content/Vocabulary Workbook Chapter Review/Test Reteaching Worksheet Challenge Worksheet

TIME ALLOCATION: 2 days

CHAPTER TOPICS

- Venezuela
- Colombia
- The Guianas

PRETEACHING CHAPTER VOCABULARY

Write the following terms on the chalkboard:

tar sands *tierra templada* *tierra helada*
tierra caliente *tierra fría*

Ask which students speak or have studied the Spanish language. Determine whether any of the students can translate the Spanish terms. Explain that the Spanish word *tierra* means "land." Ask students to hypothesize about possible definitions for the terms, using the diagram below.

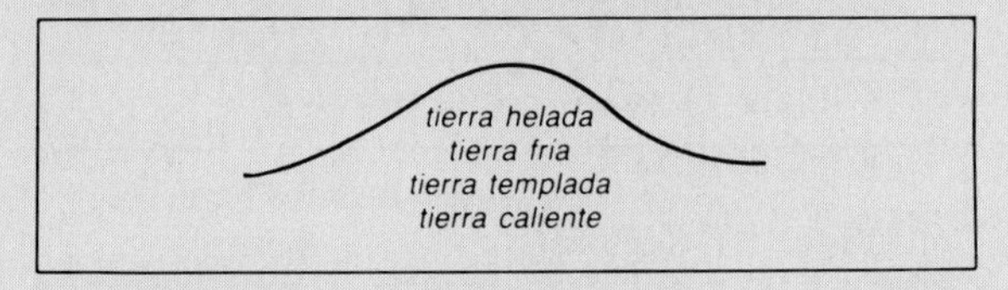

▶ LESSON 1
History and Culture

LESSON OBJECTIVES
The student should be able to
- explain geographic terminology
- locate and describe major landforms and features
- locate major natural resources
- understand the significance of the Andes in determining regions
- analyze the impact of environment on ways of life

PURPOSE
In this lesson, students will learn about Venezuela, another country that is a source of oil for the United States, as well as about Colombia and the Guianas.

MOTIVATOR
Ask students to read the primary source feature "For the Record" on page 484 of their textbook. Discuss implications of living in a country where volcanoes are prevalent.

TEACHING DESIGN
1. Present the Lecture Notes and Transparency Package to introduce Venezuela, Colombia, and the Guianas.
2. Invite students to present their "Culture Trunk" reports on the countries in this chapter. Remind students to identify a significant fact for each topic.
3. Have students complete the Chapter 40 Skills Worksheet.

CLOSURE
Ask students to review with a partner the key ideas they selected to write in their charts. Select several volunteers to present their ideas to the class.

ASSESSMENT
Have students complete the following questions from the Chapter 40 Check:

Building a Vocabulary 1, 2
Recalling and Reviewing 2
Using Geography Skills 1, 2

RETEACHING PLAN
1. Ask students to complete the Chapter 40 Reteaching Worksheet, Lesson 1.
2. Have students complete the Chapter 40 Critical Thinking Worksheet.

CHALLENGE/ENRICHMENT STRATEGY
1. Invite students to create a bulletin-board display that represents significant landforms and bodies of water in the Latin American countries studied thus far.
2. Have students research and report on the impact of the discovery of oil on Venezuela's economy.

▶ LESSON 2
Present and Future

LESSON OBJECTIVES
The student should be able to
- analyze the impact of environment on ways of life
- describe the potential for cultural interchange
- explain the economic importance of water
- describe major economic activities in the region

PURPOSE
In this lesson, students explore how geography influences the ways people live in this region.

MOTIVATOR
Draw the following on the chalkboard:

Ask students to work in groups of three to identify ways that the Andes influence ways of life in Latin America.

TEACHING DESIGN
1. Have students complete any reports not presented in the previous lesson.
2. Place students in cooperative groups of three. Write the following questions on the chalkboard.
   a. Assume that you have been asked to relocate to Northern South America. Where would you live?
   b. What occupation would you expect to practice?
   c. What freedoms or rights would you expect to have?
   d. What standard of living would you expect to achieve?

   Encourage the groups to reach a consensus on their responses. Assign one recorder and one leader.
3. Discuss the students' ideas, and continue with the following questions.
   a. Which country(ies) do you predict will become more industrialized in the future? Why?
   b. Which countries are most likely to form alliances?
   c. Which countries are most likely to remain friendly toward the United States? Why?
   d. In what ways should the industrialized nations attempt to aid these countries?
4. Lead a class discussion based on these questions. Encourage students to take notes on key ideas presented.

CLOSURE
Call upon various students to respond aloud to the following questions from the Chapter 40 Check:

Recalling and Reviewing 1, 3
Critical Thinking 4
Using Geography Skills 3, 4

ASSESSMENT

1. Ask students to complete the Chapter 40 Content/Vocabulary Worksheet in their Workbook.
2. Ask students to complete the Chapter 40 Review/Test.

RETEACHING PLAN

1. Ask students to complete the Chapter 40 Reteaching Worksheet, Lesson 2.
2. Have students create their own diagram of differences and similarities between Venezuela and Colombia.

CHALLENGE/ENRICHMENT STRATEGY

1. Invite students to report on the history of Devil's Island in French Guiana.
2. Have students complete the Chapter 40 Challenge Worksheet.

Chapter 40 Check, page 487, Answers

Building a Vocabulary

1. Rocks that contain oil. In Venezuela, along the middle Orinoco. In coming decades, after the world's more easily available oil is gone, tar sand deposits may dominate the world economy.
2. *Tierra caliente, tierra templada, tierra fría, tierra helada. Tierra helada, tierra caliente.*

Recalling and Reviewing

1. Sales of oil and petroleum products make up most of Venezuela's export income. Profits have built schools, improved living conditions, and expanded medical care.
2. A narrow tropical lowland along the coast, the northern valleys and mountains of the Andes that parallel the coast, the llanos, and the Guiana Highlands. Most people live along the Caribbean coast and in the valleys of the nearby mountains; this region has land suitable for commercial as well as subsistence agriculture.
3. Both countries are former colonies. They have similar humid tropical environments and narrow coastal plains where most of the agricultural lands are found. Their densely forested interiors are made up of rugged uplands. Rice, sugarcane, and bauxite are important to both countries, and both have mixed populations of South Asians, blacks, mulattos, Indians, and Chinese.

Critical Thinking

4. Much of the rest of Colombia is covered with thick tropical forest. This is not a productive biome because temperatures get too hot. The soil and climate of the mild coastal lowlands and Andes highlands are more productive for agriculture.

Using Geography Skills

1. 25 persons per square mile (10 per square kilometer). Humid tropical.
2. Approximately 5° N, 74° W.
3. Along the coast.
4. Guyana, Suriname, French Guiana. Venezuela.

CHAPTER 41 The Andean Countries Pages 488–495

PLANNING GUIDE		
Lesson	**Textbook**	**Teacher's Resource Binder**
1. Physical Geography and History	pp. 488–494	Lecture Notes and Transparency Package Content/Vocabulary Workbook Reteaching Worksheet Critical Thinking Worksheet Skills Worksheet
2. Economic Geography and the Future	pp. 488–494	Chapter Review/Test Reteaching Worksheet Challenge Worksheet

TIME ALLOCATION: 2 days

CHAPTER TOPICS

- Ecuador
- Peru
- Bolivia

PRETEACHING CHAPTER VOCABULARY

Write the following terms on the chalkboard:

altiplano El Niño inflation peso

Discuss various ways of classifying and dividing these terms. Ask students to find their definitions in the textbook Glossary. Challenge students to write one sentence that includes all four terms.

▶ LESSON 1

Physical Geography and History

LESSON OBJECTIVES

The student should be able to

- define geographic terms
- locate and describe landforms
- analyze the effect of environment on ways of life
- describe major economic activities in the region
- explain the causes of population patterns
- describe the agricultural base of the region

PURPOSE

This lesson will introduce students to the richness of the Inca culture and will reinforce the relationship between environment and culture.

MOTIVATOR

Locate pictures of art, crafts, and clothing produced during the Inca empire. As students study the pictures, ask them what they remember from previous studies about the Inca empire and what they deduce about the Incas from the pictures. Read a brief selection from an encyclopedia or history book about the Inca empire. Ask students to think of possible reasons that an advanced civilization could have existed in a region where little development exists today.

TEACHING DESIGN

1. Present the Lecture Notes and Transparency Package. Have students take notes. Return to the Motivator question, and complete the discussion.
2. Invite students to present their "Culture Trunks" for Ecuador, Peru, and Bolivia. Remind students to note the key ideas under the five assigned topics.
3. Ask students to work with a partner to identify the greatest similarities and differences among these countries.

CLOSURE

Invite partners to share the ideas they developed about the similarities and differences among the countries.

ASSESSMENT

1. Have students complete the following questions from the Chapter 41 Check:
   Building a Vocabulary 1, 2
   Recalling and Reviewing 1, 2, 3
   Critical Thinking 6
2. Ask students to complete the Chapter 41 Content/Vocabulary Worksheet in their Workbook.

RETEACHING PLAN

Ask students to complete the Chapter 41 Reteaching Worksheet, Lesson 1.

CHALLENGE/ENRICHMENT STRATEGY

1. Invite students to prepare a report on the specially adapted animals of the Galápagos Islands.
2. Have students complete the Chapter 41 Critical Thinking Worksheet and the Chapter 41 Skills Worksheet.

▶ LESSON 2

Economic Geography and the Future

LESSON OBJECTIVES

The student should be able to

- describe the physical setting of the region
- locate natural resources and identify their uses
- analyze the effect of environment on ways of life
- analyze population patterns

PURPOSE

In this lesson, students will examine the relationships among Ecuador, Peru, Bolivia, and the United States.

MOTIVATOR

Refer students to the map "Latin America" on page 452 in their textbook. Invite students to identify geographic features of Ecuador, Peru, and Bolivia that they predict might influence the ways of life of the people. List as many ideas as possible on the chalkboard.

TEACHING DESIGN

1. Ask students to work in groups to complete this assignment and, using a scale of 1 to 10 (highest), to rate the three countries on the following variables. Students should be prepared to explain their decisions.

Desirability as a Place for a Vacation	
Ecuador	⟵————————⟶
Peru	⟵————————⟶
Bolivia	⟵————————⟶
	1 2 3 4 5 6 7 8 9 10
Desirability as a Place to Live	
Ecuador	⟵————————⟶
Peru	⟵————————⟶
Bolivia	⟵————————⟶
	1 2 3 4 5 6 7 8 9 10
Positive Relationships with the U.S.	
Ecuador	⟵————————⟶
Peru	⟵————————⟶
Bolivia	⟵————————⟶
	1 2 3 4 5 6 7 8 9 10

   Lead a class discussion about the ideas of the groups, and try to reach a class consensus.
2. Discuss the concept of inflation. Draw a dollar sign ($) on a balloon, and blow up the balloon to represent what happens to prices when a country experiences inflation.
3. Ask students to write a scenario describing Ecuador, Peru, or Bolivia in the year 2000. It should be based on facts from the students' textbook.

CLOSURE

Write the following terms on the chalkboard:

Land Resources People

Ask students to create one sentence for each term, relating the term to the three countries studied in this chapter.

ASSESSMENT

1. Have students complete the following questions from the Chapter 41 Check:
   Building a Vocabulary 3
   Recalling and Reviewing 4, 5
   Using Geography Skills 1, 2
2. Ask students to complete the Chapter 41 Review/Test.

RETEACHING PLAN

Ask students to complete the Chapter 41 Reteaching Worksheet, Lesson 2.

CHALLENGE/ENRICHMENT STRATEGY

1. Have students complete the Chapter 41 Challenge Worksheet.
2. Select students to investigate and report the meanings of the words *Ecuador, Peru, and Bolivia*.

Chapter 41 Check, page 495, Answers

Building a Vocabulary

1. Peru, Bolivia. Lake Titicaca and Lake Poopó.
2. El Niño. Approximately every 10 years.
3. An economic situation in which prices rapidly increase because the country's money is losing its value. Bolivia's peso, or monetary unit, is nearly worthless.

Recalling and Reviewing

1. Sierra, Oriente, Costa. Because there are few roads and access to the area is difficult.
2. Controlling population growth and developing transportation systems to link the country's different regions.
3. Montaña: forests, oil, crops, livestock. Coastal region: hydroelectric power, export crops, mineral deposits (including copper), fish, guano.
4. The Andes, the altiplano between two mountain ranges of the Andes, tropical rain forests in the north and east, and a dry savanna in the southeastern region.
5. The Montaña includes the eastern slopes of the Andes and the plains of the Amazon lowland; it has humid tropical forests. The altiplano is a broad highland plain that has high-altitude grassland vegetation.

Critical Thinking

6. Answers will vary, but students should mention language and agriculture.

Using Geography Skills

1. Ecuador: Quito; Peru: Lima; Bolivia: La Paz and Sucre. 0° (Equator), 79° W.
2. Forests: in Ecuador's Oriente region, northern Peru and in the Montaña region of Peru, in northwestern and central Bolivia. Farming: in Ecuador's Sierra region, along the coastal plain and western slopes of the Andes in Peru and around Cuzco; in Bolivia: around the shores of Lake Titicaca, near Sucre and Santa Cruz.

CHAPTER 42 Brazil Pages 496–501

PLANNING GUIDE		
Lesson	Textbook	Teacher's Resource Binder
1. Physical Geography and History	pp. 496–497	Lecture Notes and Transparency Package Skills Worksheet Reteaching Worksheet
2. Regional Geography and Brazil's Future	pp. 497–500	Content/Vocabulary Workbook Chapter Review/Test Reteaching Worksheet Challenge Worksheet Critical Thinking Worksheet

TIME ALLOCATION: 2 days

CHAPTER TOPICS

- Physical Geography
- Brazil's Future
- Regions of Brazil

PRETEACHING CHAPTER VOCABULARY

Write the following terms on the chalkboard:

soil exhaustion *favelas* forward capital

Ask students to identify the term that is a Portuguese word. (*favelas*) Ask volunteers to suggest possible meanings for the three terms. Record students' suggested meanings, then ask all students to read the sentences in their textbook that apply to the terms. Ask students to compare their suggested definitions with the definitions in the textbook.

▶ LESSON 1
Geography and History

LESSON OBJECTIVES

The student should be able to

- define the significant terms of the chapter
- locate and describe major landforms and features
- identify the significance of the Amazon River to Brazil
- identify major historical events of the nation

PURPOSE

In this lesson, students will have an opportunity to learn about a nation that is considered to be one of the leading countries of the future because of its vast resources.

MOTIVATOR

Ask students to use the map "Brazil" on page 498 of their textbook and work with a partner to answer the following questions: What is the capital of Brazil? What major river empties into the Atlantic Ocean at the equator? Where are most cities located? What are the major landforms shown on the map? What countries in South America do *not* share a common border with Brazil? What is the length of the Amazon River in Brazil? Call on volunteers to share their answers.

TEACHING DESIGN

1. Present the Lecture Notes and Transparency Package to introduce Brazil. Have students take notes.
2. Invite students to present their "Culture Trunk" report on Brazil. Remind students to identify a significant fact for each topic to write on their charts.

CLOSURE

Ask volunteers to share one key idea from the topics on their charts.

ASSESSMENT

1. Have students complete the following questions from the Chapter 42 Check:
   Recalling and Reviewing 1, 3
   Using Geography Skills 1, 2
2. Ask students to complete the Chapter 42 Skills Worksheet.

RETEACHING PLAN

1. Ask students to complete the Chapter 42 Reteaching Worksheet, Lesson 1.
2. Ask students to fill in a map of Brazil noting landforms, major rivers, major cities, and bordering countries.

CHALLENGE/ENRICHMENT STRATEGY

1. Suggest that students find pictures of Brazil and bring them to class for display on a bulletin board.

2. Invite students to take an imaginary boat trip along the Amazon and describe the plant, animal, and human life they encounter on their journey.

▶ LESSON 2
Regional Geography and Brazil's Future

LESSON OBJECTIVES
The student should be able to
- identify major economic activities in regions of Brazil
- locate natural resources
- identify the agricultural base of each region
- identify patterns of urban growth in Brazil

PURPOSE
In this lesson, students will explore the relationship between Brazil's resources and economy, and they will draw conclusions about its effect on the nation's future.

MOTIVATOR
Lead a discussion comparing the westward movement in America to the movement to extend the frontier in Brazil and the effects of this extension on the native populations in both countries. Explain that modern technology allowed the Brazilian government to survey the entire nation by means of radar mapping, enabling the discovery of rich mineral deposits and an unmapped river that is 400 miles long. (Source: *National Geographic*, Vol. 152, No. 5, November, 1977, "Brazil's Wild Frontier," p. 692.)

TEACHING DESIGN
1. Have students use their notes from the lecture in Lesson 1 for the following activity. They should divide a sheet of paper into four equal sections and number them with the names of the four regions.

   Ask students to review the information in their textbook under "Regions of Brazil" on pages 497–500, writing notes on their paper about the significant facts of each region. Ask students to compare notes.
2. Have students work in small groups to create a forecast chart of the positive and negative elements of Brazil's future, using the following example.

Brazil's Future	
Positive	**Negative**
1. (Vast mineral resources)	1. (Enormous foreign debt)

CLOSURE
Ask volunteers to share information to be remembered about each region of Brazil and its future prospects.

ASSESSMENT
1. Have students complete the following questions from the Chapter 42 Check:
   Building a Vocabulary 1, 2, 3
   Recalling and Reviewing 2, 4
   Critical Thinking 5, 6
2. Ask students to complete the Chapter 42 Content/Vocabulary Worksheet in their Workbook.
3. Ask students to complete the Chapter 42 Review/Test.

RETEACHING PLAN
Ask students to complete the Chapter 42 Reteaching Worksheet, Lesson 2.

CHALLENGE/ENRICHMENT STRATEGY
1. Have students complete the Chapter 42 Challenge Worksheet and the Chapter 42 Critical Thinking Worksheet.
2. Ask students to research and report on cultural activities of Brazil such as *Carnaval*, music, art, and literature.
3. Challenge students to research and report on the ecological effects of development in the Amazon.

Chapter 42 Check, page 501, Answers

Building a Vocabulary
1. The loss in soil nutrients that results from always planting the same crop in a particular area. Many small farms are barely above subsistence level, and many farmers migrate every few years, looking for more fertile land.
2. *favelas*
3. A city built in a deliberate attempt to develop the interior region of a country. Brasília.

Recalling and Reviewing
1. The Amazon Region, the Brazilian Highlands, and the Brazilian Plateau.
2. Southeastern Brazil is the center of the coffee industry and is where most of the country's cash crops are produced. It has large mineral deposits, and most of Brazil's industry is located here.
3. Many tropical and middle-latitude crops are grown. It is famous for coffee, sugarcane, and beef production and also grows soybeans, cotton, oranges, cacao, bananas, cashew nuts, manioc, rice, maize, and tobacco.
4. Highways have resulted in the end of isolation for many Indian tribes. Many Indians either are moving to reservations or are becoming farm workers in the developing areas. Highways have made it possible for more people to settle in the Amazon Basin and for the area to be developed. The Amazon Basin has been slow to develop because of the climate, thick vegetation, and dangers of disease. Products of the Amazon rain forest are fine woods, nuts, and plants used in medicines.

Critical Thinking
5. Answers will vary, but students should mention land clearing and diseases as obstacles to settlement both in the Amazon and in the North American West. The climate and soil of the American West were more conducive to settlement than that of the Amazon region.
6. Plants would begin to establish themselves on the land that the farmer had cleared. The first plants to grow would be tough and would grow rapidly. Plants that could not survive in the open sun would grow in the shelter and shade of the first plants. Taller plants that

could get closer to sunlight might appear, such as shrubs and trees whose seeds would be blown in from a distance or carried by birds.

Using Geography Skills

1. Along the Atlantic coast. Climate, rich natural resources, industry, and ideal location for trade.
2. West of São Paulo and Salvador in the Brazilian Highlands. The area near Salvador is in the wet and dry tropical/subtropical climate region. The area near São Paulo is in a humid subtropical climate region.

CHAPTER 43 Southern South America Pages 502–509

PLANNING GUIDE		
Lesson	**Textbook**	**Teacher's Resource Binder**
1. Physical and Economic Geography	pp. 502–508	Skills Worksheet Content/Vocabulary Workbook Reteaching Worksheet
2. Political Geography	pp. 505–508	Lecture Notes and Transparency Package Chapter Review/Test Reteaching Worksheet Critical Thinking Worksheet Challenge Worksheet

TIME ALLOCATION: 2 days

CHAPTER TOPICS

- Chile
- Argentina
- Uruguay
- Paraguay

PRETEACHING CHAPTER VOCABULARY

Write the following terms on the chalkboard:

growth point cartel tannin

Ask students to suggest various strategies for defining the terms. (context, glossary, dictionary) Call on volunteers to use different strategies to define the terms for classmates. Ask volunteers to provide examples of local or state areas that would be considered growth points. Ask students to name a cartel. (Example: OPEC) Explain that tannin is used in the process of tanning leather.

▶ LESSON 1
Physical and Economic Geography

LESSON OBJECTIVES

The student should be able to

- define the significant terms of the chapter
- locate major nations of the world and their regions
- describe major landforms of southern South America
- identify major economic activities in the region
- describe the agricultural base of the region

PURPOSE

This lesson will acquaint students with the geography and resources of southern South America.

MOTIVATOR

Write the names of the countries of southern South America on the chalkboard. Divide students into groups of five, and provide each group with a set of cutout shapes of the countries. Each group should label its country cutouts and arrange them in the appropriate relationship to each other. After students have completed the task, have them check their work against the map "Southern South America" on page 504 of their textbook. Students should rearrange and relabel their cutouts if necessary.

TEACHING DESIGN

1. Present an overview of the physical geography of southern South America, pointing out features such as landforms, major rivers and bodies of water, climate, and vegetation.
2. Direct students to work with a partner to study the various regions of each country, using information in their textbook to complete the following chart.

Regions of Southern South America				
Country	**Regions**	**Climate and Vegetation**	**Resources and Economic Activity**	**Major Cities**
Chile	(Atacama Desert)			
	(Central)			
	(Southern)			
Argentina				
Uruguay				
Paraguay				

3. Invite students to present their "Culture Trunk" reports on Chile and Argentina. Remind students to identify significant fact to write on their charts.
4. Have students complete the Chapter 43 Skills Worksheet.

CLOSURE

Divide the class in half, instructing students on one side of the room to prepare country clues about the nations of this region, using information from their Regions charts and from their "Culture Trunk" charts. As clues are stated, students on the other side of the room should name the appropriate countries.

ASSESSMENT

1. Have students complete the following questions from the Chapter 43 Check:
   Building a Vocabulary 1, 2, 3
   Recalling and Reviewing 1, 2, 3, 4
   Using Geography Skills 1, 2, 3, 4
2. Ask students to complete the Chapter 43 Content/Vocabulary Worksheet in their Workbook.

RETEACHING PLAN

Ask students to complete the Chapter 43 Reteaching Worksheet, Lesson 1.

CHALLENGE/ENRICHMENT STRATEGY

1. Ask students to research and report on animals and plants that are unique to southern South America.
2. Challenge students to collect travel brochures for a bulletin-board display about resort areas in the region.

▶ LESSON 2
Political Geography

LESSON OBJECTIVES

The student should be able to
- analyze the location of cities
- identify major historical events in this region

PURPOSE

In this lesson, students will learn the history of the nations of this region and note the relationship between political unrest and economic development.

MOTIVATOR

Ask students to name the capitals of Chile, Argentina, Uruguay, and Paraguay. Place students in groups of three, and have them check the statistical charts in the back of their textbook to find the population figures of these cities. After locating the cities on maps, students should suggest possible reasons for their development as urban centers and list these reasons on a sheet of paper beside the name of each city. Call on volunteers to offer their reasons.

TEACHING DESIGN

1. Use the Lecture Notes and Transparency Package to present an overview of the history of this region. Ask students to take notes during the lecture. After the lecture, discuss the impact of political and social unrest on economic development.
2. Invite students to present their "Culture Trunk" reports on Uruguay and Paraguay. Remind students to identify a significant fact for each topic to write on their charts.

CLOSURE

Write key dates, historical figures, and cities on the chalkboard, and call on volunteers to identify them.

ASSESSMENT

1. Have students complete the following questions from the Chapter 43 Check:
   Recalling and Reviewing 5, 6
   Critical Thinking 7
2. Ask students to complete the Chapter 43 Review/Test.

RETEACHING PLAN

1. Ask students to complete the Chapter 43 Reteaching Worksheet, Lesson 2.
2. Ask students to prepare "Who Am I?" or "Where Am I?" cards for the information in this lesson. Each student should combine cards with those of a partner, draw cards, and practice identifying the correct person or place.

CHALLENGE/ENRICHMENT STRATEGY

1. Have students complete the Chapter 43 Critical Thinking Worksheet.
2. Have students complete the Chapter 43 Challenge Worksheet.
3. Ask students to report on the gauchos of Argentina and compare them to the cowboys of the United States.

Chapter 43 Check 509, Answers

Building a Vocabulary

1. A site designated by the government for growth.
2. An agreement among producers of a good to limit supplies and thereby keep prices high. To control a market. High prices.
3. A product that comes from quebracho trees, used in preparing leather. Argentina.

Recalling and Reviewing

1. Almost every landform is found in Chile—desert, fertile valleys, forests, and mountains.
2. Mining and fishing.
3. The pampas are the heartland of Argentina, a rich agricultural region and the most densely settled region.
4. Northern Argentina: grazing livestock. Production of tannin in northwestern Argentina. Western Argentina: livestock grazing. Tourism has developed around mountain resorts in the foothills of the Andes.
5. Uruguay provides general equality and many economic opportunities. It has a large middle class and a high literacy rate.
6. Eastern Paraguay is a humid subtropical lowland that rises to join the highlands of southern Brazil. It is a productive agricultural area that grows cotton, rice, tobacco, sugarcane, and tea. The western section, in the Chaco, is a less humid region of dry forests used mostly for grazing. Quebracho trees are found here.

Critical Thinking

7. Answers will vary, but students should point out the difference that government stability has made to the economic development of Uruguay.

Using Geography Skills

1. Buenos Aires, Argentina.
2. Along the coast. Because the Atacama Desert and Andes Mountains lie inland and provide fewer opportunities for trade, industry, and agriculture.
3. Four. Tropical and subtropical desert, continental steppe, Mediterranean, highland.
4. Atlantic Ocean, Straight of Magellan, Río de la Plata, Uruguay River.

UNIT 8
Review/Test

PLANNING GUIDE		
Lesson	Textbook	Teacher's Resource Binder
1. Unit 8 Review	pp. 510–511	
2. Unit 8 Test		Unit 8 Test

TIME ALLOCATION: 2 days

▶ LESSON 1
Unit 8 Review

Have students read the Unit 8 Summary. Use the Unit 8 Objectives as the basis for review, and discuss any questions that students have on the Unit 8 material.

Have students complete the questions in the textbook under the following headings:
Reading and Understanding
Mastering Geography Skills

Have students read the textbook material under the following headings and complete the appropriate or selected activities:
Applying and Extending
Linking Geography and Economics

▶ LESSON 2
Unit 8 Test

Have students complete the Unit 8 Test from the Teacher's Resource Binder.

Unit 8 Review, pages 510–511, Answers

Reading and Understanding

1. Europeans lost control of Latin America during the first quarter of the nineteenth century. Ideas of equality and independence from the American and French revolutions became well known in Latin America.
2. Variety of landforms, including high mountains, highlands, plains, plateaus, and coastal lowlands. Biomes vary from desert to humid tropical forests. Because Latin America extends across nearly 90 degrees of latitude, it includes a full range of climate regions.
3. The Andes have caused desert and continental steppe climates by blocking moisture-bearing winds from southern Argentina, northern Chile, and southern Peru. The altitude of the mountains creates a variety of climates, from the *tierra caliente* (warm) to the *tierra helada* (cold) at different elevations. Mountains have cut off countries and groups from each other, made possible the isolation and preservation of some Indian cultures, made communication and unity difficult in some countries, concentrated population along the coast in some countries, and provided mild climates for human settlement in some countries.
4. The pampas of Argentina, the llanos of Venezuela, and the plains south of the rain forest in the Amazon Basin. Mainly for cattle grazing and some agriculture.
5. For transport and to generate hydroelectric power (as in Paraguay).
6. Much of Latin America's agricultural land has been held in large estates by the wealthy and powerful. More people have desired land of their own to work, thus creating a desire for land reform. The process of land reform is still going on, so it is too early to tell what the long-range effects will be on agricultural productivity.
7. Rivers for hydroelectric power; industrial minerals such as copper, tin, iron ore, bauxite, and petroleum. Traditional ways of doing things, the isolation of settled areas from each other by rugged terrain, poor transportation and communication, political instability, a harsh climate, and little money for foreign investment.
8. Urban areas have attracted the rural unemployed who hope to find work in the cities. They bring little money and few job skills with them. This migration to the cities has created large slums.
9. Development of these areas has meant the destruction of rain forests. If the rain forests disappear, many unique plants and animal species will also disappear. The burning of trees releases large amounts of carbon dioxide into the atmosphere, which eventually could affect world climate patterns. The erosion of good soil will also result from deforestation.

Mastering Geography Skills

The cartogram should compare the GNP of Latin American countries.

Applying and Extending

Answers will vary.

Linking Geography and Economics

Answers will vary; however, students should mention that availability and cost of labor and capital would influence the choice of method that a producer would make. The plentiful supply of inexpensive labor and the lack of capital would help explain why labor-intensive industries are prevalent in Latin America.

UNIT 9
The United States and Canada
Pages 512–601

UNIT OVERVIEW

This unit presents a geographic overview of the United States and Canada. The significance of the regional landforms, history, climate, and economic activity is presented in each chapter. The unique qualities of each region are captured in regional overviews. Students will read about resources employed today in the pursuit of growth and about resources expected to be developed to ensure future growth. Similarities and differences between Canada and the United States are discussed, as well as the mutual goals and special relationships between the two countries.

CHAPTER TIME LINE

Chapter	Title	Time Allocation
44	The North American Land and People	3 days
45	New England	2 days
46	The Middle Atlantic States	2 days
47	The Midwest	2 days
48	The South	2 days
49	The Great Plains, Rockies, and Intermountain Region	2 days
50	The Pacific Coast	2 days
51	Alaska and Hawaii	2 days
52	Canada	2 days
53	Canadian Provinces and Frontier Territories	2 days
	Unit 9 Review	2 days
	Unit 9 Test	1 day
		24 days

MAJOR TOPICS

- The Influence of Landforms and Climate on Historical Development
- The Influence of Resources and People on a Nation's Standard of Living
- The Relationship Between the United States and Canada
- Ensuring Future Growth and Development, Conserving Resources, and Protecting the Environment

INTRODUCING THE UNIT

Using a large map of the United States and Canada, determine how many of the states and provinces have been visited by students in the classroom. If few students have traveled, determine how many states and provinces students have read about or seen in films. Place a pin on each state and province that has been visited by or is familiar to students. Discuss several places to determine whether or not students' impressions of them are similar. Tell students that during the study of Unit 9, they will have the opportunity to visit all 50 states and Canada without ever leaving the classroom. Ask them to suggest topics about the United States and Canada that they are interested in or curious about. List their questions on the chalkboard, and refer to them as they are answered during the study of the unit.

CHAPTER 44 The North American Land and People Pages 514–523

PLANNING GUIDE		
Lesson	Textbook	Teacher's Resource Binder
1. History and Growth	pp. 514–515	Lecture Notes and Transparency Package Content/Vocabulary Workbook Reteaching Worksheet
2. Physical Geography	pp. 515–520	Skills Worksheet Reteaching Worksheet Critical Thinking Worksheet
3. Economic Geography; Prosperity	pp. 520–522	Chapter Review/Test Reteaching Worksheet Challenge Worksheet

TIME ALLOCATION: 3 days

CHAPTER TOPICS

- Historical Geography
- Physical Geography
- Economic Geography
- Maintaining Prosperity

PRETEACHING CHAPTER VOCABULARY

Write the following terms on the chalkboard:

seaboard Gulf Stream fall line
contiguous piedmont

Indicate to students that *seaboard* and *contiguous* are terms associated with states. Allow students to identify the difference. Show examples of each type of state on a wall map. Ask students where they have heard the term *Gulf Stream*. (weather report) After making this association, ask students to hypothesize about the kind of weather that accompanies the Gulf Stream. Display pictures, and indicate sites to help students create definitions for *piedmont* and *fall line*.

▶ LESSON 1
History and Growth

LESSON OBJECTIVES

The student should be able to

- define the vocabulary of the chapter
- review and sequence key events in history
- express pride in their cultural heritage
- recognize themes used to describe the development of the United States
- project population patterns for the future

PURPOSE

In this lesson, students will learn why citizens of the United States have reason to take pride in the historical development of their country.

MOTIVATOR
Write the Pledge of Allegiance on the chalkboard. Ask students to identify this statement. Discuss their interpretations of the meaning. Indicate to students that this lesson will provide a brief review of our historical foundations.

TEACHING DESIGN
1. Present the Chapter 44 Lecture Notes and Transparency Package to review the historical development of the United States.
2. Explain to students that many themes could be used to describe the development of our country. Divide students into groups, and provide each group with one of the following pairs of words.

East	→ West	Manufacturing	→	Service
Farming	→ Manufacturing	World	→	America
Colony	→ Independence	Peace	→	War
City	→ Suburb	Rural	→	Urban

   Ask each group to prepare an explanation of the ways in which its pair of words could describe the United States, then have groups create a new pair of terms.
3. Ask each group to present its responses to the activity in Teaching Design 2.
4. Lead a class discussion comparing the development of the United States, Australia, and Latin America, using the following chart to guide the discussion.

	United States	Australia	Latin America
Time period			
Motive for colonization			
Impact on country			
Results			

CLOSURE
Ask students to write generalizations about the role of Americans in the development of the world. Invite several students to read their statements to the class.

ASSESSMENT
1. Have students complete the following questions from the Chapter 44 Check:
   Building a Vocabulary 1, 2, 3, 4
   Using Geography Skills 1
2. Ask students to complete the Chapter 44 Content/Vocabulary Worksheet in their Workbook.

RETEACHING PLAN
Ask students to complete the Chapter 44 Reteaching Worksheet, Lesson 1.

CHALLENGE/ENRICHMENT STRATEGY
1. Direct students to make a collage representing the influence of Europe on American society.
2. Ask students to conduct research to determine who settled the area where they live. Students should create an oral history tape and present it to the school library.

▶ LESSON 2
Physical Geography

LESSON OBJECTIVES
The student should be able to
- locate major cities in the United States
- describe climate characteristics
- describe the geographic regions

PURPOSE
In this lesson, students will become familiar with major U.S. cities and landform regions.

MOTIVATOR
Provide students with a United States "map of errors," a map with many major features, states, and capitals misplaced. Divide students into groups, and allot a specified time for groups to correct the maps. Then determine which group was able to correct the most errors.

TEACHING DESIGN
1. Invite students to predict what the weather will be like next month, then ask them to predict the weather six months from now. Ask students why we are able to make such predictions. Suggest that students fill in the following chart.

City	Latitude/ Longitude	Climate	Weather Description	
			January 1	July 1
New York City				
New Orleans				
Omaha				
Denver				
San Francisco				
Palm Springs				
Honolulu				
Anchorage				

2. Provide each group of students with jigsaw-puzzle cutouts of one landform region in the United States. Ask groups to select their own major cities, features, and characteristics to label. Invite one member from each group to describe its region and to help complete the jigsaw puzzle on a bulletin board or chalkboard.

CLOSURE
Ask various students to come to a wall map of the United States and identify as many landform regions as possible.

ASSESSMENT
1. Have students complete the following questions from the Chapter 44 Check:
   Recalling and Reviewing 1, 2
   Using Geography Skills 2, 3
2. Have students complete the Chapter 44 Skills Worksheet.

RETEACHING PLAN

1. Ask students to complete the Chapter 44 Reteaching Worksheet, Lesson 2.
2. Direct students to label a map of the United States showing climate and landform regions.

CHALLENGE/ENRICHMENT STRATEGY

1. Ask students to answer the following questions: Which landform region has the densest population? the least population? Which would they prefer to live in? Which do they inhabit?
2. Have students complete the Chapter 44 Critical Thinking Worksheet.

▶ LESSON 3

Economic Geography

LESSON OBJECTIVES

The student should be able to

- identify resources and their uses in the United States
- discuss the importance of conservation of resources
- cite examples of major economic activities

PURPOSE

This lesson will help students understand the importance of natural resources to the standard of living of a nation and the importance of conserving resources.

MOTIVATOR

Draw the following diagram on the chalkboard.

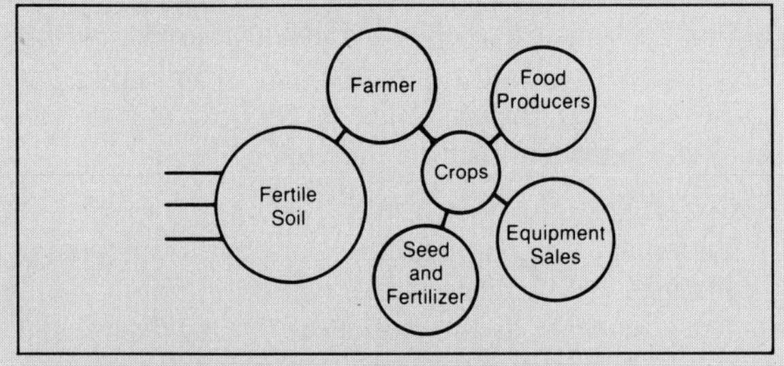

Ask students to suggest appropriate terms to add to the diagram. Think of jobs and products available as a result of the resources. Discuss the potential impact on a nation that has many valuable resources.

TEACHING DESIGN

1. Continue the Motivator activity by dividing the class into five groups and asking each group to build on one of the following clusters.

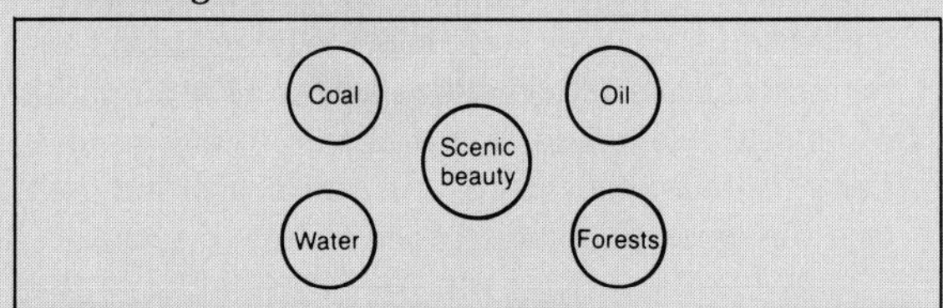

2. Ask students to imagine that they are about to embark on a train trip across the United States and Canada. As they travel, they will stop to visit several places. At each stop, they will learn about the land, people, economic activities, and significant places to tour.

Divide the class into 10 groups, and assign each group a region of the United States and Canada. Each group should plan a train trip across its region. Each group's presentation should include at least five stops and pictures to accompany the presentation sites. As the train pulls out from each site, a tour guide should provide travel time and descriptions of the land, people, special places to visit, major cities, and economic development. Give each group time to begin research.

CLOSURE

Ask students to review the resources that helped our nation prosper. Discuss the importance of resources and ways in which citizens should work to conserve resources.

ASSESSMENT

1. Have students complete the following questions from the Chapter 44 Check:
   Recalling and Reviewing 3
   Critical Thinking 4
2. Ask students to complete the Chapter 44 Review/Test.

RETEACHING PLAN

1. Ask students to complete the Chapter 44 Reteaching Worksheet, Lesson 3.
2. Direct students to create a collage that represents the vast resources of the United States.

CHALLENGE/ENRICHMENT STRATEGY

1. Have students complete the Chapter 44 Challenge Worksheet.
2. Ask students to research and report on organizations that work for conservation of resources.

Geography Skills, page 521, Answers

1. Petroleum production in millions of barrels in selected states. A bar graph makes it possible to make a quick visual comparison of economic information.
2. Texas; Montana. California: 424.2 million; Louisiana: 515 million; New Mexico: 78 million; North Dakota: 52.6 million; Oklahoma: 192 million.
3. California, Oklahoma.
4. Texas, Louisiana, California, Oklahoma, New Mexico, North Dakota, and Montana.
5. Tourism, fruit production, land use, steel production, copper mining, and wheat production.

Chapter 44 Check, Page 523, Answers

Building a Vocabulary

1. Those that border each other as a single unit. Alaska, Hawaii.
2. A warm ocean current that moves warm water northward. Summers are hot and humid; winters are mild.
3. An area at or near the foot of a mountain region. It begins in New Jersey and stretches south to Alabama. West of the Coastal Plain at the foot of the Appalachians.
4. Where the Coastal Plain meets the Piedmont. Rivers

plunge from the Piedmont to the Coastal Plain, where softer, sedimentary rocks are found. The rapids or waterfalls can be sources of electric power.

Recalling and Reviewing

1. Eleven. Alaska. The southeast.
2. Death Valley is the lowest and Mount McKinley is the highest continental point of North America.
3. Coal, oil, minerals. Oil.

Critical Thinking

4. It must maintain a balance of trade, it must find more domestic energy resources, and it must become less dependent upon other nations for raw materials. Poverty, urban decay, and environmental pollution are additional problems to solve.

Using Geography Skills

1. Southern California, east coast, region south of the Great Lakes, some parts of the southeastern United States. Mediterranean, humid continental and humid subtropical, humid continental, humid subtropical.
2. Columbia, Mackenzie, Peace, Athabasca, Slave, Nelson, St. Lawrence, Mississippi, Ohio, Red, Rio Grande, Colorado, Arkansas, Missouri, Snake. Gulf of Mexico.
3. Approximately 1,600 miles (2,560 kilometers). Central Valley, Sierra Nevada, Great Basin, Rocky Mountains.

CHAPTER 45 New England Pages 524–529

PLANNING GUIDE		
Lesson	**Textbook**	**Teacher's Resource Binder**
1. Physical Geography	pp. 524–525	Lecture Notes and Transparency Package Content/Vocabulary Workbook Reteaching Worksheet Skills Worksheet Critical Thinking Worksheet
2. Economic and Urban Geography	pp. 525–528	Chapter Review/Test Reteaching Worksheet Challenge Worksheet

TIME ALLOCATION: 2 days

CHAPTER TOPICS

- Physical Geography
- Economic Geography
- Future Prospects

PRETEACHING CHAPTER VOCABULARY

Write the following terms on the chalkboard:

granite, reforestation, moraine, biotechnology, second-growth forest

Students should be familiar with *granite*. Ask for their ideas of a definition. Suggest that students determine whether or not the school building contains any granite in its construction. Refer students to their textbook Glossary for a definition of *moraine*. Discuss places composed of moraine deposits; share pictures if available. Ask students to brainstorm about the relationship between a *second-growth forest* and *reforestation*. Help students analyze the components of the word *biotechnology* to determine a definition.

▶ LESSON 1 Physical Geography

LESSON OBJECTIVES

The student should be able to

- define significant vocabulary terms
- describe the impact of environment on ways of living
- locate regions, landforms, and features
- analyze the development of the region's landforms
- determine the agricultural base of the region
- present criteria for determining regions

PURPOSE

In this lesson, students will learn how the impact of environment on ways of life is changing in the United States.

MOTIVATOR

Display several photographs and travel brochures of New England. Ask students to suggest words to describe this region. Discuss what they expect to learn from their study of New England. Ask whether any students have visited this region, and if so, discuss their experiences. Refer students to the map "New England States" on page 527 of their textbook to identify the states in this region.

TEACHING DESIGN

1. Present the Chapter 45 Lecture Notes and Transparency Package. Ask students to take notes.
2. Refer students to the map "The United States and Canada" on page 518 of their textbook. Have students identify the climate typical of New England. Review the characteristics of this type of climate.
3. Discuss the effect of New England's climate on its natural vegetation and ways of life. Ask the following questions to guide the discussion: How long is the growing season? What crops are best suited to this environment? What adaptations must people make to live in this climate? (food, shelter, clothing) What changes in the landscape have people made to live in this environment? (irrigation, dams, reforestation) Is this region hospitable or inhospitable to human habitation?
4. Draw the following chart on the chalkboard, and use it as a springboard for a discussion of New England.

	Past	Present	Future
Economic activity			
Influence of geography			

CLOSURE

Invite students to think about a response to the following question: What characteristics of the New England states enable them to be treated as a geographic region? Create a bulletin board for Unit 9 with the following chart, and continue adding ideas of students as they study each region throughout the unit. Students may contribute pictures to accompany key words.

	Regional Characteristics
New England	

ASSESSMENT

1. Have students complete the following questions from the Chapter 45 Check:
   Building a Vocabulary 1, 2, 4
   Recalling and Reviewing 1, 2, 4
   Using Geography Skills 1, 2, 3, 4
2. Ask students to complete the Chapter 45 Content/Vocabulary Worksheet in their Workbook.

RETEACHING PLAN

1. Ask students to complete the Chapter 45 Reteaching Worksheet, Lesson 1.
2. Provide students with a map of New England with numbers rather than labels for states, capitals, large cities, and significant landforms. Give students a list of correct labels, and have them practice locating these places on their maps.

CHALLENGE/ENRICHMENT STRATEGY

1. Have students complete the Chapter 45 Skills Worksheet.
2. Have students complete the Chapter 45 Critical Thinking Worksheet.
3. Show a film or color slides featuring New England.
4. Challenge students to write and perform brief skits portraying the history of New England.

▶ LESSON 2

Economic and Urban Geography

LESSON OBJECTIVES

The student should be able to

- locate major landforms and features
- describe major economic activities of the region
- explain the importance of natural resources
- determine sources of energy available in the region
- describe cities and their functions
- describe economic, cultural, and social contributions

PURPOSE

In this lesson, students will review the physical geography of New England and study the contributions of these states.

MOTIVATOR

Refer students to the map "New England States" on page 527 of their textbook. Ask students to hypothesize about the contributions of this region to the nation. Ask students to identify products they regularly use that were produced in this region. Discuss problems that might occur if New England did not belong to the United States.

TEACHING DESIGN

1. Invite students assigned New England in Chapter 44, Lesson 3, Teaching Design 2 to present their railway tour across the region. Remind other students to take notes on the presentation. They should listen for information on the following topics: Characteristics of the Land and People, Special Places to Visit, Major Cities, and Economic Development.
2. Following the railway tour presentation, have all students take out their maps of the United States, outline the New England states, and locate the key places presented in the tour.
3. Ask students to begin a features chart for this region, noting economic activities and geographic features. Review New England's geographic features and their influence on economic activity. (Coastlines enable fishing, mountains attract tourists who like to ski, and so on.) Inform students that they will be adding to this chart as they study each region of the United States and Canada.

CLOSURE

Lead students in a discussion of contributions of New England to the standard of living in the United States.

ASSESSMENT

1. Have students complete the following questions from the Chapter 45 Check:
   Building a Vocabulary 3
   Recalling and Reviewing 3, 5
   Critical Thinking 6
2. Ask students to complete the Chapter 45 Review/Test.

RETEACHING PLAN

1. Ask students to complete the Chapter 45 Reteaching Worksheet, Lesson 2.
2. Have students use the map they created in Lesson 1 as a continuing source of review.

CHALLENGE/ENRICHMENT STRATEGY

1. Have students complete the Chapter 45 Challenge Worksheet.
2. Challenge students to create television commercials advertising products manufactured in New England.

Chapter 45 Check, page 529, Answers

Building a Vocabulary

1. A gray-to-pink speckled, hard, crystalline rock. Maine, Vermont, Massachusetts, and New Hampshire.
2. A ridge of rocks, gravel, and sand that is deposited by a glacier. Cape Cod, Nantucket, and Martha's Vineyard.

3. Because most forests were cleared for farming and for New England's early shipbuilding industry. Reforestation will ensure lumber and pulp for the future.
4. Research and products in pharmaceuticals and genetic engineering. Because it has some of the nation's leading hospitals and medical schools.

Recalling and Reviewing

1. Maine, New Hampshire, Vermont, Massachusetts, Rhode Island, and Connecticut. Northern New England (Maine, New Hampshire, and Vermont) is mountainous or hilly. It is covered by the northeastern extension of the Appalachian Mountains and includes the White Mountains, the Longfellow Mountains, and the Green Mountains. It has thousands of lakes and thin, rocky soils. Southern New England (Massachusetts, Rhode Island, Connecticut) is hilly but not mountainous. It includes the Berkshires and some plains along the Atlantic coast and in the lower Connecticut River valley. The Maine coast in the north has rocky inlets and peninsulas of granite. The southern New England coast consists of moraine deposits.
2. New England has a humid continental climate with four distinct seasons. There is a maritime influence along the coast. The winters are more severe in the north. Winter storms are common along the Atlantic coast. The rocky soil, short summers, and long, cold winters limit farming and the use of machinery.
3. They are important sources of income for the region. The waters off New England are among the world's richest fishing grounds. New Hampshire and Maine have a large lumber industry. Many New England farmers receive a second income by selling forest products. New England granite and marble are important mining resources. Asbestos production is important in Vermont.
4. New England has been modernizing its remaining traditional industries and marketing new products. The area is now concentrating on high-technology industries, such as electronics equipment and computers. Government defense programs are also important.

Critical Thinking

5. Answers will vary, but students should mention the problems that older industries in New England have when they try to compete with more technologically advanced industries. Answers to the second part of the question will vary.

Using Geography Skills

1. Eastern Massachusetts and southern Connecticut. Because of the proximity to the coast where fishing and trade occur and because most industry is there.
2. Industry and farming. These are the major economic activities in the high population-density areas.
3. New Hampshire, Vermont, and Maine.
4. Approximately 42° N, 71° W. Approximately 42° N, 73° W.

CHAPTER 46 The Middle Atlantic States Pages 530–539

PLANNING GUIDE		
Lesson	**Textbook**	**Teacher's Resource Binder**
1. Physical Geography; Washington, D.C.	pp. 530–537	Lecture Notes and Transparency Package Content/Vocabulary Workbook Reteaching Worksheet Critical Thinking Worksheet Skills Worksheet
2. Economic Geography, Important Cities, and Regional Issues	pp. 532–538	Chapter Review/Test Reteaching Worksheet Challenge Worksheet

TIME ALLOCATION: 2 days

CHAPTER TOPICS

- Physical Geography
- Economic Geography
- Urban Geography
- Regional Issues

PRETEACHING CHAPTER VOCABULARY

Write the following terms on the chalkboard:

truck farms	coke	megalopolis
bituminous coal	anthracite coal	borough

Suggest that students determine three categories for these terms. (agriculture, mining, cities) Distinguish between *bituminous* (soft) *coal* and *anthracite* (hard) *coal.* Explain that bituminous coal is used in the production of coke, which is needed to make steel. Tell students that Manhattan, Staten Island, Brooklyn, The Bronx, and Queens are the five *boroughs,* or administrative units, of New York City. Ask students if they can think of a name for divisions within other cities. On a wall map, show students the megalopolis discussed in the textbook. Remind them that *mega* means "large" and that *polis* is the Greek word for "city."

▶ LESSON 1
Physical Geography; Washington, D.C.

LESSON OBJECTIVES

The student should be able to

- define significant vocabulary terms
- describe the impact of environment on ways of life
- locate regions, landforms, and features
- analyze the development of the region's landforms
- determine the agricultural base of the region
- explain criteria for determining regions

PURPOSE
In this lesson, students will extend their familiarity with the Middle Atlantic states and learn about the nation's capital.

MOTIVATOR
Read students an excerpt from an encyclopedia about the building of the Erie Canal. Survey students to determine how many are familiar with this story. Discuss the ways in which the Erie Canal affected the development of the Middle Atlantic states. Generalize about the impact of transportation on the development of regions.

TEACHING DESIGN
1. Use the Chapter 46 Lecture Notes and Transparency Package to introduce the Middle Atlantic states. Have students take notes.
2. Refer students to the climate map "The United States and Canada" on page 518 of their textbook. Have students identify the climate types of the Middle Atlantic states and review their characteristics.
3. Discuss the effect of the Middle Atlantic region's climate on its natural vegetation and ways of life.
4. Have students read the Cities of the World feature "Washington, D.C." on pages 536–537 in their textbook. Ask students the following questions: What makes Washington, D.C., unusual as a nation's capital? What makes Washington, D.C., an ideal location for a nation's capital? What distinguishes a *capitol* from a *capital*? What are the advantages of the design of the capital? What are the disadvantages?

CLOSURE
Invite students to respond to the following question: What characteristics of the Middle Atlantic states enable these states to be treated as a geographic region? Add the Middle Atlantic states to the Regional Characteristics chart on the bulletin board.

ASSESSMENT
1. Have students complete the following questions from the Chapter 46 Check:
   Building a Vocabulary 1, 2, 3, 4
   Recalling and Reviewing 2, 5
   Using Geography Skills 2
2. Ask students to complete the Chapter 46 Content/Vocabulary Worksheet in their Workbook.

RETEACHING PLAN
Ask students to complete the Chapter 46 Reteaching Worksheet, Lesson 1.

CHALLENGE/ENRICHMENT STRATEGY
1. Have students complete the Chapter 46 Skills Worksheet.
2. Have students complete the Chapter 46 Critical Thinking Worksheet.
3. Show a movie or filmstrip pertaining to the Middle Atlantic region.

▶ LESSON 2
Economic Geography

LESSON OBJECTIVES
The student should be able to
- locate regions, landforms, and features
- describe major economic activities of the region
- explain the importance of natural resources
- describe key cities and their functions
- describe economic, cultural, and social contributions

PURPOSE
In this lesson, students will review the physical geography of the Middle Atlantic states and study the contributions of this region to our nation's standard of living.

MOTIVATOR
Ask students to imagine that an undisputable document has just been found. This document indicates that the United States territory called the Middle Atlantic states still remains the property of Great Britain. Discuss with students the immediate and future problems that this situation would create in our country.

TEACHING DESIGN
1. Invite students assigned the Middle Atlantic states in Chapter 44, Lesson 3, Teaching Design 2 to present their railway tour across the region. Remind other students to take notes on the presentation. They should listen for information on the following topics: Characteristics of the Land and People, Special Places to Visit, Major Cities, and Economic Development.
2. Following the students' railway tour presentation, have all students take out their maps of the United States, outline the Middle Atlantic states, and locate the key places presented in the tour.
3. Ask students to continue filling in their features chart for this region, noting economic activities and geographic features. Review the relationship between the geographic features of the Middle Atlantic states and their influence on economic activity.

CLOSURE
Lead students in a discussion of contributions of the Middle Atlantic states to America's standard of living.

ASSESSMENT
1. Have students complete the following questions from the Chapter 46 Check:
   Recalling and Reviewing 1, 3, 4
   Critical Thinking 6
   Using Geography Skills 1
2. Ask students to complete the Chapter 46 Review/Test.

RETEACHING PLAN

1. Ask students to complete the Chapter 46 Reteaching Worksheet, Lesson 2.
2. Have students create a map of the Middle Atlantic states, using symbols to represent key cities and capitals.

CHALLENGE/ENRICHMENT STRATEGY

1. Have students complete the Chapter 46 Challenge Worksheet.
2. Lead a student debate of the advantages and disadvantages of living within a megalopolis.

Chapter 46 Check, page 539, Answers

Building a Vocabulary

1. Their products can easily be trucked to city markets.
2. Bituminous coal is used in the production of coke, which is used to purify iron ore for making steel. It is found in western Pennsylvania, West Virginia, and eastern sections of Ohio and Kentucky. Anthracite coal is hard coal. It is found in eastern Pennsylvania.
3. A megalopolis is a continuous city. New York, Philadelphia, Baltimore, and Washington, D.C. The Los Angeles area is also a megalopolis.
4. An administrative unit of a city. Manhattan, Staten Island, Brooklyn, The Bronx, and Queens.

Recalling and Reviewing

1. New York: Albany; Pennsylvania: Harrisburg; New Jersey: Trenton; Delaware: Dover; Maryland: Annapolis; West Virginia: Charleston.
2. The Appalachian Mountains, the Piedmont, and the Atlantic Coastal Plain. There is a humid subtropical climate in the south and a humid continental in the north.
3. Fresh fruits and vegetables.
4. Transportation access to national and world markets, skilled labor force, labor pool that expanded with the growing immigrant population, new canals, roads, and railroads, which made it easy to obtain raw materials from every part of the country and the world. Coal.
5. Because of commercial and industrial development.

Critical Thinking

6. The region is making a transition from heavy manufacturing to service industries. Many of the factories are old and inefficient and cannot compete with overseas and Sunbelt industries. Many factories closed, leading to extensive unemployment. Answers to the second part of the question will vary.

Using Geography Skills

1. Mining of coal, development of oil and natural gas, farming, industry. Mining of coal, development of oil and natural gas, dairy farming, farming, industry.
2. Via the Hudson and Mohawk rivers and the Erie Canal.

CHAPTER 47 The Midwest
Pages 540–547

PLANNING GUIDE		
Lesson	**Textbook**	**Teacher's Resource Binder**
1. Physical and Economic Geography	pp. 540–545	Lecture Notes and Transparency Package Content/Vocabulary Workbook Reteaching Worksheet Critical Thinking Worksheet Skills Worksheet
2. Urban Geography	pp. 545–546	Chapter Review/Test Reteaching Worksheet Challenge Worksheet

TIME ALLOCATION: 2 days

CHAPTER TOPICS

- Physical Geography
- Agricultural Regions
- Resources and Industrial Growth
- Urban Geography
- Regional Issues

PRETEACHING CHAPTER VOCABULARY

Write the following terms on the chalkboard:

township and range system
lock
crop rotation
watershed
taconite

Ask students to study the photograph of Kansas farmland on page 542 of their textbook. Explain that the *township and range system* was used to divide land in the Midwest. Each township was divided into 36 sections, one mile square. Refer students to the photograph of *locks* on page 541 of their textbook; describe how locks are used to lower and raise ships to successive water levels through canals. Discuss how *crop rotation* was practiced by the Indians, and explain the benefits. Show students a picture of a *watershed;* discuss what could make a watershed valuable. Ask students to find the definition of the term *taconite* in their textbook Glossary.

▶ LESSON 1
Physical and Economic Geography

LESSON OBJECTIVES

The student should be able to

- define significant vocabulary terms
- describe the impact of environment on ways of life
- locate regions, landforms, and features
- analyze the development of the region's landforms
- determine the agricultural base of the region
- explain criteria for determining regions

PURPOSE

In this lesson, students will become acquainted with the contributions of the Midwest to our nation's development.

MOTIVATOR

Write the following diagram on the chalkboard:

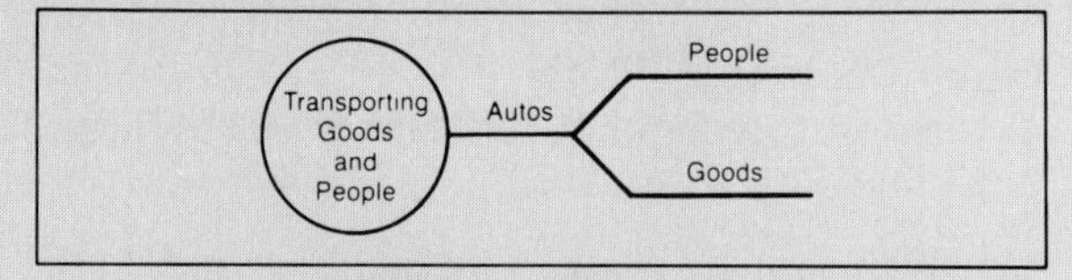

Ask students to suggest other methods we have to transport goods and people, and continue the diagram as students make their suggestions. Indicate that in their study of the Midwest, students will learn about the production and distribution of products that they use every day.

TEACHING DESIGN

1. Use the Chapter 47 Lecture Notes and Transparency Package to introduce the Midwest. Remind students to take notes.
2. Refer students to the map "The United States and Canada" on page 518 of their textbook. Have students identify the climate types of the Midwest and review their characteristics.
3. Discuss the effect of the Midwestern climate on its natural vegetation and ways of life.
4. Review the web created in the Motivator, and ask students to work in groups to complete the chart below.

Transportation					
	Automobiles	**Airplanes**	**Ships**	**Trains**	**Trucks**
Products					
Routes					

CLOSURE

Invite students to respond to the following question: What characteristics of the Midwest enable these states to be treated as a geographic region? Add the Midwest to the Regional Characteristics chart on the bulletin board.

ASSESSMENT

1. Have students complete the following questions from the Chapter 47 Check:
   Building a Vocabulary 1, 2, 3, 4
   Recalling and Reviewing 1, 2, 3, 4
2. Ask students to complete the Chapter 47 Content/Vocabulary Worksheet in their Workbook.

RETEACHING PLAN

1. Ask students to complete the Chapter 47 Reteaching Worksheet, Lesson 1.
2. Provide students with a map of the Midwest with numbers rather than labels for states, capitals, large cities, and significant landforms. Give students a list of correct labels, and have them locate these on their maps.

CHALLENGE/ENRICHMENT STRATEGY

1. Invite an investment broker to visit the class and teach students about the commodities market.
2. Have students complete the Chapter 47 Critical Thinking Worksheet.
3. Have students complete the Chapter 47 Skills Worksheet.

▶ LESSON 2
Urban Geography

LESSON OBJECTIVES

The student should be able to

- locate regions, landforms, and features
- describe major economic activities of the region
- explain the importance of natural resources
- describe key cities and their functions
- describe economic, cultural, and social contributions

PURPOSE

In this lesson, students will learn how the influence of environment on ways of life in the United States is changing.

MOTIVATOR

Indicate to students that in this chapter, there are at least four different names that apply to the Midwest. Give students time to locate these names in their textbook, and discuss the possible application of each.

TEACHING DESIGN

1. Invite students assigned the Midwest in Chapter 44, Lesson 3, Teaching Design 2 to present their railway tour across the region. Remind other students to take notes on the presentation. They should listen for information on the following topics: Characteristics of the Land and People, Special Places to Visit, Major Cities, and Economic Development.
2. Following the students' railway tour presentation, have all students take out their maps of the United States, outline the Midwest, and locate the key places presented.
3. Ask students to continue filling in their features chart for the Midwest, noting economic activities and geographic features. Review the relationship between the geographic features of the Midwest and their influence on economic activity.

CLOSURE

Lead students in a discussion of contributions of the Midwest to the standard of living within the United States.

ASSESSMENT

1. Have students complete the following questions from the Chapter 47 Check:
   Recalling and Reviewing 5
   Critical Thinking 6
   Using Geography Skills 1, 2
2. Ask students to complete the Chapter 46 Review/Test.

RETEACHING PLAN

1. Ask students to complete the Chapter 47 Reteaching Worksheet, Lesson 2.
2. Ask students to create a collage that represents the many products developed in the Midwest.

CHALLENGE/ENRICHMENT STRATEGY

1. Choose volunteers to create a map of the Midwest that shows which cities have professional football, baseball, basketball, hockey, and soccer teams. What conclusions can they draw?
2. Have students complete the Chapter 47 Challenge Worksheet.

Chapter 47 Check, page 547, Answers

Building a Vocabulary

1. The Midwest, from the air, looks like a landscape of squares. Fields, roads, and towns are all aligned. The survey system divided each township into 36 sections, each one mile square.
2. A part of a waterway enclosed by gates at each end, which is used to raise or lower boats as they pass from one water level to another. The Great Lakes.
3. It prevents the loss of minerals. Soybeans and clover.
4. A region drained by a river system. Forests are valuable wildlife and watershed areas. They provide recreational areas that are important to the Midwest's economy.

Recalling and Reviewing

1. Ohio, Indiana, Illinois, Michigan, Wisconsin, Minnesota, Iowa, Missouri, North Dakota, South Dakota, Nebraska, and Kansas. Interior Plains.
2. The rivers and lakes are used for shipping. They provide one of the largest navigation systems in the world.
3. It extends from Ohio to eastern South Dakota. Most of the corn is used to feed livestock.
4. Two types of wheat are grown to accommodate the various growing conditions in the Midwest. Spring wheat is planted in the spring and harvested in late summer. The Spring Wheat Belt is located in eastern North Dakota and South Dakota and in western Minnesota. Winter wheat is planted in the fall. It remains dormant in the winter, grows again in the spring, and is harvested in early summer. The Winter Wheat Belt is located in eastern Nebraska and Kansas but extends as far as Texas.
5. Many of the Midwest's cities are located on or near important transportation routes. Some of these industrial cities began as ports on the Great Lakes. They had access to natural resources, particularly iron ore and coal, and could ship these resources to other industrial ports with ease.

Critical Thinking

6. Factories have been refurbished; Japanese automakers have built assembly plants in the Ohio Valley and in Michigan; new industries that produce high-technology engineering products have been established. Answers to the second part of the question will vary.

Using Geography Skills

1. Michigan. Lake Superior, Lake Michigan, Lake Huron.
2. Missouri River.

CHAPTER 48 The South
Pages 548–555

PLANNING GUIDE		
Lesson	Textbook	Teacher's Resource Binder
1. Physical Geography and Cities	pp. 548–554	Lecture Notes and Transparency Package Content/Vocabulary Workbook Reteaching Worksheet Critical Thinking Worksheet Skills Worksheet
2. Economic Geography	pp. 550–553	Chapter Review/Test Reteaching Worksheet Challenge Worksheet

TIME ALLOCATION: 2 days

CHAPTER TOPICS

- Physical Geography
- Agriculture
- Resources and Industry
- Urban Geography
- Regional Issues

PRETEACHING CHAPTER VOCABULARY

Write the following terms on the chalkboard:

barrier island levee
bayou monoculture

Show a map featuring the southern coastline, and have students point out and discuss the *barrier island,* lagoons, and *bayous.* Indicate where a *levee* may be found and how it is used. Explain the relationship between the terms *monoculture* and *crop rotation.*

▶ LESSON 1
Physical Geography and Cities

LESSON OBJECTIVES

The student should be able to

- define significant vocabulary terms
- describe the impact of environment on ways of life

- locate regions, landforms, and features
- determine the agricultural base of the region
- explain criteria for determining regions

PURPOSE
Indicate to students that one purpose of this lesson is to evaluate the challenge and benefits of living in a city.

MOTIVATOR
Provide students with a physical map of the United States, and direct them to outline the South, the region they are about to study. Ask students to project the location of major cities in this area, using the skills they have learned this year. Then call on students to identify the reasons that cities are located in certain places and the services offered in cities. List their answers on the chalkboard, and save this list for use in Teaching Design 4.

TEACHING DESIGN
1. Use the Chapter 48 Lecture Notes and Transparency Package to introduce the South. Ask students to take notes.
2. Refer students to the map "The United States and Canada" on page 518 of their textbook. Have students identify the climate types of the South, and review their characteristics.
3. Discuss the effect of the South's climate on its natural vegetation and ways of life.
4. Refer students to the list created in the Motivator of services offered in cities. Ask students to create a features chart for the major cities mentioned in Chapter 48 and to list the reasons for the development of each city.

CLOSURE
Ask students to respond to the following question: What characteristics of the South enable these states to be treated as a geographic region? Add the South to the Regional Characteristics chart on the bulletin board.

ASSESSMENT
1. Have students complete the following questions from the Chapter 48 Check:
   Building a Vocabulary 1, 2, 3, 4
   Recalling and Reviewing 1, 2, 3
   Using Geography Skills 1, 2
2. Ask students to complete the Chapter 48 Content/Vocabulary Worksheet in their Workbook.

RETEACHING PLAN
1. Ask students to complete the Chapter 48 Reteaching Worksheet, Lesson 1.
2. Provide students with a map of the South with numbers rather than labels for states, capitals, large cities, and significant landforms. Give students the list of correct labels, and have them locate these places on their maps.

CHALLENGE/ENRICHMENT STRATEGY
1. Have students complete the Chapter 48 Skills Worksheet.
2. Have students complete the Chapter 48 Critical Thinking Worksheet.
3. Invite students to create a bulletin board depicting the oil industry and its products.
4. Challenge students to research and report on research centers and military bases in the South.

▶ LESSON 2
Economic Geography

LESSON OBJECTIVES
The student should be able to
- locate regions, landforms, and features
- describe major economic activities of the region
- explain the importance of natural resources
- describe key cities and their functions
- describe economic, cultural, and social contributions
- interpret a bar graph

PURPOSE
In this lesson, students will continue their study of the effects of environment on ways of life and how this relationship is changing in the United States.

MOTIVATOR
Ask students if they have heard the following word used to describe some of the southern states: Sunbelt. Ask students why this term may be applied to the region, and discuss why this region attracts people from northern states.

TEACHING DESIGN
1. Invite students assigned the South in Chapter 44, Lesson 3, Teaching Design 2 to present their railway tour across the region. Remind other students to take notes on the presentation. They should listen for information on the following topics: Characteristics of the Land and People, Special Places to Visit, Major Cities, and Economic Development.
2. Following the railway tour presentation, have all students take out their maps of the United States, outline the South, and locate the places presented in the tour.
3. Ask students to continue filling in their features chart for this region, noting economic activities and geographic features. Review the relationship between the geographic features of the southern states and their influence on economic activity.

CLOSURE
Lead students in a discussion of contributions of the South to the standard of living within the United States.

ASSESSMENT

1. Have students complete the following questions from the Chapter 48 Check:
   Recalling and Reviewing 4
   Critical Thinking 5
   Using Geography Skills 3
2. Ask students to complete the Chapter 48 Review/Test.

RETEACHING PLAN

Ask students to complete the Chapter 48 Reteaching Worksheet, Lesson 2.

CHALLENGE/ENRICHMENT STRATEGY

1. Have students complete the Chapter 48 Challenge Worksheet.
2. Challenge students to write and perform skits portraying great moments in the history of the South.

Chapter 48 Check, page 555, Answers

Building a Vocabulary

1. A long, narrow, sandy island separated from the mainland by a lagoon. Examples are Cape Hatteras, North Carolina; Miami Beach, Florida; and Padre Island, Texas.
2. In the Mississippi Delta in Louisiana, where the Mississippi River empties into the Gulf of Mexico.
3. A ridge of earth along a riverbank that prevents an area from flooding. An improved system of levees has helped prevent flooding along the Mississippi's floodplain.
4. Monoculture. It depletes soil nutrients.

Recalling and Reviewing

1. The Coastal Plain, the Piedmont, the Appalachian Mountains, the Interior Plains, and the Ozark Plateau. Major rivers: Mississippi, Ohio, Missouri, Arkansas, Red, Tennessee, Tombigbee, Rio Grande.
2. It has helped the South by depositing fertile new soil when it floods and serving as a means of transportation. It has hurt the South by causing much destruction to cities and farms when it floods its banks.
3. Oil and natural gas, petrochemical industry, coal mining, other mineral production, defense industries, textiles, and food processing.
4. Most of the population and industrial movement has been from the New England, Middle Atlantic, or Midwest states. These areas have suffered industrial and economic decline in recent years. Many immigrants have come from Latin America.

Critical Thinking

5. Answers will vary, but students should mention factors such as the development of synthetics, which might cause demand for cotton to drop. Shortages of cotton might cause demand to rise.

Using Geography Skills

1. Florida. Industry, farming, forestry, orchards, fishing, sugarcane production, livestock grazing.
2. Northern Florida and southern Georgia, parts of Mississippi and Louisiana, and much of Texas. Use of land for grazing and commercial agriculture, which require much land but few people; low coastal areas and river valleys that are subject to flooding and are poor sites for communities. Desert in some of Texas.
3. Houston: approximately 30° N, 96° W. Richmond: approximately 38° N, 78° W. Approximately 1,175 miles (1,880 kilometers).

CHAPTER 49 The Great Plains, Rockies, and Intermountain Region Pages 556–565

PLANNING GUIDE		
Lesson	Textbook	Teacher's Resource Binder
1. Physical Geography	pp. 556–564	Lecture Notes and Transparency Package Content/Vocabulary Workbook Reteaching Worksheet Critical Thinking Worksheet Skills Worksheet
2. Economic Geography, Indian Nations, and Regional Issues	pp. 557–564	Chapter Review/Test Reteaching Worksheet Challenge Worksheet

TIME ALLOCATION: 2 days

CHAPTER TOPICS

- The Great Plains
- The Rocky Mountains
- Intermountain Region
- Regional Issues

PRETEACHING CHAPTER VOCABULARY

Write the following terms on the chalkboard:

strip mining
Continental Divide
rain shadow
toponym
Mormons

Ask students if they are familiar with any of the terms on the chalkboard. Invite them to share their knowledge of the terms. Ask students in what context they have heard the use of the term *strip mining*. Discuss the relationship of strip mining to reclamation of land.

Locate the *Continental Divide* on a wall map. Ask students to indicate the significance of the Continental Divide. Explain the influence it has on the flow of rivers across the United States, and discuss how it was named.

Point out on a map a *rain shadow*. Ask students to use their knowledge of geography to describe this area. Discuss the characteristics of the land. Discuss reasons the term is characteristic of this type of land area.

List the following toponyms on the chalkboard: El Paso, Casa Grande, Mesa, and Las Cruces. Use these names to help students understand the definition of *toponym*.

Ask students what they know about the *Mormons*.

▶ LESSON 1
Physical Geography

LESSON OBJECTIVES
The student should be able to
- define significant vocabulary terms
- describe the impact of environment on ways of life
- locate regions, landforms, and features
- determine the agricultural base of the region
- explain criteria for determining regions

PURPOSE
In this lesson, students will learn how people can overcome the obstacles of an environment in order to achieve a productive way of life.

MOTIVATOR
Tell students that this chapter focuses on a region that is best studied as three subregions: the Great Plains, the Rockies, and the Intermountain region. Divide the class into small groups. Assign each group one of the subregions. Ask each group to brainstorm about terms that they would use to describe their region. List these terms on the chalkboard under the appropriate heading as follows:

Great Plains	Rockies	Intermountain Region

Discuss similarities and differences among the regions.

TEACHING DESIGN
1. Use the Chapter 49 Lecture Notes and Transparency Package to present a geographic overview of this region. Remind students to take notes.
2. Refer students to the map "The United States and Canada" on page 518 of their textbook. Have students identify the climate types of the Great Plains, Rockies, and Intermountain region. Review the characteristics of these climate types.
3. Discuss the effect of the region's climate on its natural vegetation and ways of life.

CLOSURE
Invite students to respond to the following question: What characteristics enable each area to be considered a separate subregion, and what similarities enable it to be studied as a single region called the western United States? Add this region to the Regional Characteristics chart.

ASSESSMENT
1. Have students complete the following questions from the Chapter 49 Check:
   Building a Vocabulary 1, 2, 3, 4, 5
   Recalling and Reviewing 1, 2, 3
2. Ask students to complete the Chapter 49 Content/Vocabulary Worksheet in their Workbook.

RETEACHING PLAN
1. Ask students to complete the Chapter 49 Reteaching Worksheet, Lesson 1.
2. Provide students with a map of this region with numbers rather than labels for states, capitals, large cities, and significant landforms. Give students the list of correct labels, and have them locate these places on their maps.

CHALLENGE/ENRICHMENT STRATEGY
1. Have students complete the Chapter 49 Skills Worksheet.
2. Have students complete the Chapter 49 Critical Thinking Worksheet.
3. Ask students to write and perform a skit portraying the history of Indians in this region.

▶ LESSON 2
Economic Geography and Regional Issues

LESSON OBJECTIVES
The student should be able to
- locate regions, landforms, and features
- describe major economic activities of the region
- explain the importance of natural resources
- describe key cities and their functions
- describe economic, cultural, and social contributions

PURPOSE
In this lesson, students will review the physical geography of the western United States and study the contributions of the region to the standard of living in the nation.

MOTIVATOR
Refer students to the resource map on page 559 of their textbook. Ask students to hypothesize about the contributions made by this region to the nation as a whole. Invite students to identify products that were produced in this region and that they often use. Discuss problems that would occur if this region did not belong to the United States.

TEACHING DESIGN
1. Invite students assigned the Great Plains, Rockies, and Intermountain region in Chapter 44, Lesson 3, Teaching Design 2 to present their railway tour across the region. Remind other students to take notes on the presentation. They should listen for information on the following:

Characteristics of the Land and People, Special Places to Visit, Major Cities, and Economic Development.

2. Following the railway tour presentation, have all students take out their maps of the United States, outline the states, and locate the key places presented.
3. Ask students to continue filling in their features chart with information on these states, noting economic activities and geographic features. Review the relationship between the geographic features of this region and their influence on economic activity.
4. Conduct a class discussion to complete the following:

Economic Basis			
	Past	Current	Future
Great Plains			
Rockies			
Intermountain region			

CLOSURE

Lead students in a discussion of contributions of these states to the standard of living within the United States.

ASSESSMENT

1. Have students complete the following questions from the Chapter 49 Check:
   Critical Thinking 4
   Using Geography Skills 1, 2
2. Ask students to complete the Chapter 49 Review/Test.

RETEACHING PLAN

1. Ask students to complete the Chapter 49 Reteaching Worksheet, Lesson 2.
2. Direct students to create a collage that represents the differences among the three subregions in this chapter.

CHALLENGE/ENRICHMENT STRATEGY

1. Have students complete the Chapter 49 Challenge Worksheet.
2. Encourage students to investigate careers associated with agriculture and/or national parks today.

Chapter 49 Check, page 565, Answers

Building a Vocabulary

1. Soil and rock is stripped away by large machines to get at the coal beneath the surface. It is an environmental concern. Topsoil is saved, and the land surface is returned to its original contours.
2. The Continental Divide. Eastward into the Atlantic Ocean and the Gulf of Mexico and westward and south into the Pacific Ocean and the Gulf of California.
3. A desert caused by its location on the leeward side of high mountains.
4. Names of places that reflect the history and culture of a region. El Paso, Casa Grande, Mesa, Las Cruces. Answers to the last part of the question will vary.
5. Salt Lake City. Members of the Church of Jesus Christ of Latter-day Saints.

Recalling and Reviewing

1. Coal, oil. Because of the lack of moisture, the Great Plains are better suited to grazing than to farming.
2. The Basin and Range, centered in Nevada, is a region of desert basins separated by mountains. The Colorado Plateau is centered where Utah, Colorado, Arizona, and New Mexico meet. It is a region of eroded rock layers and deep canyons, and it includes the Grand Canyon. The Columbia Plateau is centered in southern Idaho, eastern Washington, and eastern Oregon. It is mainly a series of thick layers of lava flows. Because of the wide-open spaces and sunny climate.
3. Mining, logging, livestock grazing, and tourism. Much of the region is wilderness or forest. Mining activities have decreased.

Critical Thinking

4. Answers will vary, but students should mention that coal is close to the surface, that strip mining is a relatively inexpensive method, and that strip mining removes topsoil needed to support plant life in the area.

Using Geography Skills

1. Great Plains: farming, wheat, livestock, minerals, oil, coal, and natural gas are the major economic resources. Mountain areas: concentrate primarily on forestry and minerals, some livestock, coal, and copper.
2. Montana, Wyoming, Colorado, Idaho, and New Mexico.

CHAPTER 50 The Pacific Coast Pages 566–575

PLANNING GUIDE		
Lesson	Textbook	Teacher's Resource Binder
1. Physical Geography and Regional Issues	pp. 566–574	Lecture Notes and Transparency Package Content/Vocabulary Workbook Reteaching Worksheet Critical Thinking Worksheet Skills Worksheet
2. Economic Geography, Future Growth, and Urban Geography	pp. 569–574	Chapter/Review Test Reteaching Worksheet Challenge Worksheet

TIME ALLOCATION: 2 days

CHAPTER TOPICS

- Physical Geography
- California
- Pacific Northwest
- Major Regional Issues

PRETEACHING CHAPTER VOCABULARY

Write the following terms on the chalkboard:

caldera   agribusiness

Introduce the term *caldera* with a drawing or picture of an inactive volcano. Indicate where the caldera can be found, and ask students to create a definition. Introduce the concept of *agribusiness* by comparing it to commercial farming.

▶ LESSON 1
Physical Geography and Regional Issues

LESSON OBJECTIVES

The student should be able to

- define significant vocabulary terms
- describe the impact of environment on ways of life
- locate regions, landforms, and features
- determine the agricultural base of the region
- explain criteria for determining regions
- interpret a highway map

PURPOSE

In this lesson, students will learn about one of the most valuable states in the nation in terms of the resources it provides to the entire United States. Students also will learn to use a highway map in this lesson.

MOTIVATOR

Ask students to take a few minutes to think about the past and present of California. Have students jot down a few ideas on paper using the headings Past and Present. Use the following questions to guide the discussion: How has California changed from the past to the present? Do things that attracted people to California in the past continue to attract people today? What issues now prevent people from desiring to relocate to California? What benefits does California offer residents?

TEACHING DESIGN

1. Use the Chapter 50 Lecture Notes and Transparency Package to introduce the Pacific Coast states. Remind students to take notes.
2. Refer students to the map "The United States and Canada" on page 518 of their textbook. Have students identify the climate types of the Pacific Coast states and review their characteristics.
3. Discuss the effect of the Pacific Coast region's climate on its natural vegetation and ways of life.
4. Refer students to the Themes in Geography feature "Life on the Interstate" on pages 572–573 of their textbook. Lead a class discussion on the questions in the captions to the photographs. Provide each student with a highway map of the United States. Teach students the symbols for interstate highways, state highways, and other roadways. Ask students to use this map to plan travel routes between the following locations: New York City and St. Louis; Kansas City and Detroit; Anaheim and Chicago; Dallas and Phoenix; and Houston and Denver. Teach students to use the scale to determine distances.

CLOSURE

Invite students to respond to the following question: What characteristics of the Pacific Coast enable these states to be treated as a geographic region? Add the Pacific Coast to the Regional Characteristics chart on the bulletin board.

ASSESSMENT

1. Have students complete the following questions from the Chapter 50 Check:
   Building a Vocabulary 1, 2
   Recalling and Reviewing 1, 2
   Using Geography Skills 1, 3
2. Ask students to complete the Chapter 50 Content/Vocabulary Worksheet in their Workbook.

RETEACHING PLAN

1. Ask students to complete the Chapter 50 Reteaching Worksheet, Lesson 1.
2. Provide students with a map of the Pacific Coast with numbers rather than labels for states, capitals, large cities, and significant landforms. Give students the list of correct labels, and have them practice locating these.

CHALLENGE/ENRICHMENT STRATEGY

1. Have students complete the Chapter 50 Skills Worksheet.
2. Have students complete the Chapter 50 Critical Thinking Worksheet.
3. Show a film or filmstrip about California.

▶ LESSON 2
Economic Geography

LESSON OBJECTIVES

The student should be able to

- locate regions, landforms, and features
- describe major economic activities of the region
- explain the importance of natural resources
- describe key cities and their functions
- describe economic, cultural, and social contributions

PURPOSE

At the conclusion of this lesson, students will have the advantage of knowing about all the states in the continental United States. They also will reevaluate the impact of environment on the ways people live.

MOTIVATOR

Display several pictures of Pacific Coast states, and ask students to study them carefully. Have students create a list of characteristics that make this region similar to and different from the region in which they live. (If students live in the Pacific Coast region, have them compare it to a region previously studied.)

TEACHING DESIGN

1. Invite students assigned the Pacific Coast in Chapter 44, Lesson 3, Teaching Design 2 to present their railway tour across the region. Remind other students to take notes on the presentation. They should listen for information on the following topics: Characteristics of the Land and People, Special Places to Visit, Major Cities, and Economic Development.
2. Following the railway tour presentation, have all students take out their maps of the United States, outline the Pacific Coast, and locate the key places presented.
3. Ask students to continue filling in their features checklist for this region, noting economic activities and geographic features. Review the relationship between the geographic features of the Pacific Coast and their influence on economic activity.

CLOSURE

Lead students in a discussion of contributions of the Pacific Coast to the standard of living within the United States.

ASSESSMENT

1. Have students complete the following questions from the Chapter 50 Check:
   Recalling and Reviewing 3
   Critical Thinking 4
   Using Geography Skills 2

RETEACHING PLAN

Ask students to complete the Chapter 50 Reteaching Worksheet, Lesson 2.

CHALLENGE/ENRICHMENT STRATEGY

1. Have students complete the Chapter 50 Challenge Worksheet.
2. Invite students to report on the history and current status of the film industry. Discuss why it got its start in California and why it is expanding to alternative sites.

CHAPTER 50 Check, page 575, Answers

Building a Vocabulary

1. A depression formed after a volcanic eruption or collapse of a volcanic mountain. Crater Lake.
2. Farming done on large modern farms, many of them owned by large corporations.

Recalling and Reviewing

1. Maritime and Mediterranean. Southern California has a Mediterranean climate. Very little rain falls from April to November. Hot, dry winds, called the Santa Ana winds, can cause forest and brush fires.
2. Alfalfa, almonds, grapes, olives, peaches, pears, rice, sugar beets, tomatoes, apricots, asparagus, beans, cherries, cotton, figs, melons, onions, oranges, plums, potatoes, artichokes, apples, celery, cucumbers, garlic, oranges, lemons, limes, avocados, strawberries, cattle, and dairy products. Vineyards.
3. California: minerals, oil, water, forests, fish, and industry. Pacific Northwest: fish, forests, and water.

Critical Thinking

4. Answers will vary, but students should mention uses for construction, paper and pulp products, and tourism in the region's national parks. Answers to the rest of the question will vary.

Using Geography Skills

1. Columbia River. Portland.
2. Northern California, Oregon, and Washington.
3. Approximately 38° N, 123° W. About 350 miles (560 kilometers).

CHAPTER 51 Alaska and Hawaii Pages 576–583

PLANNING GUIDE		
Lesson	Textbook	Teacher's Resource Binder
1. Physical Geography, History, and Culture	pp. 576–582	Lecture Notes and Transparency Package Content/Vocabulary Reteaching Worksheet Critical Thinking Worksheet Skills Worksheet
2. Economic Activity and Future Development	pp. 578–582	Chapter Review/Test Reteaching Worksheet Challenge Worksheet

TIME ALLOCATION: 2 days

CHAPTER TOPICS

- Alaska
- Hawaii

PRETEACHING CHAPTER VOCABULARY

Write the following terms on the chalkboard:

panhandle hot spot

Ask students how they have heard the word *panhandle* as it relates to geography. Suggest to students that they try to locate a panhandle on the map "Alaska" on page 578 of their textbook. Refer students to a wall map of the United States, and point out additional states with panhandles (Texas, Oklahoma, Florida). Indicate to students that the term *hot spot* refers to potential volcanic activity.

▶ LESSON 1

Physical Geography, History, and Culture

LESSON OBJECTIVES

The student should be able to

- define significant vocabulary terms
- describe the impact of environment on ways of life
- locate regions, landforms, and features
- determine the agricultural base of the region
- explain criteria for determining regions
- differentiate between fact and opinion

PURPOSE

This lesson concludes the study of the United States and gives students knowledge of all 50 states.

MOTIVATOR

Write the names of the states Alaska and Hawaii on the chalkboard. Ask students to work with a partner and write pairs of phrases that describe some of the differences between these two states. Allow students about four minutes to brainstorm. Collect their phrases, write them on the chalkboard, and leave them visible during the lesson.

TEACHING DESIGN

1. Use the Chapter 51 Lecture Notes and Transparency Package to introduce Alaska and Hawaii. Remind students to take notes.
2. Refer students to the map "The United States and Canada" on page 518 of their textbook. Have students identify the climate types of Alaska and Hawaii. Review the characteristics of these climate types.
3. Discuss the effect of Alaska's and Hawaii's climate on the natural vegetation and ways of life in these states.
4. Refer students to the list of differences between Alaska and Hawaii that they created in the Motivator. Ask students to classify the differences into the following categories: Landforms, Climate, History, Economy, People. Then ask students to think of similarities between the two states. List their thoughts on the chalkboard, and ask students to support their ideas with information in their textbook.

CLOSURE

Invite students to summarize the similarities and differences between the two states.

ASSESSMENT

1. Have students complete the following questions from the Chapter 51 Check:
   Building a Vocabulary 1, 2
   Recalling and Reviewing 1, 2, 3, 4
   Using Geography Skills 1, 2, 3
2. Ask students to complete the Chapter 51 Content/Vocabulary Worksheet in their Workbook.

RETEACHING PLAN

1. Ask students to complete the Chapter 51 Reteaching Worksheet, Lesson 1.
2. Provide students with a map of Alaska and Hawaii with capitals, large cities, and significant landforms indicated by numbers rather than labels. Give students the list of correct labels, and have them practice locating these.

CHALLENGE/ENRICHMENT STRATEGY

1. Have students complete the Chapter 51 Critical Thinking Worksheet.
2. Have students complete the Chapter 51 Skills Worksheet.
3. Show films pertaining to Alaska and/or Hawaii.
4. Invite a travel agent to tell students about the tourist attractions of Alaska and Hawaii.

▶ LESSON 2

Economic Geography

LESSON OBJECTIVES

The student should be able to

- locate regions, landforms, and features
- describe major economic activities of the region
- explain the importance of natural resources
- describe key cities and their functions
- describe economic, cultural, and social contributions

PURPOSE

This lesson will give students the opportunity to conclude their study of the United States and compare other regions with their own.

MOTIVATOR

Display several pictures of Alaska and Hawaii. Invite students to identify the state each picture represents and give reasons for their choices. (This task will enable students to review the previous lesson's content.)

TEACHING DESIGN

1. Invite students assigned Alaska and Hawaii to present their railway tour across these states. Remind other students to take notes on the presentations. They should listen for information on the following topics: Characteristics of the Land and People, Special Places to Visit, Major Cities, and Economic Development.

2. Following the railway tour presentation, have all students take out their maps of the United States, outline Alaska and Hawaii, and locate the key places presented.
3. Ask students to continue filling in their features chart with information on these states, noting economic activities and geographic features. Review the relationship between the geographic features of these states and their influence on economic activity.

CLOSURE

Lead students in a discussion of contributions of Alaska and Hawaii to the standard of living within the United States.

ASSESSMENT

1. Have students complete the following questions from the Chapter 51 Check:
   Critical Thinking 5, 6
2. Ask students to complete the Chapter 51 Review/Test.

RETEACHING PLAN

Ask students to complete the Chapter 51 Reteaching Worksheet, Lesson 2.

CHALLENGE/ENRICHMENT STRATEGY

1. Have students complete the Chapter 51 Challenge Worksheet.
2. Invite an adult who remembers World War II to tell students about his or her recollection of the attack on Pearl Harbor. If possible, record an interview with this guest.

Chapter 51 Check, page 583, Answers

Building a Vocabulary

1. A panhandle is a narrow arm of land attached to a larger area; southeast Alaska fits this description.
2. An area where molten material from the earth's mantle rises through the crustal plate.

Recalling and Reviewing

1. Diomede Islands: located in the middle of the Bering Strait; the Aleutian Islands: extend to the southwest from the main peninsula; Alexander Archipelago: along the Pacific coast of the panhandle.
2. Maritime in southeast Alaska and coastal south-central Alaska, and the Aleutian Islands; subarctic in interior Alaska; tundra in the extreme north between the Brooks Range and the Arctic Ocean and south along the coast of the Bering Sea.
3. In the Pacific Ocean, about 2,500 miles (4,000 kilometers) directly south of Alaska.
4. Orientals, including a large number of Japanese. Filipinos, Chinese, Caucasians, Hawaiians, part-Hawaiians. The islands were originally settled by Polynesians from the South Pacific islands. European settlers brought laborers from Japan and China to work the fields. When Hawaii became a United States territory, a large number of workers and settlers moved to Hawaii from around the world, and Japanese continued to emigrate. Most Hawaiians live in Honolulu.

Critical Thinking

5. Answers will vary, but students should mention alternative energy sources such as geothermal-, solar-, and wind-energy systems and the production of bagasse.
6. Answers will vary. Students should mention the vast, undeveloped areas still existing in Alaska and human desire for new starts in life, economic benefits, and adventure.

Using Geography Skills

1. The states are so far apart that a map including both would show mostly ocean, and the states would be too small to distinguish any important characteristics. On the Alaska map, an inch equals about 300 miles (480 kilometers), while on the Hawaiian Islands map, an inch equals about 70 miles (112 kilometers). Different scales are used because Alaska is very large and needs to be brought down to map size, while Hawaii is relatively small, and the islands need to be shown large enough for the map to be legible.
2. Juneau is in the southeast panhandle, about 58° N and 135° W. Advantages: proximity to the rest of the United States and a milder climate than the rest of Alaska; disadvantage: distant from the rest of the state.
3. Point Barrow: approximately 72° N, 157° W; Maui: approximately 21° N, 156° W.

CHAPTER 52 Canada
Pages 584–591

PLANNING GUIDE		
Lesson	Textbook	Teacher's Resource Binder
1. Physical Geography and History	pp. 584–588	Lecture Notes and Transparency Package Content/Vocabulary Workbook Reteaching Worksheet Critical Thinking Worksheet Skills Worksheet
2. Economic Geography	pp. 588–590	Chapter Review/Test Reteaching Worksheet Challenge Worksheet

TIME ALLOCATION: 2 days

CHAPTER TOPICS

- Historical Geography
- Physical Geography
- Economic Geography
- Canadian Traits

PRETEACHING CHAPTER VOCABULARY

Write the following terms on the chalkboard:

deport dominion

Ask students to complete the chart below.

Terms	Predicted Meaning Before Reading	Actual Meaning After Reading
deport		
dominion		

▶ LESSON 1

Physical Geography and History

LESSON OBJECTIVES

The student should be able to

- define significant vocabulary terms
- describe the impact of environment on ways of life
- determine the agricultural base of the region
- explain criteria for determining regions

PURPOSE

In this lesson, students will study the geography and history of Canada, their neighboring country to the north.

MOTIVATOR

Display several pictures of the Canadian landscape and cities. Ask students to estimate where these pictures were taken. If students correctly identify Canada, discuss other countries that the pictures also could depict. Discuss reasons that Canada and the United States possess many similar characteristics, and list the ideas of the students on the chalkboard.

TEACHING DESIGN

1. Use the Chapter 52 Lecture Notes and Transparency Package to introduce Canada. Remind students to take notes.
2. Refer students to the map "The United States and Canada" on page 518 of their textbook. Have students identify the climate types of Canada. Review the characteristics of these climate types.
3. Discuss the effect of Canada's climate on its natural vegetation and ways of life.
4. Provide all students with a physical map of the United States and Canada, with the landform regions of the United States outlined. Ask students to work in groups, using the maps and their knowledge of geography to predict physical boundaries for the six landform regions of Canada. Compare the students' ideas with the actual boundaries on the map " The United States and Canada" on page 516 of their textbook.

CLOSURE

Write the following chart on the chalkboard, and invite students to suggest characteristics to list.

Characteristics of Canada	
Similar to the United States	Different from the United States

ASSESSMENT

1. Have students complete the following questions from the Chapter 52 Check:
   Building a Vocabulary 1, 2
   Recalling and Reviewing 1, 2
   Using Geography Skills 1, 2, 3, 4
2. Ask students to complete the Chapter 52 Content/Vocabulary Worksheet in their Workbook.

RETEACHING PLAN

1. Ask students to complete the Chapter 52 Reteaching Worksheet, Lesson 1.
2. Provide students with a map of Canada with numbers rather than labels for provinces, territories, capitals, large cities, and significant landforms. Give students the list of correct labels, and have them practice locating these places on their maps.

CHALLENGE/ENRICHMENT STRATEGY

1. Have students complete the Chapter 52 Critical Thinking Worksheet.
2. Have students complete the Chapter 52 Skills Worksheet.
3. Show a film or filmstrip about Canada.
4. Invite students to locate articles in magazines and newspapers about current issues facing Canada.

▶ LESSON 2

Economic Geography

LESSON OBJECTIVES

The student should be able to

- locate regions, landforms, and features
- describe major economic activities of the region
- explain the importance of natural resources
- describe key cities and their functions
- describe economic, cultural, and social contributions

PURPOSE

This lesson will show the contributions that Canadians make to the United States' economy and ways of life.

MOTIVATOR

Ask students if they are familiar with the separatist controversy in Quebec. Share with students encyclopedia entries about the Quiet Revolution, the Parti Québécois, the Front de Libération du Québec (FLQ), and René Lévesque. Lead a class discussion about whether or not students think Quebec should be separate and independent from Canada.

TEACHING DESIGN

1. Solicit from students and list on the chalkboard the significant economic activities of Canada. (Examples:

agriculture, fishing, lumbering, mining, energy, manufacturing, industry, tourism) Ask students to work in groups to research and discuss answers to the following questions: Which way of living employs the most people? Which way of living contributes the most to export income? In which region is each activity found? Is more than one activity found in any region? What geographic characteristics are critical to the success of each activity? Is a particular climate necessary for each activity? Lead a class discussion to summarize students' results.

CLOSURE

Lead students in a discussion of contributions of Canada to the standard of living within the United States.

ASSESSMENT

1. Have students complete the following questions from the Chapter 52 Check:
   Recalling and Reviewing 3
   Critical Thinking 4
2. Ask students to complete the Chapter 52 Review/Test.

RETEACHING PLAN

Ask students to complete the Chapter 52 Reteaching Worksheet, Lesson 2.

CHALLENGE/ENRICHMENT STRATEGY

1. Have students complete the Chapter 52 Challenge Worksheet.
2. Direct groups of students to create commercials advertising products from Canada.

Chapter 52 Check, page 591, Answers

Building a Vocabulary

1. The French living in Nova Scotia in 1717 were deported to what was then French Louisiana. Acadians.
2. A territory or sphere of influence.

Recalling and Reviewing

1. The Labrador Current chills the shores of eastern Canada, and winds blowing across Hudson Bay make the interior of the country very cold. The ocean and wind currents along the Pacific coast give this area a mild maritime climate.
2. The Atlantic and Pacific coastal waters are among the world's richest fishing grounds. The long, indented Atlantic coastline provides good harbors. The inland lakes and streams provide a plentiful supply of freshwater fish.
3. Fish, forests, minerals, coal, petroleum, water.

Critical Thinking

4. Answers will vary, but students should mention Quebec as a symbol of Canada's French heritage and Toronto as a symbol of its English heritage.

Using Geography Skills

1. Islands: Vancouver Island, Queen Charlotte Islands, Anticosti Island, Cape Breton Island, Prince Edward Island, Newfoundland Island, Akimiski Island, Belcher Islands, Mansel Island, Coats Island, Banks Island, Victoria Island, Prince of Wales Island, Queen Elizabeth Islands, Somerset Island, Baffin Island, Prince Charles Island. Peninsulas: Boothia Peninsula, Melville Peninsula.
2. In Quebec.
3. Beaufort Sea, Amundsen Gulf, Arctic Ocean, Baffin Bay. Because the waters are frozen for much of the year.
4. Maritime, highland, continental steppe, humid continental, subarctic, tundra. To approximately 50° N.

CHAPTER 53 Canadian Provinces and Frontier Territories Pages 592–599

PLANNING GUIDE		
Lesson	**Textbook**	**Teacher's Resource Binder**
1. Regional Overview	pp. 592–598	Lecture Notes and Transparency Package Content/Vocabulary Workbook Reteaching Worksheet Critical Thinking Worksheet Skills Worksheet
2. Regional Review and Issues	pp. 592–598	Chapter Review/Test Reteaching Worksheet Challenge Worksheet

TIME ALLOCATION: 2 days

CHAPTER TOPICS

- Maritime Provinces and Newfoundland
- Saint Lawrence–Great Lakes Provinces
- Prairie Provinces
- British Columbia
- The Yukon and Northwest Territories
- Regional Issues

PRETEACHING CHAPTER VOCABULARY

Write the following terms on the chalkboard:
acid rain muskeg regionalism

acid rain Ask if any students are familiar with this term and if any have witnessed acid rain. If so, allow students the opportunity to describe it.

muskeg (grassy bog) Refer students to the Glossary in their textbook.

regionalism Ask students to use their knowledge of the suffix *-ism* to create a definition of this term. Discuss how *regionalism* differs from and is similar to *nationalism*.

▶ LESSON 1

Regional Overview

LESSON OBJECTIVES

The student should be able to

- locate regions, landforms, and features
- describe major economic activities of the region
- explain the importance of natural resources
- describe key cities and their functions
- describe economic, cultural, and social contributions

PURPOSE

Students will have the opportunity to learn more about Canada in this lesson and decide if they might like to visit this country in the future.

MOTIVATOR

Ask students to brainstorm about reasons that people choose Canada as a vacation spot. List students' ideas on the chalkboard. Show several travel brochures about Canada. Compare the students' ideas to the brochures. Discuss similarities and differences.

TEACHING DESIGN

1. Invite students assigned Canadian Provinces and Frontier Territories in Chapter 44, Lesson 3, Teaching Design 2 to present their railway tour across the region. Remind students to take notes on the presentation. They should listen for information on the following: Characteristics of the Land and People, Special Places to Visit, Major Cities, and Economic Development.
2. Following the railway tour presentation, have all students take out their maps of Canada, outline the provinces and territories, and locate the key places presented.
3. Ask students to continue filling in their features chart with information on this region. Review the relationship between the geographic features of the region and their influence on economic activity.

CLOSURE

Call on several students to identify places they would like to visit in Canada and to give their reasons.

ASSESSMENT

1. Have students complete the following questions from the Chapter 52 Check:
   Building a Vocabulary 1, 2
   Critical Thinking 5
2. Ask students to complete the Chapter 53 Content/ Vocabulary Worksheet in their Workbook.

RETEACHING PLAN

1. Ask students to complete the Chapter 53 Reteaching Worksheet, Lesson 1.
2. Have students make charts on index cards for each of the provinces and territories studied in this chapter and fill in the key ideas. A card should be created for each of the following topics: Major Economic Activity, Population Centers, and People. A sample card follows.

Major Economic Activity	
Maritime Provinces and Newfoundland	
St. Lawrence and Great Lakes	
Prairie Provinces	
Yukon and Northwest Territories	

Have students complete the cards, exchange them with a partner, and check each other's answers.

CHALLENGE/ENRICHMENT STRATEGY

1. Have students complete the Chapter 53 Critical Thinking Worksheet.
2. Have students complete the Chapter 53 Skills Worksheet.
3. Invite students to research and report on Canada's universities, museums, and symphony orchestras.

▶ LESSON 2

Regional Issues

LESSON OBJECTIVES

The student should be able to

- describe the impact of environment on ways of life
- locate regions, landforms, and features
- determine the agricultural base of the region
- use problem-solving skills

PURPOSE

In this lesson, students will review the impact that environment has on the ways people live.

MOTIVATOR

Review with students the controversy about the political status of the French in Quebec. Suggest that students examine the positive and negative factors involved in the possibility of independent status for the French in Quebec.

TEACHING DESIGN

1. Finish any reports not completed in Lesson 1.
2. Use the Chapter 53 Lecture Notes and Transparency Package to review regional and provincial boundaries. Ensure that students add only new landmarks that they failed to include previously.
3. Refer students to the map "The United States and Canada" on page 518 of their textbook. Have students identify the climate types of the Canadian provinces and frontier territories. Review their characteristics.
4. Discuss the effect of the region's climate on its natural vegetation and ways of life.

CLOSURE

Invite students to think about a response to the following question: What characteristics do the territories and provinces have in common that enable them to be treated as a geographic region? Add the Canadian Provinces and Frontier Territories to the Regional Characteristics chart.

ASSESSMENT

1. Have students complete the following questions from the Chapter 53 Check:
   Building a Vocabulary 3
   Recalling and Reviewing 1, 2, 3, 4, 5
   Using Geography Skills 1, 2
2. Ask students to complete the Chapter 53 Review/Test.

RETEACHING PLAN

1. Ask students to complete the Chapter 53 Reteaching Worksheet, Lesson 2.
2. Have students complete the following chart:

	Maritime Provinces and Newfoundland	St. Lawrence and Great Lakes	Prairie Provinces	Yukon and Northwest Territories
Location				
Population centers				
Economic activity				
Climate				
Future development				

CHALLENGE/ENRICHMENT STRATEGY

1. Have students complete the Chapter 53 Challenge Worksheet.
2. Invite students to investigate and debate the issues concerning acid rain.

Chapter 53 Check, page 599, Answers

Building a Vocabulary

1. Sulphur dioxide generated by smoke from coal-burning power plants and factories combines with water to make sulphuric acid. This acid mixes with rainwater and damages trees and lakes. Canada traces much of its acid rain to the pollution from the central United States.
2. A forested marsh that melts only during the mosquito-breeding season in the summer. The Yukon and Northwest Territories.
3. The political and emotional support for one's region before one's country. The provinces have the power to tax and much authority over local issues. Regions are suspicious of each other. Language and culture are divisive factors.

Recalling and Reviewing

1. Some farming, dairy farming, raising animals for fur, commercial fishing, manufacturing, tourism, and mining. The population is too small to support a large home market, and the markets of southern Ontario and Quebec are too far away.
2. It is Canada's most highly developed region. This area leads Canada in population, industry, commerce, wealth, and influence. It benefits from being near the St. Lawrence River and the Great Lakes. It has plentiful water power, fertile soil, and a climate that is favorable for agriculture. The Canadian Shield's mines and forests provide raw materials for industry.
3. Potash, oil, coal, natural gas.
4. Mountains and distance.

Critical Thinking

5. Students should discuss the suspicion among provinces and territories, conflict over language and culture, economic issues, and physical isolation. Answers to the second part of the question will vary.

Using Geography Skills

1. Mining iron ore. Iron ore is a raw material needed to support the industry of the St. Lawrence region.
2. Oil drilling and mining. 0-25 persons per square mile (1-10 per square kilometer). Severe climate, few settlements.

UNIT 9
Review/Test

PLANNING GUIDE		
Lesson	**Textbook**	**Teacher's Resource Binder**
1. Unit 9 Review	pp. 600–601	
2. Unit 9 Test		Unit 9 Test

TIME ALLOCATION: 2 days

▶ LESSON 1
Unit 9 Review

Have students read the Unit 9 Summary. Use the Unit 9 Objectives as the basis for review, and discuss any questions that students have on the Unit 9 material.

Have students complete the questions in the textbook under the following headings:
Reading and Understanding
Mastering Geography Skills

Have students read the textbook material under the following headings and complete the appropriate or selected activities:
Applying and Extending
Linking Geography and History

▶ LESSON 2

Unit 9 Test

Have students complete the Unit 9 Test from the Teacher's Resource Binder.

Unit 9 Review, pages 600–601, Answers

Reading and Understanding

1. In the central United States between the Appalachians and the Rocky Mountains, and in central Canada.
2. The Appalachians are older, folded mountains with peaks that have been lowered and smoothed by erosion. The Rocky Mountains have the jagged high peaks and deep valleys that characterize young mountains.
3. Michigan, Huron, Ontario, Erie, Superior. The sheet glaciers of the Ice Ages gouged the surface of the earth to form the lakes. Because they have provided a relatively inexpensive transportation system, the Great Lakes have stimulated commerce in both the United States and Canada and promoted economic development of the Great Lakes region.
4. The western United States is part of the "Ring of Fire." Throughout the ring, there are active volcanoes and earthquake faults, making living in these areas risky.
5. Agriculture in New England is limited and is focused primarily on special crops. Agriculture in the southern states centers on commercial farming.
6. United States: in New England (especially in Massachusetts and Connecticut), in the Middle Atlantic states (especially in New York, New Jersey, and Pennsylvania), and in the Great Lakes region of the Midwest. Canada: in southern Quebec and Ontario along the St. Lawrence River and the Great Lakes. Industry has developed in these areas because of easy access to trade and transportation routes, large local markets, and accessibility of energy resources and industrial raw materials.
7. Both have a maritime climate, both are covered with forests, both have abundant hydroelectric power, both are major sources of salmon and other fish, both are important for lumbering and mining, and both have easy access to Pacific trading lanes. Because they are directly in the path of prevailing westerly winds.

Mastering Geography Skills

1. To compare wheat production in millions of bushels of selected states and provinces in 1985.
2. From the graph, you can determine the actual amount of wheat produced by these states and provinces in 1985. Maps generally provide more general information: they can show where, in what area of each state and province, and under what geographic conditions (climate, soil) the wheat is grown.
3. 323.3 million bushels.
4. Saskatchewan. Kansas.
5. Montana. Manitoba.

Applying and Extending

Answers will vary.

Linking Geography and History

Answers will vary.

UNIT 10
Sharing the World's Resources
Pages 602–623

UNIT OVERVIEW

This unit focuses on concepts that have no geographic boundaries. To enjoy the benefits of the world's diverse resources and to pursue economic progress, nations actively participate in world trade. Trade, which is linked to politics as well as to resources, also means the exchange of information and ideas. The world's increasing population has placed serious demands upon the resources that serve all people. Our dependence on the world's resources for survival makes it essential that we protect the environment and conserve renewable and nonrenewable resources.

CHAPTER TIME LINE

Chapter	Title	Time Allocation
54	World Trade	4 days
55	Preserving the World's Resources	3 days
	Unit 10 Review	1 day
	Unit 10 Test	1 day
		9 days

MAJOR TOPICS

- Interdependence
- Relationship Between Resources and Trade
- Role of Trade
- Impact of People on the Environment
- Problems of Resource Depletion
- Relationship Between Population and Resources
- Current Issues Facing the People of the World

INTRODUCING THE UNIT

Bring six editions of a major daily newspaper to class, and distribute one paper to each of six groups of students. Ask students to divide the sections of the papers among group members and look for news articles that relate to world trade and to the environment. Provide each group with scissors so that students may cut out articles. After allowing adequate time, call on two representatives from each group to read the headlines of the articles they have found about the two topics. After all groups have read their headlines, call on volunteers to make generalizations about the current events and issues. Supply a large piece of butcher paper so that students may post their articles under the two topic headings on the wall. Explain that this unit will address concepts that extend beyond geographical boundaries to encompass the entire world: world trade and preserving the world's resources. Ask students to pay special attention during the next week to news broadcasts and articles about these two topics.

CHAPTER 54 World Trade Pages 604–611

PLANNING GUIDE		
Lesson	**Textbook**	**Teacher's Resource Binder**
1. Introduction	p. 604	Content/Vocabulary Workbook Reteaching Worksheet
2. Economics of Trade; History of Trade	pp. 605–610	Lecture Notes and Transparency Guide Skills Worksheet Reteaching Worksheet Challenge Worksheet
3. Trade and Politics	pp. 608–610	Critical Thinking Worksheet Reteaching Worksheet
4. Culture and Technological Exchange	p. 610	Chapter Review/Test Reteaching Worksheet

TIME ALLOCATION: 4 days

CHAPTER TOPICS
- Economics of Trade
- Trade and Politics
- Culture and Technological Exchange

PRETEACHING CHAPTER VOCABULARY
Write the following terms on the chalkboard:

barter currency cultural exchange

Define the word *barter* for students. Several students are probably familiar with the term *currency*. Display some currency, and ask students to compare it to *change*. Ask students what they believe *cultural exchange* means. Once a definition is suggested, ask students to identify examples.

▶ LESSON 1
Introduction

LESSON OBJECTIVES
The student should be able to
- define the purpose of trade
- describe trade relationships among world nations
- review the names and locations of major world nations
- suggest benefits of international trade
- define significant vocabulary
- compare country statistics

PURPOSE
This lesson will provide students with the background necessary to understanding the workings and implications of international trade.

MOTIVATOR
Ask students to suggest products that the United States trades with other nations. List these suggestions on the chalkboard. Ask students to classify the products according to export or import status or both. Emphasize that countries exchange ideas as well as products.

TEACHING DESIGN
1. Provide all students with a world map. Divide the class into eight groups, and assign each group a culture region studied during the course. Each group should identify three to five major countries in their region. Have each group identify the major trading partners and draw arrows between these countries on their map.
2. Invite a spokesperson from each group to present its findings. As each presentation is completed, label the arrows on an oversized world map.

CLOSURE
After all groups have completed their presentations, ask the following questions: Which countries appear to have the highest percentage of countries trading with them? Which countries do not serve as any country's primary trading partner? What conclusions can be drawn about the trade of developed, less developed, and developing nations?

ASSESSMENT
1. Have students complete the following questions from the Chapter 54 Check:
   Building a Vocabulary 1
   Using Geography Skills 1
2. Have students complete the Chapter 54 Content/Vocabulary Worksheet in their Workbook.

RETEACHING PLAN
1. Ask students to complete the Chapter 54 Reteaching Worksheet, Lesson 1.
2. Have students write a paragraph describing reasons that nations trade and the goods that they trade.

CHALLENGE/ENRICHMENT STRATEGY
1. Invite students to draw circles around trade organizations named on the class map. Ask students whether or not any countries trade outside their organization, and the reason if so.
2. Refer students to *Wealth of Nations* by Adam Smith to explain the advantages of trade.

▶ LESSON 2
Economics of Trade

LESSON OBJECTIVES
The student should be able to
- calculate balance of trade and balance of payments
- describe the history and purpose of international trade
- list factors that determine price of goods and services

PURPOSE
In this lesson, students will have the opportunity to understand why nations trade and the ultimate effect of their own purchases on the United States economy.

MOTIVATOR
Before beginning this lesson, ask each student to bring from home an item that he or she would be willing to barter. Review with students the definition of the word *barter*. Conduct a trading session, and give students the time and opportunity to become satisfied with their trades.

TEACHING DESIGN
1. List on the chalkboard all the satisfactory trade agreements that were reached by students in the Motivator. Discuss why some students were easily satisfied and others were not. Relate the answers to the issues regarding reasons for trade. Present the Lecture Notes and Transparency Package. Have students take notes.
2. Divide students into groups of four. Within each group, one person must take responsibility for creating a two- to three-minute "lecturette" to teach one of the following concepts: Complementarity, Prices, Balance of Trade, Balance of Payments. Allow each group ample time to review its notes and present its lecturettes.
3. Locate and share with the class a current foreign currency exchange rate between the United States and three other countries. Use the following questions to guide a discussion: What is the source of the data? What do the phrases *weak dollar* and *strong dollar* mean? In which country would U.S. dollars be most valuable? In which country would U.S. dollars be least valuable?

CLOSURE
Draw the following diagram on the chalkboard, and call on several students to present a summary of the lesson by using the diagram:

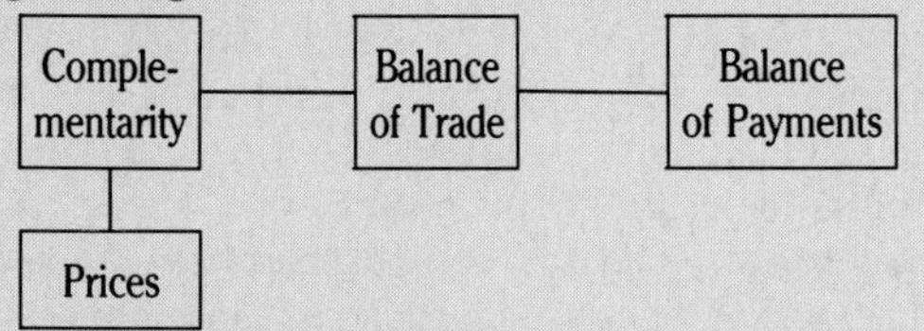

ASSESSMENT
1. Have students complete the following questions from the Chapter 54 Check:
   Building a Vocabulary 2
   Recalling and Reviewing 1, 2
   Critical Thinking 8
2. Have students complete the Chapter 54 Skills Worksheet.

RETEACHING PLAN
Ask students to complete the Chapter 54 Reteaching Worksheet, Lesson 2.

CHALLENGE/ENRICHMENT STRATEGY
1. Have students complete the Chapter 54 Challenge Worksheet.
2. Invite students to create a diagram that shows the countries from which the United States buys raw materials.

▶ LESSON 3
Trade and Politics

LESSON OBJECTIVES
The student should be able to
- explain purposes and benefits of world trade
- describe the costs and benefits of trade regulations

PURPOSE
At the conclusion of this lesson, students will be able to evaluate more carefully whether or not to purchase an item produced in a foreign country.

MOTIVATOR
Bring to class several items that were produced in countries other than the United States. Ask students to try to identify the country in which each item was produced. Ask students to describe items of theirs that were produced in other countries. Brainstorm with the class about reasons people purchase goods produced in other countries.

TEACHING DESIGN
1. Write the following phrases on the chalkboard:
   $250 billion   One out of every six American dollars

   Ask students to predict how these amounts might relate to the subject of the United States and world trade. ($250 billion is the value of U.S. imports. One out of every six American dollars is spent on import items.)
2. Use the following questions to stimulate discussion of the significant ideas related to this lesson. Why do nations trade? Why do nations regulate trade? How do nations regulate foreign trade?
3. Divide the class into four groups, and ask each group to respond to one of the following statements:
   a. The United States should increase tariffs to protect the American automobile industry.
   b. The United States should support global free trade.
   c. The EC should merge with COMECON.
   d. The United States should initiate an embargo against products of Japanese technology.

   Ask each group to select a spokesperson to share the group's response with the class. Refer students to the discussion in the Motivator to review key ideas. Ask them to assess the positive and negative results of their statements, the benefits and the costs. Give students time to prepare, then call on each group's spokesperson to share the group's response.

CLOSURE
Create a futures wheel focusing on the issue of the abolition of all tariffs on United States imports to summarize the costs and benefits of international trade.

ASSESSMENT
1. Have students complete the following questions from the Chapter 54 Check:
   Recalling and Reviewing 3, 4, 5
2. Ask students to complete the Chapter 54 Critical Thinking Worksheet.

RETEACHING PLAN
Ask students to complete the Chapter 54 Reteaching Worksheet, Lesson 3.

CHALLENGE/ENRICHMENT STRATEGY
1. Challenge students to write a letter to a local newspaper editor to support or reject higher tariffs and stiffer United States import quotas.
2. Invite students to research and report on the laws imposed in other countries to protect their nations from imports and balance of payment problems.

▶ LESSON 4
Culture and Technological Exchange

LESSON OBJECTIVES
The student should be able to
- evaluate costs and benefits of international trade
- interpret purposes for regional trading groups, multinational companies, and foreign aid
- explain the benefits of cultural and technological trade

PURPOSE
In this lesson, students will understand the value of international exchange as it affects their standard of living.

MOTIVATOR
Invite students to think of items from other countries that they have at home. Have them list five items in the chart below, and complete the other information requested:

Item	Country	Why did you purchase this item made in a foreign country rather than one made in the United States?
1.		

TEACHING DESIGN
1. Discuss with students the costs to our country every time they purchase an item produced in another country. Compare these costs with the benefits. Ask students to investigate and report on activities of American companies to rebuild damaged American markets.
2. Discuss which industries benefit and which suffer when the United States' balance of payments is negative. Reverse the question, assuming that the balance of payments is positive.
3. Divide the students in groups, and ask them to identify examples of cultural exchange that occur throughout the world. (Example: Russian ballet troupes performing in the United States.) Have students discuss how cultural exchange affects cultural diffusion and trade.

CLOSURE
Lead a closing discussion on the subject of international interdependence. What is it? Why must it occur?

ASSESSMENT
1. Have students complete the following questions from the Chapter 54 Check:
   Recalling and Reviewing 6
   Critical Thinking 7
2. Ask students to complete the Chapter 54 Review/Test.

RETEACHING PLAN
Ask students to complete the Chapter 54 Reteaching Worksheet, Lesson 4.

CHALLENGE/ENRICHMENT STRATEGY
1. Invite a former worker in the Peace Corps to visit the class and to describe his or her experiences.
2. Challenge students to research and identify the 10 countries that receive the greatest proportion of the United States' foreign aid budget and determine the interests of the United States in these countries.

Geography Skills, page 609, Answers
1. United States.
2. Calorie consumption in Iraq: 2,155; in Sweden: 3,146. Life expectancy at birth in Iraq: 56 (male), 59 (female); in Sweden: 73 (male), 79 (female).

Chapter 54 Check, Page 611, Answers

Building a Vocabulary
1. Money. Countries buy foreign currency to buy foreign goods; for example, an American company may buy French francs with American dollars to buy French goods, and French companies can use those American dollars to buy American goods.
2. Students may choose three examples from the following: musical groups, art exhibitions, dance troupes, circuses, international athletic competitions.

Recalling and Reviewing
1. Countries sell goods they can produce most efficiently, and they buy goods that others can produce more easily.

Prices fluctuate mainly because of a change in supply of or demand for the goods.
2. As demand increases, the price increases; as supply increases, the price decreases.
3. By selling more than it buys. If it buys more than it sells.
4. The General Agreement on Tariffs and Trade, an agreement to reduce tariffs and trade barriers. It was concluded in 1947 among 23 nations that controlled 80 percent of international trade. Members can protect their balances of payments by placing taxes on imports that compete unfairly with goods from local industries. Less developed countries can restrict imports in order to aid industries that are not fully established.
5. Members bargain as if they formed one country in the world market. They try to increase trade among member countries. As industrialization develops, these countries can use their raw materials and exchange them within the group for manufactured goods.

Critical Thinking

6. Aid, often in the form of a low-interest loan, enables the less developed country to purchase specific products from the developed country.
7. Answers will vary, but students should mention that currency facilitates the exchange of goods that have different value. Answers to the second part will vary.

Using Geography Skills

1. Generally, the standard of living in developed countries is higher. Answers to the second part will vary.

CHAPTER 55 Preserving the World's Resources Pages 612–621

PLANNING GUIDE		
Lesson	**Textbook**	**Teacher's Resource Binder**
1. Resources: Soil, Forests, and Water	pp. 612–615	Content/Vocabulary Workbook Reteaching Worksheet
2. Resources: Air and Energy; Population Growth	pp. 615–620	Lecture Notes and Transparency Package Reteaching Worksheet Challenge Worksheet Critical Thinking Worksheet
3. Special Features and Geography Skills	pp. 606–607, 609, and 618–619	Skills Worksheet Chapter Review/Test Reteaching Worksheet

TIME ALLOCATION: 3 days

CHAPTER TOPICS

- Soil
- Forests
- Water
- Air
- Energy
- Population Growth and Resources

PRETEACHING CHAPTER VOCABULARY

Write the following terms on the chalkboard:

renewable resources, nonrenewable resources, salinization, strip-cropping, ozone layer, greenhouse effect

Ask students to rate their knowledge of the terms in the chapter by completing the following chart before and after studying the chapter.

Words	I Can Define	I Have Seen/Heard	???

Ask students to review terms they cannot define and to underline any part(s) of the words that they are familiar with. After students have completed their ratings, call on volunteers to group the terms by suggesting categories.

▶ LESSON 1
Resources: Soil, Forests, and Water

LESSON OBJECTIVES

The student should be able to
- define the significant terms of the chapter
- identify major natural resources of the world
- explain the economic importance of natural resources

PURPOSE

This lesson will heighten student awareness of people's impact on the environment and the need for responsible use and preservation of the world's resources.

MOTIVATOR

Use the Chapter 55 Changing Environment Transparency Series to focus students' attention on people's impact on the environment and the problem of resource depletion. Project the original scene, and ask students to describe what they see. With each overlay, ask students to note changes that have occurred in the scene. Ask students to voice their reactions to the changes in the environment.

TEACHING DESIGN

1. Write the headings Renewable Resources and Nonrenewable Resources on the chalkboard. Ask students to brainstorm about resources for each heading, and record their ideas on the chalkboard. Review the lists to make sure there is a consensus on the classification.
2. Call on volunteers to name the ways in which soil fertility is reduced, mineral content is altered, and erosion can occur. Ask students to work in groups of three to

complete the following conservation checklist. Students should list additional soil conservation methods and place a check mark in the appropriate column(s).

Conservation Method	Prevents Topsoil Runoff	Prevents Wind Erosion	Replenishes Fertility
Contour plowing			

3. Ask students to continue working in their groups to create a diagram on poster board illustrating the negative effects of deforestation. For each major effect shown, subsequent negative effects should also be shown.
4. Ask each group to create organization webs to identify facts about the topics related to water, using the following examples.

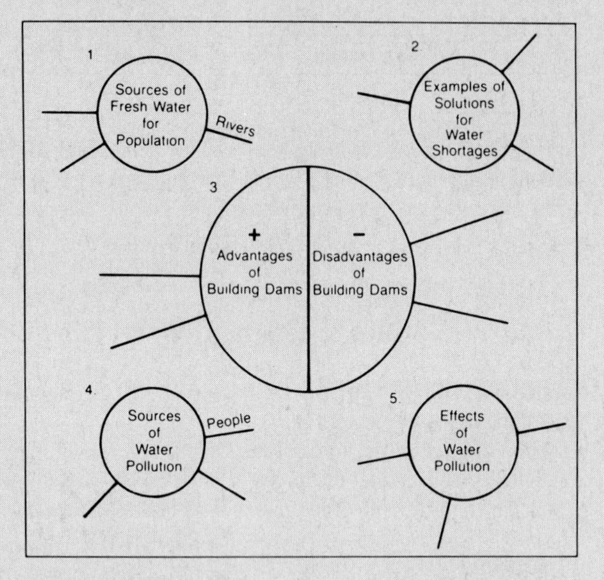

CLOSURE

Call on volunteers to present each group's information on soil conservation, deforestation, and water. Identify areas of consensus.

ASSESSMENT

1. Have students complete the following questions from the Chapter 55 Check:
   Building a Vocabulary 1, 2, 3
   Recalling and Reviewing 1, 2, 3
   Using Geography Skills 1
2. Ask students to complete the Chapter 55 Content/ Vocabulary Worksheet in their Workbook.

RETEACHING PLAN

Ask students to complete the Chapter 55 Reteaching Worksheet, Lesson 1.

CHALLENGE/ENRICHMENT STRATEGY

1. Ask students to interview a variety of people who work in the fields of environmental protection and/or conservation, then report their findings to the class.
2. Invite a member of the local water department to share information about sources of water, types of pollutants, and methods used to purify the water in your area.

► LESSON 2

Resources: Air and Energy; Population Growth

LESSON OBJECTIVES

The student should be able to

- identify major natural resources of the world
- explain the importance of natural resources
- identify environmental issues associated with cities

PURPOSE

Students will continue their study of people's impact on the environment and the need for conserving resources.

MOTIVATOR

Collect pictures showing scenes of air pollution in large cities. Ask students to describe the air quality. Show scenes of places with relatively clean air, and ask students again to describe the air quality.

TEACHING DESIGN

1. Present the Chapter 55 Lecture Notes and Transparency Package. Have students take notes.
2. Prior to class, ask the school librarian to assist in gathering resource materials directly related to the lesson topics. Divide students into six groups, and have each group investigate and report on one of the following topics: Urban and Industrial Smog, World Climate Change, Fossil Fuels, Nuclear Energy, Alternative Energy Resources, Population Distribution and Growth, and Population and Resource Balance. Inform students that each group should present a news broadcast to inform the public of the current facts and future trends about the topic. Each group will have to condense the information into a three-minute broadcast to the public. Allow students 15 minutes to gather information.

CLOSURE

Call on students to present their news broadcasts.

ASSESSMENT

1. Have students complete the following questions from the Chapter 55 Check:
   Building a Vocabulary 4, 5
   Recalling and Reviewing 4, 5
   Critical Thinking 6, 7, 8
2. Ask students to complete the Chapter 55 Content/ Vocabulary Worksheet in their Workbook.

RETEACHING PLAN

Ask students to complete the Chapter 55 Reteaching Worksheet, Lesson 2.

CHALLENGE/ENRICHMENT STRATEGY

1. Have students complete the Chapter 55 Challenge Worksheet.
2. Have students complete the Chapter 55 Critical Thinking Worksheet.
3. Conduct a poster contest with the theme Preserving the World's Resources. Ask school and community officials to be judges. Display the posters throughout the school, and award prizes to the best posters.

▶ LESSON 3
Special Features/Geography Skills

LESSON OBJECTIVES

The student should be able to

- describe major landforms and features of the earth
- analyze the site and situation of cities
- explain how geographers use national statistical data
- compare statistical information from various countries

PURPOSE

This lesson will complete the presentation of special features and skills features in the Unit 10 materials.

MOTIVATOR

Present students with the following brainteaser:

> What constitutes 10 percent of the world's land area, is claimed by seven countries, and was not visited by people until the twentieth century?

After students respond, refer them to pictures of Antarctica in their textbook, and ask them why this continent is called the last frontier. Try to reach a class consensus.

TEACHING DESIGN

1. Provide a number of globes and world atlases so that students may work in small groups to study Antarctica. Ask them to discuss the following questions, using maps and information in their textbook: What cities are located in Antarctica? What longitude lines meet at the South Pole? What is the highest point on Antarctica? What bodies of water surround Antarctica? In what area of the continent are the most features named? What features make the continent unattractive for settlement? Allow five minutes for discussion, then ask each group to develop a summarizing statement.
2. Ask students to read the Cities of the World feature "Rotterdam" on pages 606–607 of their textbook before completing the following tasks and questions with a partner: Find this city on the feature map. Describe the location of the city. What factors contributed to the growth of this city? What features does this city have in common with others that you have studied?
3. Lead students through the Geography Skills feature "Comparing Country Statistics" on page 609 of their textbook. After completing the lesson, ask students to identify the steps in reading statistical tables.

CLOSURE

Call on volunteers to share the summary statements in Teaching Design 1 and the answers to the questions in Teaching Design 2.

ASSESSMENT

1. Have students answer the questions under "Test Your Skills" on page 609 of the Geography Skills feature.
2. Have students complete the Chapter 55 Skills Worksheet.
3. Ask students to complete the Chapter 55 Review/Test.

RETEACHING PLAN

Ask students to complete the Chapter 55 Reteaching Worksheet, Lesson 3.

CHALLENGE/ENRICHMENT STRATEGY

1. Ask students to investigate and report on exploration expeditions and research activities in Antarctica.
2. Ask students to collect examples from newspapers, newsmagazines, and other print media of statistical tables to display in the classroom.

Chapter 55 Check, page 621, Answers

Building a Vocabulary

1. Renewable resources can be replenished; examples are solar energy, water, forests, and fish. Nonrenewable resources are not replaced by natural resources or are replaced at extremely slow rates; examples include oil, coal, and minerals.
2. salinization. The evaporation of irrigation water leaves behind a hard salt layer.
3. Planting alternate rows of close-growing crops and open-growing crops. To prevent wind erosion.
4. In the upper atmosphere. Ultraviolet radiation.
5. The buildup of heat caused by the buildup of carbon dioxide in the lower atmosphere, which in turn causes a long-term warming of the planet. The burning of fossil fuels and the cutting of the tropical rain forests.

Recalling and Reviewing

1. Contour farming and strip-cropping. Fertilizers, crop rotation, cover crops, and shelterbelts.
2. Soil resources must be as productive as possible to produce sufficient food for the world's people.
3. Dams, aquifers, deep wells, and desalinization plants.
4. Coal, petroleum, and natural gas. The supply of fossil fuels may not keep up with the demand much longer.
5. The world's resources are limited.

Critical Thinking

6. Answers will vary, but students should mention the by-products of manufacturing and industry and the burning of fossil fuels that create pollution.
7. Biomes are interrelated. When individuals introduce a pollutant into their own biome, it is carried by wind and/or water to biomes in other parts of the world.

8. New trees must be planted to replace the ones that are destroyed. Renewal requires foresight and planning.

Using Geography Skills

1. A tropical rain forest, which includes many different kinds of trees. The soil of this area is probably infertile. The nutrients are in the plants rather than in the soil.

UNIT 10 Review/Test

PLANNING GUIDE		
Lesson	Textbook	Teacher's Resource Binder
1. Unit 10 Review	pp. 622–623	
2. Unit 10 Test		Unit 10 Test

TIME ALLOCATION: 2 days

▶ LESSON 1
Unit 10 Review

Have students read the Unit 10 Summary. Use the Unit 10 Objectives as the basis for review, and discuss any questions that students have on the Unit 10 material.

Have students complete the questions in the textbook under the following headings:

Reading and Understanding
Mastering Geography Skills

Have students read the textbook material under the following headings and complete the appropriate or selected activities:

Applying and Extending
Linking Geography and Economics

▶ LESSON 2
Unit 10 Test

Have students complete the Unit 10 Test from the Teacher's Resource Binder.

Unit 10 Review, pages 622–623, Answers

Reading and Understanding

1. The exchange of goods and ideas. To obtain products not available at home, to improve the standard of living and quality of life. A country exports what it can produce efficiently and imports what it cannot make easily.
2. Because of changes in supply and demand. As demand increases, the price increases. As supply increases, the price decreases.
3. Large international debts, insufficient income to repay them, and the consumption of imports are signs that a country does not have a healthy balance of payments.
4. Andean Economic Community, Central American Free Trade Area, Latin American Free Trade Association, European Community, and COMECON. These groups, and GATT as well, aim to reduce tariffs and other barriers to trade; however, the regional groups reduce barriers in a small region only, while GATT includes countries that control 80 percent of international trade. Regional associations help member countries to act together as one country when they trade on the world market. This is not a function of GATT.
5. Because pollution is spread by wind and/or water to other areas of the world, and because trade has made all countries dependent upon products and resources from other environments in other parts of the world.
6. Because they are still available to the developed countries. Other sources are not widely available.
7. Balance of trade is the difference between the value of the goods a country exports and the value of the goods it imports. If a country buys more than it sells, it has an unfavorable balance of trade. Balance of payments is the difference between a country's foreign income and its foreign expenditures. If a country spends more than it earns, it has an unfavorable balance of trade.
8. Large populations put a strain on the limited resources of nations. A slower growth rate will put less strain on the limited resources of less developed countries. Some regions have surpluses of resources, while others have shortages. Conflict between regions can result.
9. The problem of water shortages has been solved to some degree by water transfer systems such as dams and aquifers, deep wells, and desalinization plants. Water is contaminated by sewage, chemical wastes, oil, thermal-waste heat, pesticides, and commercial fertilizers. Government agencies can help, but countries must act together to prevent water pollution.

Mastering Geography Skills

1. Canada. Goods can be shipped easily between the United States and Canada. Good relations.
2. Imports declined from $12,509 million (1980) to $1,907 million (1985).
3. Nigeria, the Soviet Union, Saudi Arabia, Australia. Improved: with Nigeria (exports and imports); Saudi Arabia (imports); Japan (imports and exports). Worsened: the Soviet Union (exports and imports); Australia (exports and imports).
4. Australia.

Applying and Extending

Answers will vary.

Linking Geography and Economics

Answers will vary.

TEACHER'S BIBLIOGRAPHY

UNIT 1
The Earth and Its People

BOOKS

Boyce, Ronald R. *Geographic Perspective on Global Problems: An Introduction to Geography*. New York: Wiley, 1982.

Durham, Frank. *Frame of the Universe*. New York: Columbia University Press, 1985.

Jager, Jill. *Climate and Energy Systems: A Review of Their Interactions*. New York: Wiley, 1983.

Jumper, S. et al. *Economic Growth and Disparities: A World View*. Englewood Cliffs, N. J.: Prentice- Hall, 1980.

Lewis, Richard S. *The Illustrated Encyclopedia of the Universe: Understanding and Exploring the Cosmos*. New York: Crown.

Nam, Charles B. and Susan G. Philliber. *Population: A Basic Orientation*. Englewood Cliffs, N. J.: Prentice-Hall, 1984.

Stoddard, Robert H. et al. *Geography of Humans and Their Cultures*. Englewood Cliffs: Prentice-Hall, 1986.

Audiovisual Materials

(Available for purchase or rental)

FILMS

Cosmic Zoom. National Film Board of Canada, distributed by McGraw-Hill Films, color, 8 minutes.

Earth: The Restless Planet. National Geographic Society, color, 25 minutes.

Map Projections in the Computer Age. Coronet Instructional Films, distributed by Simon and Schuster, color, 11 minutes.

RECORDINGS

Biological Catastrophies: When Nature Becomes Unbalanced. In two parts. The Center for Humanities, Inc., 160 slides, 2 tapes, script.

Man and His Environment: In Harmony and in Conflict. The Center for Humanities, Inc., 160 slides, 2 tape cassettes, 2 LP's, script.

UNIT 2
Western Europe

BOOKS

Ardagh, John. *France in the 1980s*. New York: Penguin, 1983.

Barker, Ernest. *Britain and the British People*. Westport, Connecticut: Greenwood, 1978.

Feld, Werner J. *European Community in World Affairs: Economic Power and Political Influence*. Boulder: Westview, 1985.

Hoffman, George W., et al. *Geography of Europe — Problems and Prospects*. New York: Wiley, 1983.

John, Brian. *Scandinavia: A New Geography*. Chicago: Longman, 1984.

McEvedy, Colin. *The Penguin Atlas of Recent History:* Europe Since 1815. New York: Penguin, 1982.

Meakin, W. *The New Industrial Revolution*. New York: Arno, 1977.

Riley, R. C., and Gregory Ashworth. *Benelux: An Economic* Geography *of Belgium, The Netherlands,* and *Luxembourg*. New York: Holmes & Meier, 1975.

Sawyer, P. H. *Kings and Vikings*. New York: Methuen, 1983.

Audiovisual Materials

FILMS

The Mighty Continent — A Series. BBC, distributed by Time-Life Films, Inc., color.

The Shoemaker. Indiana University, Bloomington, black and white, 34 minutes.

Scandinavia: Unique Northern Societies. International Film Foundation, color, 24 minutes.

FILMSTRIPS

European Cities — London, Paris, Brussels, Berlin — A Series. Encyclopedia Britannica Educational Corporation, color.

European Cities — Rome, Madrid, Stockholm, Vienna — A Series. Encyclopedia Britannica Educational Corporation, color.

RECORDINGS

Early English Ballads. Educational Audio-Visual, 2 LP's.

French Songs for Children. Spoken Arts, 2 LP's, text included.

UNIT 3
The Soviet Union and Eastern Europe

BOOKS

Brown, Archie, et al. *The Cambridge Encyclopedia of Russia and the Soviet Union*. New York: Cambridge University Press, 1982.

Fischer-Galati, Stephen, ed. *Eastern Europe in the Nineteen Eighties*. Boulder: Westview, 1981.

Hazard, John N. *The Soviet System of Government: 5th edition*. Chicago: University of Chicago Press, 1980.
Hollander, Paul. *Soviet and American Society: A Comparison*. Chicago: University of Chicago Press, 1978.
Mooney, Peter J. *Soviet Superpower: The Soviet Union 1945–1980*. Exeter, New Hampshire: Heinemann Educational, 1982.
Morris, L. P. *Eastern Europe Since Nineteen Forty-Five*. Exeter, New Hampshire: Heinemann Educational, 1984.
Parrott, Bruce. *Politics and Technology in the Soviet Union* (two volumes). Cambridge: MIT Press, 1985.
Richards, David, ed. *The Penguin Book of Russian Short Stories*. New York: Penguin, 1981.
Westwood, J. N. *Russia Since Nineteen Seventeen*. New York: St. Martin's, 1980.

Audiovisual Materials

FILMS

Inside the USSR — A Series. Random House, Inc., color.
Land, Features, and Climate of the USSR. Eye Gate Media, color.
The Last Year of the Tsars. Films Inc., 19 minutes.
The Mighty Volga. National Geographic Society, color, 25 minutes.
A People's Music — Soviet Style. Journal Films, Inc., color, 23 minutes.
Romanian Village Life: On the Danube Delta. International Film Foundation, color, 15 minutes.
The Russian Consumer. International Film Foundation, color, 14 minutes.
Siberia. National Geographic Society, color, 25 minutes.

FILMSTRIP

Life in the USSR. Eye Gate Media.

RECORDINGS

Russian Music. RCA, 1 LP.
A Tchaikovsky Spectacular. Angel Records, 1 cassette.

UNIT 4
The Middle East and North Africa

BOOKS

Armajani, Yahya and Thomas Ricks. *Middle East: Past and Present*. Englewood Cliffs, N. J.: Prentice-Hall, 1986.
Gray, Seymour J. *Beyond the Veil: The Adventures of an American Doctor in Saudi Arabia*. New York: Harper & Row, 1983.
Mallakh, Ragaei E. *Saudi Arabia: Rush to Development*. Baltimore: Johns Hopkins University Press, 1982.
Parker, Richard B. *North Africa: Regional Tensions and Strategic Concerns*. New York: Praeger, 1984.
Powell, William. *Saudi Arabia and Its Royal Family*. Secaucus, New Jersey: Lyle Stuart, 1982.
Quandt, William B. *Saudi Arabia in the 1980s: Foreign Policy Security, and Oil*. Washington, D.C.: Brookings, 1981.
Safran, Nadav. *Israel, The Embattled Ally*. Cambridge: Harvard University Press, 1978.
Smith, Rodney. *In the Land of Light: Israel, a Portrait of Its People*. Boston: Houghton Mifflin, 1983.

Audiovisual Materials

FILMS

The Arab Identity: Who Are the Arabs? Yorkshire Television, distributed by Learning Corporation of America, color, 26 minutes.
The Arab-Israeli Conflict. Atlantis Productions, Inc., color, 20 minutes.
Building a Dream (Israel), 2nd ed. (from The Middle East Series.) Coronet Instructional Films, color, 22 minutes.
The Changing Middle East. International Film Foundation, color, 25 minutes.
Egypt: Gift of the Nile. Coronet Instructional Films, color, 29 minutes.
Israel. National Geographic Society, color, 14 minutes.
Israel and The Arab States. Films Inc., color, 20 minutes.
A Modern Egyptian Family. International Film Foundation, color, 17 minutes.
Turkey: Nation in Transition. International Film Foundation, color, 27 minutes.

FILMSTRIPS

Ancient Egypt in the Modern World. Current Affairs Films, filmstrip with cassette, color.
The Arab World: Oil, Power, Dissension. Current Affairs Films, filmstrip with LP and cassette, color.

RECORDINGS

Hebrew Folk Songs. Folkways Scholastic Records, 1 LP.
Jan Peerce Sings Hebrew Melodies. RCA, 1 LP.

UNIT 5
Sub-Saharan Africa

BOOKS

Chaliand, Gerard. *The Struggle for Africa: Politics of the Great Powers*. New York: St. Martin's, 1982.
Gale Research Center. *Africa South of the Sahara*. New York: Europa, 1978.
Gordon, Rene. *Africa: A Continent Revealed*. New York: St. Martin's. 1981.
Hibbert, Christopher. *Africa Explored: Europeans in the Dark Continent, 1769–1889*. New York: Norton, 1983.
Riefenstahl, Leni. *Vanishing Africa*. New York: Crown, 1982.
Woods, Harold. *The Horn of Africa*. New York: Franklin Watts, 1981.

Audiovisual Materials

FILMS

Africa — Portrait of a Continent — A Series. Random House, Inc., color.

The Kalahari Desert People. National Geographic Society, color, 24 minutes.

Yonder Come Day. McGraw-Hill Films, color, 26 minutes.

RECORDINGS

The Roots of Black Music in America. Folkways Scholastic Records, 2 LP's.

UNIT 6
The Orient

BOOKS

Barnet, Richard J. *The Alliance: America, Europe, Japan — Makers of the Postwar World*. New York: Simon & Schuster, 1985.

Fishlock, Trevor. *Gandhi's Children*. New York: Universe, 1983.

Howe, Christopher. *China's Economy: A Basic Guide*. New York: Basic, 1978.

Kumar, Dharma, and Meghnad Desai. *The Cambridge Economic History of India: Vol. 2*. New York: Cambridge University Press, 1983.

Pezeu-Massabuau, Jacques. *The Japanese Islands: A Physical and Social Geography*. Rutland, Vermont: C.E. Tuttle, 1978.

Qaisar, Ahsan Jan. *The Indian Response to European Technology and Culture A.D. 1498–1707*. Delhi: Oxford University Press, 1982.

Raychaudhuri, Tapan, and Irfan Habib. *The Cambridge Economic History of India: Vol. 1*. New York: Cambridge University Press, 1982.

Schell, Orville. *In the People's Republic: An American's Firsthand View of Living and Working in China*. New York: Random House, 1978.

Welty, Thomas. *Asians: Their Evolving Heritage, 6th ed.* New York: Harper and Row, 1984.

Audiovisual Materials

FILMS

China: An Emerging Giant. National Geographic Society, color. 25 minutes.

China: A Network of Communes. Encyclopedia Britannica Educational Corporation, color, 15 minutes.

China's Changing Face. National Geographic Society, color, 25 minutes.

An Indian Worker: From Village to City. International Film Foundation, color, 17 minutes.

Japan's Living Crafts. American Educational Films, color, 21 minutes.

Transition Generation: A Third World Problem. International Film Foundation, color, 20 minutes.

FILMSTRIPS

Burma: The Golden Land. Parts 1 and 2. Current Affairs Films, color, filmstrip with cassette.

The Four Seasons in Rural Japan. Creative Learning, Inc., filmstrip, color.

India — A History. Parts 1 and 2, Educational Dimensions Group, 1 filmstrip with cassette/script, 60 frames, color.

South Korea: Tradition and the Struggle for Development. Current Affairs Films, filmstrip with cassette, color.

RECORDINGS

Music of India, Ragas and Talas. Capitol Records, Angel Records, 1 LP.

UNIT 7
The Pacific World

BOOKS

Allen, Oliver, ed. *Pacific Navigators*. New York: Time-Life, 1980.

Attenborough, David. *Journeys to the Past: Travels in New Guinea, Madagascar, and the Northern Territory of Australia*. New York: Penguin, 1983.

Broome, Richard. *Aboriginal Australians: Black Response to White Dominance, 1788 — 1980*. Boston: Allen & Unwin, 1982.

Brower, Kenneth. *A Song for Satawal: A Magical Passage to a Faraway Eden*. New York: Harper & Row, 1983.

Evans, Howard E., and Mary A. Evans. *Australia: A Natural History*. Washington, D.C.: Smithsonian, 1983.

McLauchlan, Gordon. *New Zealand Encyclopedia*. Boston: G.K. Hall, 1985.

Tyler, Charles M. *The Island World of the Pacific Ocean*. New York: Gordon, 1977.

Audiovisual Materials

FILMS

Australia: Down Under and Outback. National Geographic Society, color, 25 minutes.

Australian Marsupials. Biological Sciences Curriculum Study, University of Colorado.

Polynesia. National Geographic Society, color, 22 minutes.

Polynesian Adventure. National Geographic Society, color, 54 minutes.

Samoa I Sisifo. Journal Films, Inc., color, 26 minutes.

Tonga Royal. Journal Films, Inc., color, 20 minutes.

Way of the Nomad —Australia's Desert Aboriginals. Educational Media International, color, 28 minutes.

FILMSTRIPS

The Natural Environment. McIntyre Visual Publications, filmstrip with cassette/script, color.

RECORDING

Hovhaness — And God Created Great Whales. Eccentric Circle Cinema Workshop, 1 LP.

UNIT 8
Latin America

BOOKS

Blakemore, Harold, and Clifford Smith, eds. *Latin America: Geographical Perspectives*. Massachusetts: Methuen, 1983.

Bromley, Rosemary, and Ray Bromley. *South American Development: A Geographical Introduction, 2nd ed*. New York: Cambridge University Press, 1982.

Coe, Michael D. *Mexico*. London: Thames & Hudson, 1984.

Dickenson, John P. *Brazil, An Industrial Geography*. Boulder: Westview, 1978.

Emmerich, Andre. *Art Before Columbus*. New York: Simon & Schuster, 1983.

Goebel, Julius. *The Struggle for the Falkland Islands. A Study in Legal and Diplomatic History*. New Haven: Yale University Press, 1982.

LaFeber, Walter. *Inevitable Revolutions: The United States in Central America*. New York: Norton, 1983.

Looney, Robert E. *Development Alternatives of Mexico: Beyond the 1980's*. New York: Praeger, 1982.

Pearce, Jenny. *Under the Eagle: U.S. Intervention in Central America and the Caribbean*. Boston: South End Press, 1983.

Rollin, Sue. *The Illustrated Atlas of Archaeology*. New York: Franklin Watts, 1982.

Audiovisual Materials

FILMS

A Brazilian Family. International Film Foundation, color, 20 minutes.

Indians of the Orinoco: The Makiritare — A Series. International Film Foundation, color.

South America: Land and People. Phoenix/BFA Films and Video, Inc., color, 21 minutes.

South America: Overview. Sterling Educational Films, color, 17 minutes.

South America Today. International Film Foundation, color, 25 minutes.

FILMSTRIPS

Cuba: Balance Sheet of a Revolution. Current Affairs Films, filmstrip with cassette/script, color.

Guatemala: A Mayan Heritage. Current Affairs Films, filmstrip with record/cassette, color.

Puerto Rico: Tradition, Problems, and Prospects. Current Affairs Films, 2 filmstrips with cassettes/script, color.

The Three Worlds of Peru. Current Affairs Films, color.

Venezuela: The Challenge of New Economic Diversification — A Series. Current Affairs Films, filmstrip with cassette, color.

UNIT 9
The United States and Canada

BOOKS

Bradbury, Katherine L., and Anthony Downs. *Urban Decline and the Future of American Cities*. Washington, D.C.: Brookings, 1982.

Brookeman, Christopher. *American Culture and Society Since the Nineteen Thirties*. New York: Schocken, 1984.

Bryant, Keith L., Jr., and Henry C. Dethloff. *A History of American Business*. Englewood Cliffs, N.J.: Prentice-Hall, 1983.

Cochran, Thomas C. *Frontiers of Change: Early Industrialism in America*. New York: Oxford University Press, 1981.

Crump, Donald J., ed. *Canada's Wilderness Lands*. Washington, D.C.: National Geographic, 1982.

Doran, Charles F., and John H. Sigler. *Canada and the United States: Enduring Friendship, Persistent Stress*. Englewood Cliffs, N.J.: Prentice-Hall, 1985.

Friedland, Roger. *Power and Crisis in the City, Corporations, Unions, and Urban Policy*. New York: Schocken, 1983.

Holbrook, Sabra. *Canada's Kids*. New York: Atheneum, 1983.

Hutchison, Bruce. *The Unknown Country: Canada and Her People*. Westport, Connecticut: Greenwood, 1977.

Moon, William Least Heat. *Blue Highways: A Journey into America*. Boston: Little, Brown, 1983.

Palm, Risa. *Geography of American Cities*. New York: Oxford University Press, 1981.

Strackbein, O.R. *American Economy in a World Habitat*. New York: Vantage, 1983.

Struthers, James. *No Fault of Their Own: Unemployment and the Canadian Welfare State*. Toronto: University of Toronto Press, 1983.

Walker, David F. *Canada's Industrial Space-Economy*. Toronto: Halsted Press, 1980.

Audiovisual Materials

FILMS

The Canadians: Their Cities. Encyclopedia Britannica Educational Corporation, 16 minutes, color.

The Canadians: Their Land. Encyclopedia Britannica Educational Corporation, color, 16 minutes.

North American Indians Today. National Geographic Society, color, 25 minutes.

U.S. Specialty Cities: Manufacturing Cities. distributed by Indiana University, Bloomington, color, 26 minutes.

FILMSTRIPS

America, the Melting Pot: Myth or Reality? Current Affairs Films, filmstrip with cassette, color.

Ecology: Can Man and Nature Co-Exist? Current Affairs Films, filmstrip with record/cassette, color.

A Life Apart: A Modern-Day Frontier Family. Current Affairs Films, filmstrip with cassette/script, color.

Life in Rural America — A Series. National Geographic Society, color.

RECORDINGS

The Top Hits of 1776. Adelphi Records, Inc., 1 LP.

UNIT 10
Sharing the World's Resources

BOOKS

Allaby, Michael, and Peter Bunyard. *The Politics of Self-Sufficiency*. New York: Oxford University Press, 1980.

Fowler, John M. *Energy and the Environment*. New York: McGraw-Hill, 1984.

Fuchs, Victor R. *How We Live: An Economic Perspective on Americans from Birth to Death*. Cambridge: Harvard University Press, 1983.

Gabor, Dennis, et al. *Beyond the Age of Waste: A Report to the Club of Rome*, 2nd ed. Elmsford, New York: Pergamon, 1981.

Gaskin, D.E. *The Ecology of Whales and Dolphins*. Exeter, New Hampshire: Heinemann Educational, 1982.

George, Judith S. *Do You See What I See?* New York: Putnam, 1982.

Goldin, Augusta. *Water: Too Much, Too Little, Too Polluted*. Orlando, FL: Harcourt Brace Jovanovich, 1983.

Nussbaum, Bruce. *The World After Oil: The Shifting Axis of Power and Wealth*. New York: Simon & Schuster, 1984.

Sewell, John W. *The United States and World Development: Agenda 1980*. New York: Praeger, 1980.

Steedman, Ian. *Trade Amongst Growing Economies*. New York: Cambridge University Press, 1980.

Audiovisual Materials

FILMS

Conservation and Balance in Nature. International Film Bureau, color, 18 minutes.

Energy: The Problems and the Future. National Geographic Society, color, 23 minutes.

A World of Credit. Modern Talking Picture Service, colcr.

FILMSTRIPS

Consider the Soil First. U.S. Department of Agriculture, 1 filmstrip with cassette/script.

Critical Issues in Science and Society — A World to Feed. Center for Humanities, Inc., 160 slides, 2 tapes, 2 LP's.

Human Rights: Who Speaks for Man? Current Affairs Films, filmstrip with cassette/script, color.

Food for the World — A Series. National Geographic Society, color.

The People Problem. Current Affairs Films, 2 film strips, color.

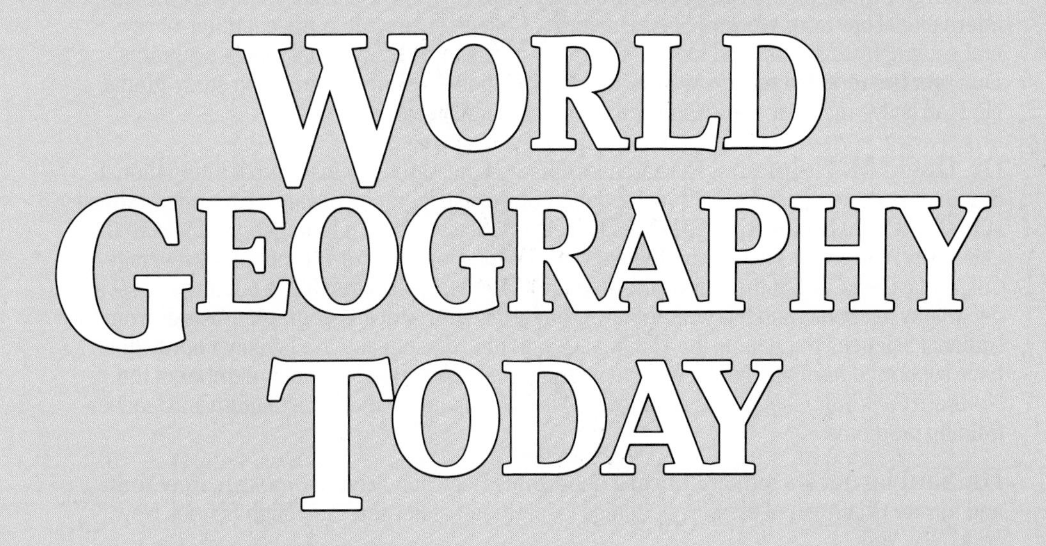

HOLT, RINEHART AND WINSTON

AUSTIN NEW YORK SAN DIEGO CHICAGO TORONTO MONTREAL

THE AUTHORS

Dr. Robert J. Sager is Professor of Earth Sciences at Los Angeles Harbor College. He also is Lecturer in geography at California State University, Fullerton, where he teaches an advanced course on the Pacific Islands and Australia. Dr. Sager received his B.S. in geology and geography and M.S. in geography from the University of Wisconsin. He holds a J.D. in international law from Western State University College of Law. He is the coauthor of several geography textbooks and has written many articles and educational media programs. Dr. Sager has received several awards, including National Science Foundation study grants. He is an active member of the California Geographic Alliance.

Dr. David M. Helgren is Research Professor at the Monterey Institute of International Studies, Monterey, California. His appreciation of world geography began as an undergraduate at Portland State University in Oregon. Dr. Helgren received his Ph.D. in geography from the University of Chicago. He has taught geography at the University of Toronto, the University of California, Davis, and at the University of Miami, Florida. Dr. Helgren is the coauthor of several geography textbooks and has written numerous articles on African geography. Awards from the National Science Foundation, the National Geographic Society, and the Leakey Foundation have supported his many field research projects. Dr. Helgren is a founding member of the Northern California Geographic Alliance. He has participated in many curriculum and teacher-training programs.

Dr. Saul Israel is a former Principal of Erasmus Hall High School, Brooklyn, New York, and former Chairman of the Social Studies Department at Seward Park High School, New York, New York.

Printed in the United States of America

For permission to reprint copyrighted material, grateful acknowledgment is made to the following sources:

Boston University Press: From "Stanley's Despatches to the New York Herald 1871-1872, 1874-1877" edited by Norman R. Bennett. Copyright© 1970 by the Trustees of Boston University.

Century-Hutchinson, London: From *Wanderings in Tasmania* by George Porter. Published by Selwyn & Blount, Ltd., Autumn Books, 1934.

Grove Press, Inc.: From *Life and Death in Shanghai,* by Nien Cheng. Copyright© 1986 by Nien Cheng.

Her Majesty's Stationery Office: From "Margaret Thatcher's Speech Delivered to a Joint Session of the U.S. Congress, Washington D.C., February 20, 1985" in *Vital Speeches.,* March 15, 1985, v. LI, no. 11.

National Geographic Society: From "Eruption in Colombia" by Bart McDowell in *National Geographic*, v. 169, no. 5, May 1986. Copyright© 1986 by National Geographic Society.

Pantheon Books; a Division of Random House, Inc.: From *Kibbutz Makom: Report from an Israeli Kibbutz* by Amia Lieblich. Copyright© 1981 by Amia Lieblich.

Times Books, a Division of Random House, Inc.: From *The Russians* by Hedrick Smith. Copyright © 1976 by Hedrick Smith.

Viking Penguin, Inc.: From *Life on the Mississippi* by Mark Twain. Copyright© 1948 by the Mark Twain Company.

ISBN 0-03-016673-X

890123456 036 987654321

Consultants and Reviewers

Robert E. Gabler
Professor of Geography
Director of International Programs
Western Illinois University
Macomb, Illinois

Stephanie Abraham Hirsh
Director of Program and Staff Development
Richardson ISD
Richardson, Texas

Basheer K. Nijim
Professor and Head
Department of Geography
University of Northern Iowa
Cedar Falls, Iowa

Thomas M. Poulsen
Head, Department of Geography
Portland State University
Portland, Oregon

Gail Riley
Social Studies Supervisor
Hurst-Euless-Bedford ISD
Bedford, Texas

Earl Price Scott
Department of Geography
University of Minnesota
Minneapolis, Minnesota

Connie S. White
Social Studies teacher
Columbus Public Schools
Columbus, Ohio

Karen Tindel Wiggins
Social Studies Consultant
Richardson ISD
Richardson, Texas

Contributing Writers

Helen Rallis-Reeder
Doctoral candidate, Education
Pennsylvania State University
University Park, Pennsylvania

Donald Rallis
Doctoral candidate, Geography
Pennsylvania State University
University Park, Pennsylvania

TABLE OF CONTENTS

Unit 5 Sub-Saharan Africa ... 290

Unit 6 The Orient ... 354

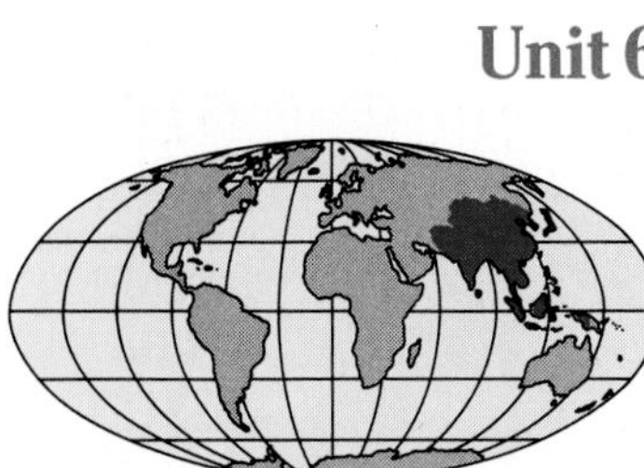

Unit 7 The Pacific World ... 420

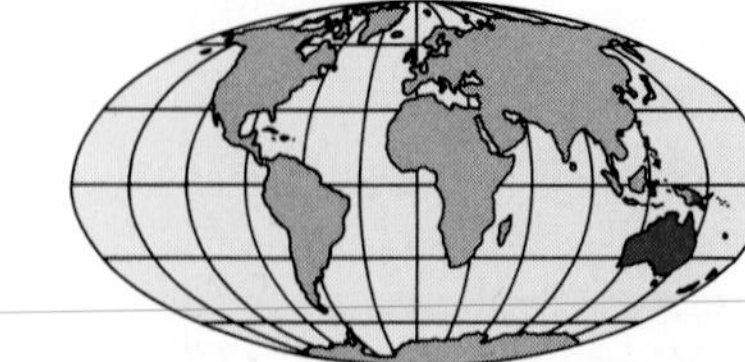

Unit 8 Latin America 448

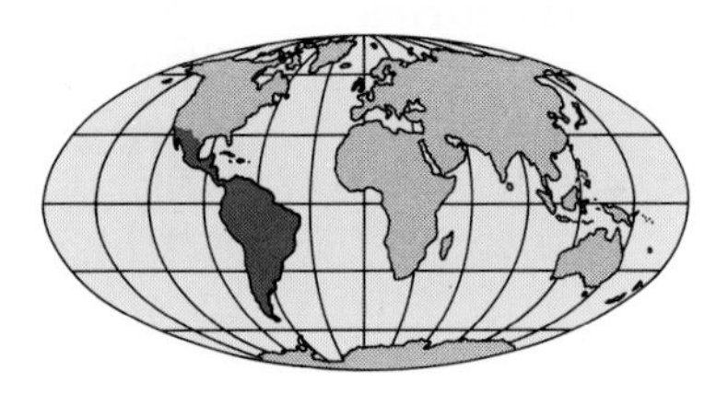

Unit 9 The United States and Canada 512

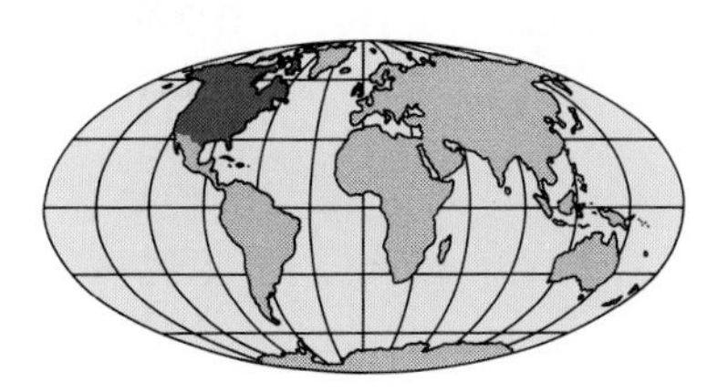

LIST OF MAPS

ATLAS

TO THE STUDENT

World Geography Today is designed to provide you with an effective tool for learning about our world. It has standard features found in all good books. It also has some special features to make your reading and studying more effective and enjoyable. The following pages describe the design of *World Geography Today* and explain how the book is organized.

Organization of *World Geography Today*

It is helpful to think of *World Geography Today* as having three parts. The first part is called the front matter. As the name implies, this is material in the front of the book. Included in the front matter is the **Table of Contents,** which lists the titles of all units, chapters, and special-feature pages throughout the book. The pages you are reading now are part of the front matter, as is the Atlas. The **Atlas** is a compilation of maps to supplement your study of world geography.

The largest part of *World Geography Today* is devoted to units and chapters describing the study of geography and the world's major regions. The first unit begins on page 30. Each unit is divided into two or more chapters.

The third part of your textbook is called the back matter. It includes an **Appendix** with information in statistical form. It also includes a **Glossary,** which lists all the important vocabulary words that appear in blue type throughout the book. Last, the back matter includes an Index. The **Index** is your guide to the hundreds of topics discussed in the book, with page references for each topic.

Each unit opens with a two-page spread like the one shown here. Three photos depicting various aspects of geography are included. Objectives for the unit are included to help guide your reading and study. For Units 2 through 9, the area of the world discussed in that unit is displayed on the world map.

UNIT TWO

WESTERN EUROPE

OBJECTIVES

- *To explain the historical development of the nations of Western Europe*
- *To explain the important role land, climate, and resources have played in the economic development of Western Europe*
- *To describe the way of life of the many cultural groups in Western Europe and how the groups interact with one another*
- *To explain how Western Europe has influenced world affairs*
- *To explain how the countries of Western Europe have attempted to develop policies of mutual political and economic cooperation*
- *To interpret a population-density map*

The Alps, the highest mountains in Europe, are famous for their beauty and are a popular tourist destination. More than 40 mountain passes make land travel possible from one country to another.

Graceful European architecture forms a backdrop for a flower market in Dinan, France. The market is a popular meeting place in France and other Western European countries.

Thousands of years ago, glaciers dug deep into Norway's river valleys. Seawater filled the valleys when the ice retreated, creating fiords. These boats float in a quiet Norwegian harbor near a fiord.

107

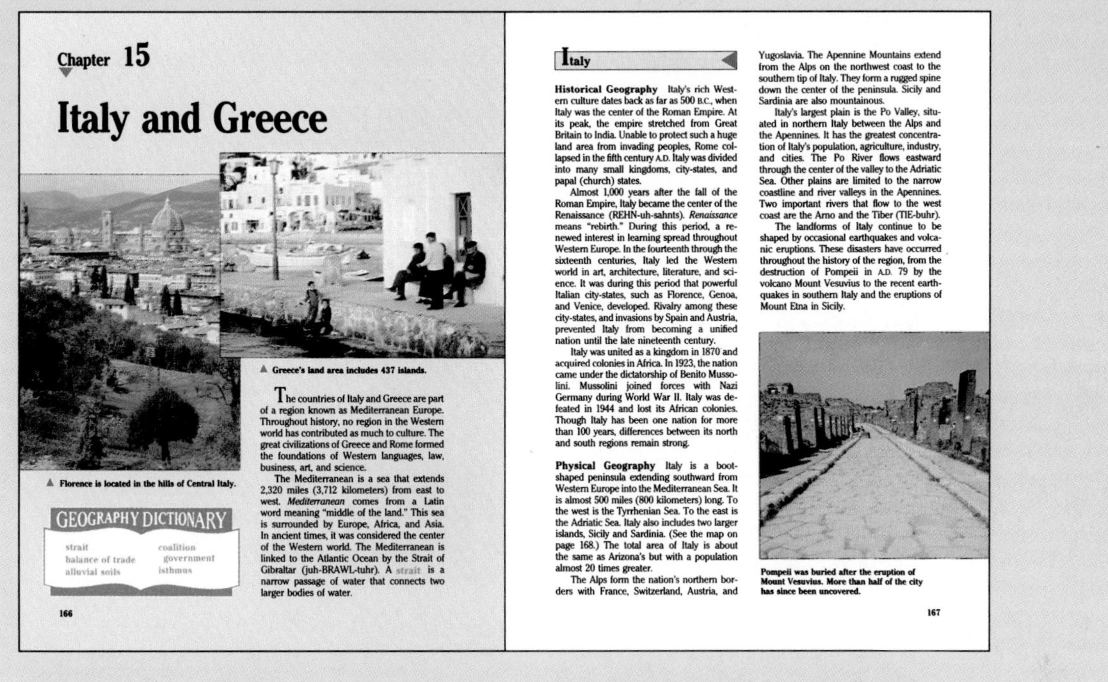

Chapter 15

Italy and Greece

▲ Greece's land area includes 437 islands.

▲ Florence is located in the hills of Central Italy.

GEOGRAPHY DICTIONARY

strait
balance of trade
alluvial soils
coalition government
isthmus

The countries of Italy and Greece are part of a region known as Mediterranean Europe. Throughout history, no region in the Western world has contributed as much to culture. The great civilizations of Greece and Rome formed the foundations of Western languages, law, business, art, and science.

The Mediterranean is a sea that extends 2,320 miles (3,712 kilometers) from east to west. *Mediterranean* comes from a Latin word meaning "middle of the land." This sea is surrounded by Europe, Africa, and Asia. In ancient times, it was considered the center of the Western world. The Mediterranean is linked to the Atlantic Ocean by the Strait of Gibraltar (juh-BRAWL-tuhr). A strait is a narrow passage of water that connects two larger bodies of water.

166

Italy

Historical Geography Italy's rich Western culture dates back as far as 500 B.C., when Italy was the center of the Roman Empire. At its peak, the empire stretched from Great Britain to India. Unable to protect such a huge land area from invading peoples, Rome collapsed in the fifth century A.D. Italy was divided into many small kingdoms, city-states, and papal (church) states.

Almost 1,000 years after the fall of the Roman Empire, Italy became the center of the Renaissance (REHN-uh-sahnts). *Renaissance* means "rebirth." During this period, a renewed interest in learning spread throughout Western Europe. In the fourteenth through the sixteenth centuries, Italy led the Western world in art, architecture, literature, and science. It was during this period that powerful Italian city-states, such as Florence, Genoa, and Venice, developed. Rivalry among these city-states, and invasions by Spain and Austria, prevented Italy from becoming a unified nation until the late nineteenth century.

Italy was united as a kingdom in 1870 and acquired colonies in Africa. In 1923, the nation came under the dictatorship of Benito Mussolini. Mussolini joined forces with Nazi Germany during World War II. Italy was defeated in 1944 and lost its African colonies. Though Italy has been one nation for more than 100 years, differences between its north and south regions remain strong.

Physical Geography Italy is a boot-shaped peninsula extending southward from Western Europe into the Mediterranean Sea. It is almost 500 miles (800 kilometers) long. To the west is the Tyrrhenian Sea. To the east is the Adriatic Sea. Italy also includes two larger islands, Sicily and Sardinia. (See the map on page 168.) The total area of Italy is about the same as Arizona's but with a population almost 20 times greater.

The Alps form the nation's northern borders with France, Switzerland, Austria, and Yugoslavia. The Apennine Mountains extend from the Alps on the northwest coast to the southern tip of Italy. They form a rugged spine down the center of the peninsula. Sicily and Sardinia are also mountainous.

Italy's largest plain is the Po Valley, situated in northern Italy between the Alps and the Apennines. It has the greatest concentration of Italy's population, agriculture, industry, and cities. The Po River flows eastward through the center of the valley to the Adriatic Sea. Other plains are limited to the narrow coastline and river valleys in the Apennines. Two important rivers that flow to the west coast are the Arno and the Tiber (TIE-buhr).

The landforms of Italy continue to be shaped by occasional earthquakes and volcanic eruptions. These disasters have occurred throughout the history of the region, from the destruction of Pompeii in A.D. 79 by the volcano Mount Vesuvius to the recent earthquakes in southern Italy and the eruptions of Mount Etna in Sicily.

Pompeii was buried after the eruption of Mount Vesuvius. More than half of the city has since been uncovered.

167

Each chapter begins like this. The "Geography Dictionary" shows you the list of important vocabulary terms for that chapter. The introduction leads you into the material that follows.

Activities that provide practice in map reading, using graphs, and other skills appear on pages titled "Geography Skills."

Each chapter ends with a "Chapter Check" so you can review the material you just read.

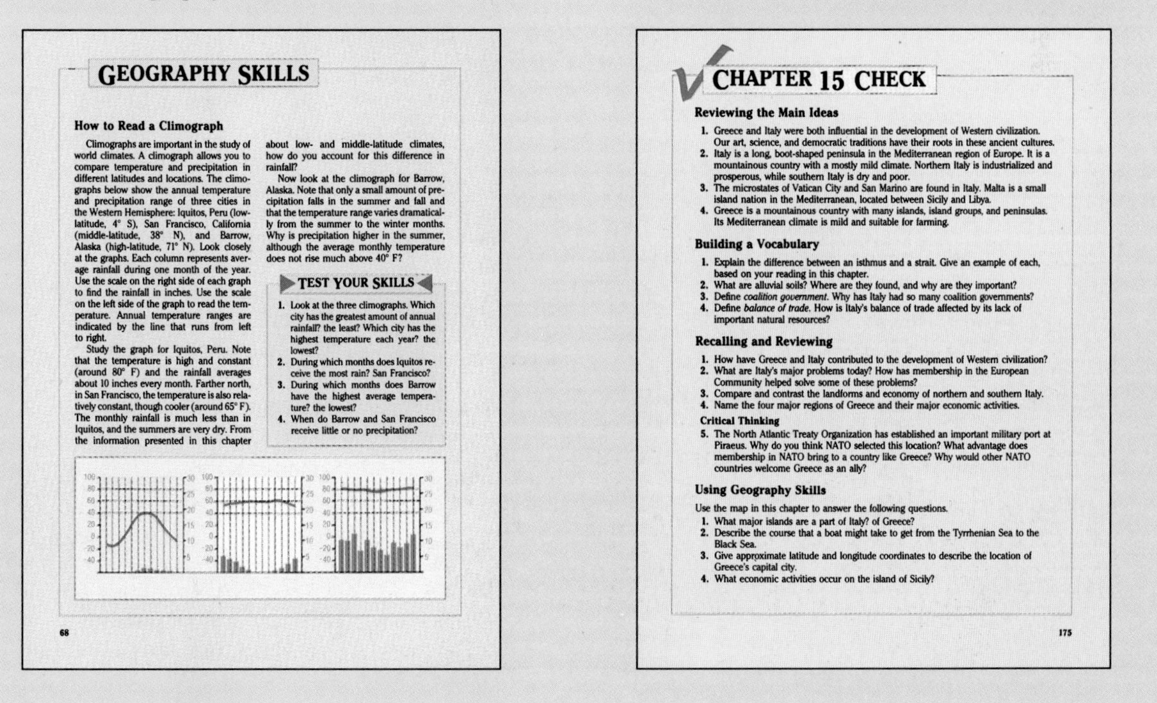

GEOGRAPHY SKILLS

How to Read a Climograph

Climographs are important in the study of world climates. A climograph allows you to compare temperature and precipitation in different latitudes and locations. The climographs below show the annual temperature and precipitation range of three cities in the Western Hemisphere: Iquitos, Peru (low-latitude, 4° S), San Francisco, California (middle-latitude, 38° N), and Barrow, Alaska (high-latitude, 71° N). Look closely at the graphs. Each column represents average rainfall during one month of the year. Use the scale on the right side of each graph to find the rainfall in inches. Use the scale on the left side of the graph to read the temperature. Annual temperature ranges are indicated by the line that runs from left to right.

Study the graph for Iquitos, Peru. Note that the temperature is high and constant (around 80° F) and the rainfall averages about 10 inches every month. Farther north, in San Francisco, the temperature is also relatively constant, though cooler (around 65° F). The monthly rainfall is much less than in Iquitos, and the summers are very dry. From the information presented in this chapter about low- and middle-latitude climates, how do you account for this difference in rainfall?

Now look at the climograph for Barrow, Alaska. Note that only a small amount of precipitation falls in the summer and fall and that the temperature range varies dramatically from the summer to the winter months. Why is precipitation higher in the summer, although the average monthly temperature does not rise much above 40° F?

TEST YOUR SKILLS

1. Look at the three climographs. Which city has the greatest amount of annual rainfall? the least? Which city has the highest temperature each year? the lowest?
2. During which months does Iquitos receive the most rain? San Francisco?
3. During which months does Barrow have the highest average temperature? the lowest?
4. When do Barrow and San Francisco receive little or no precipitation?

68

CHAPTER 15 CHECK

Reviewing the Main Ideas

1. Greece and Italy were both influential in the development of Western civilization. Our art, science, and democratic traditions have their roots in these ancient cultures.
2. Italy is a long, boot-shaped peninsula in the Mediterranean region of Europe. It is a mountainous country with a mostly mild climate. Northern Italy is industrialized and prosperous, while southern Italy is dry and poor.
3. The microstates of Vatican City and San Marino are found in Italy. Malta is a small island nation in the Mediterranean, located between Sicily and Libya.
4. Greece is a mountainous country with many islands, island groups, and peninsulas. Its Mediterranean climate is mild and suitable for farming.

Building a Vocabulary

1. Explain the difference between an isthmus and a strait. Give an example of each, based on your reading in this chapter.
2. What are alluvial soils? Where are they found, and why are they important?
3. Define *coalition government*. Why has Italy had so many coalition governments?
4. Define *balance of trade*. How is Italy's balance of trade affected by its lack of important natural resources?

Recalling and Reviewing

1. How have Greece and Italy contributed to the development of Western civilization?
2. What are Italy's major problems today? How has membership in the European Community helped solve some of these problems?
3. Compare and contrast the landforms and economy of northern and southern Italy.
4. Name the four major regions of Greece and their major economic activities.

Critical Thinking

5. The North Atlantic Treaty Organization has established an important military port at Piraeus. Why do you think NATO selected this location? What advantage does membership in NATO bring to a country like Greece? Why would other NATO countries welcome Greece as an ally?

Using Geography Skills

Use the map in this chapter to answer the following questions.

1. What major islands are a part of Italy? of Greece?
2. Describe the course that a boat might take to get from the Tyrrhenian Sea to the Black Sea.
3. Give approximate latitude and longitude coordinates to describe the location of Greece's capital city.
4. What economic activities occur on the island of Sicily?

175

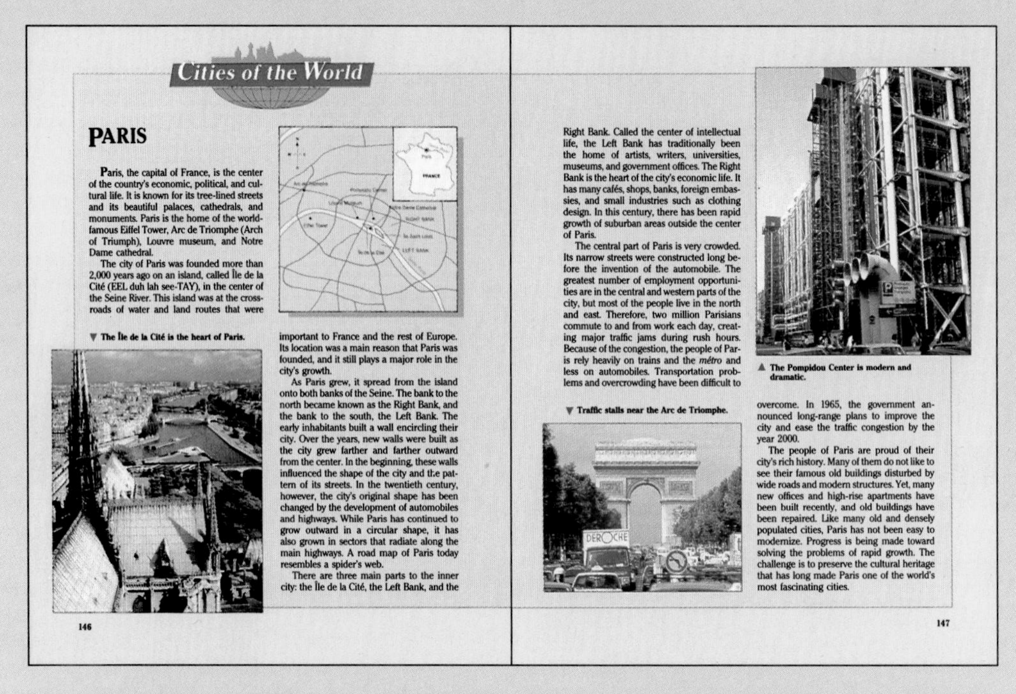

Cities of the World

PARIS

Paris, the capital of France, is the center of the country's economic, political, and cultural life. It is known for its tree-lined streets and its beautiful palaces, cathedrals, and monuments. Paris is the home of the world-famous Eiffel Tower, Arc de Triomphe (Arch of Triumph), Louvre museum, and Notre Dame cathedral.

The city of Paris was founded more than 2,000 years ago on an island, called Île de la Cité (EEL duh lah see-TAY), in the center of the Seine River. This island was at the crossroads of water and land routes that were important to France and the rest of Europe. Its location was a main reason that Paris was founded, and it still plays a major role in the city's growth.

As Paris grew, it spread from the island onto both banks of the Seine. The bank to the north became known as the Right Bank, and the bank to the south, the Left Bank. The early inhabitants built a wall encircling their city. Over the years, new walls were built as the city grew farther and farther outward from the center. In the beginning, these walls influenced the shape of the city and the pattern of its streets. In the twentieth century, however, the city's original shape has been changed by the development of automobiles and highways. While Paris has continued to grow outward in a circular shape, it has also grown in sectors that radiate along the main highways. A road map of Paris today resembles a spider's web.

There are three main parts to the inner city: the Île de la Cité, the Left Bank, and the Right Bank. Called the center of intellectual life, the Left Bank has traditionally been the home of artists, writers, universities, museums, and government offices. The Right Bank is the heart of the city's economic life. It has many cafés, shops, banks, foreign embassies, and small industries such as clothing design. In this century, there has been rapid growth of suburban areas outside the center of Paris.

The central part of Paris is very crowded. Its narrow streets were constructed long before the invention of the automobile. The greatest number of employment opportunities are in the central and western parts of the city, but most of the people live in the north and east. Therefore, two million Parisians commute to and from work each day, creating major traffic jams during rush hours. Because of the congestion, the people of Paris rely heavily on trains and the *métro* and less on automobiles. Transportation problems and overcrowding have been difficult to overcome. In 1965, the government announced long-range plans to improve the city and ease the traffic congestion by the year 2000.

The people of Paris are proud of their city's rich history. Many of them do not like to see their famous old buildings disturbed by wide roads and modern structures. Yet, many new offices and high-rise apartments have been built recently, and old buildings have been repaired. Like many old and densely populated cities, Paris has not been easy to modernize. Progress is being made toward solving the problems of rapid growth. The challenge is to preserve the cultural heritage that has long made Paris one of the world's most fascinating cities.

▼ **The Île de la Cité is the heart of Paris.**

▼ **Traffic stalls near the Arc de Triomphe.**

▲ **The Pompidou Center is modern and dramatic.**

146

147

Special-feature pages like those above appear in every unit. Some apply concepts from "Themes in Geography." Others describe the development of some of the major cities around the world.

Just as every unit begins with a two-page spread, every unit closes with two pages of "Unit Review." Like Chapter Checks, they are designed to help you recall, understand, and apply the information from the pages you just read.

UNIT 2 REVIEW

Summary

Politically, economically, and culturally, Western Europe is one of the most important regions of the world. Its importance is due partly to its favorable location. Because it is surrounded by water on three sides, Western Europe is in an excellent position for trade. Its ocean boundaries provide mild maritime and Mediterranean climates. The surrounding seas also provide a good supply of fish. The region has rich farmlands and some natural resources for industry.

Natural advantages alone do not explain Western Europe's importance, however. Much of Western Europe has a large, industrious population. Even in countries with poor supplies of natural resources, such as Switzerland and the Netherlands, people make the most of natural conditions. Western European nations today seek closer economic and political ties with each other. However, they still value and wish to retain their individual character and governments.

Reading and Understanding

1. Of the 17 nations and six microstates of Western Europe, which are landlocked? Of these, which have no water outlet to the sea? How have these landlocked nations achieved a high level of trade in Western Europe?
2. Why is Western Europe considered to be a culturally complex region? How does this complexity stand in the way of closer alliances among Western European countries?
3. Name the major rivers of Western Europe. How have the countries of Western Europe used these rivers for their economic benefit?
4. Name four Western European countries that are strong in agricultural production. How is this strength related to the landforms and climate of these four countries?
5. Name and locate the major mountain ranges of Western Europe. How are these mountain ranges important to the climates and economies of the countries of Western Europe?
6. Which Western European countries have the most important industrial areas? What factors make these areas important? What should a nation do to maintain its industrial importance?
7. Name five major cities of Western Europe. What characteristics do these cities share? Consider location, cultural activities, economic activities, historical importance, transportation, and communication.
8. Describe three environmental, social, and political issues that are of concern to Western Europeans.
9. What is the European Community? Explain how some Western European countries benefit from their membership in the European Community.
10. Which Western European countries are members of NATO? Why was it formed? How does it benefit Western Europe?

184

Mastering Geography Skills

1. Look at the climate and population-density maps on page 113. Which countries of Western Europe have the lowest population density? Why? Now look at the population-density map of the world on page 112. What other areas of the world are as densely populated as Western Europe?
2. Look at the regional map on page 110. What is the distance between Paris and Copenhagen? Paris and Rome? If you were traveling by land, which trip would be most time consuming? Why do you think so?

Applying and Extending

Use an almanac to find the following information about each of the nations of Western Europe, and record the data in a table: (1) the value of annual exports, (2) the value of annual imports, (3) the percentage of exports sent to other Western European countries, and (4) the percentage of imports purchased from other Western European nations. Study the data. Which Western European nations show a loss in their balance of payments? Do any show a gain? If so, which ones? Based on the percentages of imports and exports, how important is trade with other Western European nations to these countries? Compare your data with those found by other members of your class.

Linking Geography and History

During the nineteenth century, the British Empire included about one-fourth of the world's people as well as one-fourth of the world's land. Examine several sources, such as history texts and historical atlases, to answer the following questions: How did Britain achieve its supremacy over the seas? Which countries did it colonize, and when did they achieve their independence? Which countries and islands are still administered by Great Britain? Explore in some detail the relationship between the British Industrial Revolution and the British administration of India. What natural resources did India provide for British industry and consumption?

Reading for Enrichment

Burtenshaw, D., et al. *The City in West Europe.* New York: John Wiley & Sons, 1981.

Carrington, Richard. *The Mediterranean: Cradle of Western Culture.* New York: Viking, 1971.

Hoffman, G., ed. *A Geography of Europe: Problems and Prospects.* 5th rev. ed. New York: John Wiley & Sons, 1983.

Johnston, R., and J. Doornkamp, eds. *The Changing Geography of the United Kingdom.* London: Methuen, 1983.

Webb, R. K. *Modern England.* New York: Harper & Row, 1980.

Wright, Gordon. *France in Modern Times.* New York: W. W. Norton, 1981.

185

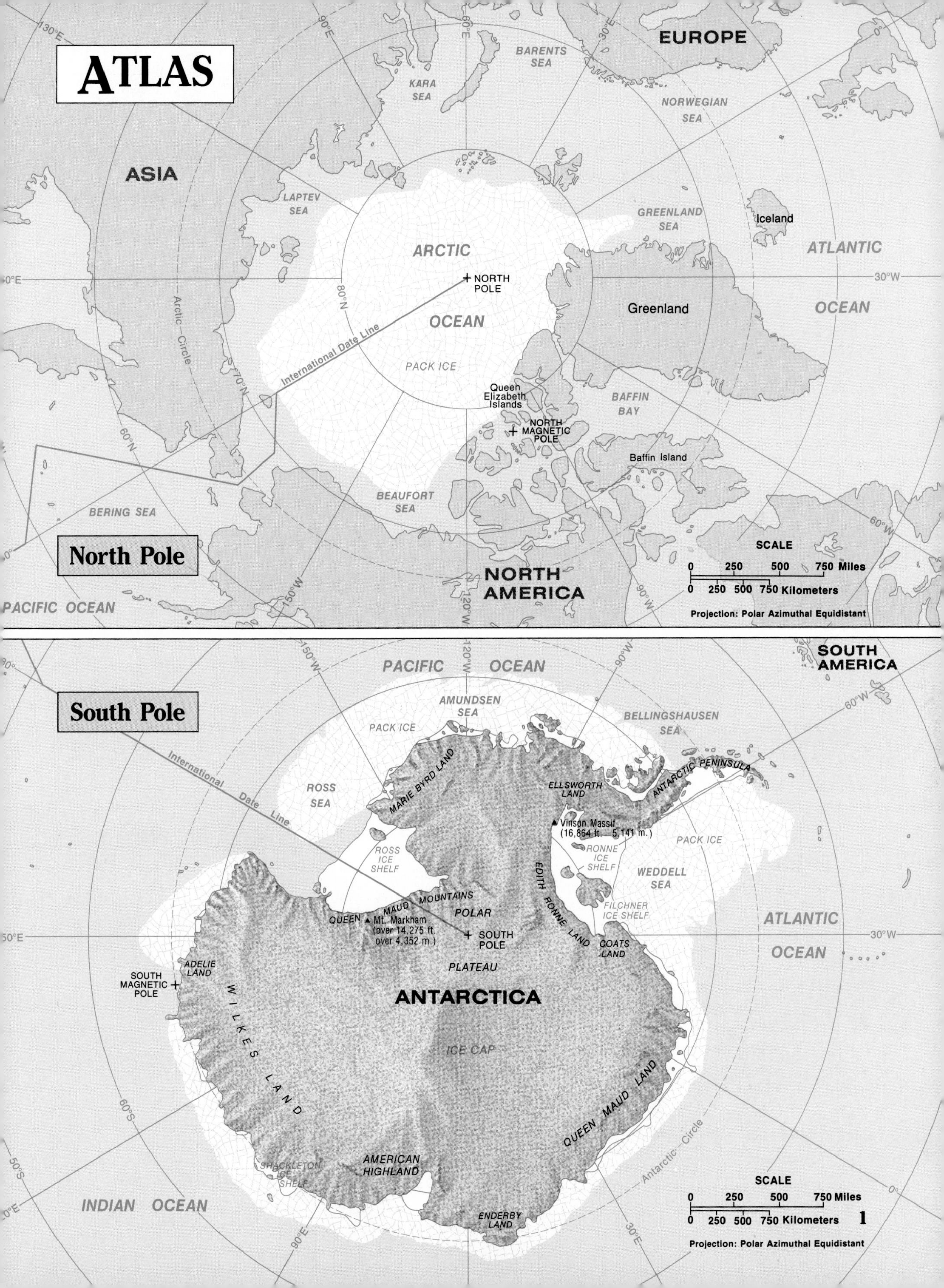
ATLAS
North Pole
EUROPE
BARENTS SEA
KARA SEA
NORWEGIAN SEA
ASIA
LAPTEV SEA
GREENLAND SEA
Iceland
ARCTIC OCEAN
NORTH POLE
ATLANTIC OCEAN
Greenland
International Date Line
Arctic Circle
PACK ICE
Queen Elizabeth Islands
NORTH MAGNETIC POLE
BAFFIN BAY
Baffin Island
BEAUFORT SEA
BERING SEA
NORTH AMERICA
PACIFIC OCEAN
SCALE
0 250 500 750 Miles
0 250 500 750 Kilometers
Projection: Polar Azimuthal Equidistant
South Pole
PACIFIC OCEAN
SOUTH AMERICA
AMUNDSEN SEA
BELLINGSHAUSEN SEA
PACK ICE
ROSS SEA
MARIE BYRD LAND
ELLSWORTH LAND
ANTARCTIC PENINSULA
International Date Line
Vinson Massif (16,864 ft. 5,141 m.)
ROSS ICE SHELF
RONNE ICE SHELF
WEDDELL SEA
EDITH RONNE LAND
QUEEN MAUD MOUNTAINS
Mt. Markham (over 14,275 ft. over 4,352 m.)
POLAR PLATEAU
SOUTH POLE
FILCHNER ICE SHELF
COATS LAND
ATLANTIC OCEAN
SOUTH MAGNETIC POLE
ADELIE LAND
ANTARCTICA
WILKES LAND
ICE CAP
QUEEN MAUD LAND
Antarctic Circle
SHACKLETON ICE SHELF
AMERICAN HIGHLAND
ENDERBY LAND
INDIAN OCEAN
SCALE
0 250 500 750 Miles
0 250 500 750 Kilometers
Projection: Polar Azimuthal Equidistant

The World: Physical

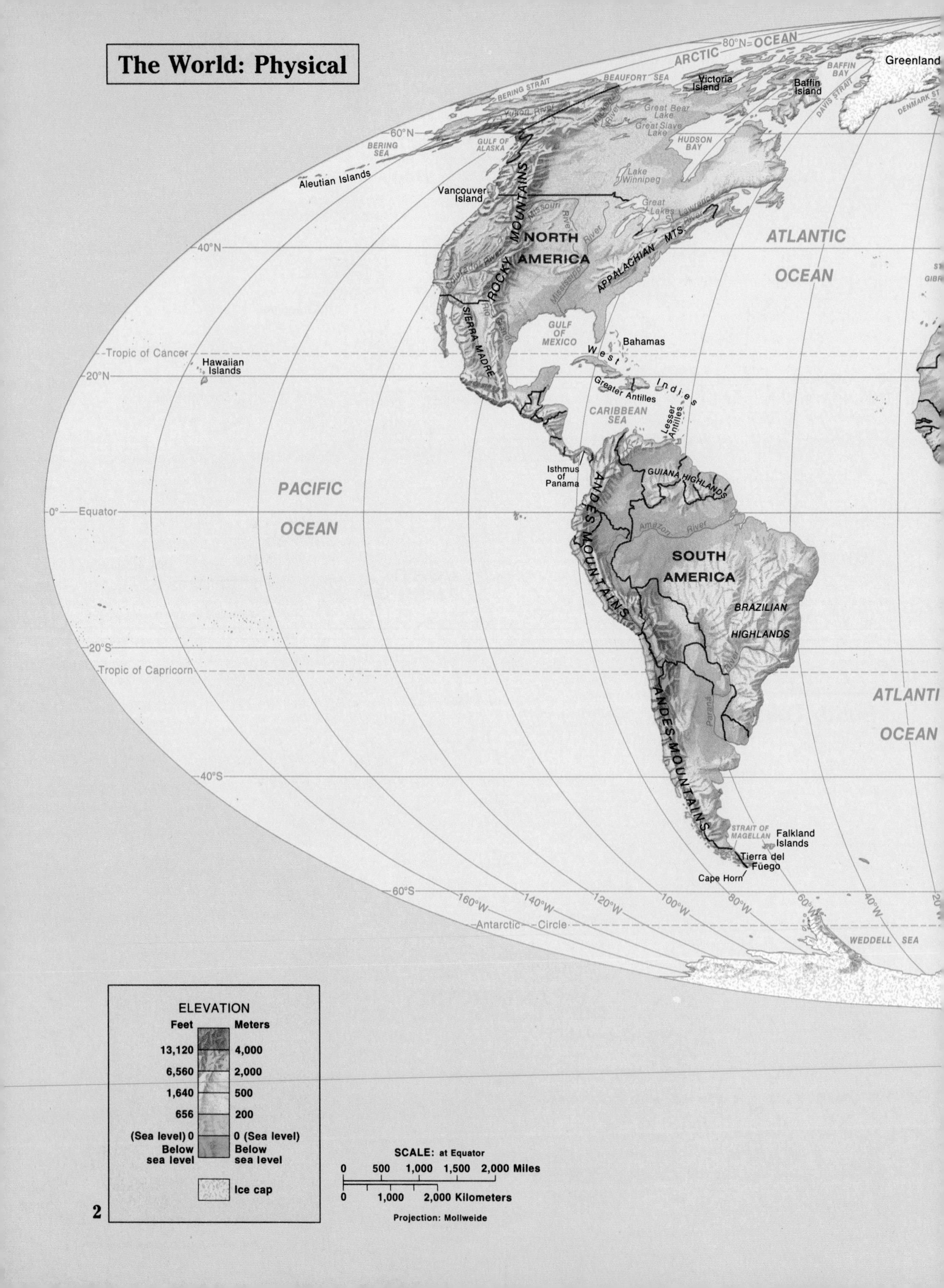

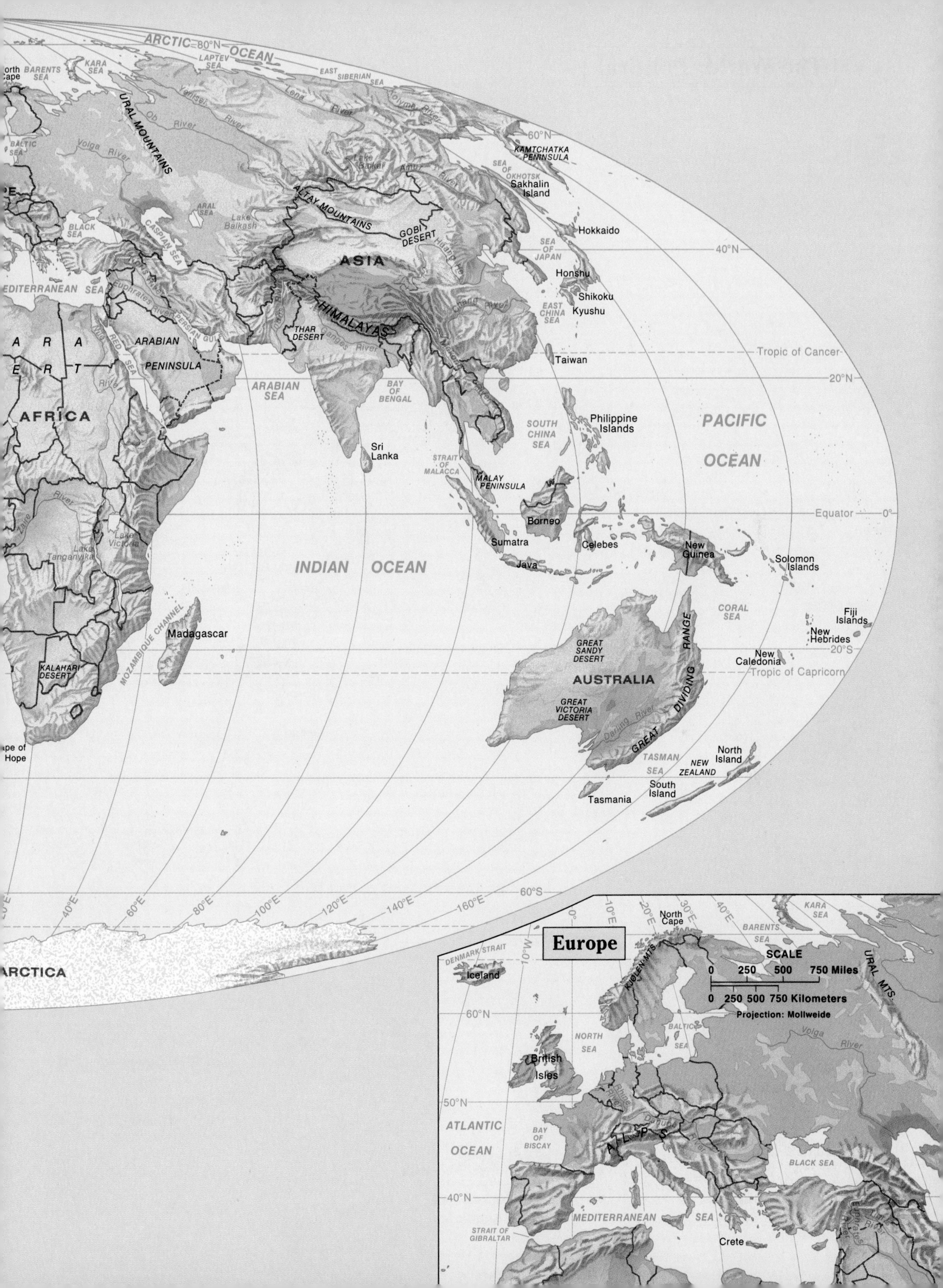
ARCTIC
OCEAN
80°N
LAPTEV SEA
KARA SEA
BARENTS SEA
EAST SIBERIAN SEA
URAL MOUNTAINS
Ob River
Yenisei River
Lena River
Kolyma River
Volga River
BALTIC SEA
Lake Baikal
Amur River
60°N
KAMTCHATKA PENINSULA
SEA OF OKHOTSK
Sakhalin Island
ALTAY MOUNTAINS
ARAL SEA
Lake Balkash
CASPIAN SEA
BLACK SEA
GOBI DESERT
Hokkaido
SEA OF JAPAN
40°N
ASIA
Honshu
Shikoku
Kyushu
EAST CHINA SEA
MEDITERRANEAN SEA
Euphrates River
PERSIAN GULF
HIMALAYAS
Chang River
THAR DESERT
Ganges River
ARABIAN PENINSULA
RED SEA
Taiwan
Tropic of Cancer
20°N
ARABIAN SEA
BAY OF BENGAL
AFRICA
SOUTH CHINA SEA
Philippine Islands
PACIFIC OCEAN
Sri Lanka
STRAIT OF MALACCA
MALAY PENINSULA
Borneo
Equator
0°
Lake Victoria
Lake Tanganyika
Zaire River
Sumatra
Celebes
New Guinea
INDIAN OCEAN
Java
Solomon Islands
CORAL SEA
Fiji Islands
New Hebrides
MOZAMBIQUE CHANNEL
Madagascar
GREAT SANDY DESERT
New Caledonia
20°S
Tropic of Capricorn
KALAHARI DESERT
AUSTRALIA
GREAT DIVIDING RANGE
GREAT VICTORIA DESERT
Darling River
Cape of Good Hope
TASMAN SEA
NEW ZEALAND
North Island
South Island
Tasmania
60°S
40°E
60°E
80°E
100°E
120°E
140°E
160°E
ANTARCTICA
Europe
North Cape
KARA SEA
BARENTS SEA
DENMARK STRAIT
Iceland
KJOLEN MTS.
SCALE
0 250 500 750 Miles
0 250 500 750 Kilometers
Projection: Mollweide
URAL MTS.
60°N
NORTH SEA
BALTIC SEA
Volga River
British Isles
50°N
ATLANTIC OCEAN
BAY OF BISCAY
ALPS
BLACK SEA
40°N
STRAIT OF GIBRALTAR
MEDITERRANEAN SEA
Crete
0°
10°E
20°E
30°E
40°E
10°W

The World: Political

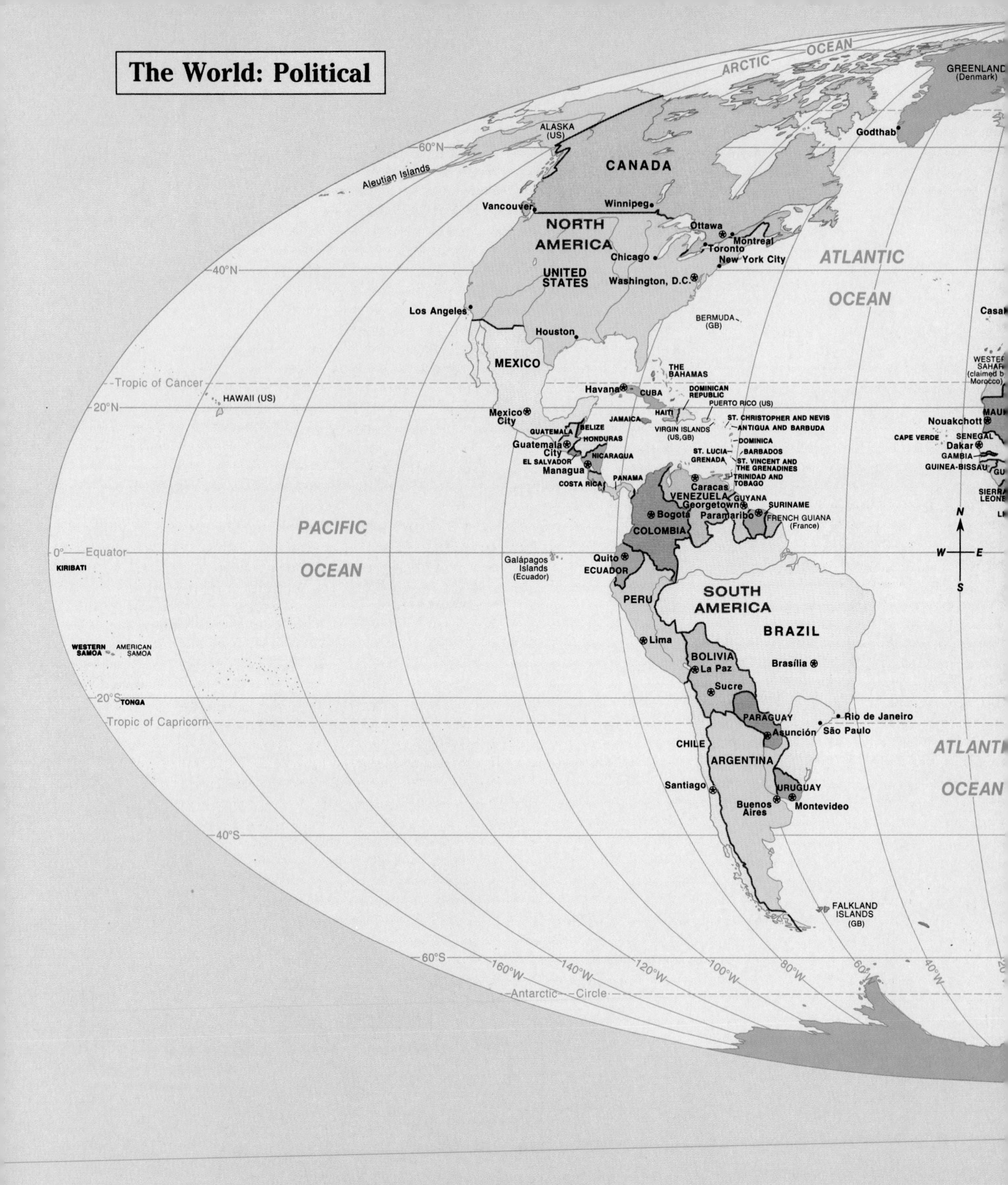

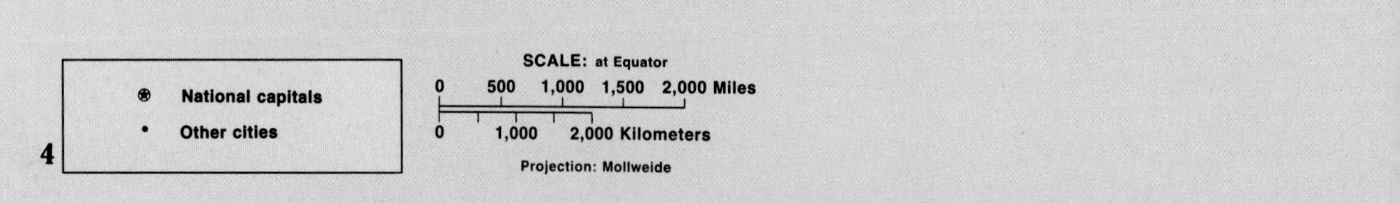

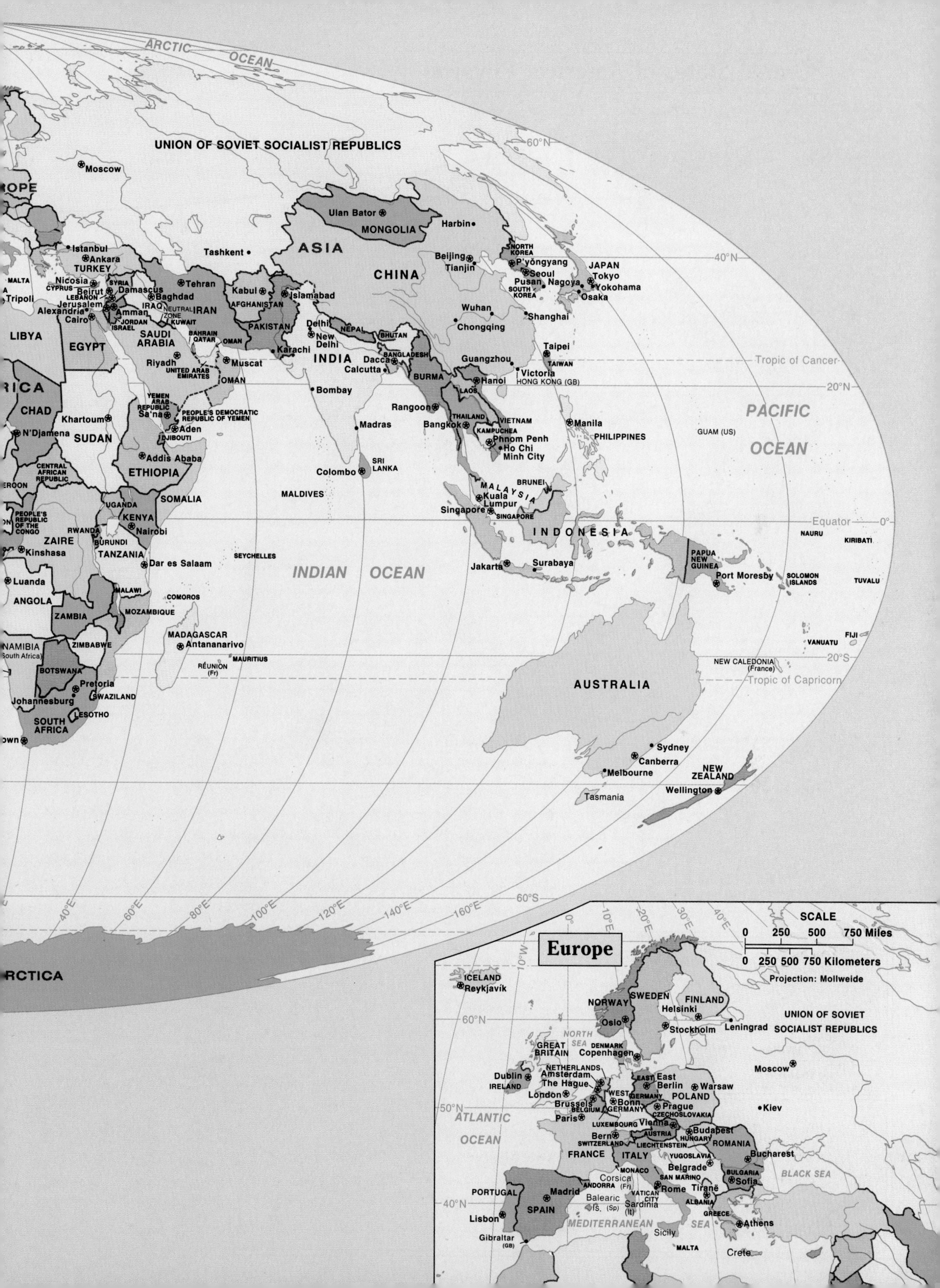

ARCTIC OCEAN
UNION OF SOVIET SOCIALIST REPUBLICS
Moscow
Ulan Bator
MONGOLIA
Harbin
ASIA
CHINA
Tashkent
Istanbul
Ankara
TURKEY
MALTA
Tripoli
Nicosia
CYPRUS
SYRIA
Beirut
LEBANON
Damascus
Jerusalem
Amman
ISRAEL
JORDAN
Alexandria
Cairo
Tehran
Baghdad
IRAQ
NEUTRAL ZONE
KUWAIT
IRAN
Kabul
AFGHANISTAN
Islamabad
PAKISTAN
Karachi
LIBYA
EGYPT
SAUDI ARABIA
BAHRAIN
QATAR
OMAN
Riyadh
UNITED ARAB EMIRATES
Muscat
YEMEN ARAB REPUBLIC
Sa'na
PEOPLE'S DEMOCRATIC REPUBLIC OF YEMEN
Aden
DJIBOUTI
CHAD
N'Djamena
Khartoum
SUDAN
Addis Ababa
ETHIOPIA
CENTRAL AFRICAN REPUBLIC
SOMALIA
UGANDA
KENYA
Nairobi
PEOPLE'S REPUBLIC OF THE CONGO
RWANDA
BURUNDI
ZAIRE
Kinshasa
TANZANIA
Dar es Salaam
Luanda
ANGOLA
ZAMBIA
MALAWI
MOZAMBIQUE
COMOROS
ZIMBABWE
NAMIBIA (South Africa)
BOTSWANA
Pretoria
Johannesburg
SWAZILAND
LESOTHO
SOUTH AFRICA
MADAGASCAR
Antananarivo
MAURITIUS
RÉUNION (Fr)
SEYCHELLES
MALDIVES
Delhi
New Delhi
NEPAL
BHUTAN
INDIA
Dacca
BANGLADESH
Calcutta
Bombay
Madras
SRI LANKA
Colombo
BURMA
Rangoon
LAOS
THAILAND
Bangkok
VIETNAM
KAMPUCHEA
Phnom Penh
Ho Chi Minh City
Hanoi
Beijing
Tianjin
Wuhan
Chongqing
Shanghai
Guangzhou
Victoria
HONG KONG (GB)
Taipei
TAIWAN
NORTH KOREA
P'yŏngyang
Seoul
SOUTH KOREA
Pusan
Nagoya
JAPAN
Tokyo
Yokohama
Osaka
Manila
PHILIPPINES
GUAM (US)
PACIFIC OCEAN
BRUNEI
MALAYSIA
Kuala Lumpur
Singapore
SINGAPORE
INDONESIA
Jakarta
Surabaya
PAPUA NEW GUINEA
Port Moresby
SOLOMON ISLANDS
NAURU
KIRIBATI
TUVALU
VANUATU
FIJI
NEW CALEDONIA (France)
INDIAN OCEAN
AUSTRALIA
Sydney
Canberra
Melbourne
Tasmania
NEW ZEALAND
Wellington
60°N
40°N
20°N
0°
20°S
60°S
Tropic of Cancer
Equator
Tropic of Capricorn
40°E
60°E
80°E
100°E
120°E
140°E
160°E
RCTICA
Europe
SCALE
0 250 500 750 Miles
0 250 500 750 Kilometers
Projection: Mollweide
ICELAND
Reykjavík
NORWAY
Oslo
SWEDEN
Stockholm
FINLAND
Helsinki
Leningrad
UNION OF SOVIET SOCIALIST REPUBLICS
NORTH SEA
GREAT BRITAIN
DENMARK
Copenhagen
Dublin
IRELAND
London
NETHERLANDS
Amsterdam
The Hague
Brussels
BELGIUM
WEST GERMANY
Bonn
EAST GERMANY
East Berlin
POLAND
Warsaw
Moscow
Kiev
Prague
CZECHOSLOVAKIA
LUXEMBOURG
Paris
Vienna
AUSTRIA
Budapest
HUNGARY
ROMANIA
Bucharest
Bern
SWITZERLAND
LIECHTENSTEIN
ATLANTIC OCEAN
FRANCE
ITALY
YUGOSLAVIA
Belgrade
MONACO
SAN MARINO
BULGARIA
Sofia
BLACK SEA
Corsica (Fr)
ANDORRA
VATICAN CITY
Rome
Tiranë
ALBANIA
PORTUGAL
Madrid
SPAIN
Balearic Is. (Sp)
Sardinia (It)
Lisbon
GREECE
Athens
MEDITERRANEAN SEA
Gibraltar (GB)
Sicily
MALTA
Crete
0°
10°E
20°E
30°E
40°E
10°W
60°N
50°N
40°N

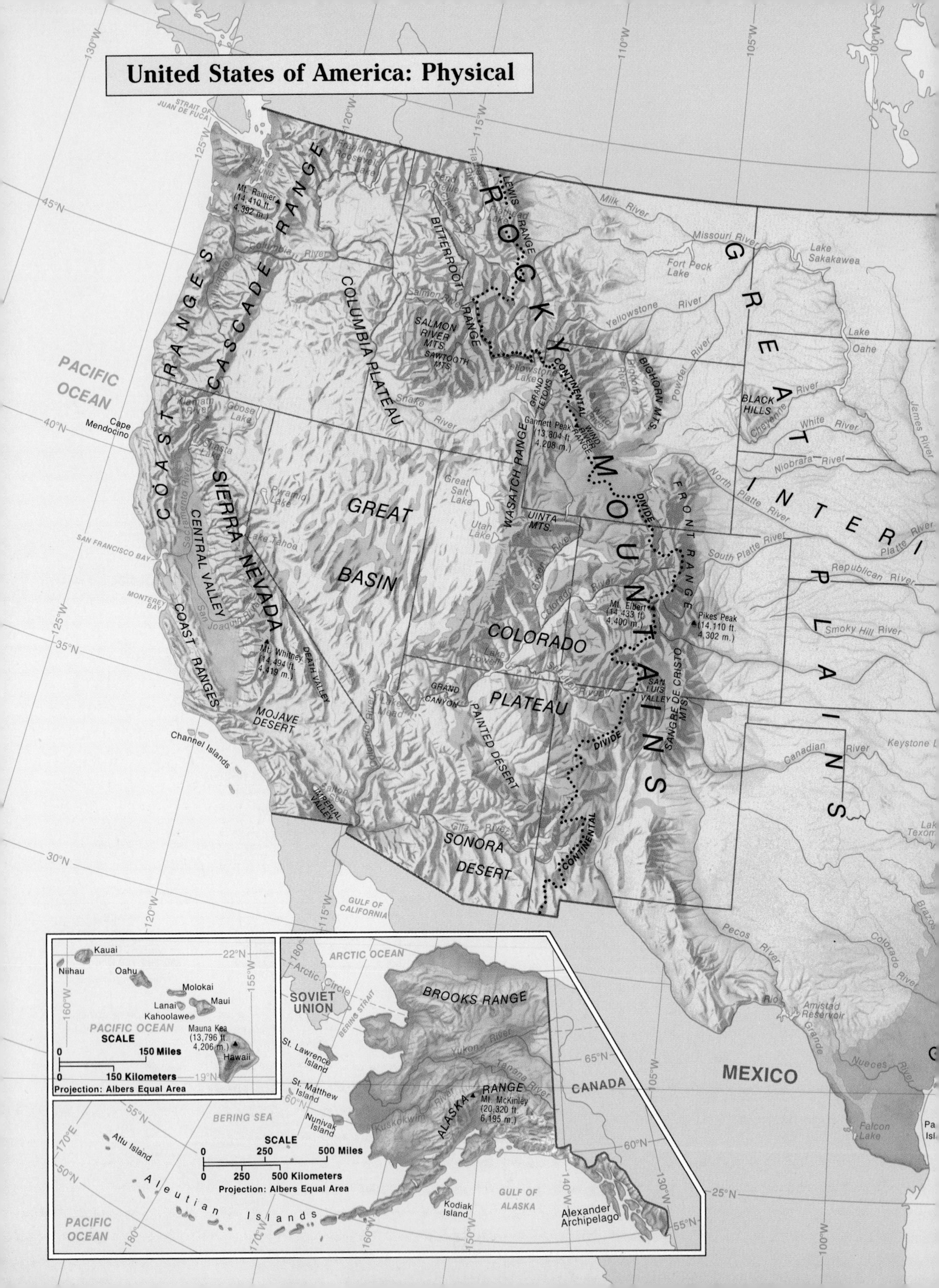
United States of America: Physical
STRAIT OF JUAN DE FUCA
PUGET SOUND
Mt. Rainier (14,410 ft. 4,392 m.)
Franklin D. Roosevelt Lake
Pend Oreille Lake
Flathead Lake
Clark Fork
LEWIS RANGE
ROCKY MOUNTAINS
BITTERROOT RANGE
CASCADE RANGE
COAST RANGES
Columbia River
Willamette River
COLUMBIA PLATEAU
Salmon River
SALMON RIVER MTS.
SAWTOOTH MTS.
Snake River
Milk River
Missouri River
Fort Peck Lake
Lake Sakakawea
GREAT PLAINS
INTERIOR PLAINS
Yellowstone River
Yellowstone Lake
Lake Oahe
Powder River
Bighorn River
BIGHORN MTS.
BLACK HILLS
Cheyenne River
White River
James River
Niobrara River
North Platte River
South Platte River
Platte River
Republican River
Smoky Hill River
PACIFIC OCEAN
Klamath River
Goose Lake
Cape Mendocino
Shasta Lake
GRAND TETONS
Gannett Peak (13,804 ft. 4,208 m.)
WIND RIVER RANGE
CONTINENTAL DIVIDE
WASATCH RANGE
Great Salt Lake
Utah Lake
UINTA MTS.
FRONT RANGE
SIERRA NEVADA
Sacramento River
CENTRAL VALLEY
Pyramid Lake
Lake Tahoe
GREAT BASIN
SAN FRANCISCO BAY
MONTEREY BAY
San Joaquin River
Green River
Colorado River
Mt. Elbert (14,433 ft. 4,400 m.)
Pikes Peak (14,110 ft. 4,302 m.)
COLORADO PLATEAU
Lake Powell
San Juan River
SAN LUIS VALLEY
SANGRE DE CRISTO MTS.
Mt. Whitney (14,494 ft. 4,419 m.)
DEATH VALLEY
COAST RANGES
MOJAVE DESERT
GRAND CANYON
Lake Mead
PAINTED DESERT
Channel Islands
Salton Sea
IMPERIAL VALLEY
Gila River
SONORA DESERT
Canadian River
Keystone Lake
GULF OF CALIFORNIA
Pecos River
Colorado River
Brazos
Rio Grande
Amistad Reservoir
Nueces River
Falcon Lake
MEXICO
130°W
125°W
120°W
115°W
110°W
105°W
100°W
45°N
40°N
35°N
30°N
25°N
Kauai
Niihau
Oahu
Molokai
Lanai
Maui
Kahoolawe
PACIFIC OCEAN
Mauna Kea (13,796 ft. 4,206 m.)
Hawaii
SCALE
0 150 Miles
0 150 Kilometers
Projection: Albers Equal Area
22°N
19°N
160°W
155°W
ARCTIC OCEAN
Arctic Circle
SOVIET UNION
BERING STRAIT
BROOKS RANGE
Yukon River
Tanana River
St. Lawrence Island
St. Matthew Island
Nunivak Island
BERING SEA
Kuskokwim River
ALASKA RANGE
Mt. McKinley (20,320 ft. 6,195 m.)
CANADA
Attu Island
SCALE
0 250 500 Miles
0 250 500 Kilometers
Projection: Albers Equal Area
Aleutian Islands
Kodiak Island
GULF OF ALASKA
Alexander Archipelago
PACIFIC OCEAN
65°N
60°N
55°N
50°N
180°
170°E
170°W
160°W
150°W
140°W
130°W

CANADA
Isle Royale
Lake Superior
MESABI RANGE
SUPERIOR UPLAND
Mississippi River
Wisconsin River
Lake Michigan
Lake Huron
Lake Ontario
Lake Erie
St. Lawrence River
St. Lawrence Seaway
St. John River
ADIRONDACK MTS.
GREEN MTS.
WHITE MTS.
LONGFELLOW MTS.
Lake Champlain
Connecticut River
Hudson R.
Finger Lakes
CATSKILL MTS.
Cape Cod
LONG ISLAND SOUND
Long Island
CENTRAL PLAINS
PLAINS
PLATEAU
Illinois River
Wabash River
Scioto River
Ohio River
Allegheny River
Monongahela River
Susquehanna River
Potomac River
James River
Kanawha River
Roanoke River
ALLEGHENY PLATEAU
APPALACHIAN MOUNTAINS
BLUE RIDGE MOUNTAINS
CUMBERLAND PLATEAU
GREAT SMOKY MTS.
PIEDMONT
ATLANTIC COASTAL PLAIN
DELAWARE BAY
CHESAPEAKE BAY
PAMLICO SOUND
Cape Hatteras
ATLANTIC OCEAN
Lake Barkley
Cumberland River
Kentucky Lake
Tennessee River
White River
Mississippi River
Tombigbee River
Coosa River
Alabama River
Pearl River
Chattahoochee River
Savannah River
Oconee River
Altamaha River
Sea Islands
Okefenokee Swamp
COASTAL PLAIN
Chandeleur Islands
Mississippi Delta
GULF OF MEXICO
FLORIDA PENINSULA
Cape Canaveral
Lake Okeechobee
The Everglades
Cape Sable
Florida Keys
STRAITS OF FLORIDA
THE BAHAMAS
CUBA
N
S
E
W
ELEVATION
Feet
Meters
13,120
4,000
6,560
2,000
1,640
500
656
200
(Sea level) 0
0 (Sea level)
Below sea level
Below sea level
Ice cap
SCALE
0
250
500 Miles
0
250
500 Kilometers
Projection: Albers Equal Area
90°W
85°W
80°W
75°W
70°W
65°W
60°W
50°N
45°N
40°N
35°N
30°N
25°N

United States of America: Political

CANADA
WISCONSIN
MICHIGAN
NEW YORK
MAINE
VT.
N.H.
MASS.
R.I.
CONN.
PENNSYLVANIA
N.J.
DELAWARE
MD.
OHIO
INDIANA
ILLINOIS
WEST VIRGINIA
VIRGINIA
KENTUCKY
TENNESSEE
NORTH CAROLINA
SOUTH CAROLINA
GEORGIA
ALABAMA
MISSISSIPPI
FLORIDA
ATLANTIC OCEAN
GULF OF MEXICO
THE BAHAMAS
CUBA
Lake Superior
Lake Michigan
Lake Huron
Lake Ontario
Lake Erie
Lake Champlain
St. Lawrence River
Connecticut River
Hudson R.
Susquehanna River
Ohio River
Illinois River
Mississippi River
Savannah River
Chattahoochee River
Kentucky Lake
Lake Barkley
Lake Okeechobee
Long Island
LONG ISLAND SOUND
DELAWARE BAY
CHESAPEAKE BAY
Cape Cod
Cape Hatteras
Cape Canaveral
Cape Sable
Sea Islands
Chandeleur Islands
Florida Keys
STRAITS OF FLORIDA
Augusta
Montpelier
Concord
Boston
Worcester
Providence
Springfield
Hartford
Waterbury
Bridgeport
New Haven
Stamford
Yonkers
Paterson
Newark
Elizabeth
New York City
Jersey City
Trenton
Albany
Syracuse
Rochester
Buffalo
Allentown
Harrisburg
Philadelphia
Pittsburgh
Erie
Cleveland
Youngstown
Akron
Toledo
Columbus
Dayton
Cincinnati
Baltimore
Dover
Annapolis
Washington, D.C.
Arlington
Alexandria
Charleston
Richmond
Roanoke
Hampton
Norfolk
Newport News
Portsmouth
Virginia Beach
Chesapeake
Frankfort
Lexington
Louisville
Evansville
Flint
Sterling Heights
Warren
Detroit
Livonia
Ann Arbor
Jackson
Lansing
Grand Rapids
Gary
South Bend
Fort Wayne
Indianapolis
Chicago
Rockford
Peoria
Springfield
Madison
Milwaukee
Minneapolis
St. Paul
Cedar Rapids
Des Moines
Davenport
St. Louis
Jefferson City
Springfield
Nashville
Knoxville
Chattanooga
Memphis
Winston-Salem
Greensboro
Durham
Raleigh
Charlotte
Columbia
Atlanta
Macon
Columbus
Savannah
Huntsville
Birmingham
Montgomery
Mobile
Jackson
Meridian
Baton Rouge
New Orleans
Tallahassee
Jacksonville
Orlando
Tampa
St. Petersburg
Fort Lauderdale
Hialeah
Miami
National capital
State capitals
Other cities
SCALE
0
250
500 Miles
0
250
500 Kilometers
Projection: Albers Equal Area
N
S
E
W
50°N
45°N
40°N
35°N
30°N
25°N
90°W
85°W
80°W
75°W
70°W
65°W
60°W

North America: Physical
ASIA
EUROP
NORTH POLE
POLAR ICE PACK
ARCTIC OCEAN
BERING SEA
St. Lawrence Island
BERING STRAIT
Nunivak Island
BROOKS RANGE
Mt. McKinley (20,320 ft. 6,195 m.)
Yukon River
ALASKA RANGE
GULF OF ALASKA
Kodiak Island
YUKON PLATEAU
BEAUFORT SEA
Banks Island
Victoria Island
Queen Elizabeth Islands
Ellesmere Island
Greenland
BAFFIN BAY
Baffin Island
DAVIS STRAIT
DENMARK STRAIT
Arctic Circle
Cape Farewell
Great Bear Lake
Mackenzie River
Great Slave Lake
Southampton Island
HUDSON STRAIT
LABRADOR SEA
Coats Island
Mansel Island
HUDSON BAY
Alexander Archipelago
Queen Charlotte Islands
Peace River
Lake Athabasca
Athabasca River
ROCKY MOUNTAINS
CANADIAN SHIELD
Nelson River
Saskatchewan River
Lake Winnipeg
PACIFIC OCEAN
Vancouver Island
Anticosti Island
Newfoundland
GULF OF ST. LAWRENCE
Prince Edward Island
Cape Breton Island
Mt. Rainier (14,410 ft. 4,392 m.)
CASCADE RANGE
COAST RANGES
Columbia River
GREAT PLAINS
Lake Superior
Lake Michigan
L. Huron
Lake Ontario
L. Erie
St. Lawrence River
Cape Mendocino
BLACK HILLS
Missouri River
Mississippi River
Cape Cod
Long Island
SIERRA NEVADA
CENTRAL VALLEY
GREAT BASIN
Platte River
CENTRAL PLAINS
APPALACHIAN MOUNTAINS
ATLANTIC OCEAN
DEATH VALLEY
COLORADO PLATEAU
Mt. Whitney (14,494 ft. 4,419 m.)
Ohio River
Cumberland R.
OZARK PLATEAU
Arkansas River
Tennessee River
PIEDMONT
ATLANTIC COASTAL PLAIN
Cape Hatteras
Bermuda
Red River
Brazos River
Rio Grande
Guadalupe Island
BAJA CALIFORNIA
GULF OF CALIFORNIA
GULF COASTAL PLAIN
FLORIDA PENINSULA
Cape Canaveral
N
W
E
S
Tropic of Cancer
SIERRA MADRE OCCIDENTAL
SIERRA MADRE ORIENTAL
GULF OF MEXICO
Florida Keys
STRAITS OF FLORIDA
Bahamas
West Indies
Cuba
Greater Antilles
Hispaniola
Puerto Rico
Lesser Antilles
Jamaica
Popocatepetl (17,887 ft. 5,453 m.)
YUCATÁN PENINSULA
SIERRA MADRE DEL SUR
CARIBBEAN SEA
Trinidad
Lake Nicaragua
CENTRAL AMERICA
ISTHMUS OF PANAMA
SOUTH AMERICA
Equator
ELEVATION
Feet
Meters
13,120
4,000
6,560
2,000
1,640
500
656
200
(Sea level) 0
0 (Sea level)
Below sea level
Below sea level
Ice cap
SCALE
0 250 500 750 1,000 Miles
0 250 500 750 1,000 Kilometers
Projection: Azimuthal Equal Area
170°E
180°
170°W
160°W
150°W
140°W
130°W
120°W
110°W
100°W
90°W
80°W
70°W
60°N
70°N
80°N
50°N
40°N
30°N
20°N
10°N
0°

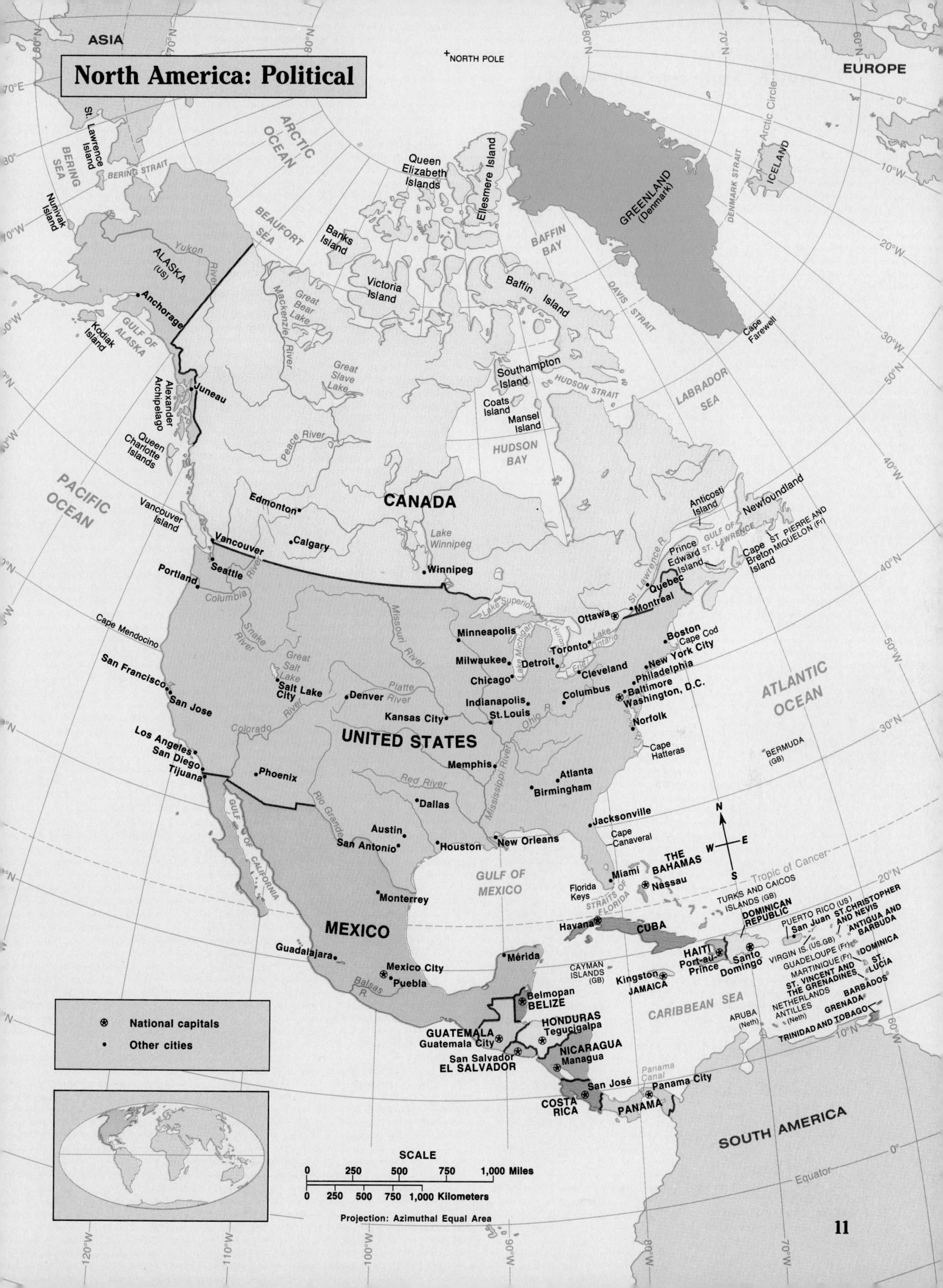
North America: Political
ASIA
NORTH POLE
EUROPE
ARCTIC OCEAN
BERING SEA
BERING STRAIT
St. Lawrence Island
Nunivak Island
BEAUFORT SEA
Queen Elizabeth Islands
Ellesmere Island
GREENLAND (Denmark)
ICELAND
DENMARK STRAIT
Arctic Circle
Banks Island
Victoria Island
BAFFIN BAY
Baffin Island
DAVIS STRAIT
Cape Farewell
ALASKA (US)
Yukon River
Anchorage
Kodiak Island
GULF OF ALASKA
Great Bear Lake
Mackenzie River
Great Slave Lake
Southampton Island
HUDSON STRAIT
LABRADOR SEA
Coats Island
Mansel Island
HUDSON BAY
Juneau
Alexander Archipelago
Queen Charlotte Islands
Peace River
PACIFIC OCEAN
Vancouver Island
Edmonton
CANADA
Calgary
Vancouver
Lake Winnipeg
Winnipeg
Anticosti Island
Newfoundland
GULF OF ST. LAWRENCE
ST PIERRE AND MIQUELON (Fr)
Prince Edward Island
Cape Breton Island
St. Lawrence R.
Quebec
Seattle
Portland
Columbia River
Lake Superior
Montreal
Ottawa
Cape Mendocino
Snake River
Missouri River
Minneapolis
Lake Michigan
L. Huron
Lake Ontario
Toronto
Boston
Cape Cod
Great Salt Lake
Milwaukee
Detroit
L. Erie
Cleveland
New York City
Philadelphia
San Francisco
San Jose
Salt Lake City
Platte River
Denver
Chicago
Baltimore
Washington, D.C.
Columbus
Indianapolis
ATLANTIC OCEAN
Kansas City
St. Louis
Ohio R.
Norfolk
Colorado River
UNITED STATES
Los Angeles
San Diego
Tijuana
Cape Hatteras
BERMUDA (GB)
Memphis
Phoenix
Atlanta
Red River
Birmingham
Mississippi River
Rio Grande
Dallas
GULF OF CALIFORNIA
Jacksonville
Austin
San Antonio
Houston
New Orleans
Cape Canaveral
N
W
E
S
THE BAHAMAS
GULF OF MEXICO
Miami
Nassau
Tropic of Cancer
Florida Keys
STRAITS OF FLORIDA
TURKS AND CAICOS ISLANDS (GB)
Monterrey
DOMINICAN REPUBLIC
Havana
CUBA
PUERTO RICO (US)
San Juan
ST. CHRISTOPHER AND NEVIS
ANTIGUA AND BARBUDA
MEXICO
VIRGIN IS. (US,GB)
HAITI
Port-au-Prince
Santo Domingo
GUADELOUPE (Fr)
DOMINICA
Guadalajara
Mérida
CAYMAN ISLANDS (GB)
MARTINIQUE (Fr)
ST. LUCIA
Mexico City
Puebla
Kingston
JAMAICA
ST. VINCENT AND THE GRENADINES
BARBADOS
Balsas R.
Belmopan
BELIZE
NETHERLANDS ANTILLES (Neth)
GRENADA
CARIBBEAN SEA
ARUBA (Neth)
TRINIDAD AND TOBAGO
National capitals
Other cities
HONDURAS
Tegucigalpa
GUATEMALA
Guatemala City
NICARAGUA
Managua
San Salvador
EL SALVADOR
Panama Canal
San José
Panama City
COSTA RICA
PANAMA
SOUTH AMERICA
Equator
SCALE
0 250 500 750 1,000 Miles
0 250 500 750 1,000 Kilometers
Projection: Azimuthal Equal Area

South America: Physical
CARIBBEAN SEA
CENTRAL AMERICA
Margarita Island
Tobago
Trinidad
Panama Canal
Orinoco River Delta
Lake Maracaibo
LLANOS
GULF OF PANAMA
Orinoco River
Meta River
Angel Falls
GUIANA
HIGHLANDS
ATLANTIC OCEAN
Devil's Island
Cape Orange
Mt. Tolima (19,049 ft. 5,808 m.)
Malpelo Island
ANDES MOUNTAINS
Amazon River Delta
Caqueta River
Rio Negro
Galápagos Islands
Equator
Mt. Chimborazo (20,561 ft. 6,269 m.)
Japura River
AMAZON BASIN
Amazon River
GULF OF GUAYAQUIL
Juruá River
Purus River
Madeira River
Tapajós River
Xingu River
Tocantins River
Parnaíba River
Mt. Huascarán (22,205 ft. 6,770 m.)
Ucayali River
BRAZILIAN HIGHLANDS
MATO GROSSO PLATEAU
PACIFIC OCEAN
Lake Titicaca
Mt. Ancohuma (21,490 ft. 6,552 m.)
Beni River
Mamoré River
São Francisco River
Lake Poopó
Pilcomayo River
CHACO
ATACAMA DESERT
ANDES
BRAZILIAN PLATEAU
Tropic of Capricorn
Paraguay River
Paraná River
Salado River
Uruguay River
San Felix Island
San Ambrosio Island
N
W
E
S
Mt. Aconcagua (22,834 ft. 6,962 m.)
Juan Fernández Islands
ATLANTIC OCEAN
RÍO DE LA PLATA
PAMPAS
MOUNTAINS
Colorado River
ELEVATION
Feet
Meters
13,120
4,000
6,560
2,000
1,640
500
656
200
(Sea level) 0
0 (Sea level)
Below sea level
Below sea level
GULF OF SAN MATIAS
Chiloé Island
Chonos Archipelago
PATAGONIA
GULF OF SAN JORGE
Cape Tres Puntas
BAHÍA GRANDE
STRAIT OF MAGELLAN
Falkland Islands
Tierra del Fuego
Cape Horn
South Georgia Islands
SCALE
0 250 500 1,000 Miles
0 250 500 1,000 Kilometers
Projection: Azimuthal Equal Area
20°N
10°N
0°
10°S
20°S
30°S
50°S
100°W
90°W
80°W
70°W
60°W
50°W
40°W
30°W
20°W

Asia: Political

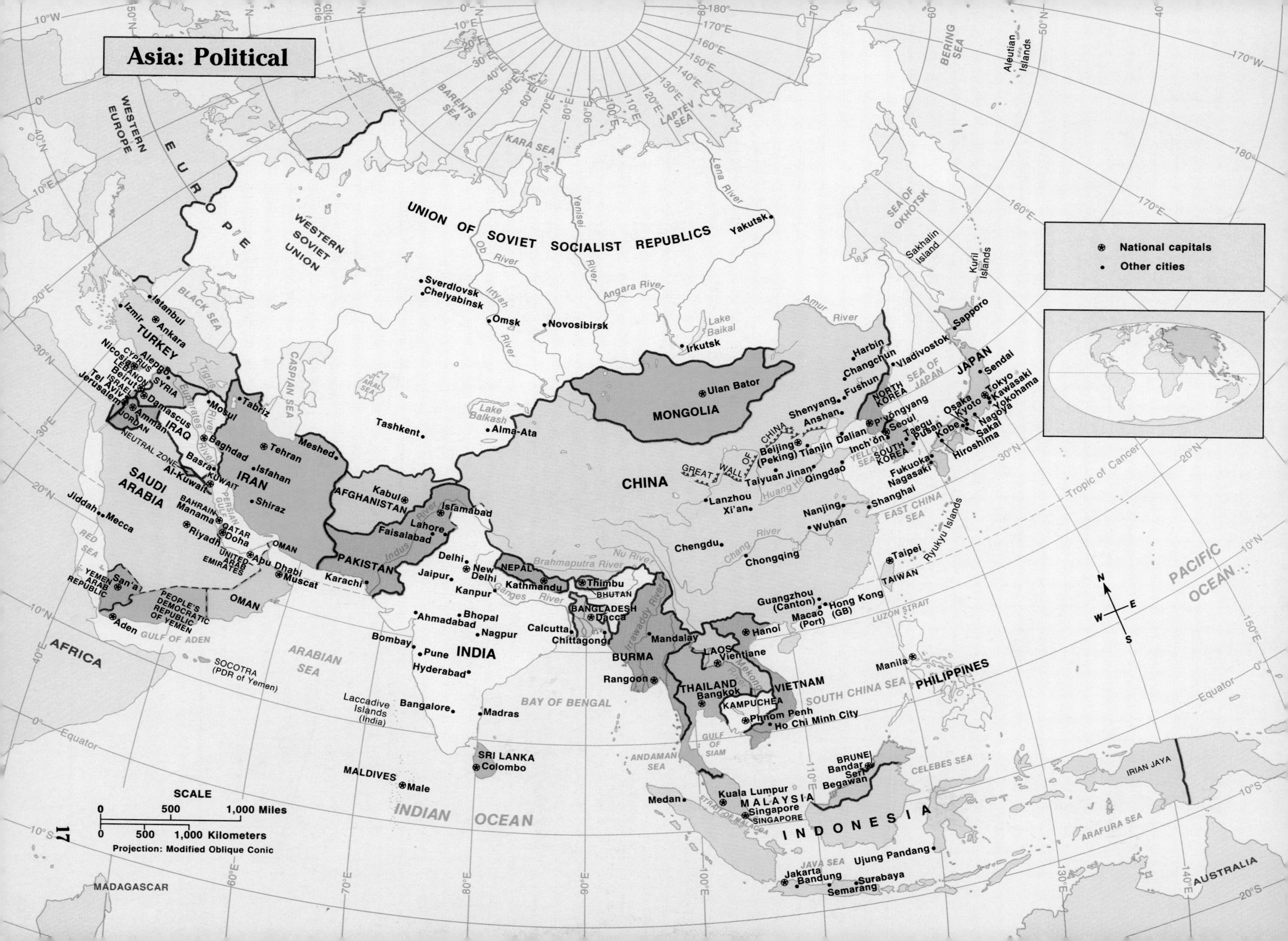

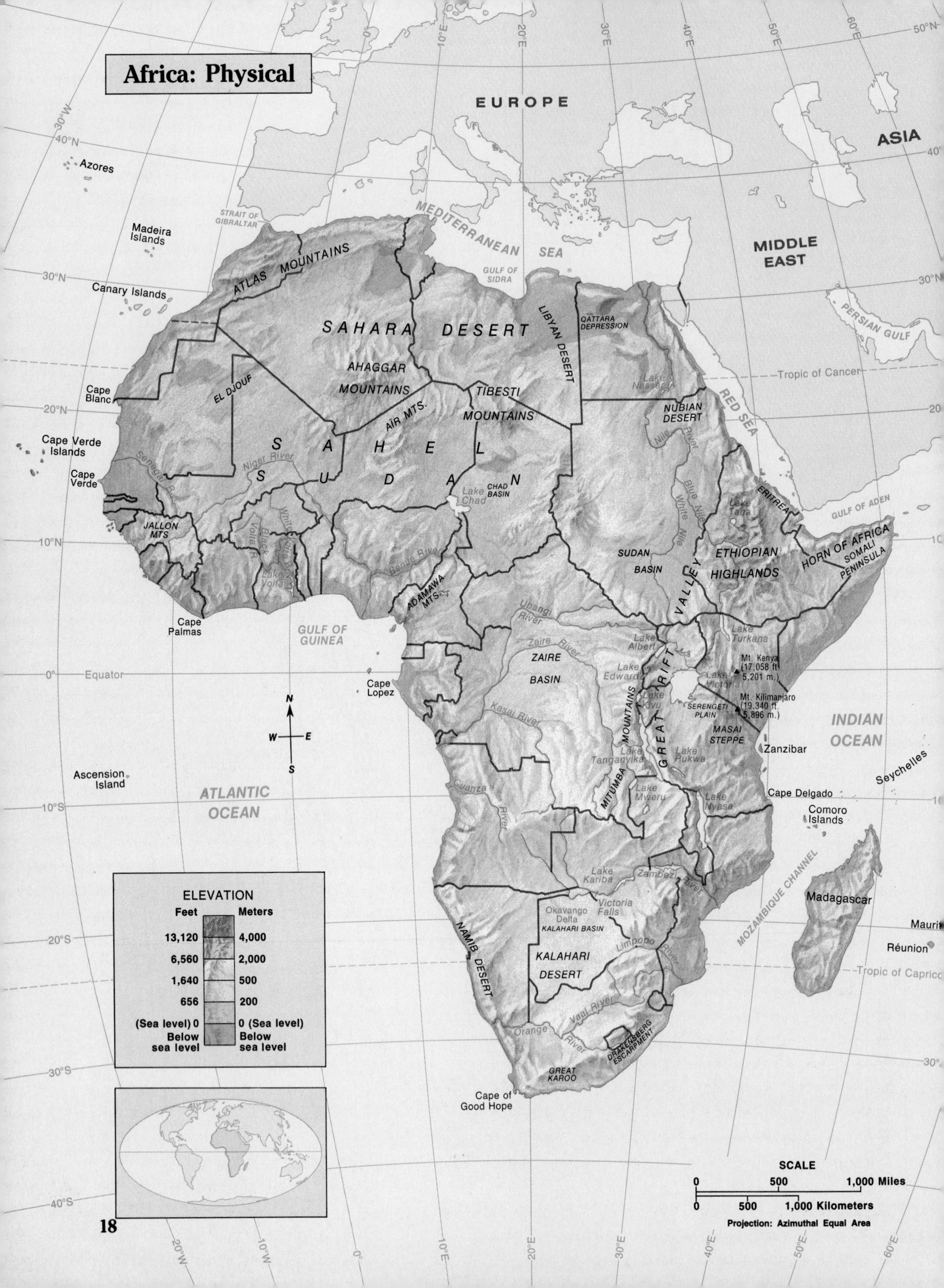
Africa: Physical
EUROPE
ASIA
MIDDLE EAST
Azores
Madeira Islands
Canary Islands
Cape Blanc
Cape Verde Islands
Cape Verde
Cape Palmas
Cape Lopez
Ascension Island
Cape of Good Hope
STRAIT OF GIBRALTAR
MEDITERRANEAN SEA
GULF OF SIDRA
ATLAS MOUNTAINS
SAHARA DESERT
LIBYAN DESERT
QATTARA DEPRESSION
AHAGGAR MOUNTAINS
TIBESTI MOUNTAINS
AÏR MTS.
EL DJOUF
Tropic of Cancer
Lake Nasser
NUBIAN DESERT
RED SEA
PERSIAN GULF
SAHEL
SUDAN
Senegal R.
Niger River
Lake Chad
CHAD BASIN
JALLON MTS
Black Volta R.
White Volta R.
Lake Volta
Benue River
ADAMAWA MTS.
GULF OF GUINEA
ATLANTIC OCEAN
Nile River
Blue Nile
White Nile
ERITREA
Lake Tana
GULF OF ADEN
ETHIOPIAN HIGHLANDS
HORN OF AFRICA
SOMALI PENINSULA
SUDAN BASIN
VALLEY
Ubangi River
Zaire River
ZAIRE BASIN
Lake Albert
Lake Edward
Lake Kivu
Lake Turkana
Lake Victoria
Mt. Kenya (17,058 ft. 5,201 m.)
Mt. Kilimanjaro (19,340 ft. 5,896 m.)
SERENGETI PLAIN
MASAI STEPPE
GREAT RIFT
MITUMBA MOUNTAINS
Lake Tanganyika
Lake Rukwa
Lake Mweru
Lake Nyasa
Kasai River
Cuanza River
Equator
Zanzibar
INDIAN OCEAN
Seychelles
Cape Delgado
Comoro Islands
MOZAMBIQUE CHANNEL
Madagascar
Réunion
Lake Kariba
Zambezi River
Victoria Falls
Okavango Delta
KALAHARI BASIN
KALAHARI DESERT
NAMIB DESERT
Limpopo River
Tropic of Capricorn
Vaal River
Orange River
DRAKENSBERG ESCARPMENT
GREAT KAROO
N
W
E
S
ELEVATION
Feet
Meters
13,120
4,000
6,560
2,000
1,640
500
656
200
(Sea level) 0
0 (Sea level)
Below sea level
Below sea level
SCALE
0
500
1,000 Miles
0
500
1,000 Kilometers
Projection: Azimuthal Equal Area
50°N
40°N
30°N
20°N
10°N
0°
10°S
20°S
30°S
40°S
30°W
20°W
10°W
0°
10°E
20°E
30°E
40°E
50°E
60°E

Africa: Political
EUROPE
ASIA
MIDDLE EAST
AZORES (Port)
MADEIRA (Port)
CANARY ISLANDS (Sp)
CAPE VERDE
MEDITERRANEAN SEA
RED SEA
GULF OF ADEN
GULF OF GUINEA
ATLANTIC OCEAN
INDIAN OCEAN
Tropic of Cancer
Equator
Tropic of Capricorn
MOROCCO
Casablanca
Rabat
Fez
Marrakesh
ALGERIA
Algiers
Oran
TUNISIA
Tunis
LIBYA
Tripoli
Benghazi
EGYPT
Alexandria
Cairo
Giza
Lake Nasser
El Aaiún
WESTERN SAHARA (occupied by Morocco)
MAURITANIA
Nouakchott
MALI
Niger River
SENEGAL
Dakar
Banjul
GAMBIA
Bamako
Bissau
GUINEA-BISSAU
GUINEA
Conakry
Freetown
SIERRA LEONE
Monrovia
LIBERIA
IVORY COAST
Abidjan
BURKINA FASO
Ouagadougou
GHANA
Accra
TOGO
Lomé
BENIN
Porto-Novo
NIGER
Niamey
NIGERIA
Kano
Ogbomosho
Ibadan
Lagos
CHAD
Lake Chad
N´Djamena
SUDAN
Omdurman
Khartoum
Nile River
Blue Nile
White Nile
ETHIOPIA
Addis Ababa
DJIBOUTI
Djibouti
SOMALIA
Mogadishu
CAMEROON
Yaoundé
CENTRAL AFRICAN REPUBLIC
Bangui
Malabo
EQUATORIAL GUINEA
SÃO TOMÉ AND PRÍNCIPE
São Tomé
Libreville
GABON
PEOPLE'S REPUBLIC OF THE CONGO
Brazzaville
Kinshasa
ZAIRE
Zaire River
Kisangani
UGANDA
Kampala
KENYA
Nairobi
Mombasa
RWANDA
Kigali
BURUNDI
Bujumbura
Lake Victoria
TANZANIA
Dodoma
Zanzibar
Dar es Salaam
Lake Tanganyika
Lake Nyasa
CABINDA (Angola)
Luanda
ANGOLA
Lubumbashi
ZAMBIA
Lusaka
MALAWI
Lilongwe
Zambezi River
MOZAMBIQUE
Harare
ZIMBABWE
Bulawayo
NAMIBIA (South Africa)
Windhoek
WALVIS BAY (South Africa)
BOTSWANA
Gaborone
Pretoria
Johannesburg
Soweto
Maputo
Mbabane
SWAZILAND
LESOTHO
Maseru
Orange River
SOUTH AFRICA
Cape Town
Port Elizabeth
ST. HELENA ISLAND (GB)
Seychelles
COMOROS
Moroni
Antananarivo
MADAGASCAR
MAURITIUS
Port Louis
RÉUNION (Fr)
N
W
E
S
National capitals
Other cities
SCALE
0
500
1,000 Miles
0
500
1,000 Kilometers
Projection: Azimuthal Equal Area
50°N
40°N
30°N
20°N
10°N
0°
10°S
20°S
30°S
40°S
30°W
20°W
10°W
0°
10°E
20°E
30°E
40°E
50°E
60°E

Australia and New Zealand

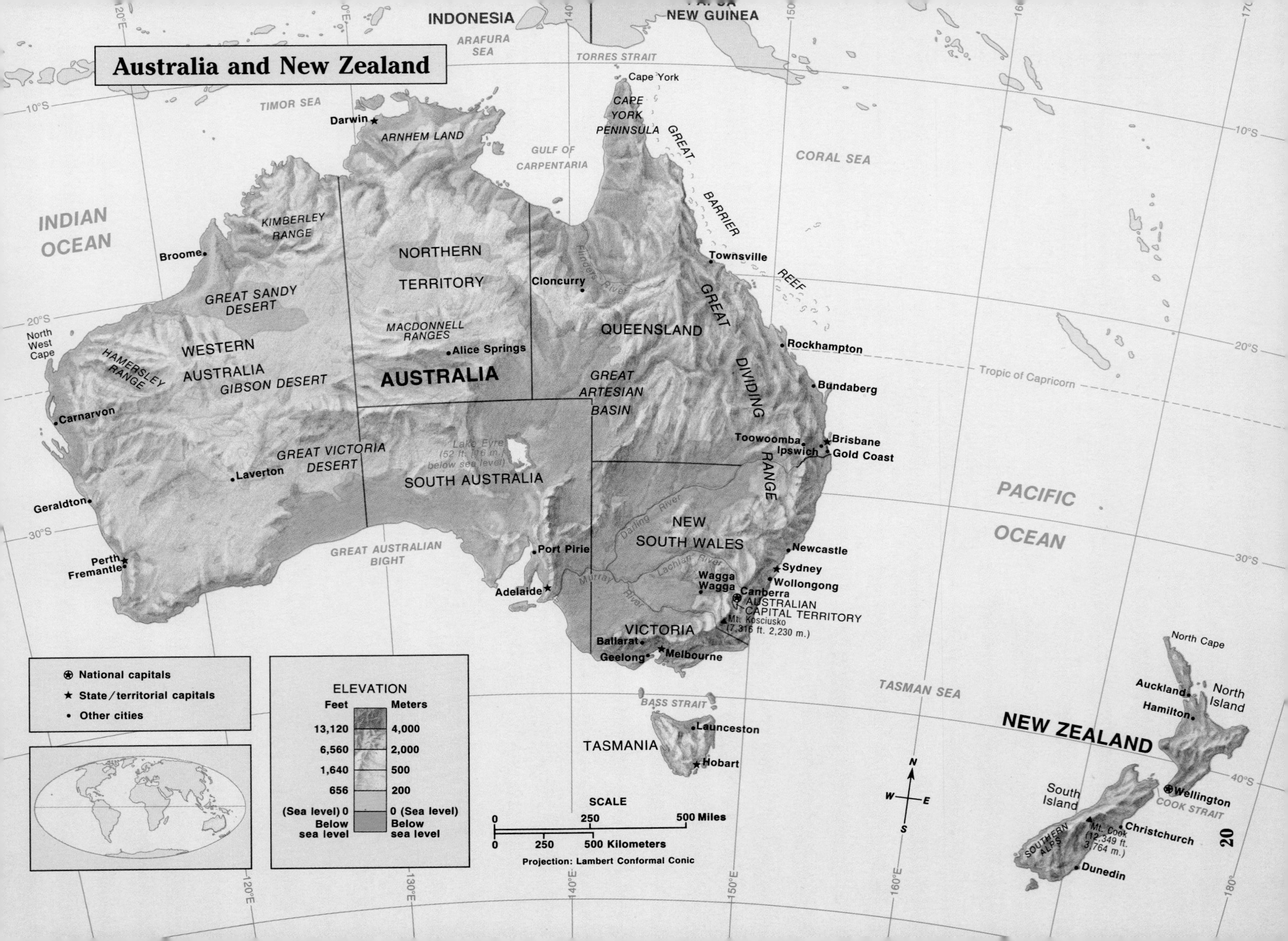

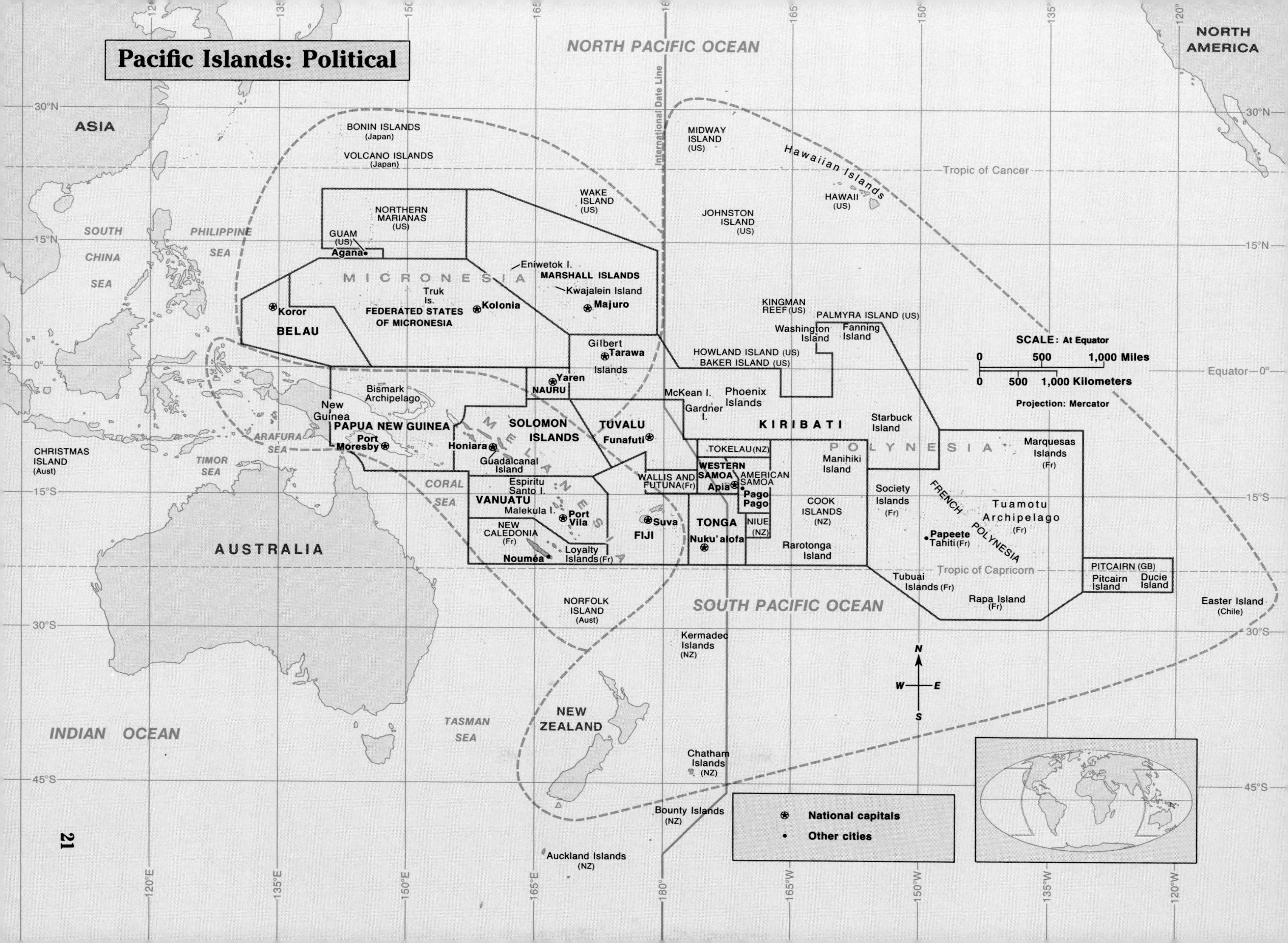
Pacific Islands: Political
NORTH PACIFIC OCEAN
SOUTH PACIFIC OCEAN
INDIAN OCEAN
NORTH AMERICA
ASIA
AUSTRALIA
NEW ZEALAND
SOUTH CHINA SEA
PHILIPPINE SEA
ARAFURA SEA
TIMOR SEA
CORAL SEA
TASMAN SEA
BONIN ISLANDS (Japan)
VOLCANO ISLANDS (Japan)
NORTHERN MARIANAS (US)
GUAM (US)
Agana
MICRONESIA
Truk Is.
FEDERATED STATES OF MICRONESIA
Kolonia
Koror
BELAU
Eniwetok I.
MARSHALL ISLANDS
Kwajalein Island
Majuro
WAKE ISLAND (US)
MIDWAY ISLAND (US)
Hawaiian Islands
HAWAII (US)
JOHNSTON ISLAND (US)
KINGMAN REEF (US)
PALMYRA ISLAND (US)
Washington Island
Fanning Island
HOWLAND ISLAND (US)
BAKER ISLAND (US)
Gilbert Islands
Tarawa
Yaren
NAURU
McKean I.
Gardner I.
Phoenix Islands
KIRIBATI
Starbuck Island
New Guinea
Bismark Archipelago
PAPUA NEW GUINEA
Port Moresby
MELANESIA
SOLOMON ISLANDS
Honiara
Guadalcanal Island
TUVALU
Funafuti
TOKELAU (NZ)
WESTERN SAMOA
Apia
AMERICAN SAMOA
Pago Pago
WALLIS AND FUTUNA (Fr)
Espiritu Santo I.
VANUATU
Malekula I.
Port Vila
NEW CALEDONIA (Fr)
Nouméa
Loyalty Islands (Fr)
Suva
FIJI
TONGA
Nuku'alofa
NIUE (NZ)
COOK ISLANDS (NZ)
Rarotonga Island
Manihiki Island
POLYNESIA
Society Islands (Fr)
FRENCH POLYNESIA
Papeete
Tahiti (Fr)
Tuamotu Archipelago (Fr)
Marquesas Islands (Fr)
Tubuai Islands (Fr)
Rapa Island (Fr)
PITCAIRN (GB)
Pitcairn Island
Ducie Island
Easter Island (Chile)
CHRISTMAS ISLAND (Aust)
NORFOLK ISLAND (Aust)
Kermadec Islands (NZ)
Chatham Islands (NZ)
Bounty Islands (NZ)
Auckland Islands (NZ)
International Date Line
Tropic of Cancer
Equator—0°
Tropic of Capricorn
30°N
15°N
0°
15°S
30°S
45°S
120°E
135°E
150°E
165°E
180°
165°W
150°W
135°W
120°W
SCALE: At Equator
0 500 1,000 Miles
0 500 1,000 Kilometers
Projection: Mercator
N
S
E
W
National capitals
Other cities

BASIC MAP AND GLOBE SKILLS

Basic Map and Globe Skills Lesson Plans and Planning Guide are found in the Teacher's Guide.

Before you begin your study of world geography, you must master some basic map and globe skills. Throughout this book, you will have the opportunity to improve these skills and build upon them.

Geographers are interested in answering such questions as these: Where is it? Why is it located there? What is it near? What influence does it have on its environment? To answer these and other questions, geographers use certain tools. The three major tools are globes, maps, and remote sensors.

The Globe

A Model of the Earth One of the best sources that geographers use to answer their questions is a **globe**. A globe is a scale model of the earth. It is especially useful for looking at the entire earth or at large areas of the earth's surface.

About 71 percent of the earth's surface is covered by water. The remainder is land. Geographers organize the earth's land surface into seven large masses of land, called **continents**. The continents are North America, South America, Europe, Asia, Africa, Australia, and Antarctica. Use a globe to locate the continents shown in Figure 1.

Notice on the globe that Europe is not really a separate continent, but part of the earth's largest landmass, called Eurasia. This landmass is traditionally divided into two continents, Europe and Asia. Landmasses smaller than continents that are completely surrounded by water are called **islands**. Locate some islands on the globe.

Like the land surface, the water surface of the earth is also organized into parts. It is divided into separate bodies of water. Study the globe to find the largest bodies of water, called **oceans**. The major oceans are the Pacific, the Atlantic, the Indian, and a much smaller ocean called the Arctic. There are other terms used to identify smaller bodies of water as well. Can you locate some of these on the globe?

Figure 1 The Continents

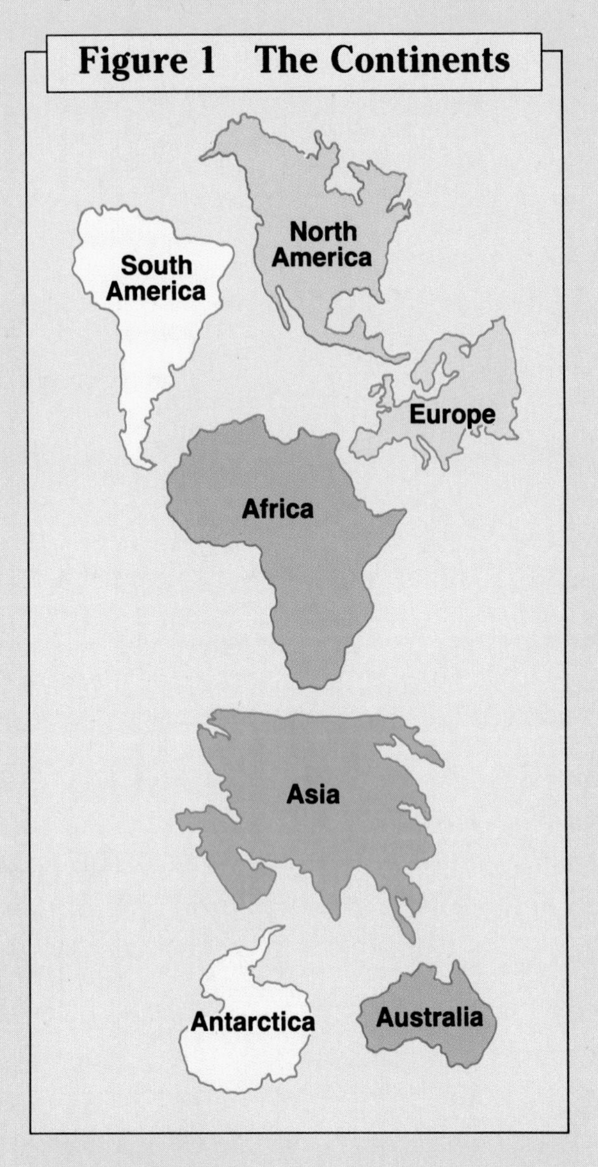

Lines on the Globe Look at Figure 2, the diagram of a globe. You will notice a pattern of lines that circle the earth in east-west and north-south directions. This pattern is called a **grid**. The two sets of lines that make up the grid do not actually exist on the surface of the earth. The purpose of these imaginary lines is to help us find the exact location of places on the earth. These lines are called **latitude** and **longitude**.

Lines of latitude are drawn in an east-west direction around the globe. Latitude is the measurement that tells you exactly how far north or south of the **equator** you are. The equator is the imaginary line that also circles the globe in an east-west direction. It is halfway between the North and South poles. In other words, latitude locates exactly where you are between the equator and either the North or the South Pole.

Because the lines of latitude are always parallel to the equator and to each other, they are called **parallels** of latitude. If you look at the diagram of the globe again, you will see that the parallels of latitude are labeled with the symbol °. This symbol is a unit of measurement called **degrees**. These parallels are also labeled with an *N* for *north* or an *S* for *south*. The letter *N* or *S* tells you whether a place is located north or south of the equator. Lines of latitude range from 0°, for locations on the equator, to 90° N or 90° S, for locations at the North or South Pole.

Latitude is only half of what you need to know in order to locate a place. There are many locations along the equator or along any parallel of latitude. How can you tell one place from another? To do this, you must use the lines of longitude. These lines circle the globe from pole to pole.

Just as we use the equator as a line to determine latitude, we use an imaginary line called the **prime meridian** to determine longitude. The prime meridian runs from the North Pole to the South Pole through Greenwich, England. Since lines of longitude are drawn in the same direction as the prime meridian, they are all called **meridians**. Imagine looking down at the globe from above the North Pole. No matter where on Earth a place is, you could draw a line that runs from pole to pole through your location.

Notice on the diagram of the globe that the meridians and parallels are similarly labeled. Each meridian indicates how many degrees of longitude a location is east or west of the prime meridian. A *W* for *west* and an *E* for *east* is added to the longitude degree. Lines of longitude range from 0° on the prime meridian to 180° at a meridian in the mid-Pacific Ocean.

By knowing the latitude and longitude of any place on Earth, you can find its exact location. No other place on Earth has this exact latitude and longitude.

Figure 2 Globe

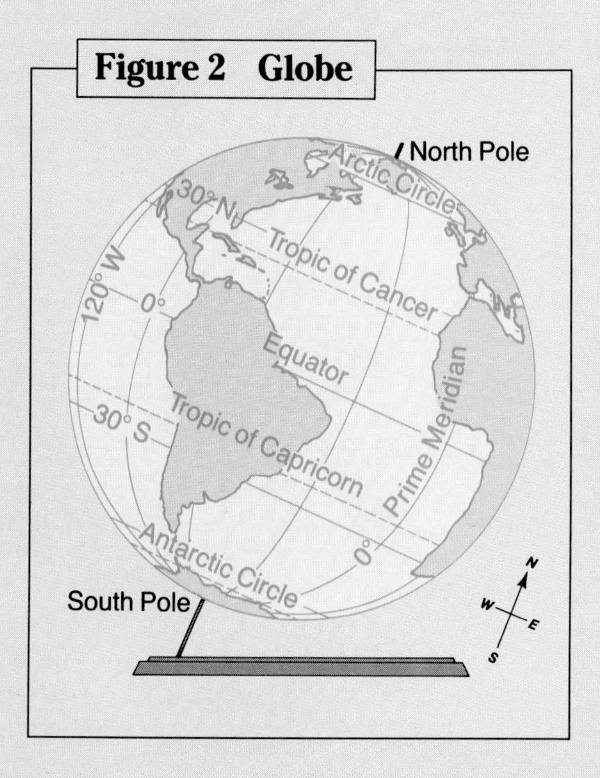

Dividing the Globe into Hemispheres Looking at the globe with the North Pole on top and the South Pole on the bottom, you can see that the equator divides the globe into two halves, or **hemispheres**. *Hemi* means "half," so *hemisphere* means "half of the sphere." The upper, or northern, half of the globe is the Northern Hemisphere. The bottom, or southern, half is the Southern Hemisphere. (See Figure 3.)

Just as the equator divides the world into two hemispheres, so do the prime meridian and the 180° meridian. The hemisphere that lies to the west of the prime meridian is the

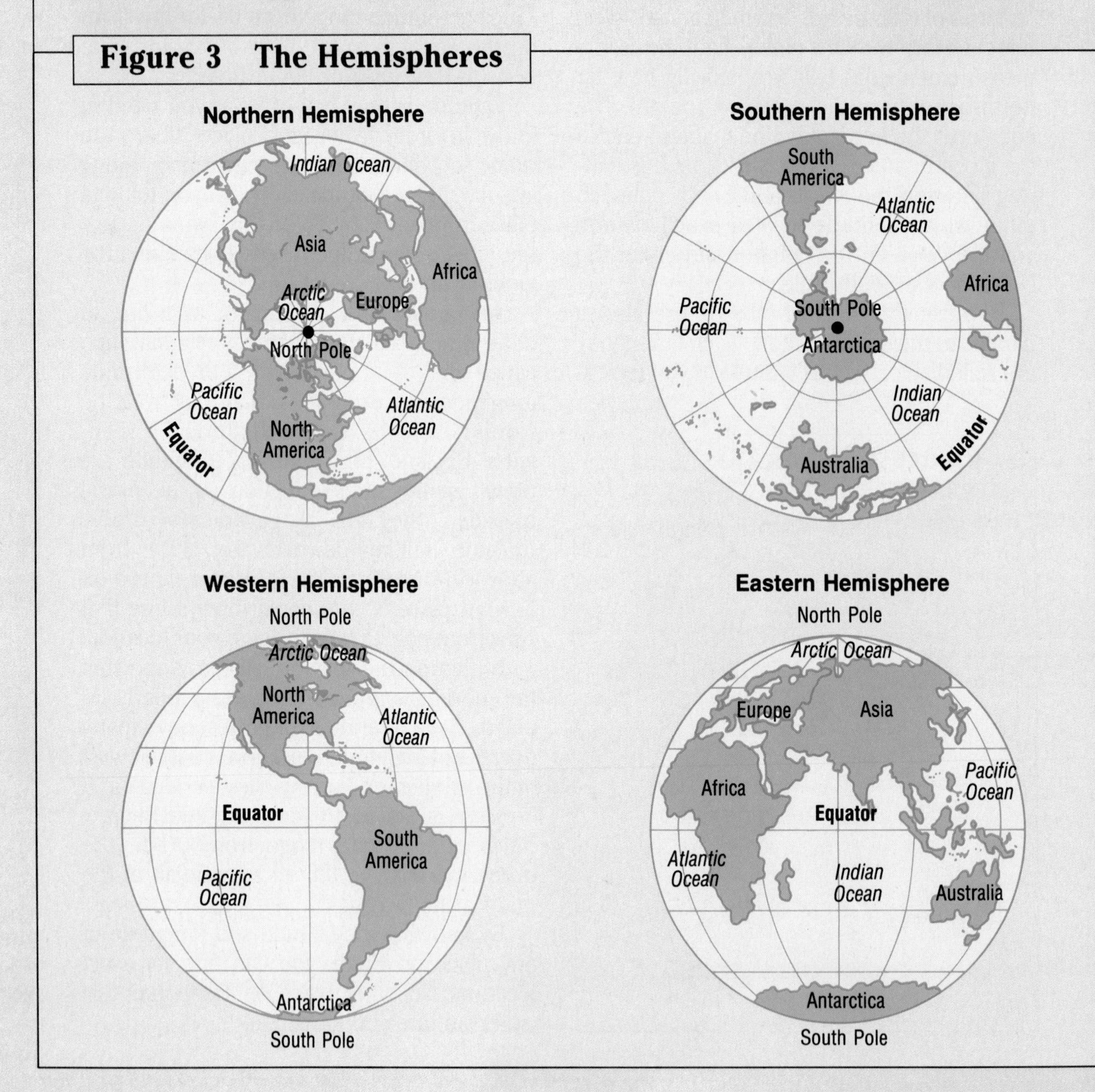

Figure 3 The Hemispheres

Western Hemisphere. The hemisphere to the east of the prime meridian is the Eastern Hemisphere. Because the prime meridian separates parts of Europe and Africa into two different hemispheres, some map makers divide the Eastern and Western hemispheres at 20° W. This places all of Europe and Africa in the Eastern Hemisphere.

Maps

The Geographer's Most Important Tool Suppose you want to plan the route to the next out-of-town football game. Would you use a globe to do that? Would it be helpful to use latitude and longitude to tell your friends where the game will be? Not really. You need to use a **map**. A map is a flat diagram of all or part of the earth's surface.

Maps are a geographer's most important tool because they show how different features of the earth are related. Maps are important to people as a source of information. Aircraft pilots, sailors, surveyors, police officers, fire fighters, soldiers, taxi drivers, hikers, and many other people rely on maps. With what kind of map are you most familiar?

Most of us have used a road map. A road map is a special-purpose map designed to help us find our way from one place to another when traveling by car. A road map is just one kind of special-purpose map. There are many other kinds of maps used for different reasons. You will learn about some of these in the chapters that follow.

An **atlas** is an organized collection of maps in one book. Some are made for a special purpose; for example, a road atlas features road maps of all the states. Turn to the Atlas at the front of this book. How will the maps in the Atlas help you in the study of world geography?

Map Projections The earth is shaped like a ball. On a globe, the shapes and sizes of the earth's geographic features are shown accurately. However, when map makers transfer information from a round globe to a flat map, problems arise. Usually, true shape and size are distorted. Map makers have different ways of presenting a round earth on flat maps. These are called **map projections**.

Look at the three major types of map projections shown on page 26. Notice that the parallel and meridian lines are curved on some projections and straight on others. Find Greenland in Figures 4a and 4c. See how its shape and size varies. Which projection shows the size and shape of Greenland most accurately?

Map projections are designed from complex numerical computations. Today, most projections are plotted by computer. The cylindrical projection is designed from a cylinder wrapped around the globe. (See Figure 4a.) The cylinder touches the globe only at the equator. The meridians are pulled apart and are parallel to each other instead of meeting at the poles. This causes a great amount of distortion near the poles.

The best-known cylindrical projection is the Mercator projection, named after its inventor, who developed it in 1569. Because it shows true direction and shape, it was useful to the sailors of that time. Although it distorts the size of areas away from the equator, it is still used quite commonly today.

The conic projection is designed from a cylinder folded into a cone and placed over the globe. (See Figure 4b.) This projection is most accurate along the latitude lines where it touches the globe. It retains almost true shape and size. The conic projection is most useful for areas that have long east-west dimensions, such as the United States.

The type of projection is identified on all maps in this text, just below the scale. Refer students to the maps on pages 110, 219, and 255 for examples.

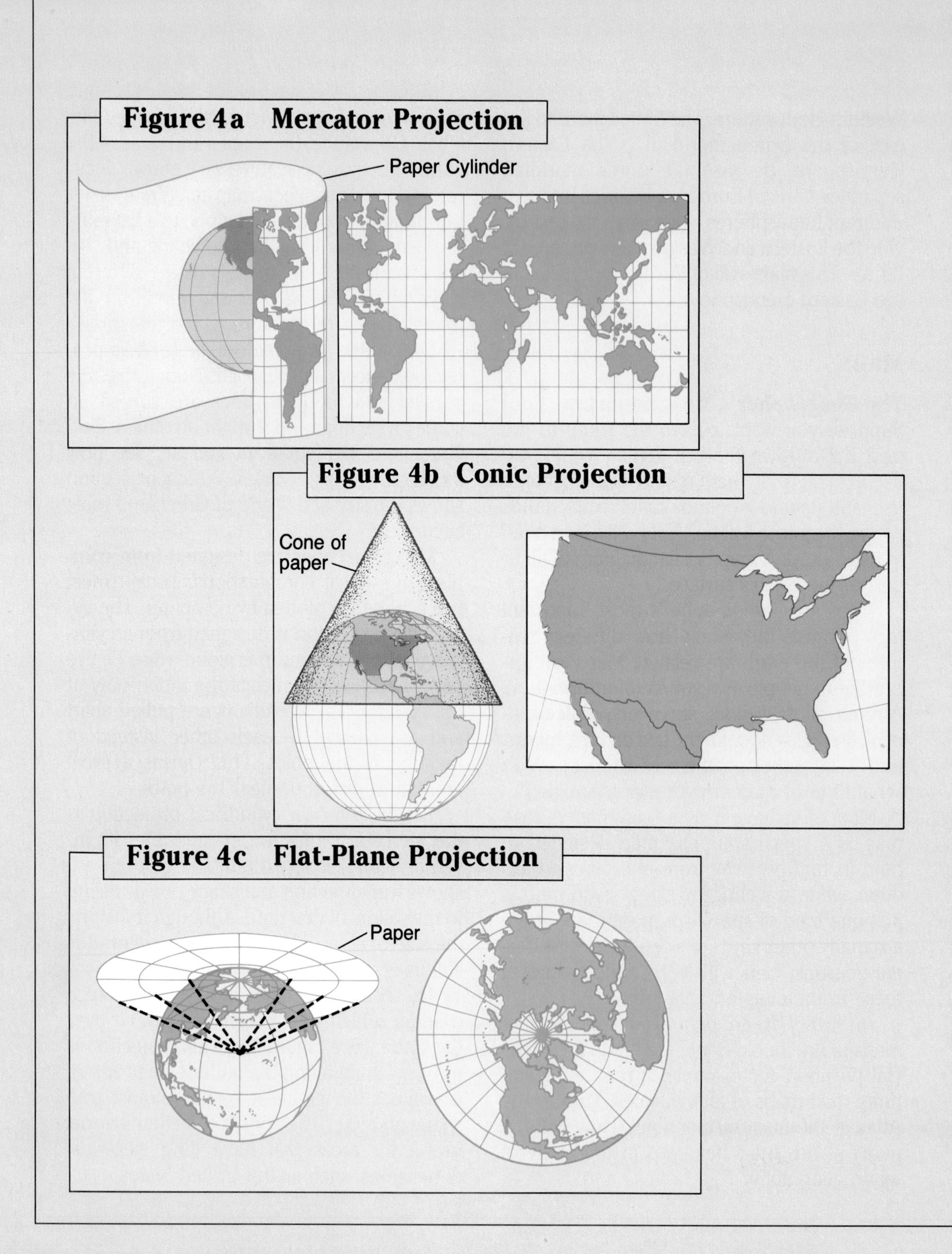
Figure 4a Mercator Projection
Paper Cylinder
Figure 4b Conic Projection
Cone of paper
Figure 4c Flat-Plane Projection
Paper

The flat-plane projection is designed from a plane touching the globe at one point, such as the North or South Pole. (See Figure 4c.) This projection is useful for showing true direction for airplane pilots and ships' navigators. It also shows true area but distorts true shape.

There are many other kinds of map projections. Each offers advantages and disadvantages. Map makers must always choose the correct projection according to the map's special purpose.

Great-Circle Routes If you were to draw a straight line between two places on a flat map, you might think you had found the shortest distance between two locations. Yet, because maps are flat, they are not true representations of the earth's surface. The shortest distance between any two points on the earth is a **great-circle route**. (See Figure 5.) Any imaginary line that divides the earth into equal parts is a great circle. For example, the equator is a great circle. Airplanes and ships are navigated along great-circle routes.

Map Essentials Study and compare the maps in the Atlas at the front of this book. You will notice that they all have several elements in common. Almost all maps have **scales**. Using a scale is the only way to compare a distance measured on the map to the actual distance on the surface of the earth. For example, without a scale, you could not determine the distance from the east coast of the United States to the west coast.

Scales appear on maps in several forms. The maps in this book provide a line scale. Look at the line scale in Figure 6. Line scales are easy to use, but you need a piece of paper and a pencil. Suppose you want to find the distance between London and Birmingham on the map of Great Britain and Ireland (Figure 6). First, mark the distance between the two cities on the map by putting two dots along the edge of a piece of paper. Then compare the two dots with the map's line scale. You can now determine the distance between the two cities. The number on top of the scale gives the distance in miles. The

Figure 5 Great-Circle Routes

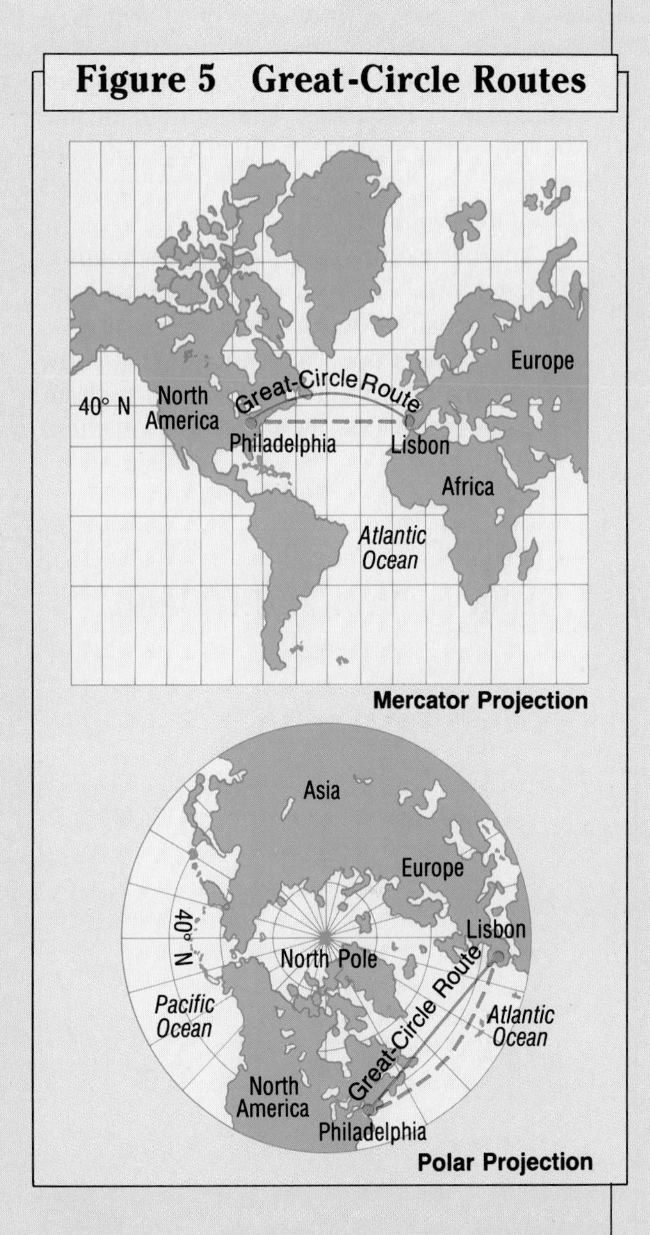

distance in miles between London and Birmingham is 100 miles. The number on the bottom of the scale gives the distance in kilometers. The distance between the two cities is about 160 kilometers.

Another feature common to most maps is a **directional indicator**. A directional indicator is usually a "north arrow." North arrows are especially important on maps that show only a small part of the earth's surface. Usually, the North Pole is not shown on these maps, and it is difficult to determine direction accurately. Map makers place on the map an arrow that points to the North Pole. By checking the north arrow, you are sure of which direction is north on the map. Other maps indicate direction by using a **compass rose**. A compass rose has arrows that point to all four cardinal directions: north, east, south, and west. All of the maps in this book have directional indicators. Find the directional indicator in Figure 6.

Figure 6
Great Britain and Ireland

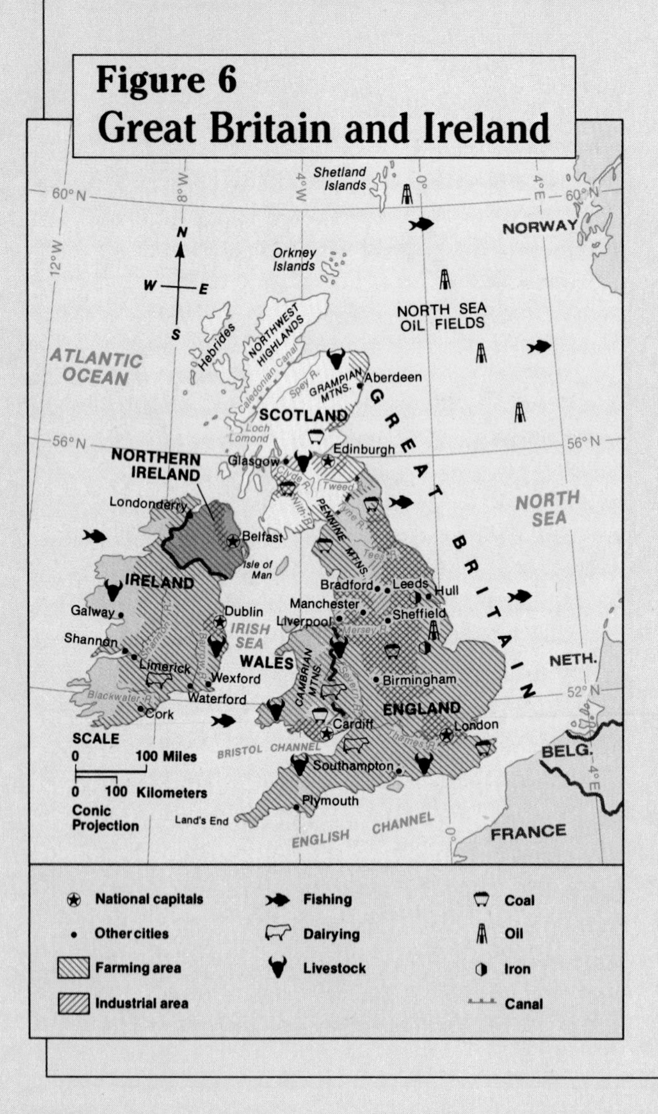

Information needed to read a map is found in the map **legend** or key. The map legend explains what the symbols on the map represent. Most of the maps you will use in this book have point symbols, line symbols, and area symbols.

Point symbols are used to specify the location of things that do not take up much space on the map. For example, the dot used to show the location and size of cities on a map is a point symbol. Find more examples of point symbols on the political map of North America on page 11.

Line symbols are used to trace the course of something on a map. They are used to show that something on the map is long but not very wide. Examples are roads, rivers, boundaries, and coastlines. Find three types of line symbols on the political map of South America on page 13.

Finally, you will see area symbols, which are used to show the location of things that take up a large amount of space on the map. Sometimes area symbols are simply colors, while at other times a pattern is used to show the areas. For example, a national-forest area on a road map might be colored green. A pattern of dots or lines might also be used. Look at the map "Midwest States," in Chapter 47. What area symbols are used on this map?

Using Map Skills Now that you are familiar with maps and their basic components, remember to use your skill and knowledge when studying maps. First, always look for and read the title of the map. The title will tell you what information can be found on the map and what area of the earth the map represents. For instance, it may be a climate map of Europe, a population-density map of Africa, or a road map of Texas.

Next, look for the directional indicator. This symbol will help you orient the map correctly. Then find the scale so that you can make accurate judgments about distances on the map. Be sure to check the legend to find information about the symbols on the map.

Remote Sensing

Our ability to make accurate maps has improved greatly in recent years. For the most part, this change is due to the use of **remote sensors**. Remote sensors are instruments that gather and record information from a distance. When used to make maps and collect data for the study of an area's geography, remote sensors are usually placed on airplanes or satellites.

A common remote sensor is a camera that records reflected light. Until recently, aerial photography was the major remote sensor used for mapping. Today, remote sensors are far more powerful than cameras or the human eye. They view energy forms other than light, such as heat, radio waves, and sound. To view vast areas of Earth from space with remote sensors mounted on satellites has allowed us to monitor changes that are occurring almost anytime and anywhere on Earth. You are probably familiar with the weather satellites that monitor our atmosphere. Their images appear on daily television weather reports. These satellites have been of tremendous help to weather forecasters, especially in tracking storms.

Figure 7

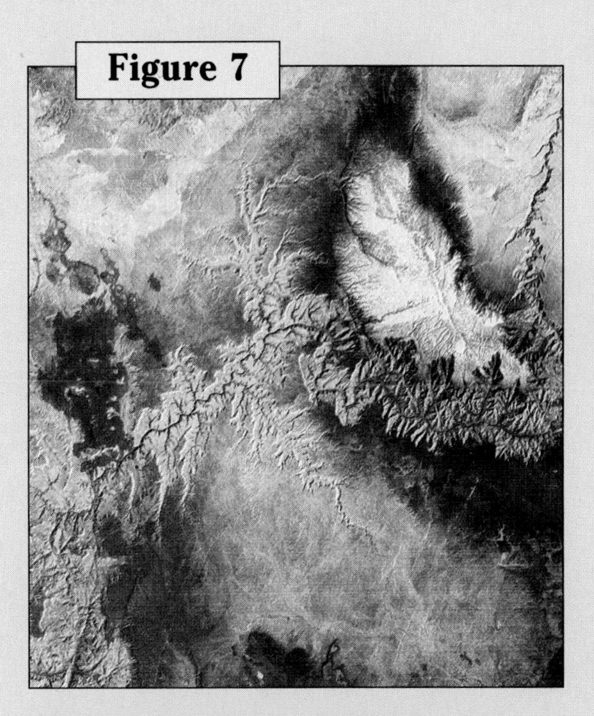

This Landsat photograph of Arizona's Grand Canyon was taken from 570 miles (912 kilometers) above the earth's surface.

Some of the photographs you will study in this book are called **Landsat** photographs. (See Figure 7.) These photographs are taken by American satellites that view the earth in both visible and infrared light. The satellites pass over the same area every 18 days, detecting changes in the earth's surface, such as vegetation and moisture. Each color of a Landsat image represents a different feature of the earth.

Though maps remain the major tool of geography, remote sensors and computers have revolutionized our ability to map the earth rapidly and accurately. Future technological advances will help us understand the earth's geography with even greater accuracy.

Additional activities are found in the Student Workbook. The Lesson Review/Test and Reteaching activities are available in the Teacher's Resource Binder.

Waves batter the southern California shoreline. Water is the most plentiful substance on Earth; without it, life could not exist. The sharing of water resources is a crucial issue for the earth's people.

It is believed that Earth is the only planet in our solar system with life. The energy that Earth needs to sustain life is generated by the sun and the heat from the earth's core.

More than five billion human beings inhabit the earth. People, such as these gathered at Wimbledon, England, have learned to adapt to their climate and physical environment.

THE EARTH AND ITS PEOPLE

OBJECTIVES

- ▷ *To describe the nature of geography and the differences between physical and cultural geography*
- ▷ *To explain why all forms of life on Earth are dependent upon the sun's energy*
- ▷ *To describe the forces at work at and beneath the earth's surface*
- ▷ *To explain what culture is, how it changes, and how cultural differences can result in conflicts between people and nations*
- ▶ *To interpret an elevation map and a cross-sectional diagram*
- ▶ *To interpret a population pyramid*
- ▶ *To interpret a climograph*

Chapter 1

Chapter 1 Lesson Plans and Planning Guide are found in the Teacher's Guide.

The Nature of Geography

Level A: Have students list the key concepts of economics, political science, and sociology. Compare these to the key concepts of geography.

▲ **The ruins of Machu Picchu are found in Peru.**

▲ **Shopping malls have become part of modern life.**

GEOGRAPHY DICTIONARY

- human geography
- physical geography
- region
- regional geography
- absolute location
- relative location
- spatial interaction
- uniform region
- nodal region
- cartography

Studying geography is discovering the world around us. Geography is everywhere. We use geography every day to find our way to school, to locate our friends, and to understand our environment. Our day-to-day travel marks our local geography. It is where we spend much of our lives. We think geographically all the time.

Geography is also about distant places. Today all of the world's peoples are linked, sharing resources, products, and ideas. Events far away influence us all. For example, a major oil discovery in Australia may mean that we can buy gasoline for less and drive our cars farther than ever. Also, the growth of the computer industry in Massachusetts might encourage workers to move there from Maine or even Mexico. We are all passengers on our spaceship Earth, and we need to know more about each other.

 Level B: Ask students to describe the environments shown in the photos above. How does each photo reflect interaction between people and their surroundings?

Have students turn to the Atlas at the front of the book to practice finding the absolute location of selected cities around the world.

What is Geography?

Kinds of Geography Geography is one of the oldest subjects for study. The term *geography* comes from the ancient language of Greece and means "earth description." It is the study of where things are located and how they got there. Geography is also the study of our interaction with the environment and the movement of materials, living things, and ideas across the earth's surface.

Human geography is the study of how the activities of people vary throughout the world. It includes the geography of economic and political activities, of cities and farms, and of religions and languages. **Physical geography** is the study of the earth's natural features, such as mountains, rivers, soil, and vegetation and of how they vary. It also is the study of world climates.

In this text, we will use both human and physical geography to study the world's **regions**. A region is an area with common characteristics that make it different from any other area in the world. Geographers use **regional geography** to discover how areas are the same or different. Regions can be areas of similar language, economy, or religion. Some regions contain a particular natural environment. Their boundaries might be rivers or mountain ranges. An area producing a common product, such as land occupied by wheat fields on the Great Plains of North America, can also be a region. Still other regions, such as the provinces of Canada, are defined by governments.

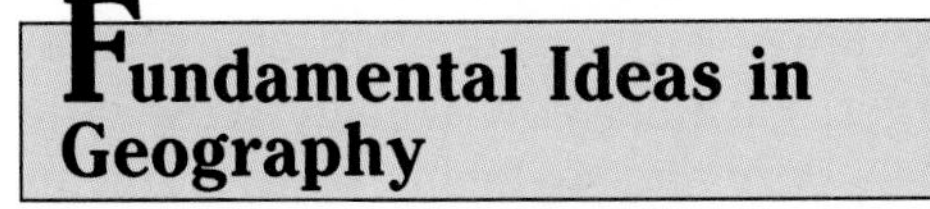

Fundamental Ideas in Geography

Location Geography begins with location. Geographers use the word *location* in two ways. **Absolute location** refers to the actual site on the earth's surface where something is found. Often the grid of latitude and longitude lines is used to describe this type of location. If you specify 40° 43′ north latitude and 74° 01′ west longitude, you are in New York City and nowhere else. Absolute locations provide the basis for recording geographic facts and for drawing maps.

Relative location is the relationship of a place to other places. For example, you might describe the location of your home by saying that you live next to the supermarket. Relative location is the situation of a place. Thus, your home is situated near the supermarket.

Place Places are keys to understanding how people interact with the world around them. Often people have special feelings about places. Poetry, such as the winter scenes of Robert Frost, often describes how a place feels, giving the place special meaning.

Geographers usually identify a place by its human or physical characteristics. Human characteristics include styles of buildings and people's activities. The tall, concrete-and-steel buildings in a large city are quite different from the houses in a suburban neighborhood. Physical characteristics of a place include its weather, rock types, and plant and animal life. A desert water hole with a cactus and a rattlesnake nearby differs greatly from a forested glen with deer and butterflies.

Cold winters are typical in the United States Midwest. Thirty inches (76 centimeters) of snow fall in Chicago every year.

Ask selected students to describe the relative location of their homes, their school, or other places.

The ancient city of Jerusalem is the capital of Israel. For centuries, Jerusalem has been a holy place for Jews, Christians, and Muslims.

Places are often perceived from different points of view. For example, Jerusalem is holy to the Jews as the location of their ancient temple. Christians think of Jerusalem as the place where Jesus taught. And for the followers of Islam, Jerusalem is a holy site where Muhammad is believed to have risen to heaven.

Relationships Within Places The relationships between people and their natural environments are important to geographers. People adapt to their environments, but they also change their environments to suit their needs. How people cause places to change is an important topic for geography.

While some changes are deliberate, others are unintended. The Spanish colonists, for example, brought horses to California to work in the fields and to provide transportation. However, seeds of a European grass called "wild oats" were in the manure dumped from the ships. Wild oats quickly spread and now are very common weeds in California. Geographic changes like this are still happening. We depend on our cars, but they change the environment by polluting the air and by using land for highways and parking lots.

Land uses change as societies progress. For example, farms in Iowa 30 years ago were often only a few hundred acres in size. Farmers grew four or five different crops. Today, there are fewer farms, but they are several times larger. Farmers now usually grow only two crops—corn and soybeans. Iowa is still an important agricultural state; its weather and soils have not changed in 30 years, but farming technology and government policies have. The changing relationships within places are important, but they can be complex and caused by events far away.

 The five basic themes of geography are location, place, relationships within places, movement, and region. These themes are emphasized throughout the text.

The concept of region is central to the study of world geography. It is the basis of Units 2 through 9.

Movement Across the Earth's Surface As natural environments vary across the earth's surface, so do the number of people, their ways of life, and their needs. One place cannot satisfy all people's needs. This leads to movement of people, goods, and ideas.

Geographers call this movement **spatial interaction.** Spatial interaction can be as simple as your trip to school today. It can be as complex as the combining of wood pulp from Canada, ink manufactured from Kentucky coal, cotton fiber from Texas, and glue from Mexican oil to make this book. Spatial interaction is most impressive on the global scale, where many types of transportation and communication move goods and information continuously.

As the amount of movement and kinds of transportation change, so do other geographic features. Because there are few cars in China, bicycles and buses are the main forms of transportation in cities. If the government of China permits more cars, the cities will need more roads and parking lots. The study of geography anticipates these changes so governments and businesses can plan for them.

Regions Regions provide manageable units for geographic study, which help us understand world events. Geographers identify two types of regions. A **uniform region** has one or more common features throughout. The country of France is a uniform region because its people share a common government, language, and culture. A **nodal region** is defined by the movement around a place. The city of Paris is at the center of a nodal region defined by the distances goods, services, and workers travel to and from the city each day. Your school is at the center of a nodal region that extends to the homes of all the students. Uniform regions have common features; nodal regions are created by movement.

How does this fast-moving traffic along El Paso's Skyline Drive illustrate the concept of spatial interaction?

Spatial interaction is the movement of people, goods, and ideas. Highways are a means of transporting people, goods, and ideas from place to place.

Geography as a Profession

Cartography To record geographic facts and ideas, geographers often use maps. Cartography is the branch of geography that studies maps and map making. Cartographers are people who make maps. Although many maps are drawn by hand, computers are revolutionizing map production. Computers store data from old maps as well as satellite scenes, photographs, and data from surveys and government reports. The cartographer then "draws" the map on a computer screen and sends it to be printed almost instantly on an electronic plotter. Cartographers work for companies that publish maps, atlases, and books and are also employed by nearly all government agencies. The largest employer of cartographers in the United States is the United States Geological Survey (USGS), which produces all the topographic maps for the United States as well as for several foreign countries.

Computers help cartographers by sorting information, then drawing the map on film, paper, or a screen.

Teaching Geography Helping students discover their world is an important profession. Many people seem to know little about their own country and even less about places far away. In a closely linked world, we must all learn as much as we can about locations, places, the relationships within places, movement, and regions.

Geographic knowledge is also necessary for good citizenship. People need to be a positive influence on their government's policies. Should we allow suburbs to be built over farmland? Where should poisonous chemicals and garbage be stored? These kinds of questions are geographic. Geography teachers help develop an informed public.

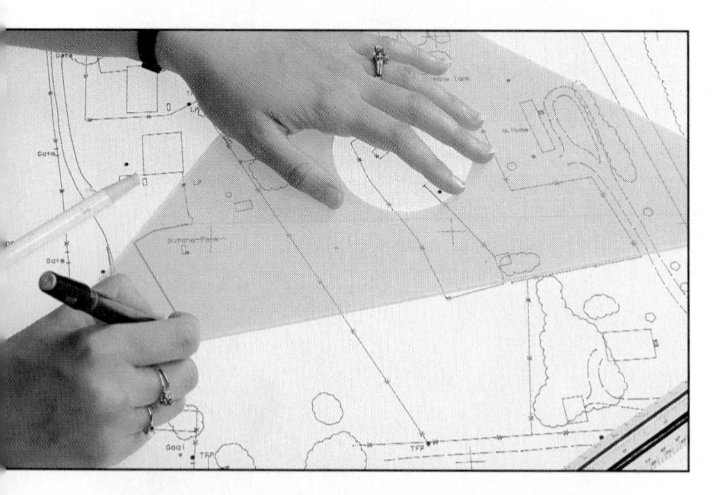

A cartographer plots the symbols and features of a map. Close attention to detail is an essential part of map making.

Applied Geography Geographers are also employed to make decisions about managing places, guiding the transport of products, and encouraging regional development. Many geographers who work in environmental management investigate land-use practices and advise how particular lands should be used. Geographers also locate natural resources and identify their possible uses.

Geographers work for businesses too. They decide where new stores should be located, schedule routes for airlines and trucking companies, and help identify new markets for businesses. Geographers also work at United Nations offices around the world. They help organizations like the Red Cross and the World Bank and serve in foreign development programs such as the Peace Corps.

Level B: Ask students to think of as many subjects or topics as possible that might be mapped. One example is world religions. List their ideas on the chalkboard.

You may want some students to read "Reviewing the Main Ideas" before they read the chapter.

CHAPTER 1 CHECK

Reviewing the Main Ideas

1. Geography is the study of where things are located and how they got there. It is also a study of how human beings interact with their environment.
2. The five basic themes of geography are location, place, relationships within places, movement across the earth's surface, and regions.
3. Regions are basic units of geographic study that can be defined in many ways. Some regions are defined by their physical environment. Others are defined by cultural, economic, or political characteristics.
4. The different professions of geography include cartography, teaching, environmental management, and consulting for businesses and nonprofit organizations.

Building a Vocabulary

1. What is physical geography? How does it differ from human geography?
2. Think of two examples of spatial interaction. Why is the concept of movement important?
3. What is the difference between absolute and relative location? Give two examples. Use a road atlas or the Atlas in the front of this book to find the absolute location of your hometown.
4. What is the difference between a uniform region and a nodal region? Is France a uniform region? Why or why not?

Recalling and Reviewing

1. Why is it important to understand the five basic themes of geography? How is geography useful to us in understanding and shaping our environment?
2. Why is it useful to study regions? How do you define a region? Name several regions in your state.

Critical Thinking

3. How do geographers use human or physical characteristics to identify places? Use the concept of place to describe your hometown. Consider the climate, natural environment, cultural diversity, building styles, and so on.

Using Geography Skills

Use the Atlas at the front of the book to answer the following questions.

1. Which of the maps are concerned with physical geography? Which maps are concerned with human geography?
2. Look at the physical map of North America. Without looking at the facing political map, tell which areas might be more populated. Why might they be?
3. Compare the physical and political maps of Europe. How many regions can you find on the maps?

Answers to Chapter Check questions are found in the Teacher's Guide.

Additional activities for Chapter 1 are found in the Student Workbook. Chapter 1 Test as well as Skills, Reteaching, Critical Thinking, and Challenge/Enrichment activities are available in the Teacher's Resource Binder.

Chapter 2 Lesson Plans and Planning Guide are found in the Teacher's Guide.

Emphasize that geography is the study of the earth; however, the position of Earth in the solar system greatly influences our planet.

Chapter 2

The Earth in Space

▲ **The Mayans studied stars from an observatory.**

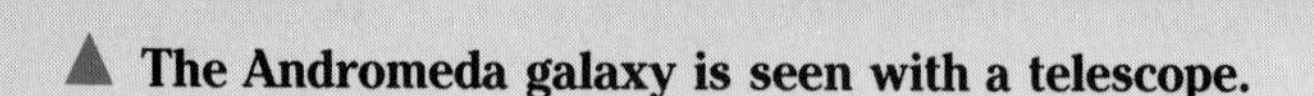

▲ **The Andromeda galaxy is seen with a telescope.**

GEOGRAPHY DICTIONARY

- universe
- galaxy
- solar system
- planets
- tropics
- polar regions
- rotation
- revolution
- equinox
- seasons
- solstice
- Tropic of Capricorn
- Arctic Circle
- Antarctic Circle
- Tropic of Cancer

If you look at the sky on a clear night, you can see thousands of stars. With a telescope, you would see millions more. All these stars are part of the **universe**. The universe includes everything that is known to exist. It is unimaginable in size. Stars are grouped together in huge, nearly circular collections called **galaxies**. Many objects that look like individual stars to the naked eye are really billions of stars in a galaxy far, far away. One galaxy is called the Milky Way. On a very clear night, you can see part of it looking like a bright streak in the sky. Many individual stars in the Milky Way are visible because it is the galaxy we live in. The sun is actually a star near the edge of the Milky Way.

You might wonder if there are places besides Earth with life like ours. Scientists also wonder this and continue to search for clues. The vast universe holds unlimited possibilities for future discovery.

Have students give specific examples of how the sun's energy affects their choice of clothes, homes, food, and outdoor activities. Ask them to think of examples from other parts of the world.

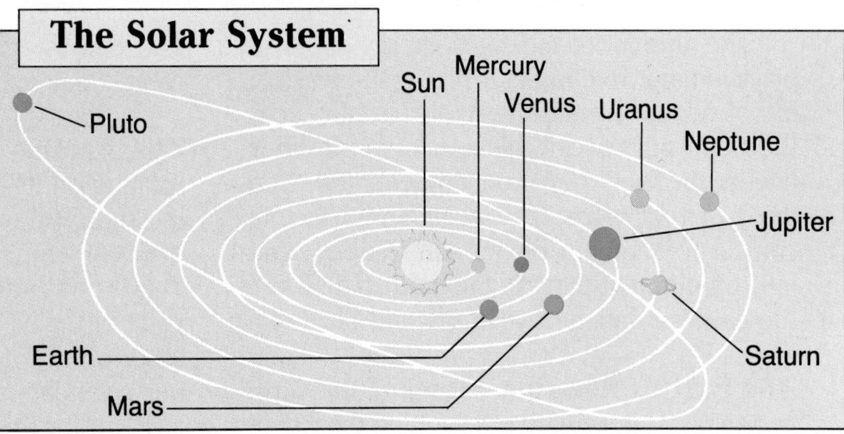

What is the distance from the earth to the sun? How large is the earth in relation to the sun?

93 million miles (149 million kilometers); The sun's diameter is more than 100 times that of the earth.

The Sun and the Earth

The Earth and the Solar System The sun is the center of the **solar system**. The nine **planets**, including Earth, that move around the sun, are part of the solar system. Planets are spheres, or ball-like objects, that circle around a central star. Some planets have many moons that circle around them, and some have none. The earth has only one moon.

The sun is small compared to some other stars. Yet, when compared to the earth, the sun is enormous. For instance, the diameter of the earth is about 8,000 miles (12,800 kilometers). The diameter of the sun is 864,000 miles (1,382,400 kilometers), more than 100 times the diameter of the earth. For a long time, people did not realize that the sun was a star. The sun looks much larger than other stars because it is so much closer to the earth. Even though the earth is 93 million miles (149 million kilometers) from the sun, the next nearest star, Alpha Centauri, is about 25 billion miles (40 billion kilometers) from the earth.

Almost all of the earth's energy comes from the rays of the sun. This energy makes life as we know it possible. Different parts of the earth receive different amounts of the sun's energy. Some parts of the earth, such as those near the equator, receive a great deal of the sun's energy. These parts are warm all the time. We call these warm areas near the equator the **tropics**. Other areas of the world get very little of the sun's energy. These areas are cold most of the time. Since they surround the earth's poles, we call these areas the **polar regions**. Some parts of the earth vary from warm to cold at different times. These lands lie between the low-latitude tropics and the high-latitude polar regions and are called the middle-latitude regions.

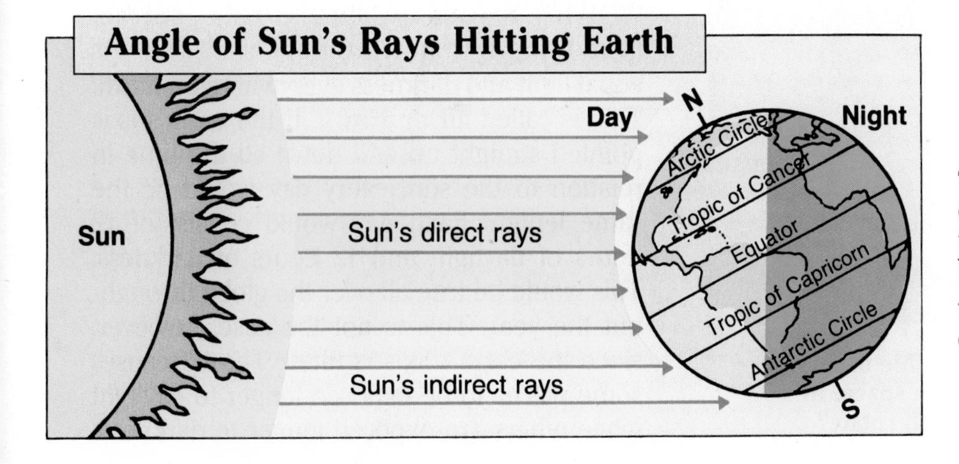

The region near the equator is warmer than the polar regions. Why?

This region receives the sun's direct rays.

The amount of the sun's energy received plays a major role in determining the ways of life of people around the world. It influences the clothes they wear, the types of homes they live in, the food they grow and eat, and even the types of sports they play. Three factors control the amount of the sun's energy that falls on different parts of the earth. These are the earth's rotation, revolution, and tilt.

The Earth's Rotation Think of the earth as having an imaginary rod running through it and sticking out at the North and South poles. This rod is called the earth's axis. The earth spins around on its axis. One complete spin of the earth on its axis is one **rotation**, equal to one day. The earth rotates eastward. We see the effects of the earth's eastward rotation as the sun rising in the east and setting in the west. To us, it appears that the sun is moving, while actually the earth is rotating on its axis.

The sun's rays of energy can strike a surface on the earth only if that surface faces the sun. If the earth did not rotate on its axis, only the part of the earth facing the sun would receive the sun's energy. That side of the earth would be extremely hot. The part of the earth facing away from the sun would always be dark and extremely cold. The earth's rotation makes it possible for all the earth's surface to be exposed for a time to the warming effects of daylight and to the cooling effects of darkness.

Day and night are clearly visible in this photo of the earth taken from space. In which direction does the earth rotate?

Eastward.

The earth's rotation, which takes 24 hours, creates day and night.

The Earth's Revolution In addition to rotating on its axis, the earth revolves around the sun. The earth makes one nearly circular orbit, or one **revolution**, every 365¼ days. This is one earth year. Each time you celebrate your birthday, you have just completed another orbit around the sun. For convenience, our calendars have 365 days in one year. To account for the remaining one-fourth day, an extra day is added to February every fourth year. This year is called a leap year.

The Earth's Tilt The earth's axis is always pointed toward the same spot in the sky. The north polar axis, for instance, always points to a star known as the North Star. The position of the axis is fixed in respect to the stars, but not to our sun. The earth's axis is tilted 23½° in relation to the sun. As the earth revolves around the sun, it leans toward or away from the sun at different positions in its orbit.

The North Pole is sometimes tilted toward the sun. When this occurs, the North Pole is in constant sunlight, and the Northern Hemisphere receives more of the sun's energy. At this time, the South Pole is in complete darkness, and the Southern Hemisphere receives less of the sun's energy. (See the diagram on page 39.) When the North Pole is tilted away from the sun and in complete darkness, the South Pole is tilted toward the sun and is in constant sunlight.

What happens when the poles are not tilted toward or away from the sun? There is equal light and darkness everywhere on Earth. This is called an **equinox**. If the earth's axis pointed straight up and down all the time in relation to the sun, every day would be the same length. Each day would consist of 12 hours of daylight and 12 hours of darkness. This would be true all over the globe throughout the year. This is not the case, however, since the earth's axis is tilted. The tilt causes some places to be exposed longer to daylight while others are exposed longer to darkness.

Explain to the class that during the last ice age, the level of the oceans was so low that many areas now separated by water, such as Asia and North America, Australia and Asia, were connected.

Forces at the Earth's Surface

Rock Weathering At the earth's surface, large rocks are broken and decayed. The process of breaking up rocks and causing them to decay is called **rock weathering**. Because wind and water are so forceful, even the hardest rock decays. Some rock weathering is like tooth decay. Chemicals in the air and water eat away at the rock. Some weathering is more like breaking or chipping a tooth. For example, roots of plants work their way into a crack and tear the rock apart. Rock weathering also results when water drips into a crack, freezes, and pushes the rock apart. Weathering is also the first step in soil formation.

Rock weathering breaks rock into smaller particles, known as **sediments**. Sediments may be mud, sand, or gravel. Weathering prepares sediments for further erosion and movement. For example, sediments may simply roll or slide downhill in a rockfall or landslide. They may also be eroded and carried by water, wind, or even moving ice.

This slope is an example of rock weathering. Even the hardest rock can be broken and decayed by air and water.

Water Whether running in streams and rivers or pounding the beach, water is a powerful force of erosion and transportation. Given enough time, water will carve through the hardest rock. Rivers sometimes carry sediments long distances. Erosion by water often begins in a tiny hillside channel narrower than a pencil. As the water flows farther downhill, the channel might grow into a deeper opening or gully. Gullies are signs that water erosion is particularly active. Eventually, water erosion will widen the gully into a valley. Over a long period of time, a river may erode the valley even further, forming a deep, narrow valley with steep sides, called a **canyon**. Even in desert regions, where there is little rainfall, the effects of water erosion are significant.

Glaciers Thick masses of ice that glide slowly across the earth's surface are called **glaciers**. There are two kinds of glaciers—sheet glaciers and mountain glaciers. Sheet glaciers, or ice sheets, cover large areas. They flow outward from great domes of ice where snow accumulates. The thick ice in the center of a glacier presses downward, and the edges of the glacier move. Glaciers have the power to level anything in their path. Yet, in some places they act more like blankets, protecting the ground beneath them from the severe cold and other eroding forces.

Today, sheet glaciers up to 10,000 feet (3,048 meters) thick cover most of Antarctica and Greenland. Scientists believe that during the last two million years, there have been periods when great sheet glaciers moved across North America and Western Europe at least a dozen times. These periods are called ice ages. During ice ages, the glaciers reached as far south as present-day St. Louis, Missouri, and Cincinnati, Ohio. Scientists think the last ice age ended about 10,000 years ago.

Mountain glaciers are smaller and much more common than sheet glaciers. Flowing like slow rivers of ice, they can be found in high mountain valleys all around the world. In such areas, temperatures are low because of high **elevations**, or height above sea level.

Level A: Have students research the three basic categories of rocks: sedimentary, igneous, and metamorphic. How are they formed? Which kinds are found in your area? Have students collect samples and identify them.

Sand dunes are caused by wind erosion. These dunes are found in Namibia, Africa, where the Namib Desert meets the Atlantic Ocean.

There are more than 1,000 mountain glaciers in the Rocky Mountains, Cascade Mountains, and Sierra Nevadas in the western United States. Mountain glaciers create sharp peaks and U-shaped valleys. Scientists believe there were several thousand mountain glaciers in North America during the last ice age.

Wind Another force that erodes and transports sediments is wind. Because plant life protects land surfaces from erosion, wind erosion is greatest in deserts, along beaches, and on dry, bare earth where people or animals have removed the plants. Wind erosion works two ways. First, wind wears down hard rocks by blowing sand and other particles against them. These particles grind down the hard rocks, often leaving smoothed and polished surfaces. Wind also changes the land by blowing sand and dust from one place to another. Hills of wind-deposited sand are called **sand dunes**. Often sand dunes are found near the places where the sand was eroded, such as along an ocean beach or beside a dry stream bed. Sometimes dust particles in deserts are lifted high into the atmosphere and carried great distances. Dust from the Sahara, a desert in northern Africa, for example, is often transported across the Atlantic Ocean to the islands of the Caribbean.

Forces Within the Earth

Inside the Earth If erosion were the only force shaping the earth, the earth's surface would have been worn flat long ago. But as erosion wears down the land, forces inside the earth continue to push up the surface and break it apart.

Scientists believe our planet is composed of several layers of different materials. (See the diagram on page 47.) At its center is the solid inner core. The next layer is the liquid outer core. The minerals of the core are rich in the elements iron and nickel. Surrounding the core is the mantle. The mantle is mostly a solid layer. A thin, rigid crust covers the mantle. If an apple-sized model of the earth were built, the crust would be only slightly thicker than the apple's skin. The gases of the atmosphere form the outermost layer of the earth. The world of people and all other living things is located in a narrow zone at the top of the crust and the bottom of the atmosphere.

Forces Beneath the Crust The earth's surface is bent and broken by heat and by rock movements in the crust and upper mantle. Heat currents travel slowly upward from the core to the crust. These heat currents partially

 Discuss with students why external forces (erosion and deposition), though gradual, are as important in creating landforms as the dramatic internal forces (volcanoes and earthquakes).

When plates slide past each other, long creases in the earth's surface are formed. The most famous example of this is the San Andreas Fault in California, where there has been plate movement many times in the last centuries. This movement has resulted in both small and large earthquakes.

The less heavy rocks of the continents ride along on the plate surfaces. When a continent is located in the quiet middle part of a plate, erosion will wear down the continent. Widespread plains are the result of this kind of erosion over a long period of time. On the other hand, when a continent is along a plate boundary, active volcanoes, faulting, and folding usually create steep hills, long valleys, and often mountains.

Most scientists believe that the theory of plate tectonics can be used to explain the history of the earth's surface. They think that the modern continents once formed a single landmass called Pangaea. About 200 million years ago, Pangaea began to split into two supercontinents. These supercontinents have been named Laurasia and Gondwanaland (gahn-DWAHN-uh-land). The supercontinent of Laurasia broke up to become the present continents of North America, Europe, and much of Asia. After breaking up, the supercontinent of Gondwanaland became India and the continents of South America, Africa, Australia, and Antarctica.

Volcanic forces created the island of Surtsey near Iceland. Wind and water are constantly eroding the face of the island.

Landform Development

Primary and Secondary Landforms

The landforms visible along a roadside or even in a photograph are the result of forces both on and within the earth's surface. Primary landforms are large masses of rock raised by volcanic eruptions and other uplifting forces. A good example of the result of these forces is the island of Surtsey, mentioned earlier. In just a few months, a piece of land appeared where only waves had washed before.

Yet, even as a primary landform is being formed, the forces of erosion are at work. No sooner had the island of Surtsey, a primary landform, appeared than the forces of erosion were evident. Ocean waves battered the growing pile of volcanic ash, washing sediments out to sea. During its first year, the island shrank and grew as the forces of the volcano and the ocean fought each other. Finally, a hard crust of lava flowed out over the cinder pile. As the lava cooled, it formed a protective rocky shield.

The island still exists, but its form has changed. The wind, rain, and pounding surf have worn its hard edges smooth. Other landforms, such as gullies and valleys, are being cut into the volcano's sides. Thus, secondary landforms develop from primary landforms as erosion works on them.

Rain and wind wear down the surface of a primary landform and remove a little more of it each year. For this reason, older secondary landforms are usually smoother, lower, and more rounded than younger ones. Younger secondary landforms, such as the island of Surtsey, have sharp edges and steep sides. Older landforms usually have less **relief** than younger ones. Relief is the difference in elevation between the top and bottom of a landform. Age is evident in hillsides that are not as steep and valleys that are wider. You might think that erosion continues until the secondary landform is worn down to a smooth plain. However, forces beneath the crust often become active before this happens.

Level B: Have students create models of various landforms and display them to the class. Ask volunteers to show the class one landform and present a brief oral report on how it is produced.

Level B: Ask students to list all the kinds of landforms they can think of and give local or well-known examples if possible.

Categories of Landforms Landforms can be further divided into two categories, based on how they are formed. One kind of landform is made of rock and has a thin layer of weathered sediments and soil at the surface. Its surface is slowly being lowered by the forces of erosion. Most mountains, valleys, plains, and **plateaus** were formed this way. A plateau is a flat-topped tableland standing above surrounding plains. Another kind of landform is formed by sediments deposited by water, wind, or ice. A sand dune in a desert is an example. Another example is a **floodplain**. A floodplain is level ground built by sediments deposited by a river or stream.

The terrain in any area is usually a jigsaw puzzle of these two kinds of landforms. For example, the sediments eroded from a mountain may be deposited by a stream as it enters a plain along the mountain's base. Such a deposit will form an **alluvial fan**. Later, the mud and sand from an alluvial fan might erode and form a **delta** at the mouth of a river. Then, waves might erode the delta and build a beach along the coast, or the sediments might be carried to the ocean floor. Today, the plains on the interior of a continent could be all that remain of a mountain range that has been eroded smooth over a long period of time.

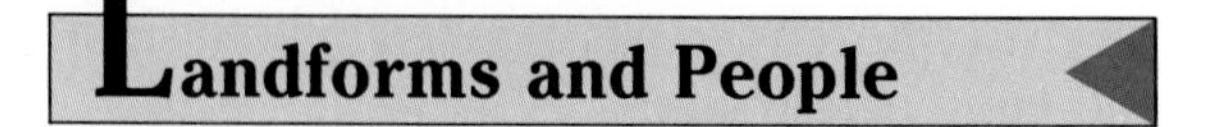

Landforms and People

People live and work on landforms. All the forces continue working as people select different ways to use the earth's surface. Some earth forces cause natural disasters. People must be aware of where and when events such as volcanic eruptions, earthquakes, floods, and dust storms might occur. Careful study of landforms will result in building better cities and highways that avoid these hazards. Soil erosion, too, must be managed so that farming can be more productive. Landforms are a key part of life on the earth's surface.

The forces that build and break down landforms are very much affected by processes in the earth's atmosphere. The next chapter examines the quickly changing geography of the atmosphere and how weather and climate affect people and their lives on Earth.

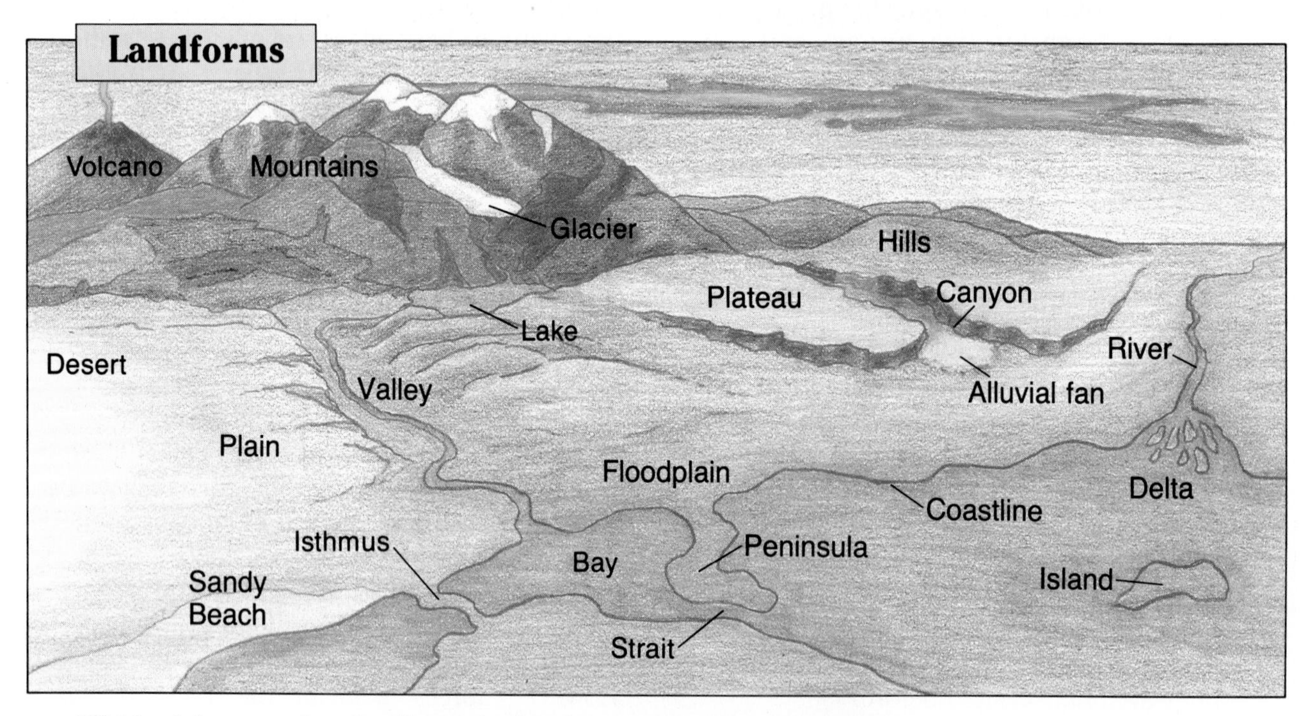

Which of the secondary landforms in this drawing are older than others?

The hills, canyon, valley, plateau, plain, and lower mountains are older. They have eroded over the years.

CHAPTER 3 CHECK

Reviewing the Main Ideas

1. The face of the earth is constantly being changed by forces inside the earth and on the earth's surface.
2. The forces of running water, waves, glaciers, and wind erode the land.
3. Other forces, such as heat and plate movement, are constantly building new landforms.
4. There are two categories of landforms—those made of rock and those made of sediments left by water, wind, or ice.

Building a Vocabulary

Students should provide an answer to Question 3 by filling in the blank, using their own paper.

1. Explain the difference between folds and faults.
2. What is the midoceanic ridge? How long is it?
3. ═══ is the difference in elevation between the top and bottom of a landform.
4. What is a canyon? How is one formed?
5. Define *alluvial fan, floodplain,* and *delta*. What do these have in common? Do any of these landforms exist in your area?
6. What is lava?

Recalling and Reviewing

1. In what ways does erosion of the earth's surface occur? Explain.
2. What are the two types of glaciers? How do they differ?
3. Explain how the forces within the earth prevent the earth's surface from becoming totally flat.
4. How do primary and secondary landforms differ?
5. Describe the theory of plate tectonics, and name the three types of plate boundaries. How were the continents and major continental landforms made?

Critical Thinking

6. What kinds of landforms are found in your state? Based on your reading of this chapter, explain how they might have been created. How have they been modified or used by people?

Using Geography Skills

1. Look at the drawing on page 52. Which landforms were formed by erosion? Which were formed by sediment deposit?
2. Using the diagram "Movement at Plate Boundaries" on page 50, describe the relationship between Plates 1 and 2. What is the result of this relationship?
3. Give two examples of erosion in your area or region. Was the erosion caused by rock weathering, glaciers, water, or wind? Explain.

Answers to Chapter Check questions are found in the Teacher's Guide.

Additional activities for Chapter 3 are found in the Student Workbook. Chapter 3 Test as well as Skills, Reteaching, Critical Thinking, and Challenge/Enrichment activities are available in the Teacher's Resource Binder.

Chapter 4

Chapter 4 Lesson Plans and Planning Guide are found in the Teacher's Guide.

Chapters 4 and 5 are closely linked, and the information in this chapter is crucial to understanding global climates in Chapter 5.

The Earth's Atmosphere

You may want to review Chapter 2 briefly with students to help them understand the relationship between the sun and the earth's atmosphere.

▲ **Clouds gather over the Galápagos Islands.**

▲ **Lightning strikes in Saskatchewan, Canada.**

GEOGRAPHY DICTIONARY

- weather
- climate
- air pressure
- trade winds
- doldrums
- westerlies
- polar easterlies
- front
- evaporation
- humidity
- condensation
- precipitation
- hurricanes or typhoons
- tornadoes
- orographic effect

The earth's atmosphere is a gas layer made up mostly of nitrogen and oxygen. It is necessary for life and provides the air we breathe. The condition of the atmosphere varies from day to day and from season to season. The **weather** is the condition of the atmosphere at a given place and time. Weather conditions in an area over a long period of time are called **climate**. Climate includes the expected, or average, weather events, as well as rare events such as floods and droughts.

Weather and climate are affected by the sun's energy, atmospheric pressure, and general earth circulation systems that you will learn about. Both are also influenced by such geographic features as oceans and mountains. The basis for all weather and climate, however, is the relationship between the earth and the sun, which was discussed in Chapter 2.

 Have students survey the photos in this chapter and list elements of weather or climate shown.

The greenhouse effect can be influenced by pollution. Explain that air pollution may raise the temperature of the earth. Carbon dioxide is the key element (it causes the earth to retain the sun's heat). What effect could this have?

Global Energy Balance

Every day, a small portion of the sun's total energy reaches the earth. Some is reflected directly back to space. A small amount stays in the atmosphere, where it is absorbed by the gases in the air. Much of the sun's energy, however, is absorbed by the earth's surface, its oceans, and continents.

Once the sunlight is absorbed, it is changed into heat energy. For example, if you stand in direct sunlight, you become warm. What you feel is the sunlight changing into heat energy. Just as you warm up by absorbing the sunlight, the earth's surface and the atmosphere also warm up when sunlight is changed into heat.

The atmosphere traps energy in much the same way that a greenhouse does. The glass allows the sunlight to pass through it. But it also traps and delays the escape of the energy after the sunlight has been changed into heat. Like the greenhouse, the earth's atmosphere allows sunlight to pass through it. When the sunlight is changed into heat energy by the earth's surface, it is trapped by the atmosphere, and the earth is kept warm. The trapped heat provides the atmosphere with the energy necessary to produce wind, rain, clouds, and storms. All weather activity is dependent upon the sun's energy.

Why does the earth not get warmer and warmer until it is so hot that we cannot survive? It is because eventually all heat energy trapped by the atmosphere escapes back into space. All the sun's energy that the earth absorbs eventually leaves the earth as lost heat, giving the earth an energy balance.

The amount of solar energy received by different regions of the earth varies from place to place. The equator receives much more solar energy than do the poles. In the tropics, more solar energy is received than is returned and lost to space. The polar regions, on the other hand, receive less energy than they lose to space. An imbalance appears to exist. We might expect the tropics to get steadily warmer and the polar regions to become colder and colder. Yet, this does not happen.

The areas near the equator do not continuously get warmer and the poles colder because major exchanges of heat energy take place between them. Excess heat energy is moved from the tropics to the polar areas via the ocean and atmosphere currents, and cold air and water are moved from the polar areas to the tropics to be reheated. This system of energy exchange balances temperature extremes. These exchanges occur both in the atmosphere and in the oceans.

Atmospheric Exchange Systems

Pressure and Wind Systems The measurement of the weight of air is called **air pressure**. Unequal heating of the earth's surface causes air-pressure differences. When air is warmed, it expands, becomes lighter, and rises. This creates low-pressure areas. Low-pressure areas tend to be unstable. That is, the air rises and cools. This rising air causes clouds, rain, and even storms. Cold air is dense and heavy and tends to sink. This action produces high-pressure areas. High-pressure areas tend to be stable. The air sinks and heats, generally causing clear, calm weather.

Wind is the motion of air between areas of different pressure. Wind always flows from high- to low-pressure areas. The seashore illustrates this point. As the land heats very quickly on a sunny day, warm air rises over the land, causing low pressure. Cool air from the higher pressure over the ocean then blows inland to replace the rising air. This cool sea breeze makes the beach very comfortable, even on a warm day. During the night, the land cools more rapidly than the ocean. The air then circulates in the opposite direction. The cool land breeze now flows from the land (high-pressure area) toward the warmer, rising air over the ocean surface (low-pressure area). (See the illustration on page 56.)

Tell students that the instrument used to measure air pressure is called a barometer.

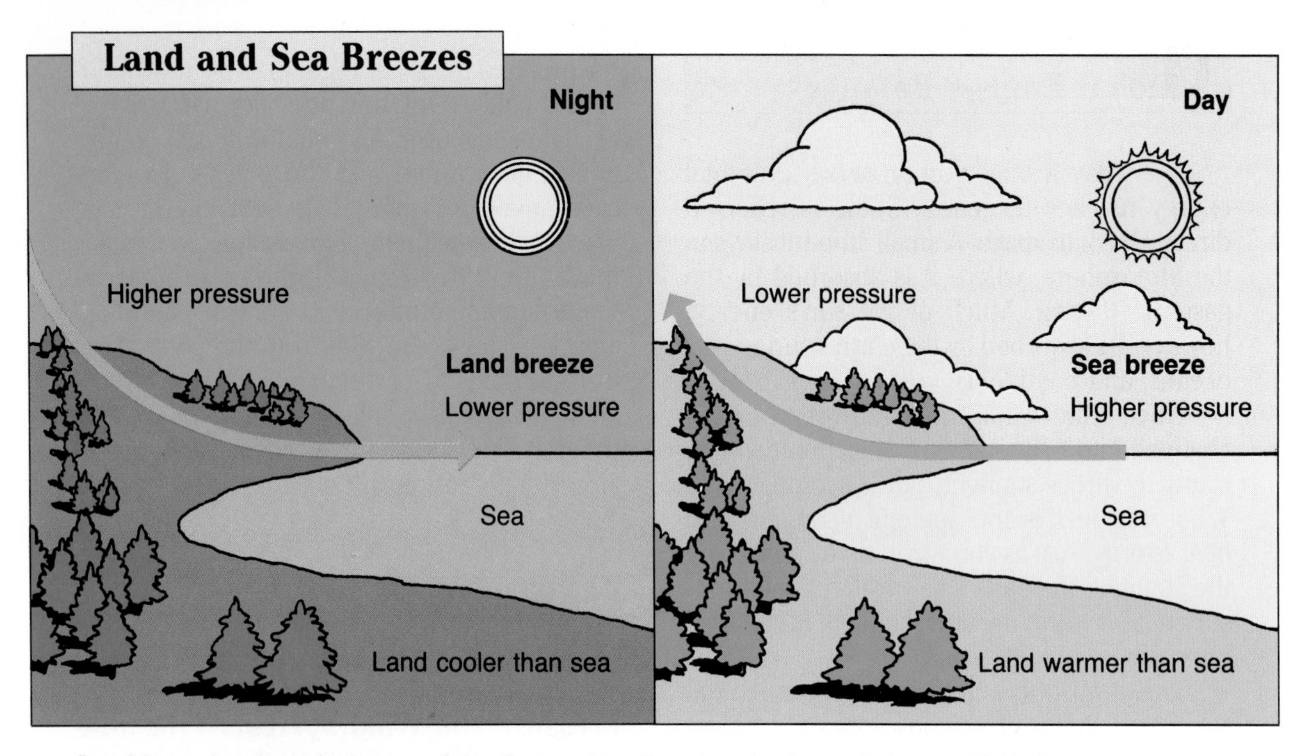

Land is cooler than the sea at night. From what direction do the night breezes blow?
from the land (high-pressure area) toward the ocean (low-pressure area)

Now look at the global pressure patterns in the diagram on page 57. Notice that warm air rising over the equator causes the equatorial low-pressure zone. This air then flows toward the poles in the upper atmosphere. Eventually, the warm air cools and sinks back to the surface at about 30° north and south latitudes. This sinking air causes the very stable subtropical high-pressure zones. At the poles, cold sinking air causes constant high pressure; the polar regions are known as the polar high-pressure zones. At about 60° north and south latitudes, air rises again, causing the subpolar low-pressure zones.

Types of Winds The earth's surface can be divided into three latitude zones in each hemisphere. The first consists of the low latitudes, which extend from 0° at the equator to 30°. The middle latitudes extend from 30° to 60°. The high latitudes extend from 60° to the poles at 90°. Knowing these general latitude zones will help you understand the various wind belts and types of climate that are discussed in this chapter.

In the low latitudes, the **trade winds** blow from the east. In the Northern Hemisphere, they blow from the northeast. In the Southern Hemisphere, they blow from the southeast. The early sailors used the northeast trade winds to travel from Europe to America. When Columbus sailed from Spain in 1492 seeking a new route to the Orient, he landed in the West Indies. Did you ever wonder why he landed there instead of at the present sites of New York or Boston? Look at the arrows showing the northeast trade winds in the diagram of global pressure and wind systems. Now, compare their flow to the world map in the Atlas at the front of the book.

Not all areas of the world are in wind belts. There are some calm areas. Those areas with no strong winds are called **doldrums**. The doldrums are centered along the equator and at about 30° north and south latitudes. Sailing ships were sometimes caught in the doldrums for long periods of time. To escape these calms, the sailors had to lighten their cargo. Because horses, along with their food and water, were the heaviest cargo, they were

Point out that the rotation of the earth causes wind deflection to the right in the Northern Hemisphere and to the left in the Southern Hemisphere. This is called the Coriolis force. It causes winds not to travel due north or south.

The trade winds blow through the islands of the low latitudes. The climate of this region is warm and rainy throughout the year.

sometimes thrown overboard. These dreaded calm seas became known as the "horse latitudes."

In the middle latitudes are the **westerlies**. These prevailing winds come from the west. After North America was settled by Europeans, sailors would travel by the northeast trade winds to America and return to Europe by the westerlies. In the high latitudes are the **polar easterlies**. Here, cold masses of air flow out of the Arctic and Antarctic regions. They bring cold conditions into the middle latitudes. Between 45° and 60° latitude, the warm westerlies come in contact with the cold polar easterlies. This meeting zone of warm and cold air is called the polar **front**. A front occurs when two different types of air meet. Fronts often cause stormy weather.

Ocean Currents The oceans are important energy storehouses. Most of the sun's energy that reaches the earth's surface is absorbed by seawater. Water heats and cools much more slowly than land. Therefore, the

Global Pressure and Wind Systems

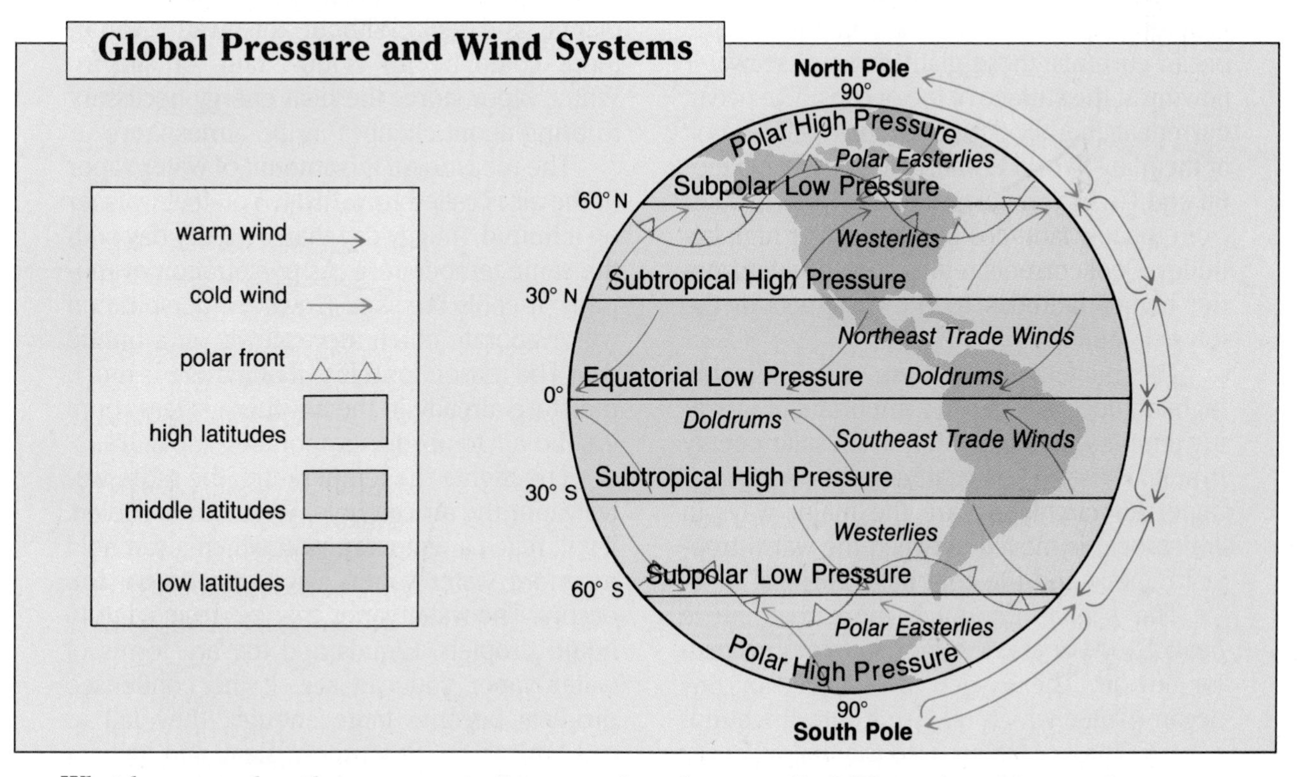

What happens when the warm westerlies meet the polar easterlies? Where does this occur?

A polar front is created; between the middle and high latitudes.

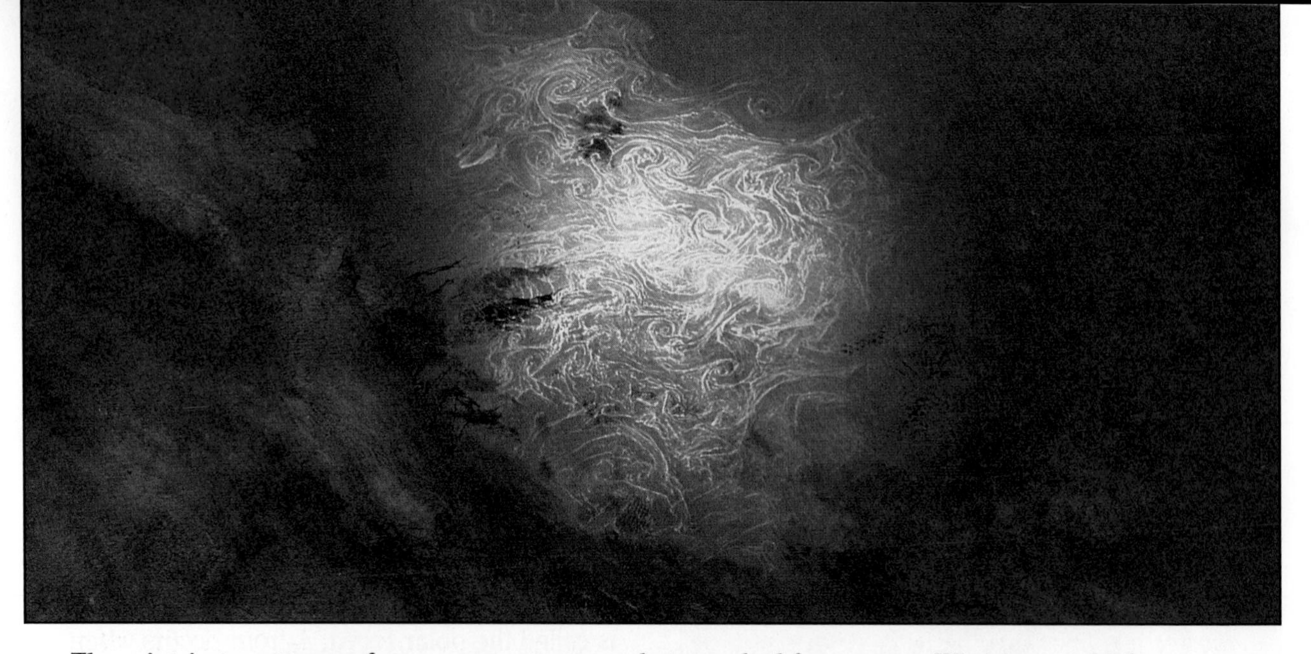

These intricate patterns of ocean currents were photographed from space. Why do you think the patterns are circular?

Wind sets ocean currents in motion. Warm and cool currents travel in different paths.

areas near oceans do not have extremes of temperature as do the areas near the centers of continents. It is the large landmasses, or continental areas, that have severe temperature ranges.

The major wind belts move across the surface of the ocean. They set in motion the ocean currents, those giant rivers of seawater flowing at the surface of the oceans. The ocean currents generally flow in circular paths. Look at the map "World Climate Regions," on pages 66 and 67. Warm currents carry heated water from the low latitudes into the cooler high latitudes. Cool currents return cooled water from the higher latitudes to be rewarmed in the lower latitudes.

The movement of warm water into the high latitudes is especially important in warming temperatures in winter, when solar energy is at its lowest. The ocean currents along with upper air circulation are the major ways in which heat is moved between the warm tropical regions and the cold polar regions.

The oceans do much more than move heat. They are also a major source of oxygen for our air. The oxygen is produced by tiny ocean plants, which release it into the atmosphere. The oceans are also the major source of moisture for the atmosphere.

Moisture in the Atmosphere A very important gas in our atmosphere is called water vapor. Most water vapor in the air is evaporated from the oceans. **Evaporation** is the process by which water is changed from a liquid to a gas. The rest comes from lakes, rivers, plants, and soil. Without this water vapor, there could be no clouds, rain, or storms. Water vapor stores the heat energy necessary to bring about changes in the atmosphere.

The measure of the amount of water vapor in the air is called **humidity**. You feel warmer on a humid, muggy day than on a dry day with the same temperature. As perspiration evaporates, it cools the skin. However, perspiration will evaporate much more slowly on a humid day. The reason for this is that if there is much moisture already in the air, there is less room for the air to absorb evaporating moisture.

The higher the temperature, the more water vapor the air can hold. When air is cooled, it will reach a temperature at which it can hold no more water vapor. Then, **condensation** occurs. The water vapor changes from a gas to liquid droplets. Clouds and fog are forms of water vapor you can see. If the condensed droplets become large enough, they fall as **precipitation**. Rain, snow, sleet, and hail are the most common types of precipitation.

Level A: Have students research El Niño, which is a warm-water ocean current usually found off the west coast of South America and which altered the climate of North and South America in 1983. Ask students to describe the usual effect of El Niño and to tell what was so unusual about it in 1983.

Precipitation is not evenly distributed. If too much precipitation falls at one time, there might be floods. If there is a long period with no precipitation, a drought might occur.

Storms The most dramatic and violent of the different weather events is a storm. A storm is a release of huge amounts of energy stored in the water vapor of the atmosphere. The vast amount of energy released from a storm is evidenced by the violence and damage that occurs.

Storms occur under low-pressure conditions in which unstable air is present. In the middle latitudes, these storms travel with the westerlies. They may be as large as 630 miles (1,008 kilometers) in diameter and travel for very long distances during their life cycle. The middle-latitude storms occur in an area along the polar front. The polar front forms a boundary line between warm, wet tropical air and cold, dry polar air.

The storm systems of the low latitudes are different from those of the middle latitudes. Since no cold air is present, tropical storms are less defined and do not occur along fronts. They move east with the trade winds, bringing an abundance of rain.

Severe tropical storms sometimes form over the warm ocean waters of the low latitudes. These storms are called **hurricanes** or **typhoons**. Hurricanes sometimes travel from the low latitudes into the middle latitudes. They bring violent winds, torrential rain, and dangerously high seas. These storms are most destructive to islands and coastal areas.

Tornadoes are the smallest but most violent of storms. These small twisting storms can destroy almost anything in their path. Tornadoes are most common in the southeastern and central United States.

The most common type of storm is a thunderstorm. A thunderstorm is any storm with lightning and thunder present. At any moment, there are about 2,000 thunderstorms taking place on Earth. Thunderstorms are most common in the hot, unstable low latitudes and along fronts in the middle latitudes.

Hurricane Elena, which struck near Biloxi, Mississippi, in 1985, caused more than one-half billion dollars in property damage.

The peak of Mount Kilimanjaro is always snowcapped. Animals graze in a milder climate region found at the base of the mountain.

Elevation and Climate

Temperature Effect The landforms of our continents, particularly high mountains, have a major effect on climate. It is elevation, the mountain's height above sea level, that has the greatest impact. An increase in elevation causes a lowering of the temperature. For every 1,000 feet (about 300 meters) up the side of the mountain, the air temperature cools about 3.6° F (2° C). The 19,340-foot (5,896-

Refer students to the map "World Climate Regions," on pages 66–67.

meter) elevation of Mount Kilimanjaro (kil-uh-muhn-JAHR-oh), in East Africa, shows us this effect. The mountain always has a snow- and glacier-covered peak, even though it is very close to the equator.

Orographic Effect Another effect of elevation on climate is called the **orographic** (ohr-uh-GRAF-ik) **effect**. When moist air flowing from the ocean meets a mountain barrier, the air is forced to rise. The higher the air rises, the more it cools. When the air cools, it is forced to give up moisture in the form of rain or snow. As a result, the side of the mountain facing the wind receives a great deal of moisture. This side of the mountain is called the windward side.

The side of the mountain facing away from the wind is called the leeward side. As the moist air moves down the leeward side, it begins to warm up and becomes drier. Since the leeward side is cut off from the moist ocean air, it has a much drier climate than the windward side. For example, because of the orographic effect of the mountains of the West Coast of the United States, so much moisture is removed from the air that deserts are located on the leeward sides of these mountains. (See the diagram below.)

Understanding climate is an important part of the study of human and physical geography. In the following chapter, you will learn more about climates throughout the world.

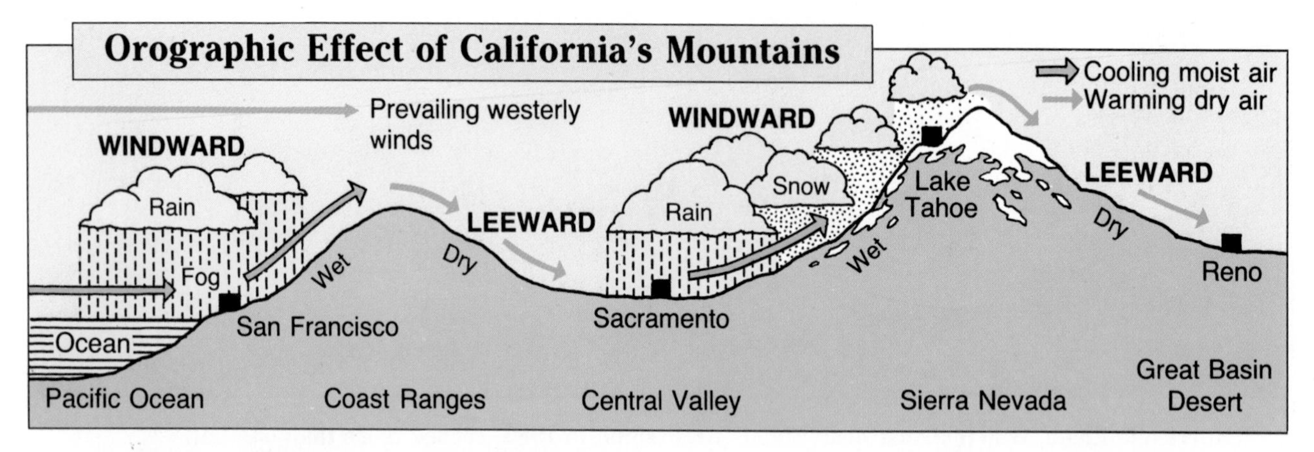

Why do the windward slopes of mountains receive more precipitation than the leeward slopes?

Moist air flowing from the ocean rises, cools, and forms precipitation when it meets a mountain barrier. Thus, the mountain slope facing the wind (windward side) receives the most moisture. As the air moves down the leeward slope, it warms and becomes drier.

CHAPTER 4 CHECK

Reviewing the Main Ideas

1. The earth maintains an energy balance by absorbing and releasing heat from the sun. When sunlight is changed into heat energy, the atmosphere traps the energy and warms the earth. This trapped heat energy provides the atmosphere with the energy necessary to produce weather activity.
2. Weather is the condition of the atmosphere at a given place and time. It is influenced by several factors, including the movement of ocean currents, global wind patterns, and differences in atmospheric pressure.
3. Most water vapor in the atmosphere is evaporated from the oceans. The rest comes from lakes, rivers, plants, and soil.
4. Storms occur under low-pressure conditions in which unstable air is present. Types of storms include hurricanes or typhoons, tornadoes, and thunderstorms.

Building a Vocabulary

1. Define *air pressure*. What causes differences in air pressure? How do changes in air pressure affect climate?
2. Rain, snow, sleet, and hail are the most common forms of what?
3. Discuss the orographic effect. How does it affect the climate of an area?
4. When the warm westerlies come into contact with the polar easterlies, what occurs? What effects does this have on weather?
5. What are the doldrums, and where are they found?

Recalling and Reviewing

1. How is an energy balance maintained on Earth?
2. What important role do the oceans play in the atmospheric exchange system?
3. Where does water vapor come from? How does its form change when the temperature rises or falls?
4. List the kinds of storms described in this chapter. What is the most common kind of storm?

Critical Thinking

5. Explain how wind and water affect weather in your area.

Using Geography Skills

Use the diagrams in this chapter to answer the following questions.

1. Between what degrees of latitude are the trade winds located? In which direction do they flow?
2. Study the diagram of land and sea breezes on page 56. Explain why the land is cooler than the sea at night and warmer than the sea during the day.

Answers to Chapter Check questions are found in the Teacher's Guide.

Additional activities for Chapter 4 are found in the Student Workbook. Chapter 4 Test as well as Skills, Reteaching, Critical Thinking, and Challenge/Enrichment activities are available in the Teacher's Resource Binder.

Chapter 5

Chapter 5 Lesson Plans and Planning Guide are found in the Teacher's Guide.

Global Climates

You may want to review Chapter 2 briefly with students to help them understand the influence of the sun and of the earth's rotation on climates.

Utah's desert region is dry and uninhabited.

Monsoons flood the streets of Bombay, India.

Understanding the weather helps us plan each day, from how we travel to what we wear. Weather events, such as rain, snow, fog, or storms, can cause many changes in our daily activities. Knowledge about the weather and climate of other parts of the world also helps us understand human activities elsewhere. People throughout the world may dress differently, eat different foods, and carry on different daily activities. Some of these differences are due to the climate where they live. Some are caused by other factors that you will study in geography.

Have students describe the climate of their area. Suggested ways of finding this information are library research or contacting a local television station, radio station, or weather bureau. Encourage students to make graphs to illustrate their description of the climate.

Climate Types

To understand the different world-climate types, think of three sets of contrasting factors: (1) warm and cold, (2) wet and dry, and (3) continental and maritime. The influence of elevation and of extreme weather events such as storms and droughts is also important. All these factors produce the different climates of the low, middle, and high latitudes.

Low-Latitude Climates The areas close to the equator have mostly warm temperatures and heavy rainfall. These wet, hot climates are called humid tropical climates. No winter or even cool weather is ever experienced by people who live in the humid tropics. Since the equator is constantly being heated by the sun's rays, warm air is always rising in the tropics. This continuous uplift of warm air brings almost daily thunderstorms and heavy rainfall. The combination of heavy rainfall and warm temperatures creates ideal conditions for plant growth. Dense tropical rain forests thrive here. The humid tropical climate can be found along the equator in South America and Africa and on the islands of the Pacific Ocean.

In some tropical areas, such as Asia and especially India, the rain is concentrated in a very wet season. During the summer months, moist air flows into Asia from the warm ocean, bringing heavy rains. During the winter, dry air flows off the continent, bringing dry conditions to the area. This change of winds, which brings both wet and dry seasons, is called the **monsoon**.

Just to the north and south of the humid tropical climate is the wet and dry tropical/subtropical climate. In this climate region, summers have heavy rainfall, while winters are very dry. This wet and dry climate is produced by the seasonal change in the way the sun's rays strike the areas north and south of the equator. For example, during the summer, or high-sun season, the sun's rays strike most directly. This high sun causes the temperature to increase. The increase in temperature causes unstable, rising air to bring heavy rainfall. During the winter, or low-sun season, the opposite occurs. Now the high-sun season has shifted to the opposite hemisphere, and the sun's rays are not striking directly. As the direct rays shift, a subtropical high-pressure zone moves into the area. This causes stable, cool, sinking air and a very dry season.

In contrast to the climates of the humid tropics is the dry tropical and subtropical desert climate. The tropical and subtropical deserts are centered at about 30° north and south of the equator. The dryness is caused by the subtropical high-pressure zones located in these areas. This pressure system is present all year. It brings stable, sinking, and dry air into the region. So, very little rain is produced in these high-pressure zones. Few plants can grow in these desert conditions.

The largest of these deserts is the Sahara, which stretches across all of northern Africa. These desert conditions also exist in southwest Asia. Most of the interior of Australia is desert as well. Small and very dry tropical and subtropical deserts are also found on the west coasts of continents. Here, the cool ocean currents cause stable conditions. It may not rain for many years. Examples of such dry coastal deserts are found along the west coasts of South America, southern Africa, and Mexico.

Middle-Latitude Climates As is the case with low-latitude climates, there are several types of middle-latitude climates. Two climates of the middle latitudes are characterized by dry conditions. These climates are the continental desert climate and the continental steppe climate, found in the interiors of large continents. They are most common in the Northern Hemisphere, where the largest continents are located.

The continental desert climate is cold in the winter and hot in the summer. This climate is often found in low areas surrounded by high mountains. The mountains block moist ocean air from reaching the areas. The Gobi Desert in Central Asia is an example of a continental desert climate.

Level A: Direct students to an almanac or similar reference to make a "That's Incredible" list of climate extremes. Students might find information on the coldest, hottest, wettest, and windiest places on Earth.

The continental steppe climate is a semiarid climate. It is a transition between the dry deserts and the more humid climates. Areas with continental steppe climates receive more moisture than the deserts but less than areas with more humid climates. Continental steppe climates are good for growing grasses that feed wild animals and cattle. In the wetter parts of the continental steppe, farming has replaced the natural grass with wheat or other grains. Examples of the continental steppe climate are the Great Plains of the central United States and the steppes of the Soviet Union.

The Mediterranean, or subtropical dry summer, climate is found on the west coasts of continents with cool ocean currents. Only 3 percent of the earth's land surface has a Mediterranean climate.

During the summer in areas with the Mediterranean climate, rain is rare. The drying influence of the stable high-pressure zone is strong at this time. However, during the winter, storms from the cooler middle latitudes enter the region and bring seasonal rains. The Mediterranean climate is found mostly in the coastal areas surrounding the Mediterranean Sea, especially in southern Europe. It is also located in such areas as southern California, central Chile, southern Australia, and a small part of southern Africa.

Much more widespread than the Mediterranean climate is the humid subtropical climate. This climate is found on the eastern side of continents with warm ocean currents. The humid subtropical climate is greatly affected by warm, moist air flowing off the warm ocean waters. Summers are hot and humid. Winters are mild, with occasional frost and little snow. Summer tends to be the wettest part of the year. Abundant rainfall and warm temperatures make this an ideal climate for agriculture and lumbering. However, the warm, wet conditions also cause areas with the humid subtropical climate to be prone to hurricanes or typhoons.

The largest areas of humid subtropical climate are found in the southeastern United States, eastern Asia, and southeastern South America. This climate is also found in small regions of Europe, eastern Australia, and southern Africa.

The coastal region of Scotland is typical of the maritime climate found in much of Western Europe.

A climate type that is influenced mainly by oceans is called maritime. The maritime climate is generally found on the west coasts of continents in the middle latitudes. Temperatures are mild all year and do not vary greatly. Storms traveling across the oceans bring a great deal of moisture to areas with this climate. Winters are foggy, cloudy, and rainy, but summers are warm and sunny. This climate supports forests and is well suited for many types of agriculture.

The maritime climate is most widespread in northwestern Europe. The absence of mountain ranges along the coast allows the cool, moist ocean air to spread far into the

Ask students to discuss the effect of a harsh climate on its inhabitants. In what ways do they adapt to it? For example, nomadic pastoralism is the way of life followed by many people living in deserts and the subarctic.

interior of Europe. In the Pacific Northwest of North America, high mountains limit the maritime climate to a narrow strip along the coast. This climate is also found at the southern tip of South America and along the mountainous coasts of Australia and New Zealand.

The humid continental climate is found in the northeastern United States, in northern Europe and the Soviet Union, and in northeastern China. Since there are no large land areas in the Southern Hemisphere in the middle latitudes, the humid continental climate is not found there.

Continental conditions are very different from maritime conditions. In the center of a large continent, there are no oceans to prevent extreme temperatures. The land heats quickly in summer and cools quickly in winter. It is this extreme range between summer and winter temperatures that is typical of humid continental climates. Moisture is usually present throughout the year. It rains in the summer and snows in the winter. This climate is wet enough to support forests. Agriculture is possible but is limited to a shorter growing season than in the subtropical climates.

High-Latitude Climates The subarctic climate is centered above 50° north latitude. As with the humid continental climate, there is no subarctic climate in the Southern Hemisphere. Lack of large land areas at these latitudes prevents the conditions necessary for the subarctic climate. The subarctic climate has long, dark, and cold winters, with temperatures staying well below freezing for almost half the year. Even in the short summer, frost can occur. For this reason, very little agriculture exists. The subarctic climate region is very large, extending across northern North America, Europe, and Asia.

Another climate with a long winter is the tundra climate. Only during the short summers are temperatures in the tundra climate warm enough for plants to grow. Sometimes frost occurs even in midsummer. People who live in these high latitudes rely on reindeer, fish, and marine mammals for food.

The tundra climate takes its name from the only vegetation that can survive there. Tundra is made up of small, hardy plants such as mosses, lichens (LIE-kuhnz), and low shrubs. Water below the tundra surface remains frozen throughout the year. This condition is called **permafrost**. Although the melting winter snow creates swamps and bogs during the summer, permafrost makes it very difficult for the water to drain away underground. Nonetheless, the wet areas can support great numbers of insects and birds during the short summer. The tundra climate stretches across the northern coasts of North America, Europe, and Asia. In the Southern Hemisphere, only a few small islands and the tip of the Antarctic Peninsula have a tundra climate.

The polar ice-cap climate is always cold. The cold is so intense and continuous that ice and snow are present all year in these areas. The Antarctic and Greenland ice caps are the largest regions of this severe climate. Plant, animal, and human life are almost impossible in polar ice-cap climates. Only animals that can feed in the polar seas survive there.

Permafrost creates a lacy pattern in the tundra region of Alaska. Water below the surface remains frozen all year.

Level B: Divide the class into three groups: low-, middle-, and high-latitude climates. Have students in each group write and present descriptions of their climate. Ask the rest of the class to guess which climates are being described.

World Climate Regions

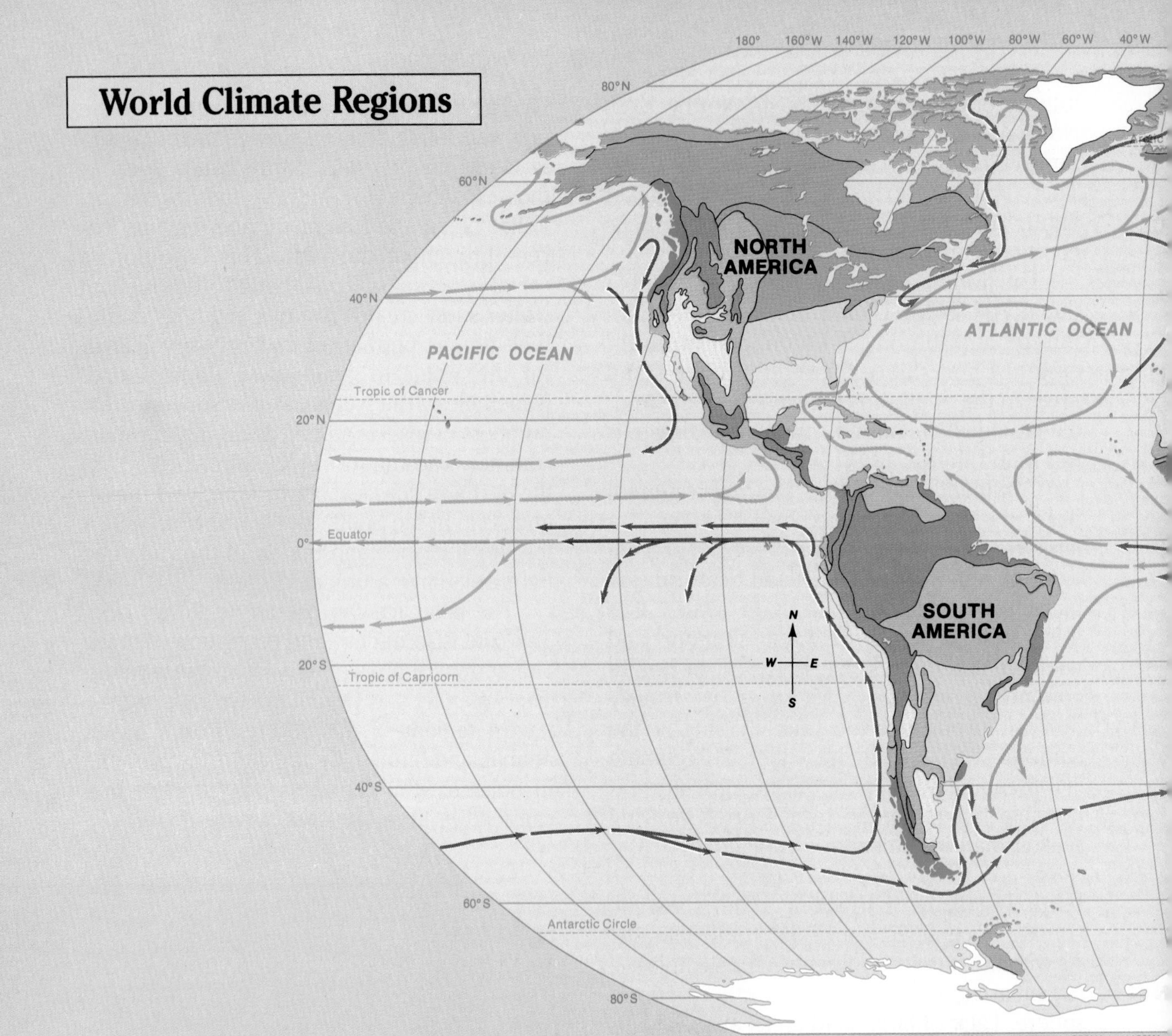

		Location (examples)	Climate	Vegetation	Human Use
LOW LATITUDES	HUMID TROPICAL	Near Equator (Amazon and Zaire river basins)	Rainy, hot	Tall trees; vines	Rubber, sugar, and banana plantations; harvesting hardwood from forest; paddy rice and shifting cultivation
LOW LATITUDES	WET AND DRY TROPICAL/ SUBTROPICAL	Between humid tropics and tropical deserts (Sahel of West Africa), with extensions into the middle latitudes	A very wet and a very dry season	Tall, tough grass; some trees	Cattle-raising; growing cotton, peanuts, wheat, barley; sheep
LOW LATITUDES	TROPICAL AND SUBTROPICAL DESERT	West side of continents (Sahara in northern Africa, Mojave Desert in California)	Hot; scarcely any rain	Desert shrubs (like cactus), scrub trees and grasses	Little unless irrigated
MIDDLE LATITUDES	MARITIME	West coasts (British Isles, Pacific Northwest)	Rainy, moderate temperatures	Mixed forest	Fruits, wheat and barley, sugar beets, wood products; grazing
MIDDLE LATITUDES	HUMID CONTINENTAL	Inland or on east coasts (northern United States, western Soviet Union)	Hot summers; cold winters; dependable rainfall	Mixed forest	Wheat, corn, soybeans; grazing
MIDDLE LATITUDES	CONTINENTAL STEPPE	Far inland, often on plateaus (North American Great Plains, southern Soviet Union)	Very hot summers; very cold winters; light rainfall	Grasses, shrubs, trees along streams	Grazing; grains, irrigated sometimes

What climate region is spread across the northern edges of Asia and North America? Tundra.

 Be sure students familiarize themselves with this map. It will be referred to throughout the book.

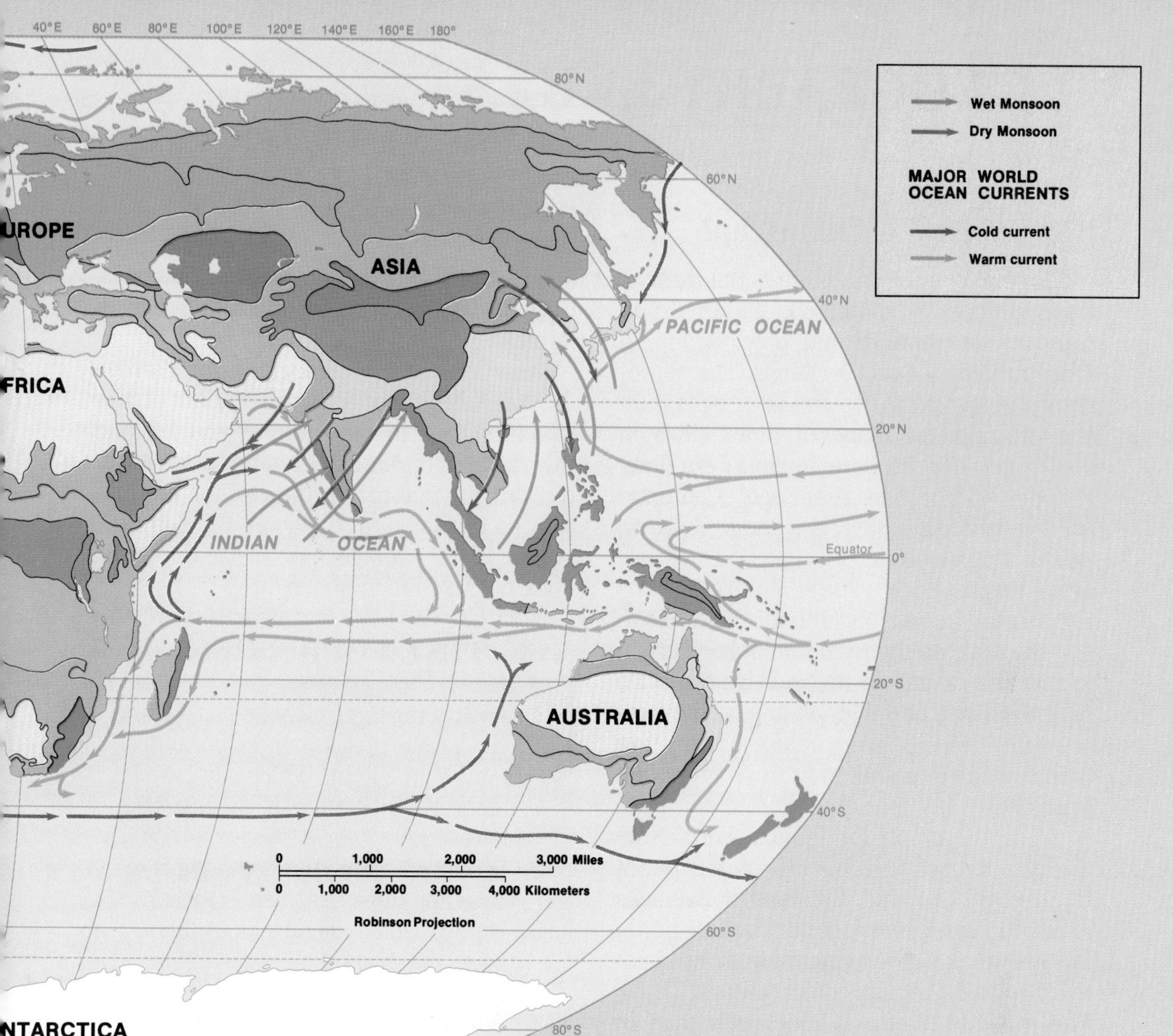

		Location (examples)	Climate	Vegetation	Human Use
MIDDLE LATITUDES	**CONTINENTAL DESERT**	Adjoining continental steppes (Gobi)	Arid; extreme temperatures	Sparse vegetation	Grazing; mining
MIDDLE LATITUDES	**MEDITERRANEAN (SUBTROPICAL DRY SUMMER)**	West side of continents, near the sea (southern Europe, central Chile, southern California)	Mild, moist winters; dry sunny summers	Scrub evergreen trees, grasses	Fruit, olives, wheat, and grapes; often irrigated
MIDDLE LATITUDES	**HUMID SUBTROPICAL**	East coast of continents (southeastern United States, south China)	Long, humid summers; short mild winters	Mixed forest	Rice, cotton, vegetables, tobacco; grazing
HIGH LATITUDES	**SUBARCTIC**	Above 50° north latitude (much of Alaska, Canada, Siberia)	Short summers; long, cold winters; humid	Forest—much of it swampy	Mining; lumbering
HIGH LATITUDES	**TUNDRA**	Northern coasts of North America, Asia, Europe	Long, cold winters; short summers	Moss, lichens, low shrubs	Fishing; trapping; hunting; oil
HIGH LATITUDES	**POLAR ICE CAP**	Greenland, Antarctica	Severe cold year-round	None	Uninhabitable
	HIGHLAND CLIMATES	Mountain areas of continents (Andes, Rockies, Himalayas)	Temperature and rainfall vary over short distances at different altitudes	Forest to tundra vegetation, depending on altitude	Lumbering, coffee, and grain at lower altitudes; limited use in higher altitudes

Why is this region colder than the climate region found at the equator?

The sun's direct rays do not fall this far north.

GEOGRAPHY SKILLS

"Geography Skills" features appear in every unit of the book.

Answers to Geography Skills questions are found in the Teacher's Guide.

How to Read a Climograph

Climographs are important in the study of world climates. A climograph allows you to compare temperature and precipitation in different latitudes and locations. The climographs below show the annual temperature and precipitation range of three cities in the Western Hemisphere: Iquitos, Peru (low-latitude, 4° S), San Francisco, California (middle-latitude, 38° N), and Barrow, Alaska (high-latitude, 71° N). Look closely at the graphs. Each column represents average rainfall during one month of the year. Use the scale on the right side of each graph to find the rainfall in inches. Use the scale on the left side of the graph to read the temperature. Annual temperature ranges are indicated by the line that runs from left to right.

Study the graph for Iquitos, Peru. Note that the temperature is high and constant (around 80° F) and the rainfall averages about 10 inches every month. Farther north, in San Francisco, the temperature is also relatively constant, though cooler (around 65° F). The monthly rainfall is much less than in Iquitos, and the summers are very dry. From the information presented in this chapter about low- and middle-latitude climates, how do you account for this difference in rainfall?

Now look at the climograph for Barrow, Alaska. Note that only a small amount of precipitation falls in the summer and fall and that the temperature range varies dramatically from the summer to the winter months. Why is precipitation higher in the summer, although the average monthly temperature does not rise much above 40° F?

TEST YOUR SKILLS

1. Look at the three climographs. Which city has the greatest amount of annual rainfall? the least? Which city has the highest temperature each year? the lowest?
2. During which months does Iquitos receive the most rain? San Francisco?
3. During which months does Barrow have the highest average temperature? the lowest?
4. When do Barrow and San Francisco receive little or no precipitation?

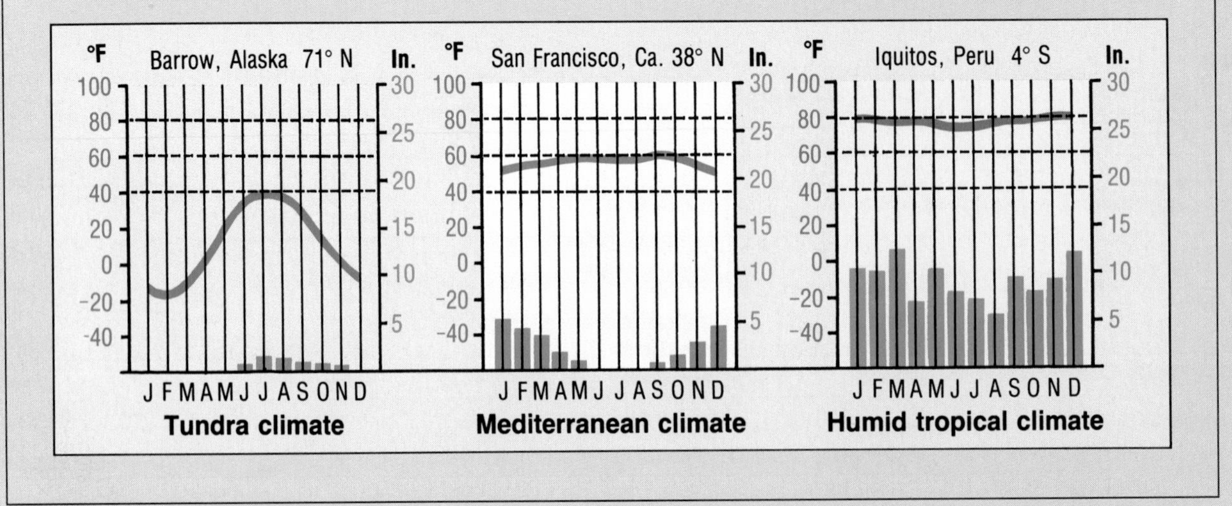

Have students use the map on pages 66–67 to identify other places where the above climates are found. What generalizations can they draw about latitude as related to these three climates?

Highland Climates As mentioned earlier, an increase in elevation causes a decrease in temperature. Unlike the other climate regions, highland climates are determined by elevation and not by latitude. High-mountain areas of the world have a variety of climatic characteristics in a small area.

The lowest elevations of a mountain will have a climate similar to that of the surrounding area. Crops grown in the surrounding area might also be grown on the mountain at the lower elevations. However, as you climb higher up the mountain, the climate conditions change. Cooler temperatures limit what can be grown. At the highest elevations, climate conditions are the most extreme. Here, temperatures may be very cold, and ice and snow will always be present. Few plants can grow, and few people can live under such conditions.

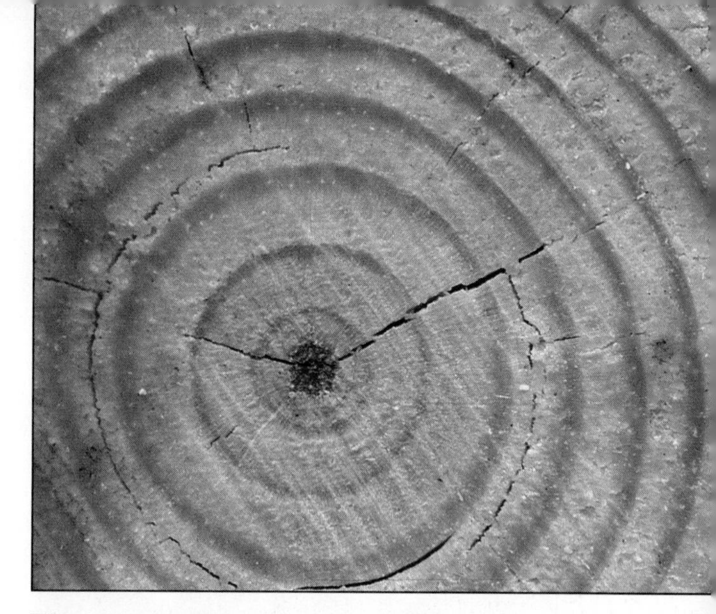

This cross section of a pine tree illustrates climate changes over time. Each ring shows one year of growth.

Climate and Human Activity

Human Activity Some of the world's climate regions, such as deserts and polar regions, have severe weather conditions, and few people can live in them. Other climate regions offer humans many opportunities and are heavily populated. People are not as limited today as they were centuries ago by weather and climate. Modern technology allows people to protect themselves from the effects of the atmosphere with indoor heating, air conditioning, and special clothing. In today's modern cities, we are able to travel, shop, and work without ever coming into contact with outside weather conditions.

Scientists must constantly monitor the atmosphere from weather stations and satellites. Statistics are continually collected to help forecast not only day-to-day weather changes, but also such events as floods and droughts or tornadoes and hurricanes. When these events happen, they can cause much damage and endanger many human lives.

Climate Change The earth's climates have not always been the same. Scientists know that there have been very warm periods, such as during the dinosaur age. There have also been very cold periods, such as the last ice age, which occurred only a few thousand years ago. Even in recent history, climate changes have taken place.

Good records of weather exist for only the last 100 years. Scientists must depend upon evidence from different sources, such as the sea bottom, glaciers, and tree trunks, to help piece together information about past climates. Evidence shows that the last 100 years were warmer than the previous several hundred years. When the trend toward a cooler climate—or even another ice age—becomes evident, it may be difficult for the world to grow enough food. Such long-term changes, both past and future, play an important role in the environment in which we live.

We must realize that the atmosphere is essential to our lives and to all life on Earth. We must continually monitor and be aware of the harm our activities can cause to our precious air. Whether humans can change the weather and climate of the whole planet is not known. But we do know that we can affect large areas, especially around cities.

Level B: Have students keep a weather log for one week. Encourage them to record the daily high and low temperatures, the amount of rain or snow, and the humidity, wind direction, wind velocity, and so on.

New Zealand's Southern Alps (left) are in a highland climate region. Scrub trees grow in the desert climate region of Arizona (below).

Icebergs float in the polar ice cap region of Antarctica (above). The tundra climate is found in the northern part of Alaska (below). Tahiti's climate is humid tropical (below right).

 Ask students to match the photos above with an appropriate climate region shown on the map on pages 66–67.

CHAPTER 5 CHECK

Reviewing the Main Ideas

1. Understanding the weather helps us plan each day and understand human activities elsewhere. Climate affects how and where people live. The study of climate also helps us understand some cultural differences.
2. Low-latitude climates are warm year-round. These are close to the equator and include the humid tropical, wet and dry tropical/subtropical, and dry tropical and subtropical desert climates.
3. Middle-latitude climates are found between 30° and 60°. These include the continental desert, continental steppe, Mediterranean, humid subtropical, maritime, and humid continental climates.
4. High-latitude climates are found between 60° and 90°. These include the subarctic, tundra, and polar ice-cap climates.
5. Highland climates are found in the high mountain areas of the world. Their characteristics depend upon elevation.

Building a Vocabulary

1. Define *monsoon*. Where do monsoons occur? Why?
2. What is permafrost? In what climate area is it found?

Recalling and Reviewing

1. Which climates are not found in the Southern Hemisphere? Why not?
2. Why are rain forests wet and hot?
3. Explain the differences between continental and maritime climate conditions. Choose two climate types to illustrate the differences.
4. Where are the tropical and subtropical deserts found? Give examples from three continents.

Critical Thinking

5. If you lived in the tropical desert, how would your life be influenced by the climate? Under which climate conditions would you prefer to live? Why?

Using Geography Skills

Use the map in this chapter to answer the following questions.

1. On the climate map, locate those areas of the world that have maritime climates. In what ways are the locations of these areas similar? Why?
2. Which climate type covers more area than the others? Which covers the least amount of area? Are these climates low-, middle-, or high-latitude climates?
3. In what ways do you think the warm and cold ocean currents affect North America's climate regions? Where is the humid subtropical climate region located in North America?

Answers to Chapter Check questions are found in the Teacher's Guide.

Additional activities for Chapter 5 are found in the Student Workbook. Chapter 5 Test as well as Skills, Reteaching, Critical Thinking, and Challenge/Enrichment activities are available in the Teacher's Resource Binder.

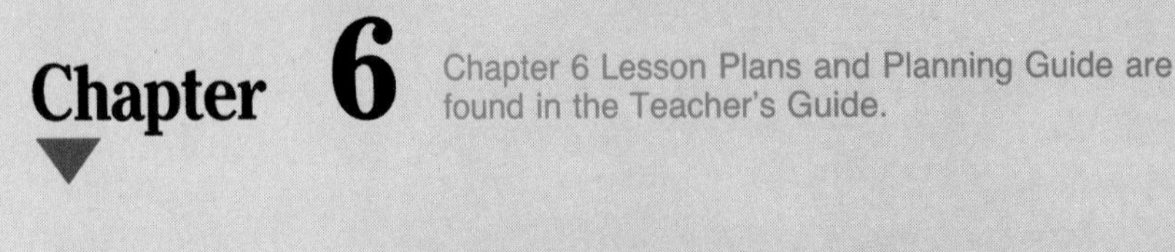

Chapter 6

Chapter 6 Lesson Plans and Planning Guide are found in the Teacher's Guide.

The Water Planet

Have students consider where the two photos below fit into the hydrologic cycle.

▲ **Venezuela's Angel Falls is 3,212 feet (979 meters) high.**

▲ **The Mackenzie Delta in Canada is a vast region.**

GEOGRAPHY DICTIONARY

hydrosphere	tributary
hydrologic cycle	estuary
transpiration	groundwater
evapotranspiration	water table
continental shelf	aquifer
headwaters	irrigation

The **hydrosphere** includes all the water of the earth. It is made up of the oceans, lakes, streams, water found underground, water in all living things, water vapor in the atmosphere, and water frozen as ice. Because it is the only planet in the solar system to have liquid water, Earth is sometimes called the "water planet."

Water is essential for all known life, especially for cell growth. All living things, plant or animal, are made up mainly of water. In fact, your body is 70 percent water. Earth is the only planet with water; we think that it is the only one with life. Not only is water important to us for survival, but it also plays an important role in human activities. For example, water is necessary for agriculture, industry, energy, and transportation.

 Have students research the amount of water that is used every day in the United States for agriculture and industry and in homes.

Explain to students that there is as much water on Earth today as there ever has been or ever will be. This is because the earth's water is constantly being recycled.

The Geography of Water

The Hydrologic Cycle Water is a unique substance. It can exist as a liquid, solid, or gas within the earth's temperature range. Other unique characteristics of water include its ability to dissolve almost anything, including the hardest rocks, and its ability to heat and cool very slowly.

The majority of the hydrosphere is in the oceans, which contain 97 percent of Earth's water. Most of the remainder is found frozen in the polar ice sheets and glaciers. Only a very small amount is found underground or in lakes and streams. Even smaller, but very essential, amounts are found in the atmosphere. Thus, more than 99 percent of the earth's hydrosphere is either salt water or ice.

The circulation of water from one part of the hydrosphere to another is known as the **hydrologic cycle**. (See the illustration below.) Water is cycled endlessly among the atmosphere, oceans, continents, ice sheets, and living things. For instance, most water enters the atmosphere by being evaporated from the ocean. It becomes water vapor and rises high into the air. As it goes up, it is cooled, and condensation occurs. That is, the water changes from a gas to tiny droplets of liquid. These collections of droplets are clouds and may fall back into the ocean as precipitation. This completes the cycle of evaporation, condensation, and precipitation.

The hydrologic cycle becomes more complex, however, when precipitation falls on the continents. Some precipitation falls on mountains or polar areas to become ice. Some flows

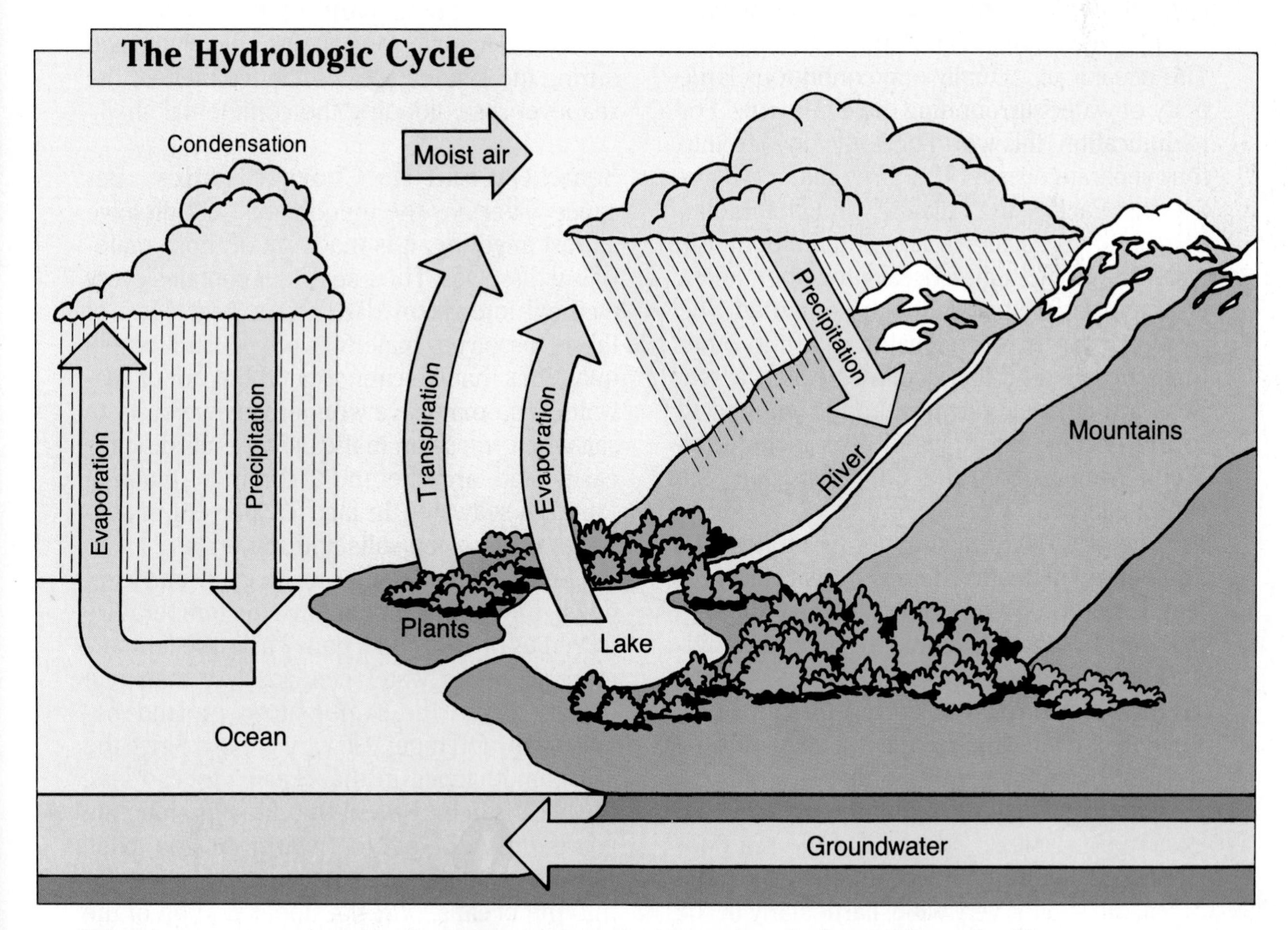

What happens to water after it evaporates into the atmosphere? after it falls on land?

Condensation occurs, and the water falls back into the ocean or onto land as precipitation. After it falls on land, it can form ice at high elevations or run off into streams and rivers. After being temporarily stored in lakes or in the ground, it eventually returns to the ocean.

Level B: Have students prepare a chart comparing the four oceans in terms of area, deepest point, average depth, and so on.

off the land through rivers and streams, possibly to be temporarily stored in lakes, and eventually back to the ocean. The ground absorbs some water by taking it into the air spaces in the soil and rock, where it can be used by plants. Here, the water is returned to the atmosphere by evaporation from the land or by a process called **transpiration**. Transpiration occurs when plants give off water vapor through their leaves. The evaporation of water from the ground combined with the transpiration of water by the plants is called **evapotranspiration**.

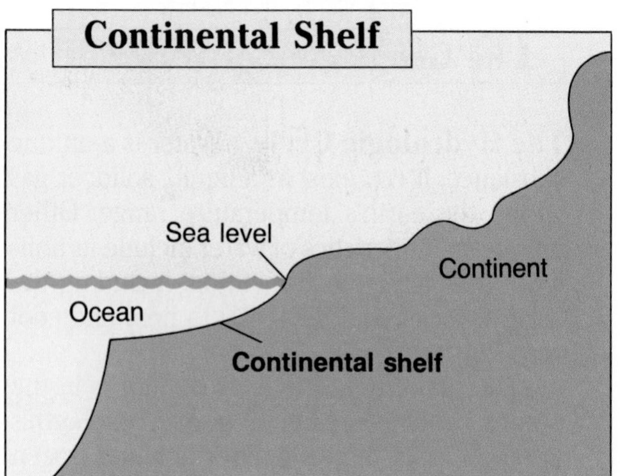

Which of the four oceans has the widest continental shelf?

the Atlantic Ocean

The Oceans

Marine Geography The oceans cover 71 percent of the earth's surface. That is why Earth is sometimes called the "blue planet." The oceans are actually one continuous large body of water surrounding the continents. For identification, this world ocean is divided into four separate oceans. The three major oceans are the Pacific, the Atlantic, and the Indian oceans. The Pacific is larger than both the Atlantic and the Indian oceans together, and it is larger than all the continents combined. The Arctic Ocean is much smaller than the other three oceans and is sometimes called a sea. Seas are smaller saltwater bodies connected to the oceans. Examples of seas include the Mediterranean Sea, the Coral Sea, and the Caribbean Sea.

The oceans average about two miles (3.2 kilometers) in depth. There is a great variety of features at the ocean bottom. One of these is the world's longest mountain range, the mid-oceanic ridge. Another feature is the Mariana Trench, located in the Pacific Ocean. More than 36,000 feet (10,976 meters) deep, it is the deepest place known on Earth.

The most shallow part of the ocean is the **continental shelf**. This part of the sea floor slopes gently from the continents. In some areas, the shelf is very wide, particularly in the Atlantic Ocean. In the Pacific, the continental shelf is very narrow. Where the continental shelf ends, there is a steep drop-off to the ocean floor. Some parts of the continental shelf were actually land thousand of years ago during the last ice age. As the ice melted, the sea level rose, flooding the continental shelf.

Seawater and Its Characteristics Because water has the unique ability to dissolve almost anything, it is made up of those materials it dissolves. Thus, seawater contains every element known on Earth, even gold. Most of these dissolved materials are in such small quantities that we cannot get them out of seawater. You may have wondered why you taste salt when you swim in the ocean. Salts dissolve easily and are the most common material found in seawater. In fact, 3.5 percent of seawater is dissolved salts and minerals.

Since water is slow to heat up and cool down, the oceans do not have the temperature extremes of land. On a daily basis, the temperature of ocean water changes less than one degree, while the temperature of land between day and night will vary greatly. Since the seasonal changes in the ocean are also less, the oceans help prevent the climates along the coasts from being too warm or too cool. Because sunlight does not penetrate very far into the oceans, only the upper portion of the water is heated.

 Ninety-seven percent of the earth's water is found in the oceans. Two percent of the earth's water is in glaciers and ice caps, and one percent is in groundwater, rivers, and lakes.

Satellite photos are among the tools that have revolutionized geography in recent years.

Like the atmosphere, the ocean is always moving and circulating. The winds push the ocean currents and cause waves to mix the surface water. Gravity causes the tides to rise and fall, and changing density causes even the deepest ocean waters to move.

Most of the ocean is deep, dark, and cold. Because marine plants need light, they are found only in the upper waters. Since the marine plants form the basis for the marine food chain, most of the marine animals also live near the surface. In fact, most marine life is found in the shallow waters of the coastal areas of the world. Though the oceans appear huge, their life-producing regions are quite limited. We still have much to explore and learn from the oceans, our "inner space."

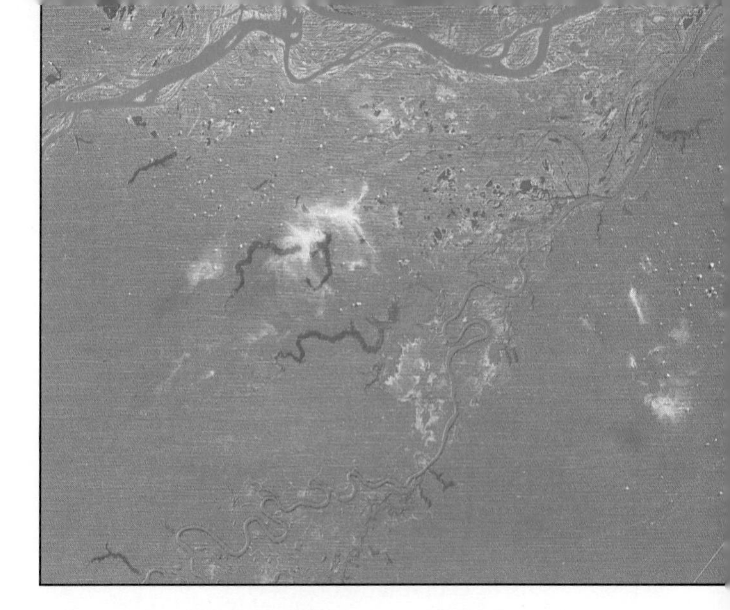

This Landsat photo of the Amazon River in Brazil shows the Amazon and its tributaries snaking through the tropical rain forest.

Rivers and Lakes

Also important in the hydrologic cycle is the formation of rivers and lakes. As precipitation falls on hills and mountains, it flows down the slopes toward the lowlands and coasts. The first streams to form from this runoff are called **headwaters**. As these headwaters join, they form larger streams and eventually large rivers. Any smaller stream that flows into a larger stream is a **tributary**. Very large river systems, such as the Amazon, Mississippi, and Volga rivers, may flow for thousands of miles and be joined by numerous tributaries.

Where rivers meet seawater, they may form a delta or an **estuary**. Estuaries are particularly important because they are rich in fish and shellfish. Good examples of estuaries are Chesapeake Bay and San Francisco Bay. As rivers carry fresh water to the coasts, they bring along minerals that were eroded from the land. These minerals are necessary to the ocean's food chain.

When runoff is slowed or stopped and water fills a depression in the land surface, a lake forms. Most lakes are freshwater, but some, such as Great Salt Lake in Utah, are salty. North America has many natural freshwater lakes, while other continents have few. The largest freshwater lake system in the world is the Great Lakes, located along the United States–Canadian border. The deepest lake in the world is Lake Baikal in the Soviet Union. Large lakes, like the oceans, have a moderating effect on the local climate.

Groundwater

Although rivers and lakes are visible to us, there is actually more fresh water beneath the surface of the continents than above it. The water found below the surface in the spaces between soil and rock grains is known as **groundwater**. Groundwater is an important resource to humans, especially in dry climates where surface waters are scarce.

The major source of groundwater is precipitation. Water from rain or melting snow sinks into the ground and slowly seeps downward. The water stops when there is no more space between the rock grains or when all the spaces are saturated with water. The top of this saturated zone is called the **water table**. To tap this groundwater, we would have to drill a well to a depth below the water table.

Direct students' attention to the Landsat photo above. The vegetation of the Amazon rain forest appears as a solid block of red. Have students research the development of Landsat photography and describe its uses in world geography.

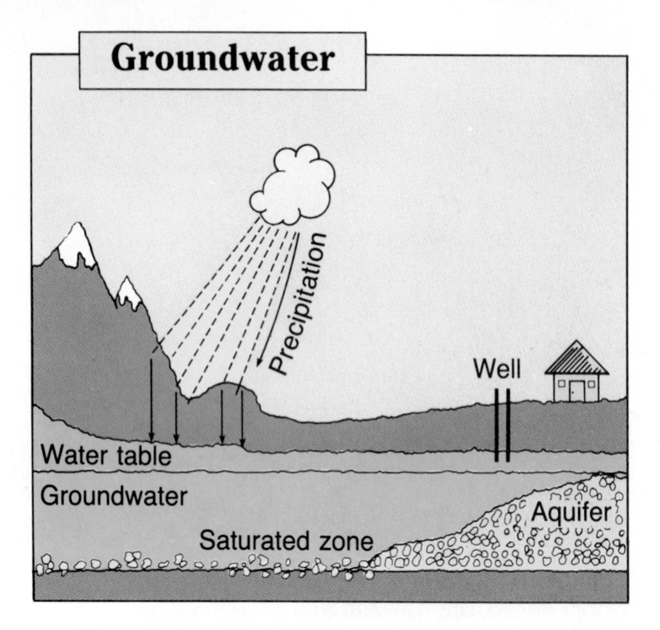

How do people gain access to groundwater?

by drilling wells to the water table

The depth of the water table varies. In areas with a wet climate, the water table will be near the surface. In desert areas, it may be hundreds of feet below the surface or may not even exist. The water table also follows the landforms and will rise under hills and drop under valleys. It will vary from year to year as rainfall patterns change. If a region suffers a drought, the water table falls. Too many wells in a region can also lower the water table.

Like rivers, groundwater can travel long distances but at a much slower rate. Generally, groundwater will flow through a rock layer called an **aquifer**. An aquifer allows space for water storage and allows water to move from space to space. Aquifers are composed mostly of sand and gravel materials. Large aquifers may be hundreds of miles across. The Ogallala Aquifer stretches from Texas to Nebraska and is a major source of water for farms in the Great Plains region.

People and Water

Water is one of our most valuable resources. Humans cannot survive for more than a day or two without it. Think about how much water you use each day in your home. In modern society, we often take our water supply for granted. In the less developed countries, people may spend hours each day walking to a community well or stream just for a supply of water.

Water is essential for agriculture. In dry climates, **irrigation** is necessary for growing crops. Irrigation is the watering of land through pipes, ditches, or canals. Few crops could be grown in areas of Texas, California, or Arizona if farmers could not get irrigation water. Industry also could not function without water. Water is an ingredient in the manufacture of many products.

Water is a valuable power source. Water stored by dams drives turbines to produce a significant portion of the world's electricity. Our oceans, rivers, and lakes also provide an invaluable world transportation network. To transport bulk cargo such as oil, wheat, or automobiles, shipping by water is a preferred method because of low cost.

As the world becomes more densely populated, our water resources become more threatened. Pollution from city, farm, and industrial wastes is dumped into our bodies of water. This affects the food chain, and our sources of fish and other seafood are threatened. In some areas, we have overused our water supplies, and water is becoming scarce. It is important to remember that our entire quality of life depends upon water.

The Anderson Ranch Dam in Idaho demonstrates how humans change their physical environment.

Level A: Have a student prepare a report detailing how a pollutant is introduced into the hydrologic cycle and its effects on water, plants, or people.

CHAPTER 6 CHECK

Reviewing the Main Ideas

1. The hydrosphere contains all the water of the earth. It includes the water in oceans, lakes, and streams, water found underground, water in all living things, water in the atmosphere, and ice. Life as we know it depends on water.
2. The circulation of water from one part of the hydrosphere to another is known as the hydrologic cycle.
3. Oceans cover 71 percent of the earth's surface. They average two miles in depth. The three major oceans are the Pacific, the Atlantic, and the Indian oceans.
4. Rivers and lakes are formed by runoff of precipitation from hills and mountains.
5. Groundwater is found beneath the surface of the earth in the spaces between the soil and rock.

Building a Vocabulary

Students should answer Question 4 by filling in the blanks, using their own paper.

1. What is the hydrosphere? the hydrologic cycle?
2. What is the continental shelf?
3. What are the differences among headwaters, tributaries, and estuaries?
4. The top of the saturated groundwater zone is called the ______ ______.
5. What is the difference between evaporation and transpiration? How does evapotranspiration occur?
6. Define *irrigation.* Why is irrigation necessary in dry climates?

Recalling and Reviewing

1. What is the deepest place known on Earth? Which areas of the ocean bottom are most shallow?
2. What are the characteristics of seawater?
3. Why are rivers and estuaries important to the ocean's food chain?
4. What is the major source of groundwater? Discuss the properties of groundwater, water tables, and aquifers. What is the relationship among the three?

Critical Thinking

5. Describe three ways life would change if the world's water resources were reduced.

Using Geography Skills

Use the diagrams in this chapter to answer the following questions.

1. Study the diagram of the hydrologic cycle on page 73. Describe the processes of evaporation, condensation, and precipitation. What happens to the precipitation that falls on continents?
2. Look at the diagram of the continental shelf. Why do you think most marine life is found in these shallow areas? Give evidence to support your answer.

Answers to Chapter Check questions are found in the Teacher's Guide.

Additional activities for Chapter 6 are found in the Student Workbook. Chapter 6 Test as well as Skills, Reteaching, Critical Thinking, and Challenge/Enrichment activities are available in the Teacher's Resource Binder.

Chapter 7

Chapter 7 Lesson Plans and Planning Guide are found in the Teacher's Guide.

World Patterns of Vegetation and Soils

▲ A plant community thrives in the South Pacific.

▲ Contour plowing prevents erosion on hilly land.

GEOGRAPHY DICTIONARY

- biosphere
- food chain
- plant community
- plant succession
- climax community
- humus
- soil horizons
- leaching
- biome
- deciduous forest
- coniferous forest
- savanna
- prairie
- steppe

All living things and the areas they inhabit on Earth are known as the **biosphere**. Plants are a major and extremely important form of life in the biosphere. In one way or another, all animals depend upon plants for their food supply. Some animals eat only plants, while others eat the animals that eat the plants. Yet others, including humans, eat both plants and animals. Plants are also a major source of oxygen for the air we breathe.

It is soil that provides the nutrients plants need to grow and to survive. Soil also supports the food we grow for our own use. This chapter will help you understand how life itself depends upon the earth's precious plant, animal, and soil resources.

Have students briefly survey the photos in this chapter and note the different patterns of vegetation and soils throughout the world. Ask them to consider the influence of climate on each.

Primary Production

Only about one percent of the sun's energy that reaches Earth is used directly by plants to make food. Each plant is like a small factory. The plant uses the sun's energy as fuel to change carbon dioxide from the air, water, and the soil's nutrients into food. Oxygen is given off as a result of this process. Some of the food produced is used to keep the plant alive. The rest is used for growth.

Plants are at the bottom of the **food chain** and support animal life. A food chain is a series of stages in which energy is passed along through living things. Some animals eat plants. These animals are eaten by other, usually larger, animals. At each stage of the food chain, the number of living things is reduced. Only a few animals, such as lions, hawks, and humans, are at the top of the food chain.

Plant Communities

Plant Growth Generally, plants do not live alone, since they are not independent. Just as you live in a society, surrounded by friends, family, and other people, plants live in groups. Such a group is called a **plant community**. In the wet tropics, these communities are quite large. Thousands of different kinds of plants are found. In addition, the many lush tropical plants in such a community support great numbers of animals, from insects to large mammals. In the desert communities, there may be only a few highly specialized plants, such as cacti, and a few specialized animals, such as lizards.

Each plant evolves in such a way that it plays a special role in its community. No two types of plants have the same needs for sunlight and nutrients. Some plants, such as grass, have shallow roots that capture water and nutrients close to the soil's surface. Other plants, such as trees, send their roots deep into the ground in search of these materials. Some

Explain why this rain forest in Australia is an example of interdependence and plant succession. Trees provide shade for ferns and support for clinging vines. This is a fully developed climax community.

plants are able to live in soil that is poor in nutrients. Others require rich soil. Large trees need a great deal of direct sunlight in order to make food. Ferns, however, can survive with very little sunlight. Only the plants whose requirements for growth are fulfilled in a community will survive there. This explains why there are forests or grasslands in some parts of the world but not in others.

All plants also depend on each other to grow. For example, trees provide the shade that keeps ferns growing on the forest floor from receiving too much sunlight. Trees also provide support for clinging vines so that they can capture needed sunlight at the top of the trees. Without trees, neither ferns nor vines would likely survive. Individual types of plants work to aid their own growth and, often, the growth of other plants in their community.

Plant Succession The cluster of plants that constitutes a community does not develop by accident. Rather, the plant community is the result of orderly change over a long period of time.

Imagine an area just after a forest fire. The first plants to grow will be very hardy. Since there are no trees to give shade, they will have to be able to endure direct sunlight. They will grow rapidly when rain falls on the ash and soil. These plants will help prevent erosion

Level A: Present these terms to students: *xerophytes, hydrophytes, tropophytes,* and *halophytes*. Explain that these terms describe four types of plants that thrive under special conditions. Have students define the special environment for each type of plant.

by holding the soil in place. They will also provide shelter for the seeds of other plants.

Grass and weeds may cover the ground for some time. But soon they will be replaced by taller plant growth. Often, these newer plants are shrubs and small trees. Their seeds may have been blown in from a distance by the wind or carried by animals such as squirrels or birds. When these seeds are dropped, a new tree may sprout. When the new tree is tall enough and has enough leaves, it will shade the ground. This shade will now make it difficult for the earlier sun-loving plants to survive. Shade-loving plants, such as ferns, will soon cover the forest floor. After many years, the area may again look as it did before the fire.

The process by which one group of plants replaces another is called **plant succession**. Each group of plants grows best in one set of conditions. When these conditions change, the growth of a new plant community begins.

The succession process can be controlled by people, but it is not easy to do. If you decide that you want only grass to grow in your yard, you can prevent other plant communities from developing. The grass must be cut frequently to keep tree seedlings from growing tall. Groups of trees might block needed sunlight from some of the grass. Fertilizer may have to be put on the grass to keep it healthy. If insects or a lack of fertilizer causes bare spots to appear on your lawn, weeds will quickly fill the open spaces. Work may be needed to keep dandelions and other weeds from taking over the lawn. Much time, effort, and money must be spent to prevent succession from changing your grassy lawn into a weed patch or a forest. In the same way, if a local plant community is destroyed, it may in time grow back.

Climax Community For each set of climate and soil conditions in an area and for

How do the plants that are able to grow after a forest fire differ from the plants in a fully recovered forest?

After a forest fire, plants that grow are those that are hardy enough to endure direct sunlight (such as grass and weeds).

Explain to students that in the United States, there are almost no climax communities because of urbanization and farming. Those that exist are preserved in national parks and forests.

any given time, there is a plant community that is best suited to use the resources there. Only after a long process of plant succession does a stable plant community develop. This last stage of succession is called the **climax community**.

Significant changes in the environment can alter a climax community. It might be destroyed by natural events such as forest fires, storms, volcanic eruptions, or plant disease. If any of these occur, the succession cycle would start over again. Humans, too, keep climax communities from continuing by burning and clearing the land for farming, by logging, by grazing their animals, or by introducing foreign plants. For these reasons, true climax communities are seldom found today in many areas of the world.

Soil Types

Factors of Soil Formation As you have learned, soil supplies the nutrients plants need to grow and survive, which helps provide the food we need to live. But what actually is soil? If you dig a shovel into the ground, what do you bring up? Soil is made up primarily of minerals from the weathering of rocks. Most of these materials are from the rocks found just below the soil layer. Sometimes the weathered rock material is brought to an area by rivers, wind, or glaciers. Soil is also composed of **humus**, which is decayed plant matter. Bacteria and insects help break down the humus, and they produce necessary space between the soil particles for gases and water. All of these conditions must exist in order for soil to provide plants with the necessities of life.

Soils differ from place to place, and there are hundreds of soil types. Climate is the major factor that controls the type of soil a location will have. Climate provides the heat, moisture, and natural vegetation that influence the soil's formation. The second major control is the rock type that provides minerals for the soil. Thus, one region may have dark, thick soils with much humus, while another has only a thin layer of sandy soil. Other factors such as the slope of the land, human activities, and time for soil development are also important.

Soil Horizons Soils develop very slowly, taking hundreds or even thousands of years to form. As they develop, they form distinct layers, called **soil horizons**. Most soils have three layers, called the A, B, and C horizons.

The top layer, or A horizon, is called topsoil. It contains humus and is where most vegetation grows. This horizon is very active and has much bacterial and insect activity as well as most of the plant roots. The middle layer, or B horizon, is called subsoil. Only deep plant roots penetrate into the B horizon. The C horizon is made up only of broken rock materials. Its lower reaches become solid rock.

Since climate is the major control of soil formation, there are similarities among soils in the same climate region. For instance, in the humid tropical climate, the warm temperatures and heavy rainfall support lush forests. These rain forests, in turn, supply great amounts of humus to the soil below. Topsoil undergoes a process called **leaching**. Leaching occurs when the nutrients necessary for plant growth are washed downward out of the

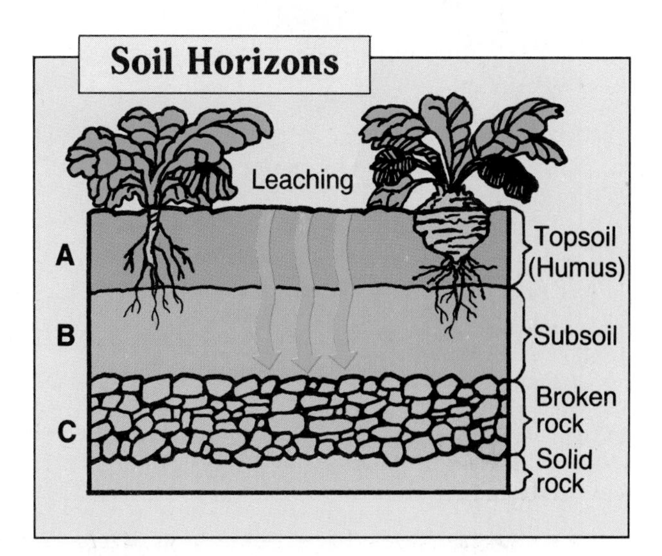

How does leaching affect the quality of topsoil?

It washes away nutrients needed for plant growth.

The distribution of soils has a major impact on a nation's agricultural potential. Have students scan Chapters 15 and 27 for examples of the relationship between soils and agriculture.

Discuss with students how the destruction of the world's rain forests can affect global climates.

topsoil by heavy rainfall. The nutrients then cannot be reached by the plant roots. Fortunately, the tropical rain forest continues to produce humus as fast as the rain washes the nutrients away. An area that supports such a lush forest would seem ideal to clear for farming. However, when the trees are removed, the soil will be depleted of its nutrients because of the constant leaching.

Biomes: World Plant Regions

What Is a Biome? A plant and animal community that covers a very large area of the earth's surface is called a **biome**. Plants are the most visible part of a biome. If you could look down on the United States from space, you would see the biomes. The forests of the East would appear green, while the deserts of the Southwest would be light brown.

The map "World Climate Regions" is found on pages 66–67.

There are five basic biomes that can be identified throughout the world: forest, savanna, grassland, desert, and tundra. Areas of the world that share the same biome type also have similar climates and soil types. (Compare the biome map below with the climate map in Chapter 5.) The differences in these large plant communities correspond to differences in climate.

Forest Biomes The forest biome is tree covered. Within this biome category, there are a number of different climax forest communities. Each forest community is determined by the types of trees growing there.

In the tropical rain forest near the equator, temperatures are warm, and water is abundant. Plants can grow all year long. Many different kinds of tall trees make up the tropical rain forest. The thick umbrella of treetops creates continuous shade on the forest floor. The constant plant growth provides much humus for the tropical soils below the forest.

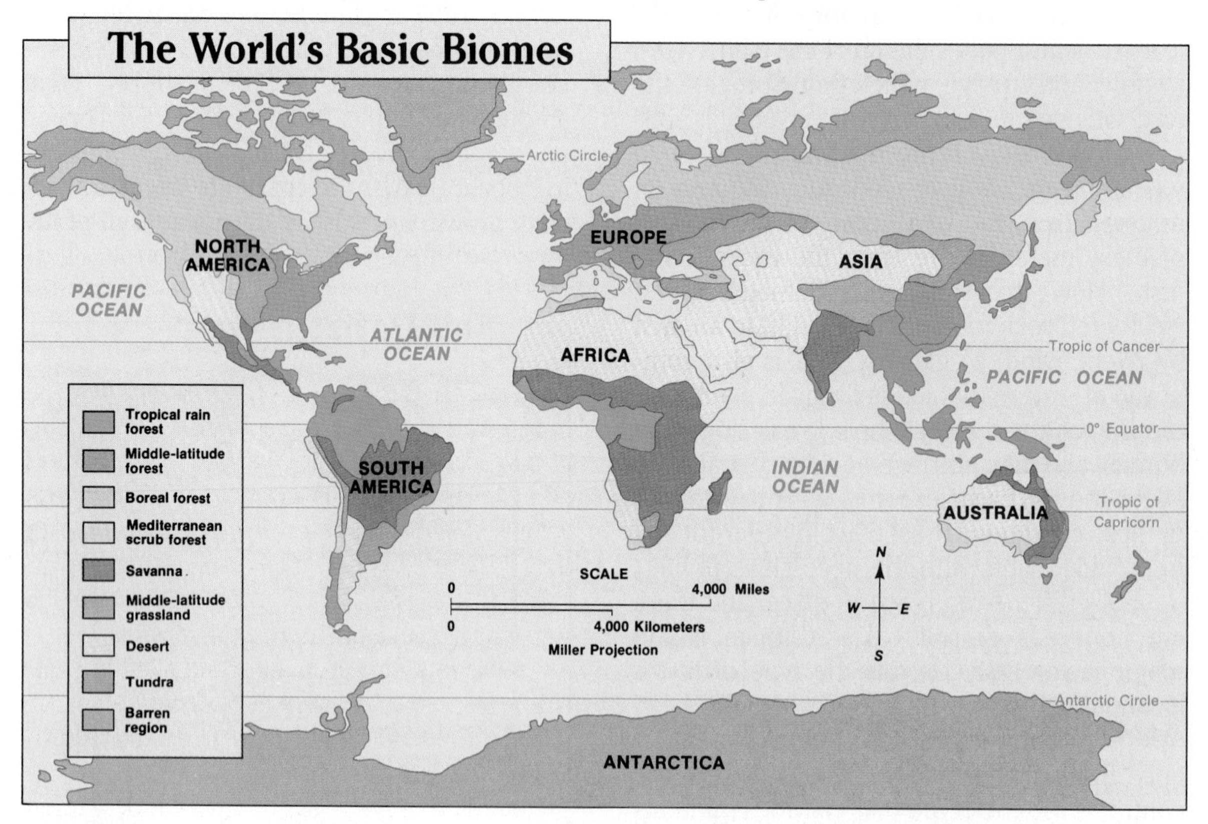

Which biomes are located at the equator? at the Arctic Circle? at the Tropic of Cancer?

equator: tropical rain forests, savanna; Arctic Circle: tundra, boreal forest, barren region; Tropic of Cancer: desert, savanna, tropical rain forest

Level B: Have students make illustrations of the vegetation regions of the earth. Have the class identify the regions.

The middle-latitude forests, such as those in the eastern United States, are less dense than those in the tropics. Enough sunlight reaches the ground to permit shrubs and other plants to grow. The middle-latitude forests can be divided into a few different types. The **deciduous** (di-SIJ-uh-wuhs) **forests** are green during the summer and lose their leaves before winter. The **coniferous** (koh-NIF-uh-ruhs) **forests** remain green year-round. Some areas are mixed where the two forest types blend.

The boreal (BOHR-ee-uhl) forests consist of needle-bearing trees. Because these trees are able to resist very cold conditions, boreal forests are found in cold climates. The thin needles cut down on the loss of water to the air and make it possible for the trees to use sunlight all year. The floors of these forests have very acid soils because of the buildup of acidic tree needles. Boreal forests stretch in a broad band across North America, northern Europe, and Asia.

Mediterranean scrub forests originally covered less than 3 percent of the earth's land surface. Most of the vegetation of this forest type is made up of short trees and shrubs. Leathery leaves and thick bark help these plants survive the long, dry summers. Forest fires are frequent during the summer season.

In Ontario, deciduous trees change color during autumn. Which of the trees in this photo are coniferous? How can you tell?

Evergreens; They do not change color or lose their leaves during winter.

The savanna biome receives little rainfall. The vegetation is grassland with scattered trees and shrubs.

Many of the Mediterranean scrub forests have been cleared for farming and housing. Clearing, along with overgrazing by animals, has destroyed much of this forest type.

Savanna Biomes On the edges of the tropical rain forests is the **savanna** biome. A savanna is a grassland with scattered trees and shrubs. Usually, savanna is located in areas with low rainfall. This rainfall generally occurs during only a few months of the year. The savanna biome occupies the areas between the tropical rain forests and the tropical deserts. Especially in Africa, savanna has increased in area at the expense of forests. Clearing forests for farming as well as overgrazing has helped spread the savanna into what were once forested regions.

Grassland Biomes Middle-latitude grasslands are found midway between the middle-latitude forest and the desert biomes. Except along rivers, grasslands lack trees. Those parts of the middle-latitude grassland that are closer to the forest biome have good supplies of rainfall. This wetter tall grassland is called **prairie**. It can be very productive once the wild grasses are removed. The deep fertile soils of the prairie are particularly important for growing grains. The world's largest wheat- and corn-growing areas are found in these prairie regions. The drier, short grassland, called **steppe**, is less productive for farming.

Desert and Tundra Biomes The remaining biomes are typical of harsh environments. These biomes include areas where temperatures are very hot or very cold, where there are very high elevations, or where it is very dry. These biomes are also found in areas with a combination of these conditions.

Desert biome plants survive by using very little water or by storing water. Cacti are examples of desert plants that store water. Other desert plants have leaves with waxy surfaces and large root systems. They can withstand high temperatures without letting much water escape into the air. Still other desert plants avoid dry conditions. These have short growing cycles and seeds that can survive underground for several years. When the rare rains come, the seeds sprout and grow into mature plants in a short time. Because there is so little moisture in the desert biomes, plants grow far apart and produce almost no humus for soil formation and little food for animals.

Like the desert biome, the tundra biome supports little plant growth. Climate conditions are too cold for anything but low shrubs, small flowering plants, and moss. Tundra communities can be found in the polar regions and in the mountains at high elevations.

Barren Regions In the far polar regions, ice and snow remain all year. This makes it impossible for plants to grow. These areas with no plant life are called barren regions. The largest barren regions are the ice caps of Antarctica and Greenland. The tops of high mountains may also be barren of plant life. Here, elevation causes polar conditions in areas far away from the polar regions. The cold temperatures, severe winds, glaciers, and snowfields make it impossible for plants to survive.

Though the desert biome is arid, it supports a small amount of vegetation throughout the year. This California desert is in bloom.

People and the Biosphere

All living things—plants, animals, and humans—are interrelated within the biosphere. That is, events in one place in the biosphere can affect plants and animals in another part far away. Natural changes in the biosphere take place all the time. For example, a volcanic eruption may destroy everything around a mountain. The plants and animals slowly return to the area and repopulate it. Changes that last for a long period of time are another matter. When an ice age replaces warmer conditions, many plants and animals die. This causes some kinds of living things to become extinct, which means that they cease to exist.

Human actions also have an effect on all parts of the biosphere. Some of the biomes have changed size as a result of human activities. The tropical rain forests, for example, are becoming smaller as the human populations increase around them. Humans cut the trees for timber and clear the land for farms. Because of the loss of the forest, the soil loses its nutrients. Although the farms may be abandoned and the loggers might move elsewhere, the forest cannot return because the necessary soil was changed.

While the forest biomes are becoming smaller, other biomes are expanding, especially the savanna and desert biomes. If our biosphere is completely changed, what will life be like in the future? Humans have a special responsibility to handle carefully the resources on which life depends.

Level A: Large-scale changes in land patterns and use can have major effects on the weather and the renewal of the world's oxygen supply. Have students write a brief science fiction scenario based on this statement.

CHAPTER 7 CHECK

Reviewing the Main Ideas

1. Life itself depends upon the earth's plant, animal, and soil resources. Plants are at the bottom of the food chain and support animal life.
2. Soil provides the nutrients plants need to grow and survive. It also supports the food humans grow for survival. Soil is composed primarily of minerals from the weathering of rocks. Climate is the major control of soil formation.
3. Different types of plants live in what are called plant communities. Plant communities evolve over a long period of time. For each set of climate and soil conditions in an area, there is a plant community best suited to use the resources in that area.
4. A plant and animal community that covers a large area of the earth's surface is a biome. There are five basic biomes in the world.

Building a Vocabulary

Students should answer Question 5 by filling in the blanks, using their own paper.

1. Define *biosphere*. Are humans part of the biosphere?
2. Explain the term *food chain*. Where are humans in the food chain?
3. Define *leaching*. Explain its effect on plant growth.
4. Define *biome*. How does climate affect a biome?
5. The ______ forests lose their leaves before winter, while the ______ forests remain green year-round.
6. Describe the process of plant succession.

Recalling and Reviewing

1. Explain how a climax community develops. How can it be destroyed?
2. Why are some plants able to survive in a desert biome? Why are polar regions called barren regions?

Critical Thinking

3. Describe the kind of plant community that grows in your state or region.
4. Contrast the biome in which you live with one you have visited or heard about.
5. If you were going to plant a vegetable garden, what would you need to know about your area to have a successful crop?

Using Geography Skills

Use the map in this chapter to answer the following questions.

1. Which biome makes up most of Western Europe?
2. Where are the middle-latitude forests located in North America?
3. If climate and vegetation influence the population of an area, which biomes would you think are the least populated?

Answers to Chapter Check questions are found in the Teacher's Guide.

Additional activities for Chapter 7 are found in the Student Workbook. Chapter 7 Test as well as Skills, Reteaching, Critical Thinking, and Challenge/Enrichment activities are available in the Teacher's Resource Binder.

Chapter 8

Chapter 8 Lesson Plans and Planning Guide are found in the Teacher's Guide.

World Cultures

This chapter explores the interaction between people and their environments.

▲ **Ancient Indians built dwellings on cliffs.**

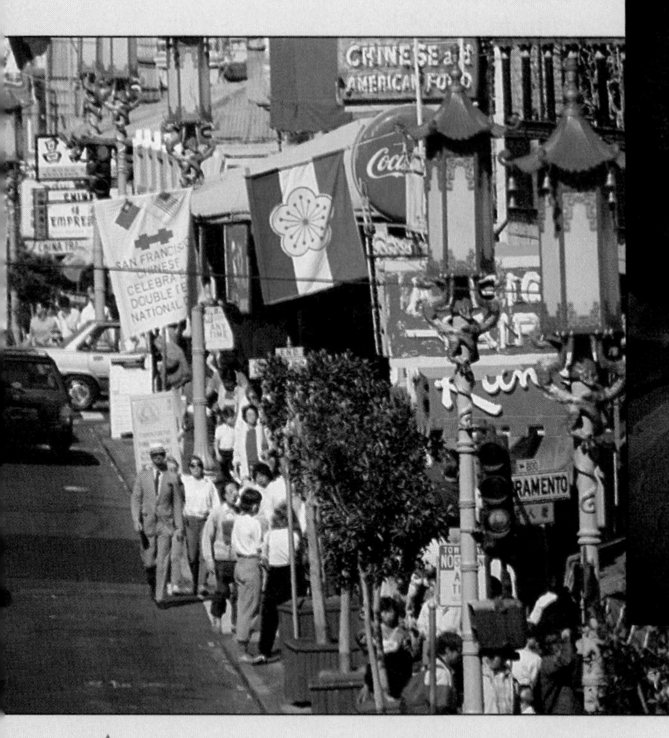

▲ **Chinese culture flourishes in San Francisco.**

GEOGRAPHY DICTIONARY

- cultural geography
- culture
- culture trait
- culture region
- innovation
- diffusion
- acculturation
- domestication
- subsistence agriculture
- urbanization
- industrialization
- nationalism
- totalitarian government
- democratic government

So far in this unit, you have learned about the forces that shape the earth's physical geography. Landforms, climates, water resources, and vegetation are important features of the earth's surface. However, the world around us also contains features obviously created by people. These include cities and villages, highways and sidewalks, and farm fields and grocery stores. People use the earth's surface in ways they believe best. For this reason, they can be viewed as another force that shapes the earth's surface.

In this chapter, you will study about people, about how people live, and about how their activities vary from place to place across the earth's surface. This study is called **cultural geography**.

 Have students compare the technology of the two societies shown in the photos above.

Encourage students to think of the many ways in which culture traits change through time. Ask about language changes (slang), technological changes (medical and scientific discoveries), and institutions (schools and the family).

Cultural Geography

What Is Culture? The word *culture* has many uses in the English language. To some people, it refers only to fine paintings, classical music, and the best manners. Scientists, however, define culture more generally. To them, **culture** is all the features of a society's way of life. It includes language, religion, government, economics, foods, clothing, architecture, and family life. Culture is all of a society's shared values and technologies.

Culture is a complex idea. One way to make culture easier to understand is to divide it into things that people normally do. Each of these things is called a **culture trait**. There are many kinds of culture traits. Explaining why they vary from place to place is seldom easy. For example, a typical American teenager eats dinner with a knife, fork, and spoon. However, a typical Chinese teenager eats with chopsticks, and Malaysian teenagers are comfortable eating with their fingers. Nonetheless, each trait is considered the best method in its own culture.

One culture trait is often closely related to another. For example, the Islamic religion forbids the eating of pork. Thus, Islamic farmers do not raise pigs. Since pigs were originally herded into forests for feeding, Islamic farmlands also have few woodlots. Culture traits are often linked to one another in this way.

Culture traits also change through time. Sometimes, these changes are as simple as a new style of dress. In other cases, they are more complex. Many American houses built before 1950 have porches or verandas at the front. But many houses built after 1950 have patios or decks at the back. To learn more about how culture traits change, talk to older people within your own culture to find out how their childhoods compare with yours.

Culture Regions An area or region where there are many shared culture traits is called a **culture region**. Culture regions usually develop slowly over centuries. In the United States, the South can be viewed as a culture region, even though its boundaries are hard to define. Political boundaries of countries often divide culture regions. France and Japan, for example, are culture regions. Yet, some countries contain several culture regions. This is particularly true in Africa. When Europeans ruled much of Africa, they combined more than one tribe, each with its own distinct culture, within the boundaries of one country. Some African countries now have people who represent more than a dozen culture regions. Culture regions can also contain collections of countries. An example is Scandinavia, which includes Sweden, Denmark, and Norway.

Culture Change Cultures develop slowly, but they are constantly changing as well. New culture traits are always being added while a few established traits are fading from memory. Two concepts are essential for understanding how these changes occur. The first is **innovation**. People are always thinking of new ways to do things. However, only those ways that are useful and valuable for that culture will last. New ideas that are accepted into a culture are called innovations. The second concept is **diffusion**. Diffusion occurs when an innovation or other culture trait spreads through a society and perhaps into another culture region.

Africans have traditionally used manual labor to process food. Here, villagers use a machine to grind manioc, an edible root.

Level A: Have students study an innovation and its diffusion. Topics might include television, a fad or fashion, or the spread of a religion. Students should describe how this innovation has changed their culture or influenced other cultures.

Culture change can be studied by looking for innovations and following their spread. Some innovations seem to have appeared only once in human history. For example, the wheeled cart first appeared about 5,000 years ago in southwestern Asia. From there, it diffused to Europe and eastern Asia. However, the wheel was unknown in America until European settlers came. Another innovation was noodles. Noodles were eaten in China, in India, and in Arab lands long ago. They were brought to Italy only 700 or 800 years ago, possibly by travelers to the East. Still other culture traits have been separate innovations in more than one region. Farming, for example, developed both in Asia and in America.

Culture traits can diffuse in two ways. First, an innovation might remain strong where it started and spread gradually to nearby areas. For example, television first became popular in the United States. It later spread throughout the world. New fashions, hairstyles, and jokes spread by expansion as well. Culture traits also diffuse when people take an innovation with them when they move to a new area. For example, Europeans and Africans brought many culture traits with them when they settled North America.

When one culture changes a great deal through its meeting with another culture, the process is called **acculturation**. The near disappearance of American Indian culture through European contact is an example of acculturation. So, too, is the adoption of blue jeans and fast foods by European teenagers. However, acculturation is a two-way street. Each culture changes the other. The European colonists learned the American Indians' way of agriculture. And Americans today enjoy wearing European designers' fashions.

People who cannot talk together cannot easily share ideas or culture traits. Language is, therefore, very important to culture change and to maintaining culture. The spread of American popular culture throughout the world is aided by the increased use of English. Today, English is spoken to some extent by one-fifth of the world's people.

People on the Land

Basic Resources All people have always had the same basic needs. Everyone has to find food and the materials to make clothing and shelter. Food, water, and the materials needed for daily life are basic resources. People spend a great deal of their time gathering resources to live. Since resources are not all found in one place, people have to journey to different places for them. Sometimes, they have other people bring resources to them. What people have thought to be necessary for daily life has changed as new innovations have been adopted.

Throughout history, three innovations have greatly affected the earth's cultural geography. These are farming, living in cities, and using machinery to make goods. As these innovations spread, places and regions changed dramatically.

Hunting and Gathering Before people knew how to grow plants and raise animals for food, they gathered plants and hunted animals in the wild. Scientists believe that from the time people first appeared on Earth until farming developed, everyone lived by hunting and gathering.

Hunter-gatherers often roamed wide areas, sometimes moving their camps with the

In some isolated regions, people still practice hunting and gathering. This hunter-gatherer lives in Africa's Kalahari Desert.

 Point out to students that the exchange of needed resources was the beginning of trade.

Mention that there are traditional agricultural groups in North America today, for example, the Amish people of southeastern Pennsylvania and other areas of the Midwest.

seasons. However, limited food resources kept populations low. Today, hunter-gatherers survive only in areas where farming has not succeeded because of poor land or climate. Among these are the Bushmen, or San, of the Kalahari Desert in Africa, and the Aborigines, the original inhabitants of the Australian deserts. Small groups of hunter-gatherers can also be found today in the humid tropical rain forests of South America, Africa, and southeastern Asia.

Even though hunter-gatherers have only simple tools and few, if any, modern products, all have complex social customs and religious beliefs. In order to survive, hunter-gatherers have to know their environments extremely well.

Agriculture Raising animals and planting crops were perhaps the most significant innovations in human history. Agriculture appeared when hunter-gatherers studied a plant or animal so closely that they found out how to grow or tame it. This innovation is called **domestication**. That is, the plants and animals were grown or raised so as to be of use to humans.

Scientists believe agriculture first appeared between 10,000 and 12,000 years ago in several areas of the Middle East and southeastern Asia. People could acquire much more food through agriculture than by hunting and gathering. They began to change the face of the earth rapidly. Forests were cleared. Fields and pastures appeared. Food resources were now grown in fields and pastures, and people no longer had to roam for food. They could live together more closely in large villages and share ideas easily. More food meant larger populations. In only a few thousand years, the farm fields and herded animals spread steadily across whole continents.

Many people still make a living by agriculture. Today, there are two types of agriculture. In traditional or **subsistence agriculture**, people grow food largely for their own use. Some surplus, or extra, crops may be traded to neighbors for other products. Some may be sold for cash. Most of the products, however, go to feed the farmers' own families. In subsistence agriculture, tools are simple, and animals are the main source of power. Most of the world's farmers practice subsistence agriculture.

In traditional agriculture, people grow food mainly for their own use. This French family is harvesting potatoes, a staple food crop.

In commercial agriculture, machinery is used a great deal. With modern fertilizers and powerful tractors, more crops are produced than a single family can use. The surplus is sold for cash. Large farms with few workers are typical in commercial agriculture.

Urbanization While hunter-gatherers usually lived in small camps, once agriculture developed, people lived either on farms or in villages surrounded by fields. Since so much more food was produced, people could now spend time doing other things. Some people made tools and clothing. Others became specialists in government and religion. These people could live together in larger settlements such as towns and cities. Growth in the proportion of people living in towns and cities is called **urbanization**. Cities first appeared more than 5,000 years ago.

Since people in cities now lived much closer to each other, communication was easier. New ideas could be tested quickly and

Japan became a powerful nation by developing an industrial base. What are the characteristics of industrialized societies?
improved standards of living; people working for companies; pollution; depletion of natural resources

adopted if they proved useful. Culture diffusion, therefore, was more rapid than before. Cities became the centers from which new techniques spread into the surrounding countryside and to neighboring cities.

Today, cities still play a major role in the creation of most innovations. They also are important in the spreading of new culture traits. Cities are centers of manufacturing, government services, education, and the arts. Innovations diffuse quickly in and around cities because cities are linked by communications facilities such as radio, television, and computers and by transportation services.

Industrialization In the early cities, tools, clothing, and other goods were made by hand. The workers started and finished each product individually. Goods could only be made slowly, and they were costly. However, in eighteenth-century Western Europe, factories began to appear in large numbers. This innovation in the organization of workers and machinery spread quickly because goods could be manufactured in large amounts at low cost. The growth of this method of production is called **industrialization**.

In factories, individual workers use machinery to make only one part of a finished product. The job of other workers is to put each of the parts together. Then, a final set of workers packages the finished product for sale. Most workers have little connection with the final manufactured product. The wages the workers earn are spent on goods and services produced by workers in other industries.

Industrialization has led to many culture changes. Many people work for companies, rather than on their own. Industrialization has improved the standard of living for many. With more goods and services, people can live in comfort. However, industrialization in some areas of the world has brought with it pollution and a depletion of natural resources.

Culture and World Events

Cultural Conflict In many cases, disagreements between people and between countries can be related to their having different culture traits. People usually believe that their own culture is best. However, if you study cultures around the world, you will find many ways to live. All seem to work. Here we will look at several causes of cultural conflict. You will find that such conflicts are often topics in world current events.

Religion Religion is a key culture trait and in many societies, the basis for many other culture traits. Generally, people who practice a particular religion receive a sense of personal fulfillment. It is a positive and life-long influence in many people's lives. Religions advise on daily behavior, diet, sacred times, and sacred places. To a large extent, the people of a religion believe theirs is the one true religion. When people do not tolerate other religions, disputes can lead to conflicts and even war. Indeed, wars over religion have been common throughout history. Because both sides believe their causes are holy, religious conflicts can be particularly bitter and long lasting.

 Level B: To reinforce the idea of culture change, have students list changes in American culture today (use of computers, space travel, and so on).

Level A: Assign a student to prepare a chart showing the number of followers of each of the major world religions. Have students develop generalizations about the distribution of each religion.

World Religions

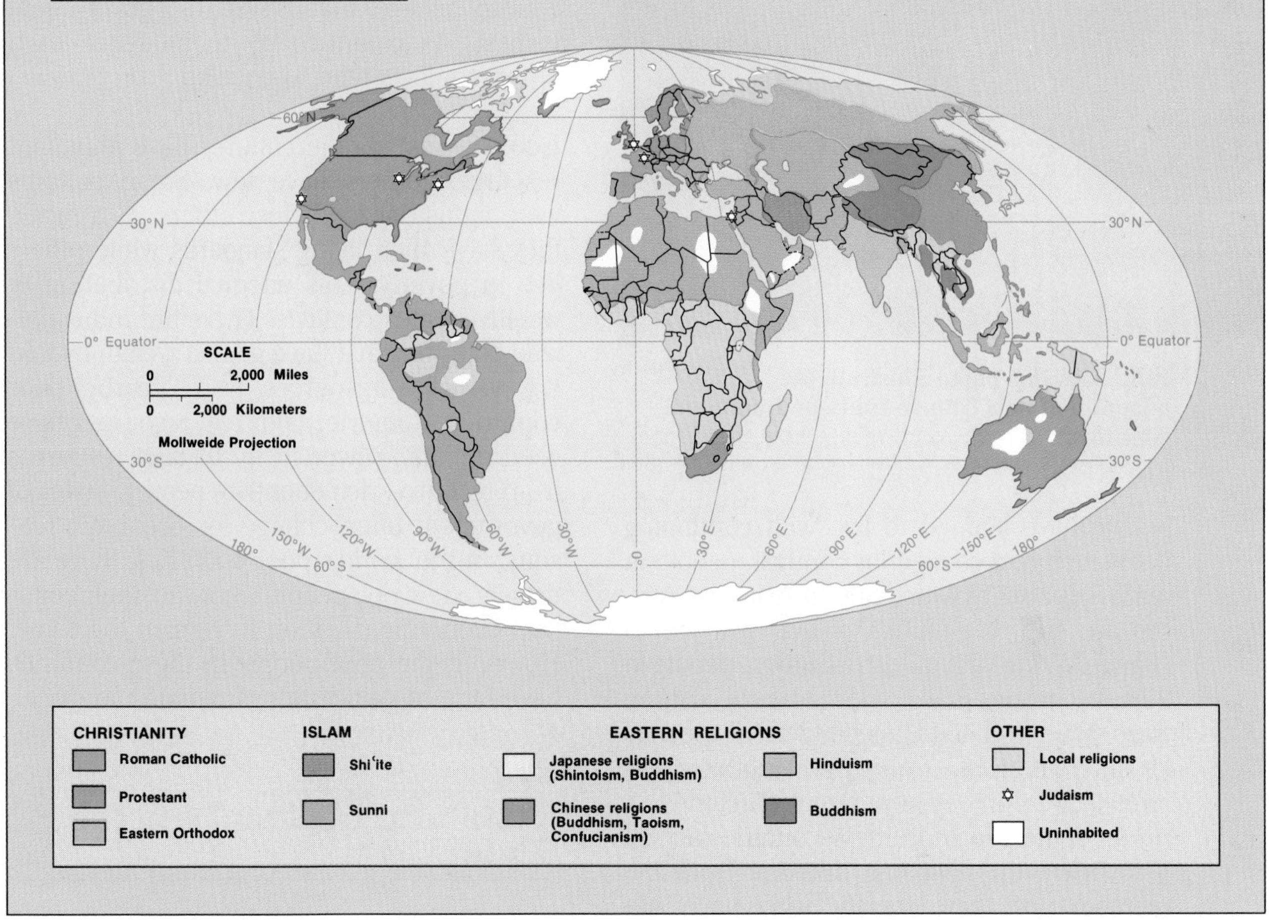

60° N
30° N
0° Equator
30° S
60° S
SCALE
0 2,000 Miles
0 2,000 Kilometers
Mollweide Projection
180°
150° W
120° W
90° W
60° W
30° W
0°
30° E
60° E
90° E
120° E
150° E
180°
CHRISTIANITY
Roman Catholic
Protestant
Eastern Orthodox
ISLAM
Shi'ite
Sunni
EASTERN RELIGIONS
Japanese religions (Shintoism, Buddhism)
Chinese religions (Buddhism, Taoism, Confucianism)
Hinduism
Buddhism
OTHER
Local religions
Judaism
Uninhabited

What is the dominant religion of the Western Hemisphere? Christianity (Roman Catholic and Protestant)

There are many modern examples of religious conflict. In Sri Lanka, Buddhists (BOO-duhsts) have battled the Hindu Tamils. In Northern Ireland, Roman Catholics and Protestants continue to fight one another. In the Middle East, religious conflicts between Jews, Muslims, and Christians have threatened major wars again and again.

Nationalism Most people feel proud of their culture and country. Feelings of loyalty toward and pride in one's country are called **nationalism**. Unfortunately, one group's pride often conflicts with the pride of another group. Some groups think they are better than others. Discrimination, or unfairness, toward different groups and even warfare can result. For example, nationalistic concerns led to civil war in Nigeria during the late 1960s. At that time, the Ibo tribe of eastern Nigeria tried to form its own country. Other Nigerians objected, and the country was torn by war. Many Ibos living in other parts of Nigeria were killed. Nigeria is now united, but people still remember the hatreds of the war.

Nationalism sometimes hides economic greed. The Ibos of eastern Nigeria knew that their part of the country had one of the richest oil deposits in the world. For this reason, they might have believed that forming a separate country would benefit them.

Traditional and Modern Values Some cultures change easily. In the United States,

Level A: Ask students to choose a conflict currently in the news, research it, and write or orally present a brief report on it.

How does this photo illustrate the adoption of new culture traits by traditional societies? These Indian women from a traditional society are adopting a Western culture trait—the radio.

people have learned to live with continuing innovation and change. In contrast, people in many countries, particularly in Asia and Africa, are just beginning to live with rapid change. Along with industrialization, people in these countries have adopted new culture traits, often from the United States. These may be different styles of music and clothing, different foods, and even different religions.

As some people in these countries accept innovation, they become different from the members of their society who have not changed. At times, the modern, or changing, people conflict with traditional, or nonchanging, people. These differences might occur between parents and children or between people who live in the cities and those who live in the countryside. Conflict might arise, too, between groups who have different views of what their country should be like. The revolution in Iran that began in 1980 was mainly a rejection of modern Western cultural values.

Politics Each culture has ideas about how its government should be organized. Some societies are governed by one person and a few advisors. These are called **totalitarian governments**. Others believe that everyone in the society should have a voice in government. These are **democratic governments**. Unfortunately, some countries want to take over other countries. This may be for economic gain or because the dominating country believes that its system of government is best. As countries try to influence each other, conflicts tend to develop.

Economics Some cultures have abundant resources; others have few. Some cultures were industrialized before others. As a result, they have high living standards while others live in poverty. This unequal distribution of wealth creates conflicts. There are industrialized countries that have gained wealth by taking valuable resources from nearby poor countries. Countries with few resources have used military power to try to take resources from resource-rich countries nearby. Styles of government often change for economic reasons. In the world's poor countries, there are usually a few rich people who want their country's economic situation to remain the same. However, the poor people in these countries have little to gain if things remain the same.

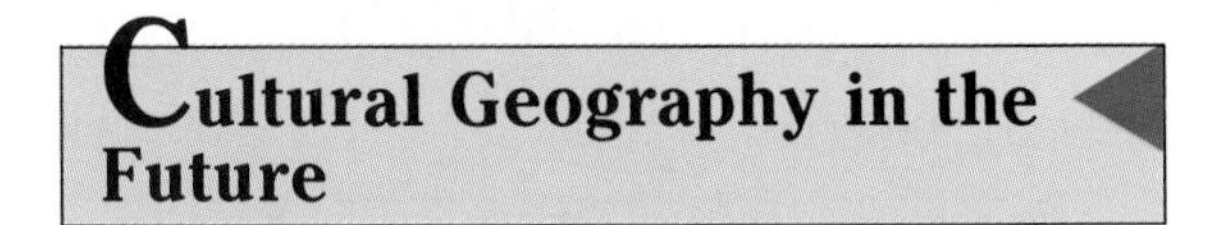

Cultural Geography in the Future

Today, culture traits spread rapidly across the world as communication and transportation become easier and easier. Students in Saudi Arabia copy the fashions they see on American television programs. Houses in Paris suburbs take on traits of houses in Houston. For better or worse, places and regions have more and more in common. The cultural geography of different areas of the world is becoming more similar.

Many geographic problems remain to be solved. People and resources are distributed unevenly over the earth's surface. The distributions of rich people and poor people and of farmers and factory workers are unequal. Understanding how different culture traits have led to conflict in the past will help people deal with conflicts in the future. Through a better understanding of our culture and other cultures, we can hope for a better future.

Discuss with students the dangers of ethnocentrism (the belief that one's own culture is best). Prejudice, intolerance, and resistance to new ideas are all outcomes of this belief. Have students give examples of ethnocentrism.

CHAPTER 8 CHECK

Reviewing the Main Ideas

1. Cultural geography is the study of how people live and how their activities vary across the earth's surface. Geographers study culture traits, culture regions, and culture change.
2. All people have the same basic needs for food, shelter, and clothing. The innovations, through history, of agriculture, urbanization, and industrialization have dramatically affected how our basic needs are met.
3. Cultural conflict is often related to a conflict between culture traits. Some of these are religion, nationalism, traditional versus modern values, and politics.

Building a Vocabulary

Students should answer Question 5 by filling in the blank, using their own paper.

1. Explain the difference between acculturation and diffusion.
2. What is a culture region? Give some examples.
3. Define *culture trait*. Give several examples. What culture traits do you see every day?
4. What word describes the major movement of people to cities?
5. The organization of workers and machinery to produce goods at a low cost is called ________.
6. Why is it important to study cultural geography?

Recalling and Reviewing

1. What basic needs do all people have? Have people always met these needs in the same way? Why or why not?
2. Explain the difference between subsistence and commercial agriculture.
3. Why do innovation and culture diffusion occur more rapidly in cities?
4. How do you gather the basic resources necessary for your own survival? Does your culture influence you?

Critical Thinking

5. Look through newspapers and magazines for examples of cultural conflict. Indicate the type of conflict and the basic causes.
6. Think of three innovations in your culture. Which ones have been retained? adapted? adopted by another culture? Which culture traits have Americans adopted from other cultures?

Using Geography Skills

Use the Atlas at the front of the book to answer the following question.

1. Look at the world political map. Imagine that you live in the uninhabited central part of Australia. Where would you need to go to be exposed to innovations and culture traits from other societies? Why?

Answers to Chapter Check questions are found in the Teacher's Guide.

Additional activities for Chapter 8 are found in the Student Workbook. Chapter 8 Test as well as Skills, Reteaching, Critical Thinking, and Challenge/Enrichment activities are available in the Teacher's Resource Binder.

Chapter 9

Chapter 9 Lesson Plans and Planning Guide are found in the Teacher's Guide.

The concepts developed in this chapter are applied throughout the book in the study of the world's regions.

Economic Development

▲ **Fishing is a main economic activity in Sri Lanka.**

▲ **Saudi Arabia benefits from its oil resources.**

GEOGRAPHY DICTIONARY

- economic geography
- primary economic activities
- secondary economic activities
- tertiary economic activities
- developed countries
- less developed countries (LDCs)
- illiteracy
- gross national product (GNP)
- capitalism
- free enterprise
- command economy
- communism
- population geography
- birthrate
- death rate
- multinational company

Economic progress varies from place to place and from country to country. Through time, the level of economic development within a country changes. Two hundred years ago, for example, the United States was a poor country, a nation of farmers who grew little more food than their families needed. Today, it is the richest country in the world. Less than 3 percent of Americans are farmers. American businesses trade throughout the world. However, even in modern America, some regions are wealthier than others. For example, decline in manufacturing in the north-central United States has caused some jobs to disappear. Although new jobs are appearing, some workers have to retrain for them. And because some of the new jobs are in different cities, workers may have to relocate. Situations similar to this exist in other countries throughout the world.

 Have students compare the level of economic activity shown in the two photos above.

Economic Geography

The branch of geography concerned with how people use the earth's resources, how they earn their living, and how products are distributed is called **economic geography**. The economic activities of people and businesses can be divided into three categories—primary, secondary, and tertiary.

Primary economic activities include agriculture, forestry, and mining. These activities make direct use of natural resources. Therefore, locations of primary economic activities are generally controlled by physical geography. Agriculture and forestry are found in regions with suitable soils and climates. Farm and forest products are often shipped long distances to factories and consumers in cities. Mines are located where rich mineral deposits are found.

Secondary economic activities are food processing and manufacturing. In manufacturing, raw materials are converted into new products. The particular locations of manufacturing industries are usually determined by three general considerations. First, some industries, such as dairies and bakeries, are located near markets so they can better serve their customers. Also, products such as bread and milk may spoil if they must be transported too far.

Second, some industries must be located where special manufacturing needs can easily be met. These may be needs for raw materials, skilled workers, or transportation facilities. The growth of the steel industry in Pittsburgh, Pennsylvania, is a good example. Iron ore, coal, and limestone are needed to make steel. Pittsburgh is located near coal fields and limestone deposits. The site of Pittsburgh, where three great rivers join, permits easy transport of iron ore from distant mines and of finished steel to distant markets.

Finally, industries are sometimes located almost by chance. Detroit became the "Motor City" mainly because Henry Ford, who grew up in nearby Dearborn, chose to build his first automobile factory there. Detroit also has the advantages of easy water transport on the Great Lakes and a central location in the north-central United States. Similarly, governments sometimes decide where industries should be located. For example, southern Italy has few industries, and many people are without jobs. The Italian government has encouraged industrial growth there by providing low taxes and generous loans.

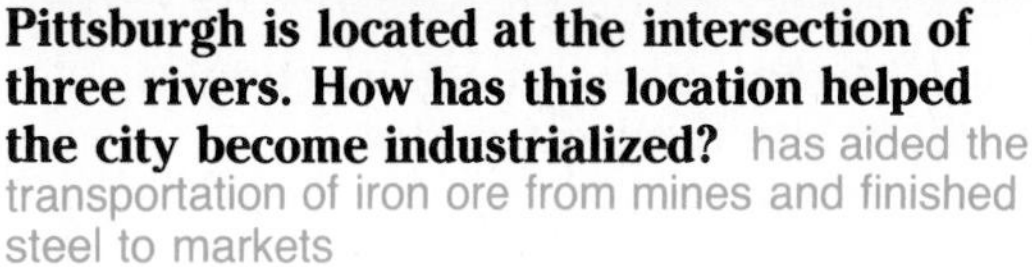

Pittsburgh is located at the intersection of three rivers. How has this location helped the city become industrialized? has aided the transportation of iron ore from mines and finished steel to markets

Tertiary economic activities are service industries. Service industries include stores, restaurants, banking and insurance, government activities, education, and transportation. Services are usually located close to where people live. Think of the location of your school and the stores where you shop. They are more than likely located near where you live. Some service industries are changing rapidly with increased use of electronic communication and computers. For example, home computers permit many office workers to do their work at home.

Level A: Have students prepare a report discussing how a site is selected for the location of a factory or store.

Examples of Economic Activities

Primary

Farming	Lumbering	Tree farming
Cattle raising	Mining	Fishing

Secondary

Processing of dairy products	Paper milling
Production of baked goods	Meat processing
Steel manufacturing	Bottling
Textile manufacturing	Canning
	Oil refining
	Automobile manufacturing

Tertiary

Selling of manufactured goods (stores)	Insurance
Food services (restaurants)	Government services
Banking	Public transportation
	Medicine
	Communications

Can you identify other economic activities that will fit into these categories?

Answers will vary.

Wealth and Poverty

Economic Development The countries of the world can be divided into two groups, developed and less developed. This division is based primarily on the countries' differing economic conditions. The world's wealthy countries are called **developed countries**. The economies of developed countries are generally based on secondary and tertiary activities. Thus, most people in developed countries work in manufacturing and service industries. These countries have abundant products and services. Generally, people in developed countries enjoy high standards of living, with abundant food, good health care, and educational opportunities. Most live in cities. These countries have a small percentage of people employed in primary economic activities, such as agriculture and mining. The industries in this category that do exist generally use modern technology.

The world's poorer countries are called **less developed countries (LDCs)**. Their economies are based mainly on primary economic activities. Unlike wealthier countries, the less developed countries have few factories and many poor people, the majority of whom practice subsistence agriculture. That is, they produce only what they need in order to live. Any manufacturing in these countries is limited to products for the local market. These include clothing, processed food, and furniture. Cars, radios, and motorcycles may be assembled with parts made in other countries. Less developed countries usually sell their natural resources to the developed countries. They then use the money they receive to buy manufactured goods from wealthier nations. Most of the world's people live in less developed countries.

The economic problems facing the less developed countries vary. Some of the countries are rich in natural resources that can be sold throughout the world. For example, people in oil-rich Saudi Arabia have incomes as high as many people in developed countries. Other less developed countries, such as Brazil, Mexico, and India, have large, complex societies, with increasing industrialization and improving educational systems. With further economic progress, some of these countries may join the developed countries.

Unfortunately, many of the less developed countries have few natural resources, rapidly increasing populations, and widespread **illiteracy**. Illiteracy is the inability to read and write. The poorest of these countries, such as Ethiopia, Afghanistan, and Laos, have little hope of developing. The governments of less developed countries tend to be unstable. This is partly the result of the countries' economic problems.

One economic measure of a country is its **gross national product (GNP)**. Gross national product is the total value of goods and services produced by a country in a year. (See the charts in the back of this book.) Per capita GNP, the gross national product divided by the number of people, is another useful measure to compare countries. A developed country usually has a per capita GNP of over $5,000. In 1985, the per capita GNP of the United States was about $16,400. In France, it was $8,450. In contrast, the per capita GNP of Mexico was $2,080, and in Nicaragua, it was only $850. Most of the poor countries in Africa have per capita GNPs of less than $300!

First, Second, and Third Worlds Countries are sometimes grouped into three broad categories based upon their economic and political systems. These categories are First World, Second World, and Third World. The First World includes the world's developed countries with democratic governments. The United States, Canada, the countries of Western Europe, Japan, Israel, Australia, and New Zealand are First World countries. The economies of many of these countries are based upon the ideas of **capitalism**. That is, resources, industries, and businesses are owned by private individuals. Some First World countries' economies are a mixture of government and private ownership.

Some capitalist countries, such as the United States, have economies based on **free enterprise**. Free enterprise gives people freedom to operate private businesses for a profit. In free enterprise, the supply and demand of products and labor determine prices and wages. Industries and merchants are free to choose the products they make and sell. Governments in these countries seek to pass laws to keep economic competition fair and encourage economic growth.

The Second World includes the Soviet Union and its allies. These countries have **command economies**. In a command economy, the prices of goods and the wages of labor are set by the government. Many Second World countries have Communist governments. **Communism** is a system in which all economic activity is controlled by a government that is ruled by a single political party. The government directs many aspects of the people's lives. Compared to those in First World countries, people in these countries have little economic freedom. In general, per capita incomes are lower in these countries than in the First World. While the Second World includes relatively developed countries, such as the Soviet Union and those in Eastern Europe, it also contains less developed countries, such as the People's Republic of China, Vietnam, and Cuba. Although Second World countries do not have the abundant consumer goods of the First World, most have good education systems and good health care.

The Third World comprises the world's less developed countries. These include most countries in Central America, South America, Africa, the Middle East, and southern and southeastern Asia. Some of these countries have much freedom. In many others, the government tries to control the people's lives. These countries are poor and generally have little influence in world affairs. To improve their economies, they need economic aid from the developed countries.

Population Growth and Economic Development

Population Explosion A major problem facing Third World countries is explosive population growth. The study of population is basic in human geography. As populations increase, so does the use of natural resources and the destruction of natural environments. **Population geography** is the branch of geography that studies how populations are distributed and how populations change in particular places and regions.

We live in a period of dramatic population growth. For much of human history, world population was low and increased slowly. The number of deaths nearly equaled the number of births. Two thousand years ago, the world had about 250 million people. Slow growth led to a doubling of world population by about A.D. 1650. Then, with better health standards and living conditions, people began to live longer. More infants survived to adulthood.

Level B: Have students gather photos from newspapers and magazines showing economic activities in developed and less developed countries.

GEOGRAPHY SKILLS

How to Use a Population Pyramid

Answers to Geography Skills questions are found in the Teacher's Guide.

Population pyramids help us understand economic, cultural, and historical patterns of a place or region. The pyramids below compare three countries' population distribution by age group and by sex.

The pyramid is a special kind of graph. Each bar represents a five-year age span. The bar on the bottom of the pyramid signifies the youngest age group. The bar at the top describes the oldest age group. The left side of the pyramid shows the percentage of males in a particular age category. The right side shows the percentage of females. The age groups are listed in the column next to the graph. The bottom, horizontal axis of the graph indicates the percentage of the total population.

Population pyramids for less developed nations such as Brazil have a much broader base, indicating a higher birthrate and death rate than in developed countries. Look at the pyramid of Brazil. You can see that 14 percent of the population is under five years of age. Twelve to 14 percent of the population is between the ages of 5 and 15. You can tell that the death rates are high because the population over 65 is small.

Now look at the pyramids of the United States and the Soviet Union. Note that the percentage of the total population under age five is similar for the two countries. Moving farther up the United States pyramid, note that the population increases in the 20–40 age range. This is the result of the post–World War II baby boom.

TEST YOUR SKILLS

1. What percentage of Soviet women are 35–39? 40–44? What are the percentages of American women in the same age brackets?
2. What is the longest bar for Brazil? the shortest? Compare the longest and shortest bars for the United States and the Soviet Union. What do your results tell you about population patterns in developed and less developed countries?
3. Find the bar for your age group on the United States pyramid. To which percentage of the general population do you belong?

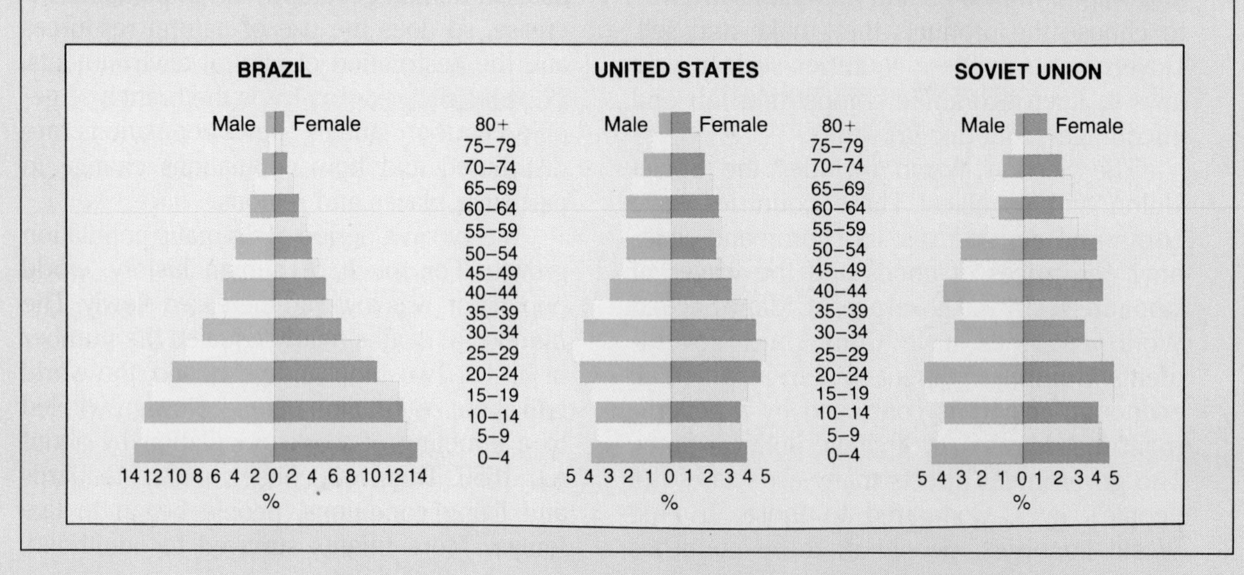

Note on the pyramid for Brazil that each age group below 79 is larger than the preceding group, indicating a continually expanding population.

The population of India is almost 800 million. Many people raise their families on the crowded sidewalks of India's large cities.

As a result, world population increased dramatically.

By 1850, the world's population passed one billion, doubling in 200 years. In 1930, world population reached two billion, doubling in 80 years. The four billion level was passed in 1971, doubling the world's population in only 41 years. In 1987, world population exceeded five billion. In the minute or so it took you to read this paragraph, the world's population increased by 162 people. Each time your heart beats, the world has 2.7 more people to feed, clothe, and educate.

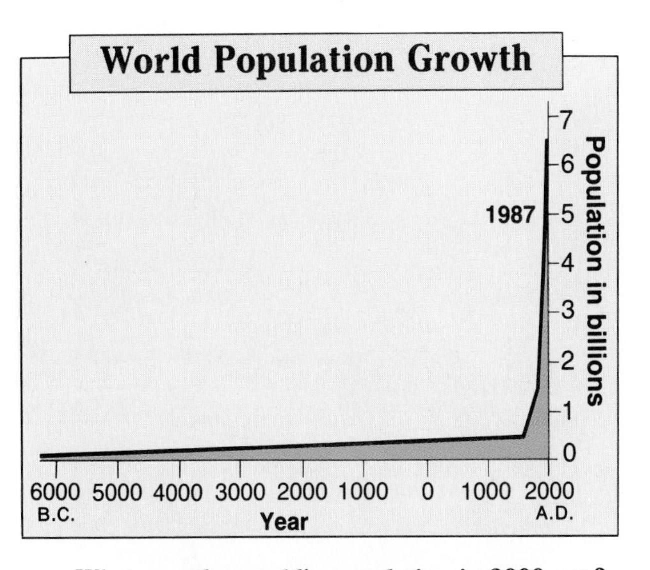

What was the world's population in 2000 B.C.? in 1987?

about 250 million; more than five billion

Population geographers explain that the world's wealthy countries have all experienced three stages of population growth as their economies developed. In a poor country, both **birthrates** and **death rates** are high. Birthrate is the number of births per 1,000 people in a given year. Death rate is the number of deaths per 1,000 people. During the first stage of growth, populations change little. Most of the country's poor people live in rural areas. As the country begins to develop economically, health care improves and people live longer. However, there is still a high birthrate. The result at this second stage is a rapidly increasing population. At the same time, industries grow and many people move to cities and towns, seeking jobs. In the final stage, population growth slows to a low level, and the country progresses economically.

Today, however, most of the world's less developed countries are only at the second stage of this sequence. That is, they are experiencing rapid population growth. This makes further economic development very difficult. The question remains whether many of today's less developed countries will complete this sequence to reach low population growth and achieve a better economic future.

Emphasize to students that the world's population is unevenly distributed. A population-density map can be found in each regional unit. See pages 112–113 for an example.

URBAN LAND USE

Look at a world or regional map, and you will see many cities and towns with large areas of countryside between them. The cities and towns are densely settled areas, called urban areas. The areas with a lower population density are called rural areas. People who live in rural areas are usually employed in primary activities such as farming or fishing. They may live miles away from their neighbors and the nearest store. People who live in urban areas are usually employed in secondary and tertiary activities such as manufacturing and banking. They live in houses or apartments that are close together.

Because land is used differently in the urban areas than in rural areas, the two areas differ in appearance. If you look at a map of a town or city, you will notice that the entire land area is developed. The streets are close together, and the areas between them are occupied by many types of buildings or other forms of urban land use, such as factories and public parks.

Throughout the world, urban areas share a common pattern. In most countries, different types of land use are arranged in the same way. The appearance of buildings in urban areas may vary greatly from one country to another. However, buildings with the same function tend to be found in the same areas within all towns and cities.

There are three categories of land use in urban areas. The first is the central business district, or CBD. This is the area people usually call "downtown." It contains banks, office buildings, government offices, hotels, restaurants, and many kinds of stores and shops.

▼ Chicago's CBD grew up around the city's lakeshore and major freeway intersections.

 Basic concepts of urban geography are discussed in this feature. These concepts are further developed in the "Cities of the World" features found throughout the text. See pages 280–281 for an example.

▲ **Suburban areas are located near cities.**

Because these businesses need to be reached easily by people from all over the urban area as well as from other cities and towns, the CBD is usually located at the intersection of the town's main roads. The CBD may start out as a small area, but as the city expands, the business activities spread.

The second major urban land-use area is the industrial area. This area, which is next to the CBD, contains industries and manufacturing and is near major transportation routes such as highways, railway lines, or a harbor. This location is important because the goods that are produced in the industrial area must be transported to and from other major areas.

The third urban land-use area, which covers the most land, is the residential area. There are several types of residential areas, and they surround the CBD and industrial areas. Usually, older and less expensive housing is located close to the industrial area. The cost of land may be lower here because of the noise and pollution from the factories. In most cases, people who work in the industrial area can use public transportation or walk to work. The more expensive residential areas are located farther from the CBD. They are often found in the scenic parts of the city, such as along rivers. Residential areas that lie on the edges of the urban area are called suburban areas. As the city expands, the area of new houses moves outward. The houses that were originally on the inner edge of this zone grow old and deteriorate. Eventually, they become part of the zone of less expensive, older housing.

Larger cities in less developed countries have another area that is not typical of developed countries. This is the area of slum housing that appears on the outskirts of the urban area. People who live here usually have come to the city in search of employment or better living conditions than they experienced in rural areas. Because they are unable to find or to afford housing in the city, they settle on open land outside the city in large shantytowns.

Try to identify the central business district, the industrial area, and the residential areas of your town or city or of the city nearest your home.

▼ **Older houses are found near the CBD.**

Ask students to categorize land-use areas in their town or city.

Choices in Economic Development

How can a poor country improve its economy? If the poor country has abundant natural resources, then development of its oil fields, mineral deposits, or valuable crops can lead to economic progress. Oil-rich countries with low populations, such as Saudi Arabia, have hired companies from Europe and North America to build whole new cities. Such countries are the fortunate few.

For a less developed country with limited natural resources, the choice is industrial development and training new workers. Yet, these answers are too simple. Poor countries usually do not have enough money or trained people to start factories or businesses.

Sometimes the governments of developed countries help less developed countries with gifts of money, equipment, and teachers. Perhaps a **multinational company** will build a factory in a poor country. Multinational companies are worldwide businesses whose activities are based in many countries. Many multinational companies have larger sales than the combined gross national products of several less developed countries. These companies are important in transferring technology to the less developed countries.

Multinational companies have invested in many less developed nations. This company is establishing its presence in Egypt.

Nigeria's low literacy rate, low per capita income and GNP, and large population make economic development difficult.

The developed countries often debate how to help the less developed countries. Some wealthy countries provide little help. They argue that there are too few resources to share and that helping poor countries will only reduce their own economic growth. Other developed countries give aid only to countries with natural resources or low population growth. Still other countries offer aid only to their political allies.

The solution to world economic progress is not to leave the peoples of some regions to lives of poverty and hardship. Instead, the people of the wealthy countries must share the world's resources and seek ways to increase economic development everywhere.

In the following chapters, you will study the geography of regions and countries all around the world. Understanding the opportunities for economic development of these regions will help you better understand the possibilities for people everywhere.

 Multinational companies account for about 15 percent of the world's goods and services. Have the class identify a multinational company to study. Ask them to find out what products or services it provides and to develop a map showing the company's locations.

CHAPTER 9 CHECK

Reviewing the Main Ideas

1. The geography and economic progress of a place or region are often closely linked. Economic geography explores how people use the earth's resources, how they earn their living, and how products are distributed.
2. Population geography is the study of how populations grow and how population is distributed.
3. The countries of the world can be grouped according to their economic conditions. These groups are developed and less developed countries. Most of the world's people live in the world's less developed countries.
4. We live in a period of dramatic population growth. In 1987, world population surpassed five billion.

Building a Vocabulary

Students should answer Question 3 by filling in the blank, using their own paper.

1. Discuss the differences among primary, secondary, and tertiary economic activities. Give an example of each.
2. Define *capitalism*. Name two countries whose economies are based upon capitalism.
3. The total value of goods and services produced by a country in a year is its ______.
4. Discuss the characteristics that distinguish developed countries from less developed countries.
5. Explain the difference between a command economy and an economic system based upon free enterprise.
6. Define *birthrate* and *death rate*. Why are these terms important to the study of population geography?

Recalling and Reviewing

1. Explain why the study of economic geography is important.
2. Discuss why rapid population growth is a serious problem for less developed countries.

Critical Thinking

3. Discuss the physical geography of the area or region in which you live and its impact upon the region's economic development. Name some of the area's primary, secondary, and tertiary economic activities.

Using Geography Skills

Use the graph in this chapter to answer the following questions.

1. What is the expected world population for the year 1990? the year 2000?
2. After what year(s) on the graph does there appear to be a dramatic rise in population growth? Why do you think this is so?

Answers to Chapter Check questions are found in the Teacher's Guide.

Additional activities for Chapter 9 are found in the Student Workbook. Chapter 9 Test as well as Skills, Reteaching, Critical Thinking, and Challenge/Enrichment activities are available in the Teacher's Resource Binder.

UNIT 1 REVIEW

Summary

When we study geography, we are exploring the world around us. Geography is a study of the earth as well as the people who inhabit it. The earth, together with our solar system, is like a speck of sand on a beach in comparison to the universe. Within the immediate environment of that speck of sand, however, much activity is taking place. The sun's energy affects the environment of the earth. It makes possible the variety of life-forms found on our planet.

Much activity is also taking place within the earth and on its surface. The face of the earth is constantly being eroded by surface forces such as water, wind, and ice. Forces within the earth, such as volcanoes, earthquakes, and plate movement, have also changed the earth's surface. All this has resulted in the variety of landforms found on our planet. These different landforms affect the way people live in various places.

The sun's energy, global pressure and wind systems, landforms, and oceans bring about a variety of climates. Climate and landforms affect how and where people live. They also affect the biosphere. The biosphere contains all plant and animal life on Earth. The type of climate found in a particular area determines the biome in that area.

The interrelationship between culture and physical geography is important. The culture of a particular group of people is influenced by the physical environment. A people's culture also affects how they use the resources of their environment to satisfy their needs. Economic development varies from region to region throughout the world. Most of the world's people live in less developed countries, where there are low standards of living. We live in a period of dramatic population growth, and many of the world's poor countries face serious population problems. The sharing of the earth's resources among all its people is a major challenge for all nations.

Reading and Understanding

1. Name the five basic themes of geography. Which theme best describes the process involved in manufacturing this book?
2. Discuss what would happen if the earth did not rotate on its axis.
3. The position of the earth in relationship to the sun influences climate. How does this relationship explain why the humid tropical climate is so much hotter than the tundra climate?
4. Explain the differences between continental and maritime climate conditions.
5. What factors can determine whether a region is wet or dry?
6. Explain the benefits human beings derive from the hydrologic cycle. Why is the earth called the "water planet"?
7. Identify the basic biomes of the world. What factor accounts for the differences among the biomes?
8. How did the development of agriculture change the way humans lived?
9. In what ways can innovation cause cultural conflicts?
10. Discuss ways in which multinational companies affect the economies of less developed nations.

 Answers to Unit Review questions are found in the Teacher's Guide.

Mastering Geography Skills

1. Look at the climographs on page 68. Given the climates of the three areas, what would you expect population and vegetation patterns to be like? Why?
2. Look at the elevation map and the cross-sectional diagram on pages 48 and 49. From what you learned in Chapter 3, how do you think the Andes Mountains were formed?
3. Refer to the population pyramids on page 98. What predictions can you make about population growth in developed and less developed countries for the next 20 years?

Applying and Extending

There are different climate regions and varying elevations throughout the world. Use an almanac to answer the questions that follow. What was the highest temperature ever recorded? What was the lowest temperature ever recorded? What were the highest and lowest temperatures ever recorded in the United States? What are the dates and places of these records? Use an atlas to find the elevations of these places in the United States. Does the elevation of an area affect the climate in any way? Why or why not?

Linking Geography and Economics

The climate, landforms, and resources of a particular area often affect the amount of industrialization and food production it will have. Developed countries of the world, such as the United States, Japan, and the countries of Western Europe, have been fortunate in being able to produce enough industrial and food products. In contrast, less developed countries in Africa, Asia, and Latin America (such countries as Chad, India, and Nicaragua) have not been as fortunate. How have climate, landforms, and resources hindered these countries in becoming economically developed? (Use atlases, almanacs, newspapers, and magazines to help you answer this question.)

This section of the Unit Review varies throughout the text. It includes links between geography and such topics as history, religion, and economics.

Reading for Enrichment

Hendrickson, Robert. *The Ocean Almanac.* Garden City, NY: Doubleday & Co., 1984.

Miller, Russell. *Continents in Collision.* Alexandria, VA: Time-Life Books, 1983.

Sagan, Carl. *Cosmos.* New York: Random House, 1980.

Switzer, Ellen. *Our Urban Planet.* New York: Atheneum, 1980.

Weiner, Jonathan. *Planet Earth.* New York: Bantam, 1986.

Wilford, John Noble. *The Mapmakers.* New York: Alfred A. Knopf, 1981.

The Unit 1 Test is available in the Teacher's Resource Binder.

The Alps, the highest mountains in Europe, are famous for their beauty and are a popular tourist destination. More than 40 mountain passes make land travel possible from one country to another.

Graceful European architecture forms a backdrop for a flower market in Dinan, France. The market is a popular meeting place in France and other Western European countries.

Thousands of years ago, glaciers dug deep into Norway's river valleys. Seawater filled the valleys when the ice retreated, creating fiords. These boats float in a quiet Norwegian harbor near a fiord.

Have students identify the aspects of physical, economic, and cultural geography found in these photos.

UNIT TWO

WESTERN EUROPE

OBJECTIVES

- ▷ *To explain the historical development of the nations of Western Europe*
- ▷ *To explain the important role land, climate, and resources have played in the economic development of Western Europe*
- ▷ *To describe the way of life of the many cultural groups in Western Europe and how the groups interact with one another*
- ▷ *To explain how Western Europe has influenced world affairs*
- ▷ *To explain how the countries of Western Europe have attempted to develop policies of mutual political and economic cooperation*
- ▶ *To interpret a population-density map*

Chapter 10

Chapter 10 Lesson Plans and Planning Guide are found in the Teacher's Guide.

The first chapter in Units 2 through 9 is an overview of the entire region. Note the regional map on page 110.

The Geography of Western Europe

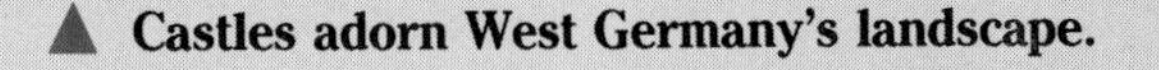

▲ Castles adorn West Germany's landscape.

▲ London's Fleet Street is a center for commerce.

GEOGRAPHY DICTIONARY

- microstate
- imperialism
- raw materials
- alliance
- peninsula
- natural boundary
- landlocked
- canal
- hydroelectricity
- economic association
- European Community (EC)
- tariffs
- terrorism

Though less than one-half the size of the United States, Western Europe is a complex region with a great variety of landscapes, climates, cultures, and nations. Western Europe consists of 17 independent nations plus six **microstates**. A microstate is a very small country that usually depends upon its larger neighbors for military protection, customs control, and foreign relations.

Western Europe is one of the most densely populated regions on Earth. Its total population is one and one-half times larger than that of the United States. It is also one of the most industrially developed and socially advanced world regions. Western Europe's languages, religions, legal systems, products, music, and art have been introduced worldwide. While there has never been a single Western European nation, the countries of Western Europe are forming closer ties.

Ask students what aspects of economic geography are shown in the photos on pages 108–109. What aspects of historical geography are shown?

How to compare a population-density map to a climate map is the subject of the Geography Skills lesson on pages 196–197.

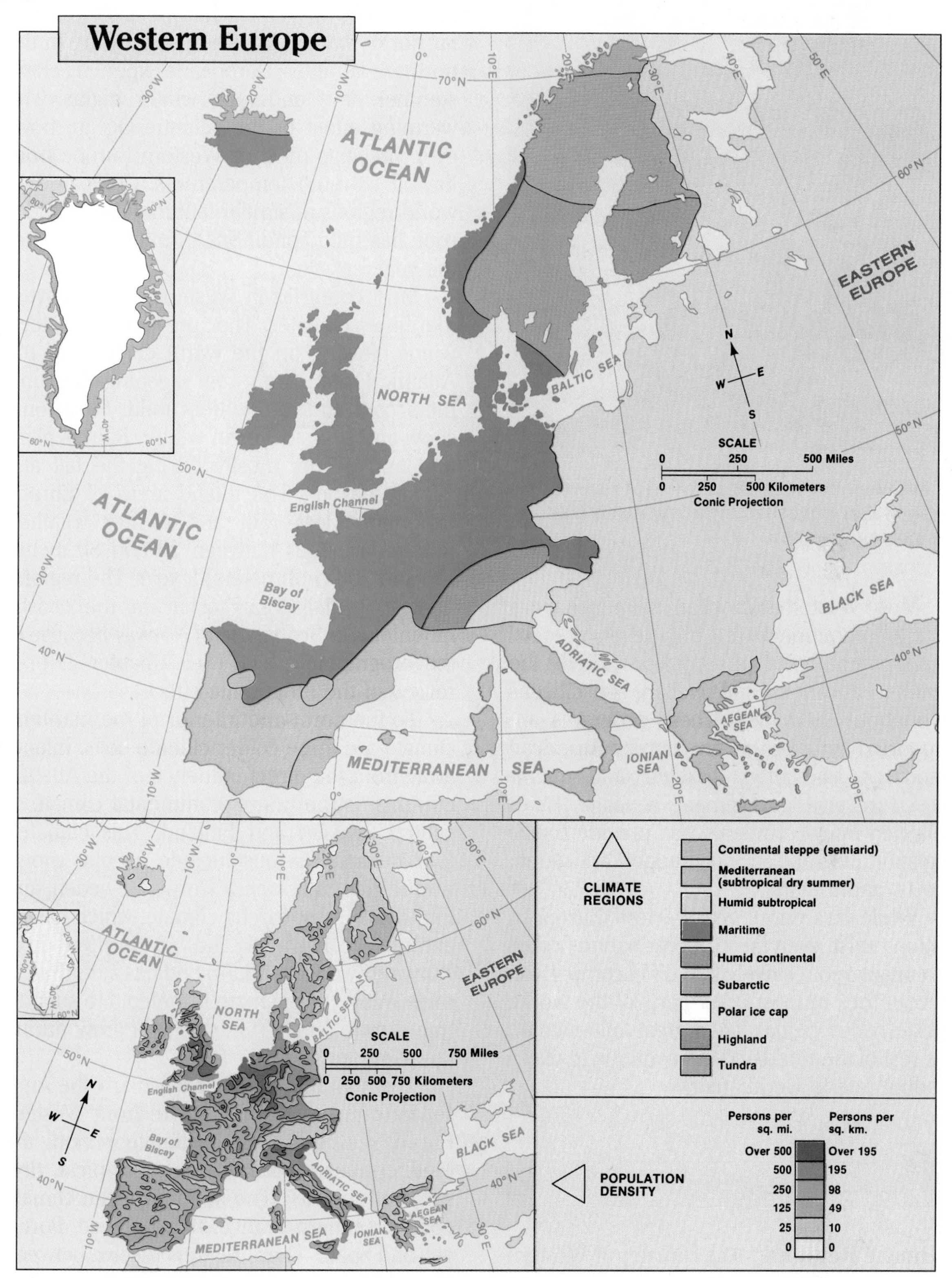

Which countries of Western Europe have the lowest population densities? the highest? Why?

Highest: Great Britain, the Netherlands, Belgium, Italy, West Germany, some parts of Spain, Portugal, France, Denmark. Lowest: Ireland, Finland, Norway, Sweden, Iceland, Switzerland, Austria. The most densely populated regions have a mild, maritime climate; the least densely populated regions have highland, subarctic, tundra, or humid continental climates.

In exchange for high taxes, Swedes receive many welfare benefits, including incentives for raising children.

Most Western European nations have a strong commitment to a high level of social development. Social programs provide for the health, education, and welfare of citizens throughout their lives. These programs are supported by taxes that are high by American standards. The aged, disabled, and unemployed receive government benefits. Taxes collected by governments also provide public transportation and strongly support education, sports, and the arts.

While less developed Western European nations exist, even these poorer nations could be considered relatively well off in comparison to countries in most other parts of the world. Western Europe has had a great influence on the rest of the world and will continue to play a leading role in world affairs.

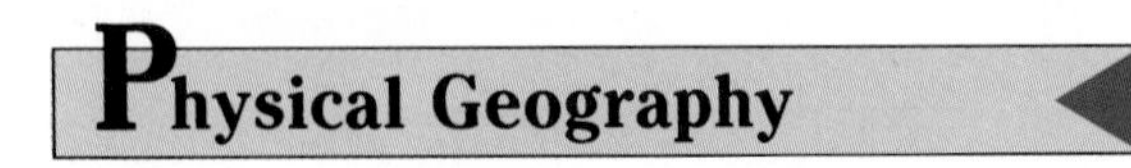

Physical Geography

Climate Regions The climates of Western Europe are influenced by the Atlantic Ocean and the high mountains of Europe, especially the Alps. Another major factor in determining the climates of Western Europe is the region's far northern location. Because they are in the northern latitudes, Europeans experience long summer days and long winter nights. The warming effect of the Atlantic Ocean, however, prevents most of Western Europe from having extreme temperatures. Compared to world regions of similar latitude, Western Europe has mild conditions throughout most of the year.

Most countries in Western Europe have a maritime climate. The prevailing westerly winds blowing off the warm currents of the Atlantic Ocean bring rain, keeping the summers cool and the winters mild. Occasional snow and frosts occur in winter. Rain is plentiful and occurs mostly during the fall and winter seasons. The inland areas of Europe experience less rain and clearer weather. Rainfall averages between 20 and 80 inches (50 and 200 centimeters) a year. The region's highest rainfall totals occur in the coastal mountains of Scotland and Norway because of the orographic effect. (See Chapter 4 for a review of the orographic effect.)

To the north and interior of the maritime climate are three colder climate belts. Inland, away from the direct influence of the Atlantic currents, is the humid continental climate of central Sweden and Finland. This climate's four distinct seasons include a cold, snowy winter. Across northern Norway, Sweden, and Finland is the subarctic climate, which has an extremely cold winter. The very northern tip of Norway, Sweden, and Finland has a tundra climate. While this climate is too cold to support tree growth, small tundra plants grow during the short summer.

The high mountains, particularly the Alps, separate the maritime climate from Western Europe's second largest climate region, the Mediterranean climate (a subtropical dry-summer climate). The Mediterranean climate covers southern France, Italy, Greece, Portugal, and Spain. These regions receive between 10 and 60 inches (25 and 150 centimeters) of rainfall a year, most of it during the winter months. A sunny, dry summer and a mild

 Level B: Have students write descriptions of the climates found in Western Europe. The descriptions should be written so that tourists would know what kind of weather to expect in the summer and winter at specific locations.

The North European Plain, formed by glaciers during the last ice age, has fertile soil and a mild, maritime climate. This river winds through fields near Hamburg, West Germany.

winter are typical. Most of the year is affected by the dry, sinking air of the subtropical high-pressure zone; therefore, little rain falls. During the winter, the subtropical high-pressure zone moves farther south. At this time, cool Atlantic storms pushed by the westerly winds bring rain into the Mediterranean region.

Within the southern part of Western Europe, two other climate types can be found. A small area of humid subtropical climate is located southeast of the Alps, mainly in the Po Valley of northern Italy. Here, summers are warm and wet. There are some dry continental steppe climate areas in parts of Spain with high elevations. These dry areas are caused by mountains that block moist ocean air from reaching farther inland.

Landform Regions Look at the map of Western Europe on page 110. Notice that the region is made up of several **peninsulas**. A peninsula is a landform that is surrounded by water on three sides. While Western Europe appears complex, it can be divided into four major landform regions. These are the Northwest Highlands, the North European Plain, the Central Uplands, and the Alpine Mountains.

The Northwest Highlands consists of a rugged, hilly landscape with some low mountains. The region includes the hills in Ireland, the Pennine Mountains in England, the Scottish Highlands, the plateau of Brittany in northern France, the Iberian Peninsula, and the Kjølen (CHUHL-uhn) Mountains in Norway and Sweden.

Much of the region was affected by the continental sheet glaciers during the last ice age. The ice removed or buried previous landforms. When the ice melted, poor drainage and thin soils were the result. Its rugged, rocky hills, poorly drained valleys, and northern location makes the Northwest Highlands the least developed of the four landform regions. Economic activity is limited to mining, fishing, forestry, and grazing.

The largest landform region is the North European Plain. It extends from the Atlantic coast of Western Europe into the Soviet Union. This broad coastal plain borders the Atlantic Ocean, the North Sea, and the Baltic Sea. In fact, during the last ice age, the North Sea, the English Channel, and the Baltic Sea were all exposed parts of this plain. As the ice melted, the sea level rose and covered these areas with water. Yet, very little of this region

Ask students to define the term *peninsula.* Then have them identify on a map of Europe the peninsulas of Iberia, Scandinavia, and Jutland (Denmark), the Balkan peninsula, the Italian peninsula, and also Europe itself.

Level A: Ask students to write travel brochures for popular tourist destinations such as the Swiss Alps or Paris. Tell them to describe the beauty, the landscape, the climate, and interesting historical sites located there.

reaches more than 500 feet (152 meters) above sea level.

The North European Plain is Western Europe's most important landform region. More than two-thirds of the population of Western Europe lives here. Nearly all of Europe's major rivers flow across this plain. Most of Western Europe's major agricultural and industrial centers are located here.

The third landform region is the Central Uplands. This ancient, hilly area of Western Europe consists of the Massif Central (ma-SEEF sehn-TRAHL) of France, the Jura (JUR-uh) Mountains on the Swiss-French border, the Black Forest and Bavarian Plateau of Germany, and the Ardennes (ahr-DEHN), a wooded plateau region in Belgium and Luxembourg. These hills and plateaus are forested and have important mineral deposits, such as coal and iron. Agriculture is quite productive in this region.

The Alpine Mountains comprise the most rugged and spectacular landform region of Western Europe. These young and active mountains are divided into several major ranges. The Alps are the highest and longest mountain range in Western Europe. They extend from southern France through Switzerland and Austria and across northern Italy. The highest peak, Mont Blanc (MOHN BLAHN), reaches to 15,771 feet (4,808 meters). Many peaks in the Alps reach more than 14,000 feet (4,268 meters). Because of their high elevations, the Alps receive much snow, and large glaciers are still present.

Another range in the Alpine Mountain region is the Pyrenees (PIR-uh-neez). These mountains form a **natural boundary** between Spain and France. When political borders are established by landforms, they are called natural boundaries. The highest peak reaches an elevation of 11,168 feet (3,404 meters). Several high mountain ranges are found in Spain, including the Sierra Nevada. South of the Alps are the Apennines (AP-uh-neenz) of Italy and the rugged Pindus (PIN-duhs)

The Pyrenees Mountains form a natural boundary between France and Spain. These are old, rugged landforms. What other European landforms form natural boundaries? Alps, Kjølen Mountains

 Ask students to name other natural features that form boundaries between countries (rivers, oceans, and so on).

The Suez Canal was constructed by a French company under the direction of Ferdinand de Lesseps.

mountains of Greece. Except for the Po Valley between the Alps and the Apennines in northern Italy, there are few large lowlands in the Alpine region.

The southern ranges of the Alpine Mountain region are the most active. Earthquakes occur in both Italy and Greece. Active volcanoes include Mount Etna and Mount Vesuvius (vuh-SOO-vee-uhs) in Italy. Since this region is very rugged, agriculture and industry are limited. Some forestry and mining are important locally. Tourism has become the major source of income for much of the Alpine Mountain region, especially for the Alps and the rugged Mediterranean coast.

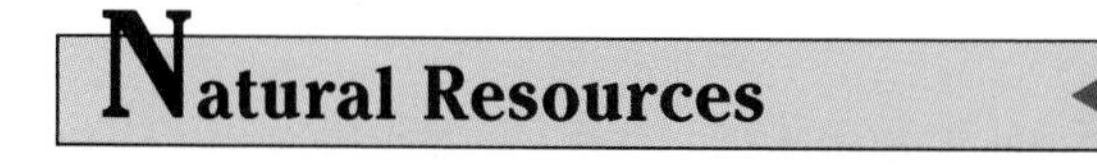

Natural Resources

Water No part of Western Europe is more than 300 miles (480 kilometers) from the sea. In fact, Switzerland, Austria, Liechtenstein, and Luxembourg are the only Western European countries that are **landlocked**. A landlocked country is one that is completely or almost completely surrounded by land. The region's long, irregular coastline is indented with hundreds of excellent ice-free natural harbors. These harbors are often located near the mouths of navigable rivers.

During the last few centuries, Europe's major sea routes were across the Atlantic to the Americas or around Africa to the Orient and the Pacific. With the opening of the Suez Canal in 1869, Western European traders were able to use the much shorter Mediterranean and Red sea route to reach the Orient and the Middle East. A **canal** is a waterway dug across land to allow ships to cross.

The Mediterranean is Western Europe's busiest sea. Italy, Greece, France, and Spain all have active ports on the Mediterranean. Although it suffers from pollution, the Mediterranean is also an important fishing and tourist area.

Two other major Western European seas are the North Sea and the Baltic Sea. The North Sea, along with its western arm, the English Channel, is important for trade between European countries. It is also one of the world's richest fishing areas. However, the North Sea fisheries have suffered in recent years because of overfishing.

About 4,500 ships use this port each year. Marseilles's location on the Mediterranean Sea has made it a major trade center.

The Baltic Sea is almost surrounded by land. Its northern arms freeze during the winter months, and icebreaking ships are needed to keep sea lanes open. The Baltic is the only important seaway in Western Europe ever blocked by ice.

The rivers of Western Europe are of great importance. Most are connected by a series of canals, making water transportation in Europe very efficient. Even cities far from the coast have access to the sea through this extensive river and canal system. The Western European nations have realized that they must work together to clean up the pollution in the rivers that flow across their national boundaries.

Forests and Land Western Europe today contains very little untouched land. Most of

Have students locate the Suez and Panama canals. Today's supertankers, because of their size, cannot use these canals. Ask students what problems this presents in the area of world trade. How can these problems be resolved?

This nuclear power plant is near Birmingham, England. How do some Western European countries meet their extensive energy needs? offshore oil and gas deposits; oil and gas imports; nuclear power plants; hydroelectricity; ocean tidal power; solar energy

the original forests have been cut for timber or cleared for farming. Wars and fires also have destroyed many of the forests that once covered the region. In recent years, air pollution has killed many trees. Only in Sweden and Finland are large areas of timber-producing forest still found. Poor grades of pine and cedar grow in the Mediterranean area. Most of the oak forests of the Mediterranean have been destroyed, leaving only a scrub-plant community covering the hillsides. Wood must be imported to satisfy the timber needs of Western Europe. Most Western European nations have initiated programs to replant trees.

Although most of the original forests are gone, Europeans have made good agricultural use of their soil. About one-third of Western Europe is farmland. Because of very productive farms, the people of Western Europe could probably grow all the food they need. However, some tropical crops do not grow well in Europe's climate. Therefore, many types of food consumed in the region must be imported.

Minerals and Energy Considering its current industrial and energy needs, Western Europe is not rich in mineral deposits. Britain, Germany, Belgium, and the Netherlands have deposits of iron and coal, which were important in the growth of these industrial countries. Sweden, France, Austria, Luxembourg, and Spain have iron deposits. Bauxite (for making aluminum) is mined in Greece, Spain, and France; Spain and Italy produce mercury. It is clear that Western Europe no longer has sufficient mineral resources to run its modern economies. Today, more than 90 percent of the necessary minerals must be imported.

Very little petroleum has been found under Western Europe's land. However, the floor of the North Sea has produced oil and natural gas. The offshore oil has been found beneath the waters claimed by Norway and Britain. The offshore gas deposits have been found mainly in areas belonging to the Netherlands, Norway, and Britain. Still, these oil and gas deposits do not satisfy Western Europe's industrial demands. Oil and gas are imported from the Middle East, Africa, and the Soviet Union.

Western Europe also depends upon other energy sources. Nuclear power plants are found in most of its nations. France, Sweden, and Belgium meet over half their energy needs with nuclear power. **Hydroelectricity** is plentiful in the mountainous nations such as Norway and Switzerland. Hydroelectricity is electricity produced by waterpower. France has been successful in producing ocean tidal power and in using solar energy.

 Note the Themes of Geography feature "Persian Gulf," on pages 270–271. Europe receives a major portion of its oil from this region.

People Its highly skilled, educated, and industrious people are probably Western Europe's greatest resource. The rapid progress of Western Europe since the devastation caused by World War II is proof of this. European workers, scientists, and technicians are continually developing new products and new manufacturing techniques. Western European farmers are some of the world's most productive. To maintain one of the highest standards of living in the world, Western Europe must continue its leadership in research, industry, and agriculture. It also must conserve its limited natural resources.

Western Europe's transportation system is modern and efficient. Travelers crowd this train station in Paris.

France's TGV is the fastest train in the world.

Economic and Political Ties

Economic Cooperation By the end of World War II, the economies of the Western European nations were destroyed. Their trade and industry had to be totally rebuilt. In addition, the Western European countries began to lose control of their colonial empires. People in the colonies in Asia and Africa were demanding independence. To Western Europe, the colonies were major sources of inexpensive raw materials that would support industrial growth.

In 1948, the United States began to send economic aid to Western Europe under the Marshall Plan. The Marshall Plan was named for United States Secretary of State George Marshall, who proposed the plan. This aid was essential in rebuilding the Western European economies. The Marshall Plan required that all the nations receiving the aid must decide as a group how the money would be used. It was an important step for a unified and successful rebuilding of Western Europe's economy. Soon the nations of Western Europe began to help themselves by forming **economic associations**. An economic association is an organization formed to break down trade barriers between member nations. The first customs union was formed by Belgium and Luxembourg in 1948.

In 1952, the European Coal and Steel Community (ECSC) was created. Its purpose was to expand steel production while lowering the costs of needed materials. The organization's six members were Belgium, the Netherlands, Luxembourg, France, West Germany, and Italy. The coal, iron, and steel industries of

Discuss with students the advantages and disadvantages of tariffs. Why do countries tax one another's imports?

these countries were operated as though they belonged to one organization rather than to six countries.

The most important and successful of these cooperative economic associations has been the European Economic Community (EEC), sometimes called the Common Market. It was formed in 1957 by the six countries of the European Coal and Steel Community. Since then, six new members have joined: Britain, Ireland, and Denmark in 1973; Greece in 1981; and Spain and Portugal in 1986. In 1975, special trade agreements were added between the Common Market and 58 nations in Africa, the Caribbean, and the Pacific. Since it not only is an economic union but also maintains close political ties, the Common Market recently renamed itself the **European Community (EC)**.

The European Community countries have done away with **tariffs**, or taxes on imported goods. This has opened trade among its member nations. The Community has combined the natural resources and industries of its members, which has resulted in impressive economic gains. With 37 percent of all global trade, it is the world's largest trading partner.

The Western European nations of Switzerland, Austria, Norway, Sweden, Finland, and Iceland belong to the European Free Trade Association (EFTA) and have not joined the European Community. However, the EFTA holds trade agreements with the European Community and is its major trade partner.

Political Ties In 1949, a military alliance called the North Atlantic Treaty Organization (NATO) was founded. The organization was formed as an alliance for the common defense of Western Europe. Its goal was to protect its members from possible invasion by the Soviet Union and its Eastern European allies. Members of NATO include 12 Western European nations plus Turkey, Canada, and the United States. The NATO members are not completely unified. They often disagree on issues such as the danger of the Soviet Union and the use of nuclear missiles.

The European headquarters of the United Nations is in Geneva. Other international headquarters also are located here.

Though most Western European nations are represented by a single culture, a few of the countries have minority groups that are seeking more independence. Some of these separatist movements, such as that of the Basques in Spain, have led to violence. Many Western European cities have large foreign populations who moved there seeking employment. Where unemployment is high and living conditions are poor, crime and violence have grown. In addition, radical political movements have led to **terrorism**. Terrorism is the use of violence as a means of political force. Terrorist groups have carried out bombings, assassinations, and kidnappings. Growing antinuclear movements and environmental political parties such as the "Greens" have also influenced the politics of Western European nations.

Western Europe's other problems are at the international level. One concern is the influence of the "superpowers," the United States and the Soviet Union, on the region. Another is economic competition from Japan and industrializing Asian nations.

Though it has a long way to go, Western Europe is seeking economic and political unity among its nations. A united Western Europe could be the most economically and politically powerful region on Earth.

 Level B: Using a key, tell students to (1) trace a political map of Europe; (2) label and color the original member nations of the EC; (3) use a second color to mark nations that joined the EC later.

Level A: Have students keep a scrapbook of newspaper articles on Western Europe for one week. At the end of the week, have students prepare summaries of their collections. Lead a discussion of current events in Europe.

CHAPTER 10 CHECK

Reviewing the Main Ideas

1. Western Europe is a complex region with a variety of landscapes, climates, cultures, religions, and languages. It consists of 17 independent nations and six microstates.
2. The climates of Western Europe are influenced by the Atlantic Ocean, as well as by the region's high mountains and far north location. Most Western European countries have a mild, maritime climate.
3. Western Europe can be divided into four major landform regions: the Northwest Highlands, the North European Plain, the Central Uplands, and the Alpine Mountains.
4. Water has always played a major role in Western Europe's economy. Italy, Greece, France, and Spain have important ports on the Mediterranean Sea. The North Sea and the Baltic Sea also provide major trade routes to Western European nations. All Western European nations have access to an extensive river and canal network.

Building a Vocabulary

Students should provide an answer to Question 2 by filling in the blank, using their own paper.

1. Define *raw materials*. During the nineteenth century, how was Britain able to obtain the raw materials it needed to become a world industrial leader?
2. A waterway dug across land to allow ships to cross is called a ____.
3. What is a natural boundary? The Pyrenees form a natural boundary between what two countries?
4. The European Community is an example of what? Define *tariffs,* and explain why the EC has done away with them.
5. What is a microstate? List the microstates of Europe. Where are they located?
6. What is a peninsula? Which Western European nations are located on peninsulas? Are any of these nations landlocked?

Recalling and Reviewing

1. In what ways is Western Europe politically and economically united? In what ways is it not united?
2. What are the major landform regions of Western Europe? Where is each located?

Critical Thinking

3. The countries of Western Europe have been able to turn from conflict to cooperation. Why was there such a long history of conflict in Western Europe?

Using Geography Skills

Use the maps in this chapter to answer the following questions.

1. Which Western European nations are island nations?
2. Name the seas that border Western Europe. Which one is almost completely surrounded by land? Which Western European countries have coasts on this sea?

Answers to Chapter Check questions are found in the Teacher's Guide.

Additional activities for Chapter 10 are found in the Student Workbook. Chapter 10 Test as well as Skills, Reteaching, Critical Thinking, and Challenge/Enrichment activities are available in the Teacher's Resource Binder.

Chapter 11

Chapter 11 Lesson Plans and Planning Guide are found in the Teacher's Guide.

Refer students to the map on page 124.

Great Britain and Ireland

▲ Ireland is famous for its green countryside.

▲ **London has many attractive public parks.**

GEOGRAPHY DICTIONARY

- lochs
- constitutional monarchy
- fossil fuels
- nationalized
- commonwealth
- bog
- peat

The two islands that lie on the continental shelf off the Western European coast are Great Britain and Ireland, sometimes called the British Isles. The larger island is Great Britain; the smaller one, to the west, is Ireland.

These islands are also divided politically into two nations, the United Kingdom and the Republic of Ireland. The United Kingdom (UK) is composed of England, Scotland, and Wales, all on the island of Great Britain, and Northern Ireland. Because the people of the United Kingdom live mainly on the island of Great Britain, they are often referred to as the British. The Republic of Ireland consists of most of the island of Ireland.

The United States and the United Kingdom share a unique relationship. Though the United States fought to gain independence from England, the United States and the United Kingdom have always maintained close ties.

 Select a student to research the term *United Kingdom.*

Ask students to research Welsh history. How did England come to control Wales? What are the prospects for Welsh independence?

Great Britain

Physical Geography Great Britain can be divided into two landform regions, lowland and highland Britain. The southeastern half of the island is lowland Britain. (See the map on page 124.) It is part of the North European Plain. Only a narrow body of water, called the English Channel, separates the British portion from the rest of the North European Plain. Since the southern part of Britain has level land with fertile soil, good farmland is found there. It is the most densely populated part of the island.

Highland Britain is in the north and west. The major highland areas are the Pennine Mountains of northern England, the Cambrian Mountains of Wales, and the Scottish Highlands. This region also includes the rugged Hebrides, Orkney, and Shetland islands. Highland Britain's only large agricultural area is the Scottish Lowland. This is a long faulted valley south of the Grampian Mountains. Most of the rest of highland Britain has limited farming because of the rocky, glaciated hills. Grazing of cattle and sheep is common. Highland Britain is where the beautiful English Lake District and the large Scottish lakes, or **lochs**, are found. Lochs are long, deep lakes carved by glaciers. Highland Britain's many rivers are a source of hydroelectric power.

Sheep grazing is common in the highland regions of Great Britain. Why are these regions not very useful for farming?

The hills are rocky and glaciated.

For centuries, the rivers of Britain have been important for trade. They connect the major ports of London, Liverpool, and Glasgow (GLAS-koh) to the sea. The Tweed, Tyne (TINE), Tees, and Thames (TEHMZ) rivers on the east coast flow into the North Sea. The Clyde, Mersey, and Severn rivers on the west coast flow into the Atlantic Ocean and the Irish Sea.

Even though Great Britain is relatively far north (50° to 60° north latitude), it has a mild, maritime climate. This is because of the prevailing westerly winds blowing off the North Atlantic Ocean. Though its temperatures are rarely extreme, Britain is known for its stormy, cloudy, and foggy weather. Much precipitation falls as drizzle; it rarely snows except in the highlands. Britain is wettest on the west coast, especially in Scotland and the Hebrides Islands, where rainfall averages more than 60 inches (150 centimeters) per year. The eastern coast is drier but still receives over 30 inches (75 centimeters) of rainfall per year.

Population Geography The United Kingdom is one of the most populous nations in Western Europe. It also has one of the region's highest population densities. Especially crowded are southeast England near London and the central part of England, known as the Midlands. Much of the hilly lands of northern England, Wales, and Scotland have low population densities.

Most of the British have their roots in five groups of people who invaded the island from continental Europe. The first people to arrive were the warlike Celts in about 700 B.C.

The Romans arrived during the first century and gave Great Britain its name. Roman soldiers occupied Great Britain for more than 400 years, leaving at the decline of the Roman Empire. The Romans built roads, walls, and cities. In the fifth century, the Anglo-Saxons invaded from present-day Germany. They

Refer students to the map on page 110. Note the location of Great Britain and Ireland.

conquered all of England, but Wales and Scotland were left to the Celts. During the Anglo-Saxon period, farming was introduced to the region, and Christianity arrived.

During the eighth and ninth centuries, the Vikings invaded the coastal areas of Great Britain and Ireland, nearly destroying the Anglo-Saxon and Celtic cultures. The Vikings eventually settled with the Anglo-Saxons. In 1066, the Normans, a Viking people who had settled in northern France, invaded England. This invasion is called the Norman Conquest. The Normans conquered the country and introduced the feudal system of land ownership and a monarchy.

The rugged Celtic areas did not come under English control for several hundred years. Wales was brought under the crown in 1275. Scotland was never conquered but was joined to England in 1707. Ireland was invaded in the twelfth century, but only the northern part remains in the United Kingdom today.

The United Kingdom is a **constitutional monarchy**. That is, it has a reigning king or queen, but a parliament serves as the lawmaking branch of government. The king or queen symbolizes unity, history, and patriotism to the British people.

Before the 1950s, many British emigrated, particularly to the United States, Canada, and Australia. Since then, the trend has reversed, and Great Britain has been receiving immigrants. Most have arrived from former colonies such as India, Pakistan, and the West Indies.

Agriculture and Fishing British farms are smaller than many in America, but farming, nevertheless, is efficient. For example, farming employs less than 2 percent of the work force yet produces more than half of the nation's food. The most important crops are wheat, barley, potatoes, and sugar beets. Two-thirds of British farmland is used for grazing cattle and sheep.

The United Kingdom's fishing industry supplies seafood for export and home use. Fishing boats travel as far as Iceland and

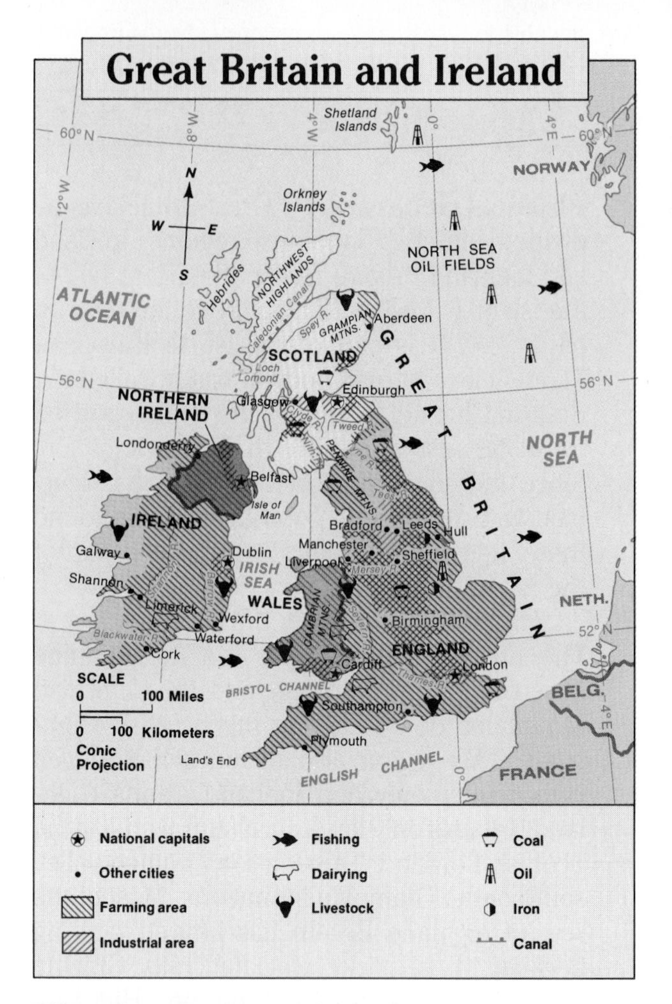

Where is coal mined in Great Britain?
In central England and southern Scotland, near the cities of Edinburgh, Birmingham, and Cardiff.

Greenland for their catch. Unfortunately, overfishing has created problems, especially in the North Sea.

British Energy Resources The United Kingdom has more **fossil fuels** than any other Western European nation. Fossil fuels are energy resources formed from plants and animals buried millions of years ago. The fossil fuels include coal, oil, and natural gas. Coal deposits are found in England, Wales, and Scotland and were extremely important in the United Kingdom's early industrialization. Though its coal deposits are not as important today, the country's reserves should last hundreds of years.

In 1969, oil was discovered under the North Sea off the Scottish east coast. Aberdeen, in eastern Scotland, has become the

One of the world's worst ecological disasters occurred in London in 1952 when 4,000 people died during a severe smog attack. Have students research ways in which the government has tried to clean the air and water.

Oil and natural-gas drilling in the North Sea oil fields has helped the British economy. These oil fields are located near Scotland.

supply center for the offshore oil industry. The United Kingdom is now a major world producer and exporter of oil. Natural gas deposits were also discovered on the continental shelf off the west coast of England. These oil and gas deposits should last well into the next century.

The United Kingdom was one of the first nations to depend upon nuclear power; about one-fifth of its electricity is nuclear generated. The United Kingdom also depends upon its supplies of local coal and hydroelectric power to generate electricity.

The Industrial Revolution In the 1700s, Britain had abundant coal and iron deposits, a large labor force, and an excellent transportation network of rivers and canals. By the early 1800s, the British had built the first railroads. These resources helped make Great Britain the world's first industrialized region.

Three of Great Britain's early industries were iron and steel, shipbuilding, and textiles. The iron and steel industry was based on deposits of iron ore and coal found in central England. This area is known as the Midlands. Birmingham, which lies in the heart of the Midlands, and nearby Sheffield became the major steel-producing cities.

During the middle of the nineteenth century, with Great Britain's steel industry already established, steel replaced wood as a material for building ships. The island's coastal cities attracted this industry. Soon, Glasgow became the world's leading shipbuilding center.

The British textile industry began mainly in northern England. It was centered primarily in Manchester, Bradford, Leeds, and the port of Liverpool. The invention of spinning machines and steam power replaced home weaving. Local wool and, later, cotton from America and India as well as wool from Australia, supplied the textile industry.

Though Great Britain was first to industrialize, it could not maintain world leadership for long. Modern foreign factories began to compete. By 1900, Germany and the United States outproduced Great Britain. They soon dominated newer industries, such as chemicals and automobiles. In recent decades, Britain's traditional industries—steel, shipbuilding, and textiles—have been almost destroyed by competition from Western Europe, the United States, and particularly Japan.

To curb its post–World War II industrial decline, the British government **nationalized** many of its industries, such as steel and shipbuilding. A nationalized industry is one that is owned and operated by the government. Because these nationalized industries were not generally successful, the government has in recent years sold some of these industries back into public ownership. Some, such as the postal service, railroads, and health service, remain nationalized.

The decline of the United Kingdom's traditional industries has created high unemployment, especially in the cities of northern England and Scotland and in the mining towns of Wales. Today, Great Britain can be divided into two economic regions by a line running from Bristol in southern England to the North Sea coast just north of Cambridge. North of this line are the decayed "smokestack" industrial cities, where agricultural land is poor. The south has prosperous farms with good soils, a milder climate, and modernized techniques.

Level B: Have students draw a line to make two columns on a sheet of paper. In the first column, they should list each of the major industries in Great Britain. In the second column, they should list the locations of these industries.

Level A: Have students research the growth of London over time. It has not grown outside the greenbelt designed to contain it. Have students brainstorm about ideas for ways to decentralize London.

Its cities have attracted growth industries in high technology, finance, and services. Today, most British are employed in service industries, especially banking, insurance, communications, and government.

London The majority of the British are urban dwellers. London, the United Kingdom's capital city and largest urban area, contains one-fifth of the nation's population.

Today, there are really "two Londons": historic London and modern, commercial London. Roman ruins can still be found beneath glass skyscrapers. It is also the home of the Houses of Parliament and Buckingham Palace, the residence of the British monarch. The United Kingdom's religious tradition can be seen at Saint Paul's Cathedral and Westminster Abbey, and its historical past is exhibited in the great museums.

Commercial London is a center of trade, industry, and services. Its early economic importance was due to its location on the Thames River, which provides easy access to the sea. London hosts a variety of industries, such as food processing, printing, aerospace, electronics, and computers. It is a center for world banking and insurance as well as a major theater, retail, and tourist center.

London's financial district, the House of Parliament, and other public buildings were built along the Thames River.

Like most other large cities, London suffers from urban decay, pollution, and traffic. The government has made effective efforts to clean up London's air and the waters of the Thames. It also maintains a "greenbelt" of parks and woodlands that circle the city.

Each day, almost one-half million people commute, or travel, to jobs in London from nearby communities. They commute on a vast rail and highway network. London today is a vibrant, important world center and by far the dominant city of the United Kingdom.

Economic Geography At its height during the nineteenth and early twentieth centuries, the British Empire controlled one-fourth of the world's land, and it ruled the seas. Prior to World War II, the British had given independence to Australia, Canada, Ireland, New Zealand, and South Africa. When World War II ended in 1945, the British Empire began to break up as other colonies sought independence. New nations appeared on the map, including India, Pakistan, Jamaica, Kenya, and Nigeria. For the most part, the British Empire no longer exists. Today, only a few small islands remain part of its holdings.

The once powerful British Empire has been replaced by a **commonwealth**. A commonwealth is an association of self-governing states with similar backgrounds and united by a common loyalty. The British Commonwealth of Nations is a voluntary association of nations that were once colonies within the British Empire. Member nations meet to discuss economic, scientific, and business matters.

The United Kingdom's joining the European Community in 1973 changed the country's economy. The British now look to other European nations for trade rather than to their distant former colonies. For instance, sugar may come from France instead of Jamaica, and dairy products come from Denmark rather than New Zealand. The United Kingdom remains a major international trading nation, exporting more than 30 percent of its goods.

Point out that Britain has had worldwide influence because of colonization. Have students do library research to show ways in which Great Britain still influences its former colonies.

Several major economic problems loom in the United Kingdom's immediate future. One is its need to increase productivity and research in order to compete in international trade. It must also close the gaps that exist between its prosperous south and poorer north areas. A high unemployment rate, especially in the northern cities, is a major concern. Finally, persuading its labor unions and management to work together is necessary if the country's industries are to be revitalized.

The United Kingdom also faces some political and social problems. There are weak, but vocal, separatist movements in Wales and Scotland. Also, the violence between Roman Catholics and Protestants in Northern Ireland continues to be a major issue.

For the Record

In 1979, Margaret Thatcher became Great Britain's first female prime minister. The following is from a speech Prime Minister Thatcher delivered in which she discussed the "new" Europe.

For five centuries that small continent [Europe] had extended its authority over islands and continents the world over. For the first 40 years of this century, there were seven great powers—the United States, Great Britain, Germany, France, Russia, Japan, and Italy. Of those seven, two now tower over the rest—the United States and the Soviet Union.

To that swift and historic change, Europe—a Europe of many different histories, many different nations has had to find a response. It has not been an easy passage—to blend this conflux of nationalism, patriotism, sovereignty into a European Community. Yet I think that our children and grandchildren may see this period—these birth pangs of a new Europe—more clearly than we do now. They will see it as a visionary chapter in the creation of a Europe able to share the load alongside you.

Adapted from: Margaret Thatcher, Speech Delivered to a Joint Session of the United States Congress, Washington, D.C., February 20, 1985. *Vital Speeches* L1, no. 11(March 15, 1985).

Ireland

A Divided Island Across the Irish Sea from Great Britain is Ireland, sometimes called the "Emerald Isle" because of its green countryside. (See the map on page 124.) The entire island has a mild, maritime climate averaging about 40 inches (100 centimeters) of rainfall per year.

Rolling green fields and pastures cover the island's central part. The Shannon River, the longest river in the British Isles, flows through the middle of the island. Low, rocky mountains form a rugged coastline. Much of Ireland is either rocky or boggy as the result of the last ice age. **Bog** is soft ground that is saturated with water. The many lakes are also products of the glaciers. For centuries, **peat** deposits from the bog areas have been used for fuel. Peat is composed of decayed vegetable matter, usually mosses. Less than 5 percent of the island is forested.

The Irish, like the Welsh and Scots, are descendants of the Celts who settled in Ireland around 400 B.C. Irish Gaelic, a Celtic language, is still considered an official language but is spoken by only a small percentage of the Irish. Today, English is generally spoken.

By the fifth century, Saint Patrick and other missionaries converted the island to Christianity. Vikings attacked and controlled Ireland's coastal towns during the ninth and tenth centuries. In the thirteenth century, the Norman rulers from Britain conquered Ireland. Eventually, the Protestant British controlled most of the land, reducing the Roman Catholic Irish to landless farmers. Lack of opportunity along with a potato famine during the nineteenth century caused many Irish to emigrate, especially to the United States.

For several hundred years, the Irish rebelled against British control. Finally, in 1922, the Irish Free State was created. In 1949, the completely independent Republic of Ireland was formed. Currently, only the northern part of the island remains under the control of the British.

Have students read the primary-source material above. Ask them to identify the countries of Europe included in the seven great powers mentioned by Thatcher. What two countries does she say have changed Europe's outlook on the world?

Northern Ireland The six northern counties of Ireland are called Northern Ireland or sometimes Ulster. Part of the United Kingdom, Northern Ireland sends representatives to Parliament in London. The majority of Northern Ireland's population is descended from Scottish and English Protestants who arrived during the seventeenth century. Irish Roman Catholics form one-third of the Northern Ireland population.

Rural Northern Ireland consists of rolling hills used for farming and cattle grazing. Early industrialization established the two main cities of Belfast and Londonderry. Belfast became an early textile center and today is Northern Ireland's largest city and capital. Londonderry is its second largest city.

In recent years, there has been bitter fighting between the Protestant majority and the Roman Catholic minority of Northern Ireland. The Roman Catholics are seeking more social, political, and economic opportunities. Many would like to join the Republic of Ireland. The Protestants fear becoming the minority and insist on remaining a part of the United Kingdom.

Because of continuing violence, British troops have been stationed in Northern Ireland since 1969. Several thousand people have been killed by the fighting and terrorist activities. The Irish Republican Army (IRA) is one of the most active of the terrorist groups.

The Republic of Ireland The Irish Republic is also known as Eire (AR-uh). It is made up of 26 counties and covers about 80 percent of the island.

Though Ireland's farming population has declined in recent decades, farm income has risen. Since joining the European Community in 1973, Irish farmers have modernized agriculture and now grow a wider variety of crops. Dairy products, beef, poultry, potatoes, and grains are produced for export and home consumption. More than half of Ireland's farm products are exported to the European Community nations. Most of Ireland's farmers own their land.

Most Irish workers are employed in manufacturing. Some of Ireland's industries are based on farm products. They include beet-sugar factories, dairies, and breweries. Other factories manufacture trucks, tractors, crystal, and numerous other products. The greatest industrial development has occurred around the seaport city of Dublin. Nearly one-third of the nation's population now lives in Dublin, the capital and center for education, banking, and shipping. The other cities are located along the coast and are much smaller. They include Cork, Wexford, Waterford, and Galway.

Though it has peat deposits, Ireland is an energy-poor island, with no oil, gas, or coal. Some hydroelectric power is produced by the Shannon River, and offshore oil or gas discoveries are future possibilities.

Ireland's future economic development depends upon more foreign investment, new industries, and increased tourism. The combination of high taxes and high unemployment rates continues to depress the Irish economy. By borrowing vast sums of money to support its unemployment, the Republic has built a huge debt. Many young people leave Ireland to seek jobs elsewhere, often in the United States, Britain, and Australia. This out-migration tends to steal Ireland's most skilled and best educated people.

Dublin is Ireland's capital, major seaport, and cultural center. Manufacturing is a major economic activity.

CHAPTER 11 CHECK

Reviewing the Main Ideas

1. Great Britain and Ireland are divided politically into two nations: the United Kingdom, which consists of Great Britain and Northern Ireland, and the Republic of Ireland.
2. Great Britain can be divided into two landform regions: lowland and highland Britain. Lowland Britain is a fertile agricultural region and the country's most densely populated area. Most of highland Britain consists of rocky, glaciated hills.
3. Although Great Britain was the world's first industrialized nation, its traditional industries—steel, shipbuilding, and textiles—have declined. Today, most British are employed in service industries.
4. Northern Ireland is part of the United Kingdom. The bitter disputes between its Protestant majority and Roman Catholic minority continue to be an important issue.
5. The Republic of Ireland is an energy-poor nation dependent upon foreign oil. Its future economic development depends upon foreign investment, new industries, and increased tourism.

Building a Vocabulary

Students should answer Question 2 by filling in the blank, using their own paper.

1. What is the role of a king or queen in a constitutional monarchy?
2. An industry owned and operated by the government is referred to as ______. Name two industries in Britain that can be included in this category.
3. Define *commonwealth*.
4. For many years, peat has been one of Ireland's major energy resources. What is peat, and where is it commonly found in Ireland?
5. What are lochs? In what part of Great Britain are they found?
6. Give three examples of fossil fuels.

Recalling and Reviewing

1. Describe the two landform regions of Britain, and tell where each is located.
2. How has London's location contributed to its economic importance?
3. What benefits does the British Commonwealth provide for its members?
4. What has caused the division of Ireland into two political units?

Critical Thinking

5. Based on what you learned in Unit 1, explain how ice ages affected the landforms of Great Britain and Ireland.

Using Geography Skills

Use the map in this chapter to answer the following questions.

1. What body of water separates Great Britain from the rest of Western Europe?
2. What resources does Great Britain have that the Republic of Ireland lacks?

Answers to Chapter Check questions are found in the Teacher's Guide.

Additional activities for Chapter 11 are found in the Student Workbook. Chapter 11 Test as well as Skills, Reteaching, Critical Thinking, and Challenge/Enrichment activities are available in the Teacher's Resource Binder.

Chapter 12

Chapter 12 Lesson Plans and Planning Guide are found in the Teacher's Guide.

Refer students to the map on page 133.

The Nordic Countries

▲ Copenhagen is the capital of Denmark.

▲ Sweden has thousands of glaciated lakes.

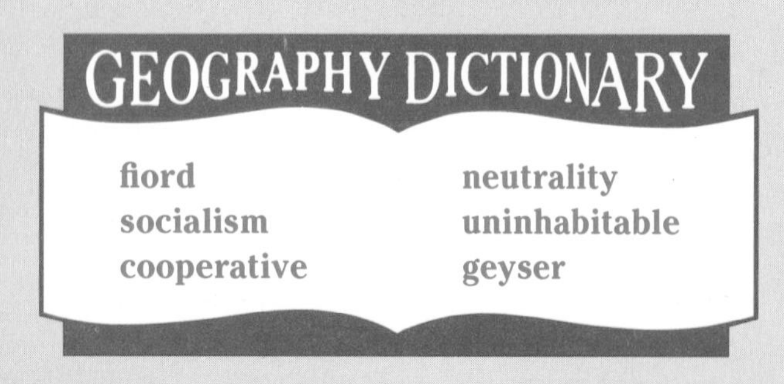

The Nordic countries include Norway, Sweden, Denmark, Iceland, and Finland. Often called Scandinavia, these nations have similar geographic, cultural, and economic characteristics. For example, they are all oriented toward the sea, each bordering a major body of water. They also share a far north location; all except Denmark have some territory above the Arctic Circle.

Once the home of fierce, warlike Vikings, today the Nordic countries are very peaceful, prosperous, industrialized nations. Generally, the people are urbanized and well educated and have high standards of living. Because of their high health standards, long life-spans, and low birthrates, they also have the oldest populations in the world.

 Have students briefly survey the photos in Chapter 12 to note physical, cultural, and economic characteristics of the Nordic countries.

Point out that Scandinavia has an area larger than that of the European core but a population of less than 25 million.

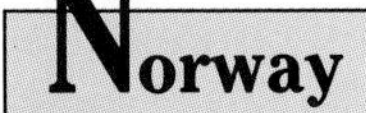

Norway

Physical Geography A long narrow country, Norway is bounded by the Atlantic Ocean and the North Sea on the north, west, and south. Along its eastern border is Sweden, which occupies the other half of the Scandinavian Peninsula. In the far northeast, Norway borders Finland and the Soviet Union. (See the map on page 133.) Norway also includes thousands of offshore islands, such as the Lofoten Islands, which have valuable fisheries. Svalbard, a group of islands in the Arctic Ocean, is also owned by Norway. However, by treaty, Svalbard's resources must be shared with other nations.

The western coastal portion of Norway is influenced by the westerly winds and the warm currents of the Atlantic Ocean. Because of the region's mild, maritime climate, the waters surrounding Norway remain ice-free all year. The mountains have cold winters with heavy snows. The far north is the thinly settled Arctic tundra and barren mountain tops. In winter, northern Norway is covered with snow and is in almost continuous darkness. At the North Cape during summer, there is almost continuous sunlight. This is why the area north of the Arctic Circle is called the "land of the midnight sun."

Norway is Western Europe's most mountainous nation. Most of the Norwegian half of the Scandinavian Peninsula is made up of the Kjølen Mountains. People in this region must live on the narrow coastal flatlands or on the slopes of the **fiords**. Fiords are narrow, deep inlets of the sea between high, rocky cliffs. They were carved out of the mountains by glaciers that extended down to the coast. The beautiful scenery of waterfalls, glaciers, and high mountain peaks rising out of the sea attracts tourists to this area in summer. Transportation by sea or over mountains on skis was once the only way to reach many Norwegian towns. Today, railroads, automobile ferries, and airlines provide transportation across this rugged land.

Fiord **in Norwegian means "a deep bay between steep, rocky walls of land." Most of Norway's people live near fiords.**

Economic Geography Norway's physical as well as human geography has always been influenced by the sea. Fishing is the oldest industry in Norway and provides a major export product. The long, rugged coastline, sheltered harbors, and the closeness of North Sea fishing grounds make Norway a leading fishing nation. Among the kinds of fish caught are cod, sardines, and mackerel. Shipping also makes up a large part of Norway's income. The country has one of the world's largest merchant fleets, including cruise ships.

Even Norway's most valuable resource, oil, comes from the sea. By the mid-1970s, Norway had developed the North Sea oil fields. It is now Western Europe's largest oil exporter, and most of the oil is exported to other Western European nations. Because of its oil, Norway has one of the world's highest standards of living. Norway's North Sea oil fields are expected to produce into the next century.

Norway is well supplied in other energy resources too. With its wet maritime climate and steep mountains, Norway generates more

Ask a volunteer to tell what a person in the merchant marine does. Then ask students to research to discover what kind of training is necessary. What are some of the advantages and disadvantages to this occupation?

Level A: Have students research the Vikings. Specific topics might include Viking raids, Norse mythology, Viking ship construction, and Viking art.

hydroelectricity than any other Western European nation. Virtually all of Norway's electricity comes from waterpower. Inexpensive electricity has attracted a number of industries to Norway, including paper, chemical, and metal industries.

While there is some farming in southern Norway, only about 3 percent of the land can be farmed. Infertile soil, steep slopes, and cool weather limit agriculture to grazing and dairy farming. Norway has become self-sufficient in meat and dairy products.

Norwegian Way of Life The Norwegians are a very prosperous people, mainly because of oil. More than half of Norway's population lives in the very southeast corner of the nation. Oslo, the capital, cultural center, and leading seaport and industrial center is located here. This modern city is situated at the end of a wide fiord. Well-planned parks, wide avenues, and houses built along the hillsides are typical. Government buildings and the royal palace are also found here. The city is close to hiking and skiing trails. Business people are often seen taking winter lunch breaks on cross-country skis through the city's parks.

Most of the rest of the country's population live near the coastal cities of Bergen, Trondheim (TRAHN-haym), and Stavanger (stuh-VAHNG-uhr). These cities provide industries, fishing fleets, and facilities to support the North Sea oil fields.

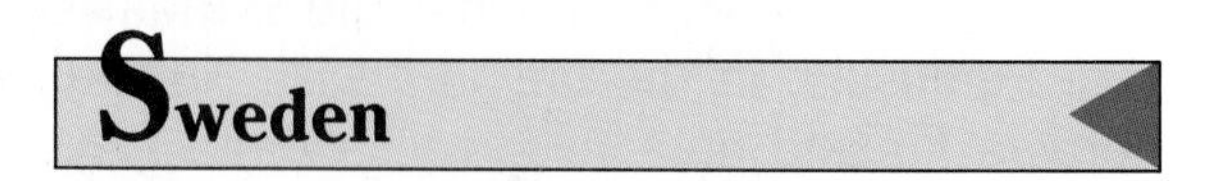

Sweden

Physical Geography Sweden is the largest Scandinavian country, both in area and in population. Except for its very southern tip, Sweden is divided between two cold climates. Its southern portion has a humid continental climate, while its central and largest portion is subarctic. The very southern coastal area has a mild, maritime climate. Unlike ice-free Norway, Sweden's east coast on the Gulf of Bothnia usually freezes during winter.

On Sweden's western border with Norway are the Kjølen Mountains. The rest of Sweden is generally a rolling, glaciated plain with thin soils and thousands of lakes. Numerous parallel rivers flow westward down into the Gulf of Bothnia. Only the very southern tip of Sweden

Oslo is Norway's leading seaport and major urban area. The city is scenic and prosperous. Find Oslo on the map on page 133.

Oslo is located at 10° E and 60° N.

Refer students to the map on page 110. Ask them to locate the Nordic countries.

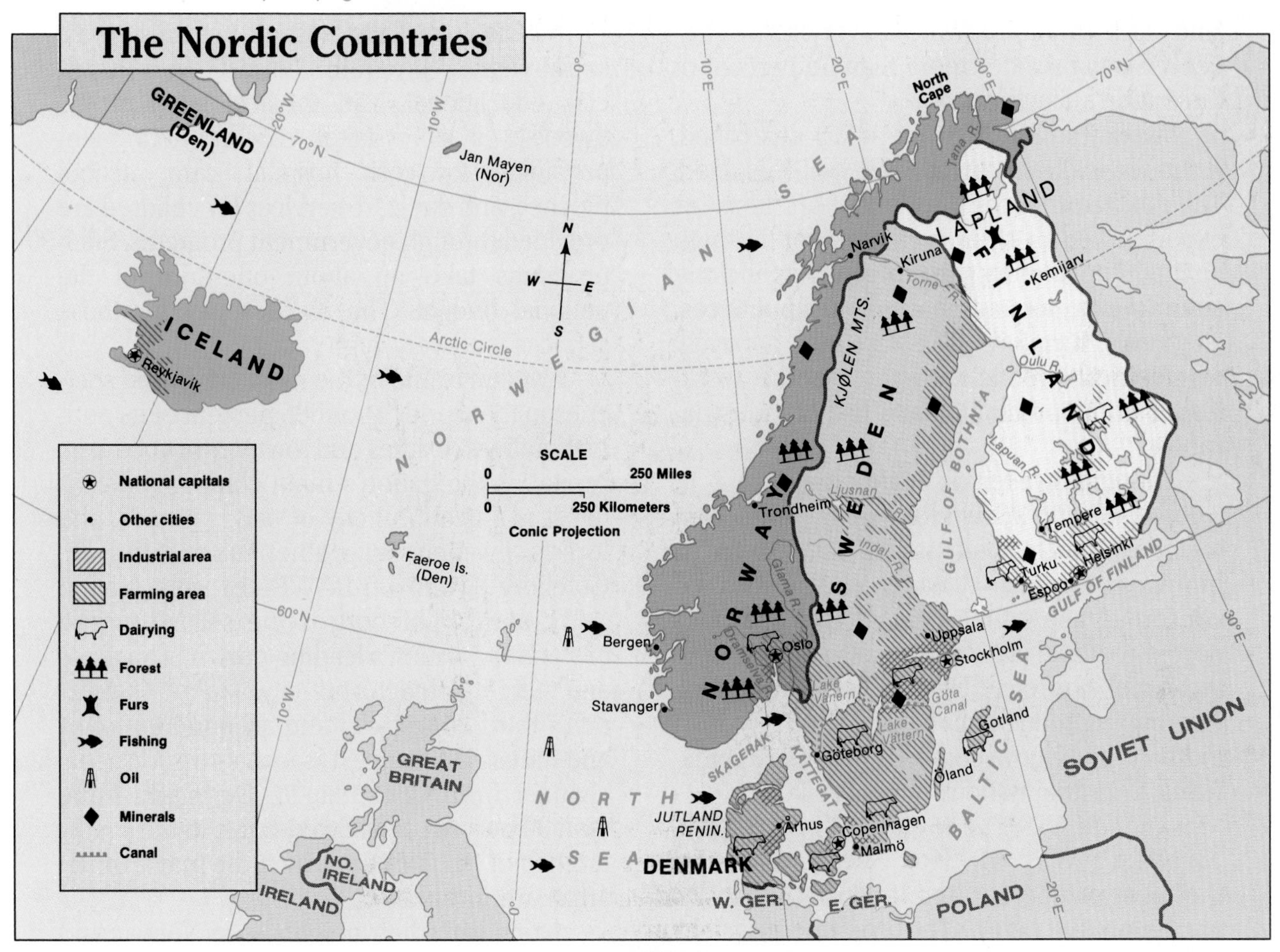

Why are the major cities of the Nordic countries located in the coastal regions?

The climate is warmer; agriculture, transportation, and shipping are more efficient along the southern coastal regions.

is part of the North European Plain. Sweden contains thousands of islands along its Baltic Sea coast. (See the map above.)

Economic Geography Sweden's four main sources of wealth are forestry, farming, mining, and manufacturing. It is manufacturing that is most important to the country's economy.

More than half of Sweden is covered with pine and spruce forests. Trees are cut in the winter. In the spring, the logs are floated downstream to sawmills. An important forest product is wood pulp for paper, matches, and building materials. One-fourth of the world's wood products are sold by Sweden. To help save the forests, the Swedish government has limited the amount of cutting allowed.

The richest farmland in Sweden is in the southern region, called Skane. Here, the soil is fertile, and the climate is mild. Wheat, rye, potatoes, sugar beets, cattle, and dairy products are sources of income. Only 5 percent of the Swedes are farmers, yet these farmers grow 90 percent of Sweden's needed food.

Being surrounded by the sea, the Swedes have traditional maritime industries such as fishing, shipbuilding, and a merchant marine fleet. These industries, however, do not play a major role in the present economy.

Northern and central Sweden have very rich iron-ore deposits. Kiruna, in the north, is the major iron-mining center. Ore is exported to the United Kingdom and West Germany from the port of Narvik (NAHR-vik) in northern Norway. Sweden also mines copper, lead,

Ask students what types of industries are publicly and privately owned in the United States. Discuss the advantages and disadvantages of public and private ownership.

zinc, and uranium. Waterpower from Sweden's rivers provides more than 50 percent of the nation's electricity.

Sweden's industries are highly diversified, technologically advanced, and well managed. The nation produces a variety of products for export. Swedish steel is famous for its high quality. The country's major products include automobiles, aircraft, household appliances, chemicals, furniture, glassware, and a variety of electrical products. Sweden is the world leader in the production and use of industrial robots.

Though Sweden is highly successful in world trade and known for quality products, it must deal with some issues that affect its future as an industrial leader. The Swedes are concerned about their dependence upon imported coal and oil. Many are also questioning the expansion of nuclear power stations for electricity production. In addition, Swedish industry must now compete with the rapidly rising industrial nations of East Asia.

Sweden's Way of Life Sweden is one of the most prosperous and democratic nations in the world today. The country practices **socialism**, an economic system by which the government owns and controls the means of producing goods. However, because Sweden's industry is both privately and publicly owned, it is sometimes called the land of the "middle way." The government controls about 10 percent of the nation's industry and services, including telephones, railroads, and airlines. The nation's electric companies may be either publicly or privately owned. The remaining 90 percent of the country's economy is in private hands.

A number of Swedish businesses are run by **cooperatives** owned by their customers. A cooperative is a business organization owned by and operated for the mutual benefit of its members. Cooperatives are particularly useful to small businesses and farmers. Those who cannot individually afford to transport products can pool their efforts to distribute their goods to the market.

Sweden's socialist system includes a vast social-welfare program. Almost every financial, educational, and medical need of the Swedish citizen is taken care of. Retirement pensions, low-cost hospital care, school lunches, and day-care services for children are provided through government programs. Such programs take up about one-third of the national budget. The Swedes pay for these services with very high taxes.

Sweden is one of the most urbanized societies in the world. About 85 percent of its population lives in cities and towns. Stockholm, its capital, is the nation's main commercial center. It is a beautiful city of parks, islands, and forests. Sweden's two other major cities are Göteborg (yuhrt-uh-BAWR-ee) and Malmö (MAL-muh). Göteborg is the second largest city and the nation's leading port. It is an auto- and steel-manufacturing city. Malmö is Sweden's third largest city and an important port and industrial center. It is just across a narrow channel from Copenhagen, Denmark. More than 90 percent of all Swedes live in the southern half of the country, where the major urban areas are located.

Along with their neighbors in Norway and Denmark, the Swedish people have a high standard of living. The government encourages recreation by providing public facilities for skiing, hiking, and sailing. Sweden has a neutral foreign policy and is not a member of NATO. **Neutrality** is a nation's policy of not taking sides in international affairs. However, Sweden maintains a strong military force. The nation is concerned about potential Soviet military threats.

Denmark

Physical Geography Denmark is the smallest of the Nordic countries. It is about the size of Massachusetts and New Hampshire combined. The majority of Denmark's land area is located on the Jutland Peninsula, which borders West Germany on the south.

Level B: Ask students to write a paragraph or two about why they think the population of Sweden is concentrated in the southern portion of the nation.

Four hundred islands make up the rest of Denmark. Just north of Denmark, across the narrow waters between the North Sea and the Baltic Sea, is Norway; to its east is Sweden. This central location between land and water routes has been an asset for Denmark. (See the map on page 133.)

Denmark has a mild, maritime climate with adequate year-round rainfall. Part of the North European Plain, Denmark's flat, rolling lowlands are excellent for farming.

Denmark is the most densely populated Nordic country. One-third of its population lives in the Copenhagen area. Copenhagen, the capital of Denmark, is situated on an island near the strait leading to the Baltic Sea. The city is a center of trade, shipping, and manufacturing.

Economic Geography Seventy percent of Denmark's land is used for agriculture. The nation's mild climate and rich pastureland have made dairy farming and hog raising the major agricultural pursuits. The Danes are famous for their fine butter, cheese, and ham. More than 90 percent of the farm income is from dairy and meat products. The remainder is derived from crops, mainly barley.

Most of Denmark's land is farmed, although the soil is not very fertile. Dairy farming is a major economic activity.

While Denmark is known as an agricultural nation, it is also modern and industrialized. Although Denmark lacks the natural resources necessary for industry, the Danes have been very successful at producing quality products for the world markets. These products include furniture, toys, glass, marine engines, medicines, and packaged foods. All of Denmark's industry is privately owned.

Denmark has a large fishing fleet, as well as one of the world's largest merchant fleets. Some offshore gas and oil deposits have been found in the Danish waters of the North Sea. These deposits may help Denmark become less dependent upon imported fuels.

Greenland: A Danish Dependency Situated between the North Atlantic and the Arctic oceans, Greenland is the world's largest island. Though located near North America, it is actually a self-governing province of Denmark. Only the southern coast of Greenland is actually green. More than 85 percent of the giant island is covered by a thick ice cap and coastal glaciers. The whole icy interior of the island is **uninhabitable**. An uninhabitable region is one that cannot support human life and settlements. Only along the southeastern and southwestern coasts are there areas of tundra climate where people can live. Even much of the coastal area is barren rock.

The Vikings discovered the island in the tenth century, naming it Greenland to encourage settlement. Most of the people of the country are Eskimo, also called Inuit (IN-yuh-wuht), or are a European-Eskimo mixture. Most Greenlanders live along the southwestern coast of the island. Fishing is the chief occupation. Livestock grazing and mining are also done on a limited scale. Eskimos along the northern coasts of Greenland still carry on traditional hunting of seals, small whales, and arctic fox.

Have students research the Mid-Atlantic Ridge.

Iceland's capital, Reykjavík, is visible on the far left of this Landsat photograph. Most of Iceland is covered by lava and volcanic ash.

Iceland

Physical Geography To the west of the Scandinavian Peninsula, in the middle of the North Atlantic Ocean, is the island country of Iceland. (See the map on page 133.) This island covers an area about the same size as the state of Kentucky. Iceland has the world's most northern capital, Reykjavík (RAYK-yuh-vik), located very near the Arctic Circle. Situated on the west coast, Reykjavík contains more than half of Iceland's population.

Iceland was originally settled by Vikings from Norway. Icelanders still speak a language very similar to that spoken by the Vikings. The Icelandic parliament is more than 1,000 years old, the oldest in the world. The country is situated on top of the volcanic Mid-Atlantic Ridge and has more than 200 volcanoes. Many of these volcanoes are active. Three-quarters of Iceland is mountainous and uninhabitable. Glacial ice covers about one-eighth of the high mountain interior. The coastline is extremely rugged and constantly battered by waves.

The name Iceland is misleading, since the majority of the island is not ice covered or extremely cold. Because of the island's location in the middle of the Atlantic, the ocean currents keep temperatures mild. The country has mainly a cool, maritime climate. It is windy, wet, and cloudy much of the year. Temperatures usually remain above freezing.

Less than one percent of Iceland supports trees. The ground is mainly barren volcanic lava or covered with grasses. Hay and potatoes are grown on the grasslands. Sheep and cattle are also grazed there. Wool is used mainly to manufacture sweaters, which are needed by the people in this cool climate.

Iceland's most important natural resource is the rich fishing waters that surround the island. Fishing and fish products are the main industries. Over 70 percent of the nation's export income comes from fishing. To protect its fishing resources, Iceland claims a 200-nautical-mile (320-kilometer) fishing limit around the island.

Because of its mountains and wet climate, Iceland produces much hydroelectricity. It also gets energy from the ground. As a result of volcanic activity just below the earth's surface, underground water rises all over the island as steam, forming **geysers**. The word *geyser* comes from an Icelandic term for a hot spring that shoots up fountains of hot water and steam into the air. In most Icelandic homes, hot water is piped from the boiling springs for heat. The hot springs are also used to heat greenhouses, where crops such as tomatoes, grapes, and cucumbers can be grown year-round. Geysers are tapped to generate electricity. This cheap electrical energy from geysers and waterpower has attracted the aluminum-refining industry to Iceland.

Despite its isolated location and barren land, Iceland has a high standard of living. The

 Level B: Have students measure distances to Oslo, Stockholm, and Copenhagen. Why is Iceland sometimes considered a part of Scandinavia?

The Icelanders have adapted to their physical environment. Heat and electricity are generated by hot springs and geysers.

country has little crime, unemployment, or illiteracy. Icelanders have made great progress in using their country's natural resources to live comfortably in a challenging environment.

Finland

Physical Geography Finland lies across the Gulf of Bothnia from Sweden. It borders the Soviet Union in the east and Sweden and Norway in the far north. (See the map on page 133.) Finland's landforms show the effects of the last ice age, during which time the whole country was covered by thick sheets of ice. Today, Finland is made up mostly of a low, glaciated plateau with more than 60,000 lakes. A narrow coastal plain is found in the west along the Gulf of Bothnia and south along the Gulf of Finland. Several thousand small islands in the Gulf of Bothnia are also part of Finland. Finland's rivers play an important role in producing hydroelectricity and in providing local transportation.

Because the Gulf of Bothnia freezes during most winters, Finland has a more severe climate than the other Nordic countries, which are nearer to open seawater. Finland can be divided into three major climate regions. In the far south of the country is a humid continental climate. The central part of the country has a subarctic climate. To the far north is a tundra climate. In the humid continental climate region, the summer growing season is very short. Despite this, the Finns grow some crops. The subarctic climate region is covered mostly by evergreen forests. In the far north, where there is only sparse tundra, the main activity is reindeer herding, which is done by people known as the Lapps. Later in this chapter, you will read more about this tundra region and the Lapps who inhabit it.

Economic Geography Less than 10 percent of Finland is suitable for farming. Glaciers

Review the effect of large bodies of water on climate. Ask students why Finland benefits little from the warm ocean currents that modify most climates of Western Europe.

A Finnish custom is the use of the sauna. Ask students what a sauna is. Ask students to speculate on why saunas are popular in Finland.

eroded most of the good soils, leaving behind either poorly drained marsh or barren rock and gravel deposits. Poor soil and a short growing season limit the Finns mainly to dairying. Hay and oats are grown to feed livestock. Potatoes, wheat, rye, and barley are other important crops.

Forests, which cover about 60 percent of the country, are Finland's greatest natural resource, and lumbering is one of the country's main industries. As a wood-producing nation, Finland ranks among the highest in the world. Wood products make up 40 percent of the country's exports. Finland's paper mills produce much of Western Europe's newsprint. Besides wood products, Finland is a producer of textiles, chemicals, ships, and electric products. The Finns are noted for their icebreakers and cruise ships. They also produce televisions and mobile telephones for Western Europe. Finland's major trade partners are other Nordic nations, the European Community, and the Soviet Union. The Soviets buy manufactured goods from Finland in trade for oil and gas.

The Finnish People Along with the other Nordic nations, Finland enjoys a high standard of living. In spite of a northern environment, the Finns have built a prosperous country.

More than half of Finland's exports are lumber and forest products. Forests cover about 60 percent of the country.

More than one-fifth of the population of Finland lives in or around Helsinki. Helsinki is the Finnish capital, leading industrial area, and major seaport. The city is situated on the Gulf of Finland, at the very southern end of the country. Tampere (TAM-puh-ray) is the second largest city and industrial center. It is located north of Helsinki, in the middle of the lake region. Turku is an important city and lumber port on the southwest coast.

Finns are not true Scandinavians. The earliest Finns came from Central Asia. Their language belongs to the same language family as Hungarian. Finnish and Hungarian are not related to languages spoken by other European nations. The second official language of Finland is Swedish. It is taught in the schools and is spoken by almost all Finns.

For most of its history, Finland was under the control of one of its neighbors. It was part of the Kingdom of Sweden from the thirteenth until the seventeenth century. Finland was under Russian control until 1917. During World War II, it was invaded by Soviet troops. Today, Finland is an independent republic. The nation follows a careful path of remaining neutral. Finland has good trade relations with the Soviet Union, but it remains a Western capitalist nation.

Lapland Across northern Finland, Sweden, and Norway is a region known as Lapland. This tundra region is populated by the Lapps, or Sami as they call themselves. These reindeer herders probably came from Central Asia. In the thirteenth century, they came into contact with Nordic fur trappers. However, they managed to maintain their culture, even with a Nordic way of life in their midst. The governments of the three nations have allowed them to retain their own language, music, and handicrafts. Today, most Lapps lead modern lives. Only about 10 percent of their population makes its living from reindeer. They sell the meat, fur, and horns to overseas markets. The reindeer herders are considered important to the Lapps as the keepers of their culture.

 Level A: Finland has had a difficult history living next to the USSR. Ask students to research the relations between Finland and the Soviet Union. Making a time line for this activity would be useful.

CHAPTER 12 CHECK

Reviewing the Main Ideas

1. Norway, Sweden, Denmark, Iceland, and Finland are the Nordic countries, often called Scandinavia. Each is a prosperous, industrialized nation. Sweden's industries are the most diversified of all the Nordic countries. Forestry, farming, mining, and manufacturing are all sources of wealth.
2. The climates and economies of these countries are influenced by the countries' nearness to the sea. Norway is bounded by the Atlantic Ocean and the North Sea. Sweden is bounded by the Gulf of Bothnia and the Baltic Sea. Most of Denmark is located on the Jutland Peninsula. Iceland rests on top of the Mid-Atlantic Ridge, in the middle of the Atlantic Ocean. Finland is bounded by the Gulf of Bothnia and by the Gulf of Finland.
3. Denmark is the smallest of the Nordic countries. Its central location between important land and water routes has been helpful in its development. Iceland is an island country. Like Norway, fishing and fish products are its main industries. Finland's poor soil limits the country mainly to dairy farming. Its greatest natural resource is its forests.

Building a Vocabulary

1. Define *cooperative*. How do businesses and small farmers benefit from cooperatives?
2. Much of Iceland is uninhabitable. How has this affected settlement patterns?
3. Define *fiord*. How were Norway's fiords formed?
4. Define *geyser*. In what Scandinavian country are geysers found?
5. Define *socialism*. Which Nordic governments practice socialism?
6. Why does Sweden have a foreign policy of neutrality yet maintain a military force?

Recalling and Reviewing

1. Why do most Norwegians live along the coast?
2. What are Sweden's main sources of wealth? How is Sweden able to grow enough food to be almost self-sufficient in food production?
3. How have landforms and climate influenced the way the Finns make their living?

Critical Thinking

4. Many sports that people enjoy today began as activities essential to everyday life. How do you think skiing has been essential to daily life in Scandinavia? What other sports originated from activities that people performed to meet their basic needs?

Using Geography Skills

Use the map in this chapter and the climate map in Chapter 10 to answer the following question.

1. What climate area is found in Finland north of 65° latitude? What economic activities are carried out in this region?

Answers to Chapter Check questions are found in the Teacher's Guide.

Additional activities for Chapter 12 are found in the Student Workbook. Chapter 12 Test as well as Skills, Reteaching, Critical Thinking, and Challenge/Enrichment activities are available in the Teacher's Resource Binder.

Chapter 13

Chapter 13 Lesson Plans and Planning Guide are found in the Teacher's Guide.

Refer students to the map on page 143.

France and the Low Countries

▲ **Sidewalk cafés line the boulevards of Paris.**

▲ **Flowers are an important Dutch export.**

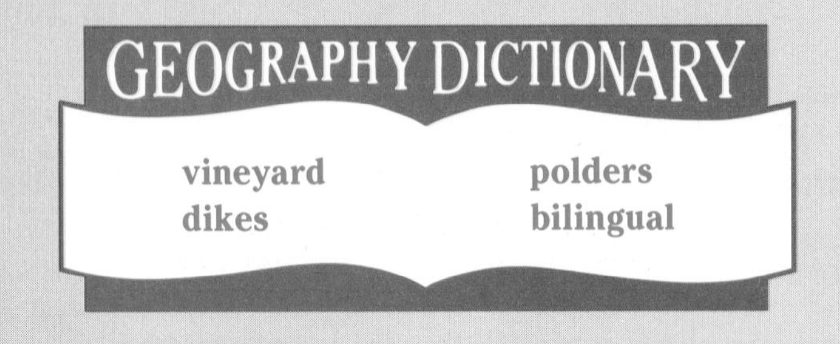

Though it is the largest nation in Western Europe, France is one of the least populated. France was once part of the Roman Empire and known as Gaul. The nation became a major European power in the seventeenth century. The French Revolution changed France from a monarchy to a republic. Napoleon spread French influence across Europe. Later, the French colonial empire extended worldwide to rival the British Empire. Today, France is a prosperous, unified nation whose people share tremendous cultural pride in their country.

To the northeast of France are three small nations known as the Low Countries. They are Belgium, the Netherlands, and Luxembourg, often called Benelux from the first letters in each country's name. The three nations have similarities besides their geographic location. They are all industrialized, modern, and prosperous nations; they are all constitutional monarchies. They also have similar histories.

 Have students briefly survey the photos in this chapter. Ask them to note the variety of economic activities shown.

France

Physical Geography France has high mountains, hills and plateaus, flat plains, and several coastlines. These features give France the greatest variety of landforms of all Western European nations. (See the map on page 143.)

The largest landform region of France is the North European Plain. This lowland area extends through most of northern France. Its level land and fertile soil make this a productive agricultural region. Only the coast of Brittany, in northwestern France, is not part of the North European Plain. Brittany is part of the Northwest Highlands.

The Central Uplands are located mainly in central and eastern France. Most of this region is hills or plateaus. The Massif Central area is in south-central France. The Ardennes, Vosges, and Jura mountain ranges form France's eastern borders with West Germany and Switzerland. The rocky and rugged island of Corsica is also part of this region.

Much of southern France is in the Alpine Mountains region. These rugged mountains form the Pyrenees, which separate France from Spain. Separating France from Italy are the Alps. Western Europe's highest peak, Mont Blanc, is in the French Alps. The rugged but beautiful French Riviera is located where the Alps meet the Mediterranean coast.

France faces three important bodies of water: the English Channel, the Atlantic Ocean, and the Mediterranean Sea. These many coastlines have helped France become a nation of international trade and a major naval power. The waters support the country's important fishing industry.

France has a number of large and important rivers. The Seine flows north to the English Channel. The Loire and Garonne (guh-RAHN) flow to the Bay of Biscay, which is part of the Atlantic Ocean. The Rhône flows south to the Mediterranean Sea. The Rhine River borders France for a part of its course, and then flows to the North Sea via West Germany and the Netherlands. The canals that connect the rivers offer an excellent system of waterways for transporting goods to neighboring Western European nations.

Except for the high mountain areas, almost all of France has a mild climate. Most of northern France has a maritime climate. Temperatures tend to be stable, and rainfall is plentiful. Inland toward the south and east, the climate becomes more continental. Summers are hotter and winters are cooler. The far southern section of France along the Mediterranean Sea has yet a different climate type. Here, the sunny and warm Mediterranean climate is found. Hot, dry summers and mild winters are typical. Only the mistral breaks the mild Mediterranean winter weather pattern. The mistral is a strong, cool wind that blows from the north toward the Mediterranean coast. These powerful winds follow the valley of the Rhône River.

Rich Farmlands Ideal climates and rich soils make France a highly productive agricultural nation. Even though France today is a modern industrial nation, agriculture remains a vital part of its economy. More than 85 percent of the land is suitable for farming. Yet,

Fertile soils are France's most important natural resource. Crops grow on more than a third of the nation's land.

Have students name well-known cheeses and wines of France. Write the names on the chalkboard. These products are usually named for the region in which they are produced. Have students locate these places on a map.

only about 8 percent of the people are now farmers. Several decades ago, farming was France's leading form of employment. In recent years, many small farming villages have lost all their young population. Mechanization has replaced much field labor, requiring only a few workers to run a large farm. The average size of French farms, particularly in the north, has doubled in the last 20 years.

Except for products grown in the tropics, France is self-sufficient in food. With more than one-fourth of its exports being food products, France is by far the largest farm exporter in the European Community.

In the north and interior of France, the leading agricultural products are wheat, oats, sugar beets, potatoes, grapes, cheese, and meat. In the Mediterranean climate of southern France, grapes, olives, vegetables, and citrus fruits are the main crops. However, irrigation is necessary for growing some crops in the southern part of France. **Vineyards** are common throughout the country. A vineyard is land cultivated for growing grapes, especially for wine production. France leads the world in the variety of wines produced and in export income from wine. Only Italy produces more wine than France. France is also a world leader in the export of wheat and cheese.

Grapes are a profitable crop in France. These vineyards near the Massif Central produce some of the world's finest wines.

Economic Geography France has not only favorable climates and fertile soils but other natural resources as well. Fishing is an important occupation on each of its coastlines, particularly the coast of Brittany in the far northwest. The mountains provide France with some waterpower and timber.

The iron-ore deposits of Lorraine, in the northeast, were the foundation of France's early steel industry. This area has been one of Western Europe's largest sources of iron ore. The iron ore was also imported by the steel industries of Belgium and Germany. The coal deposits of France are found in the north near the Belgian border. A chemical industry grew in this region to make use of the coal by-products. France today can produce only part of the high-grade coal it needs to use in steel mills and to produce electricity. Therefore, France must import coal. Today, the steel, coal, and textile industries of northeastern France are in a depressed state. Unemployment is high in this so-called rust belt, and population has been shrinking in these northeastern provinces.

Other French industries are doing well, particularly in high technology. France is one of the world's leading aerospace and military arms manufacturers and exporters. It produces satellites, missiles, jet aircraft, helicopters, and hand-held weapons. Much of this industry, which requires aluminum, is in southern France because of generous deposits of bauxite located there. Many of the provinces in southern France are gaining population because of employment offered by the aerospace, computer, nuclear-technology, and other modern industries.

About one-third of French industry and services are nationalized, or government owned. These include automobiles, aluminum, electronics, glass, engineering, and banking. Because government ownership and management of some industries has not been successful or efficient, many industries are being sold into private ownership. Much of the concern about the French economy is in the area of energy resources. French coal

 Level A: Have students research life on a typical French farm. How is it different from life on an American farm? How is it the same?

Have students locate Paris on the map on page 143.

Right Bank. Called the center of intellectual life, the Left Bank has traditionally been the home of artists, writers, universities, museums, and government offices. The Right Bank is the heart of the city's economic life. It has many cafés, shops, banks, foreign embassies, and small industries such as clothing design. In this century, there has been rapid growth of suburban areas outside the center of Paris.

The central part of Paris is very crowded. Its narrow streets were constructed long before the invention of the automobile. The greatest number of employment opportunities are in the central and western parts of the city, but most of the people live in the north and east. Therefore, two million Parisians commute to and from work each day, creating major traffic jams during rush hours. Because of the congestion, the people of Paris rely heavily on trains and the *métro* and less on automobiles. Transportation problems and overcrowding have been difficult to overcome. In 1965, the government announced long-range plans to improve the city and ease the traffic congestion by the year 2000.

▲ **The Pompidou Center is modern and dramatic.**

▼ **Traffic stalls near the Arc de Triomphe.**

The people of Paris are proud of their city's rich history. Many of them do not like to see their famous old buildings disturbed by wide roads and modern structures. Yet, many new offices and high-rise apartments have been built recently, and old buildings have been repaired. Like many old and densely populated cities, Paris has not been easy to modernize. Progress is being made toward solving the problems of rapid growth. The challenge is to preserve the cultural heritage that has long made Paris one of the world's most fascinating cities.

For the most part, the French are a prosperous people with a comfortable way of life. They enjoy excellent food, high health standards, and much cultural pride.

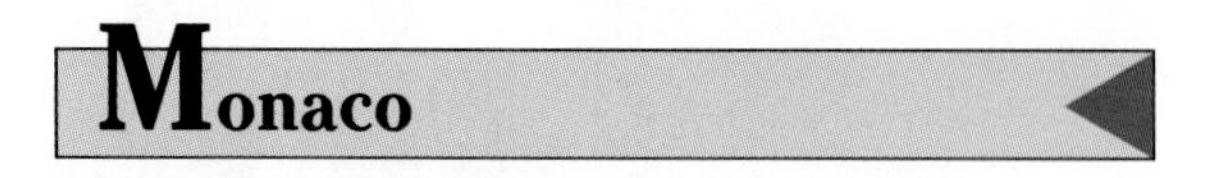

Monaco

Monaco is a microstate on the Mediterranean coast of France. It is a constitutional monarchy, officially known as the Principality of Monaco. Monaco is one of the smallest and most densely populated independent nations in the world. The majority of its people are French, with some Italian and other European residents. About one-sixth of the population is French-Italian.

This prosperous country depends mainly on tourism, banking, small industries, and postage stamps as sources of income. Many foreign companies have their headquarters in Monaco because of its low taxes. Monaco's customs regulations and defense are handled by France.

The Netherlands

Land Below Sea Level *Netherlands* means "low lands." Today, one-third of the Netherlands lies below sea level. These low-lying lands were once covered by lakes, swamps, or the North Sea. First, **dikes**, or walls to keep out water, were built. Then canals were built to remove the water. The land was eventually drained, or pumped dry, by windmills. The windmill became a symbol of the Netherlands. These lowland areas that have been drained are called **polders**. Originally, these lands were reclaimed for farmland. Today, polders are also the sites of cities, factories, and even airports. (See Themes in Geography "Polders of the Netherlands," on page 150.) The name Holland is sometimes used to refer to the Netherlands. *Holland* means "hollow land."

There are three landform regions in the Netherlands. Along the North Sea coast are high dunes, which form a barrier to the sea. Grasses are planted to stabilize this narrow wall of sand dunes from storm waves. A chain of sandy offshore islands is a continuation of the sand-dune zone. Inland from the narrow dune region are the low coastal plain and the polders. They constitute the largest region, covering about two-thirds of the nation. It is crossed by a number of major rivers, including the Rhine. Finally, in the very southern interior of the country, there is a higher area of hills covered with forest and pasture.

Economic Geography While the Netherlands is a highly industrialized nation, the country is known for its agriculture. Dutch farmers are the most productive in the world. (The people of the Netherlands are known as the Dutch.) Dairy products are major exports. Main crops include potatoes, sugar beets, and flowers. The Dutch town of Haarlem (HAHR-luhm) is the tulip-bulb center of the world.

Dutch industry is highly diversified, all privately owned, and aimed at the export market. West Germany and the other European Community nations are its major customers. The major industries are chemicals, petroleum, and metals. In fact, one of the world's largest oil refineries and the headquarters for a major world oil company are located in the Netherlands. Also important are food processing, especially the processing of dairy products and chocolate, and breweries. Electronics, optical instruments, jewelry, computers, and military arms also are produced.

With few exceptions, the Netherlands must import all its raw materials for industry. Though there are coal deposits in the hilly south, mining has been discontinued. It is cheaper for the Netherlands to import coal than to mine it.

One of the world's largest natural-gas fields is in the northeastern Netherlands. Natural-gas deposits are also located on the continental shelf of the North Sea. These deposits have been important for use in homes and for

 Level A: Students will find the bulb mania of the early 1700s in the Netherlands interesting. Ask students to research in the library the beginnings of this fad. Why did it occur?

Amsterdam is connected to other Dutch cities and international destinations by an excellent transportation network.

the manufacture of fertilizers and plastics. About half of this gas is exported by pipeline to other Western European nations. Unless more gas is found under the North Sea, the Netherlands will have to stop exporting it within the next decade. The nation will need the gas for its own use. While gas has been abundant, oil deposits are very limited in the Netherlands. Almost all of the country's oil is imported.

The Netherlands is known for its excellent transportation network of fine roads, railroads, and airports. Also, three large rivers, the Rhine, the Meuse (MYOOZ), and the Schelde (SKEHL-duh), have their outlets in the Netherlands. The rivers are connected by canals and serve as a vital link between Western Europe and the North Sea.

Urban Geography Amsterdam is the Netherlands' official capital. It is a cosmopolitan city in that people from all over the world live there. It is an old port city, with picturesque canals, museums, and art centers. Amsterdam is also the site of chocolate factories, breweries, the diamond-cutting industry, and international banking. Although Amsterdam is 15 miles (24 kilometers) from the North Sea, it is connected to it by the North Sea Canal.

The port of Rotterdam serves the trade traffic that enters and exits Western Europe on the Rhine River. Europoort, as Rotterdam is called, is the world's busiest seaport. It is also the site of oil refineries, chemical plants, and shipbuilding. The Hague is a quiet, beautiful city. It is the seat of government, where parliament meets and the royal family lives. The International Court of Justice is also here.

These three large cities have grown together to form one giant urban area known as the Randstad. The Randstad covers about one-fifth of the small country and contains more than half the population.

Dutch Society The Netherlands was formed in the sixteenth century when it gained independence from Spain. By the seventeenth century, it was one of the leading industrial and commercial nations in Europe.

Level A: Have students read *The Diary of Anne Frank* if they have not already done so. In what ways did the Dutch provide safety for Jews in World War II? Another book on the same topic is *The Hiding Place* by Corrie Ten Boom.

Themes in Geography

You may want to review the five themes of geography presented in Unit 1, pages 33–3

POLDERS OF THE NETHERLANDS

More than 50 percent of the Netherlands was once covered by shallow sea, inland lakes, and swamps. As the demand for agricultural and industrial land increased, the Dutch people found that they could obtain, or reclaim, large areas of valuable land if they drained the water from low-lying areas. This reclamation was necessary because most of the Netherlands consists of low-lying and flooded land, with very few areas of high elevation.

Location **The city of Amsterdam, the capital of the Netherlands, is built on land that was once below water. The Netherlands lies on the North Sea, along the northeast coast of Europe. Can you identify other Dutch cities that are built on reclaimed land?** Any cities that lie below sea level. These include Rotterdam and The Hague.

At first, the Dutch built walls, called dikes, around the area to be drained. Then they pumped the water in this area into canals that drained into the North Sea. In the early days, they used windmills to generate the power to operate the pumps. The windmill became a symbol of the Netherlands. Later, they used steam power and, eventually, electric power. The areas of land that were once submerged beneath the sea or inland waters are called polders. Because the polders lie below sea level, they have no natural drainage. To prevent their becoming flooded, all excess water must be pumped from them into the canals.

As a result of the creation of the polders, and because of the poor natural drainage of the land, the Dutch landscape is covered by a network of canals. These are important for drainage, for irrigating farmlands, and as waterways for transporting goods. The polders have provided very fertile farmland, and the Dutch farmers use all the space available.

The Netherlands is one of Europe's most densely populated countries. Because it is also a very small country, there is much competition among farming, industry, and cities for use of the reclaimed land. Many of the cities, including Amsterdam, the capital and largest city, have been built on polders.

Since the first construction of the dikes and creation of polders, the Dutch have struggled constantly to keep the sea from flooding their land. Storm tides from the North Sea have broken through the dikes many times in the country's history. If it were not for the dikes and sand dunes that line the shores, half the country would be flooded at high tide. Today, major projects are under way to improve the dikes along the country's

Ask students why the polders can be considered a region.

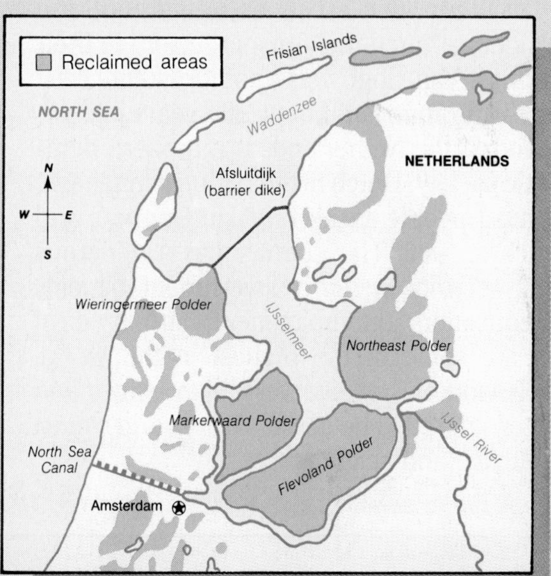

Region **Large areas of land in the Netherlands lie below sea level. This means that excess water must be drained artificially by the use of an extensive system of canals, dikes, and pumping stations. This photograph shows land after the polders were created. Do you live in a region that experiences drainage problems similar to those experienced by the Dutch people? When it rains, where does the water flow?**

No coastal area in the U.S. lies below sea level, but some cities, such as New Orleans, lie below the level of a nearby river. Some flat areas, such as Florida, have serious drainage problems. In flat areas, water collects in puddles and can cause flooding. In hilly areas, it flows downhill into streams, drainage ditches, and rivers.

coastline. These repairs will be completed by the end of this century and should provide protection from further destruction.

Relationships Within Places **Since the first polders were built, the Dutch people have fought a constant battle with the waters of the North Sea. Many times, storms have breached the dikes, causing extensive flooding and destruction. Look at the map of the Netherlands above. Can you identify the areas that would be flooded if the coastal dunes and dikes were removed?**

All areas that lie below sea level.

Dutch sailing ships and businesses were found worldwide and controlled much of the spice trade. The Dutch claimed colonies in the Caribbean and Dutch East Indies (now Indonesia). This was also a period of prosperity in Holland and a time when the great Dutch Masters painted. Such artists as Rembrandt and Vermeer captured this prosperous era in their paintings. Van Gogh was another Dutch artist of great prominence about 200 years later.

Today, the Netherlands is a modern democratic society. Dutch is the national language, but most people speak English, German, and French as well. The country has a generous social security system. Government benefits for education, unemployment, and medical care are some of the highest in the world. Dutch workers receive very high wages and are considered to be some of the world's most productive and skilled.

The major problems the Netherlands faces are housing shortages and a tremendous burden on its social security system, resulting from recent high unemployment rates. Its population is "graying"; that is, the number of elderly people is growing at a rate much faster than the number of births. This, too, places a burden on the government social security system.

Belgium

Regions of Belgium Belgium is located to the south of the Netherlands. It can be divided into three landform regions: a coastal region, a central plain, and the Ardennes Plateau. Belgium's coastal region is a plain similar to that of the Netherlands. Much of the area is land

Vincent van Gogh, a famous Dutch artist, painted *Factories at Clichy* in 1887. The painting illustrates the growth of industry in Western Europe.

reclaimed from the sea and from swampy lowlands and is used for farming. The central plain between the Schelde and Meuse rivers contains Belgium's most productive soils. In the southeast of Belgium is the Ardennes Plateau, where forested hills reach to 2,283 feet (696 meters).

Belgium was founded in 1830 when the Roman Catholic Belgians broke away from Protestant Netherlands. The country is now predominantly Roman Catholic in religion. It is divided between the French-speaking and the Dutch-speaking populations. The northern and coastal part of the country is populated by the people who speak Flemish, a language closely related to Dutch. This portion of Belgium is called Flanders and contains about 60 percent of the country's population. It is Belgium's most prosperous region. In the southern part of the country, the French-speaking people, called Walloons, make up the other 40 percent of the population. The southern area is called Wallonia. Most Belgians speak several languages, including English.

Politically, Belgium is a constitutional monarchy united by a king. In reality, however, it is a divided nation. The people have never viewed themselves as Belgians, but as either Flemish or Walloon.

Economic Geography Belgium's efficient farming provides about 80 percent of the nation's food. Most of the farms are small and produce both crops and livestock.

Belgium was one of the first industrialized nations, and its early steel industry was based on local iron and coal mining. Today, the coal deposits are almost exhausted, and the steel industry is inefficient. Many of the steel mills must either be supported by government money or be closed. The country's energy needs are supplied mainly by imported oil.

Belgium's main resource is its skilled and productive labor force. Belgian industry produces high-quality machinery, glass, pottery, and linen products. It is known for quality carpets, cut diamonds, and chocolate. A major weapons manufacturer, Belgium supplies the world with everything from bullets to jet engines. With its small population and high standard of living, Belgium is dependent upon international trade.

Flax, which is used to manufacture linen fabric, rope, thread, and linseed oil, is one of Belgium's leading crops.

Urban Geography Brussels, Belgium's capital and largest city, is located near the dividing line between Flanders and Wallonia. Although most of the people of Brussels speak French, the city is considered **bilingual**; that

Complementarity describes the acts of two or more nations who, through an exchange of raw materials and finished products, satisfy each other's requirements. Ask students to discuss how the Benelux nations are an example of this.

Brussels is a thriving city. It is headquarters for many international groups, including the European Community and NATO.

is, the people speak two languages—in this case, French and Flemish.

Brussels is an international city; more than 10 percent of its residents are foreign. It has become an important center for finance and commerce. Hundreds of organizations have their offices there, including Benelux, NATO, and the European Community. Brussels' central location in Western Europe and excellent transportation connections have helped attract many foreign businesses.

Antwerp is an important industrial city famous for its diamond-cutting business. After Rotterdam, it is Western Europe's busiest port. Antwerp is located on the Schelde River and is connected by canals to the North Sea and the Rhine River.

Luxembourg

Luxembourg covers an area smaller than Rhode Island, the United States' smallest state. This independent country is bordered by France, West Germany, and Belgium. (See the map on page 143.)

Luxembourg is a democracy with a royal family, officially called the Grand Duchy of Luxembourg. Its history can be traced back to 963. Its early independence was followed by a period of foreign rule. Though Luxembourg claimed neutrality, it was invaded by Germany during World War I and again during World War II.

The northern half of the country is the hilly and forested Ardennes. The southern half is rolling farmland with good soil. It is called Gutland (Good Land), with farms producing mainly dairy products and meat. Along the Moselle River valley, grapes are grown.

Luxembourg's iron and steel industry is located in the south near iron-ore mines. This was once the nation's single major industry. A drop in steel production has forced Luxembourg to attract other industries, such as tires and chemicals. Its most important recent economic development has been the growth of banking. Luxembourg is one of Western Europe's leading financial centers.

Luxembourg's population is mainly Roman Catholic. Many of the people are trilingual; that is, they speak three languages. The local language is called Luxembourgian, a Germanic language. German and French are also spoken, and most people speak English. Workers from Mediterranean Europe make up one-fourth of Luxembourg's population.

 Luxembourg can boast almost perfect land use. This land is both productive and picturesque. Do Americans follow the example of Luxembourg in maximizing land use?

CHAPTER 13 CHECK

Reviewing the Main Ideas

1. France, the largest nation in Western Europe, is a diverse country of high mountains, hills and plateaus, flat plains, and several coastlines. Its largest landform region is the North European Plain, which extends through most of northern France.
2. France's excellent location and its network of waterways have helped make the country a major trading nation. The country is also agriculturally productive: more than 85 percent of its land is suitable for farming.
3. The Low Countries are the Netherlands, Belgium, and Luxembourg. These three industrialized and prosperous nations share many similarities—geographic, cultural, political, and economic.

Building a Vocabulary

Students should provide an answer to Question 2 by filling in the blank, using their own paper.

1. Define *dike*. How were dikes used in the development of the polders? Why have polders been necessary in the Netherlands?
2. If you speak two languages, you are considered to be ______. If you speak three languages, what are you?
3. Vineyards are common throughout France. Explain their significance in the economy of France.

Recalling and Reviewing

1. Describe the major landform regions of France.
2. Why is the location of France ideal for trade?
3. Where is Monaco located? What are Monaco's major sources of income?
4. Describe the main landform regions of each of the Low Countries.
5. Why is industry more important than agriculture in Belgium and Luxembourg?

Critical Thinking

6. From what you have learned in this chapter, you know that the Low Countries have much in common. Would it be advantageous for the countries to unite into one political unit? Could there be any problems in unification? Explain.

Using Geography Skills

Use the map in this chapter, and maps in other chapters as indicated, to answer the following questions.

1. Where are France's vineyards located? In what climate area are they found? (Use the climate map in Chapter 10.)
2. Look at the population-density map of Western Europe in Chapter 10. What is the population density of the Low Countries? How does it compare to the population density of France? Explain.

Answers to Chapter Check questions are found in the Teacher's Guide.

Additional activities for Chapter 13 are found in the Student Workbook. Chapter 13 Test as well as Skills, Reteaching, Critical Thinking, and Challenge/Enrichment activities are available in the Teacher's Resource Binder.

Chapter 14

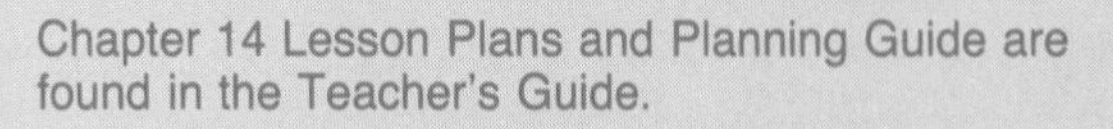
Chapter 14 Lesson Plans and Planning Guide are found in the Teacher's Guide.

Refer students to the map on page 158.

West Germany and the Alpine Countries

▲ Farms cover much of the North German Plain.

▲ The Swiss Alps contain many small villages.

GEOGRAPHY DICTIONARY

- dialect
- loess
- enclave
- confederation
- foehn
- tree line
- arable
- republic
- basin

A Germanic culture region is found throughout a large area of central Europe. The unifying factor of this region is the German language. Today, the Germanic culture region centers on West Germany, East Germany, and Austria. It also includes parts of Switzerland, Belgium, Italy, and Poland. Though there are many **dialects** of German, the language is spoken and understood by all the Germanic people. A dialect is a regional variety of a major language.

The Germanic culture has produced some of the world's greatest scientists, soldiers, and artists. It is musicians such as Bach, Handel, Beethoven, Wagner, Brahms, and Strauss who stand out as symbols of the richness of the Germanic culture.

Switzerland, Liechtenstein, and Austria are known as the Alpine Countries. Located to the south of Germany, the high peaks of the Alps run through these nations. Each of these countries maintains neutrality.

 Have students compare and contrast the photos on these two pages.

Discuss with the class the impact a unified Germany has had on world history. How might the history of the 20th century be different if Bismarck had failed? (World Wars I and II might not have occurred.)

West Germany

Divided Germany Before Germany became a nation in 1871, it was a group of more than 1,800 small kingdoms. The northern kingdoms were Protestant; the southern ones were Roman Catholic. In the nineteenth century, the German states expanded, particularly the most powerful one, Prussia. After unification in 1871, Germany soon became a major industrial power. Factories, railroads, and cities expanded rapidly. Germany began competing with Britain and France for colonies and military power. This tense competition led to World War I in 1914. Germany was defeated in 1918, and the German economy was ruined by the war. All German colonies in Africa and the Pacific were lost. The German people suffered many hardships. Under these conditions, Germany's democratic government collapsed, and Adolf Hitler and the Nazi party came to power in 1933.

By 1939, Germany had again risen to become Europe's major industrial and military power. Nazi Germany invaded neighboring countries and brought about World War II. Germany conquered most of Europe before it was defeated by the Allied forces in 1945.

After World War II, Germany was occupied by the British, Americans, French, and Soviets. Four military zones were formed. Berlin, the former capital of Germany, was divided among the four powers. Germany also lost territory to Poland and the Soviet Union.

As differences grew between the Soviet Union and the three Western powers, the chance of reunification of Germany became unlikely. In 1949, the nation was divided between east and west. The British, Americans, and French combined their zones to create the Federal Republic of Germany (FRG), known as West Germany. The city of Bonn was chosen as the capital. The Soviet zone became the German Democratic Republic (GDR). Known as East Germany, its capital is East Berlin.

During the 1950s, millions of East Germans escaped into West Germany. A majority of them were skilled workers and professionals. To prevent further loss of skilled labor, the Soviet Union built an "iron curtain" of fences and walls with armed guards to divide East and West Germany.

As a result of World War II, Germany's industries and cities were reduced to rubble. Its farms were unproductive. With help from the United States under the Marshall Plan and through the work and skill of the German people, the nation rose to become an industrial giant in less than 20 years. Today, West Germany is a modern, prosperous nation. As an industrial power, it is outproduced only by the United States, the Soviet Union, and Japan. West Germany is a member of the European Community and NATO.

Physical Geography West Germany is divided into three main landform regions: the North German Plain, the Central Uplands, and the Alps. The elevation of West Germany increases from north to south. The lowest region is the North German Plain, which is part of the North European Plain. The Central Uplands is West Germany's largest region and is made up of low mountains, forested hills, and rolling farmlands. Along the very southern

The Black Forest region is in southwestern Germany. Many wood products are made here, including toys and clocks.

Level A: Ask students to compare historical maps of Europe to trace territorial changes in Germany since 1871. They should look at maps of Germany before 1871, after World War I, and after World War II.

Have students locate West Germany and the Alpine Countries on the map on page 110.

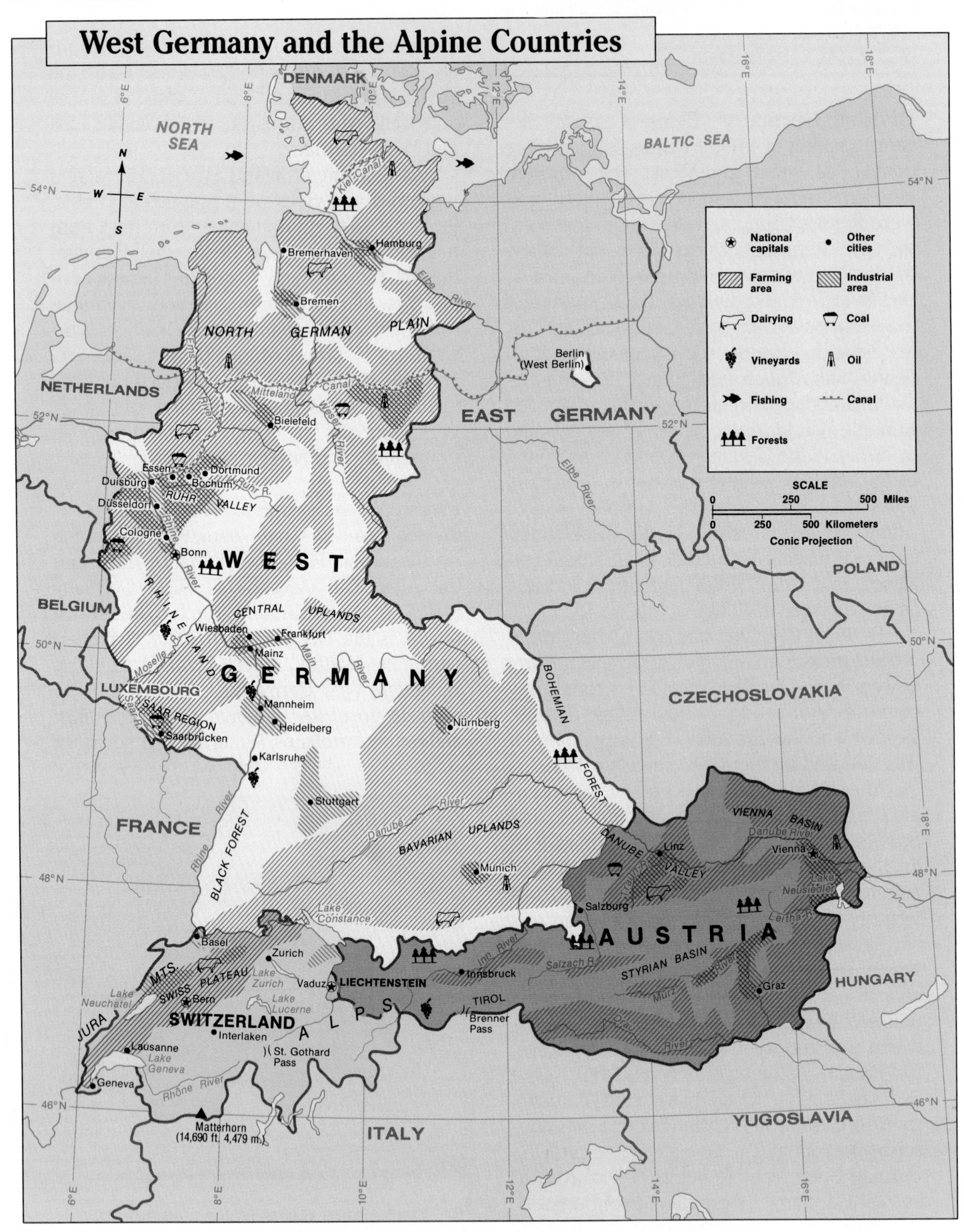

How has West Germany's river and canal system helped the nation industrialize?

Rivers and canals provide transportation for raw materials and finished goods and generate power.

West German farms produce more food than those of Great Britain but not as much as those of France. All these countries must import food, however. Ask students why this is necessary.

border of West Germany are the Alps. Here, Germany's highest peaks and most rugged landforms are found.

Germany has several large rivers. In the south, the Danube flows eastward from its headwaters in the Black Forest into Austria and Eastern Europe to the Black Sea. The Elbe, Weser (VAY-zuhr), and Rhine rivers flow to the North Sea. An excellent canal system links the rivers to form a network of waterways. The Rhine is particularly vital to West Germany's water transport system.

Most of West Germany has a maritime climate. Landforms and distance from the sea cause the temperature and rainfall to vary. The northern part of the country is affected by moist and mild ocean winds. This part of Germany has the mildest winters and the coolest summers. South, toward the interior of Germany, the continental influence is felt. Here, summers are warmer, and winters are colder. All of Germany receives adequate rainfall for farming.

Economic Geography Though an industrial nation, West Germany is not rich in natural resources. Yet, its agriculture is productive. More than 60 percent of the country is agricultural land. The main crops are grains, potatoes, sugar beets, and grapes. Though only a small percentage of the population farms, West German farmers supply the nation with three-fourths of its food. Their farms are modern, mechanized, and efficient.

West Germany's well-kept forests are mostly in the Central Uplands. Conservation and careful harvesting of timber is strictly practiced. Recent forest losses have resulted mainly from air pollution rather than from poor forestry practices. Tree loss from pollution is a national concern to West Germans, who prize their valuable forests. West Germany also has fishing resources in the North and Baltic seas.

Local coal deposits helped Germany industrialize rapidly in the last century. The coal today supplies about one-third of the nation's energy but is very expensive to mine. Since West Germany has only small deposits of iron ore, ore is imported from France and Sweden. With the use of local and imported iron ore, West Germany is the leading steel producer in Western Europe.

West Germany's major import is oil, which is used to operate the nation's industries and transportation system. Natural gas is imported by pipelines from the Netherlands and the Soviet Union.

Germany's industrious population is still its most important resource. West German workers are among the best paid and best educated in the world. They are considered world leaders in scientific, technological, and organizational skills.

West German industry is extremely diverse, producing such products as steel and machinery, automobiles, railroad equipment, consumer goods, computers, and electrical, optical, and surgical instruments. The nation's transportation network is one of the world's finest, with modern railroads, airports, and *autobahns* (freeways).

Perhaps the best way to learn about the highly developed and complex nation of West Germany is to take an imaginary trip through the country's main regions. You may want to follow along by using the map on page 158.

The West German automobile industry is world famous. How has heavy industry helped rebuild the West German economy?

West German industry is diverse and efficient.

Discuss with students the difference between the industries found in the Ruhr and those found in Saxony. Which would be less environmentally harmful? Why?

The North German Plain Along the West German coast of the North and Baltic seas, the land is poor. The weather is windy and damp. Except for yielding grasses used for cattle grazing, the soils are unproductive because of glaciation and poor drainage. Most offshore islands are covered with sand dunes. Farther to the south and inland are some of Europe's richest soils. They extend eastward into Poland and the Soviet Union. This rich agricultural belt is created by thick deposits of **loess** (LEHS), or dust-sized particles of soil deposited by the wind. Rye, potatoes, wheat, oats, and sugar beets are important crops in this region.

Hamburg, on the Elbe River, is West Germany's major seaport and second largest city. During the fifteenth and sixteenth centuries, it was one of the powerful Hanseatic ports that controlled trade in the North and Baltic seas. Bremen, West Germany's second busiest port, is on the Weser River. It is a center for shipbuilding. Downriver is Bremerhaven, West Germany's leading fishing port.

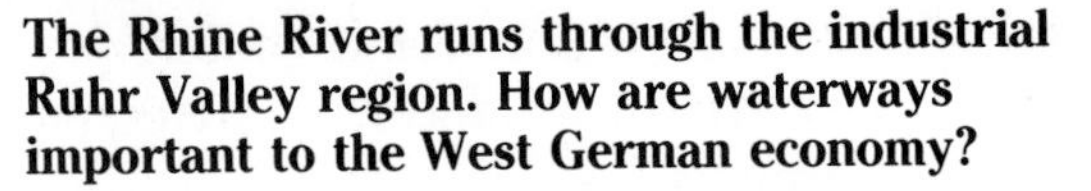

The Rhine River runs through the industrial Ruhr Valley region. How are waterways important to the West German economy?

They have made trade and transportation more efficient.

The Ruhr Valley The Ruhr Valley is the industrial heart of West Germany. This is the nation's most densely populated region. It has become almost one continuous belt of cities and industries. The major cities are Essen, Dusseldorf, and Dortmund. These cities have grown together to form the Ruhrstadt (Ruhr City). To the south, along the Rhine River, are Cologne, West Germany's fourth largest city, and Bonn, the federal capital. The Ruhr Valley is Western Europe's largest coal-producing, steel-manufacturing, and industrial area. This "smokestack region" produces a variety of other manufactured goods such as chemicals, pharmaceuticals (medicines), plastics, and automobiles.

The Rhineland South of the Ruhr is the fertile Middle Rhine Valley, also called the Rhineland. This is both an agricultural and an industrial area. The valley produces timber, barley, wheat, tobacco, and fruit. Vineyards cover the hillsides. The Rhine Valley is famous for its postcard scenery of castles, forests, and farms along the steep river gorge. The major cities along the Rhine are Wiesbaden, Mainz, Mannheim, and Heidelberg. Just to the east is Frankfurt, West Germany's main transportation and financial center. This is the second largest concentration of industries in West Germany.

The Saar West of the Rhine River, on the French border, is an old industrial area known as the Saar. Saarbrücken (zahr-BROOK-uhn) is the most important city of the region. The region grew because of rich coal deposits and the steel industry. Also an agricultural area, the Saar produces wine, vegetables, meat, and dairy products.

Bavaria and Southern West Germany The southern part of West Germany has a great variety of terrain, including the Bavarian Uplands, the Black Forest, the Bohemian Forest, and the Alps. Much of the area is forested or in farmland. Oats, barley, and hops are

Level B: Each student should design a chart to fill in about the main regions of Germany. Have students label each region and list information under these headings: Agriculture, Industry, Major Cities, Outstanding Features.

Ask students to locate Switzerland on the map on page 110. Ask why Switzerland's location is important in terms of European trade and transportation. Why don't the Alps present a major obstacle?

grown here. West Germany is the largest producer of hops, which are used by breweries.

Munich (MYOO-nik) is the nation's third largest city and the manufacturing center of Bavaria. Bavaria is the largest of 11 states in West Germany. Munich specializes in the production of electrical and scientific equipment, automobiles, textiles, and toys. Stuttgart, to the northwest of Munich, is a center of automobile and electronics industries. Munich and Stuttgart have been growing rapidly because of new industrialization, especially in high technology. The region's mountains, forests, and lakes support a major tourist industry.

West Berlin Prior to World War II, Berlin was Germany's capital and leading manufacturing center and Europe's second largest city. During the war, 60 percent of the city was destroyed. Like Germany, the city of Berlin was divided into east and west. Today, West Berlin is a geographic **enclave**. An enclave is a part of one country completely within the boundaries of another. West Berlin is surrounded by East Germany. Its border is the Berlin Wall, which was built by the East Germans to prevent people from escaping to West Berlin. The Berlin Wall is more than 100 miles (160 kilometers) long.

Munich, West Germany's third largest city, is a major industrial center. Near what international boundary is Munich located?

Austria

West Berlin is West Germany's largest city. It is a cosmopolitan city with museums, theaters, and cabarets. Its major industries are food and tobacco processing, textiles, and high technology. Tourism and shopping also add to the city's income. Yet, it is an isolated city that depends upon federal government money for support.

West Germany's Political Situation West German leaders have sought to improve relations with East Germany in recent years. The border is now more open, and West Germans are permitted to visit East Germany. Far fewer East Germans are allowed to visit West Germany, however. Although most West Germans would like to see East and West Germany reunited, most think that it will probably not happen. The present borders seem to be recognized as permanent features on today's maps of Europe.

German politics seem most divided on the subject of defense, particularly the role of NATO and the placement of nuclear weapons in West Germany. However, with its strong economy and stable government, West Germany's future looks promising.

Switzerland

Historical Geography Switzerland is a small, very prosperous country about the size of Vermont and New Hampshire combined. It was founded in 1291 when three cantons, or states, banded together. Today, Switzerland is a **confederation** of 23 cantons; its official name is the Swiss Confederation. A confederation is a group of states joined together for a common purpose. The name Switzerland comes from one of the original three founding cantons. Each canton has power to govern itself except for military defense and international relations, which are under federal control.

Level B: Ask students to trace a map of Switzerland. Have them use a map of languages of the world to make a linguistic map of Switzerland. Be sure they key the various language regions and title the map.

Switzerland has four national languages, each of which is the main language used by a part of the population. Nearly 75 percent of the Swiss speak German, 20 percent speak French, and most of the remainder Italian. In some remote Alpine villages, an Italic language called Romansh is spoken. Actually, most Swiss speak several languages. The nation is also peacefully divided by two major religions, Protestant and Roman Catholic.

By the nineteenth century, Switzerland had a large network of roads, railroads, and tunnels crossing from border to border. The country now has one of the best transportation systems in Europe. To a great extent, the rise of Switzerland to world importance is the result of its convenient geographic location at the center of Europe.

Physical Geography Switzerland can be divided into three landform regions: the Jura Mountains, the Alps, and the Swiss Plateau. Two-thirds of the country is mountainous. The Jura Mountains expand across the northwest. These forested mountains are much lower than the Alps. The Alps cover the southern half of Switzerland. The Swiss Alps reach elevations of up to 15,200 feet (4,634 meters), with many peaks above 13,120 feet (4,000 meters). The higher elevations support year-round glaciers and snowfields. Many of Europe's major rivers, including the Rhine, begin here. The important passes over the Alps are snow covered in winter, and tunnels have been dug to allow reliable transportation routes. Beautiful glaciers and lakes are located along the slopes of the Alps.

The Jura Mountains and the Alps are divided by the Swiss Plateau, or Mittelland. The Mittelland extends from Lake Geneva to Lake Constance. (See the map on page 158.) Most of Switzerland's population, along with most of its industry and farming, is located here.

Switzerland's climate varies from warm valleys to frozen mountain peaks. Most of the nation receives adequate precipitation. The Mittelland averages 50 inches (125 centimeters) per year. The surrounding mountains protect much of Switzerland from cold winter air from northern Europe. In the Alpine valleys and the Mittelland, the warm southern **foehn** (FUHRN) keeps winter temperatures mild. A foehn is a warm, dry wind that blows down the sides of mountains. On the southern slopes of the Alps, there is a small area of Mediterranean-like climate where palm trees grow within sight of the snow-covered Alps.

Resources Switzerland is a resource-poor nation. An exception is hydroelectric power, which supplies 60 percent its electricity. The lower Jura Mountains support a forest cover, while the high Swiss Alps rise above the **tree line**. The tree line is the elevation beyond which trees cannot grow; therefore, only the lower slopes of the Alps are forested. Cold, wind, and snow prevent trees from growing on the upper slopes. Alpine scenery is one of Switzerland's major resources. The snow-covered Alps are the basis of an important year-round tourist industry. Because of the country's mountainous terrain and poor soils, there is little **arable** land. Arable refers to land that is suitable for growing crops. Dairy farming and grape growing are the country's most important agricultural activities. Most Swiss farms are in the Mittelland.

Urban Geography All of Switzerland's major cities are in the Mittelland. Zurich, the largest, began as a trading and textile-weaving center. Now, it is a leading industrial and world-banking center.

Basel, on the Rhine River, connects to ocean trade through Rotterdam, in the Netherlands. Switzerland's second largest city, Basel is the home of large chemical, pharmaceutical, and engineering companies.

The medieval-walled city of Bern is the nation's capital. It is centrally located between the German-speaking and the French-speaking Swiss. Bern has a large publishing industry.

The main business of the city of Geneva is international relations. It is often the site chosen for world conferences. More than 200

 Point out that since Switzerland has no natural resources, the country must sell products whose manufacture depends upon skilled labor instead of large amounts of raw materials.

Zurich is a center for industry, international banking, and trade. Most of its people are prosperous and well educated.

international organizations maintain offices in Geneva. These include the International Red Cross (which began in Switzerland), the World Health Organization, and the United Nations, which has its European headquarters there.

Economic Geography Throughout its history, Switzerland has been in the center of trade. Even today, its economy depends upon foreign trade. More than 75 percent of all Swiss goods are exported. Most of the goods are traded with European Community nations, particularly West Germany. Switzerland is not a member of the European Community, but of the European Free Trade Association. To maintain strict neutrality, Switzerland does not belong to the United Nations or NATO.

Swiss industry is led by the machine and chemical industries. Pharmaceuticals, textiles, watches, and processed food are also very important. Swiss cheese and chocolate are world famous.

Tourism is well developed and important to the economy. However, banking brings most foreign exchange into Switzerland. The nation's peaceful stability has attracted finance and insurance industries. In fact, all of Switzerland seems to be internationalized. By maintaining their international role, the Swiss have attained the highest standard of living in the industrialized world.

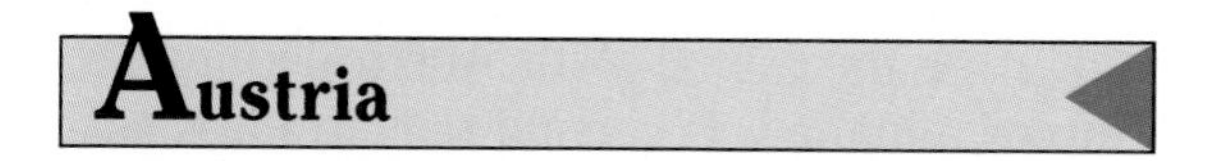

Austria

Historical Geography Throughout its history, Austria has been the crossroads of invasions and wars. Before World War I, Austria was the center of the Austro-Hungarian Empire. This empire included most of the countries that surround Austria today: Czechoslovakia, Hungary, and parts of Romania, Italy, Poland, and Yugoslavia. When World War I ended in 1918, the empire broke up. Austria shrank to a small **republic**. A republic is a form of government in which citizens elect the country's representatives and head of state.

In 1938, Nazi Germany took control of Austria. By the end of World War II in 1945, the country was occupied and divided into four military zones: American, British, French, and Soviet. The troops withdrew in 1955, and Austria was granted independent and neutral status. Today, Austria is organized into nine provinces. The population speaks German and is mainly Roman Catholic.

Physical Geography Austria is about the same size as the state of Maine and twice as large as Switzerland. (See the map on page 158.) The Alpine scenery of Austria rivals Switzerland as a tourist attraction. The Alps occupy the western two-thirds of the nation. South

Level A: Have students collect tourist literature and magazine articles on Austria. Have them write a brochure based on their research, describing what to see when visiting Austria. This could be conducted as a group activity.

Much of the music of Johann Strauss, the Waltz King, emanates from his environment in Vienna. *Tales of the Vienna Woods* and *Blue Danube* are examples. Ask students to research how music reflects the culture of an area.

of the Alps is a gentle hill country known as the Styrian Basin. A **basin** is a low area of land, generally surrounded by mountains. The Danube River flows from west to east across northern Austria. In the western part of the Danube Valley, called Upper Austria, the valley narrows. In the east, the valley widens and forms the flat Vienna Basin. This area is called Lower Austria. More than 60 percent of Austria's population lives here.

Urban Geography Along the banks of the Danube River is Vienna, Austria's capital city and center of commerce and culture. Vienna has for centuries been a city of music, drama, and art. As the capital of the Austro-Hungarian Empire, Vienna was second only to Paris as a center of culture. Today, Vienna remains the heart of Austria. It is also the center of a large agricultural region, and it is a major tourist destination.

Graz is Austria's second largest city and industrial center. It is located in the Styrian Basin. In Upper Austria along the Danube is the city of Linz, an industrial and agricultural center. Innsbruck and Salzburg are industrial and tourist centers in the Alps.

The Vienna Opera House illustrates Austria's rich cultural heritage. Austria has produced some of the world's finest composers.

Economic Geography Austria is an industrial country, but agriculture is important to its economy. Farmers make up only 6 percent of the population, yet they supply 90 percent of the nation's food. More than 40 percent of Austria is forest and supports an important forestry industry. Hydroelectricity is produced by mountain rivers. Austria is limited in other energy resources. The nation's most valuable natural resource might be its scenic beauty. Tourism is a major industry and is vital to the present economy.

Austrian industry is quite diversified. Machinery, electrical equipment, iron and steel, chemicals, glass, and paper products are produced. Most of the larger industries are nationalized. Austria trades mainly with Western European countries. It trades also with the Soviet Union and Eastern European countries for oil, gas, and coal. Like Switzerland, Austria is not a member of the European Community but of the European Free Trade Association.

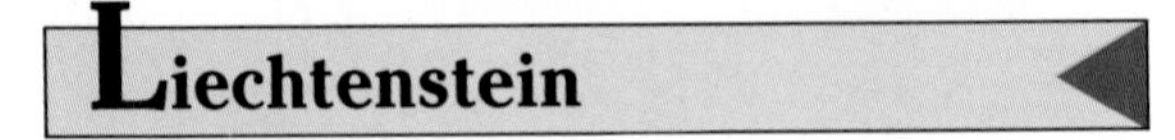

Liechtenstein

Located in the Alps on the Rhine River between Switzerland and Austria is the microstate of Liechtenstein (LIK-tuhn-s[h]tiyn). (See the map on page 158.) Liechtenstein is only 15 by 4 miles (24 by 6 kilometers) in area. Most of its people speak German and are Roman Catholic.

The economy of this prosperous country is based upon numerous industries. Small machines, ceramics, and artificial teeth are among its products. Much of the country's income is also derived from thousands of international corporations whose headquarters are there. These businesses are attracted by Liechtenstein's very low taxes. Dairy and grape farming, tourism, and the selling of postage stamps are other sources of income.

Liechtenstein has a democratic government with a ruling prince. Its customs, military defense, and foreign relations are conducted by Switzerland.

Ask several students to investigate the impact of World War II on Austria, Switzerland, and Liechtenstein. Have them hypothesize about reasons for Switzerland's success in maintaining neutrality.

minerals found in quantity are sulfur, mercury, and lead. Marble is mined in the north for building stone and sculptures. Also in the north, rivers produce hydroelectric power.

With only 20 percent of Italy's land classified as plains, farmland is limited. Most soils are rocky and eroded from past overgrazing. The Po Valley is the exception, where thick **alluvial soils** blanket the valley. Alluvial soils are those deposited by rivers. These soils are often very fertile. Italy also does not have much forest available for timber production. In addition, though Italy is a fishing nation, water pollution and overfishing have caused concern.

Northern Italy The economic heart of Italy is the Po Valley. This area contains two-fifths of the population and three-fifths of the nation's industry. Rich alluvial soil, a wet climate, and available year-round water resources make this the "breadbasket" of Italy. Po Valley farmers can grow corn, rice, wheat, grapes, sugar beets, and a variety of other crops. The area is also a poultry-raising and dairy center.

The nation's second largest city, Milan is Italy's leading industrial and banking center. It is also the transportation center for northern Italy. Its wide variety of industries include chemicals, textiles, clothing, and engineering. Milan is famous for its fashion industry. It also has a world-famous cathedral and opera house. Milan is a city in which modern skyscrapers flank old palaces.

At the western end of the Po Valley is Turin. This large industrial city produces automobiles, computers, textiles, and motorbikes. On the northwestern coast, south of the Po Valley, is Genoa. Genoa is famous as the birthplace of Christopher Columbus. Today, it is the major seaport of Italy. Shipbuilding, steel, and petroleum refining are major industries. The industrial triangle formed by Milan, Genoa, and Turin benefits from its nearness to the rest of Western Europe.

The Alps do not hinder trade or travel between Italy and the rest of Europe. Modern

Recalling what you have learned about climate and landforms, explain why the Po Valley is such a fertile agricultural region.

autostrade (freeways) and railroads pass through tunnels under the mountains. They allow northern Italy access to the markets of partners in the Economic Community, especially France and West Germany. The Alps also provide water, hydroelectric power, and some timber to the industrial areas. They are a major tourist region, with glacial lakes, ski resorts, and vineyards.

In comparison to the western end, the eastern Po Valley is less industrialized. Venice, near the mouth of the Po River, is the region's main city. Built on a group of low islands in the Adriatic Sea, this historic city is linked together by a network of bridges and canals. Boats are a major form of transportation. During the eleventh and twelfth centuries, Venice was a powerful city-state that controlled trade in the Mediterranean. Although Venice has no large factories, there is a major petroleum refinery on the nearby coast.

Central Italy Central Italy is the political and cultural heart of Italy. The Apennine Mountains dominate the region. Nearly all of the flat land is on the west coast. The main

Level A: Italy is unmatched for museums, great collections of art, and classical architecture. Have students research these topics. Why is Italy unmatched in these areas? How is Italy's history reflected in these areas?

agricultural products are olives and almonds. Tourists are attracted to the Apennines' summer and winter resorts and to Central Italy's volcanic lakes.

Rome is Italy's capital and largest city. It is located on Italy's western coastal plain. This ancient yet modern city occupies both banks of the Tiber River. Rome handles Italy's political, educational, scientific, and international affairs. Rome also has significant food-processing, fashion, and film industries. Modern Rome, like Milan, its chief rival, is well known for its air pollution and traffic jams. Ancient Rome was the center of the once vast Roman Empire. Today, millions of tourists come to visit the Forum, the Colosseum, and other Roman ruins. Each day, traffic hurries along the Appian (AP-ee-uhn) Way, an ancient Roman highway that has been in use for more than 2,000 years.

The Arno River passes through the city of Florence. One of the main centers of Renaissance Italy, Florence is still an important art, cultural, and tourist center. It has attracted industries to surrounding areas.

The Ponte Vecchio (Old Bridge) spans Florence's Arno River. The bridge, which is lined with small shops, was built in 1345.

Southern Italy Southern Italy is the poorest part of the country. This region is characterized by rural poverty and small, inefficient farms. Because the area has extremely dry summers, water is scarce for cities and agriculture. Where soils are good, especially in alluvial or volcanic soil areas, wheat, olives, and grapes are the major crops.

Southern Italy also suffers from natural disasters. Throughout history, earthquakes, volcanic eruptions, floods, and droughts have occurred. Human destruction of valuable forest and soil cover have added to the problems of the region.

Unlike northern Italy, southern Italy has few industrial centers. There is no hydroelectric power or oil and gas deposits. The Italian government has been trying to develop southern Italy's economy, mainly by bringing nationalized industries into the area. So far, this industrialization plan has not worked. Most nationalized factories in the south are unable to earn profits.

The income of southern Italians is far below that of northern Italians. In addition, unemployment rates are much higher in the south. Because of these economic hardships, many southern Italians have moved out of the area. Most have been attracted to the northern cities, while others have found work in other European Community nations. Over the last century, millions of southern Italians have emigrated to other countries, especially to the United States, Australia, Canada, and Argentina.

Naples is the largest city in southern Italy, as well as the most densely populated city in Western Europe. This crowded city is the leading manufacturing center and the major port for southern Italy. The volcanic area around Naples is southern Italy's most productive farmland. Many tourists are attracted to nearby sites such as the island of Capri, Mount Vesuvius, and the ancient city of Pompeii.

The island of Sicily is located just off the southern tip of the Italian peninsula. It is separated from the peninsula by the narrow Strait of Messina. (See the map on page 168.) Mount

 Southern Italy has not developed as rapidly as Northern Italy. Have students research the economic problems in Italy's south. Ask students why this region is economically poor. How can economic development be improved?

Sicily is the largest island in the Mediterranean Sea. Mountains and hills cover 85 percent of the island.

Etna, an active volcano, rises 10,902 feet (3,324 meters) above the eastern part of the island. Because Sicily's interior is mountainous, most of the island's population lives along the coast. Much of Sicily is not suitable for agriculture. In areas where the soil is fertile and water is available, wheat, olives, and citrus fruits are grown. Fishing and some sulfur mining are also important economic activities. Palermo and Catania are Sicily's largest cities and industrial centers.

The island of Sardinia is situated across the Tyrrhenian Sea southwest of the Italian peninsula. (See the map on page 168.) This rocky island has extremely poor soils. Sheep grazing is the main agricultural activity. There also is some mining of lead, zinc, and coal.

National Issues and Economy A rich culture, the Italian language, and strong religious ties to the Roman Catholic church bind Italians together. At the same time, the social and economic differences between northern and southern Italy divide the nation into two separate regions.

Though Italy's economy is ranked sixth in the industrial world, the country as a whole is not prosperous or greatly industrialized. Italy's agriculture is important, and its farms grow a great variety of crops. The nation is a world leader in the production of olives. However, low productivity and poor water resources and soils limit farming.

Italy's industry is very dependent upon importation of energy and raw materials. The leading industrial exports are machinery, automobiles, clothing, shoes, tires, chemicals, and refined oil. Italy has one of the largest tourist industries in Western Europe, which helps balance the economy. The economy also receives a boost from Italians who work in other European Community nations and send money home. Italy's trade is mainly with other members of the European Community, the United States, and Middle Eastern countries.

Italy is a democratic republic, but the nation has had more than 40 governments since World War II. No single political party is strong enough to gain control. As a result, Italy has been ruled by **coalition governments**. A coalition government is one in which several political parties join together to run the country. Generally, coalition governments are successful for only short periods of time. Recent Italian governments have also had to face serious challenges, including poverty in the south, poor farm productivity, terrorism, and a negative balance of trade.

Vatican City and San Marino

Italy surrounds two tiny independent states, creating countries within a country. Vatican City is a microstate located within the city of Rome. It is surrounded by medieval walls, with the Church of Saint Peter at its center. The Vatican is the headquarters of the Roman Catholic church, the world's largest Christian denomination. The Pope, who is the leader of the Roman Catholic church, also governs the Vatican. The Vatican has diplomatic relations with other nations and is policed by the Swiss Guards.

Level B: Have students list the challenges Italy faces. Discuss these challenges with students. Ask them to suggest solutions. Next, ask students to list positive developments in Italy. Does the future of Italy look promising?

Malta has a mild, sunny climate, interesting festivals, and fascinating sights. In short, Malta is an ideal vacation spot. Malta is promoting tourism because it has few natural resources.

Situated on the eastern slopes of the Apennines is the tiny country of San Marino. This microstate was founded in the fourth century and has been independent since 1631. Nearly all the citizens of San Marino, called San Marinese, are Roman Catholic and speak Italian. This mountainous country supports some farming and light manufacturing. However, most of the San Marinese make their living from tourism and the sale of postage stamps.

Malta

Malta is a small island nation located in the middle of the Mediterranean Sea between Sicily and Libya. A strategic location and good harbor have throughout history involved the country in battles over control of the Mediterranean. The Maltese people have a distinct culture and language with influences from many conquerors, especially Italians and Arabs. Maltese and English are the country's official languages; Roman Catholicism is the dominant religion. Malta was once a British colony and naval base. In 1964, it was granted independence. Malta's economy is based on some agriculture, ship repair, light manufacturing, and tourism. Its trade links are mainly with the European Community.

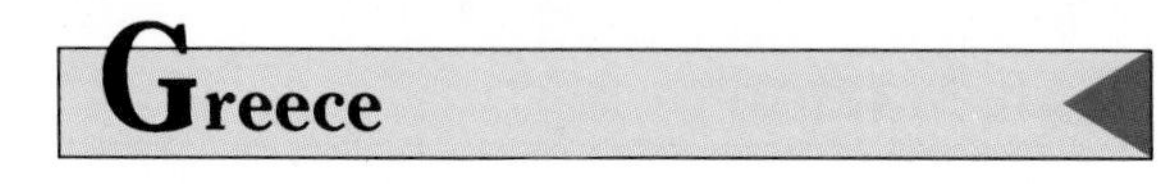

Greece

Historical Geography The ancient history of Greece was very significant in the development of Western culture. More than 2,500 years ago, the Greeks developed ideas that today are the roots of our democratic system of government. Ancient Greek art, philosophy, and science formed the basis of Western civilization.

Although it was the birthplace of democracy, Greece has had almost no democratic government in its long history. Its location at the crossroads of the Mediterranean has left the country exposed to invasions and conquerors. The two powerful Greek city-states of Athens and Sparta fought a bloody war until Athens was conquered in 404 B.C. Through the centuries, Greece was ruled by Persians, Romans, and Turks. Finally, in 1829, Greece became a free country after 400 years of Turkish rule. During World War II, the country was occupied by Germany. Immediately after the occupation ended, a bloody civil war broke out. The civil war was followed by a military dictatorship. In 1974, the Greek government finally became a parliamentary republic.

The population of Greece is united in language and religion. Greek is the national language, and most Greeks belong to the Greek Orthodox church. This church is part of the larger Eastern Orthodox church, which consists of churches derived from the Eastern Roman Empire of Constantinople. It does not include the Roman Catholic church.

Physical Geography Greece is located on the Balkan Peninsula, which extends into the Mediterranean Sea. The Ionian (ie-OH-nee-uhn) Sea is to the west; the Aegean (i-JEE-uhn) Sea is to the east. (See the map on page 168.) The coastline of Greece is formed by many bays and inlets. The strategic location of Greece is very important. The country is close to the major Mediterranean Sea routes and to entrances of the Black and Red seas.

Greece comprises mainly peninsulas, islands, and rugged mountains. About one-fifth of the area of Greece is islands. The largest of these is Crete. The Pindus and Rhodope mountains divide Greece into many isolated valleys. The highest peak in Greece is Mount Olympus, with an elevation of 9,570 feet (2,917 meters).

Except for in the high mountains, Greece has a Mediterranean climate. Summers are warm and dry, while winters are mild and rainy. The country has few rivers that flow year-round. The most valuable forest and soil resources were lost because of thousands of years of overgrazing. Today, less than 20 percent of Greece is forested. The separation of

Tell students that the first geographer was an ancient Greek, Eratosthenes. Ask students to research what Eratosthenes contributed to our knowledge of geography.

Per capita income in Greece is less than $4,000 a year. What factors might explain why Greece is Western Europe's least developed country?

Greece has poor soils, few valuable raw materials, a history of unstable government, and little industrial development.

Greece by water and mountains divides the nation into three major regions: southern Greece and the Greek Islands, central Greece, and northern Greece.

Southern Greece The Peloponnesian (pehl-uh-puh-NEE-zhuhn) Peninsula forms southern Greece. The narrow Gulf of Corinth separates it from the Greek mainland. A narrow neck of land, called an **isthmus**, joins the mainland of Greece to the peninsula. With the exception of Patras (puh-TRAS) and Corinth, there are few cities in southern Greece. The ancient cities of Olympia and Sparta were both located here. Olympia was the site of the first Olympic Games, which were held in 776 B.C. Sparta was a powerful city-state of ancient Greece. It was famous for its great athletes and warriors.

Southern Greece enjoys mild winters and sunny summers. This mild climate is ideal for growing export crops such as olives and citrus fruits. A variety of grains, fruits, and vegetables are grown for the local market. Along with agriculture, tourism and fishing are important to the economy.

In the Mediterranean Sea, directly south of the Peloponnesian Peninsula, is Crete. Greece's largest and most populated island, Crete is primarily agricultural. To the west, in the Ionian Sea, are a number of scenic islands, such as Corfu. The islands to the east, in the Aegean Sea, extend Greece's borders nearly to the mainland of Turkey. Most of these Greek islands are rocky and dry. With limited agriculture, they depend upon tourism for income. Rhodes, Mykonos, and Santorini are three well-known Aegean islands.

Central Greece Central Greece is mountainous with plateaus and coastal plains. The plains furnish fertile soils suitable for farming wheat, sugar beets, fruits, and olives.

This is the most populated region of Greece. Athens, the country's capital and largest city, is located here. Athens was a center of ancient Greek civilization. By the nineteenth century, the city was a poor village of ancient ruins. It was chosen as the capital in 1834 because of its historical importance. Athens and its seaport of Piraeus (pie-REE-uhs) have become centers of urban industrial growth. More than half of the nation's industry is located here. More than one-third of the population of Greece now lives in this smoggy urban area.

Level B: Have students map the regional divisions of Greece and label the major cities. Have them include geographically interesting details, the route of the original Marathon, the site of the first Olympics, and so on.

The modern city of Athens lies at the base of the Acropolis. The Parthenon, a gathering place for the ancient Greeks, still stands.

Millions of tourists visit Athens each year to see the famous ruins. Most of the ancient city is located in the Acropolis, on a high hilltop overlooking the modern city of Athens. The ancient marble temple, called the Parthenon, is one of the world's most photographed and famous buildings. After surviving almost 2,500 years of warfare and weather, it is now suffering from the effects of air pollution.

Northern Greece The Rhodope Mountains separate Greece from neighboring Bulgaria. The Pindus Mountains separate the northwest and northeast corners of Greece. The northwest is wetter, with forests and pasture where sheep and cattle are grazed. There is little flat land to grow crops. The northeast, known as Macedonia, has more land suitable for agriculture. On the plain of Thrace (THRAYS), tobacco, wheat, barley, and rye are grown. There is an agricultural coastal plain near the city of Salonika (suh-LAHN-i-kuh). Salonika is the second largest city in Greece and the major seaport for northern Greece.

Economic and Political Development Although Greece has some mineral deposits, it must import almost all of its oil. Offshore oil has been discovered in the Aegean Sea, but it is not a major source yet. Agriculture in Greece is limited because of the dry climate and mountainous terrain, yet it supports about 20 percent of the work force. The food-processing and food-export industries are based on such Greek agricultural products as olives, tobacco, grapes, and citrus fruits. The manufacturing of textiles, cement, metal products, and chemicals is concentrated in the Athens area.

Shipping is a vital part of the Greek economy. The world's largest merchant fleets are owned by Greek "shipping tycoons." Oil tankers, cargo ships, cruise ships, and fishing boats are all part of the Greek fleet. The ancient ruins, beautiful beaches, and sunny climate make the tourist industry an important part of the Greek economy. Also, Greeks working in other European Community nations earn needed foreign exchange to help the Greek economy.

Greece joined the European Community in 1981. While financial aid from the EC has helped improve its economic situation, Greece remains one of the poorest nations in the European Community.

Greece and Turkey have been rivals for centuries. Today, they both claim the potential offshore gas and oil sites in the Aegean. Greece is also concerned about the Turkish occupation of northern Cyprus. Although both nations are members of NATO, their relations with each other have never been good.

Many buildings on the Greek islands are painted white to deflect the hot summer sun. The village of Santorini is shown here.

Discuss with students why a seaport is important to both agriculture and industry. How are landlocked nations in Europe, such as Switzerland, Austria, and Luxembourg, able to carry on trade and industry without ports of their own?

CHAPTER 15 CHECK

Reviewing the Main Ideas

1. Greece and Italy were both influential in the development of Western civilization. Our art, science, and democratic traditions have their roots in these ancient cultures.
2. Italy is a long, boot-shaped peninsula in the Mediterranean region of Europe. It is a mountainous country with a mostly mild climate. Northern Italy is industrialized and prosperous, while southern Italy is dry and poor.
3. The microstates of Vatican City and San Marino are found in Italy. Malta is a small island nation in the Mediterranean, located between Sicily and Libya.
4. Greece is a mountainous country with many islands, island groups, and peninsulas. Its Mediterranean climate is mild and suitable for farming.

Building a Vocabulary

1. Explain the difference between an isthmus and a strait. Give an example of each, based on your reading in this chapter.
2. What are alluvial soils? Where are they found, and why are they important?
3. Define *coalition government*. Why has Italy had so many coalition governments?
4. Define *balance of trade*. How is Italy's balance of trade affected by its lack of important natural resources?

Recalling and Reviewing

1. How have Greece and Italy contributed to the development of Western civilization?
2. What are Italy's major problems today? How has membership in the European Community helped solve some of these problems?
3. Compare and contrast the landforms and economy of northern and southern Italy.
4. Name the four major regions of Greece and their major economic activities.

Critical Thinking

5. The North Atlantic Treaty Organization has established an important military port at Piraeus. Why do you think NATO selected this location? What advantage does membership in NATO bring to a country like Greece? Why would other NATO countries welcome Greece as an ally?

Using Geography Skills

Use the map in this chapter to answer the following questions.

1. What major islands are a part of Italy? of Greece?
2. Describe the course that a boat might take to get from the Tyrrhenian Sea to the Black Sea.
3. Give approximate latitude and longitude coordinates to describe the location of Greece's capital city.
4. What economic activities occur on the island of Sicily?

Answers to Chapter Check questions are found in the Teacher's Guide.

Additional activities for Chapter 15 are found in the Student Workbook. Chapter 15 Test as well as Skills, Reteaching, Critical Thinking, and Challenge/Enrichment activities are available in the Teacher's Resource Binder.

Chapter 16

Chapter 16 Lesson Plans and Planning Guide are found in the Teacher's Guide.

Refer students to the map on page 178.

Spain and Portugal

▲ Couples parade at a festival in Portugal.

▲ This Spanish town is located near the Meseta.

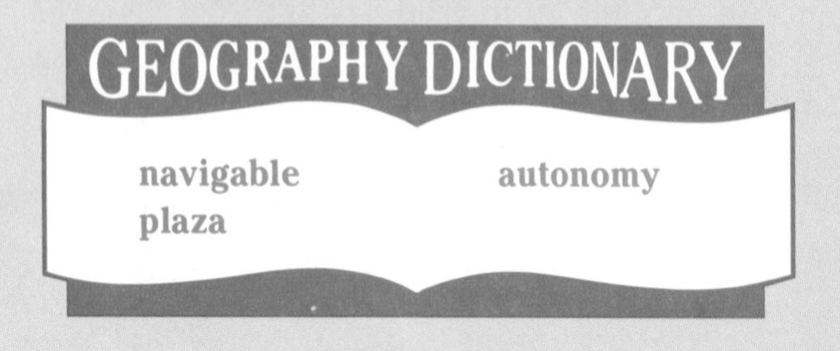

Spain and Portugal are called the Iberian (ie-BIR-ee-uhn) countries because they are located on the Iberian Peninsula. This peninsula acts as a stepping stone between Europe and Africa. Only the Strait of Gibraltar separates the two continents. The Iberian Peninsula has the Atlantic Ocean to the north and west and the Mediterranean Sea to the south. The Pyrenees Mountains separate the Iberian countries from the rest of Western Europe.

Spain and Portugal both are members of the European Community and NATO. Though not as economically prosperous as most Western European nations, both countries have experienced recent economic improvements.

 Have students compare the photos in this chapter with those found in the rest of the unit. Ask them to note differences in landforms, clothing, architecture, and so on.

Iberian History

Between the second and fifth centuries, Iberia was part of the Roman Empire. In 711, the Moors, a Muslim people from nearby North Africa, invaded the peninsula. (You will learn more about the Muslims in Unit 4.) They ruled parts of the Iberian Peninsula for more than 700 years. During those years, much of the peninsula was influenced by Muslim art, architecture, agriculture, and science. The Moors were eventually driven out of Iberia by the European peoples. Granada, the last city held by the Moors, was captured in 1492 by Spain's King Ferdinand and Queen Isabella.

From the twelfth to the sixteenth century, Spain and Portugal were the most powerful nations of Western Europe. Spanish and Portuguese explorers discovered new trade routes and new lands. Gold from the Americas flooded into the two countries.

During this period, Spain controlled much of Western Europe, including portions of France, Italy, and the Netherlands. Art, architecture, and learning flourished during this time. In 1588, Great Britain defeated the powerful Spanish Armada, Spain's fleet of armed ships. Though Spain and Portugal have never regained their world leadership roles, they have contributed their languages, religion, and other important culture traits to much of the world, especially the Americas.

Spain

Physical Geography Spain covers about four-fifths of the Iberian Peninsula. It is bounded on the west by the Atlantic Ocean and Portugal and on the south and east by the Mediterranean Sea. The Bay of Biscay forms the north coast of Spain. France borders the country on the northeast. Besides the peninsula, Spain also includes the hilly Balearic (bal-ee-AR-ik) Islands in the Mediterranean and the volcanic Canary Islands in the Atlantic

Sheep and olive orchards are major sources of income in the Andalusian region of southern Spain.

Ocean. Two small enclaves on the coast of North Africa are also claimed by Spain. (See the map on page 178.)

The interior of Spain is covered by a high plateau called the Meseta (muh-SAYT-uh). This rocky and treeless plateau averages 2,100 feet (640 meters) in elevation. It is the driest and least populated part of Spain.

Spain is one of Western Europe's most mountainous nations; mountains almost completely surround the Meseta. The Cantabrian (kan-TAY-bree-uhn) Mountains are located in the northwest. The Sierra Morena and Sierra Nevada ranges are located in the south. The highest peak on the Iberian Peninsula is found in the Sierra Nevada at 11,407 feet (3,478 meters). The Pyrenees stretch from the Bay of Biscay to the Mediterranean and are a natural boundary between France and Spain. Many peaks are at least 10,000 feet (3,048 meters) high. This mountainous terrain has isolated peoples of Spain from each other; even today, many regional differences exist, partly because of this geographical separation.

The most populated lands are the coastal plains and river valleys. The Ebro (AY-broh), the Douro (DOHR-oo), the Tagus (TAY-guhs), the Guadiana (gwahd-ee-AHN-uh), and the Guadalquivir (gwahd-l-KWIV-uhr) are the

Level B: Have students make three maps of Spain, showing climate, landforms, and population density. Have students report on ways climate and landforms affect population density of Spain.

most important rivers on the Iberian Peninsula. The Ebro empties into the Mediterranean. The other rivers flow westward into the Atlantic Ocean. The Guadalquivir is one of the few **navigable** rivers in Spain. A navigable river is one that is deep and wide enough for ships.

There are three climate types in Spain: maritime, Mediterranean, and steppe. In the northwest between the Cantabrian Mountains and the coast, there is a maritime climate. This is Spain's coolest and wettest region. Much of northwest Spain is forest and pasture. Along the southern coast of Spain, there is a Mediterranean climate. Here, the summers are sunny, hot, and dry, while the winters are mild and wet. (See the climate map of Western Europe in Chapter 10.) The Meseta of central Spain has a dry, continental steppe climate. Summers are very hot and winters are very cold. Because the surrounding mountains block moist ocean air from reaching inland, the Meseta receives little rainfall.

Agriculture Much of the country is too rugged and dry to be good farmland. For this reason, Spain is not self-sufficient in food production. However, Spanish farmers make up about 15 percent of the work force and produce a wide variety of crops. Spain is a world leader in olive, citrus-fruit, and wine exports. Because it uses greenhouses, the country supplies Western Europe with vegetables throughout the year.

Based on Spain's climates, there are also three agricultural regions. In the maritime northwest, the major agricultural activities are the raising of dairy and beef cattle as well as growing corn, potatoes, beans, and fruit. In the Mediterranean region, olives, sunflowers, almonds, citrus fruit, grapes, and vegetables are the most important products. The high and dry Meseta is not as productive, although grains, olives, and grapes can be grown where irrigation and good soils are found. Much of the rest of the Meseta is good only for grazing sheep and goats.

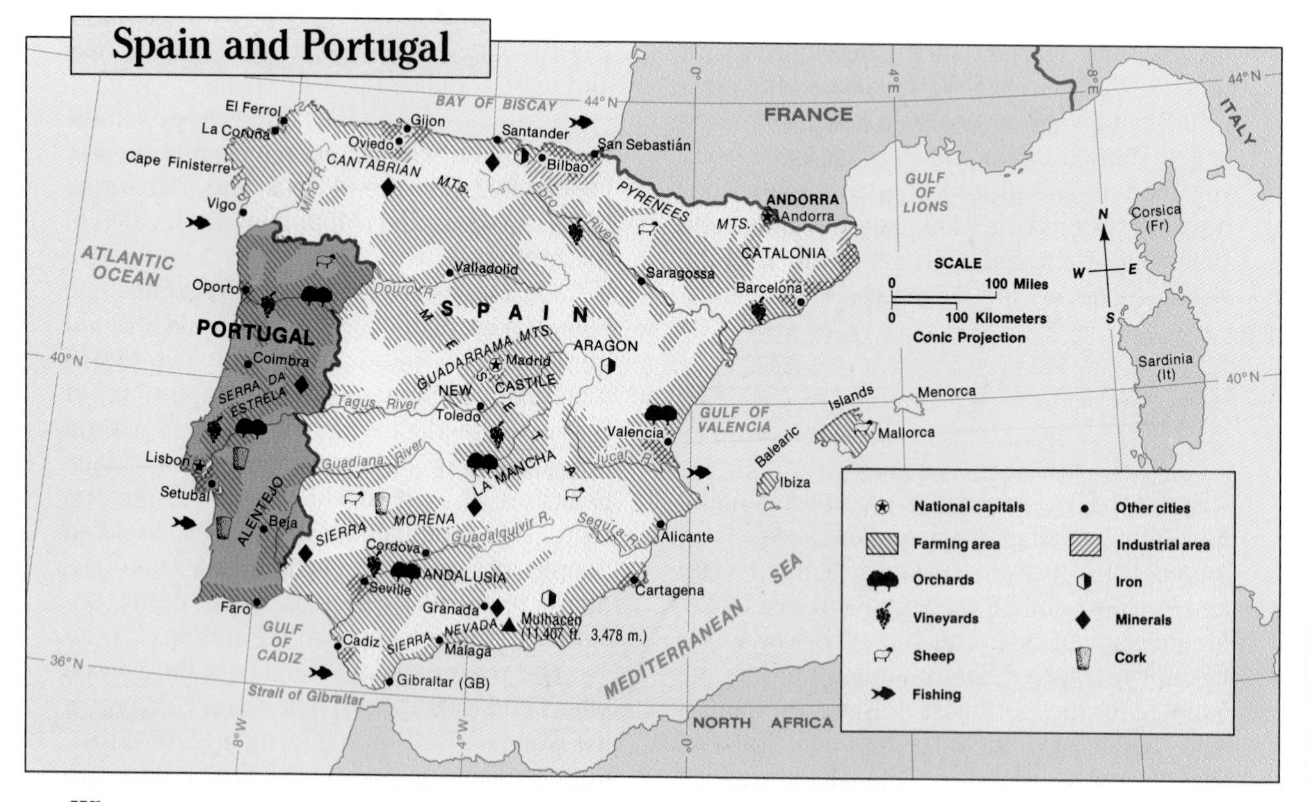

What natural boundary separates Spain from France? Pyrenees Mountains

 Have students locate Spain and Portugal on the map on page 110. Note their proximity to Africa.

The village markets are usually part of a regional network of similar markets in a preindustrial rural setting where barter is the main means of exchange. Have students research the barter system.

By Western European standards, Spanish farming is inefficient. Many of the farms in Spain need more irrigation, soil conservation, and mechanization if they are to grow more crops and earn greater profits. In northern Spain, farms are small. Available farmland is generally divided into smaller and smaller plots as land is passed on to each new generation. In contrast, in southern Spain, the farmland is organized into larger farms. Owners often employ landless workers to farm the land. On farms that are modernized, only a few farm workers are needed. This adds to rural unemployment.

Many Spanish farmers live in small villages rather than on the farmland. They travel each day from the villages to the fields. Activity in small agricultural villages centers around a **plaza**, a public square in the center of town. Often the site of a marketplace, the plaza usually serves as a gathering place for people. Many Spanish towns have lost their young people to the large cities, which offer more economic opportunities.

Economic Geography Spain has a fairly good supply of natural resources for industry. Agriculture supports a large food-processing and food-export industry. Spain is also a major fishing nation, ranking fourth in the world. Because forest resources are quite limited, timber must be imported.

Spain is rich in some minerals but lacks adequate amounts of coal and oil. The country is a major world supplier of mercury and also has good supplies of uranium, copper, zinc, lead, and tungsten. Important iron-ore mines are located in northern Spain. To support the steel industry, coal must be imported. Because Spain lacks oil, its most expensive import is petroleum. While hydroelectric power has been developed, its production rate is lower during dry years.

Some Spanish industries, such as automobiles, machine tools, and processed foods, have been very successful in the foreign-trade market. Other industries, especially steel and shipbuilding, have suffered labor cutbacks.

Spain's Mediterranean coast is a popular tourist destination. What is the climate of this region?

Mediterranean (subtropical dry summer); continental steppe

As Western Europe's leading tourist destination, Spain has one of the world's largest tourist industries. Many Western Europeans visit Spain seeking sunshine, beaches, a variety of good foods, and cultural attractions. This expanding tourist industry has led to a construction boom along the Mediterranean coast. The growth has not been all good, however. Pollution, crowding, and too many high-rise buildings are spoiling some of Western Europe's best coastal resorts.

Urban Geography Spain's largest city and capital is Madrid. Though located on the Meseta and separated from other populated areas, Madrid is the transportation, financial, and administrative center of Spain. The country's capital since the sixteenth century, it is Spain's richest city.

Barcelona (bahr-suh-LOH-nuh), located on the Mediterranean coast, is a leading port and the nation's second largest city. It is also Spain's leading industrial city. Its products include automobiles, railroad equipment, textiles, glass, tools, and machinery. Because Barcelona is an ancient Roman port with a rich cultural past, tourism is one of its major industries.

Level A: Have students use an almanac or other source of statistics to compare the economic development of Mediterranean nations (Spain, Greece, Italy) with other European nations.

A democratic-socialist government is one that is democratically elected and has socialist goals. Students can research the development of socialism and what the goals are.

The Alhambra towers over the old Moorish city of Granada. *Alhambra* means "red" in Arabic and refers to the walls' red bricks.

Spain's third largest city is Valencia. It is located on the Mediterranean coast, southwest of Barcelona. It is the center of a rich agricultural, food-processing, and resort area.

Cordova is Spain's fourth largest city. It is a port connected to the Atlantic by the navigable Guadalquivir River. The site of a 3,000-year-old trading center, Cordova features beautiful Moorish architecture from the Muslim period. Granada, at the foot of the Sierra Nevada, is the site of the Alhambra (al-HAM-bruh), Spain's largest and most beautiful Moorish palace and fortress. In northern Spain, major cities and industrial areas are Bilboa, Santander, and Oviedo with its port of Gijon.

Political and Social Development From 1936 to 1939, a bitter civil war raged in Spain. A military dictator, General Francisco Franco, ruled Spain for 36 years after the war. Today, Spain is a parliamentary democracy with a king. Its first free elections were held in 1977.

Though the Roman Catholic religion and the Spanish language are unifying forces, Spain is far from being a unified nation. About 75 percent of the population speaks Castilian, the official national language, which is a dialect of Spanish that comes from the Castile region around Madrid. Other Spaniards, however, speak a variety of other dialects and languages. In eastern Spain around Barcelona, the people speak Catalan, an Italic language. The Galician language, which is closely related to Portuguese, is spoken in the northwest. Language differences have caused problems among these various regional groups. The Spanish national government, in an attempt to solve the problems, has granted **autonomy** to the various regions. Autonomy is the right to local self-government. However, the central government maintains control of matters such as defense.

Another factor that prevents complete national unity involves the status of the Basques. The Basques live in areas primarily along the Bay of Biscay and the Pyrenees between northern Spain and France. Very little is known about the origins of the Basques. They lived in the area long before the French or the Spanish settled near them. Their language, which they call Euskara (ehws-kuh-RAH), deepens the mystery surrounding them. It is not related to any other known language in the world. Under the Franco dictatorship, the government was firmly opposed to granting rights to the Basques. The present government has granted them autonomy. There are, however, Basques who will accept only their own completely independent state. These separatist movements have sometimes turned violent.

In the past, Spain's population has consisted of a few wealthy people, with the remainder being very poor. Today, the wealth is more evenly distributed. Spain has become more industrialized and is fairly prosperous. However, the prosperity is not equally distributed throughout the nation. The areas around Madrid and Barcelona are the richest, while much of the rural Meseta remains in poverty. Yet, in recent years, the country's social, economic, and political conditions have improved drastically.

 Level B: Have students research in the library the Basque people and their culture. Why do the Basques want their own homeland? Students can tie in the idea of cultural conflict.

Mastering Geography Skills

1. Look at the climate and population-density maps on page 113. Which countries of Western Europe have the lowest population density? Why? Now look at the population-density map of the world on page 112. What other areas of the world are as densely populated as Western Europe?

2. Look at the regional map on page 110. What is the distance between Paris and Copenhagen? Paris and Rome? If you were traveling by land, which trip would be most time consuming? Why do you think so?

Applying and Extending

Use an almanac to find the following information about each of the nations of Western Europe, and record the data in a table: (1) the value of annual exports, (2) the value of annual imports, (3) the percentage of exports sent to other Western European countries, and (4) the percentage of imports purchased from other Western European nations. Study the data. Which Western European nations show a loss in their balance of payments? Do any show a gain? If so, which ones? Based on the percentages of imports and exports, how important is trade with other Western European nations to these countries? Compare your data with those found by other members of your class.

The topics of this section of the Unit Review vary throughout the text.

Linking Geography and History

During the nineteenth century, the British Empire included about one-fourth of the world's people as well as one-fourth of the world's land. Examine several sources, such as history texts and historical atlases, to answer the following questions: How did Britain achieve its supremacy over the seas? Which countries did it colonize, and when did they achieve their independence? Which countries and islands are still administered by Great Britain? Explore in some detail the relationship between the British Industrial Revolution and the British administration of India. What natural resources did India provide for British industry and consumption?

Reading for Enrichment

Burtenshaw, D., et al. *The City in West Europe.* New York: John Wiley & Sons, 1981.

Carrington, Richard. *The Mediterranean: Cradle of Western Culture.* New York: Viking, 1971.

Hoffman, G., ed. *A Geography of Europe: Problems and Prospects.* 5th rev. ed. New York: John Wiley & Sons, 1983.

Johnston, R., and J. Doornkamp, eds. *The Changing Geography of the United Kingdom.* London: Methuen, 1983.

Webb, R. K. *Modern England.* New York: Harper & Row, 1980.

Wright, Gordon. *France in Modern Times.* New York: W. W. Norton, 1981.

The Unit 2 Test is available in the Teacher's Resource Binder.

Much of the Soviet Union is covered by the rolling plains of the steppe. The Russian steppe extends from the southern Ukraine into Central Asia. The majority of the Soviet Union's farming is done in this fertile region.

Prague, the capital of Czechoslovakia, is one of the leading manufacturing centers of Eastern Europe. The countries of Eastern Europe are influenced, to varying degrees, by the Soviet Union. Each has its own distinctive style of communism.

Siberia covers about one-half of the Soviet Union. It is rich in mineral resources, but until recent years, its resources have been inaccessible to European Russia. Thirteen industrial centers are planned along the route of the new Baikal-Amur Mainline (BAM) railway.

Have students discuss the physical, cultural, and economic aspects of these photos.

UNIT THREE

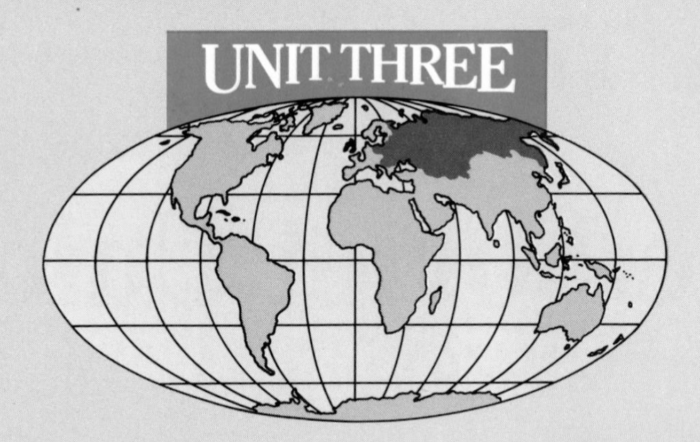

THE SOVIET UNION AND EASTERN EUROPE

OBJECTIVES

- ▷ *To describe the various factors that have led the Soviet Union to become a major influence in the world*
- ▷ *To explain how land, climate, and resources have made the western Soviet Union the heartland of the nation*
- ▷ *To explain how the Soviet Union is attempting to develop the vast resources of Siberia*
- ▷ *To explain how Eastern Europe and the Soviet Union are economically and politically tied to each other*
- ► *To determine the relationship between climate and population by comparing climate and population-density maps*

Chapter 17

Chapter 17 Lesson Plans and Planning Guide are found in the Teacher's Guide.

Refer students to the map on page 191.

The Soviet Union

▲ Kiev is one of the USSR's oldest cities.

▲ **Russians vacation at a Black Sea resort.**

GEOGRAPHY DICTIONARY

portage	collective farm
tsar	taiga
soviet	autarky
ethnic group	state farm

The Union of Soviet Socialist Republics is known as the Soviet Union or simply the USSR. Many people call it Russia because it occupies much of the same territory as the old Russian Empire. The Soviet Union covers more territory than any other country. East to west, it extends about 6,800 miles (10,880 kilometers). Its distance north to south, from the Arctic Ocean to Central Asia, is about 2,800 miles (4,480 kilometers). Despite its harsh climate, the Soviet Union has about 285 million people, making it the third most populous country, after China and India. However, only the western, or European, part of the country could be described as densely settled.

During the twentieth century, the Soviet Union has progressed from a rural, underdeveloped country to one of the world's largest industrial societies. Abundant natural resources have greatly aided Soviet economic progress. Despite this progress, living standards in the Soviet Union are still well behind those of Western Europe and the United States.

 Have students turn to the map on page 191 and locate the places shown in the photos above.

Point out that most Russian peasants were serfs tied to feudal land holdings until 1861. Students should compare and contrast slavery in the U.S. to serfdom in the USSR.

Historical Geography

Growth of the Russian Empire The history of the Soviet Union began in the grassy plain of the steppe. The steppe extends across the southwestern part of the country and reaches eastward into Central Asia. (See the map on page 191.) For thousands of years, great westward migrations of people moved along the steppe. These migrants moved their herds of sheep, cattle, and horses away from droughts and wars in regions to the east. The people who had already settled across the steppe and in nearby forests usually viewed the immigrants as invaders; battles often resulted. In most cases, however, the immigrants eventually merged with the settled population and became farmers. Each wave of immigrants brought new ways of life, languages, and religions to the region.

By A.D. 600, most of the peoples in southern Russia were Slavs. The first kingdom emerged in the ninth century at the city of Kiev (KEE-ehf), then an important trading center between the Mediterranean and Baltic seas. Goods also came from Eastern Europe and Central Asia. Kiev's first leaders were Viking traders from Scandinavia who called themselves Rus (ROOS). They eventually gave their name to Russia.

In the following centuries, many cities developed in the region between the Black and Baltic seas. Travel was easiest by water, and cities were often located at stream junctions or on easy **portages**. A portage is a low land area between lakes or rivers where boats and freight can be carried across. The people adopted Eastern Orthodox Christianity as the result of the efforts of Greek missionaries.

In 1237, Russia was invaded by Mongol tribesmen, led by Batu, the grandson of Genghis Khan, a Mongol chieftan. Their allies were the Tartars, who were the ancestors of the modern Turks. After Kiev was destroyed in 1240, Russia became the western outpost for the Mongol Empire, which also included most of China and India.

During the next two centuries, the princes of each large settlement battled each other for power and favor with the Mongol rulers. The strongest state to emerge was in the forests north of Kiev. It was called Muscovy, and its chief city was Moscow. By the late fifteenth century, a Muscovite prince, Ivan III, put an end to Mongol control and took over Russia. With this victory, Moscow became the center for the Russian Orthodox church as well.

In 1547, Ivan IV had himself crowned as first **tsar** of all of Russia. The word *tsar,* or *czar,* comes from the Latin *Caesar* and means "emperor." The Russian Empire under Ivan IV expanded from Kiev northward to the Arctic Ocean and eastward toward the Ural Mountains. Ivan became known as "the Terrible" because of some of his cruel policies and tactics. For example, he killed his oldest son with his own hands.

The collapse of Mongol rule allowed Russian fur trappers, hunters, and pioneers to expand the tsar's empire eastward into Siberia. By 1637, Russian pioneers reached the Pacific coast. Thus, the Russian tsars were able to build a land-based empire larger than any in the world. This expansion occurred at about the same time England, France, and Spain were building overseas colonial empires.

Moscow is one of the oldest cities in the Soviet Union. When did it become the center for the Russian Orthodox church?

late fifteenth century

Eventually, the Russian territory spread to what is now Alaska, in North America. In 1867, Russia sold its claims in North America to the United States for $7.2 million.

The expansion of Russia westward and southward was a slow process. Major westward expansion was delayed until the 1700s by the powerful kingdoms of Sweden and Poland. Peter the Great, who ruled Russia from 1682 to 1725, acquired coastal lands from Sweden on the Baltic Sea. In 1703, he began building a new capital city called St. Petersburg, now known as Leningrad. Considered the founder of modern Russia, Peter the Great introduced many Western European customs and ideas to the Russians.

Catherine the Great ruled from 1762 to 1796. She expanded the Russian Empire southward by warring against the Turks and their allies. Eventually, her armies took all of the northern side of the Black Sea. The Christian kingdoms of Georgia and Armenia, in the Caucasus (KAW-kuh-suhs) Mountains area, were added to the Russian Empire in the early 1800s. In the 1860s, the Russians moved east of the Caspian Sea into Central Asia. They also seized lands along the Amur and Ussuri rivers that had been part of northeastern China. Thus, many non-Russian people were brought into the empire.

Founded as St. Petersburg in 1703, Leningrad was the first Russian city to imitate the architecture of Western Europe.

Before the Communist Revolution By 1900, Russia was a country of poor farmers. Although Russia had begun to industrialize rapidly in the 1870s, there were great extremes of wealth and poverty. Only about 2 percent of the people were wealthy merchants, landowners, or royalty. A slightly larger percentage were poorly paid factory workers and crafts people. The remaining population, about 85 percent of the total, was mostly peasants. The poor economic state of these people was Russia's greatest problem.

In 1904 and 1905, there were riots in nearly every large city. People protested food shortages, low wages, and high taxes. In many rural areas, the peasants burned farm buildings and occupied wealthy estates. In answer to these protests, thousands of people were forced to migrate to Siberia. Although new social and economic programs were introduced, they were too little too late. By 1914, the lives of the peasants had seen little improvement; working conditions in factories were still terrible. Russia was a country waiting to explode. The explosion began as World War I spread across Europe.

The Communist Revolution Russia was unprepared for World War I. Sometimes four soldiers were sent to battle with only one gun among them. The losses and costs of war, combined with Russia's social and economic problems, caused the tsar to abdicate, or resign, in early 1917. A temporary republican government took over but fared no better. In the fall of 1917, the Bolshevik party overthrew the government.

The Bolsheviks wanted to remake Russia according to the ideas of Karl Marx. Marx, a nineteenth century German philosopher, saw the people of the working classes as victims of capitalism. Marx believed that the capitalists kept the workers poor to maintain their

 Have students locate the Soviet Union and Eastern Europe on the inset map of the world on page 191. Note the degrees of longitude covered by this region. Have students compare this to Western Europe on page 110.

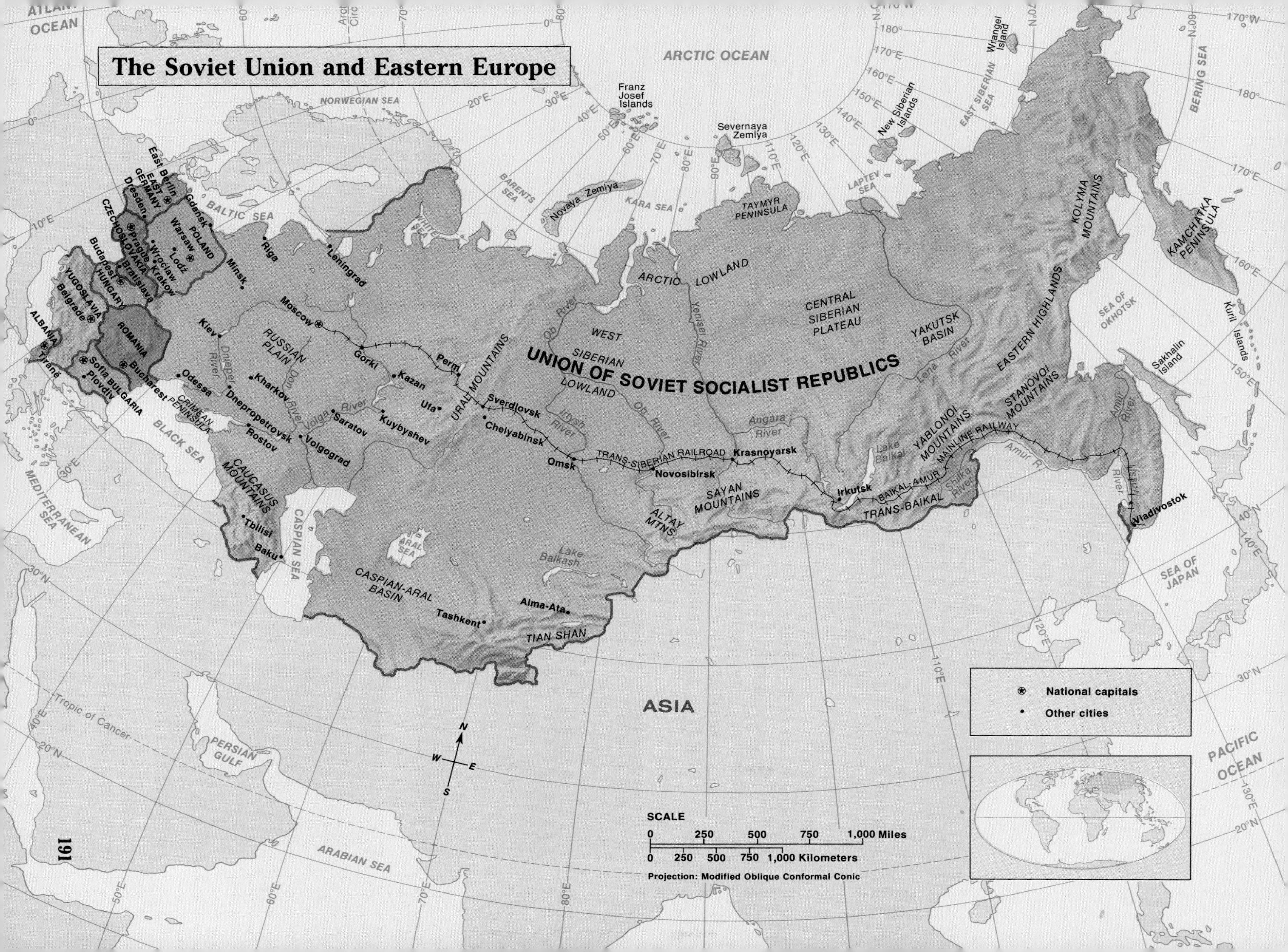
The Soviet Union and Eastern Europe
ARCTIC OCEAN
ATLANTIC OCEAN
NORWEGIAN SEA
BARENTS SEA
KARA SEA
LAPTEV SEA
EAST SIBERIAN SEA
BERING SEA
SEA OF OKHOTSK
SEA OF JAPAN
PACIFIC OCEAN
BALTIC SEA
WHITE SEA
BLACK SEA
CASPIAN SEA
ARAL SEA
MEDITERRANEAN SEA
PERSIAN GULF
ARABIAN SEA
Franz Josef Islands
Novaya Zemlya
Severnaya Zemlya
TAYMYR PENINSULA
New Siberian Islands
Wrangel Island
KAMCHATKA PENINSULA
Kuril Islands
Sakhalin Island
KOLYMA MOUNTAINS
EASTERN HIGHLANDS
STANOVOI MOUNTAINS
YABLONOI MOUNTAINS
YAKUTSK BASIN
CENTRAL SIBERIAN PLATEAU
ARCTIC LOWLAND
WEST SIBERIAN LOWLAND
UNION OF SOVIET SOCIALIST REPUBLICS
URAL MOUNTAINS
RUSSIAN PLAIN
CAUCASUS MOUNTAINS
CRIMEAN PENINSULA
CASPIAN-ARAL BASIN
TIAN SHAN
ALTAY MTNS.
SAYAN MOUNTAINS
ASIA
Ob River
Yenisei River
Lena River
Angara River
Irtysh River
Amur River
Ussuri River
Shilka River
Volga River
Don River
Dnieper River
Lake Baikal
Lake Balkash
TRANS-SIBERIAN RAILROAD
BAIKAL-AMUR MAINLINE RAILWAY
TRANS-BAIKAL
Moscow
Leningrad
Riga
Minsk
Kiev
Odessa
Kharkov
Dnepropetrovsk
Rostov
Volgograd
Saratov
Kuybyshev
Kazan
Gorki
Perm
Ufa
Sverdlovsk
Chelyabinsk
Omsk
Novosibirsk
Krasnoyarsk
Irkutsk
Vladivostok
Tbilisi
Baku
Tashkent
Alma-Ata
POLAND
Gdańsk
Warsaw
Lodź
Wrocław
Krakow
EAST GERMANY
East Berlin
Dresden
CZECHOSLOVAKIA
Prague
Bratislava
HUNGARY
Budapest
YUGOSLAVIA
Belgrade
ALBANIA
Tiranë
ROMANIA
Bucharest
BULGARIA
Sofia
Plovdiv
Tropic of Cancer
N
S
E
W
SCALE
0 250 500 750 1,000 Miles
0 250 500 750 1,000 Kilometers
Projection: Modified Oblique Conformal Conic
National capitals
Other cities

own power and wealth. This philosophy was adopted by V. I. Lenin, the leader of the Bolsheviks. To solve Russia's problems, Lenin argued that the government should own and control all the farms and factories. To carry out their plans, the Bolsheviks established a dictatorship.

The Bolsheviks had few supporters. Their Communist philosophy appealed mainly to small numbers of people living in the cities. A large group of people remained loyal to the tsar and began a civil war. The civil war ended in 1922 with Lenin's All-Union Communist Party (renamed the Communist Party of the Soviet Union in 1952) in control. Many of the empire's non-Russian peoples tried to declare their independence. Some succeeded; others, including the Georgians and Ukrainians, were reconquered by Lenin's troops.

The five years of civil war led to terrible shortages of food and fuel. Experiments in government-controlled farms and factories were often unsuccessful. Lenin believed that desperate actions were needed to save the Communist revolution. To help improve the economy, he introduced a temporary change in policy. This included returning some businesses to private control and tolerating small individual farms. This New Economic Policy resulted in increased agricultural and industrial production. The Soviet Union was repairing itself, and the Communist government gained public support.

In 1922, the country was renamed the Union of Soviet Socialist Republics (USSR). Each republic has a local governing council, or **soviet**, but political control is held by the Communist party leaders in Moscow. The republics are special regional governments that were created for the country's many **ethnic groups**. An ethnic group is a group of people who share the same language, customs, and traditions. Other republics were created for other large ethnic groups. (See the map on page 193.) The republics had few real powers, however, because all of the decisions about

The Soviet Union has many ethnic groups; all have different heritages, languages, and social traditions. How can you tell that these people are from different ethnic groups? Their coloring, clothing, and hairstyles differ.

 Have students compare Stalin's policy of genocide regarding the Ukrainians to Hitler's genocide of the Jews.

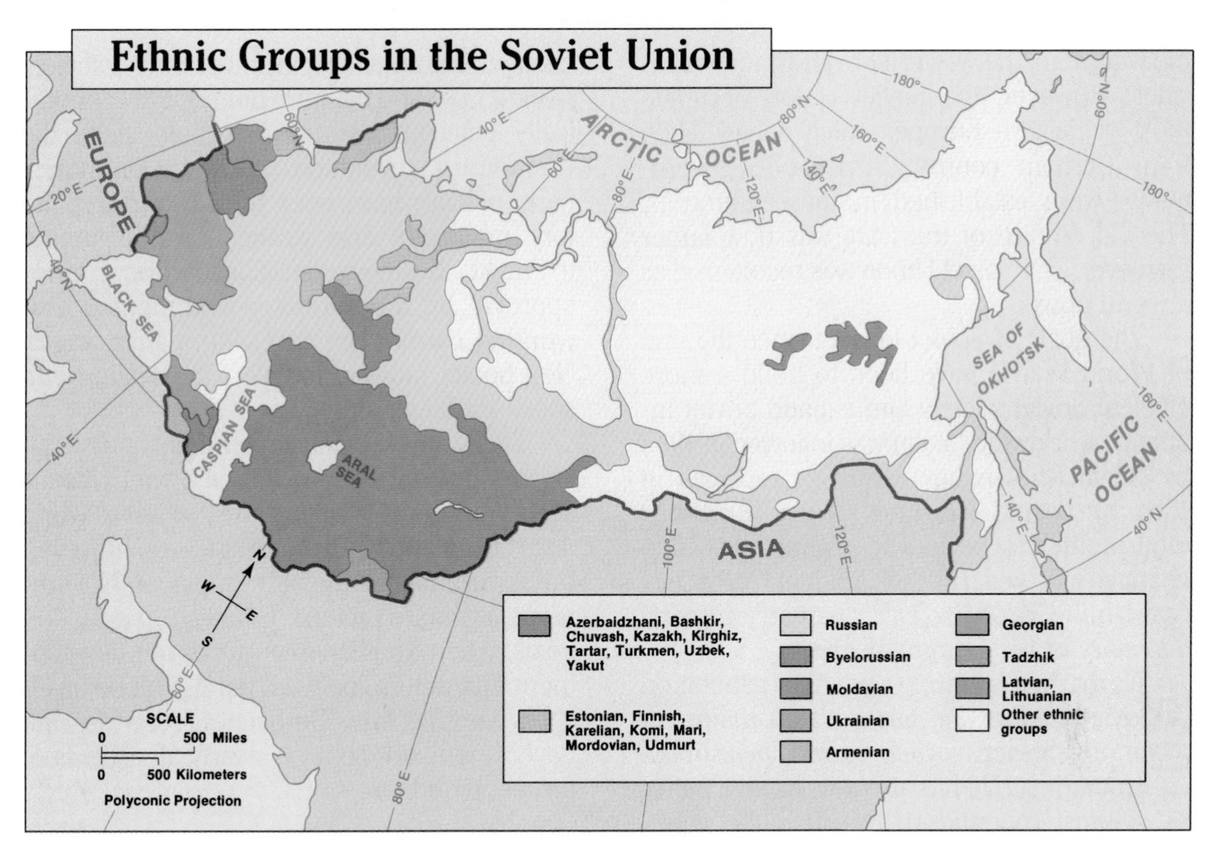

What are the dominant ethnic groups of the Soviet Union? Where are they located?

Byelorussian, Azerbaidzhani etc., Russian Ukrainian; west-central, mostly south, west

government and the economy were made in Moscow. By 1922, the USSR had become the world's first one-party, totalitarian state.

The Soviet Union Under Joseph Stalin

Many Communists began to fear that Lenin's New Economic Policy was bringing capitalism back to the Soviet Union. Lenin died in 1924. Joseph Stalin, who replaced Lenin, ended the New Economic Policy period and again organized the country on socialist principles. All farmland and factories were placed under government control. A Five Year Plan was established that set production goals for the next five years. All farmers were forced to join giant **collective farms**. On a collective farm, the workers are given a portion of the farm's profits. As Stalin began enforcing a strict Communist system, he established the basic traits of Soviet culture that are present today.

Stalin's Five Year Plans resulted in dramatic industrial growth. However, Stalin's policies were opposed by many. The Soviet people were forced to make great sacrifices in order to make industrial growth possible. Stalin's greatest opposition came from farmers. Many farmers were unhappy about losing their land and animals to collectives. In the prosperous Ukraine, farmers rebelled by attacking officials and burning farm buildings. Many were killed and others were sent to Siberian labor camps. In the middle 1930s, Stalin's farm policies caused a great famine in the farming regions. More than six million people died of starvation. By 1937, nearly all of the farmland was under government control.

The Soviet Union After World War II

During World War II, Germany invaded the Soviet Union. To defeat Germany, the USSR joined forces with the United States and its allies. The Soviet Union lost more than 20 million people before the war ended in 1945. At

Point out that the correct term for this nation is not Russia but the Union of Soviet Socialist Republics. Russia is only a part of the vast nation.

Level B: Have students create a time line showing the history of the USSR. Have them include world events as well so that they can compare Soviet history with comparable world events.

the war's end, the western Soviet Union was in ruins. However, the Soviet Union occupied most of Eastern Europe, which it had taken from German control. Soviet-style governments were established in these countries. The old empire of the tsars was now larger than ever. The Soviet Union was recognized as a world power.

The goals of Soviet leaders since the end of World War II have been to build a more efficient Soviet society and expand Soviet influence worldwide. Stalin was followed in 1953 by Nikita Khrushchev, who was replaced in 1964 by Leonid Brezhnev. When Brezhnev died in 1982, a period of several leadership changes followed. In 1985, Mikhail Gorbachev (gawr-buh-CHAWF) was named General Secretary of the Communist party.

Gorbachev is among a younger generation of Soviet leaders. He has initiated major reforms in Soviet society called *perestroika* (restructuring). He has also advocated a policy of *glasnost* (openness) by which the country's problems are discussed openly. These changes are intended to make Soviet society work harder and be more honest and economically efficient. For the first time since the Communist revolution, corrupt Communist party officials have been tried in public. Factory managers and workers are beginning to make business decisions without prior approval by the central government. Hardworking citizens are receiving higher wages. New books, movies, and plays are dealing with social problems openly.

Moscow's Red Square is the scene of the annual May Day parade. May Day is a national holiday to honor working people.

Today, the Soviet government is like an enormous corporation with all Soviet citizens as its workers. It is against the law not to work. Apart from a few private businesses, however, the government is the only employer. Government agencies act to establish production goals. They identify new areas for development and determine what goods will be available. Despite the Gorbachev reforms, the government still controls nearly all aspects of the people's lives.

Physical Geography

Location and Landforms The Soviet Union is bounded on the north by the Arctic Ocean and several smaller seas. All of these bodies of water are frozen most of the year. In the Soviet Far East are the Sea of Okhotsk (oh-KAHTSK) and the Bering Sea. Only the Sea of Japan on the Pacific Ocean has ports that allow easy access to other countries. The southern borders of the Soviet Union are marked by desert or rugged mountains and have few people. The Soviet Union's western borders lie in the heavily settled plains regions, but the Soviets are separated from their Eastern European neighbors by well-guarded military installations.

The Soviet Union can be divided into 10 major landform regions. (See the map on page 191.) Along the Arctic Ocean in the north is the broad Arctic Lowland. It extends all the way from Finland to the East Siberian Sea. South of the Arctic Lowland from the Urals westward is

Note that the Urals form the traditional natural boundary separating Europe and Asia. These two regions are actually part of the larger continent of Eurasia.

the Russian Plain. This region is where most of the Soviet people live. The low Ural Mountains form the dividing line between Europe and Asia. Though old and worn, they contain many mineral deposits.

East of the Urals is the West Siberian Lowland. Marshes and permafrost are widespread in this area. Farther to the east are the rolling uplands of the Central Siberian Plateau and the plains of the Yakutsk (yuh-KOOTSK) Basin. Both are remote and sparsely settled, but they may contain important minerals for the future. The Eastern Highlands along the Pacific coast are younger mountains than the Urals. They were created by the forces of folding and faulting. On the Kamchatka (kam-CHAT-kuh) Peninsula, there are several volcanoes.

In southern Siberia are the steep and high Central Asian ranges. These include the Tian Shan (tee-EHN SHAHN) and Altay (AL-tie) ranges on the Chinese border. Farther to the west, the Caucasus Mountains are found between the Black and Caspian seas. Finally, there are the dry plains of the basins of the Caspian and Aral seas.

The Caucasus Mountains, located between the Black and Caspian seas, form a natural boundary between the USSR and Turkey.

Rivers The vast land area of the Soviet Union is crossed by many broad rivers. These rivers form a network of waterways that is valuable for transportation and the production of hydroelectric power. The Dnieper (NEE-puhr), the Don, and the Volga rivers all start in the Russian Plain and flow toward the south. The Volga is the longest river in Europe, covering 2,293 miles (3,669 kilometers) in its journey to the Caspian Sea. Canals connect many of the rivers that flow through the plains.

The Ob, Yenisei (yehn-uh-SAY), and Lena rivers begin in the southern mountains of Siberia and flow northward to the usually frozen Arctic Ocean. Managing the rivers of western Siberia has proven difficult. In the spring, the southern, upstream parts of the rivers thaw, but the northern areas downstream remain frozen. As a result, waters from spring rains and melted snow cause great floods in many parts of western Siberia.

Climate and Vegetation Winter comes early and stays late in most of the Soviet Union. It is a land dominated by cold and bitter climates. (See the climate map in the Geography Skills feature on page 197.) The northern edge of the Soviet Union from the Norwegian border on the west to the Bering Strait on the east has a tundra climate. Temperatures here rise above freezing for only a few weeks each year. Only stunted trees, shrubs, and mosses of the tundra vegetation can grow. In the subarctic region to the south, the warmer summers allow for **taiga**, or forest, vegetation. The taiga is an enormous stand of softwood, needle-leaf trees, such as pine, fir, larch, and spruce. The taiga provides the Soviet Union with a large supply of forest products. Some agriculture is also possible in this area during the short summer.

Southwest of the taiga zone in European Russia is an area of humid continental climate. Over half of the Soviet Union's population is found in this climate region. Summers are longer here than in the subarctic region, but the winter months are very cold. Many parts of

Level A: Ask students to find the length of the growing season of the taiga (60–90 days). What crops can be grown under these conditions? (cabbages, sunflowers, and so on) Why?

GEOGRAPHY SKILLS

How to Compare a Climate Map to a Population-Density Map

Geographers use maps to display information. They use these visual presentations of data to find spatial relationships, to test ideas, and to draw conclusions. By comparing special-purpose maps of a region, for example, geographers can discover and explain many things about places.

In the past, people preferred to settle only in areas suitable for agriculture. This was true because most people farmed for a living. They depended upon a suitable climate more than people do today. The special-purpose climate and population-density maps of the Soviet Union and Eastern Europe illustrate a strong relationship between climate and settlement. Look at the climate map. Which climates in the region do you think are the most agriculturally productive? Of the climates represented in this area, the humid continental, the humid subtropical, and maritime are the most productive. These climates are found in the western part of the Soviet Union and in Eastern Europe.

Now look at the population-density map of this region. Notice that the areas of dense population are also the ones with the climates most suitable for farming. Also note that the humid continental climate region outlines the region of population density between 25 and 125 people per square mile (10 and 49 per square kilometer). In contrast, the colder subarctic climate region has a very low population density. Between 0 and 25 people per square mile (0 and 10 per square kilometer) live there. This climate region is so harsh that agricultural activity is limited.

Of course, the relationship between climate and population density is not exact. Many other factors affect settlement patterns. One of these factors is the opportunity for people to work in mines and in factories. Another is the quality of soil, which does not vary exactly with climate. For example, other areas of the world with climates like the western Soviet Union are not as agriculturally productive. These areas have soil that is rocky and less fertile than the soil of the western Soviet Union.

Today, fewer people than in the past rely upon agriculture to make a living. Therefore, settlement may be influenced by additional factors such as industrial development and nearness to natural resources or a transportation center. In order to explain the pattern of population density completely, you would have to use more than just a climate map.

Test Your Skills

1. Look at the population-density map. Which color represents the most densely populated areas? the least densely populated? Now look at the climate map. Which color represents the maritime climate? subarctic climate? humid continental climate? What conclusions can you draw about the relationship between climate and population?
2. Which is more densely populated, Eastern Europe or the Soviet Union?
3. What is a possible explanation for the low population density in the region east of the Caspian Sea?
4. Where is the densest population between the Black Sea and the Caspian Sea? To what climate region does this roughly correspond?
5. What regions of the Soviet Union are completely uninhabited? At what latitude and longitude are they located?

 Answers to Geography Skills questions are found in the Teacher's Guide.

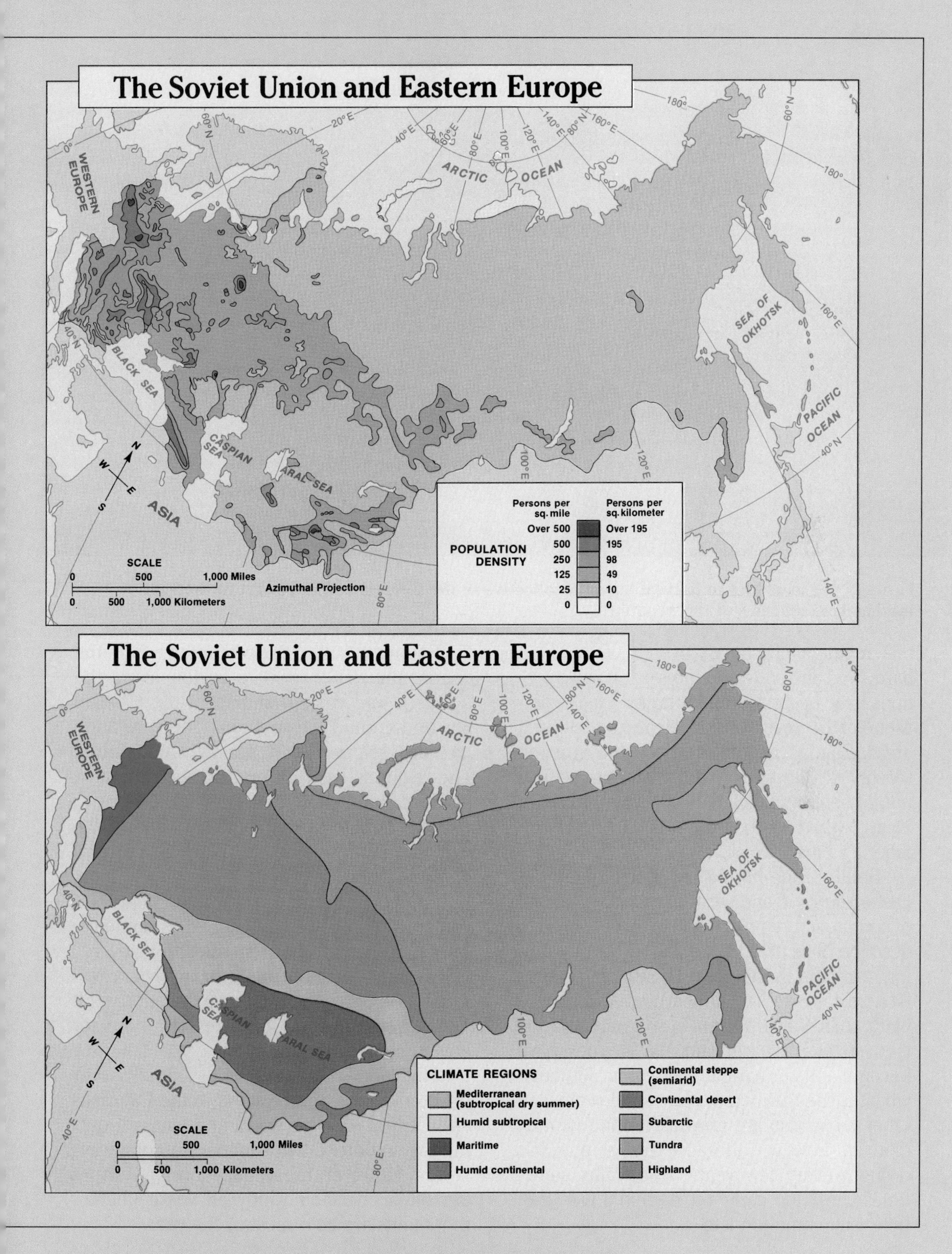

The pattern of population distribution in the Soviet Union has been compared to an arrowhead with the base in the west and the point to the east. Ask students if they can identify this pattern.

Farmers use machines to harvest tea on a collective in the Georgian SSR. Tea grows well in the highland climates of this republic.

this region were once covered with mixed hardwood and softwood forests, but large sections have been cleared for farms. This region is one of the Soviet Union's richest agricultural areas. Rainfall is sufficient, and the growing season is long enough for many crops. The humid continental climate is also found along a narrow belt south of the subarctic region on Siberia's Pacific coast.

South of the humid continental region in Eastern Europe and western Siberia is a zone of continental steppe climate. *Steppe* is used to describe both the climate and the gently rolling grassland in this area. The Russian steppe is a vast treeless plain covered with fertile black soils. Since the soil is so fertile and the land so easy to cultivate, the steppes have become a major agricultural region. Although wheat is the main crop, corn, sunflowers, and other crops also are grown. Farming is often difficult, however, because the steppe has drought every few years. Also, rains sometimes come too early, or, even in a wet year, they may come too late.

Southward, the steppes merge into the continental desert climate region along the northern and eastern shores of the Caspian Sea and around the Aral Sea. The vegetation includes sparse desert grasses and shrubs. The region is used for grazing, with irrigated farming found along rivers.

The climate of the southern edge of the Crimean Peninsula on the Black Sea is very different from that of most of the Soviet Union. It has a Mediterranean climate. Mild, wet winters are followed by hot, dry summers. There are many fruit orchards and vineyards in this area. Known as the Russian Riviera, this section of Crimea is a popular vacation spot. The government has built resorts there for Soviet workers.

Highland climates change rapidly over short distances. Some valleys in the Caucasus and Central Asia are warm and pleasant during the summers, and winters are not always severe. However, the upper mountain slopes are often cold and windy all year. Siberia's mountains are cold most of the year.

 The scientific term for brown-black soils is the Russian word *chernozem*.

Peoples of the Soviet Union

The Soviet Union has more than 150 ethnic groups. (See the map on page 193.) The Russians are the largest ethnic group in the USSR, even though they make up only about half of the total population. Twenty other groups have at least one million members. Of these, the Ukrainians of the southwestern European Russia number more than 40 million. The Uzbeks (UZ-behks) of Soviet Central Asia number about 15 million. Though many different languages are spoken in the Soviet Union, most belong to the Slavic group.

Under Soviet rule, Russians have spread throughout the Soviet Union. In many cases, Russians have brought higher living standards, modern technology, and more educational opportunities to these groups. Some minorities, particularly those in Central Asia, are increasing in number faster than the Russians. As populations rapidly increase in these less developed regions, the percentage of Russians in the Soviet Union is decreasing.

The Soviet Union is a literate society. More than 50 million students fill the nation's schools. Science and technology are stressed.

Some ethnic and religious groups, however, generally have higher living standards than the Russians. They feel that Russian control blocks their development. This is particularly true of the people of the Baltic Republics (Latvia, Lithuania, and Estonia). Russia's Jewish population of 2.5 million finds itself discriminated against in educational, social, and economic opportunities. Many have been trying to leave the Soviet Union.

Soviet Political Organization

The Soviet Union is made up of 15 Soviet Socialist Republics (SSRs). The largest and most populous of these is the Russian Soviet Federated Socialist Republic, which is usually identified by its initials (RSFSR). It extends all the way from Moscow and Leningrad eastward across Siberia to the Pacific Ocean. The Russian republic occupies three-fourths of the country's land area. (The other republics are shown on the map on page 208.)

The ruling council, or soviet, of each republic is responsible for local matters in education, agriculture, and finance. However, the governments of the republics are controlled by the central government in Moscow. Here is where the Communist party and the national Supreme Soviet meet. The Supreme Soviet is Russia's legislature. The members of the Supreme Soviet pass all laws proposed by the Communist party leaders. The Communist party has total control over the government.

In the Soviet Union, elections of representatives are important. Yet, there is seldom more than one candidate, and the real governing power rests with the Communist party. All important political decisions of the Soviet Union are made by party leaders. The party's highest authority is the Congress, which elects the Central Committee. The Central Committee determines who will be members of the important policy committees. The Politburo is

Level B: Have students review acculturation, which was discussed in Unit 1. In the USSR, the process is called Russification. Ask students how this process affects minority cultures in the Soviet Union. How might minorities feel about Russification?

Have students consider why the Soviet Union could attempt a policy of autarky. Ask if they think a country such as West Germany could have such a policy.

the most important of these committees and is the center of party power. Most government officials are members of the Communist party. Membership ensures that those who head the party also control the government. Only about 6 percent of Soviet citizens are members of the Communist party.

Economic Geography

Resources and Industry The Soviet Union is second to none in wealth of natural resources. It ranks among the world's leaders in production of oil, natural gas, coal, hydroelectric power, iron ore, gold, timber, copper, chromium, and manganese. Ores of nearly all other industrial metals are available as well. Further exploration for mineral resources is underway throughout Siberia, Central Asia, and the Arctic. The Soviet Union will have raw materials for centuries to come.

The economic system of the Soviet Union is a command economy. All aspects of the economy, especially industrial production, are planned by Soviet government managers. Preparation of the Five Year Plans continues. Since the 1960s, there has been increased emphasis on production of consumer goods. However, heavy industry and the production of military goods still make up most of the manufacturing. The Soviet Union is the world's largest producer of steel.

Soviet industrial planners have followed a policy of **autarky** (AW-tahr-kee) since the days of Stalin. Under autarky, a country strives for self-sufficiency in industrial production. As a result of this policy, the Soviet Union imports few goods and tries to meet all its needs. Rather than import goods, the Soviet people have learned to do without.

The Soviet Union is one of the world's largest oil producers. The largest oil fields are found in the middle of the Ob Valley region of western Siberia. Other oil deposits are located west of the Ural Mountains, surrounding the Caspian Sea, and on the island of Sakhalin (SAK-uh-leen), off the east coast. Though much production is used at home, substantial amounts are exported. Large amounts are also sent to Eastern Europe and to Soviet military allies around the world. A pipeline carries natural gas to Eastern and Western Europe.

Industrial development has been important to the Soviet Union. Plentiful natural resources are developed under the direction of the government.

Have students consider the viability of the Ukraine as an independent country (in terms of land area, population, and resources).

have led to the development of an iron and steel industry at Kerch. Sevastopol (suh-VAS-tuh-pohl) has a fine natural harbor and one of the most important Soviet naval bases.

The Ukraine is the second most populated republic of the Soviet Union, and Ukrainians are the country's largest minority. While their language and customs are similar to those of the Russians, the Ukrainians have long cherished a separate identity. Some Ukrainians believe that their region should be an independent country instead of a Soviet republic. Like all Soviet republics, the Ukraine enjoys some autonomy, or self-government, in local affairs, but the Ukrainians continue to press for more.

Moldavia

The tiny Soviet republic of Moldavia lies between the Ukraine and Romania. Throughout history, control of Moldavia has shifted. It has been dominated by the Turks, Polish princes, Austria, Hungary, Russia, and Romania. It also was briefly independent in the last century. The Soviet Union acquired Moldavia by force from Romania during World War II. Moldavia is almost entirely agricultural, with little industry. People there raise grain, fruit, and tobacco. Moldavia's capital is Kishinev (KISH-uh-nehf).

Moldavia, a small Soviet republic, is a rich agricultural region. These vineyards grow on the "Lenin" *kolkhoz.*

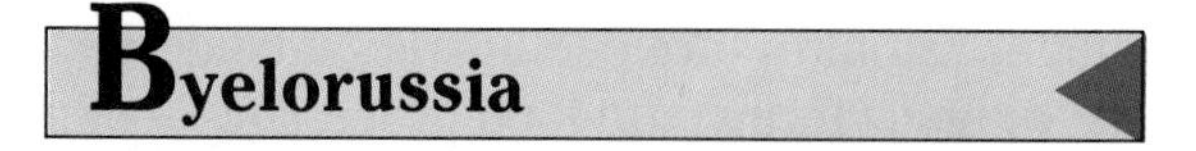

Byelorussia

Byelorussia (bee-ehl-oh-RUHSH-uh), located north of the Ukraine, is the home of another Slavic group, known as White Russians. (*Byelo* means "white" in Russian.) Byelorussians speak a language that is somewhat between Polish and Russian. Most of the people are Eastern Orthodox in religion, though some are Roman Catholic.

Byelorussia is a region with few mineral resources and poor soil. Much of the land is a large marsh. The drier parts of Byelorussia are covered with sandy soil and forests. Despite these problems, farming is one of the chief occupations. Some of the marshes have been drained to create large state farms. Grains, potatoes, sugar beets, and flax are the most important products.

Large forests in the region make lumbering a successful business. Peat is burned to provide energy for the manufacturing of cement, chemicals, and machinery. Minsk, the capital of Byelorussia, is an industrial center, as well as an important rail point connecting the Soviet Union and Poland.

The Baltic Republics

North and west of Byelorussia are Estonia, Latvia, and Lithuania, three Soviet republics bordering the Baltic Sea. Once part of tsarist Russia, they became independent after World War I. In 1940, they were absorbed again by the Soviet government and became Soviet republics. As in Byelorussia, the terrains are flat and swampy with poor soils. Russian leaders

Note the importance of the Baltic Republics in providing the USSR with access to the Baltic Sea and the Atlantic Ocean.

have long viewed control of the Baltic republics as essential for Russia's defense against European neighbors.

Estonia is the northernmost Baltic republic. The Estonian language is related to Finnish. Lumbering and dairy farming are the chief occupations. The republic has deposits of oil shale, from which petroleum products can be made. Estonia's other industries include textile manufacturing and food processing. Tallin is Estonia's capital and most important city.

While Latvia is larger than Estonia, the two republics have similar landscapes and economies. Dairy farming and lumbering are the chief industries. Riga (REE-guh), Latvia's capital, has prospered under Soviet control, partly for military reasons. Riga's harbor was developed, and its port has become one of the Soviet Union's busiest. Lutheranism is the traditional religion of Latvia and Estonia.

During the fourteenth century, Lithuania was a great kingdom extending to the Black Sea. Today, like the other Baltic republics, it is mainly agricultural. However, some manufacturing industries are found in the larger cities, such as in Vilnius (VIL-nee-uhs), the capital. Roman Catholicism is the traditional religion of Lithuania. The Lithuanian and Latvian languages are similar and very old. They are distantly related to the Slavic languages.

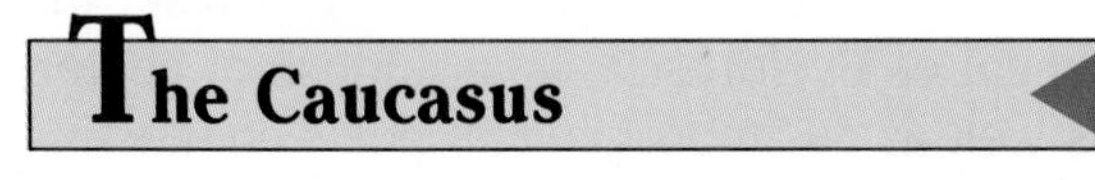

The Caucasus

The Caucasus region is the broad strip of land that separates the Caspian Sea in the east and the Black Sea and the Sea of Azov in the west. The rugged Caucasus Mountains are found in the south. The valleys and plateaus of the mountains are divided into the Soviet republics of Georgia, Armenia (ahr-MEE-nee-uh), and Azerbaijan (az-uhr-bie-JAHN). The north slopes of the mountains are part of the RSFSR.

The Caucasus region is a living museum of peoples and languages. Georgians and Armenians are the largest of the original

The population of the Caucasus region is a blend of many ethnic groups. This market in the Georgian SSR offers a variety of goods.

ethnic groups. (See the map "Ethnic Groups in the Soviet Union," on page 193.) The peoples of the Caucasus are famous for their relative independence from Moscow.

The Caucasus region is important as one of the great livestock areas in the Soviet Union. Large-scale grazing is the main occupation on the dry steppes north of the Caucasus Mountains. Meat is expensive in Soviet grocery stores. For this reason, many people eat meat only once or twice a week. The ranches of the Caucasus fill an important need.

Along the eastern shore of the Black Sea is a lowland area with a humid subtropical climate. This densely populated area produces citrus fruits and tea. Such warm-weather crops are very valuable in the Soviet Union. The lowlands bordering the southern Caspian Sea are semiarid. The chief crops along the Caspian are cotton, grains, and hay.

Russia's earliest oil fields are located on the western shore of the Caspian. Baku (bah-KOO) is a center of the oil-refining industry. Oil from the Baku area is piped to the industrial cities of the Ukraine. It is also sent by boat and rail as far as Leningrad. The Caucasus region is also rich in manganese, coal, lead, zinc, and copper.

Level B: Have students create a poster, collage, cartoon, or other visual representation illustrating some aspect of this region based on the material in this chapter. (Suggested topics: The Ukrainian Industrial Region, Dams of the Dnieper, Steel Production)

Have the students locate the Eastern Soviet Union on the map on page 191.

What are the major economic activities on the Kamchatka Peninsula? on Sakhalin Island?

forests, coal; oil, coal

can rise as high as 100° F (38° C) during this short season.

Most of Siberia is covered with the thick, softwood forests of needle-leaf trees, typical of the Russian taiga. These great forests make up one-fifth of the world's timberlands. In south-western Siberia, steppes extend from the Urals eastward for 1,000 miles (1,600 kilometers), forming a narrow triangle of fertile land between the northern taiga and dry Soviet Central Asia. The far northern lands of Siberia are icy tundra. Siberia's high latitude means long sunlight on summer days, while winter days have little sunlight.

The Ob, Yenisei, and Lena rivers flow south to north and aid transportation in the region. Navigable by barges in the summer, the rivers are used as frozen highways during the winter. The rivers divide Siberia into three natural regions: the West Siberian Lowland, the Central Siberian Plateau, and the Eastern Highlands.

The West Siberian Lowland is located between the Urals and the Yenisei River. It is drained by the mighty Ob River. The taiga part of the plain is a great swamp during spring and summer and a frozen wasteland in winter. Only the steppe part of the plain in the south is

Level B: Have students draw a map of this region showing the types of vegetation that grow here. Have students compare the region's climate and vegetation. Students can use atlases for the activity.

The great taiga forests of Siberia are covered with evergreen trees. Lumber, furs, and minerals are obtained from the forests.

arable, or suitable for farming. This is an important area for raising wheat.

The Central Siberian Plateau, located between the Yenisei and Lena rivers, is a rugged region of thick forests. The area's few settlements are found along the rivers and near mining areas. The Eastern Highlands, a mountainous and thinly populated area, extend east of the Lena River to the Bering Sea. Most people here live along the rivers and on the Pacific coast.

Resource Frontier The Russian conquest of Siberia began in 1581. Siberia's first settlers were adventurers seeking valuable furs and precious metals. Traveling along the rivers in small boats, they pushed eastward, looking for a "Northeast Passage" to China. Forts were built at river junctions and where trails crossed rivers. Some, such as Omsk, are now great cities. As the Russians spread into Siberia, they

The Russians established outposts in Alaska and as far south as California.

brought European customs and technology to people living there. As a trading relationship developed, most of the people readily accepted Russian control. Russian explorers brought the first detailed information of these little-known lands to Europe.

In the late nineteenth century, the European powers and the United States were seeking trade with China and the Pacific. Even though Russia bordered China, the easiest way into China from Moscow was by sailing through the Atlantic Ocean and then around South America or Africa! An easy overland route was needed, and in 1891, construction started on the Trans-Siberian Railroad. By 1915, the railroad connected Moscow to Vladivostock on the Sea of Japan. More than 5,700 miles (9,120 kilometers) long, the Trans-Siberian Railroad is the longest single rail line in the world. Settlement of Siberia has followed this railroad. For many towns, the train is the only way in and out.

During the twentieth century, many natural resources have been developed in Siberia. In 1984, a more direct railway was completed across eastern Siberia. This is called the Baikal-Amur Mainline (BAM) railway. The building of BAM was an impressive construc-

Workers celebrate the construction of the last link of the BAM railway. In what ways is the BAM important to the Soviets?

It has made Siberian resources more accessible to the Soviets. It will move Soviet industry away from the Chinese border. Thirteen industrial cities are planned along its route.

Historical Geography

Eastern Europe, like Russia, was settled by waves of invaders. Some were warlike; others migrated peacefully. Some wanted to farm the land and brought with them herds of horses, sheep, and cattle. Others were merchants and traders who established the towns. The new residents always brought their own language, technology, and other culture traits. Sometimes the immigrants merged with the existing people, losing much of their own culture. Other times, the settled people accepted the culture of the immigrants. By the start of the twentieth century, the cultural landscape of Eastern Europe was extremely diverse. It was common in many areas to find one set of ethnic groups living in towns while others lived in the surrounding countryside.

The languages of Eastern Europe show both the history of migrations and the many different people who settled in the region. Most of the people of Eastern Europe speak languages that belong to the Indo-European family. The most widespread languages belong to the Slavic language group. These are spoken by the Poles, Czechs, Slovaks, Yugoslavs, and Bulgarians. The Yugoslavs include several Slavic groups: the Serbs, Croats, Slovenes, Macedonians, and the Islamic Slavs of Bosnia and Herzegovina (hehrt-suh-goh-WEE-nuh). Romanian, however, is an Italic language, related to Italian, Spanish, and French. Albanian is an ancient Indo-European language, with roots that are even older than Latin. Hungarian, which is not an Indo-European language, is related to Finnish and to the ancient languages of western Siberia. Because of these many languages, Eastern Europeans have often had difficulty communicating with each other.

During the eighteenth and nineteenth centuries, Eastern Europe was controlled by neighboring empires. These included the German, Austrian, Russian, and Turkish empires. In the twentieth century, however, unhappiness with foreign control increased. By the end of World War I, all of the modern Eastern European countries had become independent. These countries included Poland, Czechoslovakia (chehk-uh-sloh-VAHK-ee-uh), Hungary, Romania, Bulgaria, Yugoslavia, and Albania.

Cultural isolation is a distinguishing feature of the mountain villages of Eastern Europe. What factors contribute to this isolation?

Many ethnic groups speak their own languages; cultural and geographic barriers make communication difficult.

When the boundaries were drawn in Eastern Europe, however, it was often impossible to separate the different groups because they were so intermixed. Thus, large numbers of Hungarians continued to live in Czechoslovakia, Romania, and Yugoslavia. Many Germans lived in Poland and Czechoslovakia. The unhappiness of minorities living on the wrong side of the border is a continuing source of political problems in Eastern Europe.

In the years between World War I and World War II, the countries of Eastern Europe enjoyed freedom from foreign control. However, during World War II, the region was caught between the Soviet Union and Germany. The end of World War II found Russian armies throughout the region. Soviet-style governments and cultural institutions were installed. East Germany—the German Democratic Republic—was created from the part of

Ask students how Eastern Europe is considered a buffer zone? What is the Warsaw Pact? How is it similar to NATO?

Have students locate Eastern Europe on the map on page 191.

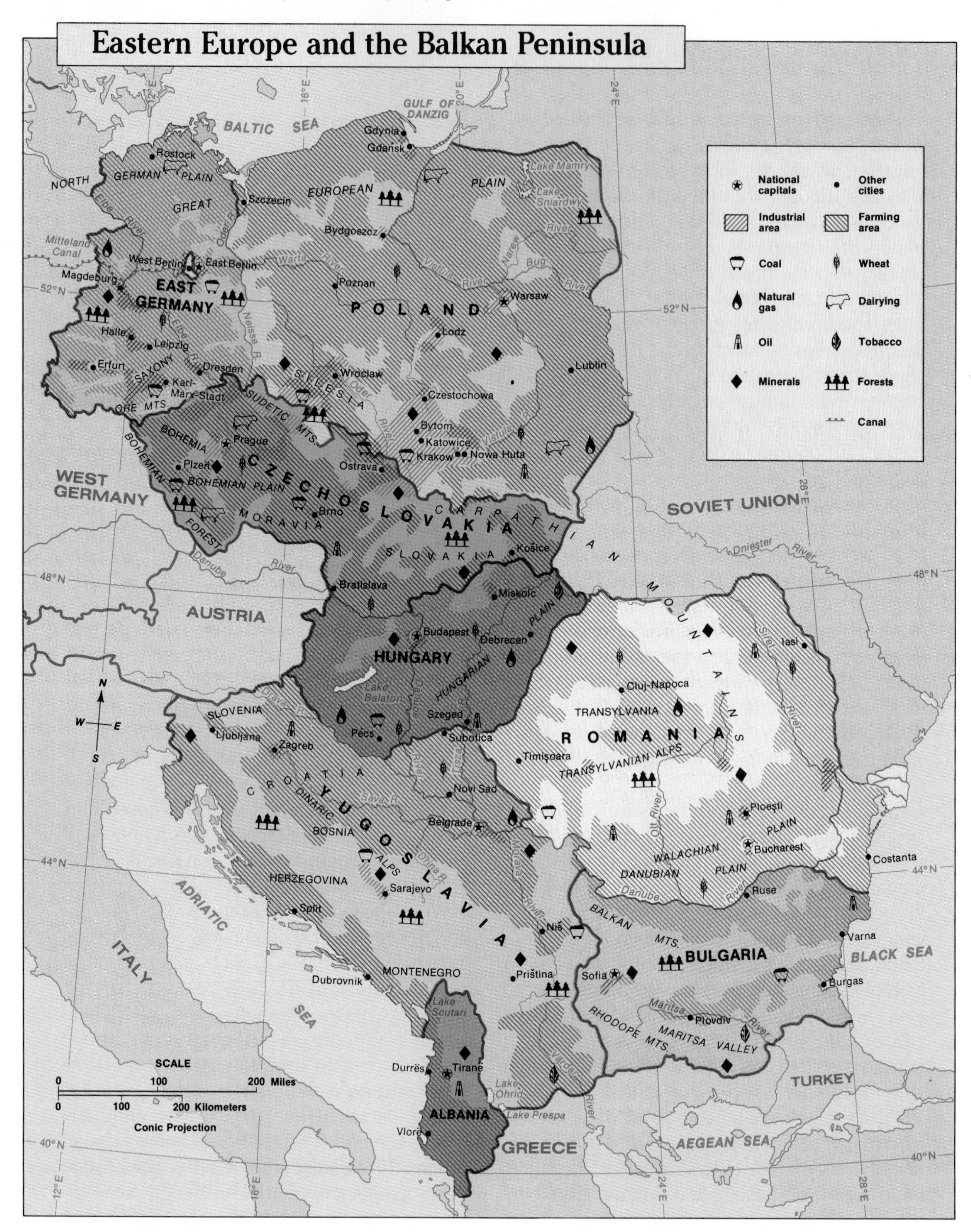

Where are the major farming areas of Eastern Europe found? the major industrial areas?

farming: throughout Poland, Czechoslovakia, East Germany, Hungary, in parts of Romania, Bulgaria, Yugoslavia, Albania

industry: East Germany, southern Poland, northwest Czechoslovakia, near large cities of Romania and Yugoslavia

Czechoslovakia exports many products to Western countries. Ask students to use almanacs to research the major products exported by Czechoslovakia.

Bohemia has long been an industrialized and economically developed region. The people of Bohemia, the Czechs, have become leaders in such fields as engineering and technology.

Moravia is east of Bohemia. Centered in a lowland gap between the Oder River of Poland and the Danube, it is a link that connects the plains of Poland and Hungary. Moravia shares part of the industrial resources of the Silesia region, which is mainly in Poland. Here, manufacturing and coal mining are of increasing importance. Moravia is a developing industrial region within Czechoslovakia. Bohemia and Moravia are usually grouped together as the "Czech" lands.

Slovakia takes up eastern Czechoslovakia. Because the landscape is mountainous, agricultural land is limited except along the Danube plains. Slovakia is a traditionally rural region that has recently industrialized. However, livestock raising, dairy farming, mining, and lumbering are still important.

Urban Geography Prague is the capital of Czechoslovakia. Although located in the far western part of the country, Prague has emerged as the country's communications center. Prague is also the traditional capital of Bohemia. It is an ancient city that is important for all of Europe. It has many beautiful old buildings and a rich cultural heritage. Prague also has a varied industrial economy. The small towns surrounding Prague specialize in making products from the minerals mined in nearby mountains.

Plzeň (PUHL-zehn) is the other large manufacturing center in Bohemia. There, iron and steel industries support Eastern Europe's largest automobile factory. Brno (BUHR-noh) is the capital of Moravia. It has textile factories, chemical plants, and machine-assembly works. Ostrava is the principal industrial city in the part of Silesia in Czechoslovakia. Bratislava, the capital of Slovakia, is on the Danube River and at the end of a natural-gas pipeline from the Soviet Union. Slovaks have been moving to Bratislava seeking high-paying jobs in the chemical industry.

Life in Czechoslovakia Czechoslovakia is a combination of Czech and Slovak territories that once had little in common with each other. However, the government's planners have been building transport and economic links to unite the country. The standard of living in Czechoslovakia is about 20 percent higher than that of Poland. Modern shops and automobiles line the streets. Imported goods are available.

Czechoslovakia, like East Germany, is closely linked to ways of life in Western Europe. With its Western outlook and prosperous industrial history, Czechoslovakia has never been attracted to Soviet communism. In 1968, however, a major reform movement was crushed by a Soviet invasion. Since then, Soviet-style policies have lasted longer than in

Wenceslas Square, a wide boulevard lined with hotels, restaurants, and shops, is the busiest street in Prague.

Level B: Have students draw a map of Czechoslovakia showing its various regions, languages, cities, and industries. How does Czechoslovakia compare to some Western European countries in a cultural sense?

the USSR itself. However, the government leaders of Czechoslovakia are now experimenting with better conditions for workers and local management of factories.

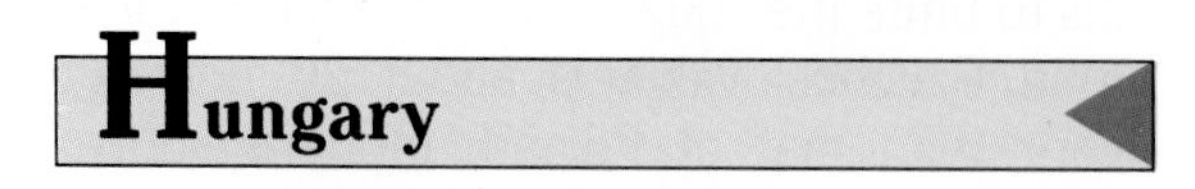

Hungary

Land and Resources Hungary is the agricultural heartland of southeastern Europe. The soil is fertile, and the land is mostly flat. The Danube and Tisza rivers provide both navigation and water resources. Important crops include wheat, corn, sugar beets, potatoes, and wine grapes. Hungary's agricultural practices are a mixture of the old and the new. The wines of the Danube River valley are well known.

Hungary is the major food exporter of Eastern Europe. Hungarian vegetables and fruits are trucked all the way to the cities of the Persian Gulf. Although nearly all of Hungary's farmland is in collectives, government planners permit farm managers to make most decisions. In addition, large lots are assigned to individual collective-farm families to cultivate as they wish. They then receive half the profits of their production.

Manufacturing accounts for about 60 percent of Hungary's income. Oil and natural gas are imported from the Soviet Union. Three modern nuclear power plants are planned to provide nearly half of the nation's electricity. There is coal in the southern part of the country near the city of Pécs (PAYCH). Three steel plants use iron ore imported from the Ukraine. Meanwhile, millions of tons of bauxite have been exported from Hungary to the Soviet Union and Czechoslovakia to make aluminum.

Hungary has a tradition of high literacy and technical progress and has become a leading producer of computer programs. Small businesses have been encouraged in farming towns. About 25 percent of Hungary's people live in the Budapest region, where manufacturing is best developed. There, machinery, precision instruments, and textiles are important products. Budapest is Hungary's capital city as well as the country's political, cultural, and economic center.

About 20 percent of all Hungarians live in Budapest, Hungary's capital and largest city. The city was built along the Danube River.

Level A: The city of Budapest is divided by the Danube River. Buda is on one side; Pest on the other. Using an atlas, ask students to list other cities in the world that are separated by rivers or other bodies of water.

Why does Hungary have a higher standard of living than Poland or Czechoslovakia? How is the standard of living related to the political climate of a country? (A higher standard of living usually reflects a stable political climate.)

Life in Hungary In 1956, many Hungarians rebelled against the Communist government. There were strikes and riots. The Soviet Union then invaded Hungary and brought back Soviet Communist rule. Many Hungarians were executed or imprisoned, and many more fled the country.

Since the 1970s, Hungary's leaders have relaxed their Soviet style of government. Today, Hungary is least like the Soviet Union of all the countries in Eastern Europe. Trade with Western Europe has increased. Factories assemble products such as computers for United States companies to sell in Western Europe. Factory managers have been given more control in running their plants. Productivity and living standards have improved. Thousands of small businesses have developed. Hungary's newspapers are the liveliest in Eastern Europe. There are often two candidates on election ballots, and government policies are often debated in public.

Romania

Land and Resources To the east of Hungary lies Romania, which is divided into four landform regions. These are the eastern part of the Hungarian Plain, the Carpathian Mountains, the hilly Transylvania plateau, and the Walachian Plain along the Black Sea. Romania was an important grain-growing region for the Roman Empire. It remains a rural, little-developed country. Old-fashioned farming methods and poor transportation are still typical of its rural areas. Although Romanian farms are not mechanized, nearly 90 percent of Romania's farmland is in state farms or collectives. Wheat, corn, barley, and oats are grown mainly on the plains. Grains, fruits, and vegetables are grown in the central highlands. Grazing is also important.

With the exception of oil, most of Romania's potential mineral resources are undeveloped. Ploeşti (plaw-YEHSHT), in the foothills of the Carpathians, is the center of the country's oil industry. The famous Ploeşti fields are the richest in Eastern Europe. New fields along the Moldavian border in eastern Romania are growing in importance. Much of Romania's oil resources have been depleted, and the country must import oil and natural gas from the Soviet Union to meet its needs.

Romania had little industry when the Communists first took over the government. Since then, industry has increased slowly, yet it remains much like that of less developed countries. A few high-value products, such as furniture and cars, are assembled for export. Other industries provide consumer goods for the local market. Most industries process primary products from the farms, oil fields, and mines. A large hydroelectric dam has been built in the "Iron Gate" area, where the Danube River crosses the Carpathian Mountains. A second dam is planned. Electricity is shared with Yugoslavia.

Romania's cultural, industrial, and commercial center is Bucharest (BOO-kuh-rehst), the capital city. Food products, textiles, and heavy machinery are manufactured here.

The Romanian government is encouraging development of the country's infrastructure. These workers are building a road.

It has been said that progress and development in Eastern Europe decline from west to east and also from north to south. Ask students for evidence to substantiate or refute this statement.

Economic and Political Ties The Romanian government has asserted its independence from Soviet control. Since 1959, Soviet troops have not been permitted to enter the country. While Romania maintains close relations with the Soviet Union, it has established political and trade relations with some Western countries, including the United States. Still, Romania remains a Soviet-style repressive society and continues to suffer from a lack of economic progress.

Bulgaria

Bulgaria occupies the eastern side of the Balkan Peninsula. (See the map on page 226.) It is crossed by two east-west mountain ranges, the Balkan and the Rhodope. To the north of the Balkan range, the land is a plateau leading to the Danube. It is important for growing corn and other crops. In the mountains, grazing and small-scale agriculture are the main activities. People with old-fashioned customs are commonly found living in isolated mountain valleys.

In the heart of the country, between the two mountain ranges, lies the valley of the Maritsa (muh-REET-suh) River. This valley, together with its Black Sea coastal lands, is Bulgaria's most important agricultural region. Wheat, corn, and other grains are grown here. The warm, Mediterranean climate in the Maritsa Valley permits the growing of fruits and vegetables. Much of this produce is canned and exported to Western Europe and the Soviet Union.

Most of Bulgaria's land is in collective farms. Farms are becoming more mechanized, and yields are increasing. Similar progress is being made in Bulgarian industry. Bulgaria is no longer one of Europe's least developed countries. Bulgaria's planners have increased educational opportunities and have encouraged Bulgarian national pride. Manufacturing in Bulgaria is geared both for the local market and for export. Important industries are concentrated near Sofia, the capital. Like the other Eastern European capitals, Sofia is the largest city in the country. It is also the historic center of Bulgaria's cultural and political life. Sofia has become a noisy, modern city with cafés and a few stylish shops.

Sheep grazing and small-scale farming are Bulgaria's main economic activities, particularly in the mountainous regions.

Have students research how communism and free enterprise exist side by side in Yugoslavia. How does the Soviet government regard Yugoslavian independence from Moscow?

Bulgaria is Eastern Europe's most loyal supporter of the Soviet Union. This close relationship gives Bulgaria an advantage in trade and development projects. Over half of Bulgaria's trade is with the Soviet Union. Three Soviet nuclear reactors generate one-third of Bulgaria's electricity. Bulgaria produces many computers for the Soviet Union and other Eastern European countries.

Yugoslavia

Land and Resources Along the western half of the Balkan Peninsula lies Yugoslavia. It can be divided into four landform regions: a narrow coastal lowland along the Adriatic Sea; the rocky and bare Dinaric (duh-NAR-ik) Alps, inland from the Adriatic coast; a portion of the Hungarian Plain, stretching across northeastern Yugoslavia; and a section of the true Alps, in the northwestern corner of the country. The Danube flows south into Yugoslavia and past Belgrade (BEHL-grayd), the capital. (See the map on page 226.)

The climate varies from cool, humid continental in the north to Mediterranean in the south. There is a narrow strip of humid subtropical climate in the northwest. Thus, while the plains along the Danube are having their summer rains, the south is experiencing its annual drought. While people are skiing in the northwestern mountains, the south is having its winter rains.

Farming is the chief occupation of the Yugoslavs. Less than one-sixth of the land is in collectives. Grapes and olives are grown along the sunny Adriatic coast. Farmers on the fertile Danubian plains grow wheat, corn, oats, and barley. Mountain pastures that are not suitable for growing crops are generally used for grazing.

Yugoslavia ranks among Europe's leading producers of lead, bauxite, and **antimony**. Antimony is a silvery white element that is used in type for printing and to make other metals. Copper, zinc, and mercury are also mined. Yugoslavia's energy resources, such as coal and waterpower, are few compared with those of Poland and Czechoslovakia.

Unlike its agriculture, much of Yugoslavia's industry is government controlled. However, in contrast to other Eastern European countries, such control is from the local community rather than from the national government. Most industrial development is centered in northern Yugoslavia. The southern part of the country has remained largely underdeveloped, although many new factories have been built there. Mining and metal industries are vital to the country. The manufacture of farm machinery and mining equipment is increasing. Automobiles are now being exported to the United States and Canada.

The warm, sunny climate of the Adriatic coast makes Dubrovnik, an ancient walled city and port, a popular tourist destination.

Cultural Variety *Yugoslavia* means "land of the southern Slavs." The main Slavic groups are the Serbians, Croatians, Bosnians, Slovenians, and Macedonians. There are also Italians, Hungarians, Albanians, Turks, and Gypsies. The country's constitution provides for six republics and two autonomous regions. There are three official languages: Serbo-Croatian, Slovenian, and Macedonian, all of which are Slavic languages. The Serbs write the Serbo-Croatian language in the Cyrillic alphabet, which is also used to write Russian. The Croats write the same language in the Roman alphabet. School is taught in

Level A: Have students research the continuing ethnic problems of Yugoslavia—for example, Serbian and Croatian nationalism and Albanian minorities. Students should draw a map of Yugoslavia showing the various ethnic groups and where they live. Newspaper and magazine articles will be helpful.

Have students research the political differences between Chinese and Soviet communism. Why did a split between the two countries occur? How and why does Albania follow the Chinese version of communism?

different languages in various regions. Most people belong to the Roman Catholic, Eastern Orthodox, or Islamic religion.

In addition to language and religion, ways of life vary in Yugoslavia. In the north, there are modern cities, while in some southern valleys, life has changed little in several hundred years. Though the country has a rich variety of cultures, the different peoples of Yugoslavia have engaged in bitter disputes. However, the notion of a united Yugoslavia seems to be growing among the people.

Political and Foreign Policy Unlike other Eastern European countries, Yugoslavia has succeeded in maintaining political independence from the Soviet Union while remaining a Communist country. This policy of political independence was developed by Joseph Tito, who led the country from World War II until his death in 1980. Tito even accepted economic and military aid from Western countries.

Politically, Yugoslavia remains somewhere between the East and West. The press is relatively free, but the Communist leaders control the government. The economy has had increasing problems, and living standards have not improved in recent years. Labor strikes in support of higher wages are common.

The Roman and Cyrillic Alphabets

Roman		Cyrillic		Pronunciation
A	a	А	а	ah
B	b	Б	б	b
C	c	В	в	v
D	d	Г	г	g
E	e	Д	д	d
F	f	Е	е	yeh
G	g	Ё	ё	yah
H	h	Ж	ж	zh
I	i	З	з	z
J	j	И	и	ee
K	k	Й	й	y
L	l	К	к	k
M	m	Л	л	l
N	n	М	м	m
O	o	Н	н	n
P	p	О	о	aw
Q	q	П	п	p
R	r	Р	р	r
S	s	С	с	s
T	t	Т	т	t
U	u	У	у	oo
V	v	Ф	ф	f
W	w	Х	х	kh
X	x	Ц	ц	ts
Y	y	Ч	ч	ch
Z	z	Ш	ш	sh
		Щ	щ	shch
		Ъ	ъ	—
		Ы	ы	i
		Ь	ь	—
		Э	э	eh
		Ю	ю	yoo
		Я	я	yah

Which Roman letters resemble Cyrillic letters?

A, a, B, b, E, e, K, k, M, m, O, o, P, p, T, t, Y, y, X, x

Albania

Albania is Europe's least developed and poorest country. Large parts of mountainous Albania seem hardly affected by the twentieth century. Yet, literacy is high. The capital city, Tiranë (ti-RAHN-uh), the site of the country's few factories, has fewer than 200,000 people. In fact, only about one-third of the Albanian people live in cities. The products of Albania are raw materials typical of a less developed country. These include mostly agricultural materials and **foodstuffs**. A foodstuff is a substance with food value, particularly the raw material of food. An example is corn. Albania has little modern manufacturing, although some oil and chrome ore are exported.

A self-reliant country, Albania is physically and culturally isolated from the rest of Eastern Europe. (See the map on page 226.) It belongs culturally to neither Eastern nor Western Europe. Albania has gained fame only for its curious political relations. The government is Communist, but the country has almost no relations with other Communist countries. Even connections with neighboring Yugoslavia are limited to a single rail line that was linked to that country in 1987. There is a freight ferry to Italy and an airline connection to Turkey. Throughout history, Albania's isolation has permitted its people to maintain their culture and identity. Today, Albania is a lonely outpost with its own variety of communism.

CHAPTER 20 CHECK

Reviewing the Main Ideas

1. Eastern Europe consists of eight countries politically tied to the Soviet Union but with various forms of Communist government: East Germany, Poland, Czechoslovakia, Hungary, Romania, Bulgaria, Yugoslavia, and Albania.
2. The countries of Eastern Europe are densely populated. They include a wide range of ethnic groups, the majority of whom are Slavic.
3. Eastern European climates vary from the moist maritime and humid continental to the warm humid subtropical and Mediterranean. The physical geography of Eastern Europe varies from plains to high mountains.
4. Since World War II, all the countries of Eastern Europe have expanded their economies. Living standards are slightly higher than those of the Soviet Union but lower than those of Western Europe.

Building a Vocabulary

1. Define *COMECON*. How has COMECON influenced the economic development of Eastern Europe?
2. Define *martial law*. Give an example of the Soviet Union's use of martial law in Eastern European countries.
3. Define *antimony*. Name one of Eastern Europe's leading producers of antimony.
4. Define *foodstuff*. Give an example of a foodstuff.

Recalling and Reviewing

1. How did the countries of Eastern Europe come under Soviet domination? Which countries are most independent from the Soviet Union? Why?
2. What factors give Poland the greatest economic potential in Eastern Europe?
3. What has been Romania's relationship with the Soviet Union? How does it differ from Bulgaria in this regard? from Czechoslovakia?
4. In what ways is Yugoslavia a diverse nation?
5. Why is Albania the poorest country in Europe?

Critical Thinking

6. Eastern Europe is an ethnically diverse area. What problems have been created by this diversity? How has the diversity benefited the region?

Using Geography Skills

Use the map in this chapter to answer the following questions.

1. Which Eastern European countries are landlocked? Is this a problem for these countries? Why or why not?
2. What major rivers does Eastern Europe share with Western Europe? How might this influence trade relations?

Answers to Chapter Check questions are found in the Teacher's Guide.

Additional activities for Chapter 20 are found in the Student Workbook. Chapter 20 Test as well as Skills, Reteaching, Critical Thinking, and Challenge/Enrichment activities are available in the Teacher's Resource Binder.

Level B: Have students write on index cards five questions dealing with information presented in this unit. Ask them to exchange cards with classmates to review the unit.

UNIT 3 REVIEW

Summary

The character of the Soviet people, like the character of the country, is complex. More than 150 different ethnic groups are found in the Soviet Union. Language, religion, and social customs vary greatly.

In 1917, when the Communists came to power, poverty and illiteracy were widespread. The Communists took action to make the Soviet Union a strong and self-sufficient nation. To do this, the Soviet government had to develop the nation's resources and industries. The Soviet people, however, have not benefited much from this buildup. They lost most of their religious and political freedom. Today, consumer goods remain in short supply and are very expensive.

Although the Soviet Union is isolated geographically, politically, and economically from much of the rest of the world, the country is a world leader. With its many resources, strong government control, and plentiful labor, the Soviet Union has become a military and industrial giant. However, compared to people in the United States, the Soviets have few freedoms.

After World War II, the Soviet Union expanded its influence over most of Eastern Europe. Communist governments were established in the countries of Eastern Europe. The Soviets separated Eastern Europe from the West. Most of the Eastern European countries, however, have established relations with the rest of the world. These countries also have tried to gain more freedom from Soviet domination.

Reading and Understanding

1. Why was Russia able to expand its territory eastward into Siberia?
2. How is government planning used in the Soviet Union today?
3. How much autonomy do the Soviet republics have? Explain.
4. What resources form the natural wealth of the Soviet Union? How have these resources influenced the nation's economic development?
5. How successful has the policy of autarky been? What are the advantages and disadvantages of this policy?
6. Describe the Soviet transportation network. How has this network contributed to the nation's economic development?
7. What problems does the Soviet Union have today? How has the Soviet government attempted to solve them?
8. How is Eastern Europe tied to the Soviet Union? Explain.
9. Why has the agricultural potential of Eastern Europe not been fully developed? Give an example of how one country is trying to improve agricultural production.
10. Which of the Eastern European countries are least industrialized? Why?
11. What term can be used to describe a group of people who have the same language and customs?

 Answers to Unit Review questions are found in the Teacher's Guide.

Mastering Geography Skills

Use the climate map and population-density map in the Geography Skills feature in Chapter 17 to answer the following questions.

1. What is the population density of the Soviet Union's continental desert climate region?
2. What is the population density at 60° north latitude, 30° east longitude? What city is located in this area?
3. Describe the population density along the route of the Trans-Siberian Railroad. What factors do you think have influenced the population density along the railway?

Applying and Extending

Use a historical atlas to do the following: (1) Develop a composite map to show the expansion of the Soviet Union to its present boundaries. Use color to show each new territorial addition. Be sure to label your map and include a legend. (2) Make a chart to show the territorial expansion, the date each territory was added, how and from whom each area was acquired. From your information, how did Russia first acquire a warm-water port? Which areas of the present Soviet Union were acquired by armed conflict? Which areas were acquired by colonization and settlement?

Linking Geography and Economics

The economies of the Soviet Union and Eastern Europe can be called command economies. Use library resources to determine the characteristics of a command economy. How does a command economy differ from the kind of economy that exists in the United States? Although the Eastern European and Soviet economies are all command economies, they differ. Select two of the countries you studied in this unit. Study the economy of each. Find out about the mix of private and public ownership, economic decision making related to prices and the use of resources, the role and organization of labor, and the role of the consumer. Write a report about the differences in these economies.

The differences between command and market economies are discussed in Unit 1, page 97.

Reading for Enrichment

Dornberg, John. *Eastern Europe: A Communist Kaleidoscope.* New York: Dial Books for Young Readers, 1980.

Kostich, Dragso. *The Land and People of the Balkans.* Philadelphia: J. B. Lippincott, 1973.

MacShane, Dennis. *Solidarity: Poland's Independent Trade Union.* Chester Springs, PA: Dufour Editions, 1982.

McDowell, Bart. *Journey Across Russia: The Soviet Union Today.* Washington, DC: National Geographic Society, 1977.

The Soviet Union. Time-Life Library of Nations series. Morristown, NJ: Silver Burdett, 1985.

▲ A stone fortress stands on an oasis at the base of the Atlas Mountains in Morocco. The Atlas Mountains extend for 1,500 miles (2,400 kilometers) across northwest Africa.

◀ Islam, which is widely practiced in the Middle East and North Africa, has about 540 million followers. These Muslims are praying toward the holy city of Mecca.

Farmers along the coast and in the mountains of the Middle East raise a variety of livestock. Most people in this region are farmers. Sheep, which are raised for meat and wool, are traded and sold at this outdoor market in Morocco. ▶

Have students compare the photos on these two pages with those found on pages 106–107 and 186–187. Ask them to note differences among the three regions.

Have students locate the Middle East and North Africa on the inset map of the world on page 244. Point out that three continents converge at this part of the world.

Saudi Arabia. Here, he worked as a merchant. When he was about the age of 40, he often went to nearby mountains to meditate about the social injustices in Mecca. Muhammad said that during one of his trips, he was told by a messenger of God to preach religion to the Arabs. He returned to Mecca as the founder of a new religion, Islam. The words that Muhammad received from God, or Allah, were later recorded by his followers in the Koran. A person who follows Islam is called a **Muslim**.

Islam shares many of the traditions of Judaism and Christianity. In fact, the Koran includes several passages directly from the Old Testament of the Bible. One of these shared traditions is **monotheism**, or belief in one god. Muslims believe in one god, Allah. Islam is a guide to daily life that is expressed in five "pillars." The first is that Muslims must accept the will of Allah and his messenger, Muhammad. The second is prayer, which is required of Muslims five times each day. The Muslim place of worship is called a **mosque**. The third pillar is charity. Muslims must give part of their wealth to the poor. Fasting during the holy month of Ramadan is the fourth Muslim duty. During Ramadan, Muslims do not eat or drink from dawn until sunset. Normal daily activities are usually reduced. The fifth pillar is a pilgrimage to Mecca at least once in a lifetime. This pilgrimage is called the *hajj*. In Mecca and the nearby holy city of Medina, the pilgrims retrace the steps of Muhammad, praying at holy sites. Medina contains the tomb of Muhammad.

After its founding, Islam spread rapidly. In 10 years, Muhammad was able to unite the tribes of western Saudi Arabia. Within a century after Muhammad's death, Muslim armies had conquered most of the Middle East and had gone as far as Morocco and Spain. In the east, Islam spread to India and Central Asia and later to Indonesia and the Philippines.

Islam is still the unifying culture trait of the Middle East. However, like Christianity, Islam has split over the years into different groups. This division has been caused by varying interpretations of the Koran and by questions of leadership throughout the Islamic world. There are two great categories of Islam, the Sunni and the Shi'ite (SHEE-ite). The Sunnis emphasize the basic Islam of the Koran and personal prayer. However, the Shi'ites believe in Imams, who are religious as well as political leaders.

Today, about 90 percent of Muslims are Sunni, and 10 percent are Shi'ite. The Shi'ites are concentrated in Iran, the Persian Gulf region, Yemen, and Lebanon. The religious revolution that began in Iran in 1979 is based on Shi'ite principles.

Colonial Empires After the conquests by the armies of Islam, the Middle East was divided into many small territories. Each territory was ruled by political and religious princes known as caliphs, emirs, or sheiks. In Turkey, a great Islamic leader named Osman founded the Ottoman Empire about 1300. In 1453, the Ottomans captured Constantinople. At its peak, the Ottoman Empire extended along the Mediterranean Sea from Tunisia to Egypt, including the land surrounding the Red Sea. In the east, the Ottomans controlled land from the Persian Gulf to the Caucasus. In the north, their empire included the Balkan Peninsula and the Crimea. As the powers of Europe expanded in the 1700s and 1800s, the Ottoman power declined.

The Ottoman Empire finally collapsed to European control at the end of World War I in 1918. Some countries of the region became colonies of France, Britain, and Italy. Others came under foreign economic control, with local leaders remaining in office. The Europeans were interested in the resources and transportation routes of the region. They tried to develop the economies of the Middle East. They used military bases in the countries to guard important shipping routes. The British and French also blocked Russian expansion into the region.

Most of the modern countries of the Middle East and North Africa achieved their independence after World War II. Since

Discuss with students how the British and French needed the Suez Canal for easy access to their colonies in the Far East and how this need eventually led to their colonizing the Middle East before World War I.

Level A: Have students research and report on the Muslim religion. Suggestions include: Details of the *Hajj;* Traditions of Ramadan; and the Spread of Islam as an Example of Diffusion (map work).

then, nationalism and modernization have been reshaping the cultures of the region. Each country has tried to maintain its independence. Yet, each has had to look to the United States, the Soviet Union, or Western Europe for economic and military support. In addition, the common traits of Islam and the Arabic language have encouraged the Arab countries to work closely together.

Many Cultures As a result of the region's long history of changing empires and religions, there are many ethnic groups in the Middle East and North Africa. The Arabs are the dominant group in all countries of the region except Turkey, Cyprus, Israel, and Iran. Nearly all Arabs are Muslims. Therefore, the many Arab countries share common cultural and religious views. The two other large ethnic groups in the region are the Turks and the Iranians. Each of these groups has its own country. While both of these peoples are Muslim, their histories and cultures differ from those of the Arabs.

The last important group is the Jews in Israel. Jews from Europe and the Middle East region began moving to Israel in large numbers after the country was formed in 1948. The people of Israel view themselves as descendants of the ancient Hebrews. The Hebrews were among the people who lived in this part of the region about 3,000 years ago. The Jewish state of Israel is located in the midst of the Arab countries and is bitterly opposed by the Arabs. The conflicts between the Arabs and the Israelis continue to be an issue in world politics. (You will study more about the formation of Israel in Chapter 22.)

There are many ethnic minorities in the region. The Berbers live in the mountains of North Africa. They are Muslim but not Arab. There are also many Christians in the Middle East, particularly in Lebanon. In Iran, there are Zoroastrians (zohr-uh-WAS-tree-uhns), who follow the religion of ancient Persia. Also Muslim but non-Arab are the Kurds. The Kurds comprise a large ethnic group living in the border regions of Iraq, Iran, and Turkey.

A Berber tends her flock in Morocco. The Berbers have lived in North Africa since about 2000 B.C. They have a unique culture.

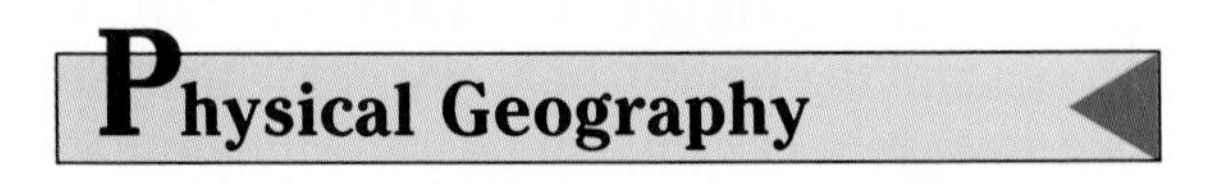

Physical Geography

Landforms and Climate The land of the Middle East and North Africa is mostly upland plains and rugged mountains. According to the theory of plate tectonics (discussed in Chapter 3), the region is at the point where the African, Asian, and European plates meet. This theory explains the region's complex mixture of mountains, valleys, and narrow gulfs and seas.

The Middle East and North Africa contain the world's largest **arid**, or dry, region. Cities and farmlands are generally located along river valleys and on other sites where water is

 Refer students to the map of world religions on page 91.

Ask students why Mecca and Medina are considered holy cities by Muslims.

The people of Saudi Arabia live where there is either water or oil. An important area of settlement is in the western mountains, where the holy cities of Mecca and Medina are located. Second in importance, and a more recent area of settlement, is near the oil fields along the Persian Gulf. One-fourth of the world's remaining oil may be buried there. Between these two areas are isolated oases. The largest oasis is in the area surrounding Riyadh (ree-YAHD), the capital. The oases of the Saudi Arabian interior have been settled for thousands of years.

Despite Saudi Arabia's oil wealth, farming and grazing are still important occupations. Dates, wheat, and barley are the main crops. In the past, nomadic Bedouins were a common feature of the desert regions. The nomads used the oases as home bases, moving into the desert during the rainy winter season. They would graze their herds of camels, sheep, and goats on the grasses that sprang up when it rained.

More and more Bedouins have settled down and taken jobs in the cities and on new, modern farms. The Saudi Arabian government encourages this settlement. Young Saudis can frequently be found behind desks as office workers, business people, and bankers. Much of the manual labor in Saudi Arabia is being done by foreign workers from Egypt, Yemen, and India.

An Islamic Holy Place Saudi Arabia has a special place within the Islamic world. The Saud family, which has ruled the country since

Much of Saudi Arabia is desert. How have oil resources helped the country develop a strong economy and position in trade?

The oil industry provides much of the country's income; Saudi Arabia can influence world oil prices.

Themes in Geography

THE PERSIAN GULF

Most of the world's oil lies beneath the earth in one small area of southwestern Asia: the Persian Gulf. Because oil has become necessary to modern life with its automobiles, aircraft, electric power, and plastics industries, it is not surprising that the Gulf region has attracted the world's attention.

The Persian Gulf is a body of water linked to the Indian Ocean by the narrow Strait of Hormuz. However, *Persian Gulf* also refers to the countries that surround the Gulf itself. The largest of these countries is the kingdom of Saudi Arabia, and the smallest is the tiny island state of Bahrain.

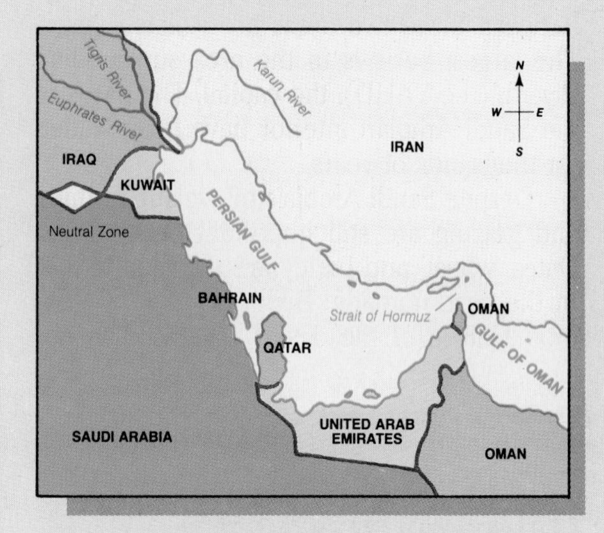

The Gulf countries may differ greatly in size and population, but they have much in common. To keep their economies running, they all rely heavily on money earned from exporting oil. The people of the region share a common religion, Islam. And almost the entire area has a dry, desert climate.

The largest customers for the oil produced in the Persian Gulf are the world's major

Region **It is because the Gulf states have so much in common—religion, climate, and dependence on oil—that we refer to the Persian Gulf as a region. Yet, despite their similarities, these countries also have important differences. The country that differs most from its neighbors is Iran. Unlike the other Gulf states, Iran is not an Arab country. Also, most Iranians belong to a branch of Islam that is different from the Islam practiced by most other Gulf residents. How do you think these differences might cause conflict between Iran and its neighbors?** Answers will vary.

 Ask students to suggest other regions in the world that may be organized around bodies of water (Pacific countries, Caribbean region, and so on).

You may want students to review the five themes of geography, discussed in Unit 1 on pages 33–35.

industrial countries, particularly the countries of Western Europe, North America, and Japan. This means that a large amount of oil must be moved from the Gulf to the places where it is needed. Some of the oil is exported overland by pipeline to neighboring countries. Most, however, travels great distances via enormous tanker ships.

Location **The Persian Gulf was an important part of the world long before oil became a valuable product. Since the time of ancient Mesopotamia, it has been one of the world's most important waterways. The Gulf's location has always encouraged trading. Long ago, Gulf traders sold food products from Mesopotamia and pearls from the Gulf; they bought porcelain wares and tea from India and China. These traders also traded products from the Gulf for African gold, spices, and slaves. What sorts of locations offer the best opportunities for trade? Why?** Areas with good transportation networks; because goods need to be moved from one place to another efficiently and quickly.

Movement **Most of the Persian Gulf's oil exports travel through the Strait of Hormuz. This makes the strait an extremely important area. If it were to be closed, for war or any other reason, the flow of oil to the industrialized world would be greatly reduced. This is why many world powers, including the United States and the Soviet Union, are concerned about what happens in the Gulf region. Study the world maps in the Atlas at the front of this book. Can you find any other narrow waterway in the world where the flow of traffic could easily be stopped?** Strait of Gibraltar, Suez Canal, Panama Canal, Bosporus, Strait of Malacca.

The Persian Gulf region has a long and impressive history that dates back to ancient times. The country called Iraq today was the site of ancient Mesopotamia. Mesopotamia was known as "the land between the two rivers," which are the Tigris and the Euphrates. Later, in another part of the Gulf, called Arabia (today called Saudi Arabia), the prophet Muhammad lived. There, he founded Islam, which is one of the world's major religions today.

its founding in 1932, is devoutly Muslim. The Sauds have organized Saudi society under strict Islamic law.

Most important to the Islamic world are the cities of Mecca and Medina. These places are where Muhammad lived and began teaching Islam. Muslims are required to make a pilgrimage to Mecca at least once in a lifetime. Today, more than two million pilgrims from all parts of the world visit the holy cities each year. In past centuries, the pilgrims came to Mecca by land in large caravans. Now, they arrive by bus, car, or plane, or by ship through the Red Sea port city of Jidda.

Before the development of the country's petroleum industry, the government of Saudi Arabia received most of its money by taxing visitors to the holy cities. Pilgrims still provide a large amount of money for the government of Saudi Arabia.

Economic Geography The oil reserves in the eastern part of Saudi Arabia are huge. Many experts believe that Saudi Arabia has more oil underground than any other country. Taxes received from oil companies and from the export of oil provide a large portion of Saudi Arabia's income. The Saudi economy falls and rises with the price of oil.

Although the oil industry produces most of Saudi Arabia's income, it provides few jobs. Because most of the oil fields are highly automated, not many workers are needed. Thus, Saudi Arabia must build industries and develop agriculture to provide more jobs for its people. Both industry and agriculture require water. Through an expensive process called **desalinization**, seawater and salty water from wells are purified by removing the salts. Riyadh alone consumes 10 percent of the world's desalinized water. The Saudi Arabian government provides money for business people to build factories and modern farms. Saudi Arabia now boasts the largest single dairy herd in the world. In addition, fields irrigated with water from deep wells and with desalinized water provide Saudi Arabia with wheat to export.

Saudi Arabia also has developed a chemical industry based on natural gas. Government-paid social services, including education, are available to all. Oil wealth is helping modernize Saudi Arabia and is improving the standard of living of many of the nation's people.

Saudi Arabia's World Role Saudi Arabia's oil wealth and devout Muslim policies give the country a special position in both the Middle East and world affairs. Since Saudi Arabia needs little oil at home, it is the world's largest oil exporter. This gives the country an influential role both in the region and in the world. The Saudis have the power to influence the price of oil. They can disrupt the world economy just by refusing to sell their oil.

As a rich Muslim state, Saudi Arabia has also given economic and military aid to the poorer states of the region. In some cases, Saudi Arabia has influenced Arab policy toward Israel. While the Saudis remain opposed to Israel, they have been patient in working for a peaceful settlement.

Small Arabian States

The Fringe of Arabia Around the southern and eastern fringe of the Arabian Peninsula is a string of small, traditional Arab states. Most consist of a strip of coastline, one or two port cities, and an isolated area of dry, undeveloped countryside. These states often appear to be relics of the last century, when the large states had not yet formed in the region. Each territory was ruled by a sultan, a sheik, or an emir. All of these small countries have significant locations in relation to world shipping. Only South Yemen does not hold important oil deposits.

Kuwait Kuwait is an oil-rich slice of desert on the Persian Gulf between Iraq and Saudi Arabia. As its rich oil deposits have been developed, its population has grown rapidly.

Kuwait has been caught in the middle of the Iran-Iraq war. The Iranians have fired missiles at Kuwaiti territory, and Kuwait's ships have been attacked.

The Suez Canal connects the Mediterranean and Red seas. How have the canal's size and location prevented it from being more useful to Egypt?

The canal is too small for most of today's supertankers; there is a continued threat of war in the region.

However, Egypt's biggest problem is its lack of cash. The total value of goods that Egypt must import is greater than the value of its exports. The country does receive important income from the tourist trade. There are also the fees ships must pay the Egyptian government to use the Suez Canal. Still, Egypt has had to borrow heavily to finance its industry. As more of these loans become due, the country may find it necessary to cut its spending for new industry. Wherever a large population cannot live on the food it produces, industry is a necessity. Until Egypt can obtain more industry, it will need aid from its oil-rich Arab neighbors as well as from the major world powers.

The future of Egypt also is tied to the country's political position. In 1979, Egypt signed a peace treaty with Israel, even though other Arab states objected. The treaty has reduced Egypt's role in Middle Eastern affairs. In 1981, Egypt's President Anwar Sadat, who signed the treaty, was assassinated. Political changes could occur in the future. With these changes, the direction of Egypt's economic progress could turn again.

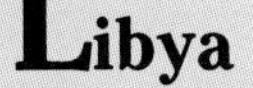

Libya

Land and People Unlike Egypt, Libya has no great river. In fact, Libya has no permanent rivers at all. The country is 93 percent desert. Libya's population of nearly four million totals only 7 percent of Egypt's. Most people in Libya live mostly along the Mediterranean shore. (See the population-density map on page 247.) Here, winter rains of the Mediterranean climate provide a yearly rainfall of 8 to 24 inches (20 to 60 centimeters). Libya's best lands are in the eastern part of the country, in the region of Cyrenaica. Traditional agriculture remains typical in rural areas. Farms produce grains, including wheat and barley, as well as citrus fruits, dates, olives, and grapes. Tobacco is also grown. Sheep and goats graze wherever pasture is available.

Resources and Economic Development For most of its existence, Libya had few known natural resources. Then, in the late 1950s, oil was discovered in the southern desert. Today, as Libyan oil flows out of the wells, money

Level A: Ask students to research the life of Anwar Sadat, the president of Egypt who was assassinated. Have them write a report on how Sadat's beliefs and policies affected the economy and political climate of Egypt.

Libya was once part of the Ottoman Empire and was a colony of Italy from 1912 until the end of World War II.

flows into the country. Libyan oil is especially in demand because it has a low sulfur content. A low sulfur content means that the oil will not pollute the air as much as oil with larger amounts of sulfur. Libya also has natural gas to export to nearby markets in Europe.

With such a small population, Libya can export most of its oil. Therefore, Libya might be able to improve its economic and social position in the years to come. Oil revenues are being spent on improved transportation, irrigation projects, public health, and new housing. Most of the money is being spent in the largest cities. These are Tripoli, the capital, and Benghazi (behn-GAHZ-ee). Libya has proposed a major pipeline that would bring fossil groundwater from the desert to the coastal cities.

Libya and World Affairs Oil wealth has given Libya political power. Colonel Muammar al-Qaddafi, Libya's leader, is a devout Muslim and an avid Arab nationalist. With the oil wealth of Libya, he supports radical causes around the world. He has tried to overthrow neighboring countries and has sent the Libyan army into Egypt and Chad. Qaddafi even organized assassinations of unfriendly Arab leaders and Libyans opposed to his policies. For these reasons, many Western governments view Libya as a troublesome country.

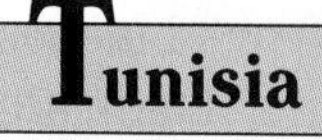

Tunisia

Physical Geography Tunisia has a strategic location between the eastern and western basins of the Mediterranean Sea. (See the map on page 279.) For thousands of years, this location made Tunisia a center of Mediterranean culture. In ancient times, the city of Carthage, now in ruins, controlled trade throughout the Mediterranean.

In the west of Tunisia, there is a plateau region that marks the beginning of the Atlas Mountains. This mountain chain extends west from Tunisia across Algeria and Morocco. Tunisia's coastal lands enjoy a mild Mediterranean climate. In these lowlands, wheat and olives are traditional crops. Cork-oak trees are also grown. More than one-quarter of Tunisia's land is suitable for agriculture. This is a large amount compared with that of other countries of the Middle East and North Africa whose territories extend into the Sahara.

Resources and Industry Although there are some industries in Tunisia, agriculture is the main occupation. The country has deposits of oil, phosphate, iron, and lead. Most of these minerals are exported. Tunisia's few industries are located in the larger towns. Tunis, the capital, is the country's leading trade and commercial center.

Tunisia's Place in the Arab World Tunisia differs from most other countries of the region. It has a broad-based economy of small farms, tilled by their owners. Although no longer a French colony, Tunisia maintains close ties to France. Tunisia is an Arab country strongly influenced by French culture. The ideals of European democracy are widely appreciated. Tunisia also has a good transportation system and several good ports. The country has become a model of stability and social progress. For example, Tunisian women

Tunisia provides many equal opportunities for women. In what other ways does Tunisia differ from other North African countries?

It has a broad-based economy, democratic ideals are appreciated, and it retains close ties with its colonial past.

Mastering Geography Skills

Use the pie graph in the Geography Skills feature on page 267 and the pie graph on this page to answer the following questions.

1. What does the pie graph on this page represent? How does it differ from the pie graph on page 267?
2. What percentage of the Middle Eastern oil reserves are held by Saudi Arabia? What does your answer tell you about the importance of Saudi Arabia to the world economy?
3. What other natural resources discussed in this unit could be represented by a pie graph? Why are pie graphs useful?

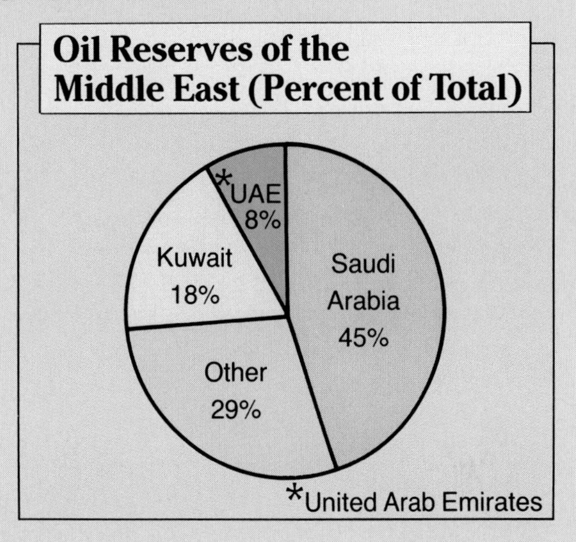

Applying and Extending

Listed to the right are some words that have entered our vocabulary from languages in the Middle East and North Africa. Look up each one in a dictionary to find out its usual meaning and the languages through which it has come to us. How do these words illustrate cultural diffusion?

carafe	kosher	salaam
bazaar	caravan	alcove
muslin	algebra	jar

Linking Geography and Religion

Jerusalem is an ancient city that is considered holy by Jews, Christians, and Muslims. It has been the site of much turmoil over the centuries because of its importance to these three major religions. Use an encyclopedia to research the history of Jerusalem. Where are the holy sites of the three religions found in Jerusalem? What do they signify to Jews, Christians, and Muslims? Present your findings in a written report.

Reading for Enrichment

Herzog, Chaim. *The Arab-Israeli Wars.* New York: Random House, 1982.

Moorehead, Alan. *The Blue Nile.* New York: Harper & Row, 1980.

Rodinson, Maxime. *The Arabs.* Chicago: University of Chicago Press, 1981.

Swift, Jeremy. *The Sahara.* New York: Time-Life International, 1975.

Worthington, E. Barton. *The Nile.* Morristown, NJ: Silver Burdett, 1978.

The Drakensberg Escarpment on South Africa's eastern coast is one of many spectacular landforms found on the African continent. Most of these landforms were formed by volcanic activity.

Sub-Saharan Africa includes hundreds of ethnic groups. Each of these groups has a different language and unique social customs. These women pass near a village in Cameroon on their way to market.

Coffee, a major cash crop for many sub-Saharan countries, is often grown on large plantations. The coffee berries are handpicked, since a way to harvest them with a machine has not been devised.

Have students identify the aspects of physical, economic, and cultural geography found in these photos.

SUB-SAHARAN AFRICA

OBJECTIVES

- ▷ *To describe the great sub-Saharan African civilizations and how Europeans affected their cultural and political development*
- ▷ *To describe how climate and landforms affect the way people live in sub-Saharan Africa*
- ▷ *To describe the economic, political, cultural, and social conditions of sub-Saharan Africa*
- ▷ *To identify available resources that may have an impact on the future economic development of sub-Saharan Africa*
- ▶ *To interpret historical events by comparing historical maps*

Chapter 25

Chapter 25 Lesson Plans and Planning Guide are found in the Teacher's Guide.

Refer students to the map on page 295.

Sub-Saharan Africa

▲ A village in Niger blends with the landscape.

▲ Senegal is a former French colony.

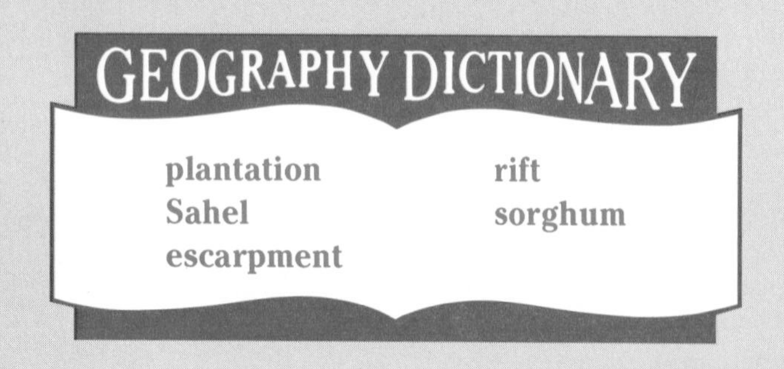

The Sahara is a desert that separates North Africa from sub-Saharan Africa. The countries of sub-Saharan Africa have many similarities. African cultures have given the world a rich tradition of folk stories, music, art, and architecture. However, the countries' differences are great. Most of the countries comprise many different ethnic groups. Religions, languages, and economic systems also separate them. These differences have often led to conflicts, both between countries and among groups within countries.

In Africa's cities, modern buildings and highways are symbols of progress. However, these cities are surrounded by slums and filled with rural migrants. The lack of economic development in most sub-Saharan countries is a long-standing issue. Creating jobs for an increasing number of people is a struggle all African governments share.

 Have students note the contrasts between urban and rural life in the photos above.

Africans are reconstructing their history by using present-day folklore, poetry, art objects, and architectural styles as clues to the past. Why do people want to learn more about their past?

Historical Geography

Early African Civilizations Little is known of the history of sub-Saharan Africa before European exploration. Instead of keeping written records, most of the people of the region kept an oral history.

We do know about a few great African civilizations, however. (See the map "Early Sub-Saharan Kingdoms," at the right.) The Kingdom of Kush was described by a Greek geographer in the fifth century B.C. The kingdom controlled much of the middle Nile River valley. The capital of Kush was Meroë (MEHR-uh-wee), in what is now Sudan. Meroë flourished from about 300 B.C. to A.D. 300. By A.D. 350, Kush had been conquered by rulers of powerful Axum. Axum, located in the highlands of present-day Ethiopia, was founded as a city of traders and merchants. Today, its ruins reveal the great innovation of dry-stone construction, by which walls were built without cement or mortar.

Other empires arose at different times in other parts of sub-Saharan Africa. Perhaps the most famous ones were founded in West Africa. These early empires grew powerful by controlling trade between tropical Africa and the area north of the Sahara along the Mediterranean Sea. Ancient Ghana was located in present-day Mali, near the headwaters of the Senegal and Niger rivers. This civilization flourished from 700 to 1200. Ghana controlled the trade routes from West Africa to North Africa. Caravans brought goods from nearby regions. They also brought gold and slaves from the coastal forests to the south. Ghana was eventually destroyed by invaders around 1200. By 1300, the most powerful empire in West Africa was Mali. It held its strong position because it had gained control of the caravan routes. Timbuktu was one of Mali's great cities.

Another West African civilization was that of Songhai (SAWNG-hie). Its capital was Gao, a city on the Niger River, east of Timbuktu. The Songhai defeated Mali in a number of wars. By 1400, the Songhai Empire extended beyond Mali territories. Because the kings of Songhai were often devout Muslims, the empire communicated with the Arab world. By the 1600s, the Songhai Empire began to decline. At about the same time, to the east, the Kanem and Bornu Empire was flourishing. This became the region's last great trading empire. It lasted until 1900.

While empires were developing in the east and west, civilizations were growing in central and southern Africa as well. These kingdoms were also involved in establishing trade routes across Africa. More than one thousand years ago, central African kingdoms were trading with Arabia, Persia, India, and China.

In southern Africa, there is a region of ruined cities. One of these cities was Zimbabwe (zim-BAHB-way), which began as a village of ironworkers during the third century A.D. A large stone structure built around the thirteenth century still stands there. The empire of Mwanamutapa (mwan-uh-MOO-tuh-puh) developed around this city. It reached its

Early Sub-Saharan Kingdoms

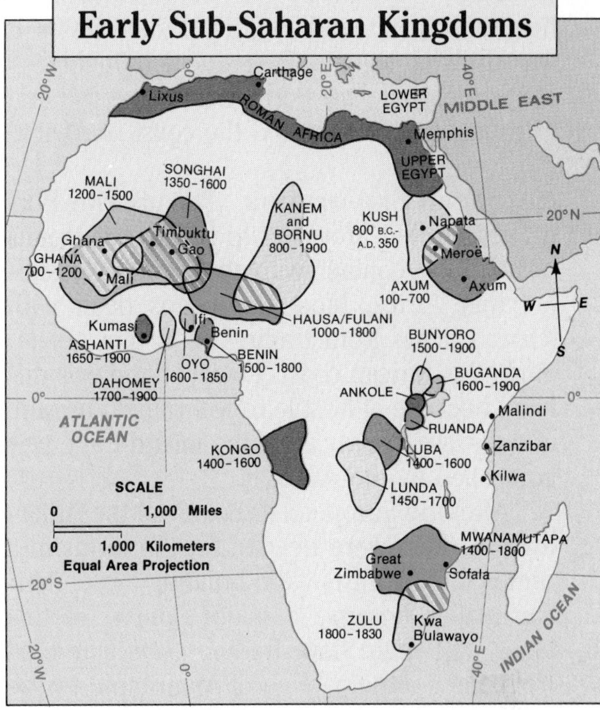

Which early sub-Saharan African kingdom was located in present-day Ethiopia? Axum

Level A: Have students research this period of European and African history and then draw a map showing the European exploratory voyages around Africa.

Have students locate sub-Saharan Africa on the inset map of the world on page 295. Point out the equator, and discuss its relationship to climate. Have students use the map scale to figure north-south and east-west distances.

height in the fifteenth century. By the 1600s, it was showing signs of Portuguese influence. A century earlier, the Portuguese were the first Europeans to sail around the coast of Africa.

European Exploration Not until the Portuguese sailed around Africa did Europeans have much contact with the people of sub-Saharan Africa. Most Europeans dealt with African tribes from trading posts they set up along the African coast. Travel inland was difficult because of hostile tribes, rugged terrain, and disease. Nearly all of the inland trade was controlled by the Africans.

After the European discovery of the Americas, workers were needed for the farms and mines the colonists were building there. African slaves became a major source of this labor. The main slave-trading ports extended from the southern edge of Mauritania to the northern edge of the Namib Desert in present-day Angola. The trading posts were located at sites where there was a harbor along with a river or trail leading inland.

Less than 2 percent of the slaves were captured by Europeans. Instead, European merchants bought the slaves from coastal African leaders. These leaders were paid with European manufactured goods, particularly guns. Most of the slaves had been captured inland. Intertribal warfare probably increased because of the slave trade.

European exploration of the African interior did not begin until a few hundred years after the slave trade began. Much of this exploration centered on the search for the sources of great rivers. Among the early explorers was Mungo Park. In 1795 and 1805, he traveled to West Africa to explore the upper parts of the Niger River. In 1857, Richard Burton and John Speke went to Africa in search of the source of the Nile River in the African interior. Four years later, Speke returned to the interior with J. A. Grant and discovered Lake Victoria, the White Nile's source.

In 1854, David Livingstone traveled from South Africa into the Zambezi River area and discovered Victoria Falls. Henry Morton Stanley, an American newspaper reporter, conducted a major expedition from 1874 to 1877. He traced the course of the Congo River in Central Africa. In Zaire (ZIE-uhr), the Congo River is known as the Zaire River.

European Colonization World events in the late nineteenth century led to dramatic changes in the European view of the African continent. Africa was the last great uncolonized area. European powers scrambled to claim territories there. To settle conflicts among themselves, the Europeans divided Africa at a conference in Berlin in 1884–1885. They took a map with little information on it and added boundaries. The boundaries were drawn with little regard for Africa's landforms, climate regions, or cultures. They divided tribes and cut through rich agricultural lands.

The Africans had little say about either these boundaries or the colonization. After a few bitter battles with the poorly armed African groups, the European colonies were quickly established. Except for Liberia and Ethiopia, all of Africa was under the rule of one European country or another by 1900.

When the Europeans arrived in Africa, they developed mines and **plantations** as sources of products for their economies at

For the Record

Days and days one may travel through primeval forests, now ascending ridges overlooking broad, well watered valleys, with belts of valuable timber crowning the banks of the rivers, and behold exquisite bits of scenery—wild, fantastic, picturesque and pretty—all within the scope of vision whichever way one may turn. And to crown the glories of this lovely portion of earth, underneath the surface but a few feet is one mass of iron ore, extending across three degrees of longitude and nearly four of latitude, cropping out at intervals, so that the traveller cannot remain ignorant of the wealth lying beneath.

From *Stanley's Despatches to the NEW YORK HERALD, 1871–1872, 1874–1877,* edited by Norman R. Bennett, p. 75.

 Have students read the primary-source material above. Ask them to cite three features of the natural environment noted by Stanley. Have them consider the significance of the iron ore mentioned in the description.

Sub-Saharan Africa
EUROPE
ASIA
MIDDLE EAST
MEDITERRANEAN SEA
PERSIAN GULF
RED SEA
GULF OF ADEN
NORTH AFRICA
MADEIRA (Port)
CANARY ISLANDS (Sp)
CAPE VERDE
Tropic of Cancer
Equator
Tropic of Capricorn
10°E
20°E
30°E
40°E
50°E
60°E
30°W
20°W
10°W
0°
50°N
40°N
30°N
20°N
10°N
10°S
30°S
40°S
MAURITANIA
Nouakchott
EL DJOUF
MALI
SAHEL
Niger River
Senegal R.
NIGER
CHAD
CHAD BASIN
Lake Chad
N'Djamena
SUDAN
SUDAN BASIN
Khartoum
NUBIAN DESERT
Nile River
Blue Nile
White Nile
ETHIOPIA
ETHIOPIAN HIGHLANDS
Addis Ababa
DJIBOUTI
Djibouti
SOMALIA
Mogadishu
SENEGAL
Dakar
Banjul
GAMBIA
Bamako
Niamey
Bissau
GUINEA-BISSAU
JALLON MTS
GUINEA
Conakry
Freetown
SIERRA LEONE
Monrovia
LIBERIA
BURKINA FASO
Ouagadougou
IVORY COAST
Bouaké
Abidjan
GHANA
Accra
TOGO
Lomé
BENIN
Porto-Novo
Lagos
Ibadan
Ogbomosho
Kano
NIGERIA
CAMEROON
Douala
Yaoundé
Malabo
EQUATORIAL GUINEA
GULF OF GUINEA
SÃO TOMÉ AND PRINCIPE
São Tomé
Libreville
GABON
PEOPLE'S REPUBLIC OF THE CONGO
Brazzaville
Kinshasa
CABINDA (Angola)
CENTRAL AFRICAN REPUBLIC
Bangui
Ubangi River
Zaire River
ZAIRE BASIN
ZAIRE
Kasai River
Mbuji-Mayi
KATANGA PLATEAU
Lubumbashi
UGANDA
Kampala
Lake Edward
Lake Turkana
Lake Victoria
KENYA
Nairobi
Mombasa
RWANDA
Kigali
BURUNDI
Bujumbura
GREAT RIFT VALLEY
TANZANIA
Lake Tanganyika
Zanzibar
Dar es Salaam
Lake Nyasa
INDIAN OCEAN
ATLANTIC OCEAN
N
W
E
S
Luanda
ANGOLA
ZAMBIA
Lusaka
MALAWI
Lilongwe
Blantyre
Zambezi River
MOZAMBIQUE
Harare
ZIMBABWE
Bulawayo
COMOROS
Moroni
MOZAMBIQUE CHANNEL
Antananarivo
MADAGASCAR
MAURITIUS
Port Louis
RÉUNION (Fr)
NAMIBIA (South Africa)
KALAHARI BASIN
BOTSWANA
KALAHARI DESERT
Gaborone
WALVIS BAY (South Africa)
NAMIB DESERT
Windhoek
Pretoria
Johannesburg
Soweto
SWAZILAND
Maputo
Mbabane
LESOTHO
Bloemfontein
Maseru
DRAKENSBERG ESCARPMENT
Orange R.
SOUTH AFRICA
Cape Town
Cape of Good Hope
National capitals
Other cities
SCALE
0
500
1,000 Miles
0
500
1,000 Kilometers
Projection: Azimuthal Equal Area

GEOGRAPHY SKILLS

How to Compare Historical Maps

Answers to Geography Skills questions are found in the Teacher's Guide.

Europeans began establishing trading posts in Africa during the fifteenth and sixteenth centuries. Full-scale European colonization of Africa began in the late nineteenth century with the Berlin Conference of 1884–1885. By 1913, almost all of Africa was under European domination. European control of Africa continued until the 1960s, when most African colonies gained independence.

The best way to trace political changes that have occurred in Africa is to compare a historical map of Africa in 1913 with a map of Africa today. A historical map gives selective information about places in the past. By comparing historical maps of a region, you can see how it has changed over a period of time. Think about the historical development of the United States. Since 1776, the United States has grown from the original 13 colonies to the present 50 states. You could study several maps of the United States at different periods in history to trace the country's development and expansion.

In the making of historical maps, a variety of methods can be used to explain events. The map of Africa in 1913 shows which areas of Africa were colonized by the various European powers as well as the colonial boundaries and names. The map "Independence in Sub-Saharan Africa" presents similar information but with additional facts. For example, you can tell from this map where the present-day boundaries are and in what year each colony gained its independence.

Both maps use color to explain historical events. If you look at the map "Independence in Sub-Saharan Africa," you can see that very few African nations were self-governing before the end of World War II in 1945. Yet, by the end of the 1960s, most African colonies had gained independence. Today, only Namibia is controlled by another country—South Africa.

There is much about the history of Africa you cannot tell from these maps. You cannot distinguish the methods used by each nation to gain independence from European control. In some colonies, independence was gained peacefully. In others, violence was used to achieve independence.

Africa is a continent that is still changing rapidly. Political boundaries between several countries are still in dispute and subject to change. Remember, when you need information about how a place has changed through time, compare historical maps. Use the two maps here and the map on page 295 to answer the following questions about African history.

Test Your Skills

1. What are the new African names of these former colonies?

Belgian Congo	Southern Rhodesia
British East Africa	German Southwest Africa
Northern Rhodesia	French Somaliland

2. Which African nations were never colonized? How many African nations were independent before 1945?
3. Which nations gained independence in the 1970s? From which European nation did each nation receive its independence?
4. Which European country controlled the most colonies in Africa? What were the names of these colonies?

World War I began in 1914. Its outcome, in 1918, affected colonization in Africa. Germany lost its colonies as a result of the war.

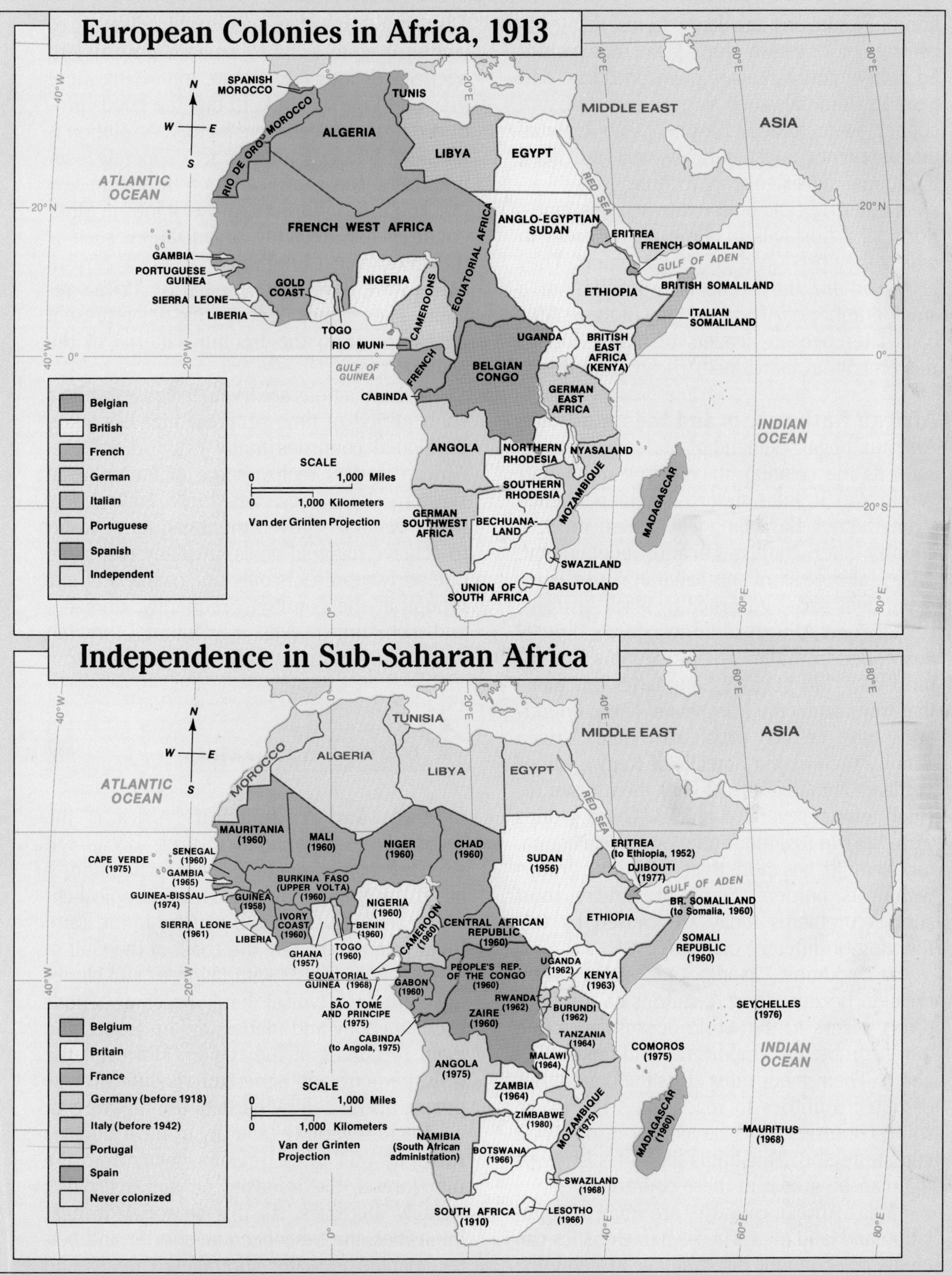

Discuss one-party rule. Contrast it to our political system. Ask students to consider why one-party rule might appeal to developing countries and its limitations for achieving consensus among a nation's citizens.

home. Plantations are large farms that concentrate on one cash crop. Cash crops included coffee, cotton, cacao, palm products, and peanuts. Minerals such as gold, iron ore, and copper were mined. Despite these colonial developments, most Africans still practiced traditional subsistence agriculture.

The Europeans built roads, schools, cities, ports, and hospitals in some of the areas in which they produced goods for export. They also paid for the European education of a small number of Africans. Many of these Africans later became leaders of independence movements in individual African countries.

African Nationalism and Independence Africans began demanding independence as soon as the colonial powers appeared. After World War II, the pressure for independence strengthened. Beginning with Sudan in 1956, country after country became independent. Today, almost all of sub-Saharan Africa is independent and is governed by black Africans.

The new African countries are not directly descended from the earlier kingdoms. For the most part, the political boundaries are those that were set by the Europeans. These boundaries have created problems. Some African people, such as the Somalis of Kenya, Ethiopia, and Somalia, have found themselves divided among several countries. Other groups were thrown together under one rule. Nigeria, for example, has more than 250 ethnic groups within its borders. Some of Africa's most serious problems today are caused by this merging of different cultures.

In addition, landlocked states were created. Because these countries do not have direct access to the world ocean, they often have problems importing and exporting goods. Their goods must cross the boundaries of other countries to reach the sea. Finally, some countries were carved out of only one climate region. This limits the kinds of crops that can be grown in these countries.

Many African countries are small in population and land area. These characteristics can create an economic disadvantage in a country. There may be too few people with advanced or technical educations to operate government agencies or industries. More important, there may be too few people to buy the goods produced by a large, efficient factory. A solution is for neighboring countries to cooperate economically. The goods can then be sold in several countries. Workers can seek jobs in other countries too. Recently, organizations such as the Economic Community of West African States have developed in each of Africa's regions. These kinds of economic ties represent an escape from the trading patterns of the colonial past.

African nations are trying to squeeze into a short period of time progress that has taken developed countries many years to achieve. African leaders realize some of the value of Western democracy. Yet, these leaders feel that a one-party government can move faster to achieve national goals. In many cases, the one party represents only one of many ethnic groups in the country. Frequently, unrest—and sometimes even revolution—threaten the stability of many African countries.

Physical Geography

Climate and Vegetation Africa is the second largest continent in area. Because it stretches across many degrees of latitude, it has many types of climate. Humid tropical climates are found near the equator in the Zaire River basin and along the coast of the Gulf of Guinea. Rain forests with tall trees and climbing vines are typical here. Precious woods such as ebony and mahogany are among the main resources of the region. However, the soils are poor, and agriculture is difficult. Few people live in central parts of the rain forests.

Regions with wet and dry tropical climates surround rain forest regions. Away from the rain forests, the intensity of winter drought steadily increases. As this seasonal drought intensifies, the trees become shorter and better adapted to water shortages. Grasses and

Have students contrast the climate map below with that of Western Europe on page 113. Note the dominance of maritime and Mediterranean climates in Western Europe. Ask students to locate these climate types on the map below.

thorny shrubs become more common. In regions with good rains, the savanna trees and grasses are tall. These regions are called tall-tree savannas. In drier regions, the savannas have low and thorny trees separated by short, drought-resistant grasses. These regions are called bush savannas. Clearing the savannas has created some of Africa's prime agricultural land. Many of Africa's famous national parks are also located in the savannas.

The dry savanna region along the southern edge of the Sahara is called the **Sahel** (suh-HAYL). Since the late 1960s, this region has been affected by long periods of drought. Many people and farm animals have died from the droughts. Many more people have moved south into the wet savanna region. This has created crowded conditions in these areas.

Regions with desert climates are found both north and south of the wet and dry tropical/subtropical climate region. To the north is the Sahara. To the south are the Namib (NAHM-ib) and Kalahari (KAL-uh-HAHR-ee) deserts. The Namib is a severe desert with only sparse plant and animal life. In contrast, the Kalahari has abundant low grasses and shrubs. Many parts of it receive up to 10 times as much rain as the Namib.

Much of southern Africa is at high altitudes and is located at latitudes as high as those in the southern United States. Many different climate regions are found here. North and east of the Kalahari Desert is a region of continental steppe climate. The desert's edge receives about 10 inches (25 centimeters) of rain each year. The result is a bush savanna. Rains increase toward the east coast, and tall-tree savannas gradually become dominant. The southern tip of Africa has a Mediterranean climate. Mild, wet winters and warm, sunny summers make this region a center for fruit growing. In summer, rainstorms move

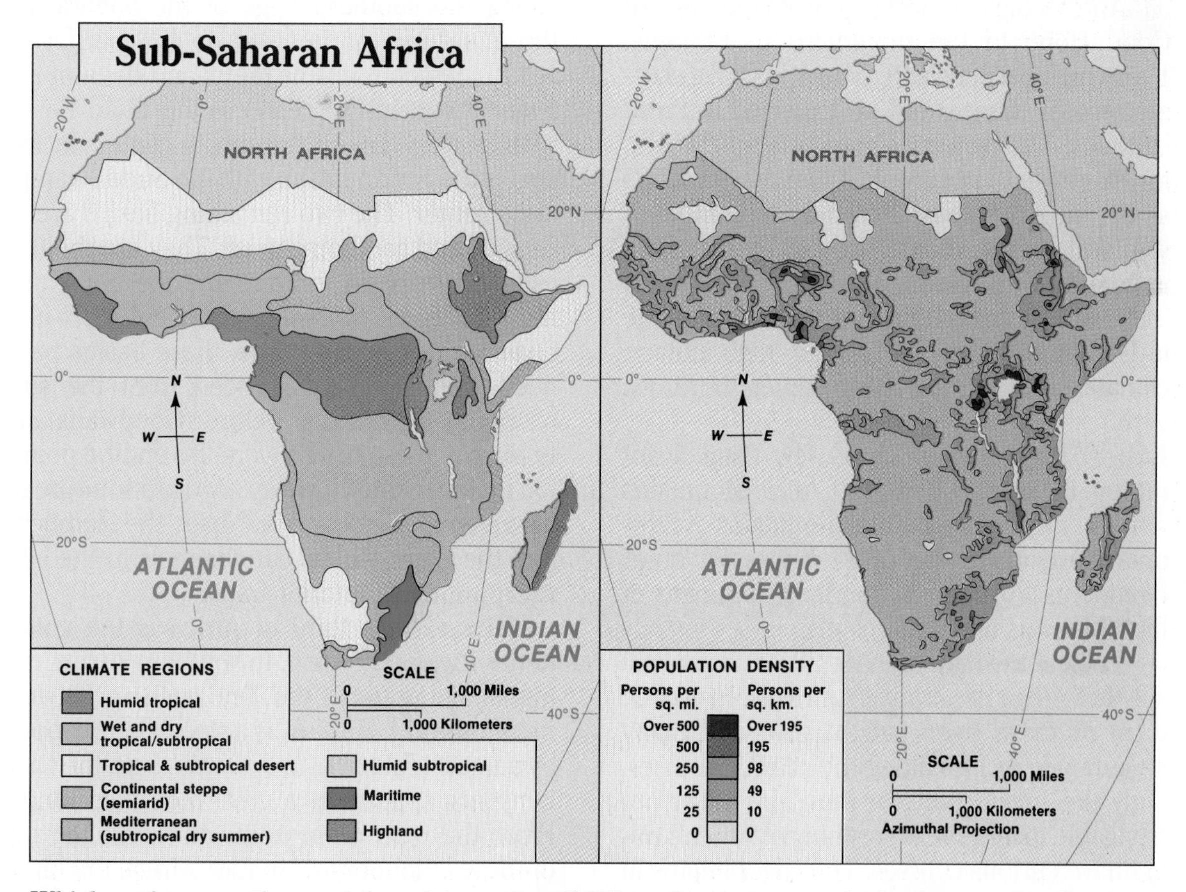

Which regions are the most densely populated? What climate types exist in these regions?

The western coast (humid tropical and wet and dry tropical/subtropical), highlands of Ethiopia and Kenya (highland), near Lake Victoria (wet and dry tropical/subtropical), eastern coast (maritime, continental steppe, and wet and dry tropical/subtropical).

Zebras gather at a water hole in Kenya. Here, thousands of exotic species of animals live in natural habitats.

into southern Africa from the Indian Ocean, producing regions of humid subtropical and maritime climate in the southeast.

Africa's highlands extend from the Cape of Good Hope to the mountains of Ethiopia. These high plateaus and mountains have cooler climates than would be expected at these latitudes. The plateau region of South Africa, for example, is not tropical like the lowlands along the east coast. Instead, subhumid to semiarid grasslands are present. Winter frost is common. In the highlands of East Africa, there are climates that resemble springtime in Europe all year. These mild, high-altitude climates attracted European settlers here.

Landforms Africa has few significant mountain ranges. Only the Atlas Mountains north of the Sahara and the mountains of Ethiopia compare to the great ranges on other continents. Africa is primarily a continent of high plateaus and wide plains.

Look at the map on page 295, and you will see that Africa has many long rivers. However, most of these rivers are narrow, and many have rapids or falls along their lower courses. Only the inland parts of the Zaire River are navigable to large vessels. Many of Africa's rivers have curious courses. The Niger begins in the Jallon Mountains along the west coast. It then flows northeastward to its inland delta in the southern edge of the Sahara before turning south toward the Atlantic. The White Nile almost dies in the great swamp of the Sudd in Sudan. The Sudd is one of the world's largest marshes. The Zambezi heads toward the Kalahari but then turns eastward over the Victoria Falls to the Indian Ocean. In southern Africa, the Orange River begins in a plateau edge called the Drakensberg near the east coast. It then flows westward to the Atlantic. Geographers think that these patterns can be traced to when Africa was the center of the supercontinent of Gondwanaland.

Recognizing that some of Africa's landforms date back to Gondwanaland helps us understand the five great depressions on the continent's surface. These depressions, or basins, are each more than 625 miles (1,000 kilometers) across and up to 5,000 feet (1,524 meters) below the surrounding highlands. Along the southern edge of the Sahara are three basins. In the west is El Djouf (ehl JOOF), which contains the inland delta of the Niger River. In the center is the Chad Basin, with shallow Lake Chad in its center. In the east is the Sudan Basin, with the Sudd swamps at its center. The two remaining basins are in Central and southern Africa. They are the forested Zaire Basin in Zaire and the arid Kalahari Basin in Botswana (baht-SWAHN-uh). Over long periods of time, these basins have filled with sediments eroded from the surrounding highlands. Before Gondwanaland broke up, these basins were the ending points for major rivers. However, Africa's four greatest rivers—the Nile, the Niger, the Zambezi, and the Zaire—have cut channels to the sea, escaping these interior traps.

The eastern third of Africa is the continent's highland region. In southern Africa, the highlands begin in the Drakensberg Escarpment. An **escarpment** is a steep slope capped by a nearly flat plateau. From the east, the Drakensberg appears as a great mountain range. From the west, wide plains dominate the top of these "mountains." In East Africa, the highlands are cut by two troughs of the great rift

 Level B: Tell students to locate the landlocked nations of Africa and list them. Which nations lie entirely within one climate region?

Level A: Ask students to research diseases found in Africa. They should use an almanac to find life expectancies for people in various African nations. Is there a correlation between the distribution of specific diseases and low life expectancy?

valleys. These troughs are the Eastern Rift Valley and the Western Rift Valley. The **rifts** begin near Lake Nyasa and continue northward into the Red Sea. In many areas, the rifts have narrow valley floors with mountainous slopes on each side. In other areas, such as Tanzania's famous Serengeti Plain, the rift is wide and not as deep. Volcanoes have been active along both rifts. The highlands of Ethiopia are made up of layer after layer of hardened lava.

Natural Disadvantages Life in Africa can be difficult. Almost every year, some regions suffer from drought while others are ruined by floods. Earthquakes and volcanic eruptions occur along the rift valleys. However, these great news-making events cannot compare with the natural disadvantages and hazards Africans must endure each day.

Despite its size, Africa has little fertile land. Much of the soil in the dry regions has too much salt or lime to be agriculturally productive. In humid regions, soils often lack plant nutrients for good crop yields. Swarms of insects such as locusts sometimes wipe out crops. Good soils are widespread only in the highlands of eastern and southern Africa. Alluvial soils along river valleys also make good agricultural land. Modern agricultural technology is needed to combat Africa's poor soil and difficult climate.

Many African villages consist of huts surrounded by farmland or dry pasture. In what ways can life in Africa be difficult?

People must endure drought, floods, earthquakes, volcanic eruptions, infertile soils, insects, and diseases.

Africa is also a continent on which diseases are a constant threat. Malaria, a disease carried by mosquitoes, and tuberculosis are widespread. In the humid tropics, sleeping sickness, carried by the tsetse (TSEHT-see) fly, attacks people and cattle. In many areas, poor diets encourage a host of related diseases. Many of Africa's diseases cause prolonged sickness but not death.

Ways of Life

The majority of people in Africa are black Africans. They are divided into several hundred ethnic groups. These groups have different languages, religions, and ways of life. One way to sort out sub-Saharan Africa's ethnic groups is to study the different language families. Most black Africans speak one of the several hundred languages in the Niger-Congo family. These languages are spread all the way from West Africa through the Congo Basin to the southeastern coast of South Africa. Along the southern Sahara are peoples who speak Semitic languages, related to those of the Berbers and Arabs living farther north. In some inland areas of East Africa and West Africa, people speak Nilo-Saharan languages.

Another language family is the ancient Khoisan, which the San (Bushmen) and Khoi (Hottentots) of southern Africa speak. These people are not black Africans. Their yellow-brown skin, short stature, and unique blood type make them unlike any other race. Madagascar, off the east coast of Africa, was settled by migrants from southeast Asia several thousand years ago. The language here is related to those of the southwestern Pacific region. The colonial period brought European languages to Africa. In South Africa, the Cape Dutch language has developed into Afrikaans. Most sub-Saharan countries have adopted French or English as an official language.

Emphasize the importance of Africa's mineral wealth to the United States. Ask students why it is in our economic and strategic interest to keep the area peaceful.

During the past 25 years, populations of people and animals have increased in Africa despite widespread diseases. The use of modern health practices is largely the reason for this growth. However, many farming communities are now overcrowded. Overgrazing by cattle and goats has caused severe soil erosion. The traditional rural cultures of Africa are undergoing dramatic change as people move to cities, seeking jobs.

Unfortunately, the number of people moving to the cities is usually greater than the number of jobs. The unemployed and those with temporary jobs live in slums that surround all African cities. Providing jobs and better housing is the most pressing issue for many African governments.

Economic Geography

Africa is a treasure-house of mineral resources. Gold, copper, chromium, manganese, uranium, and cobalt are found in abundance. Most of Africa's mineral wealth is exported to industrial countries. As a result, the resource-rich countries of Africa are often targets of the competition among powerful countries.

Oil is Africa's only mineral resource in short supply. The greatest oil deposits of Africa are found in Nigeria. Oil is also found offshore between Cameroon and Angola. Because oil is so important to modern industrial nations, all African countries have active oil-exploration programs. More oil deposits will be developed along the west coast of Africa and in the Sudan.

The rivers of Africa are a valuable source of power. Hydroelectric power stations have been built or are planned at almost every rapids or waterfall. Perhaps 30 percent of the world's future hydroelectric power exists in this region.

Despite its resources, Africa remains the least developed continent. Most Africans live by the farming and herding methods of their ancestors. The main grain crops are corn and **sorghum**. Cattle are the most important animals. In many African cultures, cattle are used to measure wealth.

Manufacturing is often limited to producing simple consumer goods and processed food. Many African countries are in economic decline. The major concerns are poor educational systems, few trained business managers, and a lack of money for investment.

Future Issues

Most countries of sub-Saharan Africa face several challenges. Fighting against often illogical colonial boundaries, African leaders must build stable countries from territories that contain peoples of different languages, religions, and cultures.

The second challenge is slow economic progress. One cause of this underdevelopment is poor government economic policies. For example, governments want their people to grow valuable export crops such as cotton and coffee. This discourages food production in rural areas. As a result, earnings from exports are spent on food imports instead of on development. Capital for investment is in short supply. Furthermore, foreign aid from the developed countries has often helped the city dwellers rather than the rural people.

Another cause of underdevelopment is the African environment. Poor soils and erratic weather make agriculture difficult. A fourth cause is Africa's dramatically increasing population. At its current rate of growth, Africa's population will double in only 24 years! Africa is becoming a continent of children who must be cared for and educated.

Improving people's lives in Africa will not be easy. Foreign aid is needed, mainly to improve agriculture and to develop small manufacturing industries. Revenues from mining and export crops are necessary to improve transportation and to provide better schools. Most important, the African countries need to cooperate more with each other.

 Point out that Africa has nearly one-third of the world's hydroelectric power potential.

Have students locate West Africa on the map of sub-Saharan Africa on page 295.

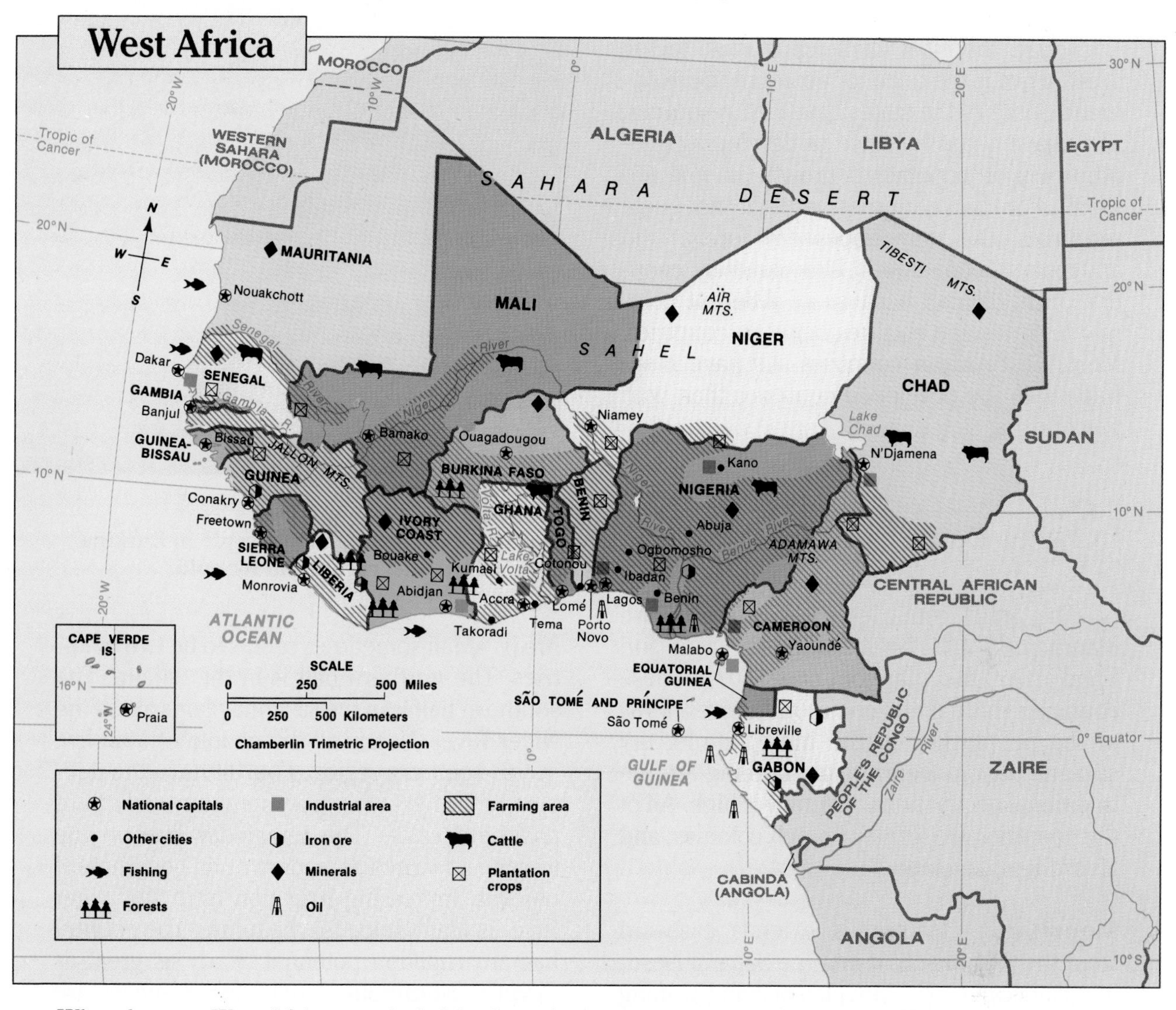

What do most West African capital cities have in common in terms of location? Why?

They are located along the coast; these areas are wetter, and they offer better opportunities for trade.

care, new agricultural methods, development of resources, and improved communications, transportation, and education. Although the Europeans often tried to be helpful, they sometimes did not understand or appreciate the local cultures. The result was frequently a disturbance of African life. Also, the profits from Africa's mineral resources went to European businesses. By the time the nations became independent, few Africans had received the training necessary to be effective leaders. This led to social unrest and even warfare. Such unrest has often discouraged investment in new businesses.

Economic Development

Many of today's African leaders have European customs. These modern leaders set the tone for life in the West African cities. West Africa's traditional crafts and industries have declined since independence. Instead, the African leaders have sought to increase the export of cash crops and minerals. This will allow the African nations to continue to buy more goods from the developed countries. The development of modern factories in the region has only just begun.

Today, the most challenging issue facing West Africa is the need for economic development. Some of the states have rich resources. They have a good chance of developing economically if population growth is not too rapid. For the countries without adequate resources, the only hopes are foreign aid and regional cooperation. Unfortunately, cooperation is difficult because of different colonial histories and rivalries between countries. Most West African countries still have more trade with the developed countries than with neighboring states.

Burkina Faso is one of Africa's least developed countries. Many people in Burkina Faso's dry Sahel region pan for gold.

Sahel Countries

The Sahel countries of West Africa are Mauritania, Mali, Niger, Burkina Faso, and Chad. (See the map on page 307.) These countries share many common features. Most of the people live in the more humid, dry savanna region. Herding and farming are the traditional occupations. Illiteracy is high. All of the countries are former French colonies, and all are less developed.

Mauritania Mauritania extends eastward from the Atlantic coast into the Sahara. Desolate landscapes of rocky plateaus and shifting sand dunes are typical. Mauritanians were once mostly nomads who followed the rains in search of better grazing lands. The people are primarily Islamic; Arabic is the official language. The capital of Nouakchott (nu-AHK-shaht) on the south coast, had 5,000 people when it was founded in 1960. The Sahel drought has forced many people to move into towns. Nouakchott now has about 600,000 people! Most of the people live in refugee camps and have little chance for work. International aid provides much of the food. Farms along the Senegal River are also important. Aid from oil-rich Arab countries is helping the country develop its mineral resources. An iron and steel factory has been built to use ores from an enormous iron deposit in the Sahara.

Mali Mali sometimes seems to be two countries. The northern half is in the Sahara. The southern half is in the savanna plains along the Niger River. Nearly all the people of Mali live in this southern region. Four-fifths of the people are farmers who grow sorghum and maize and herd cattle. The remainder lives in the towns and works as traders or as government officials. Increasing irrigation from the Niger River is Mali's hope for the future. The country has an irrigation potential nearly as great as Egypt's. Numerous foreign-aid projects are building dams and irrigation channels. Most of the country's large businesses are poorly run government corporations.

Niger Niger consists of a small savanna region along the Niger River valley and a large expanse of the Sahara and Sahel to the northeast. Only 2 percent of the country is farmed. The country has been devastated by drought. By the mid-1970s, nearly 90 percent of the farm animals had died. Perhaps a quarter of Niger's people migrated to other countries. Niger's hopes for the future are more rain and increased mining. While a large uranium deposit has been mined, much of Niger's mineral wealth remains unexplored. The people of Niger are particularly proud of their culture,

 Ask students where they would expect most people in the Sahel countries to live—north or south. Use the map on page 299 to confirm their answers.

As students read this material, have them write brief descriptions of each nation. Have students draw maps of each country, showing major landforms, climate, rivers, and bodies of water.

which blends black African traditions with Islam. The government has used this national pride to gain public support.

Burkina Faso Burkina Faso is located in the dry savannas between Mali and Niger. Cattle raising is the traditional occupation. However, the drought has reduced the number of cattle by 70 percent. Sleeping sickness and other diseases have left nearly a third of the country uninhabited. In the fertile regions, population densities are very high. There are few jobs for all these people. Many people migrate south to find work as laborers in other countries. Many others have joined the great slums surrounding the capital city of Ouagadougou (wahg-uh-DOO-goo). They have cut every tree within 25 miles (40 kilometers) of the city for fuel wood. Burkina Faso is one of the world's poorest countries. About 60 percent of the government's budget comes from foreign aid.

Burkina Faso was once called Upper Volta. The word *burkina* is common to 50 dialects in the country and means "land of honest men." *Faso* means "democratic and republican" in the language of the largest group. Nevertheless, the governments and businesses in Burkina Faso are considered corrupt. The country is run by a military dictatorship.

Chad Chad extends from the central Sahara to the tall-tree savanna region in the south. The few people in the north are mostly Islamic herders with Arabic customs. The numerous people in the south are Christian and animist farmers. In the middle of the country, the people blend these culture traits. As a result, the politics of Chad are complex. Many political groups have fought for control of the government in N'Djamena (ehn-JAHM-uh-nuh), the capital. Libya, to the north, has attempted invasions several times. The drought has reduced Lake Chad to one-third the size it was in 1950. An important fishery there has nearly collapsed. Chad has few resources, no railways, and poor roads. It has continued its dependency on France.

Atlantic Coast Countries

Senegal and Gambia Senegal is a former French colony. It surrounds the former British colony of Gambia. This unusual boundary divides similar African groups and interrupts the flow of goods and ideas. Both countries are primarily Islamic. It is possible that Senegal and Gambia will merge in the future. This union will probably not occur until the effects of different colonial governments subside.

Northern Senegal is on the edge of the Sahel and has suffered from the drought. Gambia and southern Senegal are within the savanna region. Peanuts, often grown in irrigated fields, dominate agricultural production. The national economies rise and fall with the world price of this single crop.

While Gambia has been dependent upon outside aid for years, Senegal has the best-developed economy of the Atlantic coast countries. Dakar, Senegal's capital, was the primary city of French West Africa. Today, manufacturing income is growing. Factories process peanuts, cotton, rice, and phosphate ores. Political stability has encouraged tourism from Europe.

Future progress of Senegal and Gambia depends upon improved water resources.

Gambia lies along the banks of the Gambia River. These villagers depend upon the river for food.

Both countries have attracted foreign aid to dam rivers and to build irrigation canals. The countries of the region have agreed to cooperate in developing the water resources in the Senegal River region.

Like many other less developed countries, Senegal and Gambia import more goods than they export. To do this, they have taken out loans and fallen into debt. This unfortunate situation can be ended only by reducing imports and increasing exports. Yet, this is difficult. Strict new economic rules may cause unrest and increase poverty.

Guinea Southeast of Senegal and Gambia is the Republic of Guinea. Its landscape consists of rolling plains leading to the low Jallon Mountains. The vegetation is primarily tall-tree savannas and dry forests. More than 80 percent of the people practice subsistence agriculture. Most of the people are Islamic. Cash crops include coffee, bananas, peanuts, and palm products. Palm products come from oil-palm trees. They include palm oil and palm kernels. The oil is used for making cosmetics and soap and in cooking. The kernels are processed for starch and food. Palm fiber is used for making floor mats and rope.

Despite its less developed economy, Guinea holds at least one-third of the world's bauxite reserves. In good years, the country exports about 15 percent of the world's supply of this ore. Guinea also has large iron-ore deposits as well as gold and industrial-diamond mines. These minerals provide the economic base for the country's future development.

The people of Guinea learned hard political lessons after independence. The country's first leader was a cruel dictator. More than two million people fled the country; many others died in prisons. The only foreign aid came from the USSR and the COMECON countries. Poverty increased in the rural areas, and the nation's culture suffered. These hardships continued for 25 years until the dictator died. Today, the government is making progress and has better relations with other countries throughout the world.

Guinea-Bissau and the Cape Verde Islands Guinea-Bissau and the Cape Verde Islands are former Portuguese colonies. The city of Bissau was the major port of the Portuguese West African slave trade. The nations were often administered together, and they shared common paths to establish Communist governments after independence. Portugal did little to develop either territory.

In Guinea-Bissau, there has been little industrial progress and no mineral development. The population is made up mostly of members of four African groups and a small percentage of Europeans. Subsistence agriculture based on crops such as rice and cassava is typical. Some peanuts and palm oil are exported.

The Cape Verde Islands are located 400 miles (640 kilometers) off the coast of Senegal. When the Portuguese arrived in 1456, the Cape Verde Islands were uninhabited. The 10 islands are volcanic in origin and appear stark and barren. African slaves were brought to the islands by the Portuguese in the 1460s. Intermarriage between the Africans' descendants and Europeans produced most of the present population.

Although many of the people are relatively well educated, the country is poor. Limited farmland, a dry climate, and centuries of colonial rule have made improvement of living standards difficult. Many islanders have emigrated to the United States.

Liberia and Sierra Leone Liberia and Sierra Leone are unusual in that both countries were founded as settlements for freed slaves. The name *Liberia* comes from the Latin word *liber,* which means "free." Liberia's settlers came from the United States beginning in 1822. The country became a republic in 1847 and has continued to maintain close relations with the United States.

Sierra Leone was founded in 1787 for freed slaves from the British Empire. However, it remained a British colony until 1961. In Liberia, the descendants of American slaves currently total only 3 percent of the nation's

 Level A: Have students research the differences between capitalism and Marxism. Which philosophy does each of the countries of the region follow?

population. Yet, they are the most prosperous and powerful people in the country. In Sierra Leone, the freed slaves, who call themselves Creoles, comprise about 40 percent of the population. Today, the governments of Liberia and Sierra Leone are controlled by military officers. However, these governments represent the majority of the population of both countries.

Three distinct environmental regions are found in Liberia and Sierra Leone. Along the coast are mangrove swamps that extend inland as much as 30 miles (48 kilometers). Most of the interior consists of plains covered with tall-tree savannas and humid tropical forests. Farther inland are low mountains with thick, humid tropical forests.

Neither Liberia or Sierra Leone is developed. Throughout its long history, Liberia has received little economic support from foreign countries. Sierra Leone was probably the least developed of Britain's African colonies. Subsistence agriculture is still practiced widely. Cassava and rice are the main food crops. Cash crops include coffee, cacao, and palm products. Rubber plantations, most of which were owned by a single United States company, once dominated the economy of Liberia. Today, manufacturing in Sierra Leone and Liberia is centered around processed food, textiles, and building supplies.

Liberia and Sierra Leone will focus future development on improving export crops and expanding mining projects. The countries have large reserves of iron ore, diamonds, bauxite, and **rutile** (ROO-teel). Rutile is a source of titanium, a strong, lightweight metal used in the manufacturing of jet airplanes and spacecraft.

Liberia has become a "flag of convenience" for a very large merchant fleet whose ships are owned by companies from many nations. This is because the country has low licensing fees and registration costs. In addition, Liberia does not have strict regulations and shipping rules.

Workers spread quinine bark to dry on a plantation in Guinea. Quinine is used to treat malaria and other diseases.

Level B: Have students draw an outline map of Nigeria. They should then develop a map legend and illustrate the map, showing climate, agricultural regions, major resources, and landform features.

Guinea Coast Countries

Ivory Coast The Guinea Coast countries of West Africa border the Gulf of Guinea. (See the map on page 307.) The Ivory Coast has the most prosperous economy in West Africa. Its prosperity is directly related to its political stability. It also has had continued good relations with France, its former colonial ruler. The Ivory Coast has relied on free enterprise to encourage economic growth.

The economic heartland of the Ivory Coast is the southern forested area. Here, three of the country's main exports are grown. These are lumber, coffee, and cacao beans. The coffee crop is the largest in Africa and an important part of the world market. The Ivory Coast is also one of the world's largest producers of cacao. Pineapples, bananas, and cotton are other exports.

Abidjan, the capital, is a modern city and major port. It has the largest manufacturing output of all the former French African cities. Abidjan has many small manufacturing firms, as well as an automobile-assembly plant, aluminum factories, and chemical plants.

The south has the country's best health services and educational opportunities, and it offers the most jobs. Many subsistence farmers are leaving the dry north to seek jobs in the southern cities.

Ghana Since independence in 1957, Ghana's history has been stormy. Both dictatorships and elected governments have produced bad economic policies. The economy of Ghana is worse than ever. Even the country's once vibrant culture seems to be disappearing. Many of the country's educated people now work in other countries.

Ghana was once the world's largest cocoa producer. Recent production has been only 20 percent of what it was 30 years ago. The number of people practicing subsistence agriculture seems to be increasing. When government policies falter, many Africans working on plantations return to their home regions and traditional ways of life. There, small fields of rice, sorghum, and cassava, a few animals, a vegetable garden, and fruit trees can support a good diet.

Ghana has a large aluminum plant, a steel mill, and an oil refinery. Industries are concentrated in the coastal cities. Accra, the capital, is Ghana's largest city and major seaport. One hope for the future of Ghana is continued development of the Volta River Project. This great dam on the Volta River backs up the world's largest human-made lake. The hydroelectric power generated here amounts to nearly all of Ghana's total production.

Togo and Benin Togo and Benin are small, poor countries situated between Ghana and Nigeria. Both have savanna environments with tropical forests along the coast and on the uplands. As in the other countries along the Guinea coast, coffee, cocoa, and palm oil are important products.

Togo was a German colony before World War I. After the war, it was under French control until gaining independence in 1960. Togo has phosphate reserves, which may provide important export income in the future.

Benin was formerly known as Dahomey. The French administrators developed a large school system there, but little progress was made on other development projects. Today, there are not enough jobs for the country's educated people. Independent Benin also has suffered repeated military **coups**. A coup is an overthrow of an existing government by a small group. In both Togo and Benin, the southern regions are the most developed. The capital cities, small manufacturing industries, and most commercial agriculture are located there.

Nigeria Compared to the other countries of West Africa, Nigeria (nie-JIR-ee-uh) is enormous. With more than 110 million people, it is by far Africa's most populous country. The country extends from the humid climates and mangrove coast of the Gulf of Guinea to the dry savannas along Lake Chad. In the tropical

 Have students review "African Nationalism and Independence," on page 298. Ask them to evaluate the problems described there as they apply to the Guinea Coast countries.

Environments and Cultures of East Africa

Physical Geography The rifts and volcanoes of East Africa make spectacular scenery. Mount Kenya in central Kenya is more than 17,058 feet (5,201 meters) high. Mount Elgon (EHL-gahn) on the Kenya-Uganda border is 14,178 feet (4,322 meters) high. The highest peak of Mount Kilimanjaro in Tanzania is 19,340 feet (5,896 meters) high. This is the highest peak in all of Africa. Even though this mountain is near the equator, it is so high that its twin peaks are always snow covered. The rift valleys appear as giant scars on Africa's surface. (See the photograph at the right.) The valley floors are nearly flat, while the rift walls are steep cliffs.

The Eastern Rift begins in southern Tanzania. It continues northward to Ethiopia and onto the floor of the Red Sea. The dry plains along the Eastern Rift in East Africa are sites of famous game parks. The more humid Western Rift is filled with a chain of great lakes, often located between high volcanoes. Lake Victoria, the largest lake in Africa, occupies a shallow, saucer-shaped basin on the plateau between the Eastern and Western rifts. Away from the highlands and rifts are landscapes of plains and plateaus. The greatest plains are along the Nile in the Sudan. There are many plateaus in eastern Tanzania as well.

East Africa is not only a region of magnificent landscapes, but also one of great climate contrasts. The region contains some of Africa's driest deserts and wettest forests. The deserts of northern Sudan are like those of neighboring Egypt. The Nile River forms a long oasis within the surrounding desert of bare rocks and shifting sands. Farther south, the lands along the Nile become increasingly wet. The bare desert gives way to the more common thorny-shrub desert. Eventually, a grassy-shrub savanna appears. Still farther south, vegetation changes to tall-tree savanna. The drylands of Sudan and Ethiopia also have suffered desertification by the Sahel drought.

The rift valley is volcanic. Vegetation, shown in this Landsat photograph as red or pink, is sparse here because of lack of water.

The moist climates of the Ethiopian plateau stand high above the deserts of Somalia and northern Kenya. Along the Indian Ocean coast is the wet and dry tropical climate with low-grass savannas and woodlands. Inland, dry savannas are typical up to the forested highlands. The greatest climate changes occur along the sides of the Eastern Rift Valley. The floor of the Eastern Rift is dry, with thorn scrub and dry savannas. However, the surrounding mountains have humid climates and dense forests.

Ways of Life The people of East Africa and their ways of life reflect a long and rich history. Their culture is the result of the opportunities offered by a varied physical environment. Humid highlands are intensively farmed. Dry lowland plains and plateaus are cattle-raising country. Farming is possible on irrigated lands around oases. In addition to food crops, an increasing number of cash crops are grown. Several regions of East Africa also have large commercial plantations.

Most East Africans are traditional farmers, although industrialization has begun in the region. All of the cities have factories that

Soils are partly a result of climatic conditions. Have students compare soils and climate maps of East Africa. Students should describe the relationship between climate and soils.

Level B: Have students research the Swahili language. Ask students to develop a simple vocabulary list of Swahili words. *Jambo* is "hello!"

Herding cattle is a major economic activity in Kenya. Why is Kenya's climate suitable for livestock grazing? It is too dry for farming.

produce consumer goods, processed foods, and building materials. However, many manufactured goods are imported from developed countries.

The ethnic heritage of East Africa is divided into three great groups. The Nilotic peoples live in the Nile River area on the plains of the Sudan. Several Nilotic groups migrated southward into the highlands only a few centuries ago. Among these are the Masai. The second group is the Kushites, who live in areas from the Red Sea coast through the Horn of Africa. Within this group are the Amhara of the Ethiopian highlands and the Somali of the coastal region. The third group, located in the south, is the Bantu peoples, who are similar to peoples of the West Africa coast.

Colonial Impact and Independence East Africa has always attracted the interest of outsiders. The ancient Egyptians regularly sent military expeditions up the Nile. Later, Arab traders established trading cities along the east coast. They traded gold and ivory from the interior for products from as far as the Persian Gulf and China. The great legacy of this period is **Swahili**. Swahili became a trade language along the coast. It contains many Arabic words. Through the centuries, Swahili has developed a rich literature to become the common language of the region. It is the official language of Kenya and Tanzania. The Portuguese were the first Europeans to appear in the region. They controlled several coastal cities during the 1500s.

In the nineteenth and twentieth centuries, some European colonial powers established inland cities, built railroads, and introduced commercial agriculture. European missionaries established churches and schools. A large number of European settlers arrived to farm the fertile highlands. In these areas, the black Africans either were moved out of their traditional areas or became farm workers. As elsewhere in Africa, the colonial period here had its pluses and minuses.

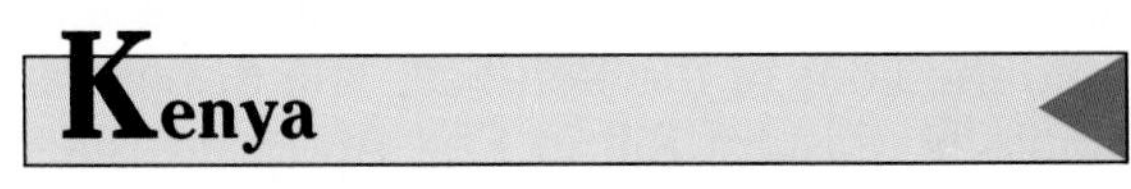

Kenya

Land and People Kenya is Africa's favorite tourist attraction. It features cool, green highlands with rich farmlands, forests, and great game parks on its dry savanna plains. There are also wide vistas of the Eastern Rift and snowcapped mountains.

Since the country gained independence from Britain in 1963, Kenya's leaders have provided a stable government. Encouraging the growth of the tourist industry and improving Kenya's small-farm agriculture have been the government's chief goals. However, Kenya's past has often been turbulent.

Kenya's first cities were founded by Arab and Persian merchants along the coast of the Indian Ocean. The merchants traded inland for ivory and slaves. Beginning in the 1500s, Portugal controlled these coastal cities for about 200 years until Arab forces recaptured them. British merchants began trading on the coast during the 1800s. In 1893, Britain extended its sphere of influence inland as far as the Western Rift Valley. This territory included modern Kenya and modern Uganda. Soon a

 Note the proximity of the Arabian Peninsula to the countries of East Africa. Refer students to the map of colonial Africa on page 297.

Level B: Have students trace both river systems in blue on an outline map. Into which ocean does each river drain?

Physical Geography

Environments The heart of Africa has dark tropical forests, sunny savanna woodlands, and pleasant forested highlands. Only in southwestern Angola along the northern edge of the Namib Desert is there a small, arid region. Most of Central Africa has a wet and dry tropical/subtropical climate. Humid and rainy tropical climates are found from the Atlantic coast inland to the mountains along the Western Rift. Central Africa is mostly plateaus and rolling plains. Along the coast are narrow lowlands that connect to several escarpments inland. High escarpments usually look like mountains when seen from below. The greatest mountain arcs in Central Africa are the Mitumba Mountains along the western side of the rift-valley lakes. Similar mountains extend through Malawi and continue south of the Zambezi River in Zimbabwe. (See the map on page 329.)

Central Africa has two great river systems—the Zaire and the Zambezi. The Zaire River drains the huge interior basin of the country of Zaire. From Kinshasa upstream, the Zaire is navigable to riverboats for about 1,085 miles (1,736 kilometers) before the great falls near Kisangani are reached. The Zambezi River drains southeastern Angola, Zambia, and northern Zimbabwe. To reach the sea, the Zambezi plunges down the gorge at Cabora Bassa to the plains of Mozambique.

The Zaire River flows for 2,716 miles (4,346 kilometers) through central Africa. Is this the longest river in Africa?

No, the Nile is the longest river in Africa.

Ways of Life In recent decades, the economies of Central Africa have been shifting from subsistence to commercial agriculture. Cash crops are produced on large commercial farms and on small plots. Many subsistence farmers have fields for both cash crops and food crops. In rural areas, people trade their products at **periodic markets**. A periodic market is an open-air trading market held regularly at a crossroads or in a town. It is also a social event where people mingle with friends and share news. (See Themes in Geography "Periodic Markets," on page 332.)

As in other sub-Saharan African countries, many people in Central Africa are migrating away from rural areas to seek jobs with cash wages. Some jobs are available in the many mines. People also look for work in the tourist industry and at game parks. However, most of the new jobs are found in factories and businesses. City governments struggle to provide basic services as slums sprawl ever outward. Most attractive houses in the city centers belong to government officials and prosperous business owners.

The countries of Central Africa import most manufactured goods from the developed countries. These goods must be paid for with cash. For this reason, the governments are always looking for more money from foreign sources. Foreign aid provides some of these funds. However, the export of cash crops and minerals is the main source of income. Sometimes, the governments have encouraged cash crops to the extent that production of food crops has decreased. In many cases, the government-controlled prices have been so low that farmers earned little for their crops.

Food production often has not kept up with the rapidly increasing populations. All countries in the region except Zimbabwe must now import food. This is particularly disappointing for Central Africa. The region has many fertile farmlands as well as income from its rich mineral deposits. These could be combined to improve food production and marketing throughout the region.

Zaire

Physical Geography A large country, Zaire is the former Belgian Congo. It occupies most of the land drained by the Zaire River and its tributaries. Landforms of Zaire's interior include the densely forested rolling plains of the Zaire Basin. The surrounding uplands are the Angolan and Shaba plateaus to the south and the Mitumba Mountains to the east. The Zaire River tumbles down steep escarpments before it reaches its mouth at the Atlantic Ocean. (See the map on page 329.)

The area of Zaire situated near the equator has a humid tropical climate. The warm temperatures and almost daily rains support Africa's greatest forests. North and south of the equatorial forests are both tall-tree and bush savanna regions. These regions have wet and dry tropical/subtropical climates.

The Belgian Congo and Independence The Berlin Conference of 1884–1885 awarded the Belgian Congo, now Zaire, to King Leopold II of Belgium. The king considered the area his personal empire. The native people of the Congo were treated badly and sometimes used as slaves.

Publicity forced King Leopold to turn over the territory to the Belgian government in 1908. Under the government, workers received better treatment and had better living conditions. Belgian businesses invested heavily in the territory. Many Belgians came to work there and soon held nearly all of the jobs in government and business. Many Africans, however, had little contact with the European customs and economic system.

After World War II, running the Congo began to cost the Belgian government more and more money. Discontent and unemployment increased dramatically when copper prices fell in the late 1950s. Adding to the discontent, the people in neighboring French and British colonies were gaining their independence. There were riots in Leopoldville (now Kinshasa) in 1959. As the unrest increased, the Belgians decided to grant the colony its independence. In June 1960, with only six months' notice, the Democratic Republic of Congo was formed.

The Belgians left behind a good school system and many hospitals. Many of the hospitals had been built by Christian missionaries. However, since Belgian workers had held nearly all the important jobs, the people of the Congo did not have the skills and experience needed to govern themselves. Furthermore, the idea of a united Congo was not widely supported. People often thought first of their own ethnic group and region.

After independence, several political parties battled to control the government. Lawlessness and rebellion were widespread. To make matters worse, the mineral-rich province of Shaba (then called Katanga) tried to withdraw from the new state. The result was civil war. The transportation system broke down, and famine threatened. After several changes of government, the commander of the army, Mobutu Sese Seko, took control. Many industries owned by Europeans were given to Mobutu's friends and relatives. In 1975, the Mobutu government changed the names of cities and provinces to local names. The country became the Republic of Zaire. This was done to remove signs of the colonial past and to unite the many African groups.

Rich Resources Zaire has many rich resources. Its humid climates encourage the growth of tropical crops. Commercial plantations produce palm products, rubber, and coffee. Numerous plantations use the navigable streams in the Zaire River system for transport. Subsistence farmers usually raise cash

 Remind students of Henry Morton Stanley's reference to iron ore. See "For the Record" on page 294.

Have students locate Central Africa on the map of sub-Saharan Africa on page 295.

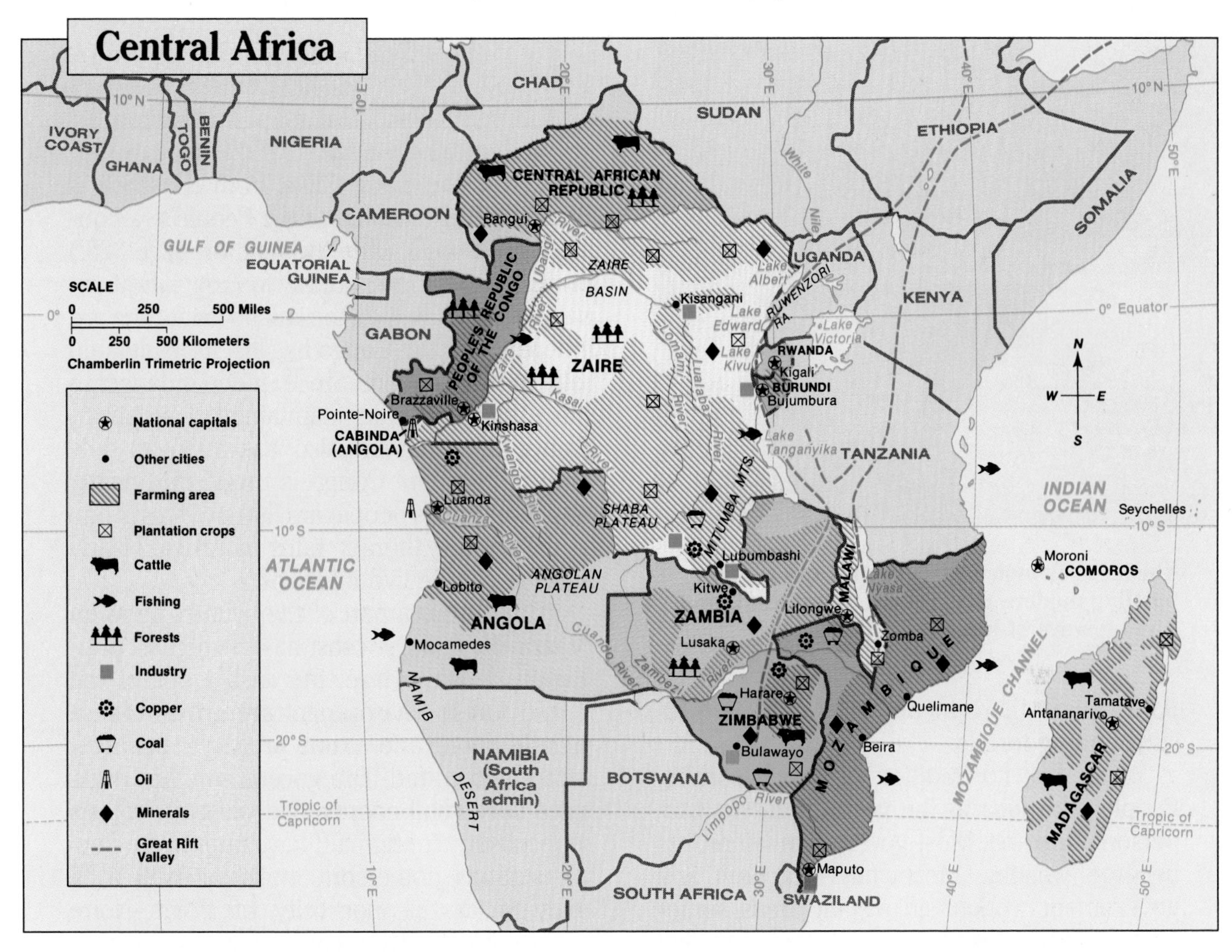

Which sub-Saharan African country has the advantage of an extensive river system? Zaire

crops too. With improved farming systems, Zaire could have food for export.

Zaire's mineral resources are even more substantial than its agricultural ones. The copper deposits in the Shaba province in the southeast are among the largest in the world. Deposits of cobalt, iron ore, manganese, gold, and tin are nearby. Combined with adjacent areas in Zambia, this is the second greatest mining region in Africa. Zaire also has enormous reserves of industrial diamonds. In addition, the rivers of Zaire have the greatest hydroelectric potential in Africa. Oil has even been found off Zaire's narrow Atlantic coast.

Life in Zaire Zaire has three great cities. The first is the capital of Kinshasa, located on the Zaire River near the Atlantic coast. The river docks in Kinshasa receive goods from throughout the Zaire Basin. Kinshasa is a congested city where traditional markets and modern stores exist side by side. Colonial buildings, high-rise apartments, and shantytowns are often found in the same neighborhood. Kinshasa is the center of political power in Zaire.

Zaire's second important city is Lubumbashi (loo-boom-BAHSH-ee). It is the industrial center of the Shaba "copper belt." The railroad from Lubumbashi carries copper to ports in South Africa. Many manufactured goods arrive on return trips. Kisangani, located on the great falls of the Zaire River, is the third important city. Industries use the electricity from the

The Themes in Geography feature on pages 332–333 tells more about life in Zaire.

Kinshasa, though Zaire's capital and a bustling modern city, retains many traditional African ways of life.

power plant there to process products from surrounding forests.

Zaire has no tradition of public service. People take jobs in the government to gain personal income. Most government ministers become wealthy. Meanwhile, teachers and government workers in distant areas sometimes wait months for their paychecks. Skilled workers often keep a small farm to ensure dependable income.

Economic and Political Future The symbols of the colonial past in Zaire are steadily disappearing. There were 80,000 miles (128,000 kilometers) of good roads at the time of independence. Now, the roads total less than 5,000 miles (8,000 kilometers). The ruins of colonial factories and farms stand silent as many lands return to subsistence agriculture. Despite this, Zaire has enormous potential. The schools are producing many educated people. The profits from cash crops and minerals are substantial. New plantations, mines, and industries are being started. The people remain hopeful. However, new political leaders are not being trained. Suspicion among the many ethnic groups continues to separate the people of Zaire.

The People's Republic of the Congo

Across the Zaire River from Kinshasa is Brazzaville, the capital of the People's Republic of the Congo. (See the map on page 329.) Brazzaville was the gateway to French Equatorial Africa. Now that the inland countries are independent, the Congo has lost its leadership role in the region. Since independence in 1960, the Congo has maintained close ties to France. However, it also has a Soviet-style government. The Congo is proud that nearly 80 percent of its people are literate. This is one of the highest literacy rates in Africa. Nearly half the people live in towns.

The southern part of the country between Brazzaville and the coast has the highest population density. Here, the cash crops of the humid forest environment are grown. These include sugarcane, palm products, bananas, cacao, and coffee. The government has organized successful cooperative villages. Projects supported by foreign aid have built hydroelectric stations and textile mills, as well as a shipyard at the port city of Pointe-Noire (pwant-nuh-WAHR). The country has oil, natural gas, and potash deposits. Oil provides much income. Cooperation among the people, the government, and foreign-aid agencies is leading to progress.

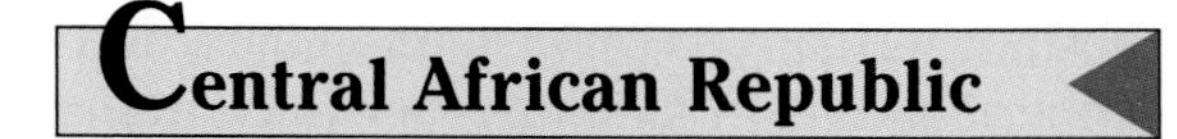

Central African Republic

North of Zaire is the Central African Republic. (See the map on page 329.) Most of the country is a rolling plateau covered with tall-tree savannas. More than 90 percent of the people are subsistence farmers with little government contact. During the colonial period, the country was exploited by development companies. These companies had exclusive licenses to rule with total authority. The local Africans came to view Europeans and their business ventures with distrust.

 In order for modern technology to be used, mineral deposits usually must be large. Ask students to think of specific ways that foreign aid could be used to develop the economy of the People's Republic of the Congo.

CHAPTER 28 CHECK

Reviewing the Main Ideas

1. Central Africa is a plains region covered with tropical forests and savannas. Most of the region has a wet and dry tropical/subtropical climate.
2. Central Africa imports many manufactured goods from developed countries. Exports of minerals and cash crops provide income to pay for these goods. Most of the countries in the region must import food.
3. The expanding population of Central Africa has forced many people to seek work in the cities. Cities are overcrowded, and many people live in poverty.
4. Politics and ethnic relations are often complex and strained in this region.

Building a Vocabulary

1. Define *periodic market*. How is the periodic market an adaptation of commercial agriculture?
2. What word describes a legal restriction placed on the movement of freight into a country? Is this an effective form of economic pressure? Why or why not? Give an example from this chapter.
3. What is vanadium? How is it used?
4. What is copra? Where in Central Africa is it grown?

Recalling and Reviewing

1. Why is Zaire a rich country? What problems must Zaire overcome?
2. Which countries developed from the Federation of Rhodesia and Nyasaland? How has their struggle for independence varied?
3. Why has the economic potential of Angola not yet been realized?
4. Why are parts of the river systems of Central Africa unnavigable?

Critical Thinking

5. Using examples from this chapter, explain how the quality and availability of transportation is related to the economic development of an area.

Using Geography Skills

Use the map in this chapter to answer the following questions.

1. Which Central African countries are landlocked? Of these, which have navigable river outlets to the sea?
2. What landform feature separates Zambia from Zimbabwe? What other countries of Central Africa have natural features for boundaries? Name the features for each.
3. What large rivers are political boundaries in Central Africa?
4. Where in Central Africa are extensive forests located? What kind of forests are these?
5. Give the latitude and longitude of the Angolan Plateau.

Answers to Chapter Check questions are found in the Teacher's Guide.

Additional activities for Chapter 28 are found in the Student Workbook. Chapter 28 Test as well as Skills, Reteaching, Critical Thinking, and Challenge/Enrichment activities are available in the Teacher's Resource Binder.

Chapter 29

Chapter 29 Lesson Plans and Planning Guide are found in the Teacher's Guide.

Refer students to the map on page 343.

Southern Africa

▲ **Mountains dominate South Africa's coast.**

▲ **Zulus are a large South African ethnic group.**

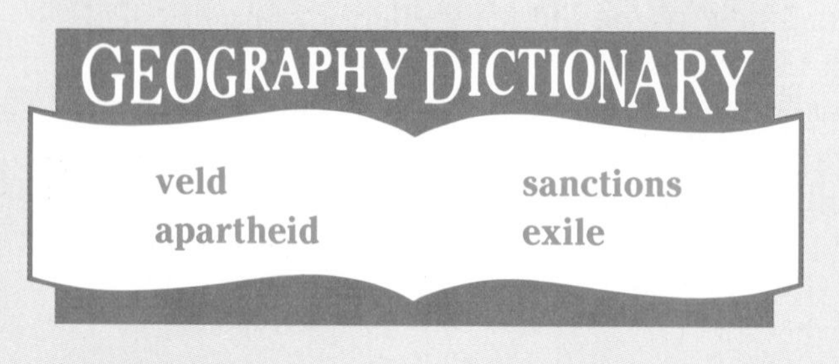

Southern Africa is a region in conflict. The proud traditions of black Africa are clashing with a modern, industrialized economy. A government run by an equally proud white minority is steadily losing power to the black majority. This struggle between black and white, modern and traditional, and Western and non-Western is being observed from all around the world.

At stake in this conflict is control over the vast mineral wealth of the region. Beneath the surface of southern Africa are some of the world's greatest mineral deposits. Access to these minerals is important for the world's developed economies. Because control of these resources means world power, all of the world's countries would like to see these struggles resolved in their favor.

 Note that unlike most other areas of sub-Saharan Africa, there is a large minority of whites in South Africa.

Republic of South Africa

Physical Geography Most of the Republic of South Africa consists of wide plains covered by dry savannas and desert shrub. (See the map on page 343.) In the western part of the country, the coastal Namib Desert and the inland Kalahari Desert combine to form a large desert region. Along the eastern and southern coasts are regions that are more humid. In the southwest, near Capetown, is a region with a Mediterranean climate. Along the east coast, near the city of Durban, the climate is humid subtropical. The high plateaus of the interior are cool and mild, and support broad grasslands.

Landforms and climates are closely related in South Africa. Along the coast are narrow, usually humid, coastal plains and lowlands that are blocked from the interior by great escarpments and mountains. Along the southern coast are mountains that rise up to 7,500 feet (2,287 meters). Along the eastern coast is the great Drakensberg Escarpment, which reaches 11,425 feet (3,483 meters). Many of these mountainsides face into the rain-bearing winds. The results are humid climates and mountain forests. However, the plains and plateaus inland are often blocked from rain-bearing winds by the mountains. As a result, the upland plains are often not only cooler but also drier.

South Africans call vegetation and landform regions **velds** (vehlts). The highveld is the vast expanse of high plains and mild grasslands in the east-central interior. Another category is bushveld. This is a savanna landscape of short trees and bushes that is found at lower elevations. Still another is lowveld, which includes the dry, tall-tree savannas and the moist forests at the base of the Drakensberg Escarpment.

The most important rivers of South Africa are the Orange and its tributary, the Vaal. The Orange River basin includes most of the highveld, as well as the western edge of the Drakensberg. The eastern and southern sides of the country are drained by shorter rivers with smaller drainage areas. There are no navigable rivers in South Africa.

African Words Used in the English Language

Words from Afrikaans

Trek
Commandeer
Commando
Apartheid

Words from Black African Languages

Banana
Yam
Voodoo
Gorilla
Banjo
Gumbo
Tote
Zombie

From what language is Afrikaans derived?

an older form of Dutch

European Settlement Europeans first settled in South Africa in 1652. They established the port of Capetown and developed farms to supply the ships of the Dutch East India Company sailing to and from Asia. Within 50 years, Dutch, French, and German farms were spread across the fertile valleys near the Cape of Good Hope. Cattle herders also were bringing the European presence into the dry mountain valleys to the east.

The Cape Dutch settlers called themselves Boers. They are the ancestors of about 70 percent of modern white South Africans, who are known as Afrikaners. Their language is Afrikaans, which is derived from an older form of Dutch. It also includes many words taken from Malayan slaves and black Africans. The Afrikaners are very loyal to Africa. They do not think of any part of Europe as their homeland.

Review the orographic effect with students to aid their understanding of the climates in South Africa. Have students compare the deserts with those of the same latitudes elsewhere in the world.

However, southern Africa in the 1600s was occupied by other people as well. Most of the western and southern parts of the region were occupied by Khoisan peoples. These included the Khoi (Hottentots), who herded cattle and goats and lived in villages. The other Khoisan people were the hunting-and-gathering San (Bushmen). Though their skin is yellow-brown, the Khoisan people are not black Africans. Archaeologists believe that the Khoisan people lived throughout Africa 10,000 years ago. Southern Africa is the only place where they survived.

Black Africans were also living in large areas of South Africa in the 1600s. Most lands along the east coast and inland north of the Orange River were settled by Bantu-speaking peoples. Like the Boers, they were herders and farmers. They knew how to work iron as well. Unlike the Boers, these people did not know of European technology. They also did not understand European concepts of landownership. The Boers eventually came to control the black Africans.

British South Africa The Cape Colony of the Boers came under British administration in 1806. The British administrators made the English language, English laws, and English education the rule. Many Boers were unhappy with the British take-over. In the 1830s, Boer families began moving farther into the interior. The migrating Boers met few black Africans on their journey. However, the lands north of the Orange River had recently lost population. Wars between different black African groups had killed many native peoples. Other people were in hiding. After the Boers arrived in the region, many blacks became workers on the Boer farms and ranches.

The Boers established two independent countries in the region. These were the Orange Free State and the Transvaal. Meanwhile, the British were establishing the colony of Natal on the Indian Ocean coast.

In 1868, diamonds were found near Kimberley. About 20 years later, the world's largest gold deposits were discovered near present-

South Africa has large mineral deposits. Some of the world's largest diamond mines are found near Kimberley.

day Johannesburg. The vast mineral wealth brought a large number of English settlers to the South African interior. Black Africans also began moving to the cities and mines to look for work.

Abundant mineral wealth and the Boers' distrust of the British led to the Anglo-Boer War (1899–1902). The Boers lost this bitter struggle. In 1910, the Transvaal and the Orange Free State were joined with the Cape and Natal colonies to form the Union of South Africa. Pretoria, in the Transvaal, became the administrative capital. Bloemfontein (BLOOM-fahn-tayn), in the Orange Free State, became the site of the judicial capital, where the Supreme Court is located. Capetown became the legislative capital, where the parliament meets. However, the government was nearly all white.

Separate Development In the 1948 elections, South Africa's government came under the control of the Nationalist party. The Nationalists have controlled the government ever since. The Nationalist party's goals have been to establish formally the separation of the races in South African law. This has been done through the policy of Separate Development. The idea of Separate Development is that each race, apart from the other races, is to improve

 Level A: Tell students that James Michener's *The Covenant* gives a popular view of the history of the region.

The South African government actively encourages the immigration of Europeans. Ask students why white South Africans feel that this policy is important to the future survival of their rule.

Urban Geography The system of cities surrounding Johannesburg is known as the Witwatersrand. This is the greatest industrial region in Africa. Most industries are South African owned. Many large European and Japanese companies are represented as well. Most of South Africa's automobiles are built near Port Elizabeth, on the southeast coast. Capetown and Durban are other great ports.

South Africa's cities are modern, but they are organized according to apartheid. White areas are reserved close to the city centers and in suburbs away from industrial districts. White neighborhoods are similar to those in American suburbs, with modern-style houses and shopping centers. Black areas are usually on the outer edges of the city, often across industrial districts away from the white neighborhoods. Often, blacks must commute long distances to work. Coloured and Indian neighborhoods are usually situated between the black and white districts. Black neighborhoods vary in style from small, government-built modern houses to slum settlements built from scrap lumber and sheet metal. South Africa's best-known black suburb is Soweto, near Johannesburg. *Soweto* is not an African word but is a shortened version of Southwestern Townships. Thousands of blacks live illegally in South Africa's cities.

South African Prospects The idea of apartheid is counter to the political and social trends of the world. Apartheid blocks progress by the black majority and causes the South African nation many practical problems. There is a shortage of skilled workers because it is difficult for blacks in South Africa to get advanced education and training. While South Africa's black population is increasing at one of the fastest rates in the world, the white population is growing very slowly. Soon, the economy will have to be open to all if the country is to prosper. As more blacks are trained and hired, the inequalities of apartheid become more unacceptable.

Capetown, where the legislative capital of South Africa is located, is an important harbor and trading center. In what province does it lie? Cape Province

Level A: Students should research SWAPO, the guerrilla movement in Namibia. What are the goals of SWAPO? How has this movement affected South Africa? Vivid photos of this nation can be found in the June 1982 issue of *National Geographic*.

South Africa's black neighborhoods have experienced serious unrest since 1985. Most people believe apartheid cannot last. Seeing the need to improve the economic position of blacks, the government claims to be ending this economic apartheid. Many more jobs with higher pay are being made available for blacks. Blacks are also receiving help in starting businesses. Many obvious signs of racial discrimination are being removed. Separate rest rooms, elevators, drinking fountains, benches, and the like are disappearing. Yet, the segregation of schools and neighborhoods continues.

More essential is ending the apartheid in the country's political affairs. Blacks want a meaningful voice in the government. How to share political power among all of South Africa's people is the great question. The possibility of civil war in South Africa is real at a time when both black and white leaders are offering possible solutions.

Events in South Africa are influencing governments and economic policies all around the world. Some governments of developed countries have placed embargoes on South African products. These restrictions are called **sanctions**. Many companies from developed countries have sold their operations in South Africa. Some countries do not permit white South Africans as tourists. Nevertheless, many South African minerals remain important for industries in the developed countries.

While governments in neighboring states also oppose apartheid, these countries often depend on South Africa's railways and harbors. They are trying to reduce their need for South African services and products. South Africa's borders are strictly defended. Some black South Africans are exiles in neighboring countries. An **exile** is one who has left or has been forced to leave his or her country.

If apartheid ended, South Africa could greatly aid other African countries with its valuable agricultural, mineral, and industrial technologies. These could improve the economies of all African nations.

Namibia

Namibia, formerly called German Southwest Africa, is located along the Atlantic coast north and west of South Africa. (See the map on page 343.) It is a former German colony. During World War I, South African forces drove out the Germans. Since then, the region has been controlled by South Africa. However, South Africa's administration has been challenged by the United Nations.

Namibia is a very dry country. Along the coast is the Namib Desert; inland is the Kalahari Desert. Most of the people live in the far north, where there is dry savanna. People also live in the central highlands, where the climates are cooler and not as arid. Ranching is the only commercial agriculture. The fishing fleet of this territory provides important income.

Namibia has enormous mineral reserves. There are whole mountains of copper, lead, zinc, and uranium. Along the southern coast are large diamond deposits. At one time, apartheid was introduced into Namibia. However, lands once reserved for whites and blacks, as well as for Khoisan peoples, have been opened to settlement by all groups. As the laws are changed, racial discrimination is being reduced. Still, a legislature has been established on the basis of race.

Beginning in the late 1970s, South Africa clashed with a Communist rebel group over plans for Namibian independence. The rebels have launched guerrilla attacks against South African troops. South African troops have attacked rebel bases inside Angola. These skirmishes continue to block Namibia's independence. With Cuban troops north of Namibia in Angola, the region also has become a focus for conflict between the United States and the Soviet Union. (Cuba has a Communist government supported by the Soviet Union.)

An interesting problem for Namibia's future is the status of Walvis Bay, the country's only good port. This small parcel of land has been part of Cape Province (South Africa)

The Namib Desert stretches along the Atlantic coast of Namibia. What resources are mined in this region? Where do most Namibians live? Diamonds; In the dry savannas of the far north and in the central highlands.

since 1884. South Africa still claims it and administers it from Capetown.

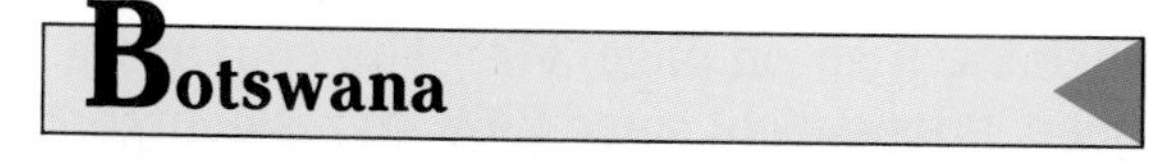

Botswana

Botswana was the former British territory of Bechuanaland. It is an arid, landlocked country occupying much of the Kalahari Desert. (See the map on page 343.) Britain took control of the area in 1886 to connect the Cape Colony with Rhodesia and to avoid the Boers in the Transvaal. Britain invested little in Botswana. At the time of independence in 1966, there were no paved roads, no electricity, and only one important factory, a meat-canning plant. A new capital was built at Gaborone. Now, mineral deposits of copper and diamonds provide important income. Botswana's diamond deposits are enormous.

Botswana has only one large ethnic group, the Tswana. Many of these people are still cattle herders and maize farmers. People live chiefly along the wetter eastern edge of the country, where dry savannas are the typical vegetation. Within the Kalahari Desert are the remaining groups of hunting-and-gathering San (Bushmen).

In the north of Botswana flows the Okavango River. This river starts in the wet highlands of Angola and flows into the Kalahari Desert. There, it forms the rich wetland forests of the inland Okavango Delta. Large game reserves are located in the region. This swamp in the desert could become a major tourist attraction in the future.

See pages 88–89 for a discussion of the lives of the San (Bushmen).

Botswana is a landlocked country. Most of its people live near the Okavango Delta in the nation's interior.

Lesotho

Lesotho (luh-SOH-toh), a landlocked country completely surrounded by the Republic of South Africa, is located in the Drakensberg Escarpment. (See the map on page 343.) Very little land in Lesotho is suitable for farming. However, the country's rough land has been an obstacle to invading tribes. A tradition of independence prevented the kingdom from being absorbed into South Africa.

Since local resources are poor, large numbers of people leave Lesotho to work in the mines of South Africa. Much internal conflict has taken place concerning what Lesotho's connection to South Africa should be. All exports of wheat, diamonds, and wool must go through South Africa to reach the world market. It would be difficult for this small state to become truly independent.

Swaziland

Like Lesotho, Swaziland is an inland country located along the Drakensberg slopes. It has access to the sea through Mozambique. (See the map on page 343.) The country has been independent since 1968. Its residents are almost entirely Swazi people, so the country has an internal unity, which makes it different from many other African countries. The Swazi government is based on Swazi tribal customs. Swaziland is fortunate in having gold, coal, asbestos, iron ore, and lumber. It also has valuable grazing lands.

Unlike Lesotho, only about 20 percent of Swaziland's trade is with South Africa. Iron ore and asbestos are leading exports. Although most of the people are subsistence farmers, cash crops are grown in the lower elevations. Sugarcane and citrus fruits are major exports.

 Namibia, Botswana, Lesotho, and Swaziland have been characterized as economic dependencies of South Africa. Ask students why this might be true.

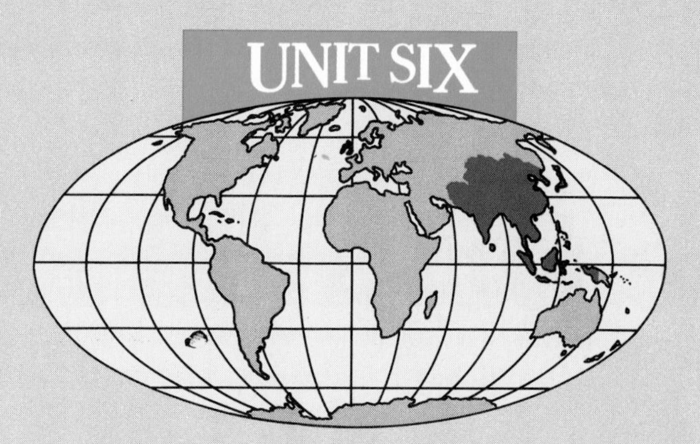

THE ORIENT

OBJECTIVES

- ▷ *To trace the history of and describe the various cultures in the Orient*
- ▷ *To describe how climate and landforms affect the way people live in the Orient*
- ▷ *To describe the political, economic, social, and cultural issues that exist in the Orient*
- ▷ *To describe the importance of the Orient as a major region of the world*
- ▷ *To describe the interchange of cultural ideas among Oriental culture groups and with the rest of the world*
- ► *To interpret a line graph*

Chapter 30

Chapter 30 Lesson Plans and Planning Guide are found in the Teacher's Guide.

Refer students to the map on page 359.

The Oriental Realm

▲ A village clings to the steep Himalayas.

▲ Old ways of life exist in modern cities.

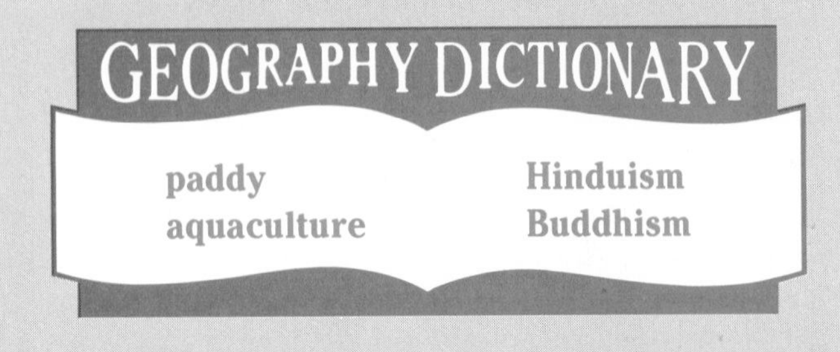

The Orient is a realm of great geographic diversity. It includes the world's highest mountains, some of the world's greatest rivers, and thousands of islands. Traditional agricultural villages exist side by side with huge, industrial cities. Economically, the world's most rapidly growing nations are located here. One of the world's richest nations, Japan, and some of the poorest, such as Bangladesh, are included in the Orient. This region also has some of the world's largest countries in population and area. The two most populated nations on Earth, China and India, are in the Orient.

Many Americans once considered the Orient exotic and remote. World War II, the Korean War, and the Vietnam War have stressed the importance of understanding the people and politics of Asia. The lives of Americans, as well as of people throughout the rest of the world, are greatly affected by the events and people of the Orient.

 Have students survey the photos in Chapters 30–34 to get a sense of the geographic diversity of the Orient.

Unlike the Orient's mainland areas, island nations, such as Japan and the Philippines, receive rain even during the winter monsoon. The reason for this rain is that the winds cross large water areas before reaching the islands. Dry winds flowing out of Asia must cross over the Sea of Japan and the South China Sea before reaching the islands of Japan and the Philippines.

Economic Geography Most of the people of the Orient are engaged in some form of agriculture. The type of farming practiced varies with the climate. In humid tropical and humid subtropical climates of the Orient, rice is the main food crop. It is generally grown in a terraced rice field known as a rice **paddy**. A paddy (sometimes spelled *padi*) is a water-covered field for growing rice and other crops. These flooded and embanked fields produce most of the food consumed by the Orient's large population. Terraced rice paddies are the dominating landscape feature of the Orient.

Commercial planters use the variations of tropical heat and moisture to produce rubber, tea, coffee, coconuts, sugarcane, and various spices. In the cooler and drier humid continental and continental steppe climates, wheat is the main food crop. In the desert and high-plateau areas, herders keep flocks of sheep or other animals.

Some of the Orient's most abundant mineral resources are tin, tungsten, coal, iron ore, and petroleum. More than 50 percent of the world's supply of tin is found in the Orient, particularly in Malaysia, Thailand, and Indonesia. More than one-third of the world's supply of tungsten is in the Orient. In fact, China is the world's leading tungsten producer. Thailand, South Korea, and North Korea also produce tungsten. Coal and iron ore, found in large quantities in China and India, have supported the industrialization of both nations.

The petroleum deposits of the Orient are of considerable importance. Indonesia and Brunei are major producers and exporters of petroleum. Although there are natural-gas and oil deposits in India, Burma, and Thailand, none of these nations appears to have large quantities that can be exported. China has developed several medium-sized petroleum fields. It may become a more important oil exporter as more oil reserves are discovered and developed.

The Orient's high mountains and great quantities of rainfall are rich sources of hydroelectric power. Japan has developed this resource to a maximum. Though China and other countries in the region also have resources for hydroelectric power, development has barely begun.

In addition to mineral resources, the Orient has fish and forest resources. Several of the region's nations, especially Japan, Taiwan, and South Korea, have large world-fishing fleets. **Aquaculture**, or water and sea farming, is an important economic activity. Seafoods, such as shrimp and oysters, are commercially farmed in protected bays and river mouths. For a local food supply, many nations in the Orient depend upon freshwater fish from lakes, rivers, and flooded rice paddies.

The forest resources of the Orient are generally divided into two types, middle-latitude forests and tropical forests. The middle-latitude forests, located mainly in Japan,

Fish is a major resource in the Orient and a major food staple. These Chinese use nets to fish from a lake in northern China.

China, and the two Koreas, have been heavily exploited over the centuries. Because their industrial and housing needs far exceed the tree growth, these nations must import timber.

The tropical-forest resources of nations such as Thailand, Indonesia, Burma, and the Philippines appear to be vast. Yet, hardwoods are being removed at a rapid and careless rate. If this rate and the destructive methods of removal continue, most of the Orient's valuable tropical-hardwood forests will be destroyed by the end of this century. The soils and wildlife dependent upon these tropical forests will disappear as well.

Cultural Geography

More than one-half of the world's people live in the Orient. Among these people is a tremendous variety of cultures. Some nations, such as Japan, comprise mainly a single culture. Others, such as India, China, and Indonesia, have a unique diversity of cultures within their borders.

There are more than 1,000 different languages in the Orient. Because of the large number of languages, some countries have adopted one official language to unify the people, as in China. English, French, and Spanish are still used in nations with colonial histories. English is the business language of many nations in the Orient.

Most of the major religious systems are found in the Orient. **Hinduism**, whose roots date to prehistoric times in India, is India's chief religion. Hindus worship many gods and believe that enlightenment is offered through *dharma,* or good conduct. They also believe in reincarnation, or rebirth after death. Islam is the main religion in Pakistan, Afghanistan,

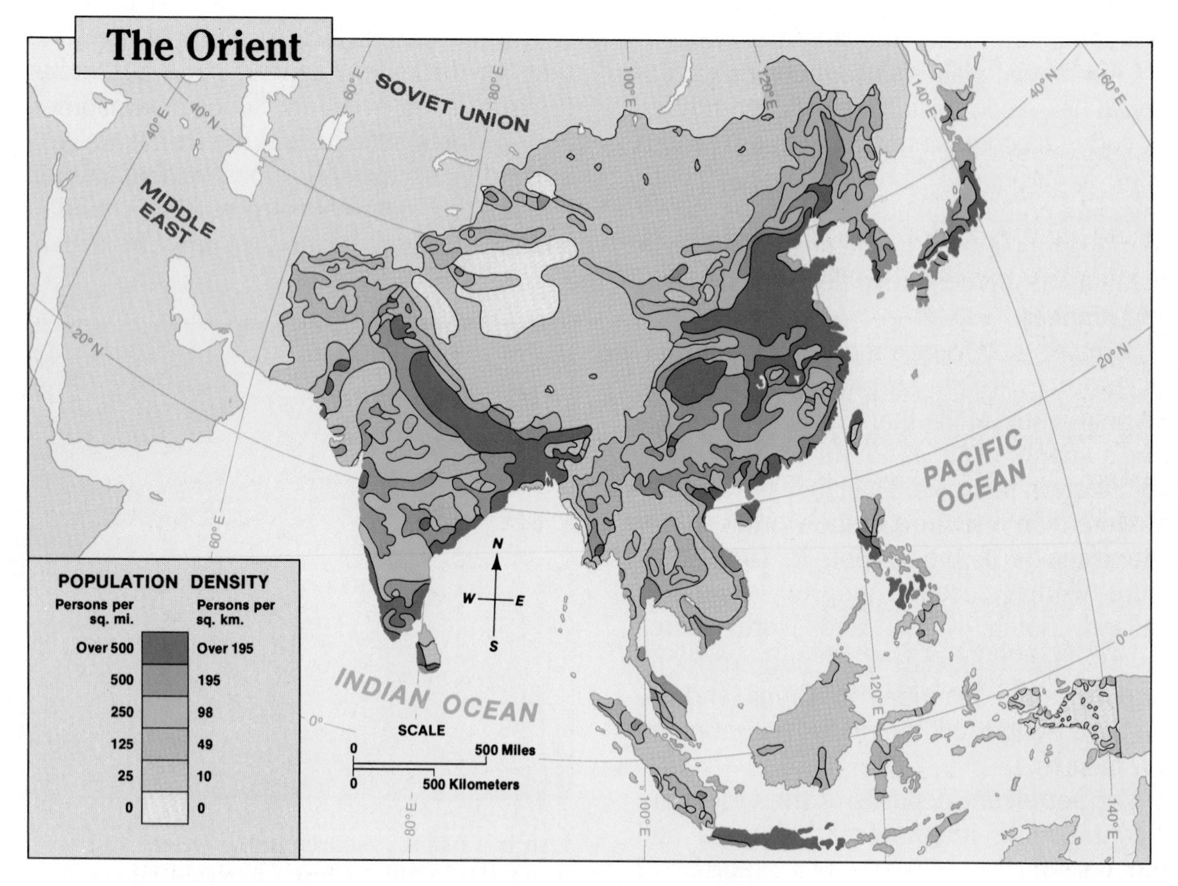

What is the population density at the southernmost tip of India?

more than 500 persons per square mile (195 persons per square kilometer)

Have students locate India and South Asia on the map on page 359.

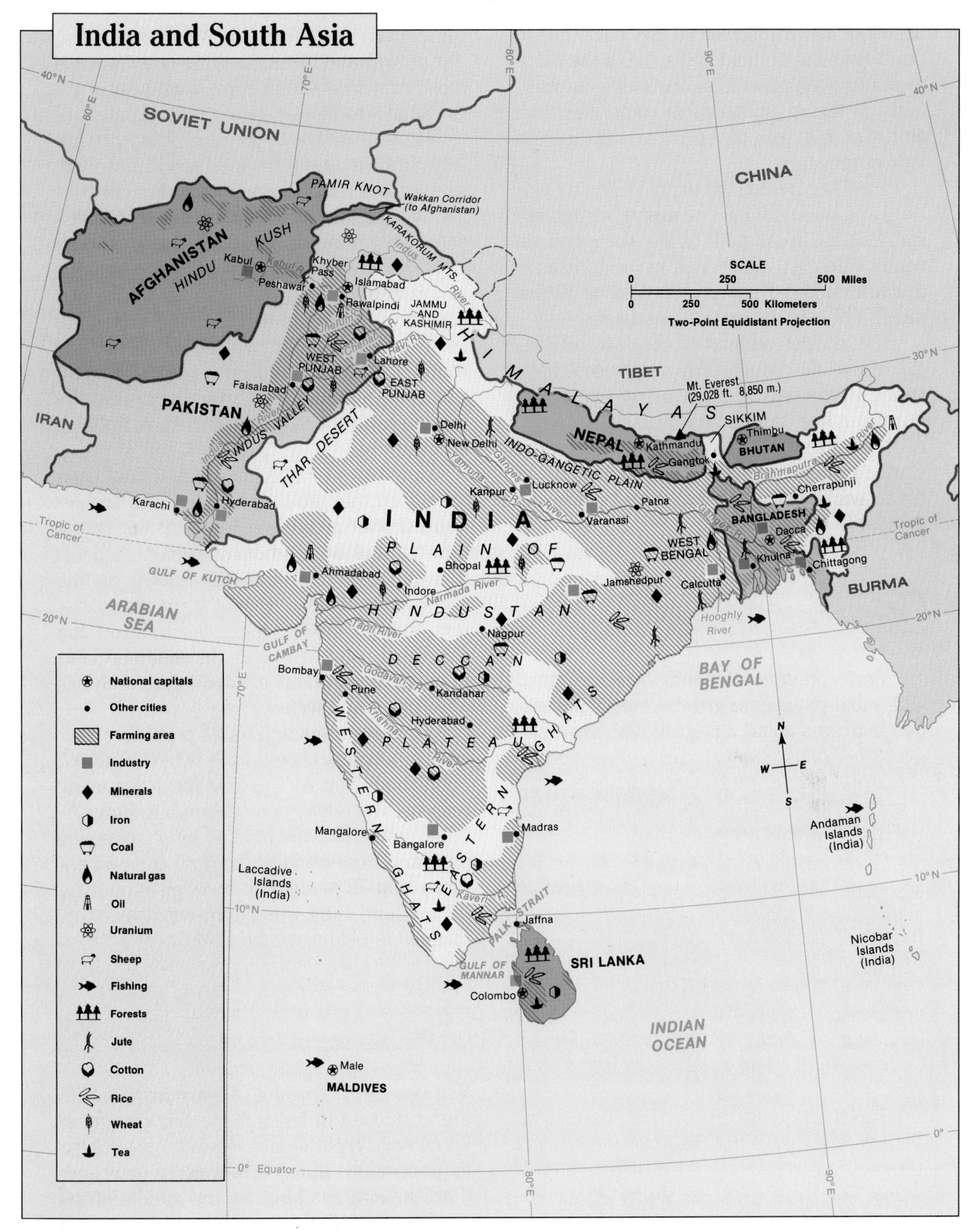

What economic activities occur along the Indus River? the Ganges River?

Indus: Forestry, coal and uranium mining, industry (including natural gas), farming. Ganges: Industry (including natural gas), farming.

Most of Bangladesh is less than 100 feet (30 meters) above sea level.

Seedlings are planted so that they have time to develop and to be held in the soil. If the monsoon rains come too soon, the seeds will wash away. If the monsoon rains come late or if almost no rain falls, the seedlings will die from lack of moisture.

The Thar Desert, the Indus Valley of Pakistan, and the mountainous desert of Afghanistan are not in the path of the wet monsoon winds. Thus, annual rainfall in these areas is much less. The Thar receives as little as five inches (13 centimeters) of rain a year.

The coastal regions of the Bay of Bengal, especially Bangladesh, are occasionally struck by severe **tropical cyclones**. These hurricane-like storms cause **storm surges**. A storm surge is a rise in sea level as the storm approaches the coast. Thousands of lives have been lost to these storms in the low delta area of Bangladesh.

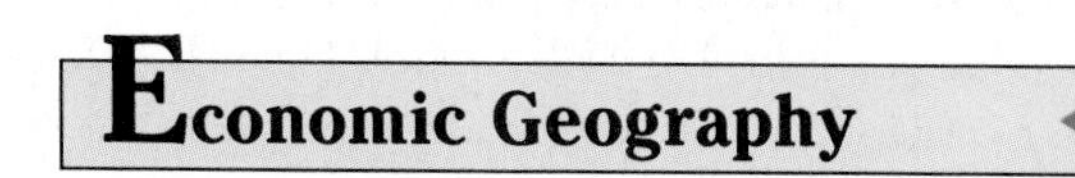

Economic Geography

Most countries of South Asia are faced with rapid population growth. India's population increases about 2 percent each year. Because India's population is near 800 million, this growth rate means that the country gains more than 15 million people each year.

Small, inefficient farms, the expense of fertilizers, and lack of mechanization prevent India from increasing its food production. Natural disasters, such as monsoon droughts and floods, cyclones, locust swarms, and crop disease, also cause setbacks and increase poverty. There is little land left to develop in India, and farmland is being lost to environmental damage. Supplying the demand for more farmland and for more wood destroys the surrounding forests. This **deforestation**, in turn, causes a loss of forest resources, destroys wildlife habitat, and creates increased flooding and soil loss.

New types of rice and wheat have been developed that produce higher yields and require shorter growing seasons. At first, these new types of grain produced such high yields that their introduction was known as the **Green Revolution**. However, natural disasters and population increases reduced the gains that had been made. In addition, these new improved grains needed larger amounts of fertilizer and water.

An answer to South Asia's problem of balancing people and resources is to control the population growth. This is difficult for many reasons. Generally, poor communication systems prevent government family-planning programs from reaching the rural areas, where more than 70 percent of the population lives. In addition, most rural farmers want large families to help work the fields.

Poverty in India is widespread. One reason is that India's birthrate is much higher than its death rate. What is India's population?

near 800 million

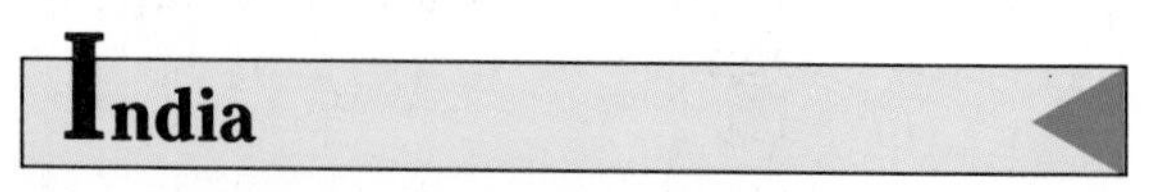

India

Village Life One out of every six people in the world lives in India, the largest country in South Asia. The nation is second only to China in population. India, however, is only one-third the size of China. By the middle of the next century, India could become the most populated nation in the world.

Most of the people in India are poor farmers who live in rural villages. Their lives have changed little over the centuries. Electricity, telephones, and running water are uncommon. Sanitation is poor. There is often one doctor for thousands of people. Sixty percent of the rural population remains illiterate.

Village houses are built of local materials. In the alluvial river valleys, straw-roofed huts are made of mud. In other areas, the houses might be made of cattle-dung plaster, brick, or bamboo.

Most Indian men wear a *dhoti* (DOHT-ee), a simple white cloth wrapped between the legs. Hindu women wear a *sari* (SAHR-ee). This is a wide piece of cloth that is draped so that it covers the body from shoulder to ankle. Indians usually can tell what part of the country a Hindu woman is from by the way she drapes her *sari*.

Most Indians wear light, loose clothing because of the hot climate. Brightly colored *saris*, worn by women, are typical in India.

Rice is the staple food of the south, wheat of the north, and millet and sorghum are eaten in the Deccan Plateau area. The Indian diet contains little meat. The Hindu religion forbids the eating of beef. The Muslim religion does not allow the eating of pork. There are some religions in India that do not allow any animal to be used as food.

Despite government attempts at land reform, more than 30 percent of India's rural population is landless field workers. Those who do own small plots of land are often forced into debt to meet farm and living expenses. Village moneylenders may charge as much as 300 percent interest on loans. Other farmers work the land of large landowners and must give a large share of the crops they grow to the landowners.

In the rural villages, the **caste system** still exists. This social system was developed thousands of years ago by the Hindu Aryans. It divided India's society into five major classes, or castes, based upon occupation. The Brahmins (BRAHM-unz) were of the highest caste. They were the priests and teachers and the only people who could read and write. The members of the lowest social caste were called untouchables. They were not allowed to enter a temple and did only the lowliest jobs. They were forbidden to have any contact with Indians of other castes. Between the Brahmins and the untouchables were the Kshatriyas (kuh-SHA-tree-uhz), Vaisyas (VISHE-yuhz), and Sudras (SOO-druhz). These people were soldiers, artisans, merchants, farmers, and laborers. A person was born into a caste and could not move into another.

Since India's independence from Britain, the government has worked hard to abolish the caste system. It declared that the poor treatment of untouchables was illegal. Mahatma Gandhi, who helped free India from British control, called the untouchables *harijans,* or "children of God." Yet, regardless of the law, the caste system has remained a part of Hindu life. As improved education and transportation reach the villages, the caste system will likely change as it has in the cities.

Discuss with students how medical advances have led to population increases. Ask them to assess both positive and negative effects of the Green Revolution.

Themes in Geography

INDIAN VILLAGES

Although India has made progress toward becoming an industrialized country, the majority of its people still lead very simple lives as rural peasant farmers. Indian peasant farmers live together in the 600,000 small villages scattered throughout the country, surrounded by farmland. These villages form the focus of life in the rural areas of India.

Each village consists of a number of small huts with only one or two rooms. These huts are clustered around a well or a stream, which forms the central point of the village. Each day, the women gather at the well with large clay pots for carrying the water. They do not have luxuries such as running water. Only one-fifth of all Indian villages have electricity. Even in villages that do have electricity, many of the huts are not connected to the electricity supply.

There are few shops in the villages. The people are mainly poor, subsistence farmers, which means that they produce almost everything they need. Some farmers own their own plots of land and grow just enough food to feed their families.

Relationships Within Places **Most of the farming methods practiced in the Indian villages today have been used for generations. Because the peasant farmers are very poor, they cannot afford modern equipment. They obtain their water from wells, using cattle to drive the pumps. There are great differences in rainfall throughout the country, so the people must live off the crop that grows best in their area. The most important crop is rice, which forms the staple diet for a large percentage of the Indian population. Because of their religious beliefs, most of the people do not eat meat. Why are there great differences in rainfall in India?** Answers will vary but should include variations in landforms, proximity to ocean, and monsoons.

 Have students compare and contrast "Indian Villages" with "*Kolkhozy* and *Sovkhozy*," on pages 202–203.

Bhutan's extreme isolation led to the idea of Shangri-La. Students may enjoy reading *Lost Horizon* by James Hilton.

Afghanistan

Afghanistan is one of the world's least developed countries. Landlocked, it is bordered by Iran on the west, Pakistan on the east and south, and the Soviet Union on the north. A narrow extension of far eastern Afghanistan, known as the Wakhan Corridor, also borders on China. Narrow mountain passes connect Afghanistan with Pakistan. The Khyber Pass has been a trade route for centuries.

All of Afghanistan is dry and mountainous. The towering Hindu Kush range cuts across northern Afghanistan. A part of the dry plateau of Iran is found in the south. The climates of Afghanistan are continental steppe and continental desert. Temperatures range from very hot in summer to below freezing in winter.

Afghanistan is a land of many tribes and many tongues. About half of its population speaks Pashto, sometimes called Pashtu, the major language of the nation. The Pashtuns have been the group with the greatest political and military power in the nation since its creation in the 1880s. About 99 percent of Afghanistan's population is Muslim.

Life in Afghanistan has changed little over the centuries. The chief occupation remains the herding of sheep and other livestock. Arable land is found along the fringes of the mountains and in a few river valleys. Cotton and sugar beets are grown, and some dried fruits are produced.

Although Afghanistan is rich in minerals, mining is, for the most part, undeveloped. A few coal mines and natural-gas wells are the only producers. Industry is limited to carpet weaving and some food processing. The largest city and capital is Kabul, in the northeast.

Great Britain and Russia began competing for control of Afghanistan during the 1800s. The British helped form a nation here in the last century as a **buffer state** between the Russians to the north and British India to the southeast. The Soviets entered Afghanistan in 1979. Thousands of Soviet troops are now stationed there to keep the Soviet-backed Communist government in power. Several Afghan resistance groups continue to attack the Soviet and Afghan government troops. The continued fighting has killed thousands and destroyed villages.

Nepal and Bhutan

Sandwiched between the giant nations of China and India are the small kingdoms of Nepal and Bhutan. Nepal is the larger of the two nations. Mount Everest, the world's highest mountain, is located on Nepal's northern border with China. Nepal's climates range from subtropical forests to ice-capped mountains. The highest population density is in the Valley of Nepal. There, the people raise wheat and corn. The valley is also the site of the capital city of Kathmandu. Most of the higher mountain areas can be used only for grazing. Rice grows on some of the lower, warmer southern slopes. Most Nepalese are subsistence farmers who practice the Hindu religion.

The Buddhist kingdom of Bhutan is a very poor and isolated country. Most of the people are subsistence farmers who grow rice, wheat, tea, and fruits.

Bangladesh

Bangladesh is located on the lower Ganges and Brahmaputra rivers. It has a coastline on the Bay of Bengal and is bordered by India on the east, north, and west. In contrast to Nepal and Bhutan, the country is almost a completely flat delta plain. The entire country has a humid tropical climate and is very dependent upon the monsoon.

Rice and jute are the most important crops grown in the rich alluvial soils. While jute is the major export, its value is declining because of competition with synthetic fibers. Bangladesh has a competitive textile industry, which is growing in importance.

The U.S. has given military aid to the Afghan resistance groups fighting against the Communist government and the Soviets.

Indian troops were invited to Sri Lanka in 1987 to help stop the terrorism of Tamil rebels that threatened to result in full-scale civil war.

Natural disasters are commonplace in Bangladesh. Especially destructive are the heavy monsoon rains, which cause flooding and tropical cyclones. Yet, the country's major challenge is overpopulation. The population growth rate of Bangladesh is greater than the growth rate of food production. Many people are malnourished most of their lives.

Since independence from Pakistan in 1971, the country has been dependent upon foreign aid and imported food. Today, Bangladesh is one of the world's poorest and most crowded nations.

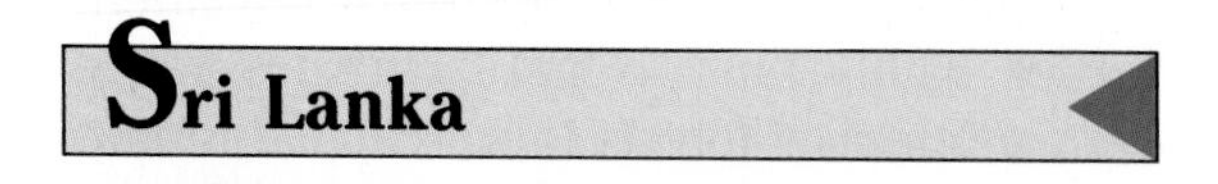

Sri Lanka

Located just off the southern tip of India, Sri Lanka is a pear-shaped tropical island in the Indian Ocean. Like India, Sri Lanka has had a democratic form of government since it received its independence from Britain.

The population of Sri Lanka is divided into two culture groups. The majority of the population is Sinhalese. The Sinhalese are Buddhist and of Aryan background. They arrived in Sri Lanka from northern India more than 2,000 years ago. About 20 percent of the population is Tamil. These Hindus migrated from southern India during recent centuries. The Tamils have complained of being discriminated against by the Sinhalese, and violence has erupted between the two groups. Some autonomy has been given to Tamils in the north and east, where they are concentrated.

Sri Lanka is influenced by the monsoons. The southwest part of the island has a humid tropical climate. The northeast half has a wet and dry tropical/subtropical climate.

The island has three landform regions. These are the low mountains in the center of the island, the fertile coastal plain, and the limestone plain in the north. Tea is grown in the mountainous interior. Along with India, Sri Lanka is a leading tea exporter. Most of the tea plantations are nationalized. The mountain area also is mined for precious gems and graphite. In addition, it supplies the nation with most of its hydroelectricity.

The coastal plain of the south and west is Sri Lanka's most populated area. Here, coconuts and rubber are the major crops. Most of the nation's rice and other food crops are also produced in this area. Colombo, the capital and largest city, is located on the west coast.

The nation's northern limestone plain is dry and less arable. New irrigation projects have helped turn some of this land into a food-growing area. Most of the minority Tamil population is concentrated here.

Sri Lanka ranks third after India and China in tea production. What other economic activities occur in Sri Lanka?

Gem and graphite mining, hydroelectric production, farming of coconuts, rubber, rice.

The Maldives

The Maldives (MAL-divez) is a group of tropical islands in the northwest Indian Ocean. Located 400 miles (640 kilometers) southwest of Sri Lanka, the nation's 2,000 tiny coral islands sit on top of an undersea mountain range. Fish products and coconuts are major exports. Tourism is the key industry.

For more than 800 years, these islands have been a Muslim **sultanate**. That is, they have been ruled by a Muslim monarch. Though never colonized, the Maldives was under British protection until 1965.

A farmer uses a buffalo to till a rice field in China's Sichuan Province. Why does rice grow so successfully in China? mild climate and summer monsoon that brings plentiful rain

method of planting is called **terracing**. The paddy is plowed, fertilized, and flooded to make it ready for the seedlings. Power for plowing is supplied by water buffalo or by small mechanical tillers. Before the harvest season, the paddies must be drained to help ripen the rice.

Southern China's other important crops include tea and cotton. China is the world's second largest producer of tea. The mountainous region between the Xi and the Chang rivers is the center of tea growing. Because most Chinese tea is used in China, the country ranks only third as a tea exporter, after India and Sri Lanka.

Cotton is the main raw material for China's textile industry. The major cotton-growing area is the Chang River valley. Since Chinese cotton has short fibers, it requires a relatively short growing season. Farmers are able to practice double cropping, alternating cotton with a food crop.

Southern China grows many other crops, including sugarcane, sweet potatoes, tobacco, peanuts, and fruits and vegetables. Silk remains important in southern China, although its role in foreign trade has lessened because of the development of artificial fibers.

Northern China

Northern China lies north of the city of Nanjing and the Qin Ling range. It includes the valley of the Huang He and the fertile North China Plain. Unlike humid southern China, northern China suffers from insufficient and uncertain rainfall. The area, with a continental steppe climate, has very cold winters, especially in the interior. Serious droughts sometimes occur. Along the more southern coastal areas of this region and on the Shandong Peninsula, the climate is humid continental. Here, winters are still severe, but rainfall is more reliable.

The Huang He begins in Tibet, flows across dry northern China, and empties into the Yellow Sea. It has changed its course many times and overflowed its banks, causing serious flood damage. For this reason, the Huang He has been called "China's sorrow." The Chinese have worked to tame the waters of this river, and it has become important for irrigation and hydroelectric power.

Deposits of fine yellowish brown soil, called loess, cover much of northern China. For this reason, the northern landscape is yellow rather than green as it is in the south. The yellow dust turns houses, clothing, and water the same shade. The Yellow Sea and the Huang He, which means "Yellow River," take their names from the yellow loess.

The continental climates of northern China influence agriculture. The short growing season and limited rain have helped make China the world's third largest wheat producer, after the Soviet Union and the United States. Yet, because of its huge population, China must import wheat, especially from the United States, Canada, and Australia. Other grains produced in this region include millet and sorghum.

Beijing is northern China's largest city, as well as China's cultural center and national capital. The city's history dates back more than 3,000 years. It has served as China's capital most of the time since the thirteenth century.

North China is much like the Great Plains of the U.S. Both regions are prone to disastrous floods and droughts.

Review the concept of heartland. Have students identify the heartland of China. (North China) Have them compare China's heartland to that of the USSR.

Beijing means "north capital." The old walled part of the city is divided into two sections, known as the Outer City and the Inner City. Within the Inner City was the Imperial City, where the government of the emperors ruled. Within the Imperial City was the Forbidden City, where the emperors lived. The city has grown beyond its original walls and is spread out across the Yellow Plain. Beijing is also a modern city with industries, subways, shops, and boulevards.

Tianjin is the second largest city and most important seaport in northern China. It is a center for the chemical, textile, and machinery industries. Most of the other large industrial cities of northern China are in the interior region near the coal and iron-ore deposits.

Northern China is the region in which Chinese culture first developed. It remains the nation's center of culture and political power. China's national language, called Mandarin, is the language of northern China. It is now spoken by a majority of people throughout China and Taiwan. More than one-half of the nation's population lives in this region.

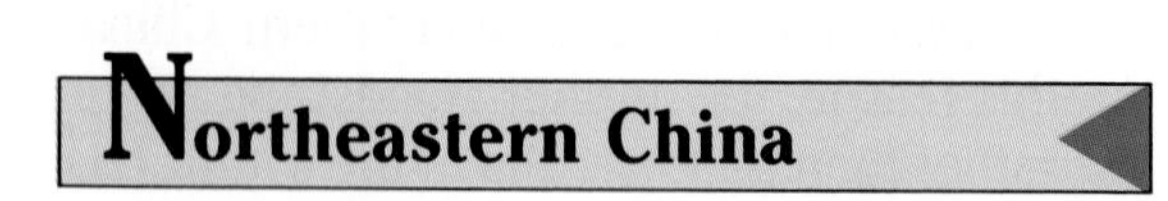

Northeastern China

Northeastern China consists of three provinces, which were once known as Manchuria, and the autonomous region of Inner Mongolia. The provinces are Heilongjiang (HAY-LUNG-jee-AHNG), Jilin, and Liaoning (lee-AU-NING). All are rich in resources.

The main landforms of the northeastern provinces are the Manchurian Plain and a surrounding rim of mountains. The region has a humid continental climate with severe winters. In the very far northern sections are forested areas of subarctic climate. The climate in these areas is similar to that of neighboring Siberia.

Most of the land in these provinces is organized into large collective farms. Because the region has a severe climate, crops are generally limited to wheat, corn, and soybeans.

Because of their mineral deposits, the northeastern provinces are among China's most important industrial regions. Oil, coal, iron ore, zinc, lead, and manganese are plentiful. The oil fields of the northeast have helped make China self-sufficient in energy. Some oil is even exported to Japan. The region also has rich forest resources. Important industries include iron and steel, chemicals, paper, textiles, food processing, and shipbuilding. The largest city and the center of the iron and steel industry is Shenyang. The standard of living in this region is higher than in most of the rest of China.

Just to the west, behind the Great Khingan Mountains, is Inner Mongolia, an autonomous region of China. The area was originally populated by Mongolians. Today, the Chinese outnumber the Mongolians many times. This dry region, at the southern fringes of the Gobi Desert, has attracted few settlers. Irrigation farming is the major activity.

Western China

Western China is made up mainly of the two large, autonomous regions of Tibet and Xinjiang (SHIN-jee-AHNG). These regions are autonomous because they have some local government and retain their own culture and language. Tibet and Xinjiang were originally populated by people who are not Han Chinese. However, Han Chinese colonists are now settling in these two regions. They probably will soon outnumber the Tibetans and other minority groups.

Western China is a rugged land of the world's highest mountains and plateaus. This region covers more than one-third of China's territory. Most of the land is too high, too dry, or too cold to support a large population. Most of the people are either nomadic herders or irrigation farmers.

Tibet occupies most of the Plateau of Tibet. It is one of the highest and most barren regions in the world. To the south, Tibet is

Level A: Ask students to assess China's chances to become a major industrial power by 2000. How might China's industrialization affect other countries, including the U.S.?

Japan's location allowed its people to borrow from China while being protected from unwanted intrusions.

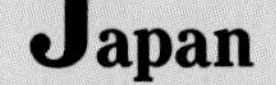

Japan

Japanese samurai warriors were hired by warring families and landowners to protect their lands and social privilege.

Historical Geography Japan's first inhabitants were the Ainu, a Caucasian people. They probably arrived from Central Asia before 5000 B.C. Today, they are a vanishing people. The Ainu were driven farther and farther north by Mongol invaders from Asia who arrived in Japan about 300 B.C. Within a few hundred years, these invaders had settled all the major islands.

Japan's oldest surviving religion is Shintoism. (See Chapter 30.) The symbol of Shintoism is a gateway called a *torii,* which is found at the entrance to Shinto shrines. In the sixth century, Buddhism was introduced into Japan from China. Although the Japanese developed a distinct culture over the centuries, much of it is related to the Chinese culture, including the Japanese writing system and the many Buddhist temples. Japanese emperors with imperial families ruled as they did in China.

Beginning in the eighth century, Japan moved away from imitating Chinese culture and developed its own political system. For hundreds of years, Japan was ruled by military leaders, or *shoguns*. They were supported by warriors, called *samurai*. Japan's island location and the shoguns' military power kept out invaders from Mongolia and China.

In the mid-1500s, the Portuguese arrived in Japan and developed a trade relationship with the Japanese. The Spanish and Dutch soon followed, resulting in the introduction of Christianity. Soon after, European traders and missionaries were either driven out of Japan or restricted to the port of Nagasaki.

From the 1600s until 1854, Japan remained isolated from the world. The arrival of American naval ships, under Commodore Matthew C. Perry, forced a commercial treaty between Japan and the United States. Many European nations soon entered into treaties with Japan. Within a few years, Japan had begun its modernization period, or Meiji Restoration. The *shoguns* were overthrown, and an emperor regained power.

Modernization took many forms. These included rapid industrialization and changes in education, law, and government. By 1890, Japan was an industrial power, had a constitution and parliament, and had won equal trade rights from Americans and Europeans.

In 1895, Japan defeated China in a war over control of Korea. In 1905, Japan defeated Russia in a war over control of Manchuria. Japan was recognized as a world power. Korea was annexed, or formally joined, to Japan in 1910. By 1920, Japan expanded its empire beyond Korea and Manchuria to include Taiwan and many of the Pacific islands.

During the Depression of the 1930s, Japanese military leaders took control of the government. Japan went to war with China in 1937. In 1940, Japan signed an alliance with Germany and Italy. Then, in 1941, it entered World War II by attacking the United States at Pearl Harbor. Japan conquered much of Southeast Asia and numerous Pacific islands.

Four years later, Japan was defeated in what proved to be an "island-hopping" war. American forces, along with Australian and British forces, slowly fought their way across the Pacific Ocean and Southeast Asia toward Japan. World War II ended, as did Japan's

The Cities of the World feature on Tokyo on pages 402–403 discusses how earthquakes have affected the city.

empire. Japanese cities and industries were destroyed by massive American bombing. The United States dropped an atomic bomb on Hiroshima and one on Nagasaki.

After World War II, Japan was occupied by United States forces until 1952. A democratic government was established in the country during this time. With United States aid, Japan again began to build itself into a major world industrial power.

Japan is a constitutional monarchy with several political parties. Members of parliament (the National Diet) and a prime minister are elected. Japan's emperor is now a symbol of the nation and has no real power.

Physical Geography Japan lies off the Pacific coast of East Asia. (See the map on page 397.) It is made up of four large "home" islands and more than 3,000 smaller ones. The four home islands from north to south are Hokkaido (hah-KIDE-oh), Honshu (HAHN-shoo), Shikoku (shi-KOH-koo), and Kyushu (kee-YOO-shoo). Honshu is the largest and most populated island. Separating the three main southern islands is the narrow Inland Sea. The Ryukyu (ree-OO-kyoo) Islands to the south are also part of Japan. Okinawa is the largest of the Ryukyus.

Mountains and hills represent 70 percent of Japan's landscape. On which island are the Japanese Alps located? Honshu

More than 70 percent of Japan is mountainous. The longest mountain range is the Japanese Alps, which form the spine of the island of Honshu. About one-fourth of Japan's 200 volcanoes are active. Mount Fuji (FOO-jee), Japan's highest peak and national symbol, is an inactive volcano.

Since Japan is situated along the Pacific "Ring of Fire," it is subject to hundreds of earthquakes each year. Where Japan is located, three large plates of the earth's crust are colliding, also causing volcanic eruptions. Thousands of people have died throughout Japan's history because of these natural disasters. Many have also been killed by large sea waves caused by earthquakes and eruptions. These waves are known as **tsunami**.

About 20 percent of Japan is plains. These are confined mainly to the coast, where the nation's agricultural and urban areas are found. Virtually all of Japan's population lives in these small coastal plain areas. Japan is a nation the size of California, with four times its population. However, the people are crowded onto one-fifth of the nation's land area.

Climate Because Japan is located at about the same latitude as the east coast of the United States, its climate is similar. Hokkaido and northern Honshu have a humid continental climate. Northern Japan's summers are cooled by the Oyashio (oh-YAH-shee-oh) Current from the north. Cold winds blowing off the Asian continent make winters in northern Japan cold, with heavy snowfall.

Southern Japan, which includes Kyushu, Shikoku, and southern Honshu, has a humid subtropical climate. There, the winters are mild, and the summers are warm and humid. The area has a long growing season. The Japanese (Kiroshio) Current, which flows from the tropical Pacific, warms southern Japan. (See the map "World Climate Regions," on page 66.) Southern Japan receives rain throughout

Refer students to the map on page 359.

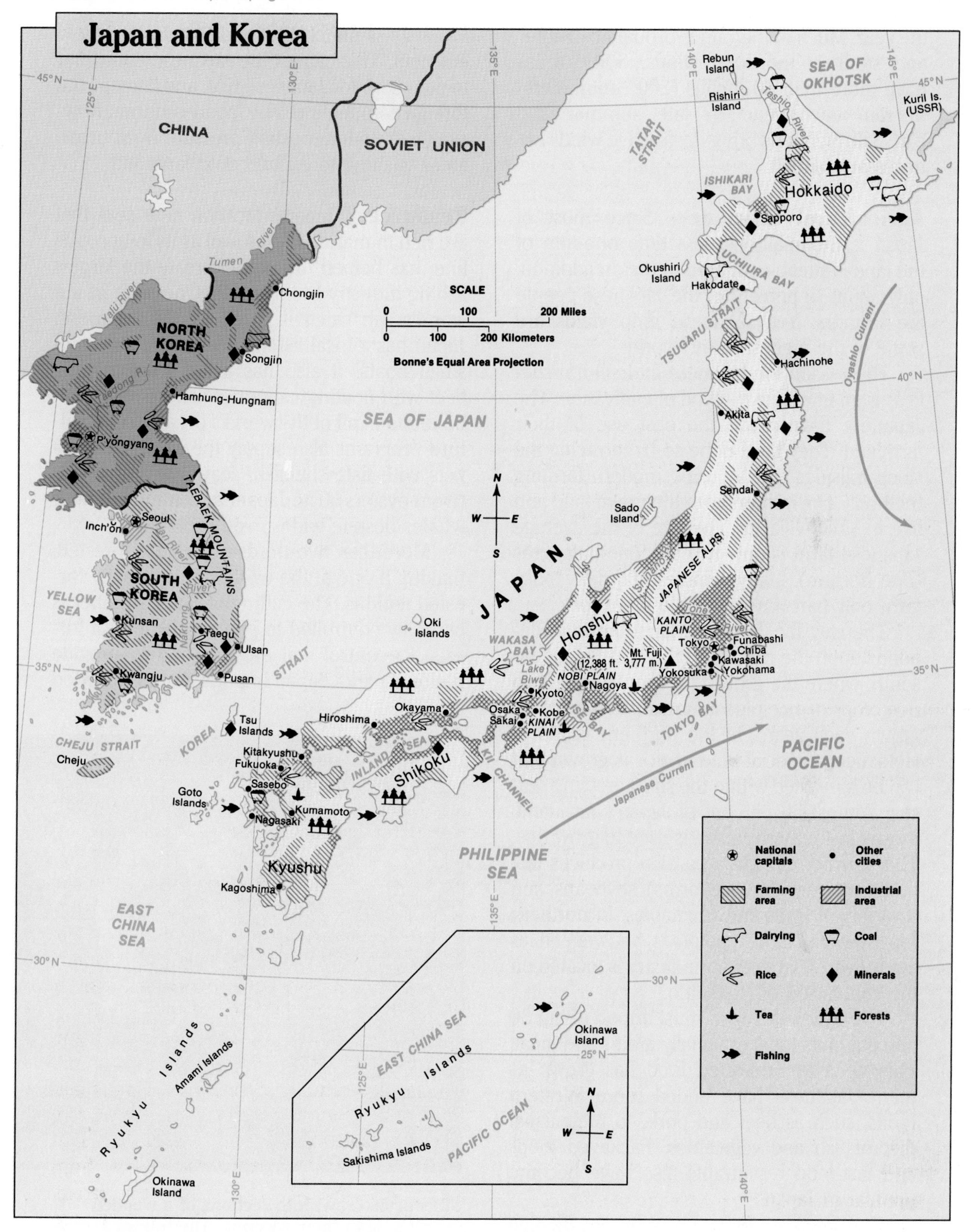

On which of the islands of Japan is the city of Tokyo located? Honshu

Level B: Have students research foods popular in Japan. How does Japanese food reflect the nation's cultural heritage and physical environment?

the year. Much of the rain is brought by winter and summer monsoon winds. Some areas have more than 80 inches (200 centimeters) of rain each year. In late summer and early fall, typhoons bring destructive winds but necessary rainfall.

Agricultural Efficiency Since most of Japan is mountainous, less than one-fifth of the land is suitable for agriculture. In addition, only about 15 percent of the Japanese people are farmers. Yet, Japanese crop yields are some of the highest in the world.

The reason for Japanese agricultural success is an intensive method of cultivation. The Japanese have made the best use of their farmland. They have done so by terracing the steep hillsides and by using modern farming methods. The farmers own their own land and live in small villages. The size of the average Japanese farm is only two and one-half acres (one hectare). Many Japanese farmers today farm only part time and hold other jobs.

The way the Japanese farm the land depends upon the climate of the area. In the far south, where the growing season is long, two rice crops or rice plus an unirrigated crop are grown each year. On the northern island of Hokkaido, a special kind of rice is grown that can be harvested within the short growing season. Almost 50 percent of Japan's farmland, mostly in southern Japan, is used to grow rice. This warmer climate area also produces tea, mulberry trees (for silkworms), soybeans, and a variety of fruits and vegetables. In northern Japan, wheat, barley, potatoes, and vegetables are grown. A large dairy industry is located on the cool island of Hokkaido.

Nevertheless, Japan must import about 30 percent of its food, especially grains and meat. This need for imported food has grown as many Japanese have added more Western foods, such as beef and pork, to their usual diet of fish and vegetables. Packaged foods and fast-food restaurants also have become popular in Japan.

The Japanese farmer is protected by high government subsidies, especially for rice. A **subsidy** is financial support given by a government. This causes the Japanese consumer to pay more for Japanese rice, and it keeps out foreign competition. The Japanese farms, however, are under constant pressure from urban areas wishing to expand onto farmland.

Resources Japan's location near seas that are rich in marine life, as well as its long coastline, has helped the nation create the largest fishing industry in the world. One-sixth of the world's fish catch is made by the Japanese. Japan has a local fishing fleet of thousands of small boats. It also has a large oceangoing fleet with floating canneries. These ships can be found in all of the world's oceans. Aquaculture programs also supply the Japanese markets with fish, shellfish, seaweed, and pearls (from oysters). The Japanese continue to hunt whales despite widespread protests.

More than two-thirds of Japan is forested, making it one of the world's most heavily forested nations. The cutting of Japan's timber is carefully controlled to prevent the loss of forests, to control soil erosion, and to provide national parks. Japan imports timber for wood

Silkworms crawl into sections of a wooden frame to spin their cocoons. The frames are placed in metal racks to give the worms air.

 Farmers are protected by the government from foreign competition. Consequently, prices for imported foods are high by American standards. Tariffs on American products such as baked goods, preserves, and ham exceed 20 percent.

Refer students to the Themes in Geography feature on the Persian Gulf on pages 270–271 to establish the connection between that region and Japan.

pulp to make paper and plywood; softwood is imported for lumber. These imports come from the United States, Canada, and the Soviet Union. Tropical hardwoods used for lumber are imported from Southeast Asia.

The largest percentage of Japan's imports are industrial raw materials and energy resources. Petroleum is its major import, mainly from the Middle East. Because Japan has small coal deposits, it must import coal from Australia and the United States. Iron ore comes mainly from Australia. Hydroelectric power plants provide about 20 percent of Japan's electricity needs; nuclear power plants provide another 25 percent. Conservation efforts have helped lower the country's dependence on imported oil.

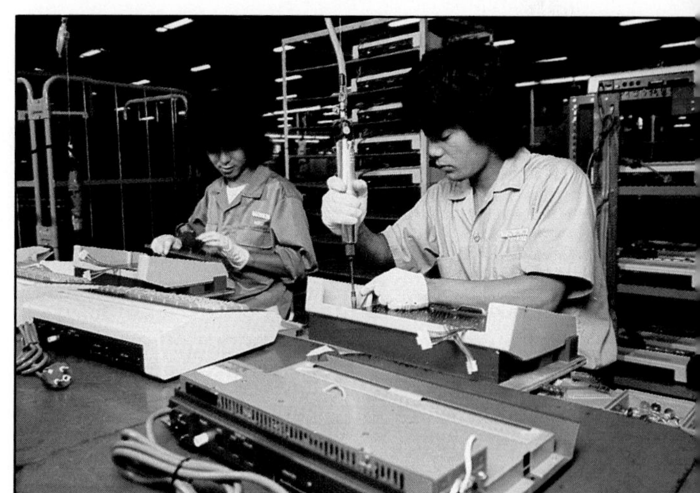

The Japanese are famous for their industrial efficiency and high technology. What factors have made Japan prominent in world trade?
shared culture, efficient industry, employee loyalty, educated people

Industrial Success Despite a tremendous dependence upon imported natural resources, Japan produces products of quality and price that are considered some of the best in the world. Today, Japan is the third largest industrial power in the world, after the United States and the Soviet Union. Its economic growth rate is the highest of the major industrialized nations. Japan ranks first in the world, surpassing the United States and Western Europe, in the production of automobiles, cameras, televisions, videocassette recorders, radios, watches, and motorcycles. The country ranks second in the production of steel, computers, telephones, and a variety of household products.

Japan is also making competitive gains on America's leadership in the computer field. At the same time, however, Japan is feeling competition from other developing Asian industrial powers, especially Taiwan, South Korea, and China. Many of Japan's older industries, such as shipbuilding and textiles, are suffering.

Many nations are looking carefully at Japan to find the reasons for Japanese technological and industrial success. Some reasons are cultural. Nearly everyone in Japan is of Japanese origin. This makes communication and the setting of common goals possible. Japanese society is efficient, and Japan's people have a strong work ethic. Japanese industry is newer and more efficiently run than many factories in the United States and Western Europe. Its workers participate in decision making in the companies that employ them. Generally, Japanese workers are loyal to their employers, and jobs are usually lifetime appointments. Though Japanese workers are unionized separately in each company, strikes are rare by Western standards. The Japanese are generally well educated and familiar with the markets, languages, and nations that buy their products.

The Japanese government works closely with industry and offers financial aid. The nation's corporations tend to have long-range goals for developing and marketing products. Its domestic market is the most competitive in the world. The Japanese are the largest consumers of Japanese industrial products. If a product can survive in Japan, it will probably compete successfully in the rest of the world. The Japanese have been able to copy Western products, upgrade and refine them, and sell them at lower prices. They are also innovative leaders in such high-technology fields as robot and computer development.

Japan's success has created problems. Japan has built up a huge trade surplus, creating

Level A: Have students use current magazine articles to research reasons for Japanese industrial success. What might the U.S. learn from Japan's example?

a **balance of payments** problem with other nations. For instance, Japanese exports to the United States are three times greater than imports from the United States. This trade imbalance could cause import nations to set up trade barriers to Japanese goods unless the Japanese import more foreign products.

Urban Industrial Core The core of Japan's cities and industry is located on three of Honshu's eastern coastal plains. This region extends southwest from Tokyo, the capital of Japan. The three areas are the Tokyo Bay area, the Kansai region, and the Nagoya area.

The Tokyo Bay area is one of the world's largest industrial and metropolitan centers. Tokyo, the nation's largest city, is Japan's center of communications, government, banking, education, and trade. It is the site of the Ginza, the world's largest and busiest shopping district. Yokohama (yoh-kuh-HAHM-uh), located just south of Tokyo, is one of the world's busiest seaports.

The Kansai region includes three large and important cities. Osaka (oh-SAHK-uh) is Japan's second largest city and industrial center. Kobe (KOH-bee) is its nearby port and also an industrial area. Kyoto (kee-OHT-oh), the ancient capital, keeps the traditions of old Japan alive by the production of handicrafts. The city's many temples attract large numbers of visitors. The Nagoya region is on Ise Bay. Nagoya is its major city and industrial center.

The port of Yokohama is located in Tokyo Bay. Japan exports large quantities of manufactured goods in exchange for raw materials.

Japan's urban areas are faced with housing shortages, air pollution, water shortages, and traffic congestion. Unlike in large American cities, however, crime rates are generally very low in Japan's cities.

Way of Life The average middle-class home in Japan is a small one-story house. Sliding wooden screens take the place of doors and windows. The rooms contain little furniture, yet 98 percent of the homes have a color television. People generally sit on cushions or mats placed on the floor. Families are used to living in close quarters. Most children do not have their own bedroom. Before entering the house, the Japanese remove their shoes to avoid dirtying the straw mats that cover the floor. Many Japanese families now also have an automobile.

As in other parts of the Orient, rice is the main food in Japan. Other foods include vegetables, seafood, and small amounts of meat. Tea is the usual beverage. Cereals, meat, dairy products, and fruits are imported mainly from the United States.

In the urban areas, Western clothing has largely replaced the traditional kimono. Business suits, blue jeans, and other Western fashions are common in the cities. Popular sports are baseball, golf, and skiing. Rock music and Western soft drinks are popular. The rural areas, however, remain traditional.

Japanese society has undergone other changes too. One is that the birthrate in Japan is now as low as that of Western Europe. The other is that Japanese women are becoming more educated and urbanized. They are seeking a larger role in the society and the work force. However, the Japanese have kept many of their traditional ways. Japanese poetry, music, painting, and theater are enjoyed by everyone, and traditional handicrafts are still

 Both Japan and India are densely populated countries, yet the contrasts between the two nations are sharp. Review pages 372–378 to have students compare and contrast these two countries of the Orient.

Have students locate Southeast Asia on the map on page 359.

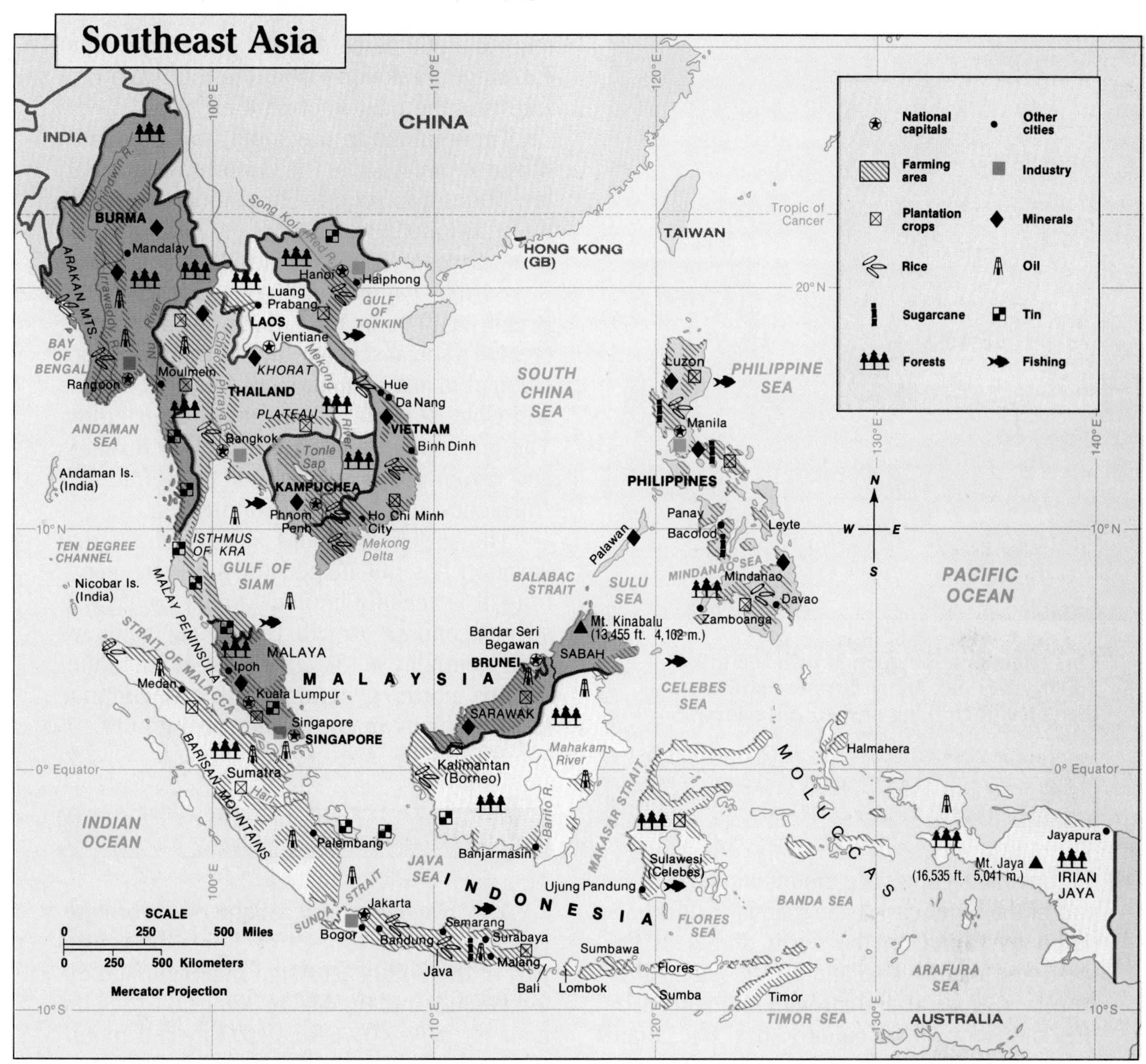

Which countries share the Malay Peninsula? What economic activities occur on the peninsula?

Singapore, Malaysia, Thailand; fishing, forestry, mining, farming, industry (including oil), plantation agriculture

Southeast Asia has large areas of valuable tropical forest, which contain sought-after hardwoods, such as mahogany, ebony, and teak. Large areas of these forests are being cut for timber. In most cases, there is little concern for future regrowth or for wildlife. In addition, population growth will probably lead to further forest clearing.

Southeast Asia has important mineral deposits. Tin and oil are by far the most valuable. The world's major tin deposits are found in Malaysia, Thailand, and Indonesia. Other important minerals include tungsten, iron ore, and manganese.

Coal is found in Vietnam and Indonesia. Petroleum is exported by Indonesia, Malaysia, and Brunei. Thailand has natural-gas deposits in the Gulf of Siam. The greatest promise for more oil and gas in Southeast Asia lies offshore. Indonesia has developed the most offshore oil fields, although several other nations have good potential supplies.

This pagoda in Rangoon is believed to be 2,500 years old. Many Burmese still wear the colorful clothing of their ancestors.

Burma

Burma occupies the mountainous western part of the Southeast Asian mainland. It is bordered by China on the north, India on the west, and mainly Thailand on the east. All of Burma is tropical and strongly influenced by the monsoon. The country has two main north-south river valleys, the Irrawaddy and the Nu. Most of Burma is a series of mountain ranges and plateaus. Its coast is mainly along the Bay of Bengal. A portion of Burma extends southward to the Isthmus of Kra along the Andaman Sea.

Burmans form more than 70 percent of the population. They are Buddhists with a culture rich in art, architecture, and literature. Many minority peoples occupy the mountains. Most of the population is crowded into the valley of the Irrawaddy River. The largest city is Rangoon, a seaport and the capital.

Rice, Burma's chief crop, is grown on the river floodplains. Rice exports account for more than 60 percent of the country's foreign exchange. Teakwood from the mountain rain forests is the next important export.

Tin is mined in the south, and oil is produced for local use in the lower Irrawaddy Valley. Mines also produce lead, zinc, silver, and the gemstone jade. Offshore gas deposits have been discovered but not developed.

Burma gained its independence from Britain in 1948. In 1962, Burma's government created a socialist economy with complete isolation from foreign influences. Resident Indians and Chinese were driven out of the country. The government took control of all business and commerce, stressing total economic self-sufficiency and Burmese culture.

The government has since realized that isolation has not helped the Burmese economy. It is seeking limited foreign contacts to speed economic growth. Burma still maintains strict neutrality in foreign affairs. In the mountainous north, government troops continue to battle many minority and rebel groups.

Thailand

The Kingdom of Thailand was formerly called Siam (sie-AM). It occupies the central part of the Southeast Asian mainland and extends south into the Malay Peninsula. (See the map on page 409.) Rugged, forested mountains form Thailand's borders with China and Burma in the north and west. The Mekong River forms its eastern boundary with Laos.

The Chao Phraya River valley, in the central part of Thailand, is the most populated and productive part of the nation. The Khorat Plateau forms the northeastern portion of the country. The climate is mostly wet and dry tropical/subtropical and is driest on the Khorat Plateau. In the south, on the narrow Malay Peninsula, the climate is humid tropical.

Thailand is the world's leading rice exporter. Sugarcane, corn, and rubber are also export crops. Many food exports go to Japan, Thailand's major trade partner. Forests

 Level A: Have students use a physical map of Thailand to suggest geographic factors that contributed to its avoidance of foreign domination.

You may want students to review the map of world religions on page 91.

provide valuable tropical hardwoods, but forest loss has caused ecological problems. Thailand is also a major fish exporter.

Thailand's mineral production includes tin, tungsten, and lead. The country also has oil resources. Large offshore natural-gas fields have been developed. A pipeline under the Gulf of Siam delivers the gas to Thailand.

Thailand has recently broadened its economy with the help of foreign investment. There has been rapid growth in textiles, electronics, and automobile assembly. There are plans for hydroelectric power plants in the north, which will provide needed electricity and irrigation water. Tourism is a major industry and is the nation's largest source of foreign income.

More than 75 percent of the population is made up of Thais, who are Buddhist. The remainder includes urban Chinese, Lao, and many mountain tribes. While most Thais live in villages, the trend is toward the cities. Bangkok is Thailand's largest city and capital. It is built on the Chao Phraya. Much of the original city is connected by canals, called *klongs*. Bangkok is famous for its *klongs,* palaces, and Buddhist monasteries. In recent years, it has become known for its traffic jams and air pollution. Also, as increasing water supplies are drawn out of wells, the city is slowly sinking.

The *klongs* of Bangkok are filled with floating markets and houseboats. More than 5 million people live in crowded Bangkok.

Thailand means "land of the free." It is the only Southeast Asian country that was not a European colony. Thailand is a constitutional monarchy with a king. However, in recent history, political power has been mainly in the hands of military leaders.

On the eastern border areas of Thailand are thousands of refugees from Kampuchea, Laos, and Vietnam. This border region has also been an area of military tension between Thailand and its neighboring countries of Kampuchea and Laos.

Kampuchea

Formerly known as Cambodia, Kampuchea is located between Vietnam and Thailand. It is the successor of the Khmer empire. (See the map on page 409.)

The majority of Kampuchea's population is found in the country's fertile, rice-growing region along the Mekong River. Most of the remainder of the country is sparsely populated. Located in the northwest is Tonle Sap, a large lake that supports a fishery. The capital is Phnom Penh, located in the rice-growing region of the south.

After independence from France in 1953, Kampuchea experienced political unrest. During the Vietnam War, the United States bombed North Vietnamese supply lines in Kampuchea. In 1975, Communist Khmer Rouge forces seized power. The Khmer Rouge government isolated Kampuchea. It launched a program to establish a "Khmer peasant nation." Cities were emptied, the educated were executed, families were separated, and citizens were forced into field labor. More than one million Kampucheans were killed.

In 1978, Vietnamese forces invaded Kampuchea. They overthrew the Khmer Rouge and set up a government friendly to Vietnam.

Chinese settlements are scattered throughout Southeast Asia. Majority groups in some countries resent the presence of the Chinese and have at times sought to drive them out. This has happened in Burma, Vietnam, and Malaysia.

Level A: Have students create a time line of the recent history of Indochina. Tell them to include details of U.S. involvement in the region.

Today, the nation is trying to rebuild after years of warfare. Thousands of Kampucheans remain in refugee camps in Thailand. Vietnamese troops are stationed in Kampuchea to prevent various guerrilla forces from taking over. Most trade is with Vietnam; most aid is from the Soviet Union.

Laos

A landlocked, mountainous nation, Laos is one of the poorest countries in the world. It was first unified in the fourteenth century under a Lao prince. From the 1890s until 1954, Laos was under French rule.

Although officially neutral, Laos served as a major supply route for the North Vietnamese during the Vietnam War. The country was heavily bombed by the United States in an attempt to block this route. In 1975, the Communist Pathet Lao forces took over the monarchy and formed the People's Democratic Republic of Laos.

The most important crop of Laos is rice. Tobacco, corn, cotton, and coffee are also grown. Electric power exported to neighboring Thailand is the major source of income for the nation. Tin and lumber are other exports. There is almost no industry, and the country requires foreign aid.

Half of the population is made up of ethnic Lao, who are closely related to the Thai people. The Lao are Buddhists and live mainly in rural valleys, especially along the Mekong River. The remainder of the population is made up of highlanders and isolated mountain tribes, such as the Hmong and Mien peoples. Actually, Laos has 68 ethnic groups. The capital and only large city is Vientiane.

The main problem the nation faces is its low standard of living. The government, though Communist, is less strict than the governments of Vietnam and Kampuchea. Private businesses and religious freedom exist. Since the Vietnam War, more than 10 percent of the Laotian population has fled the country. Many now live in the United States and Australia. Thousands more remain in refugee camps in Thailand. Within Laos are Vietnamese troops and many Soviet advisors.

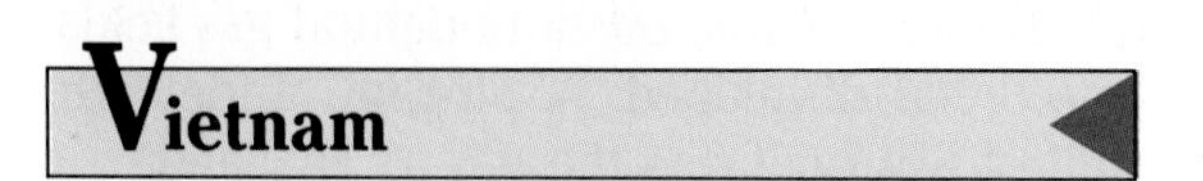

Vietnam

Vietnam is a long, narrow country that occupies the eastern portion of the Indochina Peninsula. (See the map on page 409.) The east coast is on the South China Sea. Laos and Kampuchea form its western border, and China is to the north.

The Vietnamese make up more than 85 percent of the country's population. They arrived from China more than 2,000 years ago. The largest minority groups are the Chinese, who live in the cities, and Montagnards, who live in the mountains. Vietnam, along with Laos and Cambodia (now called Kampuchea), were all part of French Indochina. When the French returned to Southeast Asia after World War II, they found that Communist forces (the Vietminh) had gained power in the north. The Vietminh, under the leadership of Ho Chi Minh, demanded independence.

Traditional and modern means of transportation exist in Hanoi. Why is Vietnam's economy underdeveloped?

The war destroyed the economy.

The U.S. supported the French and later the South Vietnamese out of fear of a Communist take-over. It was believed that the rest of Indochina would become Communist if Vietnam did so.

After eight years of fighting, the French were defeated in the north. In 1954, Vietnam was divided along the 17th parallel. North Vietnam established a Communist regime, with its capital at Hanoi (ha-NOY). South Vietnam established a government friendly to the United States, with its capital at Saigon (sie-GAHN) (now called Ho Chi Minh City). Two separate Vietnams did not work for long. Each government wanted reunification on its own terms. North Vietnam supported a guerrilla war by the Viet Cong in South Vietnam, and North Vietnamese troops invaded in 1963.

The United States supported the South Vietnamese. The Vietnam War became the longest war ever fought by the United States. More than 50,000 Americans and more than two million Vietnamese were killed. After United States troops were withdrawn, the South Vietnamese government collapsed. In 1975, Saigon was taken by North Vietnamese forces. Saigon was renamed Ho Chi Minh City, and the country was officially united as the Socialist Republic of Vietnam.

Hundreds of thousands of South Vietnamese tried to escape the Communist take-over. They became refugees and "boat people" seeking a new homeland. Many found their way to the United States, Australia, and other Western nations. Many people in Vietnam are still trying to leave the country.

In times of peace, Vietnam is a rich area. In the north, the Song Koi delta is a fertile agricultural region, and there are coal and mineral resources. The south has the great rice fields of the Mekong Delta as well as rubber plantations. The war destroyed the economy; decades later, Vietnam is still not a productive nation. The Communist government is trying to increase food production and rebuild industries.

Vietnam is now struggling to maintain its economy with the help of Soviet aid. It has a continued fear of its traditional enemy, China. The two nations have fought border conflicts in recent years. Vietnam also supports a huge military force that occupies both Laos and Kampuchea.

Today, Vietnam is one of the Orient's poorest nations. Though changes are coming slowly here, Vietnam has the potential to become a strong, self-sufficient nation.

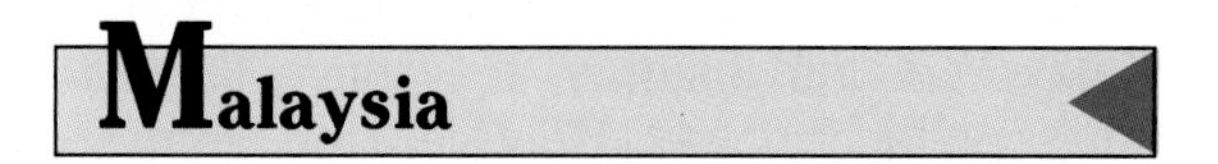

Malaysia

Malaysia is separated into two parts by the South China Sea. Western Malaysia occupies the Malay Peninsula. Eastern Malaysia is located on the northern part of the island of Borneo. (See the map on page 409.)

Malaya was governed by the British until 1957. In 1963, it was joined with Sarawak, Sabah (North Borneo), and Singapore to form a new country called the Federation of Malaysia. Singapore left the federation in 1965 and became an independent country.

All of Malaysia lies near the equator and has a humid tropical climate. The interior of the country is forested and mountainous. The greatest population concentration is along the Malay Peninsula's western coastal plain. That is where Malaysia's capital city of Kuala Lumpur (kwahl-uh LUM-pur) is located.

Kuala Lumpur holds one of Malaysia's largest Muslim populations. The city's population has more than doubled since 1970.

Ask students what effect maintaining large military forces in Laos and Kampuchea has on the Vietnamese economy.

About half the population is Malay who are Muslim. The largest minority groups are the Chinese, who make up one-third of the population, and Indians, who comprise one-tenth. In Eastern Malaysia, there are many tribal people, known as Dayaks. Since the Malays are the majority, they have the nation's political power. However, the Chinese hold the financial power of the country.

Malaysia is rich in natural resources. Its main exports are rubber, palm oil, timber, and tin. Malaysia is a world leader in all four of these commodities. Rubber plantations and tin mines are located mainly on the Malay Peninsula. Eastern Malaysia depends heavily on forestry and petroleum. Oil and gas deposits are also located in the Strait of Malacca.

Industrialization has occurred along the west coast cities of the peninsula. The major products are electronics, shoes, and textiles. Though Malaysia is fortunate in natural resources and in attracting industries, the nation faces challenges. These include removing ethnic tension between the Malay and the Chinese and increasing food production.

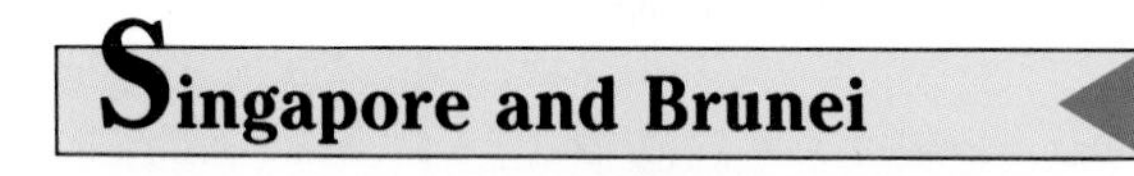

Singapore and Brunei

The two smallest nations in Southeast Asia are also the richest. Singapore is situated just off the southern tip of the Malay Peninsula. Brunei (broo-NIE) is located on the north coast of the island of Borneo. (See the map on page 409.)

Singapore is the name of the island, the city, and the independent country. It owes its prosperity to its geographic location and productive population. The port faces the busy Strait of Malacca. More than 75 percent of Singapore's population is Chinese. The minority groups are Malay and Indian.

Singapore's harbor is one of the busiest in the world. Singapore is the center of trade and banking for Southeast Asia. Its industries produce refined-oil products, ships, textiles, electronics, and many other goods.

Singapore is a modern city whose high-rise hotels, banks, and shopping malls dominate the skyline. Often called the "cleanest city in the world," Singapore fines people caught littering. Though the Republic of Singapore is strict, the population is generally content, orderly, prosperous, and well educated.

Brunei is an independent, oil-rich nation. The Sultanate of Brunei, its official name, is a Muslim country. About 75 percent of the population is Malay, while the remaining 25 percent is Chinese.

Virtually the entire economy of Brunei is based on petroleum exports, especially to Japan. All education and health care and some housing are provided to citizens. With about 30 years of oil supplies left, the government has wisely invested for the future. The government also seeks to diversify the economy in agriculture, forestry, and business.

Most of Singapore's people are Chinese, and Chinese is an official language of Singapore. How far is Singapore from China?

about 1,500 miles (2,400 kilometers)

Level A: Ask students to compile a Singapore information packet. It should contain details on cultural life, climate, local customs, money, food, tourist attractions, industrial opportunities, and so on.

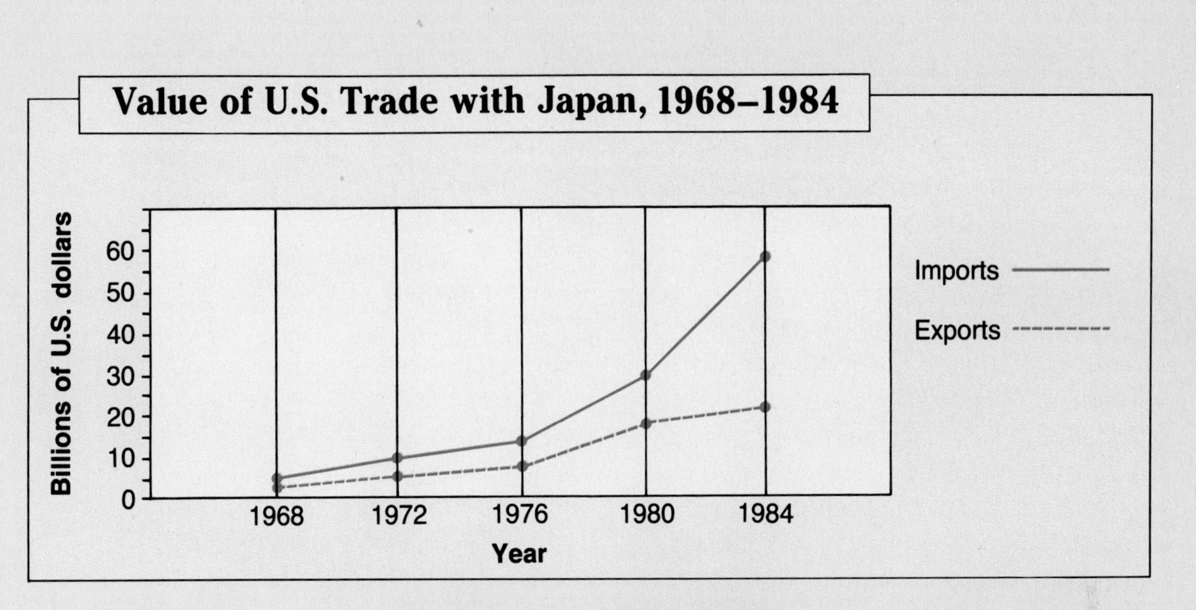

Applying and Extending

One way to find recent information about areas of the world is through the *Reader's Guide to Periodical Literature*. The *Reader's Guide* is an index of magazine articles related to a wide variety of topics. Use the *Reader's Guide* in your library to develop a bibliography of recent articles about one country in this unit. Read at least one of the articles, and summarize it.

Linking Geography and Art

Chinese art dates back to around 4000 B.C. Many paintings have been discovered on the walls of tombs in China. During four great periods in China's history, many forms of art developed and flourished. Research the development of the arts in China during the following periods: (1) the Han dynasty (202 B.C.–A.D. 220); (2) the T'ang dynasty (618–907); (3) the Sung dynasty (960–1279); and (4) the Ming dynasty (1368–1644).

Reading for Enrichment

Fairservis, Walter A. *Asia: Traditions and Treasures*. New York: Harry N. Abrams, 1981.

Hardgrave, Robert L. *India: Government and Politics in a Developing Nation*. 3rd ed. Orlando, FL: Harcourt Brace Jovanovich, 1980.

Lawson, Don. *The United States in the Vietnam War*. New York: Harper & Row, 1981.

Reischauer, Edwin O. *The Japanese*. Cambridge, MA: Harvard University Press, 1981.

Roberson, John R. *China: From Manchu to Mao (1699–1976)*. New York: Atheneum Press, 1980.

More than 25,000 islands are found in the Pacific region. Vegetation on the islands is generally dense, and the weather is warm year-round.

Soccer is one of Australia's most popular team sports. The nation's professional sports teams have large, enthusiastic followings.

Sheep graze on New Zealand's South Island. There are about 20 sheep per person in New Zealand and 10 sheep per person in Australia. Wool and wool products make up a large part of these nations' exports.

Have students identify the aspects of physical, economic, and cultural geography found in these photos.

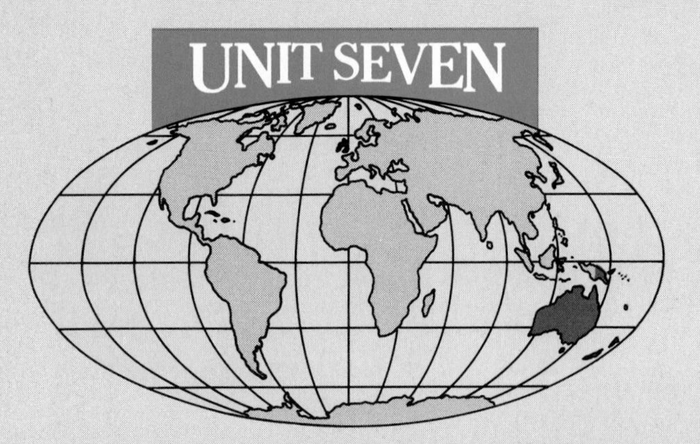

THE PACIFIC WORLD

OBJECTIVES

- ▷ *To describe how the climate and landforms of Australia have influenced where and how Australians live*
- ▷ *To explain how Australia has benefited from its vast wealth of natural resources*
- ▷ *To describe the interaction of native culture groups in Australia, New Zealand, and the South Pacific Islands with European culture*
- ▷ *To describe how the economy of New Zealand is dependent upon agriculture and foreign trade*
- ▶ *To interpret a world time-zone map*

Chapter 35

Chapter 35 Lesson Plans and Planning Guide are found in the Teacher's Guide.

Refer students to the map on page 428.

Australia

▲ **Many Australian farmers live in the Outback.**

▲ **Long summers make Sydney's beaches popular.**

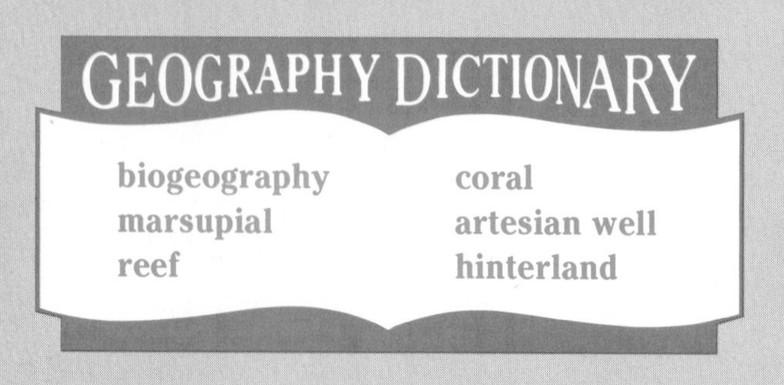

Australia, the only continent that is a single nation, is located in the Southern Hemisphere between the Indian and South Pacific oceans. Because it is situated south of the equator, Australia is sometimes called the "Land Down Under." An isolated country, its closest neighbors are New Zealand, New Guinea, and Indonesia.

While Australia is the smallest continent, it is the sixth largest country in the world in land area. With about 16 million people, however, its population is small. Most of the people are concentrated in a few large coastal cities. The rest of the continent, known as the Outback, is dry, flat, and almost uninhabited.

Because Australia is not connected to any other continent, it is also sometimes referred to as the island continent. This separation and isolation from other continents has contributed to the unique animal and plant life that is found there.

 Note physical characteristics of Australia in the photos in this chapter.

Point out that the British used America as a penal colony before 1783.

Historical Geography

The first inhabitants of Australia are called Aborigines. They were originally a nomadic hunting-and-gathering people, believed to have come to Australia from Southeast Asia more than 40,000 years ago. When the first European settlers arrived during the eighteenth century, there were about 300,000 Aborigines in Australia. Many died from diseases brought to the continent by Europeans. Today, only about one percent of the population is made up of Aborigines.

Explorer James Cook claimed Australia for Great Britain in 1770. The British settled Australia as a prison colony. The first colony was established at Sydney in 1788. Soon after, other colonies were set up at Hobart, Melbourne, and Brisbane. Adelaide was the first colony established by free settlers.

Australia gained economic importance during the 1850s when it became a leading wool producer and when gold was discovered. After 1860, other Europeans, including Germans, Scandinavians, Greeks, and Italians, settled in Australia. In 1901, the Commonwealth of Australia was formed from the six original colonies. Today, Australia comprises six states and two federal territories. The states are New South Wales, Victoria, Queensland, South Australia, Western Australia, and the island state of Tasmania. The two territories are the Northern Territory and the Australian Capital Territory, in which Canberra, the national capital, is located.

While it is a member of the British Commonwealth and recognizes the British monarchy, Australia's Prime Minister is the country's most important political figure. Australia also governs several small islands in the South Pacific and Indian oceans.

For the Record

An avenue of English trees, about 200 yards long, rises steeply up from the Penitentiary to the church. This avenue needs some judicious pruning; it has not been kept in order and the view of the church at the end is obscured by intervening foliage. As we walked along I thought of the thousands of hopeless men who were once marched this way, in clanking chains, to listen to words of peace and goodwill, with guards armed with guns watching their every movement.

From *Wanderings in Tasmania* by George Porter, pages 129–130.

Most of Australia's minerals lie in the dry inland areas far from major settlements. This mining town is in Australia's interior.

Physical Geography

Landform Regions Australia is the flattest and lowest continent. It can be divided into three major landform regions: the Western Plateau, the Central Plains, and the Eastern Highlands. (See the map on page 425.)

The Western Plateau covers more than one-half of the continent, including Western Australia, the Northern Territory, and most of South Australia. The Nullarbor Plain, a treeless plateau, extends along the southern edge of the region. The Western Plateau's flat landscape is broken by a few hilly regions. The largest of these are the Macdonnell Ranges in the Northern Territory, the Musgrave Ranges in South Australia, and the Darling, Hamersley,

Have students read For the Record and then analyze the mood of Porter's writing. What term does he use to describe the men? Why might he have used that term?

Ayers Rock is one of Australia's most unusual landforms. It is nearly 1.5 miles (2.4 kilometers) long.

and Kimberley ranges in Western Australia. Ayers Rock in the Macdonnell Ranges is the most famous landform in Australia.

To the east of the Western Plateau are the Central Plains, which extend from the Gulf of Carpentaria in the north to the Great Australian Bight in the south. They include interior Queensland and New South Wales and adjoining areas of Victoria and South Australia. The area around the dry Lake Eyre (A[UH]R) in South Australia is below sea level. Only in the far south is the region's flatness broken by the Flinders and Mount Lofty ranges.

Australia's highest elevations are found in the Eastern Highlands, sometimes called the Great Dividing Range. The Great Dividing Range extends from Cape York in the north, along the country's east coast, to Victoria in the south. The mountains appear again on the island state of Tasmania. The highest portion of the Great Dividing Range, called the Australian Alps, is located on the Victoria and New South Wales border.

Have students locate the Pacific world on the inset map on page 425.

Biogeography The study of the geographic distribution of plants and animals is called **biogeography**. Many of Australia's plants and animals are found nowhere else. The country is known for its **marsupials**. Marsupials are mammals that carry their young in pouches. The best known of these are the kangaroo and koala. Australia also is the home of the platypus and spiny anteater, the only egg-laying mammals on Earth. The plants of Australia are unusual in that they are dominated by a single plant variety, the eucalyptus. More than 90 percent of the trees in Australia are some variety of eucalyptus.

In the Coral Sea, off the northeastern coast of Queensland, is the Great Barrier Reef. More than 1,250 miles (2,000 kilometers) long, it is the world's largest coral **reef**. A reef is a ridge of **coral**, rock, or sand lying at or near the surface of the water. Coral is a rocky material (limestone) formed of the hard skeletons of the coral polyp. Coral polyps are tiny marine animals, usually smaller than the tip of a ballpoint pen. They live in colonies that are anchored to the shallow ocean floor. As old coral polyps die, their limestone skeletons serve as a platform on which new corals grow.

The Great Barrier Reef extends for more than 1,250 miles (2,000 kilometers) along Australia's northeast coast.

 Level A: Students can research the story of the first explorers to cross Australia, Robert O'Hara Burke and W. I. Willis.

CHAPTER 36 CHECK

Reviewing the Main Ideas

1. There are more than 25,000 islands in the Pacific Ocean. Some are small and uninhabited, while others, such as New Guinea and New Zealand, are large and populated.
2. The island cultures have been influenced by European and Asian explorers and settlers. The introduction of farming, tourism, and military bases has had a major impact on the region.
3. The Pacific islands are divided into three regions: Melanesia, Micronesia, and Polynesia. New Zealand comprises two large islands—North Island and South Island—as well as several smaller islands. New Zealand is a rugged, mountainous country with a maritime climate.
4. New Zealand's exports include sheep's wool and meat, dairy products, beef, apples, and kiwifruit.
5. More than 75 percent of New Zealand's population lives on North Island; 85 percent of New Zealanders live in urban areas.

Building a Vocabulary

1. What is the source of geothermal activity? Where in New Zealand does it occur?
2. What distinguishes an oceanic island from a continental island? Give two examples of each from this chapter.
3. Define *atoll*. Are the soils of atolls agriculturally productive?

Recalling and Reviewing

1. How is the sheep industry important to New Zealand?
2. Which countries are New Zealand's major trading partners?
3. Name the three main Pacific island groups.
4. How do the trade winds affect the climate of many of the Pacific islands?

Critical Thinking

5. New Zealand must import petroleum. Is this dependence upon foreign oil a problem? Why or why not? How would you suggest that New Zealand reduce its dependence on foreign oil?
6. Both cultural diffusion and cultural isolation characterize the Pacific islands. Explain how both are possible in the same region and how geographic factors are involved.

Using Geography Skills

Use the maps in this chapter to answer the following questions.

1. Name the capital of New Zealand. Is it on North Island or South Island? What are the chief economic activities around this city?
2. Where are the Southern Alps located?

Answers to Chapter Check questions are found in the Teacher's Guide.

Additional activities for Chapter 36 are found in the Student Workbook. Chapter 36 Test as well as Skills, Reteaching, Critical Thinking, and Challenge/Enrichment activities are available in the Teacher's Resource Binder.

UNIT 7 REVIEW

Summary

Australia, New Zealand, and the islands of the South Pacific are no longer isolated. The economies of Australia and New Zealand are important to world trade. The value of the Pacific islands for forest products, minerals, and tourism has increased.

Wool and meat are very important to the economies of Australia and New Zealand. The economy of New Zealand is still based mainly on agriculture. Australia, however, has become a major industrial and mining nation. The traditional trading links of Australia and New Zealand with Western Europe have weakened since the formation of the European Community. The major trading partners of Australia and New Zealand are now the nations of the Orient and the United States. Australia and New Zealand are developing new products and seeking new markets.

Most of the islands of the Pacific still depend heavily on coconut products for their incomes. However, mining and the development of tourism are bringing both prosperity and problems to some islands. Many changes are taking place in the Pacific world; however, the people of the Pacific are adapting their lives to these changes.

Reading and Understanding

1. What effects did European settlement have on the Aborigines and the Maoris?
2. Why did only small numbers of Europeans settle in Australia and New Zealand before the 1860s?
3. Why do most Australians live along the coasts?
4. How does the Great Dividing Range affect Australia's climate? How do the Southern Alps affect the climate of New Zealand?
5. Why are the Murray-Darling system and the Great Artesian Basin important?
6. How is agriculture in Australia similar to and different from agriculture in New Zealand?
7. What factors have allowed Australia to be almost self-sufficient in food production?
8. Why has Australia industrialized more extensively than New Zealand?
9. Why do hot springs and geysers occur on New Zealand's North Island?
10. What changes did Europeans bring to agriculture in the Pacific islands? How were the island people affected by these changes?
11. What are the primary economic activities of the Pacific islands? Why do many of the islands need economic aid?
12. Complete this sentence with the phrase(s) that define the term: An atoll is (a) an island that forms as a result of tectonic forces; (b) the economic association of the Pacific Islands; (c) a small coral island usually in the shape of a ring; (d) a tropical wind that brings heavy rains to the South Pacific.
13. What is the difference between a high island and a low island?
14. To which major Pacific island group does New Zealand belong? How is this reflected in the ethnic heritage of New Zealanders?

 Answers to Unit Review questions are found in the Teacher's Guide.

Mastering Geography Skills

Use the time-zone map in the Geography Skills feature on page 439 to answer the following questions.

1. How many time zones are there on the continent of Australia?
2. Through what continents does the prime meridian pass?
3. If you are traveling west, do you set your watch ahead or back each hour?
4. If it is 5 P.M. in Moscow, what time is it in Washington, D.C.?

Applying and Extending

Use a world almanac, *The Statesman's Year-Book,* or another reference source to research three of these Pacific island nations: Fiji, Kiribati, the Solomon Islands, Tonga, Tuvalu, Vanuatu, and Nauru. For each nation, gather the data in chart form with these headings: Island Group, Number of Islands, Land Area, Population, Ethnic Composition, Date of Independence, Former Colonial Power, Form of Government, Export Crops, and Industries.

Use this data to determine which island nation is most densely populated and which has the most diversified economy. How is the kind of island included in each nation related to the country's economy? What evidence is there in each country's government of the influence of non-Pacific countries?

Linking Geography and Culture

All societies depend on three types of resources to satisfy their human wants. These resources are land (natural resources), labor (human resources), and capital (tools and equipment used to produce goods).

Investigate the traditional society of the Aborigines and the modern society of Australia to determine what land, labor, and capital each uses to satisfy its wants. Write an essay comparing and contrasting the wants and resources in traditional and modern societies. Consider the following points as you plan your essay: differences in the wants of a traditional society and those of a modern society, natural resources accessible to both societies, resources that can be used only by a modern society, and ways in which the tools of modern and traditional societies differ.

Reading for Enrichment

Allen, Oliver E. *The Pacific Navigators*. New York: Time, 1980.

Cornelia, Elizabeth. *Australia: The Land and Its People*. Morristown, NJ: Silver Burdett, 1978.

Johnston, Ronald J. *The New Zealanders: How They Live and Work*. Westport, CT: Praeger Publications, 1976.

Rau, Margaret. *Red Earth, Blue Sky: The Australian Outback*. New York: Harper & Row, 1981.

Terrill, Ross. *The Australians*. New York: Simon & Shuster, 1987.

The Unit 7 Test is available in the Teacher's Resource Binder.

Iguaçu Falls, which forms part of the border between Brazil and Argentina, is about 2 miles (3.2 kilometers) wide and plunges 237 feet (72 meters).

The eastern coast of Honduras is sparsely populated. However, the nation's population is increasing at a rate ot over 3 percent a year. Many other Latin American countries are also experiencing rapid population growth.

Many Indians in Latin America practice subsistence agriculture. These small farms near Otavalo, Ecuador, are typical of this region.

Have students compare the photos here with those found on pages 186–187 and 240–241. Ask them to note differences or similarities among the three regions.

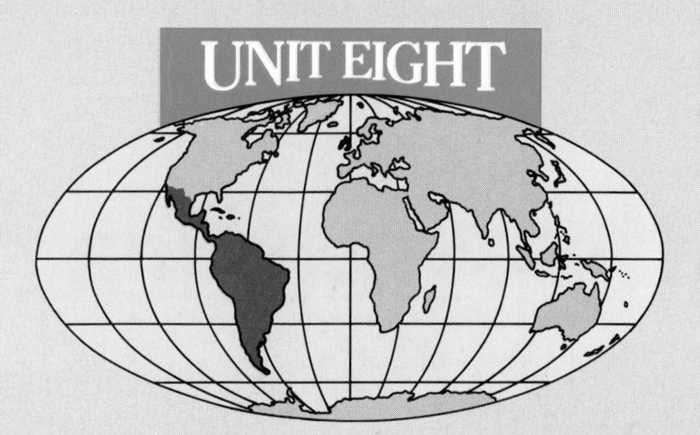

LATIN AMERICA

OBJECTIVES

- ▷ *To describe how the climate and landforms of Latin America have influenced where and how Latin Americans live*
- ▷ *To explain how the development of natural resources may affect the economies of Latin American countries in the future*
- ▷ *To describe the political, economic, social, and cultural issues of Latin America*
- ▶ *To interpret a cartogram*

Chapter 37

Chapter 37 Lesson Plans and Planning Guide are found in the Teacher's Guide.

Refer students to the map on page 452.

Latin America

▲ Latin America is famous for its colorful markets.

▲ A volcano dominates Ecuador's countryside.

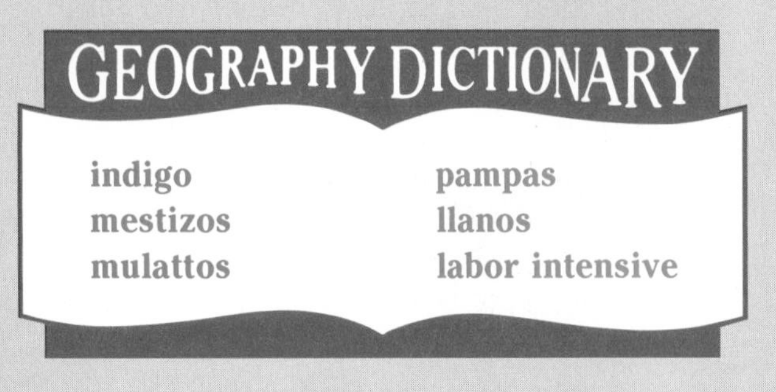

The region called Latin America extends from the northern border of Mexico to the southern tip of South America. It also includes the Caribbean islands. The region features equatorial lowlands, fertile mountain valleys, and snowcapped volcanoes. Cultural traditions merge ancient Indian ways of life with many European colonial customs.

Most of the countries of Latin America are considered less developed. Even in the region's wealthiest countries, there are many poor people. Scenes of colonial cathedrals and modern glass office buildings near slums with open sewers are common. Throughout the region, populations are increasing faster than the number of jobs. Political unrest also is common in Latin America as the people seek new ways to organize their societies.

 Have students use the map on page 452 to find the locations of places shown in the photos in this chapter.

Historical Geography

Indian Heritage In 1492, Christopher Columbus landed at an island in the Bahamas southeast of Florida, which he named San Salvador. He claimed this island for Spain. Within a few years, there were thousands of European colonists on the larger Caribbean islands. However, the New World of the European colonists was not uninhabited. Because Columbus mistakenly thought he had reached the islands off the coast of Asia, he named the islands' inhabitants Indians. These Native Americans had lived in the New World thousands of years.

The Indian civilizations produced great monuments and cities. The first great civilization was that of the Maya (MAH-yuh), whose society grew and declined between A.D. 100 and 900. The ruins of Mayan cities now lie in the forests of the Yucatán Peninsula of Mexico, Belize, and Guatemala.

The empires of the Aztecs of Mexico and the Incas of Peru developed around 1300. The Aztecs followed a series of little-known civilizations that developed after the fall of the Maya. The Aztec empire was centered at Tenochtitlán (tay-nawch-tee-TLAHN) on the present site of Mexico City. Tenochtitlán and the Aztec empire were conquered in 1521 by the Spanish explorer Hernando Cortés (kawr-TEHZ). The Inca empire of western South America extended along the Andes from Ecuador to Chile. The Incas began their conquests of neighboring tribes in about 1200. The Incas were conquered by a small army of Spanish soldiers led by Francisco Pizarro in 1532.

Besides conquering armies, other factors helped bring an end to the Indian civilizations of Latin America. For example, Europeans brought with them diseases which killed millions of Indians. Yellow fever and malaria, brought by African slaves, appeared a few decades later. Today, the Indian populations are rapidly increasing. However, they remain among the poorest people in Latin America.

Ruins such as Chichén Itzá, an old Mayan city on the Yucatán Peninsula, are evidence that the Mayans once ruled this region.

European Colonization Spain and Portugal, both countries with Latin cultures, were rich and powerful in 1500. Each country was eager for the profits and prestige of establishing overseas empires. A treaty in 1494 divided the Western Hemisphere into Spanish and Portuguese territories. Portugal received a large triangle of eastern South America, including present-day Brazil. Spain was awarded the Caribbean and the remainder of what is now Latin America.

Latin America was quickly settled by Spanish and Portuguese colonists. The colonists extended the Indian's gold and silver mines and opened new ones. However, ranching and farming were the key economic activities. The colonists brought European crops, livestock, and farming methods to the New World. They also adapted Indian crops to their needs and adopted some Indian farming practices. Spanish colonists were awarded large land grants by the king. Indians living on this land became farm workers.

By the mid-1600s, the French, British, and Dutch began to acquire lands that the Spanish had neglected. These included the smaller islands of the Caribbean, as well as Jamaica and Haiti. Their only mainland settlements were British Honduras (now Belize), British Guiana

Level A: Ask students to report on the Mayans, Toltecs, and Aztecs. If possible, have them use sources containing accounts of the civilizations given by the Spanish conquistadores.

Latin America
UNITED STATES
ATLANTIC OCEAN
GULF OF MEXICO
BERMUDA (GB)
Tijuana
Ciudad Juárez
Rio Grande
GULF OF CALIFORNIA
SIERRA MADRE OCCIDENTAL
SIERRA MADRE ORIENTAL
Monterrey
San Luis Potosí
León
Guadalajara
MEXICO
Mexico City
Puebla
Balsas River
Mérida
YUCATÁN PENINSULA
GULF OF HONDURAS
THE BAHAMAS
Nassau
Havana
CUBA
TURKS AND CAICOS ISLANDS (GB)
Greater Antilles
DOMINICAN REPUBLIC
Santo Domingo
HAITI
Port-au-Prince
Hispaniola
PUERTO RICO (US)
San Juan
VIRGIN ISLANDS (US, GB)
ANTIGUA AND BARBUDA
Lesser Antilles
JAMAICA
Kingston
Basseterre
St. Johns
ST. CHRISTOPHER AND NEVIS
GUADELOUPE (Fr)
Roseau DOMINICA
MARTINIQUE (Fr)
Castries ST. LUCIA
Kingstown
Bridgetown BARBADOS
ST. VINCENT AND THE GRENADINES
St. George's GRENADA
TRINIDAD AND TOBAGO
Port of Spain
CARIBBEAN SEA
NETHERLANDS ANTILLES (Neth)
ARUBA (Neth)
BELIZE
Belmopan
GUATEMALA
Guatemala City
HONDURAS
Tegucigalpa
San Salvador
EL SALVADOR
NICARAGUA
Managua
Lake Nicaragua
Panama Canal
San José
COSTA RICA
Panama City
PANAMA
Barranquilla
Cartagena
Maracaibo
Lake Maracaibo
Caracas
Valencia
Barquisimeto
Orinoco River
VENEZUELA
LLANOS
Georgetown
GUYANA
Paramaribo
SURINAME
Cayenne
FRENCH GUIANA (Fr)
GUIANA HIGHLANDS
Médellin
Bogotá
COLOMBIA
Cali
PACIFIC OCEAN
Galápagos Islands (Ecuador)
Equator
Quito
ECUADOR
Guayaquil
Río Negro
Amazon River
Belém
Trujillo
PERU
BRAZIL
BRAZILIAN HIGHLANDS
Recife
Callao
Lima
ANDES
Cuzco
MATO GROSSO PLATEAU
Salvador
São Francisco River
Arequipa
Lake Titicaca
La Paz
BOLIVIA
Brasília
Lake Poopó
Sucre
MOUNTAINS
Paraguay River
Belo Horizonte
Tropic of Capricorn
Tropic of Cancer
CHACO
PARAGUAY
Paraná River
BRAZILIAN PLATEAU
Santa Cruz
Rio de Janeiro
São Paulo
ATACAMA DESERT
Asunción
Iguaçu Falls
Curitiba
National capitals
Other cities
Uruguay River
Porto Alegre
Mt. Aconcagua (22,834 ft 6,962 m)
Córdoba
Valparaiso
Santiago
Rosario
ARGENTINA
URUGUAY
Buenos Aires
Morón
San Justo
Montevideo
RIO DE LA PLATA
CHILE
PAMPAS
ATLANTIC OCEAN
PATAGONIA
SCALE
0 500 1,000 Miles
0 500 1,000 Kilometers
Projection: Azimuthal Equal Area
FALKLAND ISLANDS (GB)
Stanley
Tierra del Fuego
Cape Horn
SOUTH GEORGIA (GB)
40°N
30°N
20°N
10°N
0°
10°S
20°S
50°S
130°W
120°W
110°W
100°W
90°W
80°W
70°W
60°W
50°W
40°W
30°W
20°W
N
S
E
W

Have students locate Latin America on the inset map of the world on page 452. Point out Mexico, Central America, the Caribbean, and South America.

Languages of Latin America

Language	Country
Spanish	Mexico, Argentina, and most other Latin American countries
Portuguese	Brazil
French	Haiti, French Guiana
Dutch	Suriname, Aruba
English	Belize, Guyana, some West Indies countries and territories
American Indian	Throughout Latin America
Nahuatl	Mexico, El Salvador
Quechua	Peru, Bolivia, Ecuador, Guatemala
Guaraní	Paraguay, Argentina, Brazil

Where is Quechua spoken? Peru, Bolivia, Ecuador, Guatemala

(now Guyana), French Guiana, and Suriname. These colonies were all based on plantation agriculture and used African slave labor. The British also brought many Asians from India, and the Dutch brought Indonesians to work the plantations. Major plantation crops were sugarcane, tobacco, and **indigo**. Indigo is a plant used to make a blue dye.

Because there were few women among the colonists, settlers often married Indian or African women. Many Latin Americans are **mestizos**, or people with both Indian and European ancestors. The descendants of plantation colonists and Africans are known as **mulattos**. Today, the geography of the peoples, languages, and cultures in the region is a result of the colonization process.

Independence By 1800, ideas of independence introduced by the American and French revolutions were well known throughout Latin America. Between 1808 and 1825, many colonies gained independence. Yet, the economies of the various countries changed little after independence. Rich landowners, merchants, and generals merely replaced the colonial government. The life of the rural mestizos, Indians, and blacks remained the same.

Latin America's trade shifted mainly to Great Britain and the United States. Great Britain protected the new Latin American states with its navy. The United States offered diplomatic protection under the Monroe Doctrine, which declared the countries of Latin America to be off limits to new European colonization.

Physical Geography

Landforms According to the theory of plate tectonics, South America and Africa broke apart from the supercontinent Gondwanaland more than 100 million years ago. The South Atlantic Ocean opened between them. As South America moved westward, other old continental fragments from the Pacific region collided with the west coast. These fragments may have formed the first Andes Mountains. A few million years ago, South America began to override the Pacific plate. This movement created an ocean trench off the coasts of Peru and Chile. Since then, mountains have continued to be built in the Andes as the result of folding, faulting, and volcanic activity.

The other mountains of South America are the Guiana Highlands and the highlands of eastern Brazil. These ancient highlands were formed much earlier than the Andes, when South America and Africa collided and created Gondwanaland.

Plains cover most of South America. The largest plains are in the Amazon River basin. The two other large plains are the **pampas** of Argentina and the **llanos** (LAN-ohz) of Venezuela.

Plate tectonics are also used to explain the creation of Caribbean islands. As the South American plate drifted westward from Africa, the smaller Caribbean plate cruised eastward. This Caribbean plate carried only a few continental fragments. These pieces are said to have become the mountainous islands of the Greater Antilles—Cuba, Jamaica, Hispaniola, and Puerto Rico. At the eastern edge of the

Level B: Ask students to: (1) Construct a time line of Latin American history; (2) Draw historical maps related to the time line.

Caribbean plate was another ocean trench and a chain of islands—the Lesser Antilles.

Central America and Mexico are believed to have been formed from a jumble of small plates attached to the rim of the North American and Caribbean plates. Mexico consists mostly of a high central plateau with mountains along each side. Central America, the region from Panama to Guatemala, has many short mountain ranges and great volcanoes. The presence of these mountains and of a related ocean trench off the coast of southern Mexico and Central America indicates plate collision. Scientists believe that Central America and South America were connected only a few million years ago.

Climate Because Latin America extends across nearly 90 degrees of latitude, it includes a full range of middle-latitude as well as tropical climate regions. South America contains the largest humid tropical region in the world. This is the huge forested area centered on the Amazon River. It rains almost every day. The other areas of humid tropical climate in Latin America are along coasts where the moist trade winds move onto shore.

Many areas of Latin America have a wet and dry tropical/subtropical climate. These include the wide, savanna-covered plains of Venezuela and central Brazil. The many areas of dry forests on the Caribbean islands and the Pacific coast of Central America also have a wet and dry tropical/subtropical climate.

Southern South America has a variety of middle-latitude climates. Southern Chile is influenced by the moist westerlies. The maritime climate found here is cool and wet. Snow and rainfall are particularly heavy in the southern Andes. However, east of the Andes in southern Argentina, the mountains block the rain from reaching the area. As a result, semiarid continental steppe and continental desert climates are found. Farther north, the central valley of Chile has a Mediterranean climate, with winter rains and summer drought. In southeastern South America is a large humid subtropical region.

The Atacama Desert is a mining region in northern Chile and southern Peru. It extends for about 600 miles (960 kilometers).

The largest and driest region of Latin America is the Atacama Desert of northern Chile and southern Peru. Subtropical high pressure brings clear skies and dry weather to this area throughout the year. In addition, the high Andes block rainy weather. Thus, the western slopes of the Andes are very dry. High pressure also produces the deserts of northwest Mexico.

The mild climates on mountain slopes between the elevations of 3,000 and 6,000 feet (915 and 1,829 meters) are particularly important in tropical and subtropical Latin America. In the mountain valleys, there are fertile soils and water from mountain streams. These are some of the densest areas of settlement in the region.

River Systems Three great river systems drain the eastern side of the South American continent. These are the Amazon, the Paraná, and the Orinoco rivers. The Amazon River is 4,000 miles (6,400 kilometers) long. Ocean ships can navigate upstream nearly 2,000 miles (3,200 kilometers), all the way to the city of Iquitos (i-KEET-ohs), Peru. The Orinoco and Amazon river basins are linked by the Río

 Point out Latin America's highest peaks, active volcanoes, and earthquake-prone areas. Remind students that these are part of the "Ring of Fire" around the Pacific Ocean. Review pages 47–51.

CHAPTER 37 CHECK

Reviewing the Main Ideas

1. Latin America includes Mexico, Central America, the Caribbean islands, and South America. Most Latin American countries are considered less developed, though many have extensive resources. Increasing populations, the contrast between rich and poor, and political instability hinder economic development.
2. Ancient Indian civilizations flourished in Latin America. The cultures of their descendents have been influenced and modified by European colonization.
3. Latin America includes a wide range of tropical and middle-latitude climate regions. Its environments include the great Andes Mountains of South America and the mountains of Mexico, plains, desert, savannas, and tropical rain forests.

Building a Vocabulary

Students should answer Question 4 by filling in the blanks, using their own paper.

1. What is indigo? What other plantation crops were grown by the European colonists?
2. From which groups are mestizos and mulattos descended?
3. What are pampas and llanos? Where are they located?
4. Industries that use hand labor to assemble products are often called ______ ______. What impact can these industries have on a country's economy?

Recalling and Reviewing

1. Why did few Indians survive the European settlement of Latin America?
2. How did independence change the countries of Latin America?
3. Describe the four types of farming systems found in Latin America. Which type provides the least cash income?
4. Name the three great river systems of South America. Where is each located?
5. Why is the theory of plate tectonics particularly applicable to South America?

Critical Thinking

6. How have political instability and population growth made economic development difficult for Latin America?

Using Geography Skills

Use the maps in this chapter to answer the following questions.

1. What climate regions are found in Latin America? Why are there so many?
2. Why do you think the population of South America is concentrated in the coastal regions?
3. What natural features might affect transportation between the capitals of Peru and Brazil?
4. Through what countries do the Andes Mountains run?

Answers to Chapter Check questions are found in the Teacher's Guide.

Additional activities for Chapter 37 are found in the Student Workbook. Chapter 37 Test as well as Skills, Reteaching, Critical Thinking, and Challenge/Enrichment activities are available in the Teacher's Resource Binder.

Chapter 38 Lesson Plans and Planning Guide are found in the Teacher's Guide.

Refer students to the map on page 465.

Mexico

▲ **Mexico City has more than 18 million people.**

▲ **A volcano towers over Mexico's corn fields.**

Mexico is a vibrant and complex nation. Its land area is the third largest in Latin America, after Brazil and Argentina. The country's environments vary from deserts in the north to tropical forests in the south. Cultural contrasts between wealth and poverty and between modern cities and ancient villages are everywhere. With more than 82 million people, Mexico has more than twice the population of Spain and is the largest Spanish-speaking country in the world. The works of Mexican artists and writers are displayed and read around the world.

Mexico's oil resources are important to the United States economy. Mexicans who live and work in the United States have added vital new traits to American culture. Political and social change is a continuing feature of Mexico. Today, this change has an impact on the political, economic, and social issues of all of North America.

 Have students survey the photos in this chapter to note physical, historical, and economic aspects of Mexico's geography.

provide schools and jobs. Many people have left the country. Continued political unrest and violence have held back the country's economic and social progress.

Nicaragua The physical geography of Nicaragua is much like that of neighboring Honduras. There is a hot, humid plain along the Caribbean coast, a rugged inland landscape of low mountains, and a volcanic plain along the Pacific coast. However, in Nicaragua, a rift valley with large lakes is found inland from the Pacific coastal plains. The fertile volcanic soils of the rift valley make this area the most densely settled part of the country. The capital city of Managua is located here.

Outside its rift valley, Nicaragua is thinly settled. The country's population is less than 3.5 million. Economic progress has been slow throughout Nicaragua's history, and political unrest has been typical. Most of the people continue to farm. Primary exports are coffee, cotton, sugarcane, and bananas. Copper, gold, silver, and lead have been mined in the central highlands.

In 1979, Nicaragua began a period of dramatic political change. The result has been a new revolutionary government designed with socialist goals. Close relations have been established with Cuba and the Soviet Union. The governing party calls itself Sandinista, named after César Sandino, a Nicaraguan patriot of the 1930s. Many wealthy and well-educated Nicaraguans have fled the country. Large estates and businesses have been nationalized.

Unfortunately, the people of Nicaragua have no more freedom than they had before. The Nicaraguan government has given aid to guerrilla rebels in nearby El Salvador. The United States has supported guerrillas inside Nicaragua who are fighting the Sandinista government. The Nicaraguan economy is near collapse.

Costa Rica Costa Rica's central highlands with fertile soils and mild climates are the nation's heartland. Located here is the capital city of San José. Many coffee farms, which provide the country's main exports, also are found in this area.

How has political instability made daily life difficult for many Nicaraguans, particularly those in the rural areas? Economic progress has been slow.

Unlike most Central Americans, Costa Rica's people are primarily of Spanish descent. When the European colonists first arrived, the local Indians quickly disappeared through disease and migration. As a result, the colonists did not have local labor to develop large estates. Instead, small family farms became the custom. This development created a large middle class that has supported stable democratic governments. Costa Rica has the highest standard of living in Central America.

Like the other countries of the region, Costa Rica has a high population growth rate. In the past, the opening of new lands closer to the coasts often relieved population pressure. Only recently have landless people and suburban slums become a serious concern. These problems have become more urgent because of the revolution taking place in nearby Nicaragua. Costa Rica senses that its democratic traditions are being threatened.

Panama Made up mostly of low hills and mountains, Panama has only narrow coastal plains. In the western part of the country are high volcanic peaks. Most Panamanians live along the southern (Pacific) side of the country where the tropical climate has a three-month dry season.

Level A: Have students gather newspaper and magazine articles about the debate over support for the Contras in Nicaragua. Have them prepare summaries of the arguments in support of, and in opposition to, Contra aid.

To some, modern Panama seems like three separate countries. In the east, toward South America, is a densely forested region that is nearly uninhabited. In the middle of the country is a prosperous area surrounding the Panama Canal. This area has an international flavor, with large cities. In the west are the more rural areas, where traditional, mixed-crop farming is common.

Panama gained its independence from Colombia in 1903 with the aid of the United States. The Americans intervened so that the Panama Canal could be constructed. The canal was built to carry trade between the east and west coasts of the United States. The Panama Canal also had military importance for the United States. The canal's importance for the United States has decreased in recent years. Railroads and trucks now carry much of the freight across the United States. Furthermore, many modern ships are too large for the canal.

Canal fees and canal-side industries provide nearly half of Panama's gross national product. The cities at either end of the canal—Colón on the Caribbean and Panama City on the Pacific—are major industrial centers.

The Panama Canal was completed in 1914. Why has the canal become less important to the United States recently? Railroads and trucks carry most of the freight across the United States.

The Islands

Physical Geography To the east of Central America lie the islands Columbus called the West Indies. They extend in a wide arc from south of Florida to the coast of Venezuela. Separate island groups included in the West Indies are the Greater Antilles, the Lesser Antilles, and the Bahamas. (See the map on page 472.)

The islands of the Greater Antilles are Cuba, Hispaniola, Puerto Rico, and Jamaica. Hispaniola is divided between the countries of Haiti and the Dominican Republic. The Greater Antilles are made of rocks like those of the continents. The Lesser Antilles include the Virgin Islands, Barbados, Trinidad and Tobago, and many other small islands. The Bahamas, north of the Greater Antilles and entirely in the Atlantic Ocean, include more than 700 islands. The small islands, like those in the Pacific, can be divided into low and high islands. The low islands are raised coral reefs with limestone rocks; they have poor soils and drier climates. The high islands are the remains of volcanoes; they have richer soils and wetter climates.

The islands of the West Indies have pleasant humid tropical and wet and dry tropical/subtropical climates. The islands receive a yearly average of 40 to 60 inches (100 to 150 centimeters) of rainfall. Even on rainy days, the sun shines most of the time. Some of the islands depend almost entirely on the tourist trade for their income.

Crops, particularly fruits and vegetables, can be grown throughout the year because of the region's favorable climate. However, soils are usually fertile only in the valleys of the higher islands. On the low islands as well as on Jamaica and Puerto Rico, there are many limestone rocks. These rocks and their soils

Where are the Bahamas located? How does the islands' location explain their close relationship with the United States?
southeast of Florida; they are influenced by the U.S. economy and culture

are very well drained. The rainwater quickly sinks downward below the plant roots. The result is frequent drought conditions, even though it may rain often.

Social Issues Nearly every island in the West Indies still shows the effects of colonialism and slavery. Much of the best land is part of large plantations or estates, many of which are controlled by foreign firms. Because of rapid population growth, most natural forests have been replaced by small, mixed-crop farms with poor soil. Most of the islands must import food. When people give up farming, they travel to the islands' capital cities, only to live in growing slums.

Even though most of the islands are politically independent, they remain economically dependent on foreign countries. The islands export their raw materials and import most of their manufactured goods. In this environment of poverty, inequality of wealth, and foreign dependence, political unrest is common. Social revolution is also a constant threat. The countries of the West Indies are discovering that they have many common issues. Regional cooperation is a hope for the future.

Cuba Cuba, the largest island of the Greater Antilles, lies just 90 miles (144 kilometers) off the southern coast of Florida. Mountains and hills cover about one-fourth of the island. The remainder of the land consists of gently rolling hills and fertile valleys. Unlike the rest of the West Indies, Cuba still has much little-used land.

The United States was once Cuba's most important market for exports. American companies owned almost all of Cuba's industries. However, the Cuban economy was corrupt and was controlled by a few wealthy people. A revolution in 1959 brought Fidel Castro to power. By 1961, Castro's government had taken over most privately owned property. It established close relations with the Soviet Union. Now, Cuban society is organized along Communist lines. The economy is centrally planned. Farmlands have been put into cooperatives and large state sugarcane plantations.

One-half of the farmland is used to grow sugarcane, Cuba's most important crop. There are many reasons Cuba is a sugar-producing country. Its soil is fertile and holds moisture well. The wet and dry tropical/subtropical climate changes little throughout the year. Finally, because the land is level, machinery can be operated on the sugarcane plantations.

Cuba's tobacco industry is famous, but it accounts for only a small portion of the country's exports. Tobacco is grown primarily in the western part of the island, mostly for the manufacture of cigars.

Besides sugarcane and tobacco, Cuba produces rice, coffee, vegetables, and tropical fruits. Most of Cuba's industry is based on the processing of agricultural products. Since Castro's revolution, more and more of the economy has depended on sugarcane. As a result, lands that once grew food now grow sugarcane. Therefore, Cuba must import much of its food. As in other Soviet-style countries, consumer goods are limited.

Havana, the capital of Cuba, is the largest city of the West Indies. Once the major port of the Spanish colonies, Havana is the country's most important trade, transportation, and

Note on a world map the proximity of Africa to this region. About 10 million black slaves were brought to the Americas from Africa.

Nearly two million people live in Havana, Cuba's capital and largest city. What kind of government does Cuba have? Communist

manufacturing center. It is a mixture of the old and the new. Its more than 450-year history can be seen in the old forts and churches and in the narrow streets. Havana also has wide avenues, large parks, and modern buildings.

The Cuban economy is dependent on aid and trade with the Soviet Union and other COMECON countries. However, Cuba maintains limited trading relations with other countries. Besides sugar and tobacco products, its exports include fish, nickel, and citrus fruits. Unfortunately, heavy military spending, often overseas, continues to weaken Cuba's economy. Cuban troops are based in several distant countries, including Angola and Ethiopia. Cubans in these countries support governments fighting anti-Communist guerrillas.

The Castro government is a dictatorship, and personal freedoms are limited. Since the revolution, large numbers of people who opposed Castro's government have fled Cuba. Many have made their new homes in the United States.

Haiti Haiti occupies the mountainous western third of the island of Hispaniola. It is the poorest and most densely populated country in the Americas. Haiti was the major French sugar-supplying colony until a slave uprising in 1791 brought independence. Since then, the country has experienced repeated corrupt governments and financial disorder. A collection of mulatto families has controlled the nation's wealth, while the black peasantry has gained little in two centuries. Civil disorder in 1986 forced the end of a strong dictatorship. There has been no effective government since then.

Most of the people farm tiny plots. Garden crops include cassava, yams, and **plantains**, a type of banana used in cooking. Coffee, cotton, and sugarcane provide some export income. Coffee also comes from wild stands of trees that were once coffee plantations.

The official language of Haiti is French, although most Haitians speak Haitian creole. This language is mostly French, with some Spanish, English, and African words. While Roman Catholicism is the recognized religion, **voodoo** is important throughout Haitian culture. Voodoo is a Haitian version of African animism. Followers believe that spirits of good and evil play an important part in daily life.

The capital city of Haiti, Port-au-Prince (pohrt-oh-PRINS), is the country's manufacturing center. In a country with low literacy, industries often depend on hand labor and simple machines. Clothing and baseballs are exported to the United States. Because of few economic opportunities and many political problems, a large number of Haitians are trying to flee their country. Most have become refugees in the United States.

Dominican Republic Haiti's neighbor, the Dominican Republic, occupies the eastern part of Hispaniola. The people of the Dominican Republic are of mixed Spanish and African descent. Because the country was one of Spain's earliest colonies, Spanish culture remains strong. Santo Domingo (duh-MING-goh), the capital, was the first permanent European settlement in the Western Hemisphere.

The Dominican Republic is a land of steep mountains and fertile valleys. Its forests con-

 Level B: Challenge students to find out what famous explorer is entombed in Santo Domingo. (Christopher Columbus)

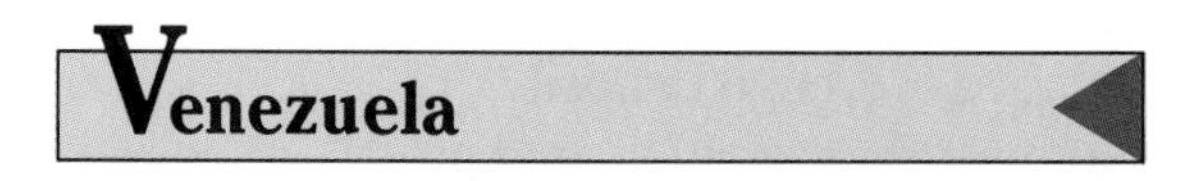

Venezuela

Physical Geography Venezuela is located on South America's northern Caribbean coast. The country is divided into four landform regions: (1) a narrow tropical-lowland coast that is surprisingly arid; (2) the northern mountains and valleys of the Andes that parallel the coast; (3) the llanos, the great savanna plains between the Andes and the Orinoco River; and (4) the Guiana Highlands in the southeast. (See the map on page 482.)

The Orinoco River and its tributaries drain the llanos as well as the Guiana Highlands. When streams plunge over the steep escarpments in the highlands, they create many of the world's highest waterfalls. Angel Falls, with a drop of 3,212 feet (979 meters), is the world's highest. In contrast, parts of the llanos are so flat that the water from summer floods stands on the plains for months.

Climates in the region are mostly wet and dry tropical/subtropical. The temperatures are warm to hot all year. The rains come mostly in the summer months when the equatorial low-pressure zone moves over the region. However, the highlands of Venezuela cause many climate variations. Mountains along the north coast can receive rain throughout the year as the trade winds blow moist air inland from the Caribbean.

Vegetation and soils follow the patterns of climates and landforms. The coastal region has humid air but little rain. Scrub savannas with cacti are common. Mountain slopes receive abundant rain and are often cloud covered during the rainy season. Thick forests were once widespread here, but 400 years of clearing the land for firewood and lumber has opened many areas to grasses. The llanos are one of the world's greatest short-tree and grass savannas. Huge expanses of grassland are flooded during rainy seasons. The plains in the far south are covered with humid tropical forests. The vegetation of the Guiana Highlands includes tall-tree savannas as well as dense forests.

Economic Geography Most of the people of Venezuela live along the Caribbean coast and in the valleys of the nearby mountains. These areas were the focus for early Spanish, as well as Indian, settlement. Most Venezuelans are mestizos. Small mixed-crop farms and large commercial farms are spread throughout this northern part of the country. The main crops are sugarcane and coffee. Corn and wheat are grown on the cooler mountain slopes.

The llanos are a land of large cattle ranches. Until recently, few people have lived in the region. However, as irrigation projects are being developed, many mixed-crop farmers are moving there. The Guiana Highlands are nearly empty of people. Any large settlements there are around mines. There also are a few communities of traditional Indians who still practice slash-and-burn agriculture.

Although agriculture is important, Venezuela's mineral wealth provides its chief source of income. As recently as 1920, Venezuela was a poor country with few opportunities. Then, the vast oil deposits surrounding Lake Maracaibo (mar-uh-KIE-boh) were developed. Since then, Venezuela has become a leading oil-exporting country. Though Venezuela has several oil refineries, much of the

Venezuela's llanos, which lie between the Andes Mountains and the Guiana Highlands, support many large cattle ranches.

Level A: Ask students to use the library to find five books on each country in this chapter. Have them record from the card catalog the author's name, the title, and the year of copyright for each book.

Have students locate northern South America on the map on page 452.

production is refined outside the country on offshore islands such as Curaçao and Trinidad. A large amount is also shipped to refineries in the United States.

The Venezuelan government has used oil profits to build schools, improve living conditions, and expand medical care. However, Venezuela's economy has become dependent on oil revenues. The country suffers when oil prices drop. For this reason, the government of Venezuela is encouraging new industries that process food and minerals.

The Guiana Highlands are rich in minerals, including large deposits of iron ore. There are also copper, tin, and manganese deposits. Large dams have been built on the Orinoco River. Steel mills using oil and hydroelectric power are producing a steel surplus.

Venezuela will have metal and oil resources for years to come. Along the middle Orinoco are huge deposits of **tar sands**. Tar sands are rocks that contain oil. However, because the oil has to be cooked out of the rocks, this oil is expensive. Nonetheless, these deposits may dominate the world economy in coming decades after the world's easily available oil is gone.

Caracas (kuh-RAK-uhs) is the capital of Venezuela's increasingly democratic government. The city is also the center of Venezuelan culture. It has a new subway system and a ring of modern expressways. There are modern office buildings and many technical and professional job opportunities. Venezuela's per capita income is the highest in Latin America. Skilled workers have migrated to Caracas from throughout Latin America.

As in other Latin American cities, however, slums encircle Caracas. There also are many poor people in rural areas. Improving

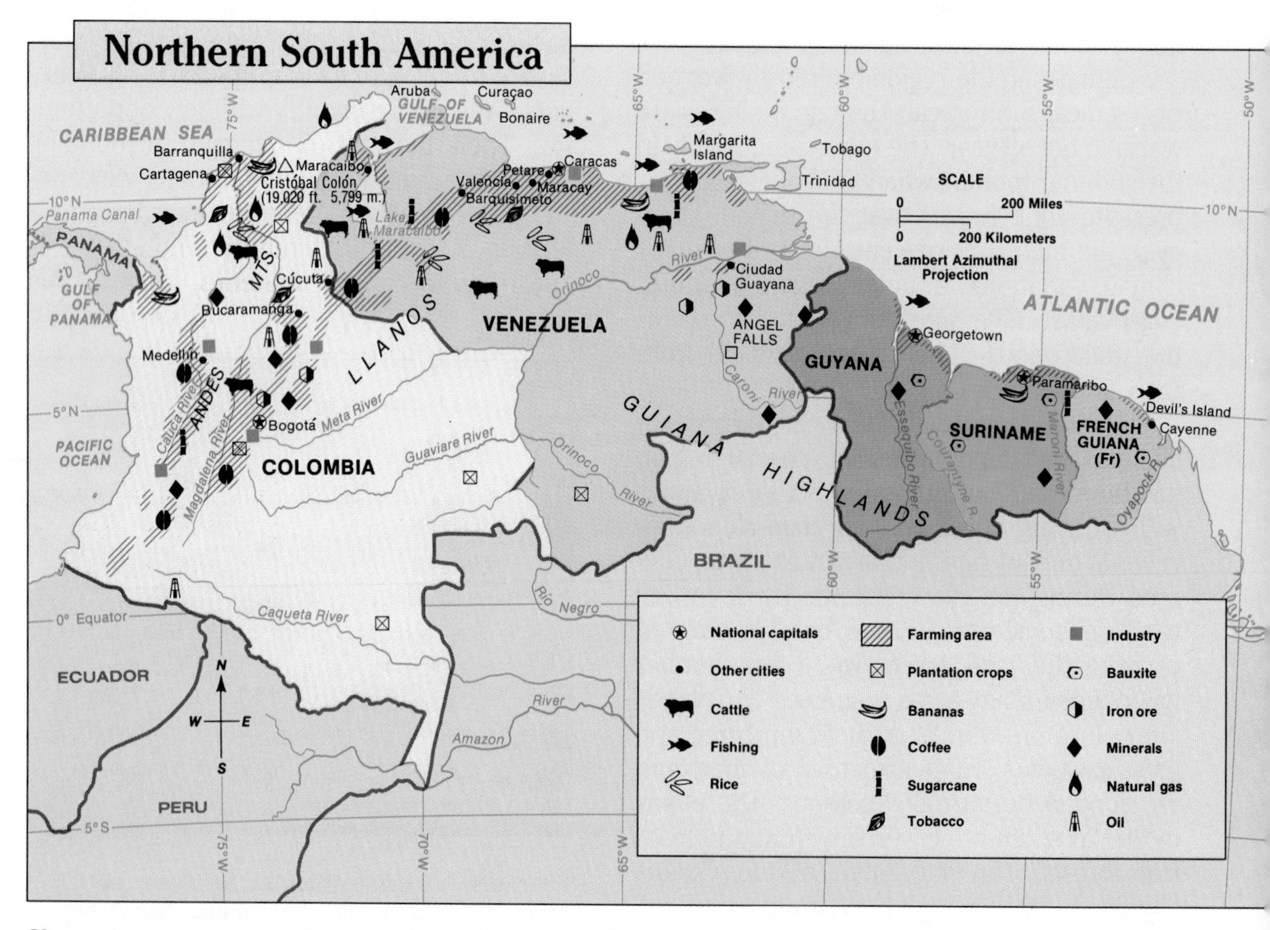

Name the major economic activities of Venezuela. Where do most of these occur? industry, fishing, farming, cattle raising, the growing of sugarcane, rice, tobacco, and bananas, mining, development of natural gas, oil, and iron ore deposits; along the northern coast

Use the map on page 452 to note areas of high elevation in northern South America.

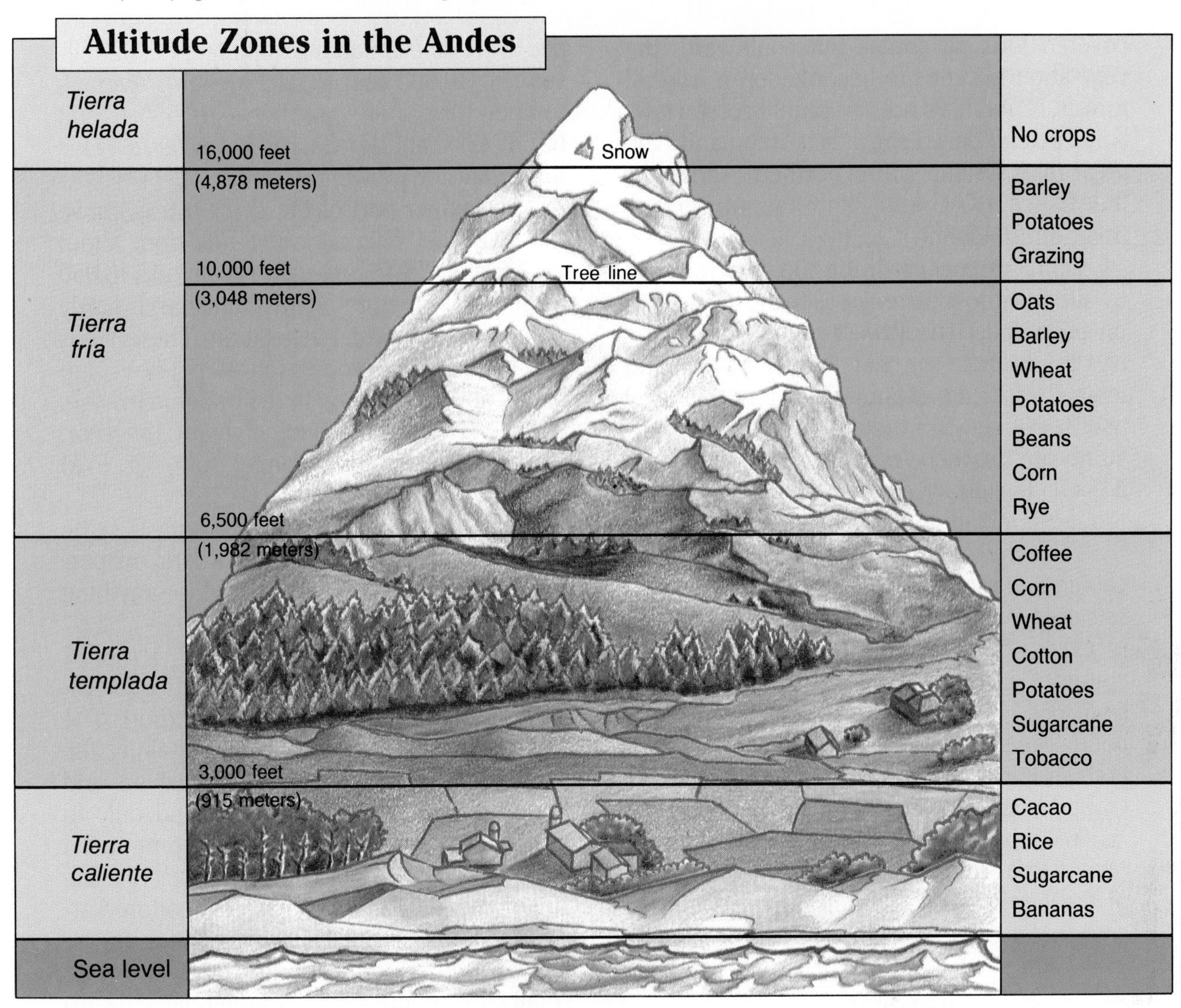

Why are so few crops grown in the upper *tierra fría*? It is too cold for most crops to grow.

the lives of the poor people living in these areas is Venezuela's main task. Oil wealth and new lands to develop in the south suggest that Venezuela can continue as an economic and political leader in the region. Venezuelans are proud of their country's successes and feel confident about the future.

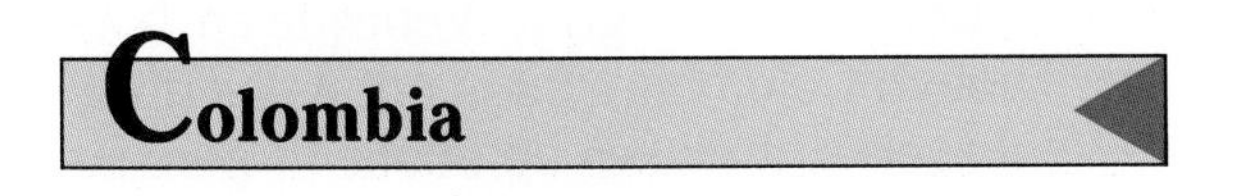

Colombia

Physical Geography Colombia is the only South American country with coastlines on both the Caribbean Sea and the Pacific Ocean. (See the map on page 482.) However, in Colombia, the Andes Mountains are divided into four chains, making travel difficult between the coasts. The rugged, hilly western coast used to be almost impossible to reach from the Caribbean coast. With the opening of the Panama Canal, road construction, and the development of air transportation, communications have improved.

Colombia has two major landform regions: the Andes Mountains and valleys in the west and the plains of the llanos and the Amazon basin to the east. In the north, the plains are

Level B: Ask students to compile a list of crops that can be grown in the region where they live.

covered with savannas, but southward, the vegetation thickens to become dense tropical forests. Colombia's hot and humid coastal lowlands extend only a short distance inland. The great river systems of the northern Andes are the Cauca (KAU-kuh) and the Magdalena (mag-duh-LAY-nuh).

Latin Americans divide the Andes slopes by altitude into four zones. (See the illustration on page 483.) The lowest zone is called the ***tierra caliente,*** or "hot country." It is found in the warm atmosphere between sea level and 3,000 feet (915 meters). Average temperatures are between 75° F and 80° F (22° C and 24° C). The *tierra caliente* produces tropical products, including bananas and cacao.

Farther up the mountains, the air becomes cooler, and moist climates with mountain forests are typical. This zone of pleasant climates is called ***tierra templada,*** or "temperate country." Here, average temperatures range from 65° F to 75° F (17° C to 22° C). Many Latin Americans live in this zone, where both tropical and middle-latitude types of crops are grown.

The next zone, as you move upslope, is the ***tierra fría,*** or "cold country." The lower part of the *tierra fría* has a natural vegetation of cool forests and grasslands. This lower *tierra fría* extends from about 6,500 feet (1,982 meters) to 10,000 feet (3,048 meters). Average temperatures here are between 55° F and 65° F (12° C and 17° C). The lower *tierra fría* is important for growing grains.

The upper part of the *tierra fría* is above the tree line. This zone extends from about 10,000 feet (3,048 meters) to as high as 16,000 feet (4,878 meters). Grasslands and hardy shrubs are the usual vegetation. These lands can be used for crops such as potatoes and barley and for grazing. In the upper *tierra fría*, frosts may occur on any night of the year. Average temperatures range from 42° F to 55° F (6° C to 12° C).

Tierra helada, or "frozen country," is the area of highest altitude. It is a land of permanent snow. Here, it is too cold for anything to grow.

For the Record

On November 13, 1985, Colombia's volcano Nevado del Ruiz erupted. Sixteen-year-old Slaye Molina remembers this:

"People screamed, 'The world is ending!' We ran upstairs to our terrace, but we saw another house collapse. So we rushed outside, though my grandmother had just had an operation and could not run. A friend took my hand and dragged me faster toward a hill. I looked back and saw my grandmother and aunt and uncle embracing each other. I do not want to be selfish, but I only thought about saving myself. I ran. The mud would catch up, and we would run faster. . . ."

From "Eruption in Colombia," *National Geographic,* Vol. 169, No. 5, May 1986, p. 648

Economic Geography Coffee is Colombia's leading crop and its major export. The most famous coffee area is found along the Andes ridges between the Magdalena and Cauca rivers. The country is second only to Brazil in coffee production. Other commercial crops grown in Colombia include rice, corn, sugarcane, tobacco, bananas, and cotton. Cattle grazing is an important occupation in the llanos. Most Colombians are farmers whose small plots produce both commercial and subsistence crops. One of Colombia's newest industries is marketing cut flowers, which are flown every night to markets in the United States. Clothing manufacturing is also an expanding industry.

The mountains of Colombia are rich in minerals. Ninety percent of the world's emeralds come from Colombia. However, Colombia's iron ore is a more valuable mineral resource. Gold, silver, copper, salt, and coal are mined as well. Oil is found in the eastern part of Colombia along the Venezuelan border. Though Colombia's oil is not as plentiful as Venezuela's, it makes up about 15 percent of Colombia's national income.

A good trade location, plentiful natural resources, and improved interior transportation

 Have students read the primary-source account above. Ask what emotions Slaye Molina might have experienced. Ask if students think there are universal feelings that people share, regardless of culture.

Point out that Ecuador is the world's major exporter of bananas.

Ecuador

Physical Geography The equator passes directly through Ecuador, which is the Spanish word for "equator." Ecuador can be divided into three landform and climate regions. (See the map on page 491.) In the east are humid tropical lowlands of the Amazon Basin. This forested region is known as the Oriente, which means "east." Latin Americans call the thick rain-forest vegetation the *selva,* or "forest." Less than 5 percent of Ecuador's 10 million people live in the Oriente.

In the middle of the country are two high mountain ranges that form the Andes. The valleys between these mountain ranges are at elevations up to 10,000 feet (3,048 meters). The Andes rise up to 10,000 feet higher. There are more than two dozen high volcanic peaks that have permanent snowcaps, even at the equator. As a result, the high valleys have mild climates all year. About 45 percent of Ecuador's population lives in this central mountain region, known as the Sierra.

Ecuador's third region is the lowlands along the Pacific coast, known as the Costa, which means "coast." The climates here vary from humid tropical to a narrow fringe of desert on the coast. Most of the region has a wet and dry tropical/subtropical climate. This region, with rich agricultural potential, has the fastest growing population and economy in Ecuador. About half of Ecuador's people live in the Costa.

Economic Geography Ecuador's economy closely reflects its physical geography. The Sierra is a historic region of dense Indian settlement. Many people still speak Quechua, the ancient Inca language. The Sierra contains many small subsistence farms where corn, potatoes, and barley are grown. The fields are often on steep slopes where soil erosion is severe. More than 60 percent of the farms are smaller than 12 acres (five hectares), and many are too small to support a family. The Sierra is also a region of large haciendas.

Indians work in a field near a village in Ecuador. Why is productive agriculture difficult throughout much of this country?
Much of Ecuador is forest and mountains.

Though the Sierra's rock masses have provided gold and silver since colonial times, production today is low. Other large mineral reserves may be discovered in the future. Many mountain streams have been dammed to provide hydroelectricity.

The Oriente is a frontier land. Landless farmers from the Sierra are moving down the Andes slopes to clear small farms from the dense forests. Because there are few roads, access to this area is difficult. The Oriente's greatest resource is oil, which is piped over the Andes to the Pacific port city of Esmeraldas. Oil, which accounts for more than 50 percent of Ecuador's export income, is the financial basis for development of the entire country.

The Costa has emerged during the twentieth century as Ecuador's most important agricultural region. The key export crops are bananas, coffee, and cacao. Rice and sugarcane are grown for local markets. Many of the Costa's farms are medium sized, large enough for prosperous family farming. The government has been successful in encouraging cooperatives and in providing technical advice. There are still some large haciendas, where cattle ranching is the major activity. Offshore

Level B: Ask why farmers would be reluctant to move from the Sierra to the Oriente. Have students think of five questions a Sierra farmer would ask before moving to the Oriente.

For every increase of 1,000 feet (300 meters) in elevation, temperature drops about 3.6°F (2°C). Have students estimate the impact of altitude on Quito's climate.

from the Costa are rich fishing grounds that are just beginning to be fished. Natural-gas deposits have been found beneath the Costa's lowlands.

The great oil wealth of the Oriente and the agricultural progress of the Costa have moved Ecuador to world prominence. Ecuador does have some challenges ahead. One is its dramatic population growth. Ecuador has the second highest population growth rate in South America. If the current rate continues, the country's population will double by the year 2012. Providing jobs, schools, and other needs of this growing nation will be difficult.

Ecuador's second serious challenge is transportation. Despite a railroad stretching from the coast to Quito and many roads, the country's regions are not linked. The Andes make construction and maintenance of transportation systems both difficult and expensive.

After decades of corrupt governments, Ecuadorians are having more political success. With continued resource development, Ecuador could become one of Latin America's prosperous countries in the next century.

Population and Cities Ecuador's population is about 40 percent mestizo and 40 percent Indian. There are also descendants of Spanish and European colonists and of African slaves. The Indians live primarily in the Sierra, and blacks and mulattos live primarily in the Costa. After centuries of separation, cooperation between the several ethnic groups is now common.

Much of Quito is a reminder of Spanish colonial times. Ecuador became independent in 1822.

Guayaquil (gwie-uh-KEEL), on the Pacific coast, is Ecuador's largest city. Guayaquil is also Ecuador's most important port and manufacturing center. Guayaquil's modern buildings, banks, and factories are symbols of a modern Ecuador.

Ecuador's capital city is Quito (KEE-toh), in the Sierra. Quito is situated in an Andean valley on the equator at an altitude of 9,350 feet (2,850 meters)! It was a center of Indian culture for centuries before the Spanish conquest. A capital of the Inca empire, it became the center for colonial government. Quito is the focus of Ecuador's traditional life of great haciendas and Indian subsistence farming. Despite repeated earthquakes over the centuries, Quito is still a classic example of a Spanish colonial city. It has the old churches, monuments, and colorful markets that are tourist favorites.

Galápagos Islands The arid Galápagos (guh-LAHP-uh-guhs) Islands are among the world's most unusual regions. They are owned by Ecuador and are located in the Pacific Ocean about 650 miles (1,040 kilometers) off its coast. About 4,000 people live on the islands. Much of the plant and animal life in the Galápagos has special adaptations because of the islands' long isolation from the mainland. These special animals and plants have been studied intensively by scientists.

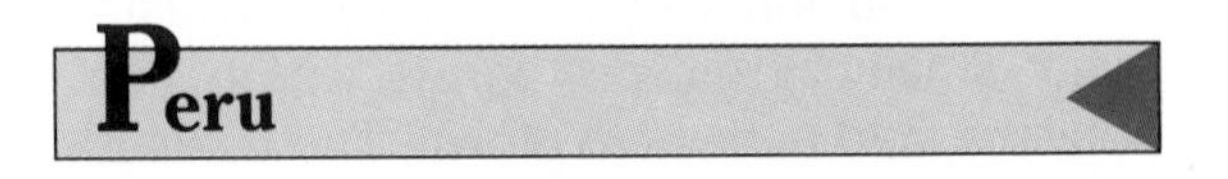

Peru

Physical Geography East and south of Ecuador is Peru. (See the map on page 491.) The Andes divide Peru into three regions. The first is the eastern humid tropical region, often

UNIT NINE

THE UNITED STATES AND CANADA

OBJECTIVES

- *To describe how the climate and landforms of the United States and Canada have influenced where and how Americans and Canadians live*
- *To explain how the development of natural resources found in North America has benefited both the United States and Canada*
- *To describe the economic, social, and cultural issues that exist in the United States and Canada and the ways in which both nations are addressing these issues*
- *To interpret a bar graph*

Chapter 44 Lesson Plans and Planning Guide are found in the Teacher's Guide.

Refer students to the map on page 516.

The North American Land and People

▲ **This wheat field is in Wyoming.**

▲ **A sailboat cruises on Lake Michigan.**

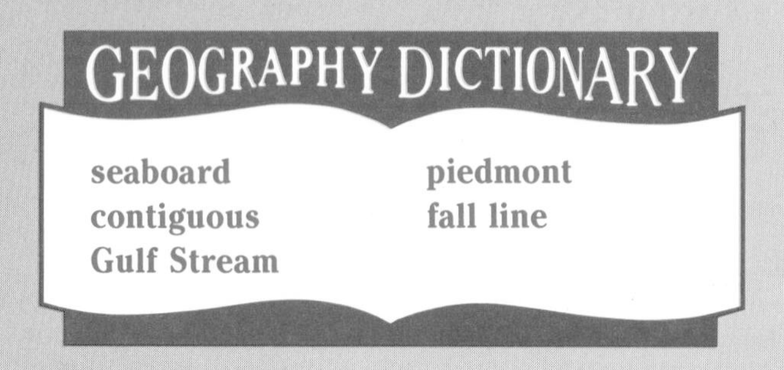

The United States and Canada share the majority of the North American continent, as well as the world's longest peaceful boundary. The United States ranks fourth in the world in land area and in population. The country faces oceans and seas on three sides, making its location ideal for international trade. The United States consists of 50 states.

Canada is the world's second largest nation in land area but has a population of about one-tenth that of the United States. Most Canadians live near the United States border, while the rest of Canada has rich natural resources and few people. Like the United States, Canada is well situated for trade. The country consists of 10 provinces and two territories.

In this unit, Chapters 44 through 51 cover the geography of the United States. Chapters 52 and 53 explore the geography of Canada.

 Have students survey the chapter photos for evidence of economic development.

Historical Geography

Settlement and Growth The first inhabitants of America were Indians, whose ancestors probably arrived from Asia more than 30,000 years ago. By the time of European exploration, there were several million Indians in the present-day United States.

The first Europeans to arrive were Viking explorers about A.D. 1100. The Spanish arrived about 500 years later, settling mainly in the southern part of the United States from Florida to California. They were soon followed by the French, who explored major river systems in the interior of the present United States. The British, however, became the nation's major influence, with settlements all along the Atlantic **seaboard**. A seaboard is the land area near the sea. The British eventually gained all French and Spanish territory east of the Mississippi River.

After gaining independence from Britain, the United States grew with immigrants from all over the world. Most of the people came from Europe, especially Britain, Germany, and Ireland. Between 1870 and 1930, more than 30 million European immigrants arrived in America. With the development of plantation agriculture in the south, millions of Africans were brought to the United States and sold as slaves.

As the nation's population gradually moved westward, more states were added to the original 13. Diseases and warfare with settlers killed many Native Americans. The surviving Indian populations were driven farther west, where they were eventually forced onto reservations.

Today, the United States has the most diverse population in the world. It is about 78 percent European, 12 percent black, 7 percent Hispanic (of Latin American descent), 2 percent Asian, and 1 percent Native American (Indian, Eskimo, and Hawaiian). Recent immigration has been mainly from Latin America, especially from Mexico. Hispanics now make up the fastest-growing sector of the nation's

Ethnic groups in the United States originated from almost every part of the world. How does this photograph show ethnic diversity?

It shows different groups.

population. There also have been great numbers of people arriving from Asia, particularly from Vietnam, and from the Middle East. In recent years, many people have immigrated to the United States illegally. Most come to seek employment or to escape political instability in their home countries.

The United States is predominantly Christian in religion. The population is about one-half Protestant and one-third Roman Catholic. There are also significant numbers of Jews, Mormons, and Muslims, as well as numerous other religious groups.

While the United States is ranked fourth in world population, it comprises a very small percentage of the total world population. With low birthrates by world standards, the population of the United States will form an even smaller percentage in the future. Low birthrates combined with high health standards means that the nation's population is also aging. More than 12 percent of the population is now over age 65.

Physical Geography

Climate The 48 **contiguous** states are located in the middle latitudes, where a variety

Refer students to the population pyramid of the U.S. on page 98.

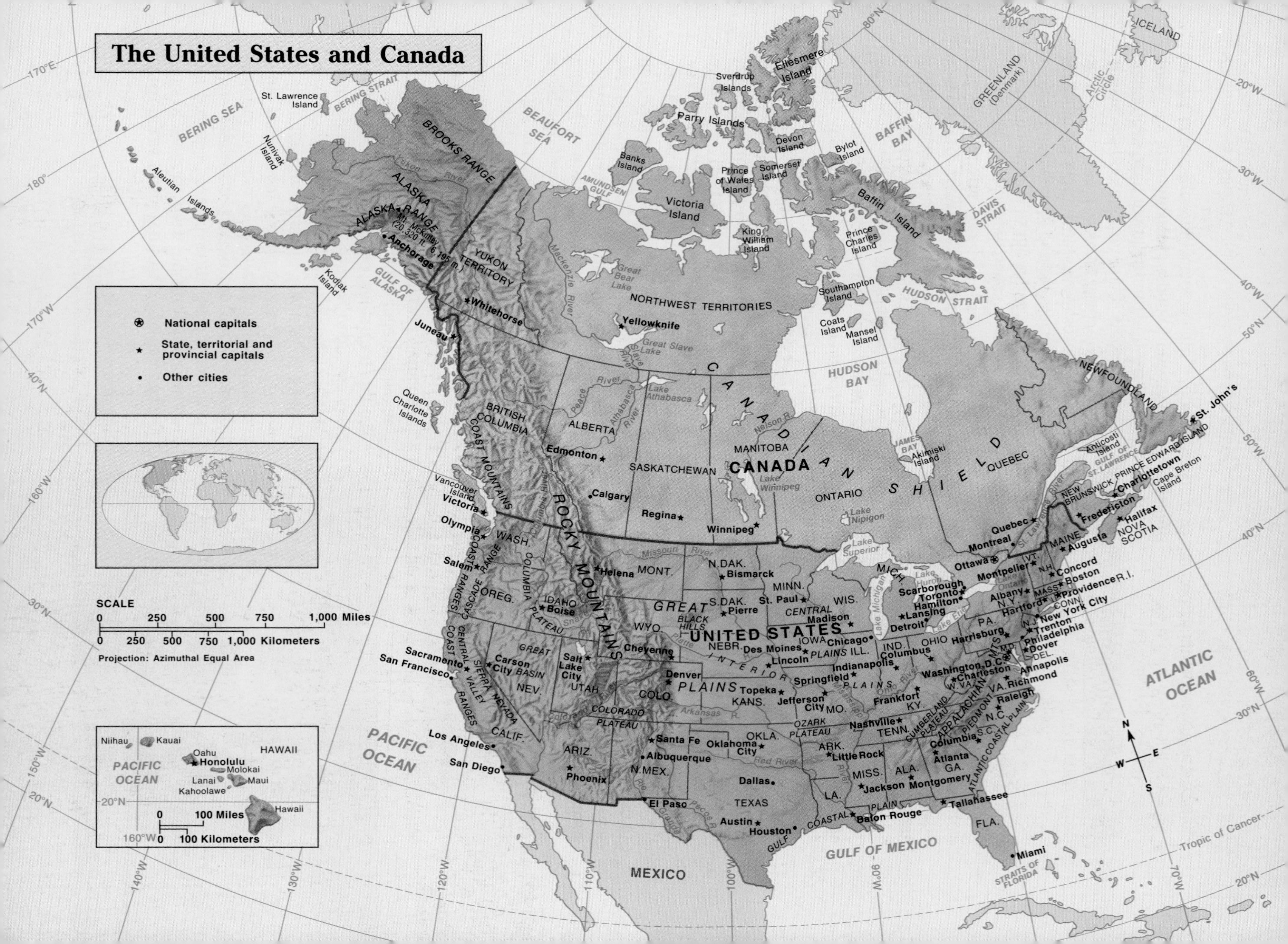

The United States and Canada
National capitals
State, territorial and provincial capitals
Other cities
SCALE
0 250 500 750 1,000 Miles
0 250 500 750 1,000 Kilometers
Projection: Azimuthal Equal Area
Niihau
Kauai
Oahu
Honolulu
Molokai
Lanai
Maui
Kahoolawe
Hawaii
HAWAII
PACIFIC OCEAN
0 100 Miles
0 100 Kilometers
BERING SEA
BERING STRAIT
St. Lawrence Island
Nunivak Island
Aleutian Islands
Kodiak Island
GULF OF ALASKA
BROOKS RANGE
ALASKA RANGE
Mt. McKinley (20,320 ft. 6,195 m.)
Anchorage
Juneau
YUKON TERRITORY
Whitehorse
BEAUFORT SEA
AMUNDSEN GULF
Banks Island
Victoria Island
Parry Islands
Sverdrup Islands
Ellesmere Island
Devon Island
Somerset Island
Prince of Wales Island
King William Island
Bylot Island
Baffin Island
Prince Charles Island
Southampton Island
Coats Island
Mansel Island
BAFFIN BAY
DAVIS STRAIT
HUDSON STRAIT
HUDSON BAY
JAMES BAY
Akimiski Island
GREENLAND (Denmark)
ICELAND
Arctic Circle
NORTHWEST TERRITORIES
Yellowknife
Great Bear Lake
Great Slave Lake
Lake Athabasca
Mackenzie River
Peace River
Athabasca River
Slave River
Nelson R.
BRITISH COLUMBIA
Queen Charlotte Islands
Vancouver Island
Victoria
COAST MOUNTAINS
ALBERTA
Edmonton
Calgary
SASKATCHEWAN
Regina
MANITOBA
Winnipeg
Lake Winnipeg
CANADA
CANADIAN SHIELD
ONTARIO
Lake Nipigon
QUEBEC
Quebec
Montreal
Ottawa
NEWFOUNDLAND
St. John's
Anticosti Island
GULF OF ST. LAWRENCE
St. Lawrence River
NEW BRUNSWICK
Fredericton
PRINCE EDWARD ISLAND
Charlottetown
Cape Breton Island
NOVA SCOTIA
Halifax
Toronto
Scarborough
Hamilton
ROCKY MOUNTAINS
Columbia River
Missouri River
WASH.
Olympia
COAST RANGE
CASCADE RANGE
OREG.
Salem
COLUMBIA PLATEAU
IDAHO
Boise
Snake R.
MONT.
Helena
WYO.
Cheyenne
N.DAK.
Bismarck
S.DAK.
Pierre
BLACK HILLS
GREAT PLAINS
MINN.
St. Paul
WIS.
Madison
CENTRAL
Lake Superior
Lake Michigan
Lake Huron
Lake Erie
Lake Ontario
MICH.
Lansing
Detroit
UNITED STATES
IOWA
Des Moines
Chicago
ILL.
INTERIOR PLAINS
NEBR.
Lincoln
Platte R.
CALIF.
Sacramento
San Francisco
COAST RANGES
CENTRAL VALLEY
SIERRA NEVADA
Los Angeles
San Diego
NEV.
Carson City
GREAT BASIN
UTAH
Salt Lake City
COLO.
Denver
COLORADO PLATEAU
Colorado River
ARIZ.
Phoenix
N.MEX.
Santa Fe
Albuquerque
El Paso
Rio Grande
Pecos R.
KANS.
Topeka
Arkansas R.
MO.
Jefferson City
Springfield
OZARK PLATEAU
OKLA.
Oklahoma City
Red River
TEXAS
Dallas
Austin
Houston
GULF COASTAL PLAIN
ARK.
Little Rock
LA.
Baton Rouge
Mississippi River
MISS.
Jackson
ALA.
Montgomery
TENN.
Nashville
KY.
Frankfort
Ohio River
IND.
Indianapolis
OHIO
Columbus
GA.
Atlanta
FLA.
Tallahassee
Miami
STRAITS OF FLORIDA
S.C.
Columbia
N.C.
Raleigh
PIEDMONT
ATLANTIC COASTAL PLAIN
CUMBERLAND PLATEAU
APPALACHIAN MTS.
VA.
Richmond
W. VA.
Charleston
Washington, D.C.
MD.
Annapolis
DEL.
Dover
PA.
Harrisburg
Philadelphia
N.J.
Trenton
N.Y.
Albany
New York City
CONN.
Hartford
R.I.
Providence
MASS.
Boston
VT.
Montpelier
N.H.
Concord
MAINE
Augusta
PACIFIC OCEAN
ATLANTIC OCEAN
GULF OF MEXICO
MEXICO
Tropic of Cancer
N
S
E
W

Have students find pictures that illustrate each climate type.

of climates occur. Contiguous states are those that border each other as a single unit. Only Alaska and Hawaii are not contiguous states. The climate of the United States is controlled mainly by a large continental area, high mountain ranges, westerly winds, and ocean currents. The country has 11 climate types, the greatest variety of climates of any nation.

A humid continental climate is found in the northeastern quarter of the nation. It stretches westward from the Atlantic coast to approximately 100° west longitude. This climate region has four distinct seasons, including a warm, humid summer and a cold, snowy winter. The local effect of the Great Lakes and the Atlantic Ocean makes temperatures somewhat milder, but it increases precipitation.

A humid subtropical climate is found in the southeastern quarter of the United States. This region extends from the Atlantic coast to central Texas. The climate is influenced by the warm waters of the Gulf of Mexico and by the **Gulf Stream** current in the Atlantic Ocean. The Gulf Stream is a warm ocean current that moves warm water northward. Summers are hot and humid, and winters are mild. Rainfall is distributed fairly evenly throughout the year. Some snow and frost may occur in winter. The very southern tip of Florida extends beyond the humid subtropical region. This area has a wet and dry tropical/subtropical climate.

West of 100° west longitude is the continental steppe climate of the Great Plains. The steppe climate is semiarid and supports grassland but is not forested. Summers are hot, while winters can be very cold. The Great Plains region is subject to violent weather, including thunderstorms, hailstorms, snowstorms, and tornadoes. Drought is also a concern here.

The interior western states have a variety of climates because of the mountainous terrain. The Rocky Mountains have highland climates, with temperatures and precipitation dependant upon elevation and exposure. The Intermountain region, located between the major mountain areas of the west, has mainly desert and steppe climates.

Hundreds of tornadoes strike the Great Plains every year. What climate is found in this region? continental steppe

The Pacific Coast region has two climates. The Pacific Northwest is dominated by a mild maritime climate. It has cool, cloudy, wet winters and mild, sunny summers. The maritime climate region is a narrow band between the coast and mountains. It extends from northern California through Oregon and Washington to as far north as Alaska.

The Mediterranean climate is found in southern and central California. This climate is known for its long, sunny, dry summers and its mild, wetter winters. In most years, no rain falls in the Mediterranean climate region of California between April and November.

Hawaii, the southernmost state, has a humid tropical climate. Because they are located in the easterly trade wind belt, the Hawaiian islands are wetter on the windward, eastern sides and drier on the leeward, western slopes.

Alaska is the only state that extends into the subarctic and tundra climates. It actually has three climates: maritime along the southeast coast, subarctic in the interior and along the west coast, and tundra along the north and east coasts. The subarctic climate area is forested and has severe seasonal changes. Winters are cold and long, while summers are short and quite warm. The tundra climate area is treeless and supports only small plants.

Have students find the U.S. and Canada on the inset map of the world on page 516. Note the position of this region in terms of latitude.

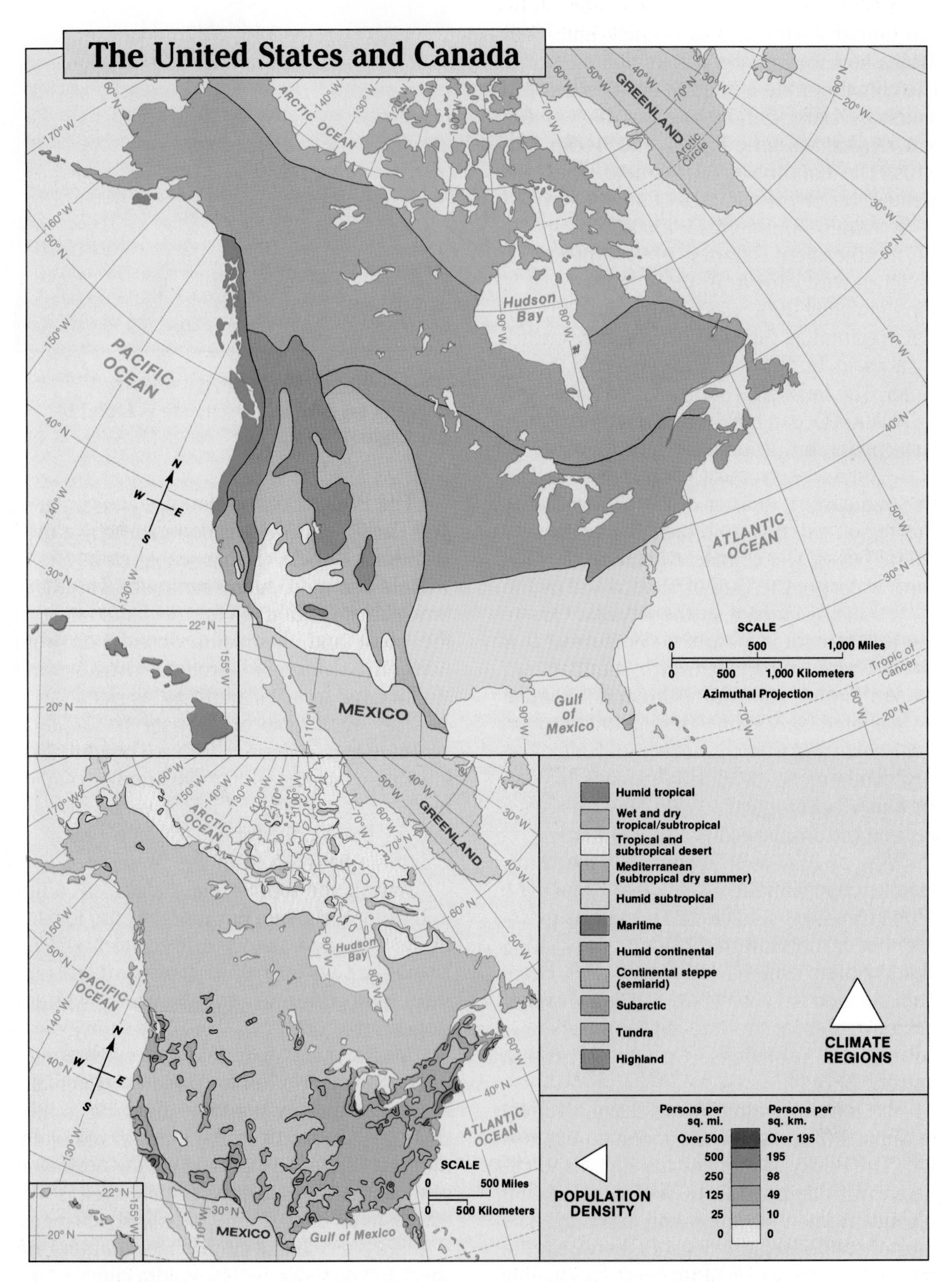

What is the population density of the subarctic climate region? the maritime region?

subarctic: 25 persons per square mile (10 persons per square kilometer); maritime: 25 persons per square mile (10 persons per square kilometer), 125 persons per square mile (49 persons per square kilometer)

CHAPTER 44 CHECK

Reviewing the Main Ideas

1. The United States and Canada share most of the North American continent as well as the world's longest peaceful border. Both of these countries, because of their access to sea routes, are well situated for trade.
2. The United States comprises 50 states. Canada has 10 provinces and two territories. The United States has the most diverse ethnic population in the world. Its population is also one of the world's most urbanized.
3. The climate and landform regions of the United States are varied. The nation's climates are controlled by ocean currents, high mountain ranges, westerly winds, and a large continental area.

Building a Vocabulary

1. What are contiguous states? Which states discussed in this chapter are not contiguous?
2. Define *Gulf Stream*. What effect does the Gulf Stream have on the climate of the southeastern United States?
3. What is a piedmont? Where is the North American Piedmont region located?
4. Where is the fall line located? What are the advantages of building a city along the fall line?

Recalling and Reviewing

1. How many climate regions are found in the United States? In what state are the subarctic and tundra climate regions found? Which region is dominated by the humid subtropical climate?
2. What is significant about Death Valley, California, and Mount McKinley, Alaska?
3. What energy resources are found in the United States? What energy resources must the United States import?

Critical Thinking

4. What problems must the United States solve in order to maintain a high standard of living?

Using Geography Skills

Use the maps in this chapter to answer the following questions.

1. Which regions of the United States are most densely populated? What is the climate of each of these regions?
2. Name the major rivers of North America. Into what body of water does the Mississippi River flow?
3. What is the distance between Sacramento, California, and Denver, Colorado? What major landforms lie between the two cities?

Answers to Chapter Check questions are found in the Teacher's Guide.

Additional activities for Chapter 44 are found in the Student Workbook. Chapter 44 Test as well as Skills, Reteaching, Critical Thinking, and Challenge/Enrichment activities are available in the Teacher's Resource Binder.

Chapter 45

Chapter 45 Lesson Plans and Planning Guide are found in the Teacher's Guide.

New England

Refer students to the map on page 527.

▲ Bostonians are surrounded by a colonial past.

▲ Forests cover much of New England.

The New England states are located in the far northeastern corner of the United States. They include Maine, New Hampshire, Vermont, Massachusetts, Rhode Island, and Connecticut. The total population of this region is 13 million, about 5 percent of the nation's total.

New England has played a major role in United States history. It was first settled at Plymouth, on Massachusetts Bay, in 1620 by the Pilgrims. They were soon followed by other colonists, who farmed, fished, or obtained lumber and furs from the forests. In the eighteenth century, New England became the center of the American Revolution.

During the nineteenth century, New England became the first industrialized region in North America and developed major textile and leather industries. These industries began to decline in the twentieth century. Today, New England is undergoing a second industrialization. Its new industries are centered around high technology and finance.

GEOGRAPHY DICTIONARY

- granite
- moraine
- second-growth forest
- reforestation
- biotechnology

Ask students if they can make a generalization about New England from the photos in this chapter.

Physical Geography

Landforms New England's variety of landforms includes seashore, mountains, hills, and plains. (See the map on page 527.) Northern New England is covered by a northeastern extension of the Appalachian Mountains. Its major mountain ranges are the White Mountains in New Hampshire, the Green Mountains in Vermont, and the Longfellow Mountains in Maine. The region's highest peak is Mount Washington in the White Mountains, which reaches an elevation of 6,288 feet (1,917 meters). Because of glacial erosion during the last ice age, northern New England has thousands of lakes and thin, rocky soils.

Southern New England is hilly but not mountainous. Many of the area's rugged landforms were caused by glaciers. As the great ice sheets pushed their way to the Atlantic coast, they piled up rock material, forming hundreds of hills. The Berkshire Hills of western Massachusetts, usually called the Berkshires, form the highest hill region in southern New England. The only plains in New England are along some sections of the Atlantic coast and in the lower Connecticut River valley.

Even New England's coastline has great variety. In the north, the coast of Maine has rocky inlets and peninsulas of **granite**, such as those found in Acadia National Park. Granite is a gray-to-pink, speckled, hard, crystalline rock formed deep in the earth's crust. Because of the narrow inlets and rugged shoreline, some of the world's greatest tides occur along the northern Maine coast. Glacial deposits are responsible for the shoreline features on the southern New England coast. For example, Cape Cod and the islands of Nantucket and Martha's Vineyard are glacial **moraine** materials piled up by the great ice sheets. A moraine is a ridge of rocks, gravel, and sand that is deposited by a glacier.

Climate New England has a humid continental climate, with four distinct seasons. The region is famous for its colorful fall. Along the coast, there is a maritime influence. Coastal fog is common in New England, especially in summer, when the warm Gulf Stream waters meet the cool waters of the Labrador Current. Occasionally during summer, hurricanes strike coastal New England. More common are winter storms from the North Atlantic, called northeasters, which bring cold snowy weather with very strong winds and large ocean waves.

The Atlantic coast of New England is a rich source of fish and shellfish. Clam digging is a popular activity in Massachusetts.

Economic Geography

Agriculture Because of the region's rocky terrain and short growing season, agriculture in New England is quite limited. Only about 12 percent of New England's land is farmed.

The Connecticut River valley is the leading agricultural area in New England. It produces a variety of crops, including tobacco, apples, and other fruit. The swampy lowlands of Cape Cod grow 60 percent of all the cranberries produced in the United States. The Aroostook (uh-ROOS-tuhk) Valley in northern Maine produces about 10 percent of the country's

Point out to students that many New England farmers went to the Midwest during the nineteenth century because the soil in the Midwest was more productive for farming.

potatoes. Washington County, in Maine, specializes in blueberries.

New England's cool, moist climate and hilly pasture make dairy farming the region's most important agricultural activity. Vermont is one of America's leading dairy states. Its cheese, ice cream, and butter are famous for their high quality. There are also poultry farms throughout New England.

Farming in the region has been in decline for more than a century. Many abandoned farms have been bought by urban dwellers to use as summer homes. Nearer to the cities, many commuters have become part-time farmers.

Fishing, Forestry, and Mining Coastal New England was the center of the American whaling fleet during the last century. Whale oil made New Bedford, Massachusetts, the richest city in the United States during the whaling period. Today, New England whaling boats are used to take tourists to see and photograph these large marine mammals.

The shallow waters off New England's coast are among the world's most productive fishing grounds. Where the cool Labrador Current flows over the shallow New England continental shelf, there is a plentiful supply of fish. The most productive waters are around Georges Bank, a shallow underwater plateau off the Massachusetts coast. Haddock, mackerel, and many shellfish, especially lobsters, are caught off New England's shores.

Boston, Portland, New Bedford, and Gloucester (GLAHS-tuhr) are the chief ports for New England's fishing fleet. The New England fisheries supply more than 10 percent of the total United States fish catch. Recent oil and gas discoveries on the continental shelf off New England have attracted the interest of major oil companies. This is a major concern to New Englanders, since an oil spill could be devastating to the region's fishing industry.

Softwood forests, including many valuable white pines, were once an important New England resource. Much of the forest, however, was cleared for farming and for New England's early shipbuilding industry. Today, New Hampshire and Maine have lumbering operations, mainly to supply the pulp and paper industry. Virtually all forests in New England are **second-growth forests**. These are the trees that cover an area after the original stands have been removed.

Lobster is a major catch along New England's coast. Almost every small community has its own fishing fleet.

Reforestation projects have been set up to ensure a good supply of lumber and pulp for the future. Reforestation is the planting of new trees on land that was once forested. New England, especially Maine, is now one of the most heavily forested regions in the United States. Almost 75 percent of the region is forest covered as a result of reforestation. Today, many New England farmers receive additional income by selling firewood and maple syrup from forests and woodlots on their land.

New England's most abundant mining resource is granite. It is found in Massachusetts, Vermont, New Hampshire, and Maine. Vermont is also noted for a high grade of marble, and it is the leading state in the mining of asbestos.

 Level A: Have students research oil-resource development in New England. How can oil-resource development be[nefit] the New England economy? Does oil development pose hazards for the region?

Have students locate the New England States on the map on page 516. Note the proximity of Canada and the Atlantic Ocean.

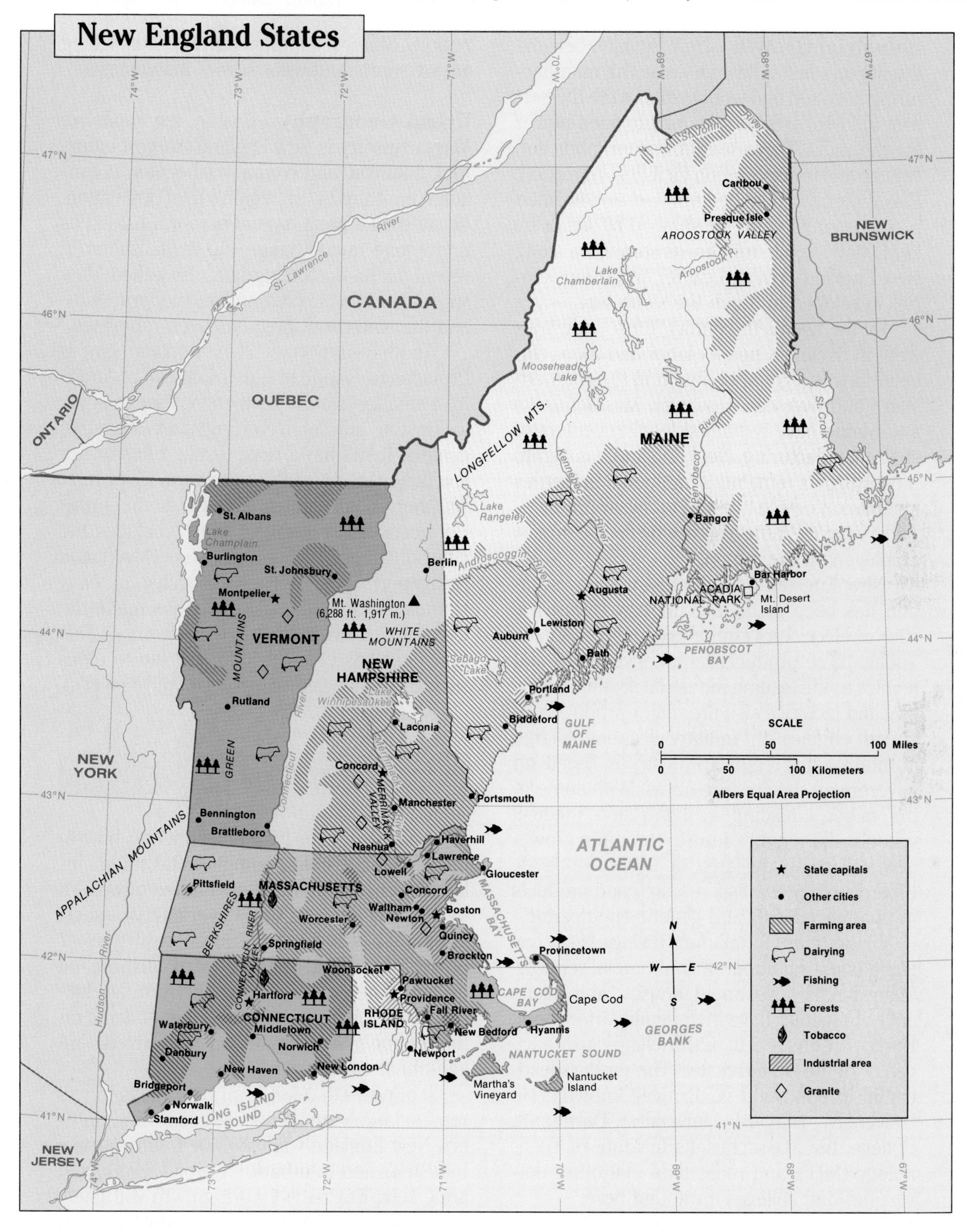

Where is tobacco grown in this region? Massachusetts, Connecticut

Have students look at a landforms map that shows the Hudson and Mohawk River valleys. Ask students why New York City has better access to the interior of the country than New England.

Point out that the first women's college in the U.S., Mount Holyoke, was founded in South Hadley, Massachusetts, in 1837.

Industrial Growth New England became the nation's first industrial area. The manufacturing of cotton and wool textiles were the earliest industries. Swift streams provided power for the mills. There was abundant labor and capital, and transportation facilities were good. Raw materials could be imported, and finished products could easily be shipped by sea or by land. Many textile-mill towns sprang up along New England's rivers. Leather and shoe factories developed in much the same way.

By the twentieth century, these industries were in decline. Cheaper labor and more efficient factories could be found in the southern states and overseas. However, Massachusetts and Maine have remained leading centers of shoe manufacturing. New England has had to modernize its remaining traditional industries and market new textile and footwear products. Today, service industries are by far the largest employers in the region.

New England's new industrial concentration has been in high-technology industries, such as electronics equipment and computers. Today, these products rank above shoes and textiles as the leading industrial products. New England factories also are major producers of aircraft engines and military weapons. A large portion of the region's industry is based on government defense programs. With some of the nation's leading hospitals and medical schools, the region's future economic growth also will be in the area of **biotechnology**. Biotechnology involves research and products in pharmaceuticals and genetic engineering.

In addition to manufacturing, New England offers a number of commercial services. Many banks, investment houses, and insurance companies have their headquarters here. Hartford, Connecticut, is the leading insurance center of the United States. The region's leadership in education is also well known. Yale University, Harvard University, Dartmouth College, the Massachusetts Institute of Technology (MIT), and more than 250 other universities and colleges are found here.

A cool climate, beautiful scenery, and historic sites have made tourism important to the region. Skiing, hunting, fishing, and sailing attract sports enthusiasts from many states.

Urban Geography Boston, the capital of Massachusetts, is New England's largest industrial, financial, and commercial center. It contains one-third of the region's total population. Located on a great natural harbor, Boston has undergone rapid change and revitalization in recent years. Boston Harbor, also called Massport, can now accommodate modern ships, and the waterfront area has been renovated.

Another important New England city is Providence, Rhode Island, the capital of America's smallest state. Providence is second only to Boston among New England's industrial centers. It still has a large textile industry and is a major producer of jewelry and silverware. Another major industrial area is the lower Connecticut River valley, which includes the cities of Hartford, Waterbury, New Haven, and Bridgeport. The Merrimack Valley of New Hampshire and Massachusetts has made the transition from textiles to electronics. This transition has brought industrial life to cities such as Manchester, Lawrence, and Lowell.

Future Prospects

A challenge for New Englanders is how to have adequate energy supplies and still maintain the environment. There has been concern in the New England states over energy issues. These issues include the location of nuclear power plants and the effects of offshore oil drilling. The pollution from oil-burning electric power plants and the building of dams on the region's few remaining wild rivers are additional concerns.

It appears that northern New England will remain largely rural and forested, while southern New England will continue to attract new industries and to urbanize beyond the Boston area. New England's future growth will likely be based on high-technology industries and tourism.

 Have students make an economic map of Maine illustrating the major economic activities that take place in the state. Students should use a variety of economic maps found in the library.

CHAPTER 45 CHECK

Reviewing the Main Ideas

1. The New England states include Maine, New Hampshire, Vermont, Massachusetts, Rhode Island, and Connecticut. This region has played an important role in United States history.
2. All of New England has a humid continental climate, with a maritime influence along the coast. The landforms of this region vary from mountains and hills to plains and seashore.
3. Dairy farming is New England's most important agricultural activity. However, farming in this region has declined during the last century.
4. Fishing, forestry, and mining are traditional economic activities in New England. High-technology industries and services are replacing manufacturing in the region.

Building a Vocabulary

1. Define *granite*. Where is granite mined in New England?
2. What is moraine? Which shoreline features of the New England coast are the result of moraine deposits?
3. Why are most of the forests in New England second-growth forests? Why is reforestation important to this region?
4. Define *biotechnology*. Why is New England likely to experience economic growth in this area?

Recalling and Reviewing

1. Name the six New England states. Describe the various landforms found in each.
2. Describe the climate of New England. How do landforms and climate affect agriculture in New England?
3. How are fishing, forestry, and mining important to New England?
4. How is industry changing in New England today? Why is it changing?

Critical Thinking

5. Why do older industrialized regions such as New England find it hard to compete in today's markets? What steps could be taken to improve New England's competitive position?

Using Geography Skills

Use the map in this chapter or the maps in Chapter 44 to answer the following questions.

1. Where in New England is the highest population density found? Why?
2. What economic activity is most important in New England? How can you tell?
3. Which New England states share a border with Canada?
4. At what latitude and longitude is Boston located? Hartford?

Answers to Chapter Check questions are found in the Teacher's Guide.

Additional activities for Chapter 45 are found in the Student Workbook. Chapter 45 Test as well as Skills, Reteaching, Critical Thinking, and Challenge/Enrichment activities are available in the Teacher's Resource Binder.

Chapter 46

Chapter 46 Lesson Plans and Planning Guide are found in the Teacher's Guide.

Refer students to the map on page 533.

The Middle Atlantic States

▲ **Delaware's wildlife is protected in this refuge.**

▲ **Cities are home to 74 percent of all Americans.**

GEOGRAPHY DICTIONARY

- truck farms
- bituminous coal
- coke
- anthracite coal
- megalopolis
- borough

The Middle Atlantic states include New York, Pennsylvania, New Jersey, Delaware, Maryland, and West Virginia. The region also contains the United States' capital, Washington, District of Columbia. Approximately one-fifth of the nation's population lives here in what is one of the world's most industrialized and urbanized areas.

The Middle Atlantic states have played an important role in the development of the United States. For example, significant iron-ore and coal deposits, together with important land and water routes, helped the region become an industrial giant during the nineteenth and early twentieth centuries. This industrialization drew thousands of immigrants to the United States. Today, this region is where many of the nation's economic and political decisions are made.

 Point out urban scenes in the photos in this chapter and relate them to the map on page 533.

Physical Geography

Landform Regions Three major landform regions cross the Middle Atlantic states: the Atlantic Coastal Plain, the Piedmont region, and the Appalachian Mountains. (See the map on page 533.)

The Atlantic Coastal Plain extends through all of the Middle Atlantic states except West Virginia. It is flat and remains close to sea level, even inland. Interesting geographic features of the Atlantic Coastal Plain include Long Island and the coastal estuaries. Long Island was deposited off the New York coast by ice sheets during the last ice age. It is mainly a series of moraine deposits that are separated from the mainland by Long Island Sound.

Chesapeake Bay, which runs north from the coast of Virginia, is the largest and most productive estuary in the United States. It is a major supplier of fish and shellfish. Delaware Bay, which separates Delaware and New Jersey, forms another large estuary. A smaller, but still important, estuary is at the mouth of the Hudson River at New York City. All three estuaries provide excellent natural harbors and remain ice-free all winter.

Inland from the Atlantic Coastal Plain is the Piedmont region. The Piedmont gently slopes down from the Appalachians to meet the Coastal Plain. Where the two regions meet, rivers plunge over the fall line. Because of the available water power, many cities developed in this region.

The Appalachian Mountains extend through the Middle Atlantic states region in a northeast-southwest direction. Here, the largest part of the Appalachians is made up of a series of folded parallel ridges and valleys known as the ridge and valley section. It was formed more than 300 million years ago when, according to the theory of plate tectonics, eastern North America collided with Africa.

Major rivers of the region cut through some of the Appalachian ranges on their way to the Atlantic. These include the Potomac, Susquehanna, Delaware, and Hudson rivers. The Susquehanna, which enters the Atlantic in Chesapeake Bay, is the largest North American river flowing into the Atlantic.

The far northern part of this region was glaciated during the last ice age. The ancient Adirondack Mountains in northern New York were eroded by ice sheets. New York's Finger Lakes also were carved by the ice. Two of the Great Lakes, Lake Ontario and Lake Erie, also were formed by glaciers.

Climate The Middle Atlantic states have two major climate types: a humid subtropical climate in the south and a humid continental climate in the north. The dividing line between the two climates extends westward from the Atlantic coast south of New York City.

Summers in both climate regions can be very hot and humid. Air flows in from over the warm Gulf Stream current in the Atlantic and from over the warm Gulf of Mexico. Though winters are generally not as severe as in regions farther inland, there can be cold periods when arctic air from Canada enters the region. Snowstorms occasionally occur, causing extreme transportation problems for the major Middle Atlantic cities.

Harsh storms strike the Middle Atlantic region during winter, shutting down entire metropolitan areas.

Level A: James Michener's *Chesapeake* gives an interesting view of the geography and history of the region. Assign students different chapters to read and report on. How has the geography of the area influenced its historical and economic development?

Emphasize that fresh vegetables need to be grown close to the urban markets of the region.

Economic Geography

Agriculture Agriculture in the Middle Atlantic states is practiced primarily to feed the people of the region's many large cities. Fresh fruits and vegetables are produced in Long Island, New Jersey, Delaware, and Maryland. These farms are sometimes called **truck farms** because their products can easily be trucked to the big-city markets.

Some areas of the region specialize in certain crops. Potatoes are grown in New Jersey and Long Island. Vineyards are found in the Finger Lakes area and in Long Island. Other areas specialize in orchards and berries. Large farms in Pennsylvania produce grains as well as beef and pork. The poultry farms of Maryland and Delaware supply chickens and fresh eggs for the region. Pennsylvania and New York are leading milk-producing states.

During recent years, there has been a decrease in farmland in the Middle Atlantic states. Part of this loss has been the result of urban expansion into rural areas. Valuable farmland is being bought for housing and industry. Other farms, particularly smaller ones, have not been profitable.

Natural Resources The Middle Atlantic states are highly industrialized. Coal has been the region's key natural resource. One of the largest **bituminous coal**, or soft coal, areas in the world is found in western Pennsylvania, in West Virginia, and in the bordering eastern sections of Ohio and Kentucky. Bituminous coal has a wide variety of industrial uses. The most important is the production of **coke**, or coking coal, which is used in blast furnaces to purify iron ore for making steel. Because it was relatively low in cost, plentiful, and mined locally, bituminous coal became the basis for the industrialization of the Middle Atlantic states and the Midwest.

Deposits of **anthracite coal**, or hard coal, are found in eastern Pennsylvania. Almost all of the anthracite mined in the United States comes from this small area on the upper Susquehanna River between Wilkes-Barre and Scranton, Pennsylvania. Anthracite mining has declined over the last several decades.

The coal-producing region of Appalachia has suffered economic decline. What effect has this had on its population? unemployment, poverty, migration

The coal-producing region in the Appalachians also has suffered decline in recent years. This coal region, sometimes called Appalachia, suffers some of the highest unemployment and poverty rates in the nation. Thousands of people have left the region, seeking jobs elsewhere. Coal mining also has left scars on the land. Though Appalachia is centered on West Virginia and Pennsylvania, it actually extends into adjoining Midwest and southern states.

Although America's first oil wells were in Pennsylvania, this region has never been a major petroleum producer. There are small petroleum and natural-gas deposits in the Appalachian Mountains.

While this region is usually not associated with forests and wilderness areas, there are several areas that should be mentioned. The Adirondack Mountains and the Catskill Mountains in New York, the Pine Barrens in New Jersey, and numerous mountain ranges in Pennsylvania and West Virginia have remained unpopulated. Though these forested regions are not major wood producers, they serve as important areas for wildlife preservation and for recreation for the huge urban populations nearby.

Point out that coke is made by heating coal, in the absence of air, in order to force out gases and liquids. This process leaves mostly carbon.

Have students locate the Middle Atlantic States on the map on page 516. Note that Washington, D.C. is part of this region.

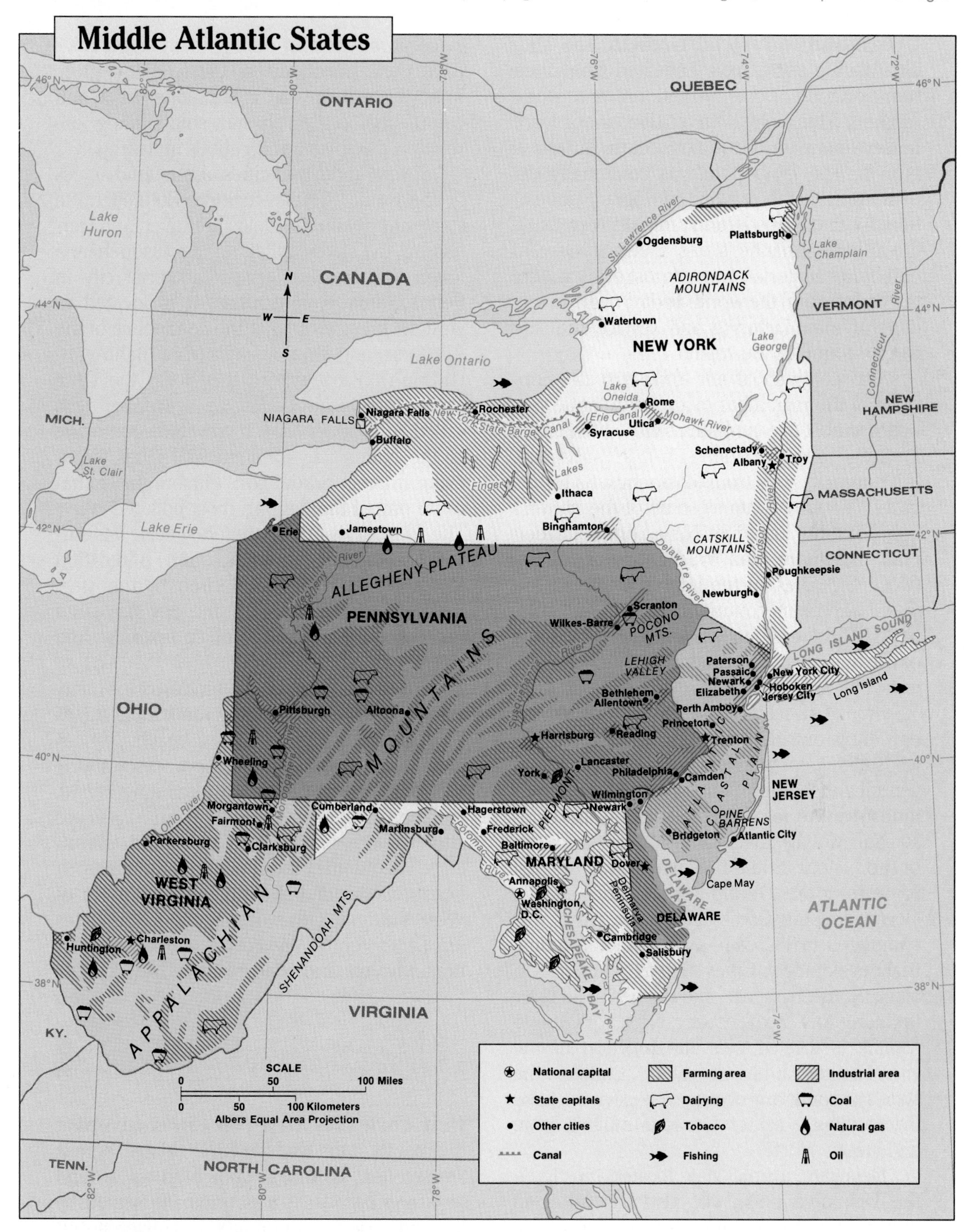

What waterway connects Lake Erie with Utica, New York? New York State Barge Canal (Erie Canal)

Urban and Industrial Growth The Middle Atlantic states have benefited from their transportation access to national and world markets. This along with a skilled labor force helped maintain the region's early industrial growth. The labor pool was continually expanding with the growing immigrant population. As canals, roads, and railroads were built, the Middle Atlantic states found it easy to obtain raw materials from almost every part of the country and the world. Today, nearly every kind of manufacturing and service industry can be found in the region.

The commercial and industrial development of the region has made the Atlantic seaboard almost one continuous city. Often called a **megalopolis** (mehg-uh-LAHP-uh-luhs), this supercity runs from Massachusetts to Virginia. The area contains some of the nation's largest cities, including New York, Philadelphia, Baltimore, and Washington, D.C. Between the cities, in the interurban areas, are many suburban communities and industrial parks. The megalopolis is connected by some of the busiest airports and interstate highways in the nation.

Industrial activity also has spread to several areas outside the major megalopolis cities, such as upstate New York, Delaware, and central and western Pennsylvania. New York's industries are located chiefly along the Hudson and Mohawk river valleys. This is the route of the historic Erie Canal, which was the first route to connect trade between the Atlantic Ocean and the Great Lakes. The canal was completed in 1825 and soon attracted industrial development. Cities such as Schenectady, Utica, Rome, Syracuse, and Rochester all grew up along this canal. Albany, New York's state capital, is located near the junction of the Hudson and Mohawk rivers. Buffalo, the state's second largest city, is a port on Lake Erie and is a center for flour milling, iron, and steel.

Delaware, a small state located mostly on the Delmarva Peninsula, also has attracted industrial and commercial development. Because Delaware's tax laws are favorable to corporations and banks, many companies have their headquarters there. Wilmington, in northern Delaware, is the major urban and commercial center. It has shipbuilding, oil-refining, chemical, and rubber industries.

Several industrial cities also extend across Pennsylvania. These include Bethlehem, a steel-mill city on the Delaware River, and Harrisburg, the state capital, on the Susquehanna River. The second largest industrial city in Pennsylvania is Pittsburgh. It is located in western Pennsylvania at the confluence of the Allegheny and Monongahela (muh-nahn-guh-HEE-luh) rivers, where they form the Ohio River. Pittsburgh began as a frontier fort, called Fort Pitt, in 1754. It grew to become one of the nation's major industrial cities, especially in steel production. Pittsburgh also became one of the nation's most polluted cities. The Pittsburgh urban area, however, has suffered economic problems and population losses because of the decline of the steel industry. In recent years, the city has been revitalized with towering office buildings and cleaner air.

Tourism is an important industry for many cities and towns throughout the Middle Atlantic region. The major tourist beach destinations in the region are the New Jersey shore, especially Atlantic City, eastern Long Island, and the Delmarva Peninsula. Tourists also are attracted to resorts in the Catskill and Adirondack mountains, the Pocono Mountains in Pennsylvania, and the rugged mountains of West Virginia. Other important tourist sites are Lancaster, Valley Forge, and Gettysburg, Pennsylvania, and Niagara Falls, New York.

Urban Geography

New York City New York City is situated at the mouth of the Hudson River, where there is an excellent natural harbor built up around bays and islands. It was originally settled by the Dutch, who named it New Amsterdam. It was seized by the British in 1664 and renamed

Level B: Have students use *The New Yorker* magazine or the Sunday *New York Times* Entertainment section to find the wide range of events and activities taking place in New York City.

Have students locate the Midwest on the map on page 516. Note the Great Lakes and the proximity of Canada.

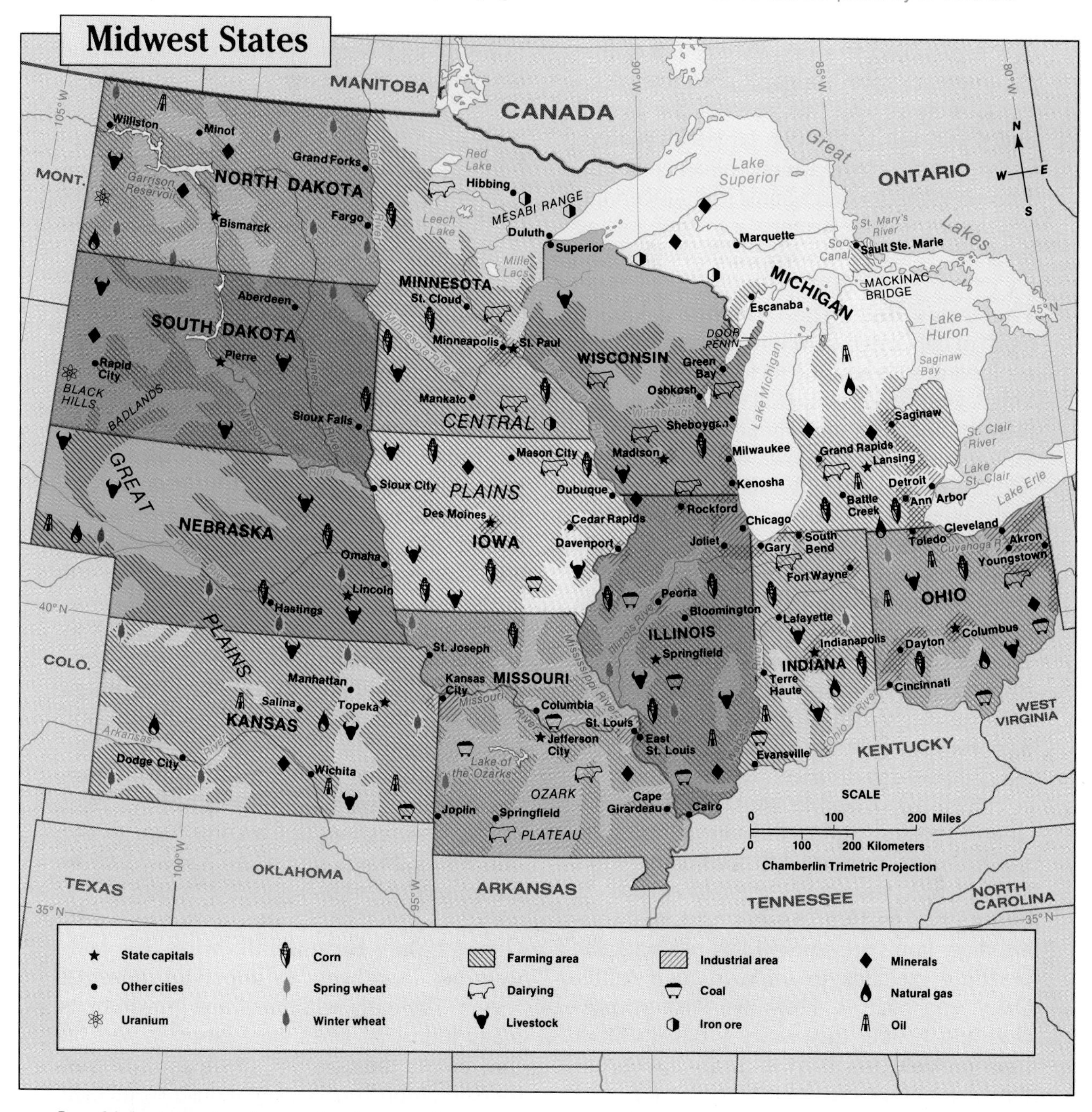

In which states is winter wheat grown? winter wheat: Kansas, Nebraska, Missouri, Illinois, Indiana, Ohio

occurred with the development of large-scale irrigation.

Rainfall in the wheat belts often varies. Even a slight reduction in rainfall can mean crop failure or that expensive irrigation systems must be installed. Hailstorms, insects, and plant disease also can be a threat. Soil erosion is another major concern. In windy areas, rows of trees, called shelterbelts, have been planted to protect the soil. To prevent soil erosion caused by runoff in the more hilly and rolling countryside, contour plowing, or plowing along natural contour lines, has been used.

Though farmers in the wheat belts make the United States the leading wheat exporter, they are subject to world market prices and government price supports. Political decisions, such as how much wheat the United States will sell to the two largest importers, China and the Soviet Union, influence prices. To protect themselves against such uncertainties, more wheat farmers are now raising livestock.

The Dairy Belt The Dairy Belt is located north of the Corn Belt, where the summers are cooler and soils become rocky and less fertile. This region covers almost all of Wisconsin and most of Michigan and Minnesota. The Dairy Belt also extends into Iowa on the southwest and into New York and Vermont on the northeast. Often called "America's Dairyland," Wisconsin produces more milk, butter, and cheese than any other state. Products from the Dairy Belt are sold in the nearby midwestern urban areas. With refrigerated trucking, however, fresh Wisconsin dairy products are shipped to other regions of the country. Other products, such as condensed, evaporated, and powdered milk, are not as limited by time, temperature, and distance.

Much of the land in the Dairy Belt is devoted to pasture and to the cultivation of the hay, oats, and corn used to feed dairy herds. Many farms also grow vegetables such as beans, peas, beets, and sweet corn. Midwestern dairy farms are efficient and use scientific breeding methods to improve their herds. Dairy cooperatives help the farmers process and market their dairy products. Strict laws regulate the dairy industry to ensure that dairy products are safe and pure. Like wheat farmers, the dairy farmers of the Midwest have had problems in recent years because of large dairy product surpluses and low prices.

Many dairy farms have woodlots, which can be used to earn extra income from firewood, maple syrup, Christmas trees, and wood pulp. Because of the warming effect of the Great Lakes, specialized fruit-growing areas are located around the Great Lakes, especially in western Michigan and the Door Peninsula in Wisconsin. Cherries, cranberries, apples, and other fruits are grown.

Resources and Industrial Growth

Forests After the major forests of New England and the Middle Atlantic states were cut, the lumberjacks moved into Wisconsin and Michigan. Logging began in the Midwest during the 1860s and lasted until just after the turn of the century.

Today, the region's virgin pine forests are gone, although many second-growth forests have appeared. Lumbering is no longer a major industry. The forests of northern Wisconsin and Minnesota and of Michigan's Upper Peninsula now serve as major recreational areas. The region's forests, lakes, streams, and hills attract millions of tourists, who come to camp, hunt, fish, and ski. The Ozarks of southern Missouri also remain mainly forested and serve as a tourist center. These forests also serve as valuable wildlife and **watershed** areas. A watershed is a region drained by a river system. Recreation has become vital to the Midwest and joins agriculture and industry as a cornerstone to the region's economy.

Great Lakes Industrialization The Midwest has long been an important industrial region. The early settlement and growth of its many industrial cities were based in part on the cities' location on or near important transportation routes. Key industries include food processing, automobile assembly, machinery, and the manufacturing of iron and steel.

Many industrial cities of the Midwest grew as ports on the Great Lakes. Here, they had access to the coal deposits of the Appalachians, local coal mines in Illinois, Indiana, and Ohio, and the iron-ore deposits of upper Michigan and northern Wisconsin and Minnesota. The iron ore was shipped out of the ports of Duluth and Superior on Lake Superior to

 Level B: Ask students to list and locate on an outline map the leading industrial cities of the Midwest.

Physical Geography

Climate Except for the higher peaks of the Appalachian highlands, where temperatures are cooler, the South has a humid subtropical climate. Throughout the region, summers are long, hot, and humid, and winters are mild.

For most of the region, rainfall averages 40 to 60 inches (100 to 150 centimeters) per year and is evenly distributed. In the Appalachians, annual totals may reach more than 100 inches (250 centimeters). Portions of the South experience occasional winter snowstorms, especially in the Appalachians. During winter, cold arctic air may cause crop damage as far south as Florida and southern Texas.

During late summer and early fall, hurricanes may strike from the warm Atlantic and Gulf of Mexico. The region is also subject to severe thunderstorms and tornadoes.

Landform Regions The South's landform regions include the Coastal Plain, the Piedmont, the Appalachian Mountains and Ozark Plateau, and the Interior Plains. (See the map on page 551.) The largest of these is the Coastal Plain, which lies along the coast of the Atlantic Ocean and the Gulf of Mexico. Much of this region is low and has a shoreline bordered by **barrier islands**. A barrier island is a long, narrow, sandy island separated from the mainland by a lagoon. Examples are Cape Hatteras, North Carolina, Miami Beach, Florida, and Padre Island, Texas.

The Coastal Plain contains the Mississippi Delta, the lower Mississippi Valley, and the low Florida Peninsula. It also includes coral islands at the tip of Florida, known as the Florida Keys. Much of the Coastal Plain is swampy, especially the Everglades in Florida, the Okefenokee (oh-kuh-fuh-NOH-kee) Swamp in Georgia, and the Mississippi Delta in Louisiana. The low Coastal Plain extends inland to the fall line, where the plain meets the hard rock of the Piedmont. The rolling hills of the Piedmont rise inland in central Virginia and the Carolinas and in northern Georgia.

Farther inland are the Appalachian Mountains, which consist of several major ranges, including the Blue Ridge, Great Smoky, and Cumberland mountains. The Appalachians extend through the western part of Virginia and the Carolinas, the eastern portions of Kentucky and Tennessee, and into northern Georgia and Alabama. The Ozark Plateau is located in Arkansas and Oklahoma and extends northward into Missouri. This ancient highland region is forested and rugged. Much of the South lies in the Interior Plains. This region includes central Kentucky and Tennessee and portions of Texas and Oklahoma.

River Systems The Mississippi is one of the largest river systems in the world. Along with its tributaries, which include the Ohio and Missouri rivers in the north and the Arkansas and Red rivers in the south, it drains nearly half of the United States.

Great amounts of sediment are carried by the river and deposited at its mouth, forming the large Mississippi Delta. The delta is located in Louisiana, where the river empties into the Gulf of Mexico. In the delta, the land is low

For the Record

Once, in one of those lovely island chutes, we found our course completely bridged by a great fallen tree. This will serve to show how narrow some of the chutes were. The passengers had an hour's recreation in a virgin wilderness, while the boat-hands chopped the bridge away; for there was no such thing as turning back, you comprehend.

From Cairo to Baton Rouge, when the river is over its banks, you have no particular trouble in the night . . . but from Baton Rouge to New Orleans it is a different matter. The river is more than a mile wide, and very deep—as much as two hundred feet, in places. Both banks, for a good deal over a hundred miles, are shorn of their timber and bordered by continuous sugar plantations . . . .

From *Life on the Mississippi* by Mark Twain, page 105.

Life on the Mississippi was first published in 1883. Ask students to read what Twain says about the differences in various parts of the river. Note that Cairo is near the junction of the Mississippi and Ohio rivers.

Note that Texas and Oklahoma are sometimes grouped into a region called the Southwest. Both states share physical features of the Great Plains as well.

and swampy and cut through by small, sluggish streams called **bayous**.

The Mississippi has often flooded its banks and caused much destruction to the areas along its floodplain. Yet, the river serves the South as a means of transportation, and the floods are a source of fertile new soil. Flood danger and damage from the Mississippi have been lessened by better flood forecasting, by dams along its upper tributaries, and by an improved system of levees (LEHV-eez). A **levee** is a ridge of earth along a riverbank that prevents an area from flooding. The first Mississippi River levee was built at New Orleans in the early eighteenth century.

The Tennessee River originates in the Appalachians and flows westward through several southern states. With its large tributary, the Cumberland, it joins the Ohio River near its junction with the Mississippi River. In 1933, the federal government formed the Tennessee Valley Authority (TVA). The TVA built dams, reservoirs, and canals in the Tennessee River valley. The program was designed to control flooding, slow soil erosion, allow barge navigation, and produce electricity.

The newest river project is the Tenn-Tom Waterway. As the result of dredging and the construction of locks on the Tombigbee River, the Tennessee River now connects the Ohio River and the port of Mobile, Alabama, on the Gulf of Mexico.

West of the Mississippi River, the Arkansas (ahr-KAN-zuhs) River also has been dredged, and locks have been built. This has made Tulsa, Oklahoma, into a port by connecting it with the Mississippi River. These channel and lock systems provide barge transportation for iron, oil, coal, and wheat to the Gulf ports. These systems also provide flood control and hydroelectricity.

The lower Rio Grande forms the international boundary between the United States and Mexico. It runs from El Paso, where the river enters Texas, to Brownsville, where the Rio Grande meets the Gulf of Mexico. The irrigation waters from this river are important to the productive farms along this borderland.

Agriculture

Traditional Crop Regions The first farmers in the South were the Native Americans who lived in small villages. They grew corn, tobacco, beans, and squash, and they hunted in the forests. The largest tribes were the Choctaw, Cherokee, Creek, and Seminole. When settlers moved into this region, the Indians were driven off their lands. Many were sent to reservations in Oklahoma.

The climate, rainfall, and soil on the Coastal Plain were well suited to growing cotton. Cotton became the South's leading commercial crop. The Cotton Belt followed the Coastal Plain and the Piedmont from southern Virginia to Texas. African slaves were brought into the South to work the large plantations. In the years before the Civil War, the Cotton Belt became the leading cotton-growing area in the world. After the Civil War, with slavery ended, the cotton plantations were broken up into smaller units.

During the last several decades, cotton production in the South has drastically declined. Pests, such as the boll weevil (an insect whose larvae destroy the immature cotton boll), competition, expensive mechanization, and synthetic fibers have all contributed to this decline. Today, most of the United States' cotton is grown on large irrigated and mechanized farms farther west. The leading cotton states are now Texas, California, and Arizona. During the mid-1980s, the South's cotton production increased. This was partly a response to the demand for more cotton clothing.

Tobacco is the South's other traditional crop. Like cotton, tobacco is a labor-intensive crop, and before the Civil War the industry also made use of slave labor. Tobacco is grown mainly on the Coastal Plain and the Piedmont between Virginia and Georgia and in Kentucky and Tennessee west of the Appalachians. North Carolina, Virginia, and Kentucky remain the core of the Tobacco Belt. Cities especially noted for the manufacture of cigarettes and pipe tobacco are Durham and

Level A: The TVA project was a response to unique economic conditions of the times. Have students write reports on the TVA and how it has changed the region's physical, economic, and cultural geography.

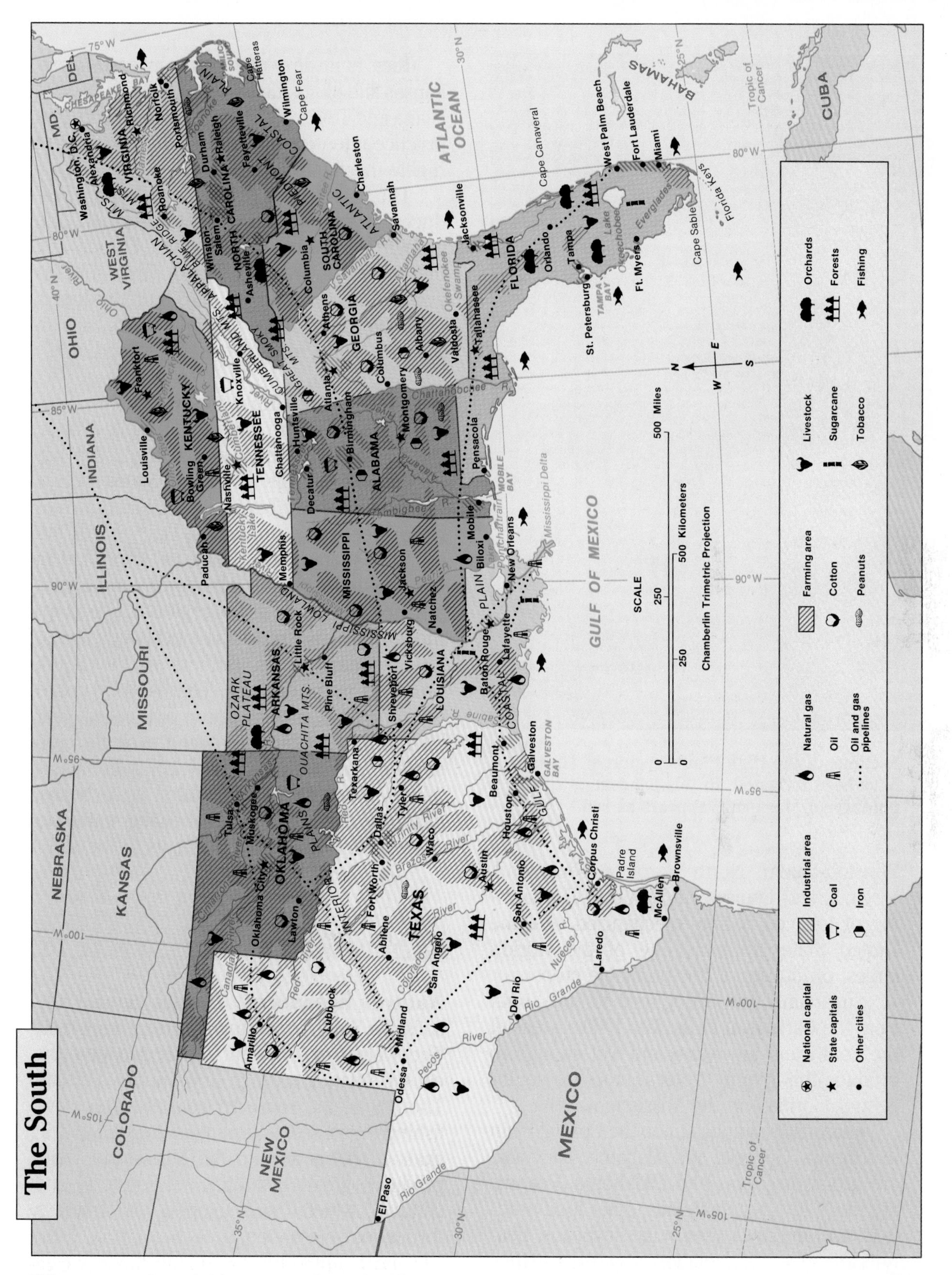

What economic activities occur along the Mississippi River? industry, oil production, farming, growing of cotton and sugarcane, livestock grazing

Have students locate the South on the map on page 516. Note the proximity of the Atlantic Ocean, the Gulf of Mexico, and Mexico.

Point out to students that the U.S. government subsidizes the tobacco industry.

Sections of the High Plains region of Texas have been irrigated. In this Landsat photograph, farmland appears as red.

Winston-Salem, North Carolina, Richmond, Virginia, and Louisville, Kentucky. The farming of tobacco is now mechanized, and the United States remains one of the world's largest producers of this profitable crop.

Sugarcane was introduced to the South from the Caribbean in the late 1700s. Areas of the South that are warm and wet enough to produce this mainly tropical crop are south-central Florida and the Mississippi Delta.

Peanuts are important on the Coastal Plain in Virginia, Georgia, and Alabama. They also are grown in Texas and Oklahoma. Peanuts are made into vegetable oils, other food products, and livestock feed. Like soybeans, peanuts are grown as a rotation crop because they add nutrients to the soil.

Rice is an important crop in eastern Arkansas and along the wet Coastal Plain of Louisiana and Texas. Unlike rice farms of Asia that practice intensive agriculture, these farms are highly mechanized.

The South is one of America's leading producers of citrus fruits. Oranges, lemons, and grapefruits are grown in central Florida and the lower Rio Grande valley of Texas. Frozen and bottled fruit juices also are produced, especially in Florida. Georgia specializes in peaches, and Virginia is an apple producer.

Changing Southern Agriculture Soybeans have become the most valuable crop in the South. Corn, for livestock feed, also has become increasingly important. States such as Florida, Georgia, Texas, and Oklahoma are major beef-cattle producers. Dairy herds are found in many states. Hogs are important in Georgia and Alabama. The poultry industry is a major enterprise in the region. Large commercial farms are replacing most family farms.

For many years, farmers in the South practiced **monoculture**. That is, they raised only one crop, such as cotton or tobacco, year after year. Constantly growing a single crop depletes soil nutrients. Today, commercial fertilizers, along with crop rotation of soybeans, peanuts, and corn, and allowing cattle to graze, return important nutrients to the soil.

Resources and Industry

Natural Resources Besides climate and soils, the South boasts many natural resources. The waters of the Gulf of Mexico and the coastal Atlantic are rich in marine resources. Though all the southern coastal states have fishing industries, Louisiana is the leader. The shallow waters around the Mississippi Delta produce great quantities of oysters and shrimp, which are shipped to markets throughout the nation.

The South's forests provide many useful products. Southern pine is plentiful on the

 Level A: Peanuts were developed as an important crop due partly to the research of George Washington Carver. Have students prepare reports on Carver and his research.

The Great Plains

Physical Geography The Great Plains are located in the middle of the United States between the Central Plains on the east and the Rocky Mountains on the west. They extend from the Mexican border to the Canadian border. The Great Plains include the western parts of Texas, Oklahoma, Kansas, Nebraska, North Dakota, and South Dakota and the eastern parts of New Mexico, Colorado, Wyoming, and Montana. (See the map on page 559.)

The Great Plains were formed by sediments eroded from the mountains and deposited by rivers. They rise gradually from east to west. By the time they reach the foot of the Rocky Mountains, they are more than 4,000 feet (1,220 meters) above sea level.

The continental steppe climate dominates the region. The Great Plains are semiarid and become drier toward the west. Temperatures are extreme, with winters reaching to −40° F (−40° C) and summers reaching above 100° F (38° C). In winter, blizzards occur, bringing cold winds filled with blowing snow. In summer, hot, moist air from the Gulf of Mexico brings severe thunderstorms. Strong, dry foehn winds, called chinooks, descend from the Rockies and affect the western portions of the Great Plains. Droughts, however, are the major climate hazard of the region and occur every few decades. During drought periods, huge dust storms may cover large areas.

The vegetation of the Great Plains is steppe grasses, shrubs, and sagebrush. The region is treeless except along the riverbanks. Early settlers considered the region a wasteland because there were no trees there. They called it the "Great American Desert."

Agriculture Because of the lack of moisture, most of the Great Plains has been better suited to grazing than to farming. In fact, wheat farming is a quite recent development in this area.

During the mid-1800s, huge herds of cattle and flocks of sheep roamed the open range of the Great Plains. They replaced the huge buffalo herds hunted by the Plains Indians prior to the arrival of the railroads and settlers. Eventually, barbed wire, steel plows that cut through the tough sod, and windmill pumps for wells ended the era of open range. The plains were divided into fenced ranches and farms. Today, livestock raising is still a major economic activity, but it is generally combined with wheat farming.

About 80 percent of Wyoming's land is used for grazing. These sheep are grazing on the Great Plains of eastern Wyoming.

The Great Plains region is cut across by the 100th meridian. West of this longitude, the region becomes drier. Generally, less than 20 inches (50 centimeters) of rain fall per year, making irrigation necessary for crop growth. To the east of the 100th meridian, as the rainfall amounts increase, the farmers face less risk of crop failure.

Underground water supplies are pumped for use in the fields. The largest source of water under the Great Plains is the Ogallala Aquifer. This huge underground storage system stretches from South Dakota to Texas. For many years, this aquifer could supply all the farmers' needs. With increased irrigation and wasteful water practices, the water table began to drop during the 1950s. Today, the depletion rate has slowed with more efficient irrigation

Point out that because the Great Plains were largely treeless, sod was used as a building material. Sod houses were cool in the summer and warm in the winter.

Have students research the causes and effects of the Dust Bowl. Students should discuss how modern farming methods can prevent future dust bowls.

methods. A decline in farm acreage also has lessened the demand for water.

The Great Plains is crossed by a number of rivers, including the Yellowstone, Missouri, Platte (PLAT), Arkansas, and Pecos. These rivers originate in the Rockies and flow eastward. In the spring, the rivers are full, while in the summer, the water slows to a trickle. Today, major dam-building projects have made farming less of a risk. The largest project was the Missouri River basin, which drains nearly all of the northern Great Plains. The series of dams and reservoirs has allowed better irrigation, flood control, and land management.

A productive irrigated farm area just to the east of the Colorado Rockies, called the Colorado Piedmont, is a major producer of sugar beets, corn, beans, and cattle feed. On the Colorado Piedmont, cattle are raised in huge feedlots. Other large irrigated farm areas are found along the Platte River of western Nebraska and the High Plains area centered around Lubbock, Texas. In the latter, cotton and cattle feed are the main crops.

Some areas that do not produce wheat or have irrigated farmland are used mainly as beef-cattle ranches. The largest are found in the Wyoming Basin, the Edwards Plateau in West Texas, and the grass-covered dunes of the Nebraska Sand Hills.

Though this is a productive agricultural region, the farmers here are always challenged by climate and world markets. During the 1930s, the combination of a drought and the Great Depression meant disaster for thousands of farmers on the plains. Many farmers migrated to California and other states to seek jobs. During this catastrophic era of plains farming, this area was known as the Dust Bowl because of the dust storms created by the bare abandoned land. Today, most farmers are as concerned about the price of wheat on world markets, which also can be disastrous to them, as they are about drought.

Fair distribution of water resources is important. Dams and irrigation projects help make water available to this region.

Resources and Industry Both coal and oil have been found beneath the Great Plains. Oil production on the plains has been mainly from West Texas, Wyoming, Colorado, and eastern New Mexico. Major coal deposits are mined in Montana and Wyoming. The coal, which is buried just beneath the surface, can be removed by **strip mining**. In strip mining, soil and rock is stripped away by large machines to get at the coal beneath the surface. While strip mining is an environmental concern, if proper reclamation methods are practiced, the land remains usable after the mining is finished. To reclaim the land, the topsoil is saved, and the land surface is returned to its original contours. If this is not practiced, serious erosion occurs.

Denver is the dominant city and largest urban area of the Great Plains and of the adjoining Rocky Mountain region. It is the capital of Colorado and the leading industrial, transportation, and financial center of the Great Plains region. Located at the foothills of the Rockies, it serves as the economic center for both regions.

 Ask why many mining towns are no longer inhabited. One reason is that the mines are no longer usable.

producing region, supplies of oil may be mined in the future from deposits of oil shale found in western Colorado and Utah.

The Intermountain region is an important tourist destination, particularly Arizona and Utah. Because of the spectacular beauty of this part of the nation, much of the area has been set aside as national parks. Grand Canyon National Park in northwestern Arizona is one of nature's most impressive sights and attracts almost three million tourists per year. The canyon walls of rock layers that vary in color have been carved by the Colorado River over millions of years. The canyon is about one mile (1.6 kilometers) deep and 217 miles (347 kilometers) long.

Other national parks in the Intermountain area are Zion and Bryce Canyon in Utah, Mesa Verde (MAY-suh VUHRD) in southwestern Colorado, and Great Basin in Nevada. Las Vegas and Reno, Nevada, are the major tourist cities in the region.

Boise, with just over 100,000 people, is the capital and largest city of Idaho. What is Boise's absolute location?

approximately 44°N, 116°W

Population Geography The Intermountain region began to gain population after World War II when the federal government built dams and water projects, military bases, and major highways. The widespread use of air conditioning made the region even more attractive.

The southern part of the Intermountain region, especially Arizona and New Mexico, has a very distinct cultural geography because of strong Indian and Spanish influences. Both states have large Indian populations and have been inhabited by Indians for centuries.

The Spanish arrived in this area prior to the Pilgrims' landing in New England. Many Spanish **toponyms** can be found on maps of the southwest United States. Toponyms are names of places and often reflect the history and culture of a region. Examples are El Paso, Casa Grande, Mesa, and Las Cruces. Spanish architecture and language reflect this cultural influence. Today, the majority of the population of New Mexico speaks Spanish. Most of the farms in these states are dependent upon migrant labor from Mexico to work in the fields. The continuing migration from south of the border ensures that Latin culture will remain important in this region.

Although it is still one of the least densely populated regions in the United States, the Intermountain region is also one of the fastest growing. People are attracted to the region because of the wide-open spaces and warm, sunny climate. Cities here have grown rapidly, especially Phoenix, Tucson, Las Vegas, Albuquerque, El Paso, Salt Lake City, Boise, and Spokane.

Phoenix, the capital of Arizona, is the largest city in the Intermountain region. It is a major industrial city and retirement community. Phoenix contains more than half of the population of Arizona in its expanding urban area. Prior to the 1970s, the economy of Arizona was based mainly on agriculture. Today, the state has an industry-based economy.

The second largest urban area in the region is Salt Lake City. It is the state capital of

Point out that in many of the region's "boomtowns," city services cannot keep up with the rapid population growth.

Point out that the Mormons settled the Salt Lake region in an effort to find religious freedom.

Utah and the headquarters of the Mormon church. **Mormons** are members of The Church of Jesus Christ of Latter-day Saints. They are called Mormons because they follow the Book of Mormon. Salt Lake City is a transportation center and the site of nearby irrigation farming, refineries, and steel mills. It is located near a major ski-resort region in the Wasatch Mountains.

More than 65 percent of Utah's population lives in Salt Lake City and other nearby cities between the foot of the Wasatch Mountains and the Great Salt Lake. Mormon pioneers, led by Brigham Young, settled here in 1847. They were the first settlers to use large-scale irrigation in this arid region.

The Indian Nations The earliest inhabitants in the Intermountain region were the Indians who arrived in this area more than 12,000 years ago. The first major agricultural Indian cultures were established by A.D. 600. When the Spanish arrived in the fourteenth century, there were more than 100,000 Indians living in the region. Most were farmers, but some, such as the Apaches, Utes (YOOTS), and Comanches, were nomadic hunter-gatherers. The name Utah comes from the Ute Indians.

The largest American Indian population is found on the Colorado Plateau of Arizona and New Mexico. The Navajos (NAV-uh-hohz) and the Hopis (HOH-peez) are the largest nations, or tribes. For many years, the Navajos lived as nomadic herders, moving over the plateau with their sheep, goats, and horses in search of pasture. Herding is still an important part of Navajo life, but now many Navajos are farmers as well as herders. The Navajos are famous for their artistic skill in making beautiful baskets, blankets, and jewelry. Some industry and tourism also has been developed by the Navajos. The Hopis are mainly farmers and crafts workers. They live in small farming villages on high plateaus. They also graze sheep and goats on the surrounding land.

The mining of coal and oil on the Indian lands has created as many problems as it has solved. Coal mining and coal-burning electric power plants are causing air pollution. Some Indian religious beliefs are against mining because it desecrates, or violates, the ground. The wages and royalties earned by the tribes from coal and oil are changing traditional ways of life. Even so, poverty and unemployment remain high among the Indian nations.

Regional Issues

The Great Plains, Rockies, and Intermountain region of the western United States are sometimes collectively called the "Empty Quarter." Because this is the nation's most sparsely populated region and occupies almost one-fourth of the area of the United States, this title seems fitting.

The major limiting factor for this region is water. Without continued water supplies, the region cannot develop cities, farms, or mines. Irrigation agriculture is the major consumer of this precious resource. However, rapid urban development and increased mining have made competitive claims to the water. In some areas, underground water supplies have become dangerously low. While future dam building is possible, new dams would threaten wild rivers and flood beautiful canyons.

Rapid development of the region threatens the environment. Cities such as Phoenix and Denver already have major air-pollution problems. Other cities have grown too rapidly and lack adequate traffic management and housing. Many mining communities have gone from boomtowns to ghost towns in a few years. Mining development affects the water and air quality. It also conflicts with tourism as the scenery is blighted by mines, smoke, and rail and truck traffic.

The arid and mountainous environment of the region will limit settlement. Most growth will be in the already large cities with adequate water supplies. Any future changes will require a balance between preservation of the environment and economic growth.

 Level B: Have students research the differences between the Indian culture of the Southwest and mainstream American culture.

CHAPTER 49 CHECK

Reviewing the Main Ideas

1. Much of the western United States can be divided into three regions: the Great Plains, the Rocky Mountains, and the Intermountain region.
2. The Great Plains are dominated by the continental steppe climate. Although several rivers cross the plains, much of the region is dry. The land is irrigated or used for grazing. Coal and oil are important resources in this region.
3. The Rocky Mountains stretch from Canada through New Mexico. Most of this region is rugged and remains wilderness or forested. Tourism is an important industry in the Rocky Mountains.
4. An arid region with a variable climate, the Intermountain region is growing rapidly. Mining and tourism are important to the region.

Building a Vocabulary

1. What is strip mining? Why is it necessary to reclaim the land after strip mining, and how is this done?
2. What divides the major river systems of North America? Into which bodies of water do these river systems flow?
3. Define *rain shadow*.
4. Define *toponym*. Give some examples of toponyms from your reading in this chapter. Can you identify any other toponyms that reflect the history of a region?
5. What city is the headquarters of the Mormon church? What is another name for the Mormons?

Recalling and Reviewing

1. What mineral resources are found in the Great Plains? Why is grazing usually preferable to farming in this region?
2. Describe the three landform regions of the Intermountain region. Why has this region been growing at such a fast rate?
3. What are the major economic activities in the Rocky Mountain region? Why is this region sparsely populated?

Critical Thinking

4. Strip mining is an issue in the states of Montana and Wyoming. Why is coal strip-mined in these states? What danger does strip mining pose to the environment?

Using Geography Skills

Use the map in this chapter to answer the following questions.

1. How are the economic activities of the Great Plains different from the economic activities of the mountain areas?
2. Through which states does the Continental Divide pass?

Answers to Chapter Check questions are found in the Teacher's Guide.

Additional activities for Chapter 49 are found in the Student Workbook. Chapter 49 Test as well as Skills, Reteaching, Critical Thinking, and Challenge/Enrichment activities are available in the Teacher's Resource Binder.

Chapter 50

Chapter 50 Lesson Plans and Planning Guide are found in the Teacher's Guide.

Refer students to the map on page 568.

The Pacific Coast

▲ **About 82,000 Chinese live in San Francisco.**

▲ **Yosemite's El Capitan rises over a green valley.**

The region of the United States known as the Pacific Coast includes most of California and the western parts of Oregon and Washington. Because of differences in climate and economy, the region is usually divided into California and the Pacific Northwest.

If California were a country, its economy would rank sixth in the world, outproducing such nations as the United Kingdom, India, and Brazil. Californians total 10 percent of the United States' population. California is a state of great geographic diversity. Its farms, factories, and industries produce goods sold throughout the world.

The Pacific Northwest states of Washington and Oregon have been somewhat isolated from the rest of the country. They have been producers of fishing, farming, and forestry products and are known for their natural beauty.

 This region is famous for its physical beauty. Have students survey the photos in the chapter for evidence of this.

Physical Geography

Landform Regions The Pacific Coast region can be divided into three major landform areas: the Coast Ranges, the Sierra Nevada and the Cascade Range, and the valleys between the mountain systems. (See the map on page 568.)

The Coast Ranges extend southward along the Pacific Ocean from Washington to Mexico. They are made up of numerous parallel mountain ranges. In California, they are separated by major earthquake-fault systems, such as the San Andreas Fault. Although most of the Coast Ranges form a rugged coastline, there are a few lowland areas between the mountains and the sea. The largest lowland area is the Los Angeles Basin. The Coast Ranges also form the eight Channel Islands off the southern California coast. The Coast Ranges include two mountain ranges. Located in northern California and southern Oregon are the forested Klamath Mountains. On Washington's Olympic Peninsula are the glaciated Olympic Mountains.

Inland and east of the Coast Ranges are the Sierra Nevada and the Cascade mountains. The Sierra Nevada, a giant range of faulted and glaciated granite, is in California. It is one of the longest and highest mountain ranges in the United States. Mount Whitney in the Sierra Nevada is the highest peak in the 48 contiguous states. It reaches an elevation of 14,494 feet (4,419 meters). The Sierra Nevada is dotted by hundreds of lakes, including giant Lake Tahoe. The spectacular beauty of the Sierra Nevada can be best seen in Yosemite (yoh-SEHM-uht-ee) and Sequoia (si-KWOY-uh)-Kings Canyon national parks.

The Cascades are a volcanic mountain chain extending from Mount Baker in northern Washington to Mount Lassen in northern California. The largest of these volcanoes are Mount Rainier, Mount Hood, and Mount Shasta. The Cascades also include the active Mount St. Helens in Washington and Crater Lake in Oregon. Crater Lake is the deepest

Mount Rainier is one of several high peaks on the Olympic Peninsula. Why is this region geologically active? It is volcanic.

lake in the United States. It fills the huge **caldera** of Mount Mazama, an extinct volcano. A caldera is a huge depression formed after a volcanic eruption or collapse of a volcanic mountain.

Between the Cascades and the Sierra Nevada ranges and the Coast Ranges is a lowland region that contains three valleys. Just to the south of Puget Sound, in the northwest corner of Washington, is the fertile Puget Sound Lowland. The Willamette Valley is in Oregon, and the large Central Valley dominates the center of California. All three valleys are centers of agriculture for the Pacific Coast states.

Natural Hazards The geologic structure of the Pacific Coast has caused earthquake and volcanic activity. The San Andreas Fault system occurs where the Pacific and North American plates meet. This active plate boundary extends along the northern California coast through the San Francisco peninsula, then southeast through southern California into the Gulf of California.

The Pacific plate is slowly moving northward along the San Andreas Fault. Scientists believe that millions of years from now the Los Angeles region and parts of coastal California will move past the coast of northern California.

Level B: Have students make maps of this region's landforms.

Have students locate the Pacific Coast states on the map on page 516.

As the Pacific plate moves past the North American plate, an increasing level of stress is built up between the plates. The shock waves caused by moving rock along the fault line may create a major earthquake.

In recent years, scientists have warned that Washington and Oregon also could be hit by severe earthquakes. An undersea ridge just off the shore of these states is at an active plate boundary and could cause severe tremors.

All three Pacific Coast states share the volcanic Cascade Range. The most recent severe eruption was of Mount St. Helens in 1980. However, volcanic peaks in all three states have the potential for future eruptions.

Climate The Pacific Coast has two dominant climate types. A maritime climate exists along the coasts of Washington, Oregon, and northern California. A Mediterranean climate is found in southern and central California.

Rainfall and cloudy weather are common in the maritime climate region, especially in winter. The Pacific Northwest lies directly in the path of prevailing westerly winds. These moist winds moderate summer temperatures and warm the land in winter. They are blocked by the Sierra Nevada and the Cascade mountains. The moisture that the winds bring from Pacific storms falls on the western slopes of the mountains. The Olympic Mountains in Washington receive more than 142 inches (360 centimeters) of rain per year. The Sierra Nevada and the Cascade mountains receive record amounts of snowfall each winter.

The western slopes of the Cascade Range and the Sierra Nevada act as a watershed for the agricultural valleys and cities of the Pacific Coast. The mountains store up rainwater and snow in high lakes and snowfields. The construction of dams, reservoirs, and aqueducts has allowed the water to be stored and distributed to farmlands and cities below.

Southern and central California, with their Mediterranean climate, have long, dry, sunny summers and wetter winters. Most rainfall comes from Pacific storms between December and March. Very little rain falls from April

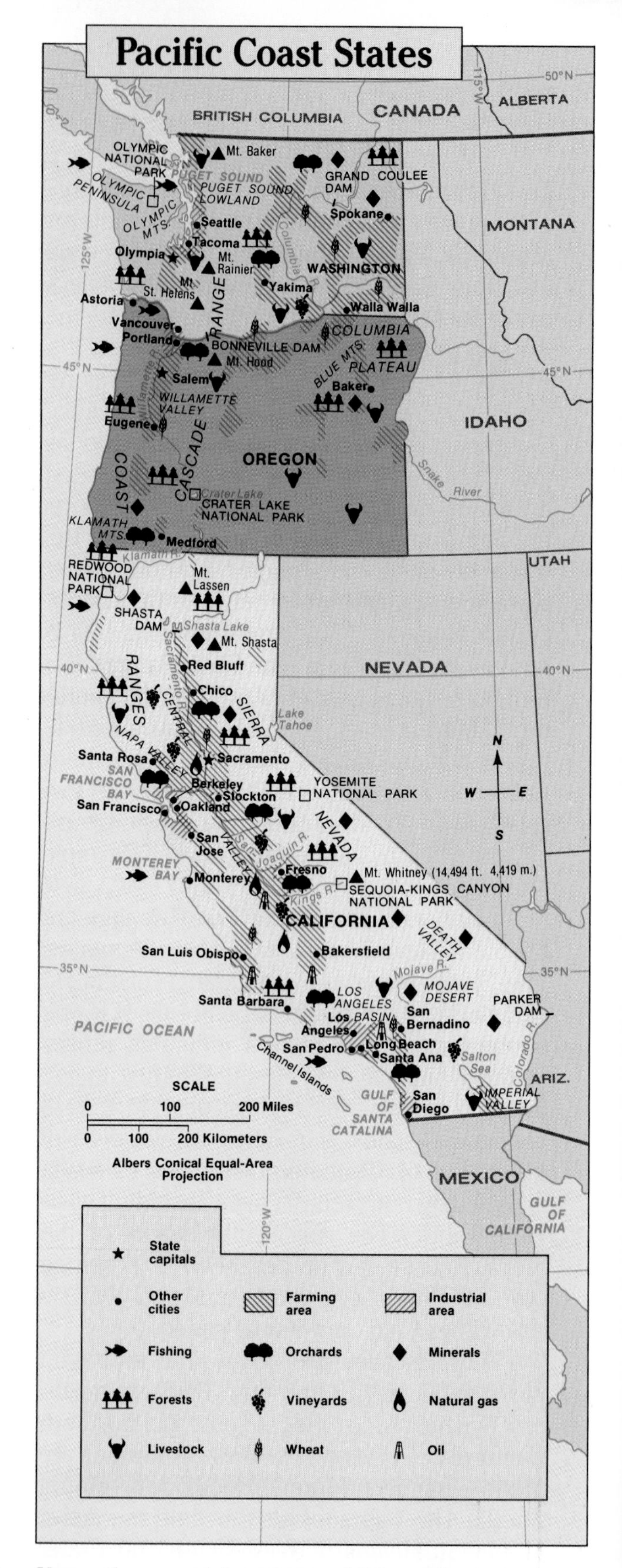

Name the mountain ranges of this region.

Cascade Range, Coast Ranges, Sierra Nevada, Blue Mountains, Olympic Mountains, Klamath Mountains

CHAPTER 50 CHECK

Reviewing the Main Ideas

1. The Pacific Coast region includes the states of California, Oregon, and Washington. The major landforms of this region are the Coast Ranges, the Sierra Nevada, and the Cascade Range. This region is subject to volcanic and earthquake activity.
2. California's Central Valley is the most productive farming region in the United States. California is rich in minerals, especially oil. Forests, fishing, high technology, and other industries make this one of the nation's most prosperous states.
3. California is a highly urbanized state. Its most populated cities are Los Angeles, San Francisco, and San Diego.
4. The major resources of Washington and Oregon are fish, forests, and water. Portland, Oregon, and Seattle, Washington, are important ports and major cities.

Building a Vocabulary

1. Define *caldera*. What lake discussed in this chapter fills a caldera of an extinct volcano?
2. What is agribusiness? Why do you think agribusiness has developed into a profitable industry in California?

Recalling and Reviewing

1. What are the dominant climates of this region? Why is southern California subject to brush and forest fires?
2. Name some agricultural products of California. For what is Napa Valley famous?
3. What are the major resources of the Pacific Coast states?

Critical Thinking

4. The forests of the Pacific Coast are among the region's most valuable resources. Why are the forests valuable? What kinds of lumbering techniques do you think should be used to ensure that this resource lasts long into the future? What role do you think the governments of the Pacific Coast states should take in the management of its forests? Why?

Using Geography Skills

Use the map in this chapter to answer the following questions.

1. What river forms part of the boundary between Washington and Oregon? What major city is located on this river?
2. Where in the Pacific Coast states is lumbering an important economic activity?
3. At what approximate latitude and longitude is San Francisco located? How far is it from Los Angeles?

Answers to Chapter Check questions are found in the Teacher's Guide.

Additional activities for Chapter 50 are found in the Student Workbook. Chapter 50 Test as well as Skills, Reteaching, Critical Thinking, and Challenge/Enrichment activities are available in the Teacher's Resource Binder.

Chapter 51

Chapter 51 Lesson Plans and Planning Guide are found in the Teacher's Guide.

Refer students to the maps on pages 578 and 581.

Alaska and Hawaii

▲ Much of Alaska has a tundra climate.

▲ Hawaii is a state of beauty and contrast.

Alaska and Hawaii, the youngest states, were admitted into the Union in 1959. They are the only states that are not part of the contiguous United States. That is, they do not have borders with any other states.

Alaska, the largest state, is separated from the rest of the nation by great distances across western Canada and the North Pacific Ocean. It is the country's northernmost state and the only one that has part of its area in the tundra and subarctic climate regions. Scenic Alaska is a vast storehouse of natural resources.

Hawaii, the fourth smallest state, is separated from the rest of the United States by almost 2,500 miles (4,000 kilometers) of the North Pacific Ocean. It is the nation's southernmost state and the only one that is located within the tropics. While each of the major Hawaiian islands is unique, they all share a tropical climate, steep volcanic mountains, and beautiful beaches.

 Note the sharp contrasts between these two states as reflected in the photos in this chapter.

Historical Geography

New France The first Europeans to visit Canada's eastern shores were Viking adventurers, who arrived between A.D. 1100 and 1300. However, they left no permanent settlements or other imprint on the landscape. The formal exploration of Canada by Europeans began in 1497 when John Cabot landed on Newfoundland and other islands. Cabot, like Columbus, was from Genoa, Italy, though he was employed by Britain. During the first part of the 1500s, Canada's eastern coast was visited by people from many European nations.

The first great explorer of the St. Lawrence River was Jacques Cartier (zhahk kahr-TYAY) of France. He traveled upstream as far as the site of Montreal in 1535, nearly a century before the American colonies in New England were established. Cartier and the French explorers who followed him had three goals. The first was a search for a northwest passage to China and India. The second was development of the fur trade. The third goal was conversion of the Indians to Roman Catholicism. By 1608, the French, under the leadership of explorer and geographer Samuel de Champlain, started their first permanent settlement in New France at Quebec City.

The French quickly realized that the St. Lawrence River and the Great Lakes were a great avenue into the interior of North America. The lakes and rivers were connected by easy portages. French settlers came to farm on the fertile lowlands along the St. Lawrence and in Nova Scotia. This region was then known as Acadia (uh-KAYD-ee-uh). Fur trappers and trading posts spread into central North America. New France became the largest French colonial settlement in the New World.

The population density of northern Canada is low. Many homesteads like this lie many miles from the nearest town.

British North America In 1717, England acquired Nova Scotia for its fishing ports and farmlands. The French who were living in Nova Scotia were deported to what was then French Louisiana. To **deport** means to send out of a country. The descendants of these French settlers are the Acadians of Louisiana and Texas. In the early 1700s, England's expanding American colonies and France's interests in North America came into conflict. The population of the British colonies was rapidly increasing and expanding into lands along the Ohio and Mississippi valleys claimed by France. Both sides acquired Indian allies, and in 1756, France and England went to war. In North America, this war ended in 1759 when Quebec City was captured. The British, however, recognized the French traditions and laws of Quebec. French and English became official languages.

British settlement of Canada increased dramatically during the American Revolution. Many United Empire Loyalists, or American colonists who remained loyal to Britain, left the United States in order to remain under British rule. Most went to live in southern Ontario. Their descendants have been economically and politically powerful throughout Canada's history.

In 1791, Canada was divided into Upper Canada, which is modern Ontario, and Lower Canada, which is modern Quebec. Lower Canada was so named because it was at the downstream end of the St. Lawrence drainage

Level B: Students should make models of the landforms of Canada, draw maps showing the country's climate and vegetation, or draw a cross-sectional diagram of the nation.

system. At this time, Canada's Maritime Provinces along the eastern coast were separate British colonies.

In the first half of the nineteenth century, Canada grew steadily. The English and French communities squabbled over language and immigration policy. French Canadians have been a minority since the early 1800s. The cities of Quebec, Montreal, and Toronto grew into major commercial centers. Farther west, the Hudson's Bay Company and the North West Company extended their fur-trading posts to the Pacific coast and the subarctic. In 1846, the United States and Great Britain agreed that their western border should follow the 49th parallel.

Dominion of Canada In 1867, the British Parliament passed the British North America Act, which created the Dominion of Canada. A **dominion** is a territory or sphere of influence. The dominion joined the provinces of Ontario, Quebec, Nova Scotia, and New Brunswick into a confederation governed by a parliament and prime minister. The city of Ottawa, in Ontario but on the Quebec border, became the capital. Manitoba, British Columbia, and Prince Edward Island joined the dominion in the 1870s. Alberta and Saskatchewan, however, were not carved out of the Northwest Territories until 1905. Newfoundland joined Canada only in 1949. Canadian provinces each have their own parliaments, which are headed by a provincial prime minister. Provincial governments have more authority than do state governments in the United States. Within the province, they have more direct control over taxes, education, and civil rights.

Most of Canada's west was settled after the country's transcontinental railroad was finished in 1885. Western Canada was homesteaded by thousands of immigrants who rode the rails to their new homes. Canada has continued to be a land of immigrants. Today, many of the nation's citizens are immigrants from Europe, the Caribbean, and Asia.

Canada's connections to Britain have remained close. Canada's national government has a parliament like Britain's, and Britain's queen also is Canada's queen. Canada joined World War I with Britain in 1914, rather than with the United States in 1917. Canadians were fighting in World War II more than two years before the United States joined the war. In foreign relations, Canada has become a moderating force. Canadians are proud of their international policies, which they feel stress peaceful relations with all countries.

Today, the Canadian and United States economies and cultures are closely linked. The countries are each other's most important trading partner. Canadians are the most frequent foreign visitors to the United States, while Americans are the most frequent foreign tourists to Canada.

Physical Geography

Landforms Canada can be divided into six landform regions: the Appalachian Mountains, the St. Lawrence and Great Lakes Lowlands, the Canadian Interior Plains, the Canadian Shield, the mountains of the Canadian Pacific coast, and the mountains of the northeast Arctic. All of these regions except the Arctic mountains are connected to similar landform regions in the United States. (See the map on page 587.)

Nearly all of Canada was covered by great ice sheets during the ice ages. Canada's Appalachian Mountains are in southeastern Quebec and the Maritime Provinces of New Brunswick, Nova Scotia, Prince Edward Island, and Newfoundland. These highlands were heavily eroded during the ice ages. The ridges and mountains are lower than those in the United States. The St. Lawrence River valley and Great Lakes areas were major routes of waters from melting glaciers. Today, these parts of Ontario and Quebec contain some of Canada's most fertile soils.

The Canadian Interior Plains extend across Manitoba, Saskatchewan, and eastern Alberta. In the north, they extend all the way to

 Level A: Have a class representative write for educational materials on Canada that are available at the Canadian Embassy (1771 North St., NW, Washington, D.C. 20036).

and is in great demand. Canada now produces a large part of the world's supply of potash. Regina, Saskatchewan, is the center of the potash industry as well as of the province's food-processing industry.

The development of oil, coal, and natural-gas deposits in western Saskatchewan and Alberta has brought the western prairies to world attention. More than 85 percent of Canada's coal and oil comes from the province of Alberta. Billions of dollars have been spent developing oil reserves in this area. Oil production is growing every year. Alberta's oil travels in pipelines to both the St. Lawrence and the Pacific coast. Income from oil exceeds income from wheat. Alberta has two rapidly growing cities, Calgary and Edmonton. Each is an important oil and agricultural center and is among the most vibrant cities in North America. The glass office buildings of Calgary and Edmonton stand as striking monuments on the Canadian prairie.

British Columbia

The province of British Columbia lies on the Pacific coast of Canada. It is a land of rugged mountain ranges, intermountain plateaus, and fertile river valleys. Large mountain ranges extend the length and width of the province. The mountains descend steeply to the Pacific Ocean, leaving few areas of level land along the coastline. These mountain ranges have long isolated British Columbia's coastal cities from the rest of Canada.

British Columbia is rich in natural resources. Like the Pacific Northwest of the United States, much of British Columbia is covered with forests of fir, spruce, and cedar trees. More than half of the income of the province comes from its forests. In addition, the salmon and mining industries contribute to British Columbia's wealth. Lead, zinc, gold, and coal are the leading minerals in the province. Because of its location on the west coast, British Columbia can trade easily with the Orient. Much of its mineral production and forest products are sold to Japan.

Water resources play a large role in British Columbia's economy. Mountain streams provide a good supply of inexpensive electric power. At Kitimat (KIT-uh-mat), 400 miles (640 kilometers) north of Vancouver, a dam 16 times the height of Niagara Falls generates huge amounts of electricity. Abundant hydroelectricity has encouraged industries from other parts of Canada to relocate to British Columbia.

Vancouver, with a population of more than one million, is British Columbia's largest city and the third largest city in Canada. It grew to become western Canada's greatest city after the Trans-Continental Railway built its terminus at the mouth of the Fraser River. Vancouver has Canada's major ice-free harbor. It is the nation's outlet to the Pacific Ocean as well as the largest cargo port on the Pacific. Salmon canning and the manufacture of wood and paper products are the major industries of Vancouver.

Vancouver is the commercial center of British Columbia. It is also the busiest port on North America's Pacific coast.

Level A: Have students research and report on the Hudson's Bay Company, and its control of the fur trade in British Columbia during the nineteenth century.

Remind students that in the far north of Canada, crops that require a short growing season are raised. Examples are flax and sunflowers.

Victoria, the capital of British Columbia, is located on the southern tip of Vancouver Island. It is a shipping center and the home port for a large fishing fleet. The city cultivates an old-English charm that attracts a large number of tourists every year.

The Yukon and Northwest Territories

The Canadian north stretches from Alaska to the Atlantic coast. Along the shore of the Arctic Ocean are tundra grasslands with permafrost beneath. Farther south is an almost continuous band of forested hills and plateaus cut by millions of lakes and thousands of rivers. The most important of these are the Mackenzie (muh-KEHN-zee) River and the Great Bear and Great Slave lakes. **Muskegs** are among the most common landforms. A muskeg is a forested marsh that melts only during the mosquito breeding season in the summer.

The Yukon and the Northwest Territories occupy more than one-third of Canada's territory, yet their combined population is only about 75,000. It is a land of few people, of isolated towns and villages, and of only a few, usually gravel-surfaced, highways. Airplanes and satellite communications have brought this vast region into the national sphere only in the last 20 years. In spite of this region's severe climate, the north has important promise for Canada's future. It is one of the world's great modern frontiers. Rich deposits of metals and fossil fuels have been discovered. The supplies of fresh water and potential hydroelectricity are enormous.

Today, the people of the Canadian north are a mixture of hardy frontier folk. Permanent settlements such as Whitehorse and Yellowknife are classic frontier towns. The residents include people who have left Canada's cities in the south as well as many native peoples. The native peoples are Inuits (Eskimos) and Indians. Years ago, many of the native peoples lived by hunting and gathering. While some still live this way during the summers, many are now employed by the region's mining and transportation companies or military bases.

Regional Issues

Canada's culture is much like that of the United States. The United States and Canada are friends. However, Canadians are very nationalistic. They fear that their culture will be overwhelmed by the United States.

When Canadians deal with each other, they often show considerable **regionalism**. Regionalism is the political and emotional support for one's region before one's country. The threat to divide Canada into several countries has appeared repeatedly in Canada's history. While Canada's national government has international duties similar to those of the United States federal government, the 10 Canadian provinces are much more powerful than states in the United States. Each province has much greater taxing powers and authority over local issues than does the government in Ottawa.

This situation has kept regionalism alive. People in the Maritime Provinces and in the Canadian west have long been suspicious of events in Ottawa, where Ontario and Quebec seem to dominate. People in Ontario and Quebec have been battling over language and culture for more than two centuries. Oil-rich Alberta would rather not share its large tax revenues from the petroleum industry with Ottawa. British Columbia's Pacific viewpoint and distant location have allowed the prospect of separation to be raised again and again. To make matters worse, many Canadians have often had closer cultural and economic ties to the United States than to neighboring provinces. Nonetheless, these disputes are often more like family squabbles. Most Canadians continue to support a united Canada rather than several small countries.

 Level A: Students can organize a Canada-America Friendship Day. Ask students to create activities that can help everyone learn more about Canada.

UNIT TEN

SHARING THE WORLD'S RESOURCES

OBJECTIVES

- ▷ *To describe how trade has increased in complexity since its beginnings and how the economies of each nation in the world are closely linked*
- ▷ *To describe how the natural resources of the world are vital to human existence and ways in which humans can conserve natural resources*
- ▷ *To describe the ways in which the quality of human life has improved and progressed during this century and the challenges that will face humans in the future*
- ▶ *To interpret information presented in a table to compare statistics among countries*

Chapter 54

Chapter 54 Lesson Plans and Planning Guide are found in the Teacher's Guide.

World Trade

▲ **The Tokyo Stock Exchange is active.**

▲ **The port of New Orleans is a busy trade center.**

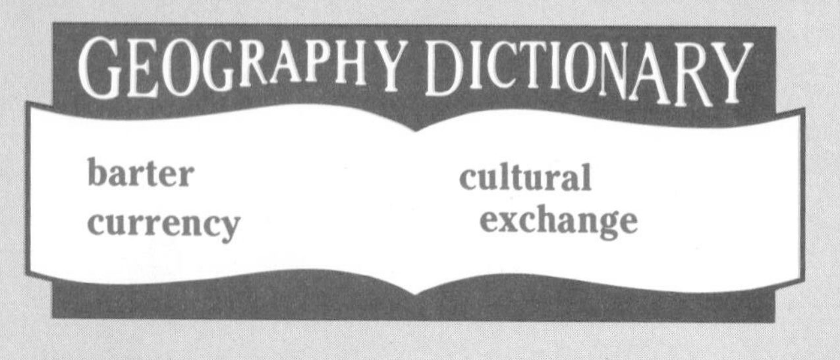

The world's great ports link mines, factories, farms, and homes around the world. Goods arrive and depart, transported by the barge, truck, and train systems of the continents and the great ships of the oceans. The world's docks, warehouses, and factories at ports are symbols of the economic advantages of trade.

Ports also are places where the world's resources can be seen. Petroleum, coal, minerals, lumber, and agricultural products have their own transportation requirements. All of these resources demonstrate the diversity of the world's geography. Trade is necessary for economic progress.

Ports, however, are only part of world trade. For example, trade also includes computers sharing data on an international telephone link. The goods here are information and ideas. Neither bulky nor heavy, they can be stored by the millions on computer disks.

Level B: Ask students to label on an outline map of the world the distribution of a specific resource (oil, iron ore, natural gas, and so on).

Level B: Ask students to list animals that depend upon forests for survival.

Soil

Throughout history, people have developed ways to conserve and enrich the soil. "Return to the soil what you get from it" has been the rule for thousands of years in China. The world's farmers know that the soil is like a bank. It is impossible to continue making withdrawals without making deposits.

Today, most farmers are aware of the importance of soil conservation. Farmers use contour plowing, by which they plow and plant crops in rows that follow the contour of the land. Furrows run across the slope of the land, preventing runoff from eroding topsoil. Farmers also prevent wind erosion by practicing **strip-cropping**. They plant alternating rows of close-growing crops, such as peas and beans. These plants alternate with open-growing crops, such as corn, which stand high and expose the bare soil to wind and rain.

The fertility of the soil can be preserved in several ways. Use of natural or chemical fertilizers is one; another is crop rotation. A field that is planted with wheat one year may be planted with alfalfa or some other soil-enriching crop the next. Use of cover crops is yet another way to preserve fertility and to prevent erosion. Wind erosion also is lessened by the planting of shelterbelts, which are rows of trees along the edge of fields that block the wind.

Many dry regions that depend upon irrigation are suffering from **salinization**, or salt buildup in the soil. The evaporation of irrigation water leaves behind a hard salt layer. High salt content becomes destructive to crops and eventually destroys the soil's productivity.

In developed countries, such as the United States, suburban and urban areas have expanded across some of the best farmland. In less developed countries, overgrazing is one of the worst enemies of topsoil. Both reduce the amount of land available for food production. Today, because the world's growing population must be fed, our soil resource is more valuable than ever.

Once fertile soil in Nebraska has been destroyed by erosion. What soil conservation methods are most effective? contour plowing, strip-cropping, crop rotation, fertilizers, shelterbelts

Forests

Forests, if properly used and cared for, are valuable renewable resources. The first European settlers in the United States found a land of seemingly endless forests. As the nation's population grew, many forests were cut to provide timber for fuel, homes, and industry. New trees were not planted to replace those that were destroyed.

It was not until the early 1900s that the federal government became aware of the need to conserve our forests. Now, in many regions, only a certain number of trees may be cut each year. National forests have been created, and lumbering in these areas is controlled. Other nations, such as Sweden, West Germany, Israel, Finland, New Zealand, and the Soviet Union also practice reforestation.

In contrast, the tropical rain forests of Southeast Asia, Africa, and Latin America are disappearing at a rapid rate. These nations generally do little to reforest the regions that have been cleared. One-third of the remaining rain forests are located in Brazil, Zaire, and

Level B: Ask students to obtain information on soil conservation from the county agricultural extension service in their region.

Level A: Have students research pollution and/or flood disasters in recent times.

Renewing the world's resources is a growing concern. On this tree farm in New Zealand, new trees will be planted to replace those cut.

Indonesia. It is hoped that countries will realize the consequences to soil, climate, wild life, and human life before the rain forests are completely gone.

Water

Water Shortages With the rapidly growing world population, maintaining water quantity and quality are serious problems in many areas of the world. In the western United States, expensive water transfer systems with dams and aquifers must send water long distances. In many parts of the world, deep wells are the only source of water. The water table, however, is dropping in these areas as water demands increase, forcing the drilling of deeper wells or even drying up the aquifer.

In order to solve the problem of inadequate water supplies, some countries have built desalinization plants. Fresh water is made from salty seawater. The major problem with desalinization is expense. Energy is required to produce fresh water, and the corrosive salt water destroys equipment rapidly. Thus, most countries cannot afford to desalinize. Presently, two-thirds of the desalinization plants are on the Arabian Peninsula, where water is expensive and energy is cheap. Desalinization also is used where there is no alternative, such as at Key West, Florida, and in the Virgin Islands.

As weather and climate conditions change, people must not only find new water sources but also conserve water. In recent years, drought has caused serious water shortages in Africa's Sahel region, Australia, and the southwestern United States.

Flood Control Although water shortages affect many countries, floods also create great environmental problems. As natural disasters, floods rank high in the taking of human life. Floods have killed millions of people over the centuries. The Huang He (Yellow River) of northern China is known as "China's sorrow." The Huang He has drowned one million people in a single flood.

Dams and levees are built to control flood waters. In a positive way, dams provide flood protection, hydroelectricity, stored water for farms and cities, and a place for fishing and recreation. The negative effects are that productive farmland or forests may be flooded behind the dam.

After the building of the Aswān High Dam in Egypt, the Nile's floods were minimized. However, the riverbanks were no longer enriched by sediments deposited by the floodwaters. Egyptian farmers now must use chemical fertilizers to replenish their soils.

Water Pollution As the population of the world continues to grow along with the expansion of agriculture and industry, the by-products of human activity also increase. Water cannot naturally remove pollutants as fast as the pollutants are being added to the water.

Human life is endangered by impure water supplies. This is particularly true in less developed nations that are densely populated. There, sewage comes into contact with the drinking-water supply, and diseases spread.

In industrialized nations with sewage-treatment plants, sewage is not the major form

 Point out to students that dams can fill up with silt and change rivers into lakes. Discuss with students the pros and cons of dam construction.

of water pollution. Here, water is contaminated by industries' chemical wastes, oil, and thermal-waste heat. Farms also pollute rivers and streams with runoff of pesticides and chemical fertilizers.

The oceans are the final dumping ground for human wastes. Some is by accident, as when an oil tanker sinks. Some is by direct dumping of dangerous toxic-chemical wastes and radioactive materials. Most pollution in the seas is from rivers, which wash all types of human waste to the sea. Such pollution can contaminate fish and shellfish and, thus, make people sick or even poison them.

Some water, such as that of rapidly flowing rivers, recovers rapidly from pollution. Other water, such as the water in deep lakes and groundwater, which circulates slowly, may stay polluted for centuries.

Air pollution is a problem shared by most of the world's large cities. What are some primary sources of air pollution? automobiles, factories

Air

Urban and Industrial Smog Though it is often taken for granted, air also is a natural resource. It is an essential gas that we breathe, it protects us from harmful radiation, and it helps maintain the temperature balance of the planet. There appears to be an endless supply of air; however, our atmosphere is subject to change by human pollution.

Air pollution, or smog, is a serious problem facing many large urban and industrial areas. *Smog* originally meant "smoke plus fog," but it now refers to both industrial air pollution and photochemical smog. Photochemical smog is produced when sunlight interacts with exhaust gases to produce haze.

Automobiles release exhaust fumes, and factories pour smoke and fumes into the atmosphere. Some cities, such as Los Angeles, New York, Rome, Tokyo, and Athens, have become infamous for their smog. In their rush to industrialize, many less developed nations now also face smog problems in their cities. Examples include Mexico City, Istanbul, Cairo, São Paulo, and Bombay.

Acid particles released by industries cause damage hundreds of miles downwind. The acid combines with water vapor in the atmosphere, causing acid rain and acid fog. These acid droplets can harm lakes, forests, and human health.

World Climate Change Air pollution is a worldwide problem and may have the potential to change the earth's atmosphere. In the upper atmosphere is a region known as the **ozone layer**. Ozone (O_3) is a gas formed from an interaction between oxygen and sunlight. This layer of gas protects life from dangerous ultraviolet radiation. Recent evidence indicates that the ozone layer has become depleted, especially over the Antarctic, where there appears to be a "hole" in the ozone.

A group of chemicals called CFCs (chlorofluorocarbons) apparently rise in the atmosphere and set off chemical reactions that destroy the ozone layer. CFCs are released from aerosol spray cans, air conditioners, refrigerators, liquid cleaners, and plastic foams. Concerned nations have agreed to cut back on production of products that release CFCs. The United States and Scandinavian countries have already banned aerosol spray cans.

Point out to students that most of the world's oxygen is produced by plant life in oceans. If the oceans became seriously polluted, what would be the effect on the earth's atmosphere?

Another concern is the **greenhouse effect**. This is caused by an apparent buildup of carbon dioxide (CO_2) in the lower atmosphere, which traps heat, causing a long-term warming of the planet. During the 1890s, a Swedish chemist warned that massive coal burning would cause increased heat to be trapped in the atmosphere. Recent studies indicate that this theory was probably correct. When the earth has natural fluctuations in carbon dioxide, the climate changes. Ice ages occurred when carbon dioxide content was low. Warmer periods of the earth's history occurred when carbon dioxide had increased in the atmosphere. Natural conditions such as volcanic activity and forest fires release great amounts of this gas. Humans also produce great quantities by burning fossil fuels. There has been an increase in carbon dioxide in the atmosphere over the last two centuries.

Two ways to protect ourselves from a continued heat buildup are to cut back on the burning of fossil fuels and to stop cutting the tropical rain forests. The forests produce oxygen, which helps keep the proper ratio of gases in the lower atmosphere.

Refineries that process petroleum and other chemicals contribute to the world's water and air pollution.

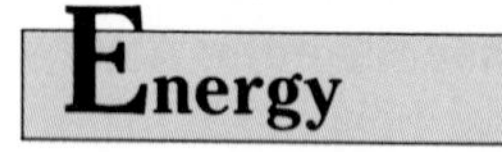

Energy

Fossil Fuels Coal, petroleum, and natural gas are known as fossil fuels because they are believed to have been formed slowly from the remains of prehistoric plants and animals. Since the Industrial Revolution, great amounts of energy have been needed for factories, transportation, and farms. During the last two centuries, coal was the major fuel. In this century, petroleum and natural gas have replaced coal as the preferred energy sources. In the next century, as petroleum and gas supplies decline, what will be the major fuel?

Coal is still an important source of energy. It is needed, along with iron ore, to make steel, a basic material for all industrial nations. Coal also is the source of many chemical products, such as synthetic rubber, synthetic fabrics, plastics, and paints. Coal also is burned to produce electricity.

Because coal is a solid fuel, it has certain disadvantages, and burning it causes acidic air pollution. However, through expensive liquification and gasification processes, coal can be changed into a cleaner-burning liquid fuel or gas.

Petroleum is the source of gasoline, which runs our transportation systems. It also provides fuel oils for heat, chemical fertilizers, pesticides, and plastics. No nation can be an industrial power without a dependable source of petroleum. Petroleum is a nonrenewable resource. The world's supply may last only another 50 years or so.

The largest petroleum reserves are found in Saudi Arabia and in the neighboring Persian Gulf countries. Oil is shipped easily by sea in large tankers, and it is the leading commodity traded in the world today. It also is a major cause of air pollution.

Natural gas, which is often found along with oil, also is in limited supply. The largest

 Tell students that although the U.S. is a major petroleum producer, it is also the world's greatest consumer. The U.S. must import one-third of its petroleum.

Point out that conservation is the cheapest way to extend energy supplies into the future.

reserves of natural gas are in the Soviet Union and the Persian Gulf. Natural gas is the cleanest-burning fossil fuel and can be shipped through pipelines across land. For overseas export, it is liquified into a smaller volume.

Nuclear Energy The world's first nuclear power plant began operation in Britain in 1956. Today, there are hundreds of nuclear power plants scattered across the United States, Europe, the Soviet Union, Japan, and many other countries.

Nuclear power was seen as a clean, inexpensive power source. However, there are two serious problems that the nuclear industry faces: nuclear accidents and nuclear waste. In 1986, a nuclear accident occurred at Chernobyl, in the Soviet Union. Some people died from radiation, and others who were contaminated will have shorter lives. Many square miles of territory around the plant had to be evacuated, and people lost homes and farms.

Chernobyl, in the Soviet Union, was the site of a nuclear accident in 1986. What problems have been associated with nuclear power plants?
disposal of nuclear wastes, potential for accidents

Contamination spread over much of Western Europe. This so far has been the world's worst nuclear accident.

All nuclear power plants produce nuclear waste. Exposure to these radioactive substances is dangerous. Radioactive wastes decay slowly, which means that they must be stored and monitored for thousands of years. There also is the danger of accidents in transporting these waste products to storage facilities.

Since Chernobyl, many countries are reevaluating their nuclear-energy programs. Other nations had noted concern before 1986. The nuclear industry in the United States is at a standstill; Sweden is going to phase out its nuclear plants; and numerous countries, including Austria, Denmark, and New Zealand, have nonnuclear policies.

Other Energy Resources Flowing water is an important source of energy. Hydroelectric power produced by the force of running water is a relatively inexpensive and pollution-free source of electrical energy. Also, it is a renewable energy source. Water power offers hope to less developed countries that do not have coal and petroleum reserves.

Two other sources of renewable clean energy are underground heat and the sun. Geothermal energy is created by steam from underground sources in volcanic regions. Geothermal energy is now being produced in the United States, Japan, New Zealand, Italy, and Iceland. Solar energy involves capturing the sun's light energy and converting it into heat or electricity. Experimental solar electric plants are now operating in France and the United States. If these prove successful, many more may be built in the future.

Population and Resources

Population Distribution and Growth

The world's population is not evenly distributed. Extreme climates, rugged terrain, and

ANTARCTICA: THE LAST FRONTIER

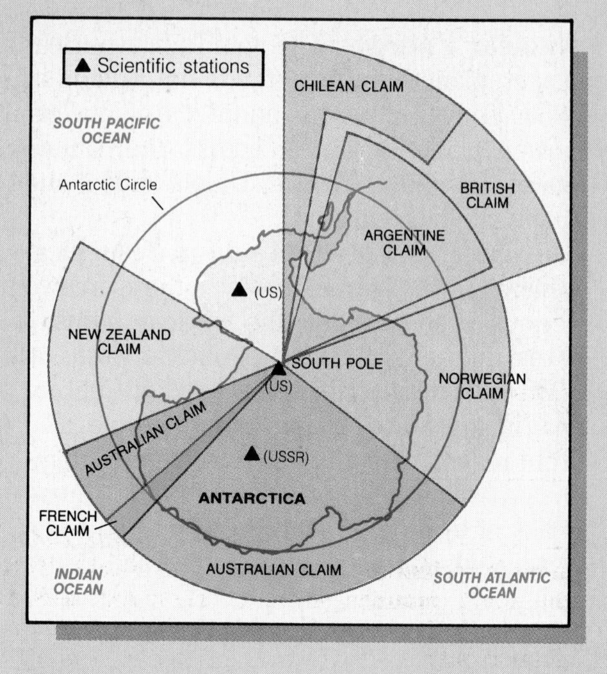

The continent of Antarctica lies in the southernmost part of the world, surrounding the South Pole. It was first discovered in 1772, but it was not until 1911 that the first human being reached the South Pole. Even today, much of this empty, icy land remains unexplored and uninhabited. The only people living in Antarctica are researchers in scientific research stations.

The icy wastes of Antarctica offer little to attract explorers or settlers. For most of the year, temperatures remain bitterly cold, dropping to −120° F (−84° C) or lower! Six months of the year are spent in the darkness of the long polar night, when the sun never rises. This means that most exploration must take place during the continent's short summer. Scientists live all year in research stations located at various points on the ice cap. Outdoor work is often made impossible by howling winds and Antarctic storms.

Although Antarctica makes up 10 percent of the world's land, it is not owned by any one nation. When Antarctica was discovered, there were no native inhabitants who could claim the land as their own. The continent has been divided into pie-shaped claims by Argentina, Australia, Chile, Great Britain, New Zealand, Norway, and France. Neither the United States nor the Soviet Union has made claims, but neither nation honors the claims of others. Officials of the USSR and the United States reason that no claim is valid until permanent settlement has taken place.

In 1959, a very successful international agreement, called the Antarctic Treaty, was drawn up by 12 countries. The countries agreed to prevent any new claims to the territory. The treaty banned all military activity in the region and made Antarctica a nuclear-free zone. It also allowed all countries the opportunity to undertake scientific exploration, provided that they share the results of their discoveries. The signers of the treaty agreed that the treaty would be reviewed in 1991, but most experts doubt that it will be changed. More than 100 developing countries have condemned the treaty organization as unfair.

As the world's natural resources are used up, countries are looking south to Antarctica as a possible area of untapped resources. So far, the Antarctic Treaty has prevented any

Level B: Have students list the various resources of the earth. Ask students to classify these resources as renewable, nonrenewable, or recyclable.

Place **The vast ice cap that covers Antarctica provides a challenging environment for scientists from all over the world. Much of their research is conducted during the short summer, when the sun never sets. For six months of the year, the region remains in darkness. Can you explain what causes the long polar day and the long polar night?** The revolution of Earth around the sun and the tilt of the earth's axis. Antarctica faces the sun continually in summer and is hidden from the sun in winter.

prospecting. Another factor protecting the continent is the massive, moving ice cap that covers it. The ice is as much as three miles (4.8 kilometers) thick in some places! Even if minerals were discovered under the ice, it would be too costly to reach them. The ice itself, however, is a potential resource. Three quarters of the world's fresh water is trapped in the ice. Drought-stricken countries have considered the possibility of towing icebergs from Antarctica to help ease their water shortages.

Region **The only animals in Antarctica live along the coastline. The most common are seals, penguins, and whales. Whaling used to be Antarctica's only industry, but uncontrolled hunting has reduced the numbers of whales considerably. To prevent further destruction, conventions were adopted in 1978 and 1982 to protect the seals and other marine resources. Why do no animals live in the interior of Antarctica?** Because of the sub-zero temperatures and the lack of food and water in the interior.

Ask students why nations might be especially concerned about preserving and protecting Antarctica and its resources.

lack of water and fertile soil prevent many areas from being populated. Most productive regions of the world today, however, are populated.

Population growth rates will affect the future of the world. Some nations, such as Mexico, Kenya, Turkey, and Pakistan, have rapid growth rates. Many industrial nations have very slow growth rates. These include the United States, Canada, the Soviet Union, Australia, and the countries of Western Europe. Some nations, such as China and Indonesia, are trying hard to control population growth.

Why would a nation want to control population growth? Mainly to raise the standard of living for its citizens. A slower growth rate will put less strain on the limited resources of that nation. The nations with the highest growth rates are generally the poorest nations per capita.

In contrast, some nations, such as Singapore, Japan, and several European countries, have been successful at controlling population. These nations now are concerned about their success. With low birthrates, there are fewer young people coming into the work force, which may lower productivity and consumer spending. At the same time, the percentage of elderly people grows, causing an economic strain because of the increased cost of health and welfare care. It is probably "population balance" that a nation seeks, rather than population control.

Population and Resource Balance The present population of the world is more than five billion. If current growth rates continue, it is estimated that the world population will be eight billion by 2020. More than one million people are added to the world every five days. The question is whether the earth can support this many people.

Natural resources, like people, are not distributed evenly. Some regions have surplus water, food, or energy resources. Other areas have shortages. This, of course, leads to world trade, but also to world conflict as nations try to secure needed resources. The people of the world must share and use the planet's resources more wisely.

Many developed countries, such as Japan, are experiencing a low death rate. The number of older people is increasing in these countries.

Many people are optimistic about the future. As the earth's population increases, they believe, scientists will discover ways to produce new kinds of foods. Farmers will increase crop yields through new methods, more food and water will be taken from the sea, and high-rise buildings will permit more people to live in less space.

Yet, others wonder whether answers to continued population growth can be found. They ask if we have not reached the point at which important resources are already in short supply. Will substitutes be found for many of our dwindling natural resources? Can we continue to pollute the environment without harming our planet, our nation, and ourselves? Will countries continue to trade valuable resources? Will the richer nations share their wealth and resources with the less fortunate nations? A better understanding of the world's geography will help us face these challenging questions.

Country	Land Area	Population	Language	GNP	Type of Government
Greece	50,949 sq. mi. (130,429 sq. km.)	9,987,000	Greek (official), Macedonian	369.9 billion (1984)	presidential parliamentary government
Greenland	840,000 sq. mi. (2,150,400 sq. km.)	54,000	Danish and Greenlandic (official)	380 million (1984)	province of Denmark
Iceland	39,769 sq. mi. (101,809 sq. km.)	246,000	Icelandic	2.3 billion (1984)	republic
Ireland	27,137 sq. mi. (69,471 sq. km.)	3,547,000	English and Irish Gaelic (official)	17.5 billion (1984)	republic
Italy	116,324 sq. mi. (297,789 sq. km.)	57,298,000	Italian (official)	367 billion (1984)	republic
Liechtenstein	62 sq. mi. (159 sq. km.)	27,000	German (official)	524 million (1980)	constitutional monarchy
Luxembourg	999 sq. mi. (2,557 sq. km.)	367,000	French and German (official), Luxembourgian, English	5 billion (1984)	constitutional monarchy
Malta	122 sq. mi. (312 sq. km.)	336,000	English and Maltese (official)	1.2 billion (1984)	parliamentary democracy
Monaco	.7 sq. mi. (1.8 sq. km.)	29,000	French (official), Italian, Monegasque, English	158.3 billion (1984)	constitutional monarchy
Netherlands	16,133 sq. mi. (41,300 sq. km.)	14,561,000	Dutch (official), English, German, French	135.8 billion (1984)	constitutional monarchy
Norway	125,050 sq. mi. (320,128 sq. km.)	4,166,000	Norwegian (official), English, Finnish, Lapp	57.1 billion (1984)	constitutional monarchy
Portugal	35,672 sq. mi. (91,320 sq. km.)	10,250,000	Portuguese (official)	20.1 billion (1984)	republic
San Marino	24 sq. mi. (61 sq. km.)	23,000	Italian (official)	177 million (1980)	republic
Spain	194,885 sq. mi. (498,906 sq. km.)	38,818,000	Castilian Spanish (official), Catalan, Galician, Basque	172.4 billion (1984)	parliamentary monarchy
Sweden	187,901 sq. mi. (481,027 sq. km.)	8,358,000	Swedish (official), Finnish, Lapp	99.1 billion (1984)	constitutional monarchy
Switzerland	15,943 sq. mi. (40,814 sq. km.)	6,556,000	German, French, Italian (all official), Romansh	105.1 billion (1984)	federal republic

The Soviet Union and Eastern Europe

Country	Land Area	Population (1986)	Language	GNP	Type of Government
Albania	11,100 sq. mi. (28,416 sq. km.)	3,023,000	Albanian (official), Greek	2.6 billion (1982)	Communist state
Bulgaria	42,823 sq. mi. (109, 627 sq. km.)	8,974,000	Bulgarian (official), Turkish	25.1 billion (1984)	Communist state
Czechoslovakia	49,381 sq. mi. (126,415 sq. km.)	15,552,000	Czech and Slovak (official), Hungarian	84 billion (1984)	Communist state
German Democratic Republic (East Germany)	41,827 sq. mi. (107,077 sq. km.)	16,636,000	German (official)	93.6 billion (1984)	Communist state
Hungary	35,921 sq. mi. (91,958 sq. km.)	10,624,000	Hungarian (official)	18.3 billion (1984)	Communist state
Poland	120,727 sq. mi. (309,061 sq. km.)	37,456,000	Polish (official), Ukranian, Byelorussian	143 billion (1984)	Communist state
Romania	91,700 sq. mi. (234,752 sq. km.)	22,809,000	Romanian (official), Hungarian, German	45.6 billion (1984)	Communist state

Soviet Union	8,649,500 sq. mi. (22,142,720 sq. km.)	80,038,000	Russian (official), Slavic, other Indo-European, Altaic, Uralian, and Caucasian languages	1,925 billion (1984)	Communist state
Yugoslavia	98,766 sq. mi. (252,841 sq. km.)	23,289,000	Serbo-Croatian, Slovenian, Macedonian (all official), Albanian, Hungarian	39 billion (1983)	Communist state, federal republic in form

Middle East and North Africa

Country	Land Area	Population (1986)	Language	GNP	Type of Government
Algeria	919,565 sq. mi. (2,354,086 sq. km.)	22,564,000	Arabic (official), Berber, French	50.7 billion (1984)	republic
Bahrain	264 sq. mi. (676 sq. km.)	435,000	Arabic (official)	4.3 billion (1984)	traditional monarchy (emirate)
Cyprus	3,572 sq. mi. (9,144 sq. km.)	674,000	Greek and Turkish (official)	2.4 billion (1984)	republic
Egypt	385,229 sq. mi. (986,186 sq. km.)	48,007,000	Arabic (official)	33.3 billion (1984)	republic
Iran	636,443 sq. mi. (1,629,294 sq. km.)	46,097,000	Farsi (official), Azerbaijani, Kurdish	159.1 billion (1983)	Islamic republic
Iraq	169,235 sq. mi. (433,242 sq. km.)	15,946,000	Arabic (official), Kurdish	31.3 billion (1981)	republic
Israel	7,992 sq. mi. (20,460 sq. km.)	4,381,000	Hebrew and Arabic (official), Yiddish	21.3 billion (1984)	republic
Jordan	34,443 sq. mi. (88,174 sq. km.)	2,749,000	Arabic (official)	4.3 billion (1984)	constitutional monarchy
Kuwait	6,880 sq. mi. (17,612 sq. km.)	1,791,000	Arabic (official), Kurdish	27.6 billion (1984)	constitutional monarchy (emirate)
Lebanon	3,950 sq. mi. (10,112 sq. km.)	2,707,000	Arabic (official)	4.6 billion (1983)	republic
Libya	675,000 sq. mi. (1,728,000 sq. km.)	3,953,000	Arabic (official)	29.8 billion (1984)	socialist republic
Morocco (including Western Sahara)	280,117 sq. mi. (717,100 sq. km.)	22,605,000	Arabic (official), Berber	14.3 billion (1984)	constitutional monarchy
Oman	120,000 sq. mi. (307,200 sq. km.)	1,288,000	Arabic (official), Indian	7.4 billion (1984)	absolute monarchy (sultanate)
People's Democratic Republic of Yemen	130,066 sq. mi. (332,969 sq. km.)	2,365,000	Arabic (official)	1.1 billion (1984)	people's republic
Qatar	4,400 sq. mi. (11,264 sq. km.)	311,000	Arabic (official)	6 billion (1984)	traditional monarchy
Saudi Arabia	865,000 sq. mi. (2,214,400 sq. km.)	11,670,000	Arabic (official)	116.4 billion (1984)	monarchy
Syria	71,498 sq. mi. (183,035 sq. km.)	10,612,000	Arabic (official), Kurdish, Armenian	18.5 billion (1984)	republic
Tunisia	59,664 sq. mi. (152,740 sq. km.)	7,327,000	Arabic (official), French	8.8 billion (1984)	republic
Turkey	300,948 sq. mi. (770,427 sq. km.)	52,419,000	Turkish (official), Kurdish, Arabic	57.8 billion (1984)	republic
United Arab Emirates	30,000 sq. mi. (76,800 sq. km.)	1,700,000	Arabic (official)	28.5 billion (1984)	federation of emirates
Yemen Arab Republic	52,213 sq. mi. (133,665 sq. km.)	7,046,000	Arabic (official)	3.9 billion (1984)	republic

Sub-Saharan Africa

Country	Land Area	Population (1986)	Language	GNP	Type of Government
Angola	481,350 sq. mi. (1,232,256 sq. km.)	8,823,000	Portuguese (official), Niger-Congo languages	7.6 billion (1982)	people's republic
Benin	43,450 sq. mi. (111,232 sq. km.)	4,126,000	French (official), Fon, Yoruba, other Niger-Congo languages	1.1 billion (1984)	republic
Botswana	224,607 sq. mi. (574,994 sq. km.)	1,126,000	English (official), Tswana, Khoisan	940 million (1984)	parliamentary republic
Burkina Faso	105,869 sq. mi. (271,025 sq. km.)	8,126,000	French (official), Sudanic languages	1 billion (1984)	military rule
Burundi	10,745 sq. mi. (27,507 sq. km.)	4,830,000	Rundi and French (official), Swahili	1 billion (1984)	republic
Cameroon	179,714 sq. mi. (460,068 sq. km.)	9,873,000	English and French (official), Niger-Congo languages	8 billion (1984)	republic
Cape Verde	1,557 sq. mi. (3,986 sq. km.)	342,000	Portuguese (official), Crioulo	100 million (1984)	republic
Central African Republic	240,324 sq. mi. (615,229 sq. km.)	2,706,000	French (official), Banda, other Niger-Congo languages	680 million (1984)	republic under military control
Chad	495,755 sq. mi. (1,269,133 sq. km.)	5,139,000	French (official), Arabic, Sara, other Nilo-Saharan languages	425 million (1984)	republic
Comoros	719 sq. mi. (1,841 sq. km.)	409,000	French and Arabic (official), Comorian	106 million (1983)	federal Islamic republic
Djibouti	8,950 sq. mi. (22,912 sq. km.)	456,000	French and Arabic (official), Issa, Afar	302 million (1984)	republic
Equatorial Guinea	10,831 sq. mi. (27,727 sq. km.)	322,000	Spanish (official), Fang, Bubi	60 million (1983)	republic
Ethiopia	472,400 sq. mi. (1,209,344 sq. km.)	48,850,000	Amharic (official), other Semitic languages	4.8 billion (1984)	military rule
Gabon	103,347 sq. mi. (264,568 sq. km.)	1,187,000	French (official), Fang, other Niger-Congo languages	2.8 billion (1984)	republic
Gambia	4,127 sq. mi. (10,565 sq. km.)	765,000	English (official), Malinke, other Niger-Congo languages	180 million (1984)	republic
Ghana	92,098 sq. mi. (235,771 sq. km.)	13,144,000	English (official), Akan, other Niger-Congo languages	4.7 billion (1984)	republic
Guinea	94,926 sq. mi. (243,011 sq. km.)	6,225,000	French (official), Niger-Congo languages	1.8 billion (1984)	republic
Guinea-Bissau	13,948 sq. mi. (35,707 sq. km.)	891,000	Portuguese (official), Niger-Congo languages	160 million (1984)	republic
Ivory Coast	123,847 sq. mi. (317,048 sq. km.)	10,694,000	French (official), Akan, other Niger-Congo languages	6 billion (1984)	republic
Kenya	224,961 sq. mi. (575,900 sq. km.)	21,148,000	Swahili (official), other Niger-Congo languages, Semitic and Nilo-Saharan languages	5.9 billion (1984)	republic
Lesotho	11,720 sq. mi. (30,003 sq. km.)	1,586,000	English and Sesotho (official)	790 million (1984)	constitutional monarchy
Liberia	38,250 sq. mi. (97,920 sq. km.)	2,303,000	English (official), Niger-Congo languages	990 million (1984)	republic

Madagascar	226,658 sq. mi. (580,244 sq. km.)	10,294,000	French (official), Malagasy	2.6 billion (1984)	republic
Malawi	45,747 sq. mi. (117,112 sq. km.)	7,279,000	English and Chewa (official), other Niger-Congo languages	1.4 billion (1984)	one-party republic
Mali	478,841 sq. mi. (1,225,833 sq. km.)	8,457,000	French (official), Niger-Congo and Semitic languages	1.1 billion (1984)	republic
Mauritania	397,700 sq. mi. (1,018,112 sq. km.)	1,689,000	Arabic (official), French	750 million (1984)	republic
Mauritius	788 sq. mi. (2,017 sq. km.)	1,034,000	English (official), French, Hindi, Bhojpurī	1.1 billion (1984)	independent state
Mozambique	308,642 sq. mi. (790,124 sq. km.)	14,143,000	Portuguese (official), Niger-Congo languages	4.5 billion (1981)	people's republic
Namibia	318,818 sq. mi. (816,174 sq. km.)	1,204,000	Afrikaans and English (official), Niger-Congo and Khoisan languages	1.7 billion (1984)	administrative control by South Africa
Niger	458,074 sq. mi. (1,172,669 sq. km.)	6,423,000	French (official), Hausa, Niger-Congo languages	1.2 billion (1984)	republic
Nigeria	356,669 sq. mi. (913,073 sq. km.)	98,112,000	English (official), Hausa, Yoruba, Ibo, other Niger-Congo languages	74.1 billion (1984)	federal republic
People's Republic of the Congo	132,047 sq. mi. (338,040 sq. km.)	2,097,000	French (official), Niger-Congo languages	2 billion (1984)	people's republic
Republic of South Africa	470,413 sq. mi. (1,204,257 sq. km.)	33,704,000	Afrikaans and English (official), Niger-Congo and Khoisan languages	74 billion (1984)	republic
Rwanda	10,169 sq. mi. (26,033 sq. km.)	6,336,000	Rwanda and French (official)	1.6 billion (1984)	republic
São Tomé and Príncipe	386 sq. mi. (988 sq. km.)	110,000	Portuguese (official), Fang	30 million (1984)	republic
Senegal	75,955 sq. mi. (194,445 sq. km.)	6,699,000	French (official), Wolof, other Niger-Congo languages	2.4 billion (1984)	republic
Seychelles	175 sq. mi. (448 sq. km.)	66,000	English and French (official), Creole patois	145 million (1984)	republic
Sierra Leone	27,699 sq. mi. (70,909 sq. km.)	3,733,000	English (official), Mende, Temne, other Niger-Congo languages	1.1 billion (1984)	republic
Somalia	246,000 sq. mi. (629,760 sq. km.)	5,992,000	Somali and Arabic (official)	1.4 billion (1984)	republic
Sudan	966,757 sq. mi. (2,474,898 sq. km.)	24,603,000	Arabic (official), Dinka, other Semitic and Nilo-Saharan langauges	7.4 billion (1984)	republic under military control
Swaziland	6,704 sq. mi. (17,162 sq. km.)	682,000	English and Swazi (official)	590 million (1984)	monarchy
Tanzania	364,881 sq. mi. (934,095 sq. km.)	22,463,000	Swahili and English (official), other Niger-Congo languages	4.5 billion (1984)	republic
Togo	21,925 sq. mi. (56,128 sq. km.)	3,072,000	French (official), Ewe, other Niger-Congo languages	730 million (1984)	republic
Uganda	93,100 sq. mi. (238,336 sq. km.)	15,638,000	English (official), Ganda, other Niger-Congo languages, Nilo-Saharan languages	3.3 billion (1984)	republic
Zaire	905,365 sq. mi. (743,900 sq. km.)	31,079,000	French (official), Luba, Kongo, other Niger-Congo languages	4.2 billion (1984)	republic

Zambia	290,586 sq. mi. (743,900 sq. km.)	6,896,000	English (official), Bemba, other Niger-Congo languages	3 billion (1984)	republic
Zimbabwe	150,873 sq. mi. (386,235 sq. km.)	8,553,000	English (official), Shona, other Niger-Congo languages	6 billion (1984)	republic

The Orient

Country	Land Area	Population (1986)	Language	GNP	Type of Government
Afghanistan	251,825 sq. mi. (644,672 sq. km.)	16,892,000	Pashto and Dari (official), Afghan, Turkish languages	3.5 billion (1982)	Communist regime backed by Soviet force
Bangladesh	55,598 sq. mi. (142,331 sq. km.)	103,084,000	Bengali (official)	12.4 billion (1984)	republic
Bhutan	18,150 sq. mi. (46,464 sq. km.)	1,446,000	Dzongkha (official), other Tibetan languages	135 million (1984)	constitutional monarchy
Brunei	2,226 sq. mi. (5,699 sq. km.)	233,000	Malay and English (official), Chinese	4.3 billion (1983)	monarchy (sultanate)
Burma	261,228 sq. mi. (668,744 sq. km.)	38,493,000	Burmese (official)	6.6 billion (1984)	people's republic
China	3,696,100 sq. mi. (9,462,016 sq. km.)	1,053,703,000	Chinese (official), other Chinese dialects	318.3 billion (1984)	Communist state
Hong Kong	400 sq. mi. (1,024 sq. km.)	5,533,000	Chinese and English (official)	34 billion (1984)	British colony
India	1,183,427 sq. mi. (3,029,573 sq. km.)	777,230,000	Hindi and English (official), more than 800 Indic languages	197.2 billion (1984)	federal republic
Indonesia	741,101 sq. mi. (1,897,218 sq. km.)	168,662,000	Bahasa Indonesia (official), Javanese, Sundanese	85.4 billion (1984)	republic
Japan	145,870 sq. mi. (373,427 sq. km.)	121,470,000	Japanese (official)	1,248.1 billion (1984)	constitutional monarchy
Kampuchea	68,898 sq. mi. (178,939 sq. km.)	7,469,000	Khmer (official), Vietnamese, Chinese	1.1 billion (1977)	Communist state
Laos	91,400 sq. mi. (233,984 sq. km.)	3,703,000	Lao (official), Khmu, Tai	602 million (1975)	Communist state
Macao	6 sq. mi. (15 sq. km.)	433,000	Portuguese (official), Chinese	780 million (1983)	overseas province of Portugal
Malaysia	127,581 sq. mi. (326,607 sq. km.)	16,090,000	Malay (official), Chinese, Tamil, English	30.3 billion (1984)	federal constitutional monarchy
Maldives	115 sq. mi. (294 sq. km.)	189,000	Diyehi (dialect of Sinhala)	56 million (1984)	republic
Mongolia	604,000 sq. mi. (1,546,240 sq. km.)	1,938,000	Khalkha (Mongolian) (official), Kazakh	1.8 billion (1984)	Communist state
Nepal	56,827 sq. mi. (145,477 sq. km.)	16,863,000	Nepali (official), Maithilī, Bhojpurī	2.6 billion (1984)	constitutional monarchy
North Korea	47,250 sq. mi. (120,960 sq. km.)	20,543,000	Korean (official)	14.7 billion (1984)	Communist state
Pakistan	307,374 sq. mi. (786,877 sq. km.)	102,878,000	Urdu (official), Punjabi, Sindhi, Pashto	35.4 billion (1984)	federal Islamic republic
Republic of the Philippines	115,800 sq. mi. (296,448 sq. km.)	56,004,000	Pilipino and English (official), Tagalog, Cebuano	35 billion (1984)	republic
Singapore	239 sq. mi. (612 sq. km.)	2,588,000	Bahusa Malaysia, Tamil, Mandarin Chinese, English (all official), Chinese	18.4 billion (1984)	parliamentary republic

South Korea	38,259 sq. mi. (97,943 sq. km.)	41,569,000	Korean (official)	84.9 billion (1984)	republic
Sri Lanka	25,332 sq. mi. (64,850 sq. km.)	16,087,000	Sinhalese (official), Tamil, English	5.7 billion (1984)	parliamentary republic
Taiwan	13,900 sq. mi. (35,584 sq. km.)	19,439,000	Mandarin Chinese (official), South Fukien Chinese	57.5 billion (1984)	one-party republic
Thailand	198,115 sq. mi. (507,174 sq. km.)	52,654,000	Thai (official), Chinese, Malay, Khmer	42.8 billion (1984)	constitutional monarchy
Vietnam	128,052 sq. mi. (327,813 sq. km.)	61,218,000	Vietnamese (official), Tay, Khmer, Thai	10 billion (1984)	Communist state

The Pacific World

Country	Land Area	Population (1986)	Language	GNP	Type of Government
American Samoa	77 sq. mi. (197 sq. km.)	36,000	Samoan and English (official)	160 million (1984)	territory of U.S. with bicameral legislature under Interior Dept.
Australia	2,966,200 sq. mi. (7,593,472 sq. km.)	15,912,000	English (official), Aboriginal languages	185 billion (1984)	federal parliamentary
Cook Islands	91 sq. mi. (233 sq. km.)	17,000	English and Maori (official)	20 million (1980)	self-governing in free association with New Zealand
Fiji	7,056 sq, mi. (18,063 sq. km.)	710,000	English (official), Hindi, Fijian	1.3 billion (1984)	independent parliamentary state
French Polynesia	1,359 sq. mi. (3,479 sq. km.)	180,000	French (official), Tahitian	1.3 billion (1984)	overseas territory of France
Guam	209 sq. mi. (535 sq. km.)	121,000	English (official), Chamorro, Philippine languages	760 million (1984)	under jurisdiction of U.S. Interior Dept.
Kiribati	328 sq. mi. (840 sq. km.)	65,000	English (official), Kiribati	30 million (1984)	republic
Nauru	8 sq. mi. (20 sq. km.)	8,000	Nauruan (official)	20 million (1984)	republic
New Caledonia	7,233 sq. mi. (18,516 sq. km.)	151,000	French (official), Melanesian, Wallisian	920 million (1984)	French overseas territory
New Zealand	103,493 sq. mi. (264,942 sq. km.)	3,288,000	English (official), Maori	23.5 billion (1984)	parliamentary republic
Papua New Guinea	178,704 sq. mi. (457,482 sq. km.)	3,400,000	English (official), Pidgin English, Papuan and Melanesian languages	2.5 billion (1984)	independent parliamentary state
Solomon Islands	10,640 sq. mi. (22,238 sq. km.)	277,000	English (official), Pidgin English, Kwara'ae	160 million (1983)	independent parliamentary state
Tokelau Islands	4.7 sq. mi. (12 sq. km.)	1,600	English (official), Tokelauan	1 million (1980)	administered by New Zealand
Tonga	288 sq. mi. (737 sq. km.)	98,000	Tongan and English (official)	80 million (1983)	constitutional monarchy
Tuvalu	9 sq. mi. (23 sq. km.)	8,000	Tuvaluan, Kiribati, English	5 million (1981)	constitutional monarchy
Vanuatu	4,706 sq. mi. (12,047 sq. km.)	137,000	French and English (official), Bislama	64 million (1982)	republic
Western Samoa	1,093 sq. mi. (2,798 sq. km.)	160,000	Samoan and English (official)	127 million (1981)	constitutional monarchy under chief

Latin America

Country	Land Area	Population (1986)	Language	GNP	Type of Government
Antigua and Barbuda	171 sq. mi. (438 sq. km.)	81,000	English (official)	150 million (1984)	republic
Argentina	1,073,399 sq. mi. (2,747,901 sq. km.)	31,030,000	Spanish (official), Italian	67.2 billion (1984)	federal republic
Aruba	75 sq. mi. (192 sq. km.)	61,000	Dutch (official), Papiamento	Not available	republic
Bahamas	5,382 sq. mi. (13,778 sq. km.)	235,000	English (official)	960 million (1984)	independent commonwealth
Barbados	166 sq. mi. (425 sq. km.)	253,000	English (official), English creole	1.1 billion (1984)	independent state within commonwealth
Belize	8,867 sq. mi. (22,700 sq. km.)	171,000	English (official), Spanish, Carib	180 million (1984)	republic
Bermuda	21 sq. mi. (54 sq. km.)	57,000	English (official)	920 million (1984)	British colony
Bolivia	424,164 sq. mi. (1,085,860 sq. km.)	6,611,000	Spanish, Quechua, Aymara (all official)	2.6 billion (1984)	republic
Brazil	3,286,500 sq. mi. (8,413,440 sq. km.)	138,403,000	Portuguese (official)	227.3 billion (1984)	federal republic
Chile	284,521 sq. mi. (728,374 sq. km.)	12,278,000	Spanish (official), Araucanian	20.3 billion (1984)	republic
Colombia	440,831 sq. mi. (1,128,527 sq. km.)	28,231,000	Spanish (official)	38.4 billion (1984)	republic
Costa Rica	19,730 sq. mi. (50,509 sq. km.)	2,534,000	Spanish (official)	2.9 billion (1984)	republic
Cuba	42,803 sq. mi. (109,576 sq. km.)	10,194,000	Spanish (official)	15.8 billion (1983)	Communist state
Dominican Republic	18,704 sq. mi. (47,882 sq. km.)	6,390,000	Spanish (official), French patois	6 billion (1984)	republic
Ecuador	103,930 sq. mi. (226,061 sq. km.)	9,651,000	Spanish (official), Quechua	10.3 billion (1984)	republic
El Salvador	8,124 sq. mi. (20,797 sq. km.)	5,461,000	Spanish (official), Nahuatl	3.8 billion (1984)	republic
French Guiana	35,900 sq. mi. (91,904 sq. km.)	86,000	French (official), Creole	180 million (1983)	overseas department of France
Grenada	133 sq. mi. (340 sq. km.)	97,000	English (official)	80 million (1984)	independent state
Guadeloupe	687 sq. mi. (1,759 sq. km.)	334,000	French (official), Creole	1.2 billion (1983)	overseas department of France
Guatemala	42,042 sq. mi. (107,628 sq. km.)	8,191,000	Spanish (official), Quiché, Mam, other American Indian languages	9.1 billion (1984)	republic
Guyana	83,000 sq. mi. (212,480 sq. km.)	796,000	English (official)	470 million (1984)	cooperative republic
Haiti	10,579 sq. mi. (27,082 sq. km.)	5,427,000	French (official), French Creole	1.7 billion (1984)	republic
Honduras	43,277 sq. mi. (110,789 sq. km.)	3,938,000	Spanish (official), American Indian languages	3 billion (1984)	republic
Jamaica	4,244 sq. mi. (10,865 sq. km.)	2,351,000	English (official), English Creole	2.5 billion (1984)	independent state within commonwealth
Martinique	421 sq. mi. (1,077 sq. km.)	328,000	French (official), Creole	1.3 billion (1983)	overseas department of France

Mexico	756,066 sq. mi. (1,935,529 sq. km.)	80,472,000	Spanish (official), Nahuatl, other American Indian languages	158.3 billion (1984)	federal republic
Netherlands Antilles	308 sq. mi. (788 sq. km.)	176,000	Dutch (official), Papiamento, English	1.4 billion	territory within the Netherlands
Nicaragua	49,363 sq. mi. (126,369 sq. km.)	3,384,000	Spanish (official), American Indian languages	2.7 billion (1984)	republic
Panama	29,762 sq. mi. (76,191 sq. km.)	2,227,000	Spanish (official), Guaymí, Cuna	4.2 billion (1984)	republic
Paraguay	157,048 sq. mi. (402,043 sq. km.)	3,531,000	Spanish (official), Guaraní	4.1 billion (1984)	republic
Peru	496,225 sq. mi. (1,270,336 sq. km.)	20,207,000	Spanish and Quechua (official), Aymara	18 billion (1984)	republic
Puerto Rico	3,515 sq. mi. (8,998 sq. km.)	3,286,000	Spanish and English (official)	14 billion (1984)	commonwealth within United States
St. Lucia	238 sq. mi. (609 sq. km.)	140,000	English (official), French patois	150 million (1984)	independent state within commonwealth
Suriname	63,251 sq. mi. (161,923 sq. km.)	395,000	Dutch and English (official), Hindi, Javanese, Creole	1.4 billion (1984)	military-civilian rule
Trinidad and Tobago	1,978 sq. mi. (5,064 sq. km.)	1,202,000	English (official), Hindi	8.4 billion (1984)	parliamentary republic
Uruguay	68,037 sq. mi. (174,175 sq. km.)	3,035,000	Spanish (official)	5.9 billion (1984)	republic
Venezuela	352,144 sq. mi. (901,489 sq. km.)	17,791,000	Spanish (official)	57.4 billion (1984)	federal republic
Virgin Islands (U.S. and GB)	195 sq. mi. (499 sq. km.)	126,000	English (official), Spanish	900 million (1984)	U.S. islands under the jurisdiction of the Interior Dept; GB islands administered by commonwealth

United States

State	Land Area	Population (1986 estimate)	Per Capita Income (1986)	Entered Union	Capital
Alabama	51,705 sq. mi. (132,365 sq. km.)	4,053,000	$11,115	1819	Montgomery
Alaska	591,004 sq. mi. (1,512,970 sq. km.)	534,000	$17,744	1959	Juneau
Arizona	114,000 sq. mi. (291,840 sq. km.)	3,317,000	$13,220	1912	Phoenix
Arkansas	53,187 sq. mi. (136,159 sq. km.)	2,372,000	$10,773	1836	Little Rock
California	158,706 sq. mi. (406,287 sq. km.)	26,981,000	$16,778	1850	Sacramento
Colorado	104,091 sq. mi. (266,473 sq. km.)	3,267,000	$15,113	1876	Denver
Connecticut	5,018 sq. mi. (12,846 sq. km.)	3,189,000	$19,208	1788	Hartford
Delaware	2,044 sq. mi. (5,233 sq. km.)	633,000	$15,010	1787	Dover

Florida	58,664 sq. mi. (150,180 sq. km.)	11,675,000	$14,281	1845	Tallahassee
Georgia	58,910 sq. mi. (150,810 sq. km.)	6,104,000	$13,224	1788	Atlanta
Hawaii	6,471 sq. mi. (16,566 sq. km.)	1,062,000	$14,691	1959	Honolulu
Idaho	83,564 sq. mi. (213,924 sq. km.)	1,003,000	$11,432	1890	Boise
Illinois	57,871 sq. mi. (148,150 sq. km.)	11,553,000	$15,420	1818	Springfield
Indiana	36,413 sq. mi. (93,217 sq. km.)	5,504,000	$12,944	1816	Indianapolis
Iowa	56,275 sq. mi. (144,064 sq. km.)	2,851,000	$13,222	1846	Des Moines
Kansas	82,277 sq. mi. (210,629 sq. km.)	2,461,000	$14,379	1861	Topeka
Kentucky	40,409 sq. mi. (103,447 sq. km.)	3,728,000	$11,129	1792	Frankfort
Louisiana	47,752 sq. mi. (122,245 sq. km.)	4,501,000	$11,227	1812	Baton Rouge
Maine	33,265 sq. mi. (85,158 sq. km.)	1,174,000	$12,709	1820	Augusta
Maryland	10,460 sq. mi. (26,778 sq. km.)	4,463,000	$16,588	1788	Annapolis
Massachusetts	8,284 sq. mi. (21,207 sq. km.)	5,832,000	$17,516	1788	Boston
Michigan	97,102 sq. mi. (248,581 sq. km.)	9,145,000	$14,064	1837	Lansing
Minnesota	86,614 sq. mi. (221,732 sq. km.)	4,214,000	$14,737	1858	St. Paul
Mississippi	47,689 sq. mi. (122,084 sq. km.)	2,625,000	$9,552	1817	Jackson
Missouri	69,697 sq. mi. (178,424 sq. km.)	5,066,000	$13,657	1821	Jefferson City
Montana	147,046 sq. mi. (376,438 sq. km.)	819,000	$11,904	1889	Helena
Nebraska	77,355 sq. mi. (198,029 sq. km.)	1,598,000	$13,777	1867	Lincoln
Nevada	110,561 sq. mi. (283,036 sq. km.)	963,000	$15,074	1864	Carson City
New Hampshire	9,279 sq. mi. (23,754 sq. km.)	1,027,000	$15,922	1788	Concord
New Jersey	7,787 sq. mi. (19,935 sq. km.)	7,620,000	$18,284	1787	Trenton
New Mexico	121,593 sq. mi. (311,278 sq. km.)	1,479,000	$11,037	1912	Santa Fe
New York	52,735 sq. mi. (135,002 sq. km.)	17,772,000	$17,118	1788	Albany
North Carolina	52,669 sq. mi. (134,833 sq. km.)	6,331,000	$12,245	1789	Raleigh
North Dakota	70,702 sq. mi. (180,997 sq. km.)	679,000	$12,284	1889	Bismarck
Ohio	44,787 sq. mi. (114,655 sq. km.)	10,752,000	$13,743	1803	Columbus
Oklahoma	69,956 sq. mi. (179,087 sq. km.)	3,305,000	$12,368	1907	Oklahoma City
Oregon	97,073 sq. mi. (248,507 sq. km.)	2,698,000	$13,217	1859	Salem

Pennsylvania	46,043 sq. mi. (117,870 sq. km.)	11,889,000	$13,944	1787	Harrisburg
Rhode Island	1,212 sq. mi. (3,103 sq. km.)	975,000	$14,670	1790	Providence
South Carolina	31,113 sq. mi. (79,649 sq. km.)	3,378,000	$11,096	1788	Columbia
South Dakota	77,116 sq. mi. (197,417 sq. km.)	708,000	$11,850	1889	Pierre
Tennessee	42,144 sq. mi. (107,889 sq. km.)	4,803,000	$11,831	1796	Nashville
Texas	266,807 sq. mi. (683,026 sq. km.)	16,682,000	$13,523	1845	Austin
Utah	84,899 sq. mi. (217,341 sq. km.)	1,665,000	$10,743	1896	Salt Lake City
Vermont	9,614 sq. mi. (24,612 sq. km.)	541,000	$12,845	1791	Montpelier
Virginia	40,767 sq. mi. (104,364 sq. km.)	5,787,000	$15,374	1788	Richmond
Washington	68,139 sq. mi. (174,436 sq. km.)	4,463,000	$14,498	1889	Olympia
West Virginia	24,231 sq. mi. (62,031 sq. km.)	1,919,000	$10,530	1863	Charleston
Wisconsin	66,215 sq. mi. (69,510 sq. km.)	4,785,000	$13,796	1848	Madison
Wyoming	97,809 sq. mi. (250,391 sq. km.)	507,000	$13,230	1890	Cheyenne
District of Columbia	69 sq. mi. (177 sq. km.)	626,000	$18,980	—	Washington, D.C. (national capital)
Total U.S.	3,623,461 sq. mi. (9,276,060 sq. km.)	241,489,000	$3,670.5 billion (1984 GNP)	—	Washington, D.C.

Canada

Province/ Territory	Land Area	Population (1986)	Per Capita Income in U.S. $ (1986)	Entered Confederation	Capital
Alberta	248,800 sq. mi. (636,928 sq. km.)	2,348,800	$ 9,508	1905	Edmonton
British Columbia	358,971 sq. mi. (918,966 sq. km.)	2,892,500	$10,203	1871	Victoria
Manitoba	211,723 sq. mi. (542,011 sq. km.)	1,069,600	$10,108	1870	Winnipeg
New Brunswick	27,834 sq. mi. (71,255 sq. km.)	719,200	$ 7,637	1867	Fredericton
Newfoundland	143,510 sq. mi. (367,386 sq. km.)	580,400	$ 7,713	1949	St. John's
Nova Scotia	20,402 sq. mi. (52,229 sq. km.)	880,700	$ 7,701	1867	Halifax
Ontario	344,090 sq. mi. (880,870 sq. km.)	9,066,000	$10,731	1867	Toronto
Prince Edward Island	2,185 sq. mi. (5,594 sq. km.)	127,100	$ 7,626	1873	Charlottetown

Quebec	523,859 sq. mi. (1,341,079 sq. km.)	6,580,000	$ 3,270	1867	Quebec
Saskatchewan	220,348 sq. mi. (564,091 sq. km.)	1,019,500	$ 8,773	1905	Regina
Northwest Territories	1,271,442 sq. mi. (3,254,891 sq. km.)	50,900	$ 9,965 (average figure for both territories)	1869 (date established)	Yellowknife
Yukon Territory	184,931 sq. mi. (473,423 sq. km.)	22,800		1898 (date established)	Whitehorse
Total Canada	3,849,675 sq. mi. (9,855,168 sq. km.)	25,640,000	$330.9 billion (1984 GNP)	—	Ottawa

World's Longest Rivers

River	Location	Length
Nile	Africa	4,187 mi. (6,699 km.)
Amazon	South America	4,000 mi. (6,400 km.)
Mississippi-Missouri	United States	3,860 mi. (6,176 km.)
Ob-Irtysh	Soviet Union	3,461 mi. (5,538 km.)
Chang (Yangtze)	China	3,434 mi. (5,494 km.)
Huang He	China	2,903 mi. (4,645 km.)
Zaire	Africa	2,716 mi. (4,346 km.)
Amur	Asia	2,705 mi. (4,328 km.)
Lena	Soviet Union	2,653 mi. (4,245 km.)
Mackenzie	Canada	2,635 mi. (4,216 km.)
Mekong	Asia	2,600 mi. (4,160 km.)
Niger	Africa	2,600 mi. (4,160 km.)
Yenisei	Soviet Union	2,566 mi. (4,106 km.)
Murray-Darling	Australia	2,310 mi. (3,696 km.)
Volga	Soviet Union	2,293 mi. (3,669 km.)
Paraná	South America	1,827 mi. (2,923 km.)

World's Largest Islands

Island	Area
Greenland	840,000 sq. mi. (2,150,400 sq. km.)
New Guinea	344,927 sq. mi. (883,013 sq. km.)
Borneo	290,320 sq. mi. (743,219 sq. km.)
Madagascar	226,657 sq. mi. (580,242 sq. km.)
Baffin	183,810 sq. mi. (470,554 sq. km.)
Sumatra	182,561 sq. mi. (467,356 sq. km.)
Great Britain	88,146 sq. mi. (225,654 sq. km.)
Honshu	86,246 sq. mi. (220,790 sq. km.)
Ellesmere	82,119 sq. mi. (210,225 sq. km.)
Victoria	81,930 sq. mi. (209,741 sq. km.)

World's Largest Bodies of Water

Body of Water	Location	Area
Oceans		
Pacific Ocean		63,800,000 sq. mi. (163,328,000 sq. km.)
Atlantic Ocean		31,800,000 sq. mi. (81,408,000 sq. km.)
Indian Ocean		28,900,000 sq. mi. (73,984,000 sq. km.)
Arctic Ocean		5,400,000 sq. mi. (13,824,000 sq. km.)
Major seas, bays, and gulfs		
Caribbean Sea		1,063,000 sq. mi. (2,721,280 sq. km.)
Mediterranean Sea		967,000 sq. mi. (2,475,520 sq. km.)
South China Sea		895,400 sq. mi. (2,292,224 sq. km.)
Bering Sea		876,000 sq. mi. (2,242,560 sq. km.)
Sea of Okhotsk		610,000 sq. mi. (1,561,600 sq. km.)
Gulf of Mexico		596,000 sq. mi. (1,525,760 sq. km.)
East China Sea		482,000 sq. mi. (1,233,920 sq. km.)
Hudson Bay		475,000 sq. mi. (1,216,000 sq. km.)
Sea of Japan		389,000 sq. mi. (995,840 sq. km.)
North Sea		222,000 sq. mi. (568,320 sq. km.)
Andaman Sea		218,100 sq. mi. (558,336 sq. km.)
Black Sea		178,000 sq. mi. (455,680 sq. km.)
Red Sea		169,000 sq. mi. (432,640 sq. km.)
Baltic Sea		163,000 sq. mi. (417,280 sq. km.)
Largest lakes	Location	
Caspian Sea	Asia-Europe	143,240 sq. mi. (366,694 sq. km.)
Lake Superior	North America	31,700 sq. mi. (81,152 sq. km.)
Lake Victoria	Africa	26,820 sq. mi. (68,659 sq. km.)
Aral Sea	Asia	24,909 sq. mi. (63,767 sq. km.)
Lake Huron	North America	23,000 sq. mi. (58,880 sq. km.)
Lake Michigan	North America	22,300 sq. mi. (57,088 sq. km.)
Lake Tanganyika	Africa	12,350 sq. mi. (31,616 sq. km.)
Lake Baikal	Asia	12,160 sq. mi. (31,130 sq. km.)
Great Bear Lake	North America	12,028 sq. mi. (30,792 sq. km.)
Lake Nyasa	Africa	11,150 sq. mi. (28,544 sq. km.)

Highest and Lowest Continental Points

Highest Point	Height Above Sea Level	Lowest Point	Depth Below Sea Level
Africa Mount Kilimanjaro, Tanzania	19,340 ft. (5,896 m.)	Lake Assal, Djibouti	512 ft. (156 m.)
North America Mount McKinley, Alaska	20,320 ft. (6,195 m.)	Death Valley, California	282 ft. (86 m.)
South America Mount Aconcagua, Argentina	22,834 ft. (6,962 m.)	Valdés Peninsula Argentina	131 ft. (40 m.)
Antarctica Vinson Massif	16,864 ft. (5,141 m.)	Unknown	— —
Asia Mount Everest, Nepal-Tibet	29,028 ft. (8,850 m.)	Dead Sea Israel-Jordan	1,302 ft. (397 m.)
Australia Mount Kosciusko, New South Wales	7,316 ft. (2,230 m.)	Lake Eyre, South Australia	52 ft. (16 m.)
Europe Mount Elbrus, Soviet Union	18,481 ft. (5,634 m.)	Caspian Sea, Soviet Union	92 ft. (28 m.)

GLOSSARY

Pronunciation Key

This key will help you understand the pronunciations of difficult words given in the textbook.

Key Word	Example	Key Word	Example
map	Alaska (uh-LASS-kuh)	book	Cooktown (KOOHCK-toun)
day	Malaya (muh-LAY-uh)	humid	Utah (YOO-taw)
date	Maine (MANE)	sun	Kentucky (ken-TUCK-ee)
air	Canary (kuh-NAIR-ee)	deep	Armenia (ahr-MEE-nee-uh)
father	Baja (BAH-HAH)	tip	Britain (BRIT´-n)
law	Bengal (ben-GAWL)	pie	Thailand (TIE-land)
wet	Genoa (JENH-uh-wuh)	tide	Kuwait (koo-WITE)
boy	Detroit (di-TROIT)	top	Bosporus (BAHS-puh-ruhs)
now	Moscow (MOSS-kow)	toe	Ohio (oe-HIE-oe)
moon	Rangoon (rang-GOON)	note	Gabon (guh-BONE)

A

absolute location the actual site on the earth's surface where something is found, 33

acculturation the process by which one culture changes through contact with another, 88

acid rain pollution caused by sulphur dioxide from coal-burning plants mixing with water in the atmosphere to make a sulphuric acid that then mixes with rainwater, 592

agribusiness farming done on large modern farms, many of which are owned by large corporations, 569

air pressure the measurement of the weight of air, 55

alliance an agreement between countries to support one another against enemies, 111

allspice a spice, or seasoning, produced in Jamaica, 478

alluvial fan a landform created by sediment eroded from a mountain and deposited by a stream as it enters a plain along the mountain's base, 52

alluvial soils soils deposited by rivers, 169

altiplano a broad highland plain between the eastern and western mountain chains of the Andes, 491

animism a traditional belief in nature and spiritual beings, 306

Antarctic Circle the line of latitude located at 66 1/2° south of the equator, 42

anthracite coal hard coal, 532

antimony a silvery white element used in printing type and to make other metals, 235

apartheid (uh-PAHR-tayt) the South African policy of separation of the races, 343

aquaculture water and sea farming, 361

aquifer a rock layer composed mostly of sand and gravel materials through which groundwater can flow, 76

arable land that is suitable for growing crops, 162

archipelago (ahr-kuh-PEHL-uh-goh) a large group of islands, 415

Arctic Circle the line of latitude located at 66 1/2° north of the equator, 42

arid dry, 246

artesian well one in which water rises toward the surface without being pumped, 427
atlas an organized collection of maps in one book, 25
atoll a ring of coral islands surrounding a shallow lagoon, 437
autarky (AW-tahr-kee) a policy to develop self-sufficiency in industrial production, 200
autonomy the right to self-government, 180

B

balance of payments the difference between a nation's total payments to other countries and the total receipts from them over a given period of time, 400
balance of trade the difference between the value of the goods a country exports and the value of the goods it imports, 168
barrier island a long, narrow, sandy island separated from the mainland by a lagoon, 549
barter to trade in goods without using money, 605
basin a low area of land, generally surrounded by mountains, 164
bayou a small, sluggish stream, 550
Bedouins migrating desert herders, 248
bilingual able to speak two languages, 153
biogeography the study of the geographic distribution of plants and animals, 424
biome a plant and animal community that covers a very large area of the earth's surface, 82
biosphere all living things and the areas they inhabit on Earth, 78
biotechnology involves research and products in pharmaceuticals and genetic engineering, 528
birthrate the number of births per 1,000 people in a given year, 99
bituminous coal soft coal, 532
bog soft ground saturated with water, 127
borough one of five administrative units into which New York City is divided, 535
Buddhism a leading religion in parts of the Orient, founded in India and based on the belief of enlightenment through meditation, 363
buffer state a neutral area separating rival powers, 379

C

caldera a huge depression formed after a volcanic eruption or a collapse of a volcanic mountain, 567
canal a waterway dug across land to allow ships to cross, 117
canyon a deep, narrow valley with steep sides, 45
capital any source of wealth used to produce more wealth, 286
capitalism an economic system in which resources,industries, and businesses are owned by private individuals, 97
cartel an agreement among producers of a product to limit supplies to keep prices high, 503
cartography the branch of geography that studies maps and map making, 36
cash crop a crop produced for direct sale in a market, 282
cassava a fleshy root crop that provides a nutritious starch, 306
caste system a social system that divided India's society into five major classes based on occupation, 373
Christianity the religion founded on the teachings of Jesus Christ, 243
chromium a blue-white metal used to plate metal, 249
climate weather conditions in an area over a long period of time, 54
climax community the stable plant group that forms the last stage of plant succession, 81
coalition government a government in which several political parties join together to run the country, 171
coke coking coal, used in blast furnaces to purify iron ore for making steel, 532
collective farm state-owned farms whose workers receive a portion of the profits, 193
columbite a metal used to harden stainless steel, 313
COMECON the economic association organized by the Soviet Union to bind the countries of Eastern Europe together and to the Soviet Union, 227
command economy an economic system in which the prices of goods and the wages of labor are set by the government, 97
commodity a valuable economic good, 249
commonwealth an association of self-governing states with similar political and cultural backgrounds and united by a common loyalty, 126
commune a cooperative group of farmers who pool their labor, 385
communism a social system in which all economic activity is controlled by a government ruled by a single political party, 97
compass rose a directional indicator with arrows that point to north, south, east, and west, 28
condensation the process by which water vapor changes from a gas to liquid droplets, 58
confederation a group of states joined together for a common purpose, 161
coniferous forest (koh-NIF-uh-ruhs) a middle-latitude forest that remains green year-round, 83
constitutional monarchy government that has a reigning king or queen, but with a parliament as the law-making branch of government, 124
contiguous bordering each other as a single unit, 515
continents the seven large masses of land on Earth, 22
Continental Divide the crest of the Rocky Mountains that divides the major river systems of North America into those that flow eastward into the Atlantic Ocean and Gulf of Mexico and those that flow westward into the Pacific Ocean and Gulf of California, 560
continental island island that was formed from sections of ancient continents, 437
continental shelf the part of the sea floor that slopes gently from the continents and is the most shallow part of the ocean, 74
cooperative a business organization owned by and operated for the mutual benefit of its members, 134
copra the dried meat of the coconut from which coconut oil is obtained, 336
coral a rocky limestone material formed of the hard skeletons of the coral polyp, 424
cottage industries the production of small consumer items by workers in homes and small workshops, 377
coup an overthrow of an existing government by a small group, 312
crop rotation substituting a different crop in some years for the regularly grown crops to prevent mineral loss from the soil, 542
cultural exchange the trading of ideas between countries, 610
cultural geography the study of how people and their activities vary from place to place, 86
culture all the features of a society's way of life, 87
culture region an area with many shared culture traits, 87
culture trait something that people in a culture normally do that distinguishes that culture from others, 87
currency money, 605

D

death rate the numbers of deaths per 1,000 people in a given year, 99
deciduous forest (di-SIJ-uh-wuhs) a middle-latitude forest that is green during the summer and loses its leaves before winter, 83
deforestation the destruction of forests for more farmland or wood, 372
degrees a unit of measurement, 23
delta the landform at the mouth of a river created by an alluvial deposit, 52
democratic government government in which everyone in a society has a voice, 92
deport to send out of a country, 585
desalinization the process that purifies seawater and salty water from wells by removing the salts, 272
desertification the process by which the desert has been expanded through overgrazing and the cutting of trees for wood, 305
developed countries the world's wealthiest countries, in which most people work in manufacturing and service industries, 96
dialect a regional variety of a major language, 156
diffusion the spread of an innovation or culture trait through a society or into another culture region, 87
dikes walls built to keep out water, 148

directional indicator a symbol on a map which always points north, 28
doldrums calm areas centered along the equator and at about 30° north and south latitudes, 56
domestication the innovation of raising animals and growing crops so they are of use to humans, 89
dominion a territory or sphere of influence, 586
double cropping raising two crops in one year, 388
dryland agriculture farming that depends upon rainfall rather than irrigation, 218
dynasty a government ruled by a family whose power lasts for a long period of history, 383

E

earthquake the shock waves caused by moving rock above a fault, 47
economic association an organization formed to break down trade barriers between member nations, 119
economic geography the study of how people use resources, earn a living, and distribute products, 95
El Niño a weather disturbance that brings destructive torrential rains to the dry coastal lands of Peru and Ecuador, 492
elevation the height of land above sea level, 45
embargo a legal restriction placed on the movement of freight into a country, 334
enclave a part of one country completely within the boundaries of another, 161
Equator an imaginary line that circles the globe in an east-west direction exactly halfway between the North and South poles, 23
equinox a time of equal light and darkness everywhere on the earth when the poles are not tilted toward or away from the sun, 40
erg a shifting sea of sand found in the Sahara, 277
erosion the wearing away of the earth's surface, 44
escarpment a steep slope capped by a nearly flat plateau, 300
estuary a water passage formed where rivers meet seawater, 75
ethnic group a group of people who share the same language, customs, and traditions, 192
European Community (EC) an economic association of European nations that maintains close political ties (also called the Common Market), 120
evaporation the process by which water is changed from a liquid to a gas, 58
evapotranspiration the evaporation of water from the ground combined with the transpiration of water by plants, 74
exile a person who has left or has been forced to leave his or her country, 348
exotic river rivers that begin in humid regions and cross into desert areas, 248
export economy an economy in which most products are for sale to other countries, 404

F

fall line where the Atlantic Coastal Plain meets the Piedmont, 519
fallow land on which crops are not planted, in order to conserve moisture and nutrients in the soil, 263
fault the result of rock layers that break and move apart, 47
favelas (vah-VEH-lahs) huge slum areas in Brazil, 498
fiord a narrow, deep inlet of the sea between high, rocky cliffs, 131
floodplain level ground built by sediments deposited by a river or stream, 52
foehn (FUHRN) a warm, dry wind that blows down the side of a mountain, 162
fold the result of rock layers that become bent, 47
food chain a series of stages in which energy is passed along through living things, 79
foodstuff a substance with food value, particularly the raw material of food, 236
forward capital a city that is built from the ground up in a deliberate attempt to develop the interior region of a country, 499
fossil fuels energy sources formed from plants and animals buried millions of years ago, 124
fossil groundwater water trapped for many years beneath the desert, 278
free enterprise the type of economy that gives people freedom to operate private businesses for a profit, 97
free port a port city in which there are almost no taxes on goods unloaded there, 286
front the meeting zone of two different types of air, 57

G

galaxy a huge, nearly circular collection of stars, 38
geothermal refers to heat from the earth's interior, 435
geyser a hot spring that shoots up fountains of hot water and steam into the air, 136
glaciers thick masses of ice that glide slowly across the earth's surface, 45
globe a scale model of the earth, 22
granite a gray-to-pink, speckled, hard, crystalline rock formed deep in the earth's crust, 525
great-circle route the shortest distance between any two points on Earth, 27
Green Revolution the introduction of new types of high-yield grains, 372
greenhouse effect a condition caused by the apparent buildup of carbon dioxide in the lower atmosphere, which traps heat, causing a long-term warming of the planet, 616
grid pattern of lines used as a reference for locating points, 23
gross national product (GNP) the total value of goods and services produced by a country in a year, 96
groundwater the water found below the earth's surface in the spaces between soil and rock grains, 75
growth point an area designated by a government for industrial development, 502
guerrilla one who takes part in irregular warfare, especially through harassment and sabotage, 258
Gulf Stream a warm ocean current that moves warm water northward, 517
gum arabic a substance from the sap of acacia trees, used in some medicines and candies, 322

H

habitation fog a weather condition caused by the warmth, dust, and water vapor created by cities, 218
hacienda a large estate in Mexico, 464
headwaters streams formed from runoff from precipitation falling on hills and mountains, 75
heartland central region important to a nation's economy or defense, 206
hemisphere half a sphere, the globe divided into halves, 24
Hinduism India's chief religion, based on the worship of many gods and the belief that enlightenment is offered through good conduct, 362
hinterland a region surrounding and serviced by an urban area, 432
Holocaust the persecution and death of millions of Jews by the Nazis during World War II, 256
hot spot an area where molten material from the earth's mantle rises through the crustal plate, 580
human geography the study of how the activities of people vary throughout the world, 33
humidity the measure of the amount of water vapor in the air, 58
humus decayed plant matter found in soil, 81
hurricanes severe tropical storms that sometimes move from the low latitudes to the middle latitudes bringing violent winds, torrential rain, and high seas (also called typhoons), 59
hydroelectricity electricity produced by water, 118
hydrologic cycle the circulation of water from one part of the hydrosphere to another, 73
hydrosphere all the water of the earth, including oceans, lakes, streams, and underground, in all living things, in the atmosphere, and in ice, 72

I

illiteracy the inability to read and write, 96
imperialism the policy of gaining control over territory outside a nation, 111

indigo a plant used to make a blue dye, 453
industrialization the growth in the method of producing goods in large amounts at low cost with the use of machines, 90
inflation an economic situation in which prices rapidly increase because the country's money is losing value, 494
infrastructure basic residential and transportation facilities that must be built before new industries can be started, 207
innovation a new way of doing things, 87
intensive agriculture farming that requires much human labor, 408
irrigation the watering of land through pipes, ditches, or canals, 76
Islam the religion founded by Muhammad that is based on the belief in one god, 243
island landmass that is smaller than a continent and completely surrounded by water, 22
isthmus a narrow neck of land connecting two larger areas of land, 173

J

Judaism the first religion based on the belief in a single god, 243

K

kibbutz an Israeli collective farm, 257

L

labor intensive refers to industries that make products that must be assembled by hand, 456
land reform the breaking up of large estates, 464
landforms shapes on the earth's surface, 47
landlocked completely or almost completely surrounded by land, 117
LANDSAT photographs taken by American satellites, 29
latitude the distance north or south of the equator; lines of latitude are drawn in an east-west direction around the globe, 23
lava melted, liquid rock from within the earth that is brought to the earth's surface, 47
leaching the process by which nutrients are washed out of the topsoil by heavy rainfall, 81
legend a key that explains what the symbols on a map represent, 28
less developed countries (LDCs) the world's poorer countries in which most people practice subsistence agriculture, 96
levee (LEHV-ee) a ridge of earth along a riverbank that prevents flooding, 550
llanos (LAN-ohz) large, grassy plains found in South America, 453
lochs long, deep lakes in Scotland carved by glaciers, 123
lock a part of a waterway enclosed by gates at each end, used to raise or lower boats from one water level to another, 541
loess (LEHS) dust-sized particles of soil deposited by the wind, 160
longitude the distance east or west from the prime meridian; lines of longitude are drawn in a north-south direction around the globe, 23

M

map a flat diagram of all or part of the earth's surface, 25
map projection different ways of presenting a round globe on a flat surface, 25
marsupial a mammal that carries its young in a pouch, 424
martial law law administered by the military forces of a country when civilian law is unable to keep public order, 230
megalopolis (mehg-uh-LAHP-uh-luhs) a number of commercially and industrially developed cities and suburban areas so close to one another as to appear as one continuous city, 534
meridians lines of longitude, 23
mestizos people with both Indian and European ancestors, 453
microstate a very small country usually dependent upon its larger neighbors, 108
midoceanic ridge a chain of mountains along the ocean floor, 50
millet a grass grown for its small edible seeds, often used to make flour or to cook in soup, 322
minaret the tower of a mosque, 259
monoculture the practice of raising only one crop year after year, 552
monotheism the belief in one god, 245
monsoon a change of winds that brings wet and dry seasons, 63
moraine a ridge of rocks, gravel, and sand that is deposited by a glacier, 525
Mormons members of the church of Jesus Christ of Latter-day Saints, 564
mosque the Muslim place of worship, 245
mulattos people of mixed black and white ancestry, 453
multinational company a business with activities based in many countries, 102
muskeg a forested marsh that melts only during the summer, 598
Muslim a person who follows Islam, 245

N

nationalism feelings of loyalty toward and pride in one's country, 91
nationalized owned and operated by the government, 125
natural boundary a political boundary established by landforms, 116
navigable deep and wide enough for ships, 178
neutrality a nation's policy of not taking sides in international affairs, 134
nodal region an area defined by the movement around it, 35
nonalignment political position of not siding with the major world powers, 378
nonrenewable resources resources that are not replaced by natural processes or are replaced at extremely low rates, 612

O

oasis a desert site that has springs or wells or water beneath dry riverbeds, 247
oceans the world's largest bodies of water, 22
oceanic island island that is formed of volcanic material from the ocean floor, 437
orographic effect (ohr-uh-GRAF-ik) an effect of mountains on climate which produces moisture on the windward side of the mountains and dry air on the leeward side, 60
ozone layer a region in the upper atmosphere made up of ozone (O_3) gas formed from an interaction between oxygen and sunlight that protects life from dangerous ultraviolet radiation, 615

P

paddy a water-covered field for growing rice and other crops, 361
pampas large, grassy plains in South America, 453
panhandle a narrow arm of land attached to a larger area, 577
parallels lines of latitude drawn around the globe that are always the same distance from the equator, 23
peat substance found in bogs that is composed of decayed vegetable matter, 127
peninsula a landform surrounded by water on three sides, 115
periodic market an open-air trading market held regularly at a crossroads or in a town, 327
periphery region surrounding the core, 206
permafrost water below the tundra surface that is permanently frozen, 65
peso the monetary unit of many Latin American countries, 494
physical geography the study of the earth's natural features, vegetation, climate, and how they vary, 33
piedmont (PEED-mahnt) an area at or near the foot of a mountain region, 519
plains nearly flat areas of land, 47
planets spheres, or ball-like objects, that revolve around a central star, 39
plant community a group of plants that live and grow together, 79
plant succession the process by which one group of plants replaces another, 80

plantain a type of banana usually used in cooking, 476
plantation a large farm that concentrates on one cash crop, 294
plate tectonics (tehk-TAHN-iks) the theory that the earth's surface is made up of rigid, moving plates, 50
plateau a flat-topped tableland that rises above surrounding plains, 52
plaza a public square in the center of town, 179
polar easterlies cold winds in high latitudes that blow from the Arctic or Antarctic, 57
polar regions cold areas that surround the earth's poles, 39
polders lowland areas that have been drained or reclaimed, 148
population geography the study of how populations change in particular places and regions, 97
portage a low land area between lakes or rivers where boats and freight can be carried across, 189
prairie wet, tall grassland on fertile soil, 83
precipitation condensed droplets of water vapor that appear as rain, snow, sleet, and hail, 58
primary economic activities agriculture, forestry, mining, and other actions that make direct use of natural resources, 95
prime meridian an imaginary line that runs from the North Pole to the South Pole, through Greenwich, England, 23

R

rain shadow a desert caused by its location on the leeward side of high mountains, 561
raw materials materials in their natural state that are used to produce finished goods, 111
reef a ridge of coral, rock, or sand lying at or near the surface of the water, 424
reforestation the planting of new trees on land that was once forested, 526
reg a gravel-covered plain found in the Sahara, 277
region an area with common characteristics, making it different from any other area in the world, 33
regional geography the study of how areas are the same or different, 33
regionalism the political and emotional support for one's region before one's country, 598
relative location the relationship of a place to other places, 33
relief the difference in elevation between the top and bottom of a landform, 51
remote sensors instruments that gather and record information from a distance, 29
renewable resources those resources that can be replenished, 612
republic a form of government in which citizens elect the country's representatives and head of state, 163
revolution the circular orbit of the earth around the sun, 40
rift a narrow trough, or valley, with mountainous slopes on each side, 301
rock weathering the process of breaking up rocks and causing them to decay, 45
rotation one complete spin of the earth on its axis, 40
rutile a source of titanium, a metal used to manufacture jet airplanes and spacecraft, 311

S

Sahel (suh-HAYL) the dry savanna region along the southern edge of the Sahara, 299
salinization salt buildup in the soil, 613
sanctions trade and other restrictions placed on one country by other countries, 348
sand dunes hills of wind-deposited sand, 46
savanna a grassland with scattered trees and shrubs, 83
scale the line on a map that represents distance, 27
seaboard the land area near the sea, 515
seasons times of the year when varying amounts of the sun's energy reach the earth, causing greater or lesser heat, 41
second-growth forest the trees that cover an area after the original stands have been removed, 526
secondary economic activities food processing and manufacturing, 95
sediments small particles of mud, sand, or gravel caused by rock weathering, 45
Sikhism a religion in India that is a blend of Hindu and Muslim teachings, 369
silt sediment coarser than clay yet finer than sand deposited by rivers, 278
sisal a strong durable fiber used to make rope and twine, 319
slash-and-burn farming a type of agriculture in which forests are cut and burned for fields, 408
socialism an economic system by which the government owns and controls the means of producing goods, 134
soil exhaustion refers to the loss in soil nutrients that results from always planting the same crop in a particular area, 497
soil horizons distinct layers formed by developing soils, 81
solar system nine planets and all other objects revolving around the sun, 39
solstice when the earth's axis is tilted at its greatest angle toward or away from the sun, 41
sorghum a grain crop, 302
soviet a local governing council in the Soviet Union, 192
spatial interaction the movement of people, goods, and ideas across the earth's surface, 35
state farm a government-operated farm whose workers receive wages, 201
steppe dry, short grassland, 83
storm surge a rise in sea level as a storm approaches the coast, 372
strait a narrow passage of water that connects two larger bodies of water, 166
strip-cropping the planting of alternating rows of close-growing crops and open-growing crops to prevent wind erosion, 613
strip mining the practice of using large machines to strip away soil and rock to get the coal beneath the surface, 558
subcontinent a large landmass smaller than a continent, 368
subsidy financial support given by a government, 398
subsistence agriculture growing only enough crops for one's family, 89
sultanate a country ruled by a Muslim monarch, 380
Swahili the official language of Kenya and Tanzania and the common language of East Africa, 318

T

taconite (TAK-uh-nite) low-grade iron ore refined into iron-rich pellets, 545
taiga an enormous stand of softwood, needleleaf trees, 195
tannin a product of the quebracho tree, used in preparing leather, 506
tar sands rocks that contain oil, 482
tariffs taxes on imported goods, 120
terracing a method of planting crops in hilly country on a series of flat steps built at different levels, 389
terrorism the use of violence as a means of political force, 120
tertiary economic activities service industries, 95
theocracy a country governed by religious law, 268
tierra caliente the "hot" zone of the Andes found in the warm atmosphere between sea level and 3,000 feet (915 meters), 484
tierra fría the "cold" country zone of the Andes from about 6,500 feet (1,982 meters) to about 16,000 feet (4,878 meters), 484
tierra helada the "frozen" zone that is the area of highest altitude in the Andes, 484
tierra templada the "temperate" zone of pleasant climates in the Andes, 484
toponym the name of a place that often reflects the history and culture of a region, 563
tornadoes small, twisting, violent storms, 59
totalitarian government a government ruled by one person and a few advisors, 92
township and range system the settlement pattern of the United States Midwest that was started as a survey system by the federal government to divide the land for sale, 540
trade winds winds in low latitudes that blow from the east, 56
transpiration the process by which plants give off water vapor through their leaves, 74
tree line the elevation beyond which trees cannot grow, 162
tributary any smaller stream that flows into a larger stream, 75
tropical cyclone hurricane-like storms, 372

Tropic of Cancer the latitude line located at 23 1/2° north of the equator, 42
Tropic of Capricorn the latitude line located at 23 1/2° south of the equator, 42
tropics warm areas of the earth near the equator, 39
truck farms farms whose products can easily be trucked to the big-city markets, 532
trust territory a region that is placed under the control of another nation until it can govern itself, 314
tsar emperor of Russia, 189
tsunami large sea waves caused by earthquakes and eruptions, 396
typhoons severe tropical storms that sometimes move from the low latitudes to the middle latitudes bringing violent winds, torrential rain, and high seas (also called hurricanes), 59

U

uniform region an area that has one or more common features throughout, 35
uninhabitable unable to support human life and settlements, 135
universe includes everything that is known to exist, 38
urbanization the growth in the proportion of people living in towns and cities, 89

V

vanadium a gray mineral that combines with iron to make very strong steel, 335
veld (vehlt) vegetation and landform regions in South Africa, 341
vineyard land cultivated for growing grapes, especially for wine production, 142
volcano a mountain of lava, 47
voodoo a Haitian version of African animism, 476

W

water table the top of the saturated zone of groundwater, 75
watershed a region drained by a river system, 544
weather the condition of the atmosphere at a given place and time, 54
westerlies prevailing winds in middle latitudes that come from the west, 57

Z

Zionism the movement to establish a Jewish state in Palestine, 256

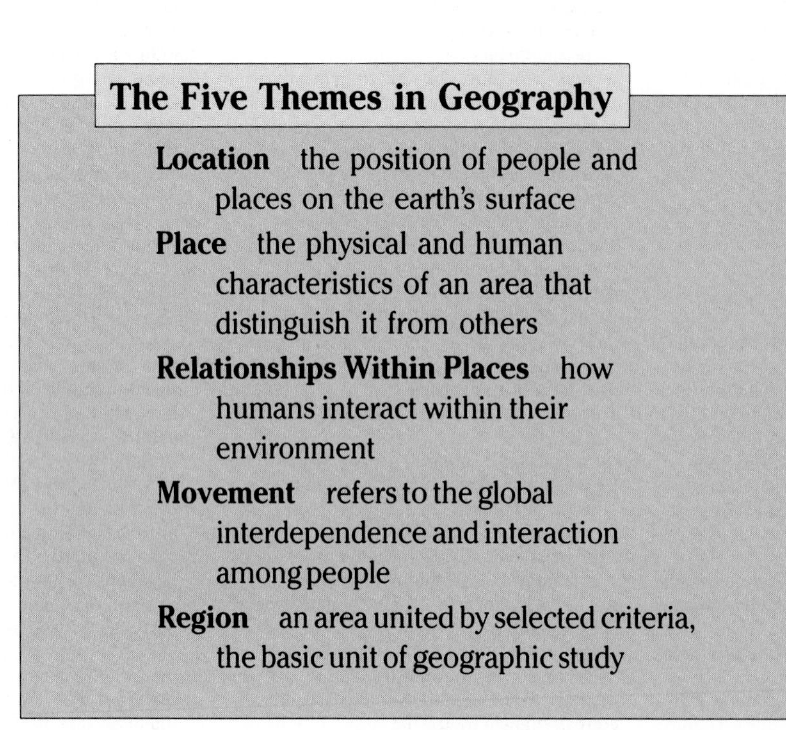

The Five Themes in Geography

Location the position of people and places on the earth's surface

Place the physical and human characteristics of an area that distinguish it from others

Relationships Within Places how humans interact within their environment

Movement refers to the global interdependence and interaction among people

Region an area united by selected criteria, the basic unit of geographic study

INDEX

d indicates diagram
g indicates graph
m indicates map
p indicates photo

A

B

C

D

E

F

G

H

I

J

K

L

M

N

O

P

Q

R

S

T

U

PHOTO CREDITS

Abbreviations used: (t) top; (c) center; (b) bottom; (l) left; (r) right.

Basic Map and Globe Skills: Page 29, Earth Observation Satellite Corporation.
Unit 1: Page 30(t), David Muench Photography; 30(b), NASA; 31, Judy White/Berg and Associates.
Chapter 1: Page 32(l), Sheryl S. McNee/Tom Stack and Associates; 32(r), Payne Johnson/FPG International; 33, Cameramann International, Ltd.; 34, FPG International; 35, Don Klumpp/Image Bank; 36(t), Steve Dunwell/Image Bank; 36(b), HRW Photo by Eric Beggs.
Chapter 2: Page 38(l), Robert Frerck/Odyssey Productions; 38(r), California Institute of Technology and Carnegie Institute of Washington; 40, NASA; 42, Gary W. Guippen/Earth Scenes.
Chapter 3: Page 44(l), Breck P. Kent/Earth Scenes; 44(r), James Sugar/Black Star; 45, Manuel Rodriequez; 46, Jim Brandenburg; 51, Stella Sneed/Bruce Coleman Inc.
Chapter 4: Page 54(l), Ed Nagele/Photo Source; 54(r), Elliott Varner Smith; 57, Peter French Photography; 58, 59, NASA; 60, Lynn McLaren/Picture Cube.
Chapter 5: Page 62(l), Dave Brown/Journalism Services; 62(r), Bruno Barbey/Magnum Photos; 64, Jerry Derbyshire/Berg and Associates; 65, Carol Brown/Center for Creative Photography; 69, Nancy Christensen/Berg and Associates; 70(tl), Robert Frerck/Odyssey Productions; 70(cl), M.P. Kahl/Tom Stack and Associates; 70(cr), Tom Algire Photography; 70(bl), HRW Photo by Dennis Cowals; 70(br), Cameramann International, Ltd.
Chapter 6: Page 72(l), Michael Nichols/Magnum Photos; 72(r), Paul von Baich; 75, Earth Observation Satellite Corporation; 76, David R. Frazier.
Chapter 7: Page 78(l), M. P. Kahl/DRK Photo; 78(r), Grant Heilman/Grant Heilman Photography; 79, Michael Guest/Berg and Associates; 83(t), Nadine Zuber/Photo Researchers; 83(b), D. Cavagnaro/DRK Photo; 84, Warren Jacobi/Berg and Associates.
Chapter 8: Page 86(l), Guido Alberto Rossi/Image Bank; 86(r), David Muench Photography; 87, Robin White/Fotolex Associates; 88, Carol Jopp/Robert Harding Picture Library; 89, Cameramann International, Ltd.; 90, Four By Five; 92, Carl Purcell/Words and Pictures.
Chapter 9: Page 94(l), Robert Frerck/Odyssey Productions; 94(r), Dirck Halstead/Gamma-Liaison; 95, John Neubauer/International Stock Photo; 99, Susan Malis/Marilyn Gartman Agency; 100, AP/Wide World Photos; 101(t), Phil Degginger/Bruce Coleman Inc.; 101(b), Mark Antman/Image Works; 102(t), HRW Photo by Owen Franken; 102(b), Alex Webb/Magnum Photos.
Unit 2: Page 106(t), Guido Alberto Rossi/Image Bank; 106(b), Norma Watts; 107, Eric Meola/Image Bank.
Chapter 10: Page 108(l), P. J. Griffiths/Magnum Photos; 108(r), 109, Bob and Ira Spring Photography; 114, George Mars Cassidy/Picture Cube; 115, Dieter Blum/Peter Arnold Inc.; 116, Luis Casteneda/Image Bank; 117, Ellis Herwig/Picture Cube; 118, Annie Griffiths/DRK Photos; 119, Harry Gruyaert/Magnum Photos; 120, Frank Lambrecht/Berg and Associates.
Chapter 11: Page 122(l), HRW Photo by Russell Dian; 122(r), Four By Five; 123, Chris Thomson/Image Bank; 125, Peter Paz/Image Bank; 126, Peter Jordan/Gamma Liaison; 128, Ellis Herwig/Picture Cube.
Chapter 12: Page 130(l), Ron Sanford/F-Stop Pictures; 130(r), Henry D. Meyer/Berg and Associates; 131, M. Desjardins/Photo Researchers; 132, Hubertus Kanus/Shostal and Associates; 135, Thomas Nebbia; 136, Earth Observation Satellite Corporation; 137, GEO Heat Center/Oregon Institute of Technology; 138, Guido Mangold/Image Bank.
Chapter 13: Page 140(l), Lisl Dennis/Image Bank; 140(r), 141, HRW Photo by Russell Dian; 142, Bruno Barbey/Magnum Photos; 144, Stuart Cohen; 145, Rick Berkowitz/Picture Cube; 146, Marc Romanelli/Image Bank; 147(t), R. L. Goodard/Berg and Associates, 147(b), Robert Harding Picture Library; 149, Cameramann International, Ltd.; 150, Bob and Ira Spring Photography; 151(t), Dr. Georg Gerster/Photo Researchers; 151(b), Carol Lee/Picture Cube; 152, The St. Louis Art Museum/Gift of Mrs. Mark C. Steinberg; 153, Robert Davis/Photo Researchers; 154, Robert Harding Picture Library.
Chapter 14: Page 156(l), Bob and Ira Spring Photography; 156(r), Eric Lessing/Magnum Photos; 157, F. Hidalgo/Image Bank; 159, Sepp Seitz/Woodfin Camp and Associates; 160, Julie O'Neil/Picture Cube; 161, Cameramann International, Ltd.; 163, Wallace Garrison/Index Stock International; 164, Kay Reese/Berlitz.
Chapter 15: Page 166(l), Henri Cartier-Bresson/Magnum Photos; 166(r), Tom Stack and Associates; 167, S. Vidler/Four By Five; 169, Four By Five; 170, Larry Dale Gordon/Image Bank; 171, Michael Pasdzior/Image Bank; 173, Four By Five; 174(t), Robert Frerck/Odyssey Productions; 174(b), Bob West/Berg and Associates.
Chapter 16: Page 176(l), Paula M. Lerner/Picture Cube; 176(r), Harry Gruyaert/Magnum Photos; 177, Charlotte Kahler; 179, Carl Purcell/Words and Pictures; 180, HRW Photo by Jan Lukas; 181, John L. Stagl/Image Bank; 182, Robert Frerck/Odyssey Productions.
Unit 3: Page 186(t), Burt Glynn/Magnum Photos; 186(b), Norma Watts; 187, TASS/Sovfoto.
Chapter 17: Page 188(l), Peter Marlow/Magnum Photos; 188(r), Inge Morath/Magnum Photos; 189, TASS/Sovfoto; 190, HRW Photo by Russell Dian; 193(tl), Burt Glynn/Magnum Photos; 193(tc), Eve Arnold/Magnum Photos; 193(tr), 193(bl), Burt Glynn/Magnum Photos; 193(bc),

193(br), Eve Arnold/Magnum Photos; 194, Novosti/Gamma-Liaison; 195, Burt Glynn/Magnum Photos; 198, TASS/Sovfoto; 199, Elliott Erwitt/Magnum Photos; 200, 202, TASS/Sovfoto; 203, Shostal Associates; 204, HRW Photo by Russell Dian.
Chapter 18: Page 206(l), HRW Photo by Russell Dian; 206(r), Wolfgang Kaehler Photography; 207, TASS/Sovfoto; 210(b), Patrick Morvan/Sipa Press; 211, Pete Turner/Image Bank; 213, TASS/Sovfoto; 214, Inge Morath/Magnum Photos.
Chapter 19: Page 216(l) Elliott Erwitt/Magnum Photos; 216(r), TASS/Sovfoto; 217, Earth Observation Satellite Corporation; 218, Robert Harding Picture Library; 220(t), 220(b), 221, 222, TASS/Sovfoto.
Chapter 20: Page 224(l), Robert Harding Picture Library; 224(r), Cameramann International, Ltd.; 225, P. Slaughter/Image Bank; 228, Leonard Freed/Magnum Photos; 229, Cameramann International, Ltd.; 231, Bruce Rosenblum/Picture Cube; 232, Christian Vioujard/Gamma-Liaison; 233, Owen Franken/Sygma; 234, David Burnett/Woodfin Camp and Associates; 235, William Kennedy/Image Bank.
Unit 4: Page 240(t) K. Kummels/Shostal Associates; 240(b), Henri Bureau/Sygma; 241, Bruno Barbey/Magnum Photos.
Chapter 21: Page 242(l), Shostal Associates; 242(r), Bob and Ira Spring Photography; 243, Art Resources; 246, Robin White/Fotolex Associates; 248, Cameramann International, Ltd.; 249, Robert Azzi/Woodfin Camp and Associates; 250, Don Smetzer/Click/Chicago.
Chapter 22: Page 252(l), Sygma; 252(r), Stanley Rowin/Picture Cube; 254, Photri; 256, Camermann International, Ltd.; 259, Wolf Achtner/Sygma; 260, Frank Folwell/Click/ Chicago.
Chapter 23: Page 262(l), Michael Fenton; 262(r), Sygma; 264, Jonathan Wright/Bruce Coleman Inc.; 266(t), Neil Hart, 266(b), HRW Photo by Richard Weiss; 269, M. Biber/Photri; 270, Alain Nogues/Sygma; 271(t), Peter Kemp/Wide World Photos; 271(b), Alan McWeeney/Archive Pictures Inc.; 273, Kurt Scholz/Shostal Associates; 274, John Elk III.
Chapter 24: Page 276(l), Four By Five; 276(r), Bernard Pierre Wolf/Alpha; 277, Robin White/Fotolex Associates; 278, Earth Observation Satellite Corporation; 280, Jake Rajs/Image Bank; 281(t), Robert Harding Picture Library; 281(b), Robert Uzzi/Woodfin Camp and Associates; 283, Photri; 284, Mark Antman/Image Works; 285, Bruno Barbey/ Magnum Photos; 286, Four By Five.
Unit 5: Page 290(t), Gerald Cubitt; 290(b), Marcel Isy-Schwart/Image Bank; 291, Sassoon/Robert Harding Picture Library.
Chapter 25: Page 292(l), Four By Five; 292(r), Richard Wood/Picture Cube; 300, Gerald Cubitt; 301, Philip Jon Bailey/Picture Cube.
Chapter 26: Page 304(l), John Elk III/Bruce Coleman Inc.; 304(r), Norman Myers/Bruce Coleman Inc.; 305, Jack Couffer; Bruce Coleman Inc.; 306, Richard Wood/Picture Cube; 308, L. Gunez/Sipa; 309, Wolfgang Kaehler Photography; 311, A. Tessore/Shostal Associates; 313, J. Kraay/Bruce Coleman Inc.
Chapter 27: Page 316(l), Gerald Cubitt; 316(r), Robin White/Fotolex Associates; 317, Earth Observation Satellite Corporation; 318, Four By Five; 320, Gerald Cubitt; 321, Lynn McLaren/Picture Cube; 322, F. Jackson/Bruce Coleman Inc.; 324, Carl Purcell/Words and Pictures.
Chapter 28: Page 326(l), Lauré Communications; 326(r), Robin White/Fotolex Associates; 327, M. Rousseall/Shostal Associates; 330, P. Schmid/Shostal Associates; 331, 332, 333(t), 333(b), David M. Helgren; 334, Gerald Cubitt; 335, Lauré Communications; 336, Wernher Krutein/Photovault; 338, Gerald Cubitt.
Chapter 29: Page 340(l), Gerald Cubitt; 340(r), William E. Ferguson; 342, 344, Gerald Cubitt; 345(t), Louise Gubb/Gamma-Liaison; 345(b), Gerald Cubitt; 346, Four by Five; 347, Gerald Cubitt; 349, Tony Pupkewitz; 350, E. R. Degginger/Bruce Coleman Inc.
Unit 6: Page 354(t), Robert Harding Picture Library; 354(b), Bruno Barbey/Magnum Photos; 355, Four By Five.
Chapter 30: Page 356(l), Robin White/Fotolex Associates; 356(r), Four By Five; 357, John Elk III; 361, George Corrigan/Photri; 363, Photri; 364, Geoff Thompkinson/Click/Chicago; 366, David Wells/Image Works.
Chapter 31: Page 368(l), 368(r), Robert Frerck/Odyssey Productions; 369, Four By Five; 372, Wolfgang Kaehler Photography; 373, R. A. Acharya/Dinodia Picture Library; 374(l), J. Joshi/Bruce Coleman Inc.; 374(r), Robert Harding Picture Library; 375, Mike Lichter/International Stock Photo; 376, 377, 378, Robert Frerck/Odyssey Productions; 380, Four By Five.
Chapter 32: Page 382(l), Carl Purcell/Words and Pictures; 382(r), Four By Five; 383, Weng nai giang/Mark McLaren Inc.; 384, Thomas Nebbia/Click/Chicago; 387, David Wells/Image Works; 388, Earth Observation Satellite Corporation; 389, Bruno Barbey/Magnum Photos; 391, Kirkendall-Spring Photographers; 392, A. Grace/Sygma.
Chapter 33: Page 394(l), Kirkendall-Spring Photographers; 394(r), Dennis Cox/Click/Chicago; 395, Art Resource; 396, Cameramann International, Ltd.; 398, Kirkendall-Spring Photographers; 399, 400, 402, Cameramann International, Ltd.; 403(t), 403(b), Four By Five; 404, Camermann International, Ltd.
Chapter 34: Page 406(l), Four By Five; 406(r), Harald Sund/Image Bank; 408, Cameramann International, Ltd.; 410, 411, Four By Five; 412, Dirck Halstead/Gamma-Liaison; 413, Camermann International, Ltd.; 414, Berlitz/Click/Chicago; 415, John Elk III; 416, Photri.
Unit 7: Page 420(t), Peter French Photography; 420(b), Carl Purcell/Words and Pictures; 421, Dennis Stock/Magnum Photos.
Chapter 35: Page 422(l), Fritz Prenzel/Peter Arnold Inc.; 422(r), Robin Smith/Shostal Associates; 423, Four By Five; 424(t), Jack Riesland/Berg and Associates; 424(b), P. Saloustrous/Shostal Associates; 426, NASA/Earth Observation Satellite Corportaion; 430, 431(t), Four By Five; 431(b), Cameramann International, Ltd.; 432, Steve McCurry/Magnum Photos.
Chapter 36: Page 434(l), Cameramann International, Ltd.;

434(r), Bob and Ira Spring Photography; 435, Warren Jacobs/Shostal Associates; 437, Allen Russell; 441, Dilip Mehta/Woodfin Camp and Associates; 443(t), Paul Chesley/Aspen Photographers; 443(b), Nicholas DeVore/Aspen Photographers; 444, George Holton/Photo Researchers.
Unit 8: Page 448(t), G. A. Mathers/Robert Harding Picture Library; 448(b), Vivienne Moos/The Stock Market; 449, Stephanie S. Ferguson/William E. Furguson.
Chapter 37: Page 450(l), Four By Five; 450(r), Martha Cooper/Peter Arnold Inc.; 451, Four By Five; 453, The Stock Market; 454, K. Scholz/Shostal Associates; 458, Chip and Rosa María Peterson.
Chapter 38: Page 460(l), Chip and Rosa María Peterson; 460(r), Four by Five; 461, Alberto Guatti; 462, Robin White/Fotolex Associates; 463(t), Beryl Goldberg; 463(b), Steve Allen/Peter Arnold Inc.; 464, Alberto Guatti; 446, Eric Carle/Shostal Associates; 467(t), Chip and Rose María Peterson; 467(b), Four By Five.
Chapter 39: Page 470(l), Chip and Rosa María Peterson; 470(r), David Brownell Photography; 471, Suzanne L. Murphy/D. Donne Bryant Photography; 473, Luis Villota/Stock Market; 474, Joe Viesti; 475, Four By Five; 476, Fred Ward/Black Star; 477, Jack Riesland/Berg and Associates.
Chapter 40: Page 480(l), 480(r), Chip and Rosa María Peterson; 481, Hubertus Kanus/Shostal Associates; 485, J. Ianiszewski/Shostal Associates; 486, Chip and Rosa María Peterson.
Chapter 41: Page 488(l), Jean S. Buldain/Berg and Associates; 488(r), Chip and Rosa María Peterson; 489, Rick Merron/Magnum Photos; 492, Thomas Nebbia; 493, Four By Five; 494, Luis Villota/Stock Market.
Chapter 42: Page 496(l), Don Klein; 496(r), Bruno Barbey/Magnum Photos; 499, Karl Kummels/Shostal Associates; 500, Robert Perron.
Chapter 43: Page 502(l), Joe Viesti; 502(r), Jaques Jangous/Peter Arnold Inc.; 503, Gary Milburn/Tom Stack and Associates; 505, Robin White/Fotolex Associates; 506, Earth Observation Satellite Corporation; 507, Cameramann International, Ltd.; 508, The Picture Cube.
Unit 9: Page 512(t), Steve Satusher/Image Bank; 512(b), HRW Photo by Russell Dian; 513, G. R. Roberts.
Chapter 44: Page 514(l), William D. Adams; 514(r), Grant Heilman Photography; 515, John Coletti/Stock Boston; 517, Edi Ann Otto; 519, Clyde H. Smith/F-Stop Pictures; 520, Grant Heilman Photography.
Chapter 45: Page 524(l), Carol Palmer/Picture Cube; 524(r), Isaac Geib/Grant Heilman Photography; 525, Grant Heilman Photography; 526, Robert De Gast/After Image.
Chapter 46: Page 530(l), HRW Photo by Russell Dian; 530(r), Larry Lefever/Grant Heilman Photography; 531, Alex Webb/Magnum Photos; 432, Lee Balterman/Marilyn Gartman Agency; 435, Four By Five; 437(t), James T. Lehrner/Picture Cube; 437(b), Mary Altier/Picture Cube; 438, Ulricke Welsch.
Chapter 47: Page 540(l), Four By Five; 540(r), J. C. Allen and Son; 541, Cameramann International, Ltd.; 542, Grant Heilman Photography; 545, Four By Five.
Chapter 48: Page 548(l), Larry Lefever/Grant Heilman Photography; 548(r), Four By Five; 552, Earth Observation Satellite Corporation; 554, Robin White/Fotolex Associates.
Chapter 49: Page 556(l), Stephen Trimble/DRK Photo; 556(r), Grant Heilman Photography; 557, HRW Photo by Claude Poulet; 558, David R. Frazier; 560, Four By Five; 461, Frank J. Staub/Picture Cube; 562, 563, David R. Frazier.
Chapter 50: Page 556(l), Alan Pitcairn/Grant Heilman Photography; 556(r), Robin White/Fotolex Associates; 567, Ed Cooper Photography; 569, Alan Pitcairn/Grant Heilman Photography; 570, Lewis Kemper/DRK Photo; 571, Four By Five; 572, M. Lewy/Shostal Associates; 573(t), Gus Schonefeld/Berg and Associates; 573(b), R. Pauli/Shostal Associates; 574, Bob and Ira Spring Photography.
Chapter 51: Page 576(l), David R. Frazier; 576(r), Sharon Cummings/Dembinsky and Associates; 577, Steve Kraseman/DRK Photo; 579, Grant Heilman Photography; 580, Thomas Nebbia; 582, Dirk Gallian/Journalism Services.
Chapter 52: Page 584(l), Joseph R. Pearce/DRK Photo; 584(r), Wayne Lankinen/DRK Photo; 585, HRW Photo by Russell Dian; 588, G. Hunter/Shostal Associates; 589, Philip John Bailey/Picture Cube; 590, Susan Van Etten/Picture Cube.
Chapter 53: Page 592(l), Ron Goulet/Dembinsky and Associates; 592(r), Ken Kaminsky/Picture Cube; 594, 595, 597, Four By Five.
Unit 10: Page 602(t), Tom Bean; 602(b), Four By Five; 603, Steve Niedorfy/Image Bank.
Chapter 54: Page 604(l), Cameramann International, Ltd.; 604(r), Shizuo Kambayashi/Wide World Photos; 605, Ellis Herwig/Picture Cube; 607(t), Lee Foster; 607(b), V. Lefteroff/Shostal Associates; 608, Carl Purcell/Words and Pictures; 610, TASS/Sovfoto.
Chapter 55: Page 612(l), Larry Lefever/Grant Heilman Photography; 612(r), Cameramann International, Ltd.; 613, Grant Heilman Photography; 614, David R. Frazier; 615, Budd Titlow/F-Stop Pictures; 616, Cameramann International, Ltd.; 617, TASS/Sovfoto; 169(t), 619(b), Four By Five; 620, James H. Simon/Picture Cube.

Cover: Cover image part of "Worlds of Color" copyright 1987, Allan Cartography. Available from Raven Maps & Images, Medford, Oregon.

Text maps: Mapping Specialists, Ltd.
Atlas and regional maps: R. R. Donnelley & Sons